D1717062

Controlled Drug Delivery

Challenges and Strategies

Kinam Park, Editor
Purdue University

ACS Professional Reference Book

AMERICAN CHEMICAL SOCIETY
WASHINGTON, DC

Library of Congress Cataloging-in-Publication Data

Controlled drug delivery : challenges and strategies / Kinam Park, editor.

p. cm.—(ACS professional reference book)

Includes bibliographical references and index.

ISBN 0-8412-3418-3

1. Drug delivery systems. 2. Drugs—Dosage forms.

I. Park, Kinam. II. Series.

RS199.5.C66 1997

615′.6—dc21 96-51016

CIP

The paper used in this publication meets the minimum requirements of American National Standard for Information Sciences—Permanence of Paper for Printed Library Materials, ANSI Z39.48-1984.

Copyright © 1997 American Chemical Society

All Rights Reserved. Reprographic copying beyond that permitted by Sections 107 or 108 of the U.S. Copyright Act is allowed for internal use only, provided that the per-chapter fee of $17.00 base + $0.25/page is paid to the Copyright Clearance Center, Inc., 222 Rosewood Drive, Danvers, MA 01923, USA. Republication or reproduction for sale of pages in this book is permitted only under license from ACS. Direct these and other permission requests to ACS Copyright Office, Publications Division, 1155 16th St., N.W., Washington, DC 20036.

The citation of trade names and/or names of manufacturers in this publication is not to be construed as an endorsement or as approval by ACS of the commercial products or services referenced herein; nor should the mere reference herein to any drawing, specification, chemical process, or other data be regarded as a license or as a conveyance of any right or permission to the holder, reader, or any other person or corporation, to manufacture, reproduce, use, or sell any patented invention or copyrighted work that may in any way be related thereto. Registered names, trademarks, etc., used in this publication, even without specific indication thereof, are not to be considered unprotected by law.

PRINTED IN THE UNITED STATES OF AMERICA

About the Editor

Kinam Park is a professor in the School of Pharmacy of Purdue University. He received his B.S. degree in pharmacy from Seoul National University and then served in the Korean army as a lieutenant. After receiving his Ph.D. degree in pharmaceutics from the University of Wisconsin in 1983, he had postdoctoral training in the Department of Chemical Engineering of the same university for two years. He joined the faculty of Purdue University in 1986 and was promoted to full professor in 1994.

He is an associate editor and the book review editor of the journal *Pharmaceutical Research* and a member of the editorial boards of the *Journal of Biomaterials Science—Polymer Edition* and the *Journal of Bioactive and Compatible Polymers*. He received the NIH New Investigator Research Award in 1986 and the Young Investigator Award from the Controlled Release Society in 1992. He was elected to the Board of Governors of the Controlled Release Society in 1993. He was elected as a fellow of the American Association for Pharmaceutical Scientists in 1993 and of the American Institute for Medical and Biological Engineering in 1996.

His research focuses on the surface modification of biomaterials to prevent protein adsorption and cell adhesion. His research also involves the use of smart hydrogels and superporous hydrogels in the design of controlled drug delivery systems. He has published more than 120 papers and presented more than 130 papers and invited lectures at national and international meetings. He coauthored the book *Biodegradable Hydrogels for Drug Delivery* (with Waleed Shalaby and Haesun Park) and coedited the book *Hydrogels and Biodegradable Polymers for Bioapplications,* (with Raphael Ottenbrite and Samuel Huang), which was based on the International Symposium on Biorelated Polymers sponsored by the Division of Polymer Chemistry, Inc., of the American Chemical Society.

Contents

SELF-REGULATED DRUG DELIVERY

DELIVERY OF PEPTIDE AND PROTEIN DRUGS

TISSUE ENGINEERING AND GENE THERAPY

NEW BIOMATERIALS FOR DRUG DELIVERY

MODELING OF CONTROLLED DRUG DELIVERY

REGULATORY ISSUES

Contributors

Julia E. Babensee *page 311*
Department of Chemical Engineering and Applied Chemistry, and Centre for Biomaterials, University of Toronto, Toronto, ON, M5S 3E5, Canada

You Han Bae *page 147*
Department of Materials Science and Engineering, Kwangju Institute of Science and Technology, 572 Sangam-dong, Kwangsan-gu, Kwangju, Korea

Ronald T. Borchardt *page 205*
School of Pharmacy, Department of Pharmaceutical Chemistry, The University of Kansas, 2095 Constant Avenue, Lawrence, KS 66047

Terry L. Bowersock *page 269*
Department of Veterinary Pathobiology, Purdue University, 1243 Veterinary Pathology Building, West Lafayette, IN 47907–1243
Current address: Pharmacia & Upjohn, 7000 Portage Road, Kalamazoo, MI 49001–0199

Joseph Cappello *page 439*
Protein Polymer Technologies, Inc., 10655 Sorrento Valley Road, San Diego, CA 92121

A. G. A. Coombes *page 73*
Department of Pharmaceutical Sciences, The University of Nottingham, University Park, Nottingham, United Kingdom NG7 2RD

Henry R. Costantino *page 229*
Department of Chemical Engineering, Massachusetts Institute of Technology, Cambridge, MA 02139
Current address: Department of Pharmaceutical Research and Development, Genentech, Inc., 460 Point San Bruno Boulevard, South San Francisco, CA 94080–4990

Basil I. Dahiyat *page 469*
School of Medicine, Department of Biomedical Engineering, The Johns Hopkins University, Baltimore, MD 21205

M. C. Davies *page 73*
Department of Pharmaceutical Sciences, The University of Nottingham, University Park, Nottingham, United Kingdom NG7 2RD

S. S. Davis *page 73*
Department of Pharmaceutical Sciences, The University of Nottingham, University Park, Nottingham, United Kingdom NG7 2RD

Richard L. Eckert *page 347*
Departments of Physiology/Biophysics, Dermatology, Reproductive Biology, and Biochemistry, Case Western Reserve University School of Medicine, 2109 Adelbert Road, Cleveland, OH 44106–4970

Thomas H. Ferguson *page 289*
Animal Science Product Development, Elanco Animal Health, A Division of Eli Lilly and Company, P.O. Box 708, Greenfield, IN 46140

D. Channe Gowda *page 405*
School of Medicine, Laboratory of Molecular Biophysics, University of Alabama at Birmingham, VH300, Birmingham, AL 35294–0091, and Bioelastics Research, Ltd., 1075 South Thirteenth Street, Birmingham, AL 35205

Rajesh K. Gupta *page 229*
Massachusetts Public Health Biologic Laboratories, Boston, MA 02130

Michael J. Hageman *page 205*
Drug Delivery Research and Development, The Upjohn Company, Kalamazoo, MI 49007

Cynthia M. Harris *page 405*
Bioelastics Research, Ltd., 1075 South Thirteenth Street, Birmingham, AL 35205

Jorge Heller *page 127*
Advanced Polymer Systems Research Institute, 3696 Haven Avenue, Redwood City, CA 94063

Allan S. Hoffman *page 485*
Center for Bioengineering, University of Washington, Box 35–2255, Seattle, WA 98195

Harm HogenEsch *page 269*
Department of Veterinary Pathobiology, Purdue University, 1243 Veterinary Pathology Building, West Lafayette, IN 47907–1243

L. Illum *page 73*
Department of Pharmaceutical Sciences, The University of Nottingham, University Park, Nottingham, United Kingdom NG7 2RD

Kenneth James *page 389*
Department of Chemistry, Rutgers—The State University of New Jersey, New Brunswick, NJ 08903

Kazunori Kataoka *page 49*
Department of Materials Science and Research Institute for Biosciences, Science University of Tokyo, 2641 Yamazaki, Noda, Chiba 278, Japan

Tetsuhiko Koguma *page 357*
Institute of Applied Biochemistry, University of Tsukuba, Tennoudai 1-1-1, Tsukuba Science City 305, Japan; and National Institute of Bioscience and Human Technology, Agency of Industrial Science & Technology, MITI, Tsukuba Science City 305, Japan

Joachim Kohn *page 389*
Department of Chemistry, Rutgers—The State University of New Jersey, New Brunswick, NJ 08903

Joel M. Kupfer *page 27*
Department of Medicine/Division of Cardiology, UCLA School of Medicine Cedars-Sinai Medical Center, Los Angeles, CA 90048

Robert Langer *page 229*
Department of Chemical Engineering, Massachusetts Institute of Technology, Cambridge, MA 02139

Christopher P. Leamon *page 11*
Glaxo Inc., 5 Moore Drive 2.2154, Research Triangle Park, NC 27709

Kam W. Leong *page 469*
School of Medicine, Department of Biomedical Engineering, The Johns Hopkins University, Baltimore, MD 21205

Jun Liao *page 455*
Department of Chemistry, High Technology Materials Center, Virginia Commonwealth University, Richmond, VA 23284

Philip S. Low *page 11*
Department of Chemistry, Purdue University, West Lafayette, IN 47907

D. Robert Lu *page 559*
Department of Pharmaceutics, College of Pharmacy, University of Georgia, Athens, GA 30602

Chi-Hao Luan *page 405*
School of Medicine, Laboratory of Molecular Biophysics, University of Alabama at Birmingham, VH300, Birmingham, AL 35294–0091

Chi Xiang Luan *page 405*
School of Medicine, Laboratory of Molecular Biophysics, University of Alabama at Birmingham, VH300, Birmingham, AL 35294–0091

Philip R. Mayer *page 589*
Clinical Research and Development, Wyeth-Ayerst Research, Philadelphia, PA 19101–8299

D. J. Mooney *page 333*
Departments of Biomedical, Chemical Engineering, and Biologic and Materials Sciences, University of Michigan, Ann Arbor, MI 48109–2136

Randall J. Mrsny *page 107*
Department of Pharmaceutical Research and Development, Genentech, Inc., 460 Point San Bruno Boulevard, South San Francisco, CA 94080–4990

Steven L. Nail *page 185*
Department of Industrial and Physical Pharmacy, Purdue University, West Lafayette, IN 47907

Balaji Narasimhan *page 529*
School of Chemical Engineering, Purdue University, West Lafayette, IN 47907

Michael R. Neuman *page 163*
Department of Obstetrics/Gynecology, MetroHealth Medical Center, Cleveland, OH 44109

Jun Ohkawa *page 357*
Institute of Applied Biochemistry, University of Tsukuba, Tennoudai 1-1-1, Tsukuba Science City 305, Japan; and National Institute of Bioscience and Human Technology, Agency of Industrial Science & Technology, MITI, Tsukuba Science City 305, Japan

Masaya Orita *page 357*
Institute of Applied Biochemistry, University of Tsukuba, Tennoudai 1-1-1, Tsukuba Science City 305, Japan; and National Institute of Bioscience and Human Technology, Agency of Industrial Science & Technology, MITI, Tsukuba Science City 305, Japan

Raphael M. Ottenbrite *page 455*
Department of Chemistry, High Technology Materials Center, Virginia Commonwealth University, Richmond, VA 23284

Timothy M. Parker *page 405*
Bioelastics Research, Ltd., 1075 South Thirteenth Street, Birmingham, AL 35205

Garnet E. Peck *page 577*
School of Pharmacy and Pharmaceutical Sciences, Purdue University, West Lafayette, IN 47907

Shao Qing Peng *page 405*
School of Medicine, Laboratory of Molecular Biophysics, University of Alabama at Birmingham, VH300, Birmingham, AL 35294–0091

Nicholas A. Peppas *page 529*
School of Chemical Engineering, Purdue University, West Lafayette, IN 47907

William M. Reichert *page 163*
Center for Emerging Cardiovascular Technologies (NSF/ERC) and Department of Biomedical Engineering, Duke University, Durham, NC 27708

Joseph R. Robinson *page 1*
School of Pharmacy, University of Wisconsin, 425 North Charter Street, Madison, WI 53706

J. A. Rowley *page 333*
Departments of Biomedical, Chemical Engineering, and Biologic and Materials Sciences, University of Michigan, Ann Arbor, MI 48109–2136

Irwin A. Schafer *page 347*
Department of Pediatrics at MetroHealth Medical Center, Case Western Reserve University, Cleveland, OH 44109

P. D. Scholes *page 73*
Department of Pharmaceutical Sciences, The University of Nottingham, University Park, Nottingham, United Kingdom NG7 2RD
Current address: 3M Healthcare, Loughborough, United Kingdom

Christian Schöneich *page 205*
School of Pharmacy, Department of Pharmaceutical Chemistry, The University of Kansas, 2095 Constant Avenue, Lawrence, KS 66047

Steven P. Schwendeman *page 229*
Department of Chemical Engineering, Massachusetts Institute of Technology, Cambridge, MA 02139
Current address: College of Pharmacy, Division of Pharmaceutics and Pharmaceutical Chemistry, The Ohio State University, Columbus, OH 43210

Michael V. Sefton *page 311*
Department of Chemical Engineering and Applied Chemistry, and Centre for Biomaterials, University of Toronto, Toronto, ON, M5S 3E5, Canada

A. Adam Sharkawy *page 163*
Center for Emerging Cardiovascular Technologies (NSF/ERC) and Department of Biomedical Engineering, Duke University, Durham, NC 27708

Takashi Shimayama *page 357*
Institute of Applied Biochemistry, University of Tsukuba, Tennoudai 1-1-1, Tsukuba Science City 305, Japan; and National Institute of Bioscience and Human Technology, Agency of Industrial Science & Technology, MITI, Tsukuba Science City 305, Japan

Ronald A. Siegel *page 501*
School of Pharmacy, Departments of Biopharmaceutical Sciences and Pharmaceutical Chemistry, University of California, San Francisco, CA 94143–0446

Daniel J. Smith *page 347*
Departments of Chemistry and Biomedical Engineering, University of Akron, Akron, OH 44325

Kazunari Taira *page 357*
Institute of Applied Biochemistry, University of Tsukuba, Tennoudai 1-1-1, Tsukuba Science City 305, Japan; and National Institute of Bioscience and Human Technology, Agency of Industrial Science & Technology, MITI, Tsukuba Science City 305, Japan

Dan W. Urry *page 405*
School of Medicine, Laboratory of Molecular Biophysics, University of Alabama at Birmingham, VH300, Birmingham, AL 35294–0091

Akira Wada *page 357*
Institute of Applied Biochemistry, University of Tsukuba, Tennoudai 1-1-1, Tsukuba Science City 305, Japan; and National Institute of Bioscience and Human Technology, Agency of Industrial Science & Technology, MITI, Tsukuba Science City 305, Japan

Masaki Warashina *page 357*
Institute of Applied Biochemistry, University of Tsukuba, Tennoudai 1-1-1, Tsukuba Science City 305, Japan; and National Institute of Bioscience and Human Technology, Agency of Industrial Science & Technology, MITI, Tsukuba Science City 305, Japan

Jie Xu *page 405*
School of Medicine, Laboratory of Molecular Biophysics, University of Alabama at Birmingham, VH300, Birmingham, AL 35294–0091, and Bioelastics Research, Ltd., 1075 South Thirteenth Street, Birmingham, AL 35205

Zhong Zhao *page 469*
School of Medicine, Department of Biomedical Engineering, The Johns Hopkins University, Baltimore, MD 21205

De-Min Zhou *page 357*
Institute of Applied Biochemistry, University of Tsukuba, Tennoudai 1-1-1, Tsukuba Science City 305, Japan; and National Institute of Bioscience and Human Technology, Agency of Industrial Science & Technology, MITI, Tsukuba Science City 305, Japan

Preface

Significant advances in controlled-release drug delivery have been made in the past 30 years. Of the many achievements made during this period, probably the most significant is the ability to deliver drugs for extended periods of time, up to years at a constant rate. In the early stage of developing controlled-release drug delivery systems, the main goal was to design the devices with zero-order release properties. With the concerted efforts of scientists in different disciplines, the task of preparing devices for zero-order release of small molecules was rather easily conquered. It turned out, however, that much of the technology developed for traditional low-molecular-weight drugs might not be applicable to the delivery of large molecules such as peptide and protein drugs. Furthermore, zero-order delivery is not necessarily the best mode of delivery for all drugs. For controlled drug delivery to be truly useful in the treatment of various diseases, we have to overcome many new challenges in drug delivery. We are entering a new era of controlled-release technology. This new era can be called "the next generation of controlled release". Dealing with the challenges we will face in this new era is the focus of this book.

A brief overview on the history of controlled-release drug delivery in Chapter 1 is followed by chapters dealing with intracellular delivery and targeting. Noninvasive intracellular delivery of high-molecular-weight therapeutic agents (which are essentially unwanted molecules by the body) is in a sense a challenge to the barrier that nature has established to protect itself. Partial solutions to this challenge are discussed in Chapters 2 and 3. Intracellular delivery is useful and desirable only if drugs are delivered to the target sites. Targeting to various tissues in the body or specific regions in the gastrointestinal tract will make many drugs more useful. The targeting aspect of drug delivery is discussed in Chapters 4–6.

Controlled delivery would be the optimal way of administering drugs if the drug delivery could be precisely matched with physiological needs. To make such a delivery possible, development of self-regulated drug delivery systems is essential. These delivery systems are gaining more importance because many of the newly developed protein drugs need to be delivered in a pulsatile fashion. Chapters 7 and 8 describe how drug delivery is coupled with environmental signals, either chemical or physical, to produce modulated drug delivery systems. Chapter 9 focuses on glucose sensors and the importance of biocompatibility of the implanted sensors in applications for long-term feedback-controlled insulin delivery.

Recent advances in genetic engineering have made it possible to produce once-rare protein drugs in large quantities. Successful long-term treatment of chronic diseases by the protein drugs requires nonconventional drug delivery systems. Protein drugs are difficult to produce, but they are even more difficult to deliver. Development of suitable delivery systems for protein drugs is the bottleneck for their successful application in the treatment of chronic diseases. It is necessary to understand what is possible and what is not for the development of protein delivery systems. Of the many routes of drug administration, oral delivery is most desirable for obvious reasons. Oral delivery of protein drugs, however, is most difficult to achieve. Recent successes in oral vaccination, albeit limited, provide hope for further development of oral protein delivery systems. Topics on the delivery of peptide and protein drugs are discussed in Chapters 10–14.

Tissue engineering and gene therapy occupy a monumental position on the frontiers of controlled drug delivery. Tissue engineering offers the possibility of replacing metabolic and structural functions of virtually any lost or deficient tissues. Gene therapy is essentially the redesign of cells by introducing therapeutic genes into genetically disabled cells to deliver therapeutic agents made by the gene. Microencapsulated cells, whether normal or genetically altered, are capable of functioning as a source for chronic systemic delivery of agents with desirable bioactivity. Chapters 15–18 deal with applications of tissue engineering, gene therapy, and microencapsulation in drug delivery.

We are currently in the middle of the so-called "biomaterials availability crisis". The lack of biocompatible polymers will also influence in a negative way the development of new controlled drug delivery systems. Development of biocompatible materials is critical for long-term drug delivery from implantable devices. Due to the failure of some biomaterials and subsequent legal problems, efforts in developing new biomaterials have been diminishing. It is time to reenergize research efforts in biomaterials. Chapters 19–24 describe new polymeric materials that can be used for drug delivery. They include pseudo-poly(amino acid)s derived from natural metabolites, transductional protein-based polymers, protein polymers mimicking the properties of body materials, polymers degradable to non-toxic compounds, bioactive polymers, and intelligent polymers.

As we have witnessed during the development of the first generation of controlled drug delivery systems, modeling study is very helpful, and perhaps essential, in the design of new controlled-release devices. As emphasis is given to developing self-regulated drug delivery systems, modeling studies will have to focus on this aspect, too. In addition, as computer simulation becomes more affordable as a result of drastic increases in the performance of personal computers and workstations, simulation of controlled drug delivery on a computer is expected to flourish. These are the topics of Chapters 25–27.

The final two chapters (Chapters 28 and 29) deal with regulations and pharmacokinetic aspects important in the development of controlled-release dosage forms. Although regulations change as time passes, it is necessary to understand them at one point.

The goal of this book is to combine in one place new information on diversified subjects related to the development of future controlled drug delivery systems. It is hoped that this book can save a lot of time for many researchers who may otherwise have to make numerous trips to the library. This book is intended to serve as a reference book for students, both industrial and academic scientists, engineers engaged in basic research and dosage form development, and others whose jobs require understanding of the next generation of controlled drug delivery systems. This book will serve its purpose if it stimulates new ideas by young scientists and triggers development of better controlled drug delivery systems that can benefit society.

I thank all the authors, who took time out of their busy schedules to write the chapters, and the reviewers, who provided helpful suggestions. My thanks also go to Anne Wilson, who guided me in the beginning of this project, and Marc Fitzgerald, who undertook the most difficult task of handling manuscripts for reviews. This book was possible only because of all those contributors.

KINAM PARK
Purdue University
West Lafayette, IN 47907–1336

1

Controlled Drug Delivery

Past, Present, and Future

Joseph R. Robinson

Delivery of drugs to humans has evolved from primitive extracts and inhalants to more reliable conventional dosage forms, such as tablets and capsules. These conventional systems were expected to be benign insofar as drug stability and bioavailability are concerned, whereas more modern and future delivery systems are expected to increase activity of the drug through greatly improved spatial and temporal drug delivery. Fundamental to future delivery systems is the availability of new polymeric materials.

From the earliest times, humans have found ways to introduce drugs into the body. Undoubtedly, this process began with the chewing of leaves and roots of medicinal plants and inhaling soot from the burning of medicinal substances. It may seem incongruous today, but as recently as the 1950s, it was routine to treat asthma by smoking medicinal cigarettes. These primitive approaches to delivering drugs, which lacked consistency and uniformity, let alone specificity, gave way in the latter part of the last century, and more specifically in this century, to uniformity of drug delivery systems (nee dosage forms). Common systems included elixirs, solutions, syrups, extracts, pills, capsules, tablets, emulsions, suspensions, cachets, troches, lozenges, nebulizers, and many other "traditional" delivery systems. Many of these delivery systems resulted from the fact that most drugs in the last century were derived from plant extracts.

The modern era of medicine probably began in the 1920s with the introduction of vaccines, followed by the penicillin and steroid eras. Potent drugs were discovered on a routine basis, but the potency and activity of these agents could have been, and often were, thwarted by the wrong or improperly constructed delivery system. The historic view of delivery systems, which continued up to the 1960s, was "do no harm". Clearly, the delivery system was viewed as a benign carrier of drugs that at its best did not destroy the integrity of the drug and permitted its full availability to the

Copyright © 1997 American Chemical Society

body or site of application. As analytical capabilities improved, the modern era of drug development began. It became possible to follow the time course of drugs in the body, explore the metabolic fate and mechanistic activity of drugs, introduce rational design into new drugs, specify an acceptable drug concentration range in tissues and fluids of the body, etc. Through all of this evolution, the formulation-development scientist attempted to construct delivery systems that protected drugs, to devise sustained release products if prolonged action was needed, and to make every effort to insure that full drug bioavailability took place.

Beginning in the 1960s the modern view of drug delivery began to unfold: namely, that the delivery system could substantially contribute to performance of drugs through proper spatial placement or temporal delivery. The philosophy and product performance of the ALZA Corporation were major catalysts to this movement. From this period onward began the serious effort of understanding the disposition of drugs and delivery systems in the various routes of administration and devising strategies for drug delivery to overcome biological barriers and the physical–chemical properties of modern drugs. The relative importance of drug delivery at present and in the future can be better appreciated by examining the changing economic and therapeutic marketplace.

Economics

By the year 2000, the worldwide value of pharmaceuticals will be $250 billion. About 10% of this value will be attributable to biotechnology drugs, and 10% will be from drug delivery systems (*1*). A $25 billion market for drug delivery systems represents an enormous growth from comparable benchmarks in 1990 and 1980. Closer inspection of the U.S. market alone, as shown in Table 1.1, gives a much better sense of the expected economic impact of drug delivery systems. The production and sale of pharmaceuticals is influenced not only by health-care needs and associated advances in medicine but also by economic conditions, including national health-expenditure levels, government policies, and medical-provider resources. Nevertheless, interest in improved drug delivery systems will continue and will grow accordingly as the worldwide economy becomes more robust.

TABLE 1.1. U.S. Market Estimates for Drug-Delivery Systems

Route	*1998 ($ billion)*	*Yearly Growth from Base Year (%)*
Oral	5.8	9.7
Implantable and parenteral	5.9	14.0
Topical	1.5	7.0
Total	14.3	11.0

NOTE: Base year is 1993.
SOURCE: Data are from reference 8.

Drugs

The development and production of pharmaceuticals includes classical synthesis of small molecules and a number of newer technologies. Recombinant DNA is an example and involves the genetic modification of microorganisms to transform them into drug-producing factories. A number of important pharmaceuticals already employ this technique, such as human insulin, interferon, interleukin-2, erythropoietin, and tissue plasminogen activator. Table 1.2 provides a listing of approved recombinant therapeutics.

Other approaches include development of chiral compounds, monoclonal antibodies, and gene-based therapies. Naturally, there is also an ever-expanding group of peptide and protein drugs. Each of these agents, by virtue of size, stability, or the need for targeting, will require a specialized drug delivery system. It would seem reasonable, based on these needs, that future drug delivery systems will either be extremely complex with multifunctional components or that new polymers with specialized properties will be needed.

TABLE 1.2. Approved Recombinant Therapeutics

Year	*Product*	*Indication*	*Company and Trade Name*
1982	Human insulin	Diabetes	Eli Lilly/Genentech (Humulin)
1985	Somatrem for injection	hGH deficiency in children	Genentech (Protropin)
1986	Interferon-α-2a	Hairy cell leukemia	Hoffmann-La Roche (Roferon-A)
	Hepatitis B vaccine, MSD	Hepatitis B prevention	Merck (Recombivax HB); Chiron
	Muromonab-CD3	Reversal of acute kidney transplant rejection	Ortho Biotech (Orthoclone OKT3)
	Interferon-α-2b	Hairy cell leukemia	Schering-Plough/Biogen (Intron A)
1987	Somatropin for injection	hGH deficiency in children	Eli Lilly (Humatrope)
	Alteplase (TPA)	Acute myocardial infarction	Genentech (Activase)
1988	Interferon-α-2a	AIDS-related Kaposi's sarcoma	Hoffmann-La Roche (Roferon-A)
	Interferon-α-2b	AIDS-related Kaposi's sarcoma Genital warts	Schering-Plough/Biogen (Intron A)
1989	Erythropoietin	Anemia associated with chronic renal failure	Amgen (Epogen); Johnson & Johnson; Kirin
	Interferon-α-n3	Genital warts	Interferon Sciences (Alferon N Injection)
	Hepatitis B vaccine	Hepatitis B prevention	SmithKline Beecham (Engerix-B); Biogen
1990	Erythropoietin	Anemia associated with AIDS/AZT	Amgen (Procrit); Ortho Biotech
	PEG-adenosine	ADA-deficient SCID	Enzon (Adagen); Eastman Kodak
	Interferon-γ-1b	Management of chronic granulomatous disease	Genentech (Actimmune)
	Alteplase (TPA)	Acute pulmonary embolism	Genentech (Activase)
	CMV immune globulin	CMV prevention in kidney transplant patients	MedImmune (CytoGam)
	Erythropoietin	Anemia associated with chronic renal failure	Ortho Biotech (Procrit)
1991	Filgrastim (G-CSF)	Chemotherapy-induced neutropenia	Amgen (Neupogen)
	β-Glucocerebrosidase	Type I Gaucher's disease	Genzyme (Ceredase)
	Sargramostim (GM-CSF)	Autologous bone marrow transplantation	Hoechst-Roussel (Prokine); Immunex
	Sargramostim (GM-CSF)	Neutrophil recovery following bone marrow transplantation	Immunex (Leukine); Hoechst-Roussel
	Interferon-α-2b	Hepatitis C	Schering-Plough/Biogen (Intron A)
1992	Antihemophilic factor	Hemophilia B	Armour (Mononine)
	Aldesleukin (interleukin-2)	Renal cell carcinoma	Chiron (Proleukin)
	Indium-111 labeled antibody	Detection, staging, and follow-up of colorectal cancer	Cytogen (OncoScint CR103); Knoll
	Indium-111 labeled antibody	Detection, staging, and follow-up of ovarian cancer	Cytogen (OncoScint OV103); Knoll
	Antihemophilic factor	Hemophilia A	Genetics Institute; Baxter Healthcare (Recombinate)
	Interferon-α-2b	Hepatitis B	Schering-Plough/Biogen (Intron A)
1993	Erythropoietin	Chemotherapy-associated anemia in nonmyloid malignancy patients	Amgen (Procrit); Ortho Biotech
	Interferon-β	Relapsing/remitting multiple sclerosis	Chiron; Berlex (Betaseron)
	DNase	Cystic fibrosis	Genentech (Pulmozyme)
	Factor VIII	Hemophilia A	Genentech; Miles (Kogenate)
	Erythropoietin	Anemia associated with cancer and chemotherapy	Ortho Biotech (Procrit)
1994	Filgrastim (G-CSF)	Bone marrow transplant	Amgen (Neupogen)
	Enzyme (PEG-L-asparaginase)	Refractory childhood acute lymphoblastic leukemia	Enzon (Oncaspar)
	Human growth hormone	Short stature caused by human growth hormone deficiency	Genentech (Nutropin)
	Glucocerebrosidase	Type I Gaucher's disease	Genzyme (Cerezyme)

NOTES: AZT is azidothymidine; ADA is adenosine deaminase; SCID is severe combined immunodeficiency disease; CMV is cytomegalovirus; and G-CSF is granulocyte colony stimulating factor.
SOURCE: Data are from reference 9.

Drug Delivery Systems: Historical Perspective

In the late 1940s *sustained release* products were commercialized successfully first by Smith, Kline & French (Spansules) and then by a variety of other pharmaceutical firms using several different technologies. The primary reason for these systems was to reduce the frequency of dosing, and on occasion, the local or systemic side effects. These systems enjoyed modest therapeutic and economic success. Table 1.3 gives a short list of major attributes of these modified-release dosage forms.

A major change occurred in the late 1960s, heavily influenced by the ALZA Corporation, where drug delivery systems were thought of as economically viable investments that could contribute to therapy by improving existing and projected new drugs. The early systems from the ALZA Corporation were referred to as *controlled release products*, where the controlled release aspect pertained to a reliable, reproducible system whose rate of drug release was often essentially independent of the environment in which it was placed. The reliable, reproducible nature of drug release emphasized the need for precision of control and attempted to minimize any contribution to intra- and intersubject variability associated with the drug delivery system. Simultaneous with the issue of precision of control was the nature of the kinetics of drug delivery. It is easy to show that the rate of drug input, from a declining plasma drug level, must be zero order if a constant, invariant blood drug level is to be maintained. Thus, we moved from the period of sustained release, where simple prolongation of the blood level in the therapeutic range was the endpoint, to controlled drug release rate as a critical component of successful drug therapy. Clearly, the emphasis on zero-order release was, and to a certain extent still, is excessive; but irrespective of whatever kinetic pattern is used, it can and does influence successful therapy.

A significant challenge in drug delivery is to prepare delivery systems that release drug in a non-zero-order fashion such as a pulsatile or ramp or some other pattern and, to add an additional constraint, in many cases, that the release of drug must be immediate. Of course, the ideal system is a feedback controlled device that releases the appropriate amount of drug in response to a therapeutic marker. One of the more interesting drug delivery systems that has evolved out of recent advances in cell and molecular biology is to implant normal or mutant cells that secrete drugs in response to a local or circulating chemical stimulant. Pancreatic islet cells, which secrete insulin in response to circulating glucose levels, im-

TABLE 1.3. Major Characteristics of Controlled Drug Delivery: Past, Present, and Future

Period	*Characteristics*	*Primary Technology*
1950–1970	Sustained release	Waxes
	Small molecular weight drugs	Plastics
	Primary purpose is patient compliance	Oils
		Few conventional polymers, primarily cellulosics
1970–1990	Controlled release	Skin patches
	Emphasis on zero-order release kinetics	Iontophoresis
	Low and high molecular weight drugs	Expanded polymer base
	Early stages of cellular and subcellular targeting	Early development of biocompatible polymers
1990–2010	Delivery of genes	New polymers with designed properties
	Nonparenteral delivery of proteins and carbohydrates	Range of bioerodible polymers
	Modulated, multipulsatile delivery from a single dose	Chemical and electrical regulated systems
	Self-regulated drug delivery	

planted in the peritoneal cavity illustrate this approach. Other cell types include hepatocytes. In all cases, a biocompatible implantable container that permits diffusion of therapeutic agent and marker is needed as is a suitable polymeric matrix to hold the cells. The field of biomaterials is essential in this area.

One last issue that is a component of controlled drug delivery is spatial placement of the drug. Targeting of drugs to specific organs, cells, subcellular content, and specific substrates is a critical, yet largely unmet, objective of drug delivery. It is easy to appreciate direct injection into a joint as a desirable alternative to drug supply from the vascular circulation. More difficult is placement of drug to internal organs or cells of the body unless it is through surgical intervention. At times surgical intervention for drug delivery purposes is the most attractive approach. Thus, treating brain tumors with various cytotoxic/cytostatic agents by placement of these drugs in a suitable long-acting erodible polymer minimizes systemic toxicity of these agents while maximizing localized response. However, there is not a substantial menu of erodible polymers to accommodate the range of solubility and other physical chemical properties of the large and growing list of antitumor agents. Suppose that surgical intervention for purposes of drug delivery system placement is not an option. The alternative is to place the drug delivery system into the bloodstream, which perfuses most tissues of the body, and to try to direct the system to an organ or cell type. There are many problems with this approach including avoiding loss of the systems from the bloodstream to the reticuloendothelial system, escaping the bloodstream to the tissue, targeting to a receptor on the cell, and internalizing into the cell. Each of these barriers must be overcome for a successful system, and although significant progress has been made in these areas, we are far from having a successful system or systems.

In summary, controlled drug delivery implies that the drug delivery system can influence performance of a drug by manipulation of its concentration, location, and duration.

Controlled Drug Delivery: The "Early Years" (1970–1990)

In the early time period of sustained release (i.e., 1950–1970), these products employed fairly traditional, well-accepted ingredients and technology (*2*). Thus, waxes, a few insoluble polymers, and a small number of water-soluble polymers constituted the bulk of ingredients used in sustained release. As the discipline moved to controlled release of drugs, our lack of understanding of the pharmacodynamics of most drugs, as well as a primitive understanding of the physiological and anatomical constraints of all routes of administration, became serious impediments to well-designed drug delivery system. It is reasonable to categorize this period of time as attempting *to understand the biological factors that influence design and performance* of drug delivery system. For all routes of administration this remains an ongoing challenge at an organ, cell, and biochemical level. Thus, for the oral route we can describe transit time, some elements of local pH, bacterial population, certain receptors, and a general understanding of the absorption process across the various tissues of the gastrointestinal tract; however, our understanding is far from complete. Of course, as an alternative to physiological and biochemically based delivery systems, we can design systems that are self-contained and independent of the environment. However, this design will not allow good local delivery and will not allow full control over the time, location, and drug concentration profile of the oral drug delivery system.

The technological base during this period moved forward. Introduction of transdermal patches, new and improved oral inhalation products, a number of erodible injectable polymers, and various oral products are illustrative of improved dosage forms. The polymer area and associated production technology moved ahead but at a slow pace. Clearly, most companies were unwilling to invest in new polymer technologies, given the regulatory hurdles and the cost of establishing safety, unless the market for application was large. This very limited list of available polymers significantly constrained the evolution of new delivery systems.

At the heart of this issue of investing in new polymers is whether or not the polymer–drug delivery system materially contributes to the therapeutic outcome of the treatment. Historically, justification for sustained/controlled drug delivery system has been improved patient compliance to treatment regimens and isolated examples of reduced local and systemic side effects. Few examples exist where the therapeutic benefit of the drug delivery system was reported over and above the issues of compliance and side effects. Clearly, the pharmaceutical–medical community is at a point where

- large and sensitive drugs must be delivered in a consistent reproducible fashion in a noninjectable delivery system (*3*);
- localized and regional delivery, such as to the brain and the eye, is needed;
- site-specific delivery is needed to reduce side effects;
- prolonged delivery from a single dose is needed for effective chronic treatment; and
- thorough pharmacodynamic and toxicokinetic analysis of drugs is necessary if the value of an effective controlled drug delivery system is to be realized.

Controlled Drug Delivery: Modern and Future Years (1990–2010)

The year 1990 is an arbitrary year in terms of judging trends in drug delivery, but it is important in the sense that improvements are part of a continuous rather than a discrete step.

There is little doubt that newer drugs, including gene therapy, are intended to treat the cause of a disease rather than the symptoms. These newer drugs have inherent disadvantages of size, usually sufficiently large to slow or inhibit membrane permeability, sensitivity in dose and dosing regimen, commonly potent drugs needing short exposure time to the target receptor, and often being quite unstable to the external environment. One can develop a short list of needed technologies over this 20-year period.

Carriers To Assist Membrane Permeability

These carriers can be tissue friendly, classic penetration enhancers, or preferably modeled after the physiologically based chaperones, e.g., macromolecules (*4*).

Biocompatible–Bioerodible Injectable Polymers

A sizable number of nontoxic, erodible polymers possessing a range of physicochemical properties are needed to embrace the diverse properties of new and emerging drugs. These drugs include microencapsulation and transplantation of endocrine tissue (*5*).

Stimuli-Sensitive Polymers

These polymers respond to pH, temperature, and electrochemical stimuli as well as to more subtle specific biochemical triggers. Such intelligent polymers are necessary for feedback controlled drug delivery systems (*6*).

Kinetic- and Equilibrium-Modulated Polymers

The need for cycling in release rate is essential for many drugs. Such flexibility in polymers is currently at a primitive stage.

Platforms for Tissue Engineering

According to Hubbell and Langer (*7*), there is need for "biocompatible casings for all transplants, polymer composites for patching wounds, scaffolds that guide and encourage cells to form tissue bioreactors for large-scale production of therapeutic cells." From this list, a number of issues become instantly apparent. First, drug delivery is a multidisciplinary activity involving polymer scientists, pharmaceutical scientists, chemical engineers, and a variety of biologically oriented scientists. Second, the trend to produce polymers possessing multiple properties, depending on the environment, and highly specialized functions is apparent. Third, the driving force for most of these changes is an expanding understanding of biology as it pertains to drug delivery systems. If the decades 1970–1990 represent the period to define what is needed in controlled drug delivery and understand, even at an organ level, disposition issues for the various routes of administration, the period 1990–2010 will represent the true biomedical polymer period. Of course success of this period is dependent on continued economic success of drug delivery systems and the willingness of certain companies or entrepreneurs to invest in this future.

References

1. *The U.S. Pharmaceutical Market Year in Review, 1994;* IMS International: New York, 1994; Eino Nelson Conference, Bahamas, 1994; Second Annual Drug Delivery Research Conference, Mabon Securities Corporation, New York, 1993.
2. *Sustained and Controlled Release Drug Delivery Systems,* 2nd ed.; Robinson, J. R.; Lee, V. H., Eds.; Marcel Dekker: New York, 1987.
3. *Peptides: Theoretical and Practical Approaches to Their Delivery;* Capsugel Proceedings, NJ, 1991.
4. Milstein, S. J.; Leipold, H.; Sarubbi, D.; Leone-Bay, A.; Freire, E.; Robinson, J.; Mlynek, G. "Oral Bioavailability of Partially Folded Proteins," accepted by *Science (Washington, D.C.).*
5. Kossovsky, N. *CHEMTECH* **1995,** *25*(11), 20–26.
6. Yoshida, R.; Uchida, K.; Kaneko, Y.; Sakai, K.; Kikuchi, A.; Sakurai, Y.; Okano, T. *Nature (London)* **1995,** *374,* 240.
7. Hubbell, J. A.; Langer, R. *Chem. Eng. News* **1996,** *73,* 42–54.
8. *Drug Delivery Systems;* The Freedonia Group: Cleveland, OH, 1994.
9. Damm, J. B. L. *BioPharm (Eugene, Oreg.)* **1995,** 43–47.

Intracellular Delivery and Targeting

2

Intracellular Delivery of Peptides and Proteins

Philip S. Low and Christopher P. Leamon

Proteins and peptides with biological activities could be employed to cure a plethora of human diseases if delivery of the polypeptide drugs to their target cells were readily achievable. The major obstacles still to be surmounted routinely include cell-specific targeting, entry of the therapeutic protein into the target cell, and escape of the protein drug from endosomal/degradative compartments into the cytosol. This review describes the major advances that have led to at least partial solutions to each of these delivery obstacles. The review also points out additional impediments that may arise in selected protein delivery cases and their possible solutions. Although proteinaceous drugs do not currently command a major share of the pharmaceutical market, with a few key advances, their utility as therapeutic agents could greatly expand.

Nature establishes barriers to protect itself against entry of unwanted molecules. Tight junctions between cells of epithelial origin, close associations of endothelial cells lining the blood vessels, the plasma membranes surrounding cells, and organellar membranes within cells all represent barriers designed to prevent movement of undesired molecules from one biological compartment to another. Unfortunately, when a biochemical process within a cell malfunctions and must be corrected by pharmaceutical methods, many of these barriers become obstacles to delivery of useful drugs to their sites of action. In fact, numerous otherwise potent remedies for a plethora of diseases remain idle on research laboratory shelves solely because of difficulties with delivery. With the advent of molecular biology methods that allow production of protein-based pharmaceuticals, this delivery problem is especially aggravated, because most macromolecular drugs cannot penetrate the aforementioned barriers by the well-established pathways of organic molecule delivery.

In general, only three pathways have been successfully exploited for delivery of drugs into cells (*1*). First, most commercial pharmaceuticals cross membranes by passive diffusion through the lipid bilayer. Unfortunately, this pathway is limited to low molecular weight solutes that are peculiarly soluble (or transportable) both in the extracellular aqueous environment and within the intrabilayer

Copyright © 1997 American Chemical Society

hydrocarbon milieu. Because peptides larger than a few amino acids are not generally hydrocarbon-soluble, this route of entry cannot be exploited by most proteinaceous drugs. Second, some pharmaceutical agents enter cells via the natural transporters for cellular nutrients and fuels. Whereas these pathways can be highly efficient and ubiquitously expressed, they simultaneously suffer because of high selectivity (i.e., the glucose transporter, GLUT 1, transports D-glucose but not L-glucose). Consequently, these routes of entry are also not useful for transporting peptides or proteins. Finally, larger molecules can pass membrane barriers by endocytosis or transcytosis, especially if receptors are present to concentrate the macromolecules at sites of entry. Because this pathway constitutes the predominant route exploited by nature for transport of macromolecules across compartmental barriers, this pathway may also be the most promising for delivery of exogenous peptides and proteins to their sites of action. In this review, we focus on progress that has been made in delivering peptides and proteins into cells by endocytosis-related routes.

Receptor-Mediated Endocytosis of Peptides and Peptide Conjugates

Background on Receptor-Mediated Endocytosis

The term endocytosis refers to the process whereby a membrane invaginates and pinches off to form an endocytic vesicle within the membrane-bounded compartment. With few if any exceptions, the process occurs in all eukaryotic organisms (*2–4*). Localized membrane proteins, lipids, and extracellular solutes also internalize during this event. In many cells, endocytosis at the plasma membrane is so active that the entire membrane surface is internalized and replaced in less than half an hour (*4, 5*).

A special class of membrane proteins, called receptors, uses endocytosis to physically mediate the cellular uptake of exogenous ligands. This "receptor-mediated endocytosis" can be broken down into specific events. Initially, exogenous ligands bind specifically to externally oriented receptors in the cell membrane. Binding is rapidly (usually <2 min) and is followed by membrane invagination at the receptor–ligand locus until a distinct internal vesicle, called an early endosome or receptosome, forms within the cell (*6*). Proton pumps located in the endosomal membrane then quickly lower the lumenal pH to about 5.5, triggering dissociation of the receptor–ligand complex (*7*). Endosomes may then move within cells randomly or by saltatory motion along tracks of microtubules (*8a*) and eventually come into contact with the trans Golgi reticulum. Here, they are believed to interact or fuse with Golgi elements or other membranous compartments and then convert into tubulovesicular complexes and multivesicular bodies. In these sorting compartments the fates of the receptor and ligand are believed to be determined. Some ligands and receptors are returned via vesicular recycling to the surface of the cell, where the ligand is released back into the extracellular milieu and the receptor is reused (e.g., transferrin). Some ligands are directed to the lysosomes for destruction while their receptors are returned to the cell surface for recycling (e.g., asialoglycoprotein). Also, some ligands and their receptors are collectively sent to the lysosomes to be degraded (e.g., epidermal growth factor) (*6*).

There are many advantages to exploiting receptor-mediated endocytosis for protein delivery into living cells. First, ligands generally bind to their receptors with high affinity, thus permitting effective sequestration of the ligand from low concentrations in the surrounding medium. Second, many receptors recycle back to the cell surface, thereby permitting multiple rounds of ligand capture. Third, different cell types express receptors to varying degrees (if at all), a phenomenon that can be exploited in the design of cell-type-specific drug delivery systems. Finally, different intracellular compartments can be targeted by carefully choosing the appropriate targeting ligand (*8b*).

Examples of the use of receptor-mediated endocytosis in drug delivery are now widespread.

To date, the majority of ligand–macromolecule conjugates were prepared by chemically coupling the two species together (*9*); however, genetic coupling via the production of fusion proteins also was described (*10*). A variety of ligands were used to introduce functionally active proteins into target cells, including peptide hormones, asialoglycoproteins, vitamins, monoclonal antibodies, viruses, and transferrin molecules. Following internalization by their target cells, the intracellular itinerary of the ligand–peptide conjugates appears to vary with the ligand and cell type involved, and thus the intracellular fates of the complexes cannot be generalized. Nevertheless, some common features do exist and these will be elaborated (Figure 2.1). In Figure 2.1, similar to the free ligand, endocytosis

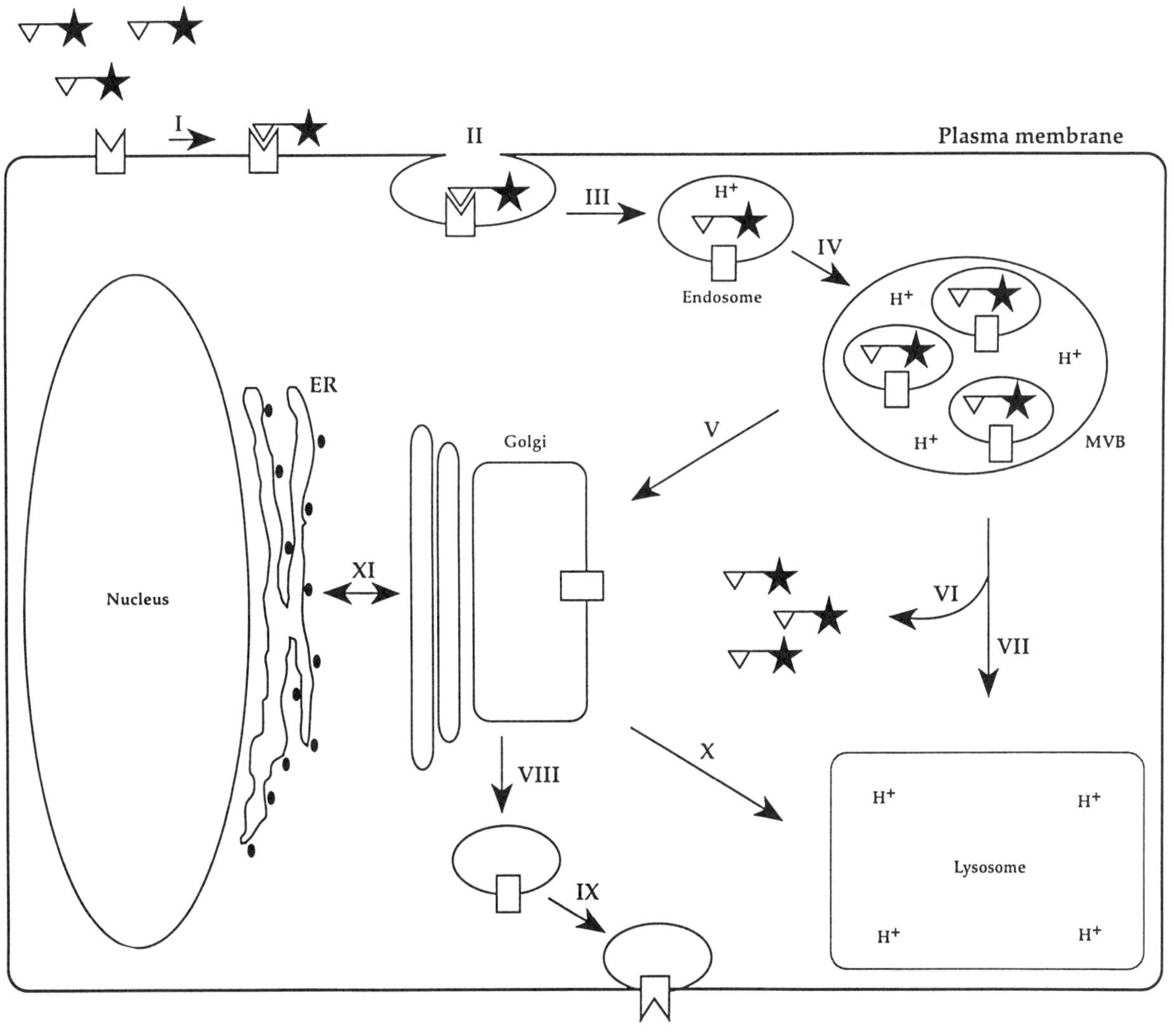

FIGURE 2.1. Receptor-mediated endocytosis of ligand–macromolecule conjugates. The conjugate consists of a cell-surface receptor binding ligand (triangle) covalently attached to a normally impermeable macromolecule (star).

of the ligand–macromolecule conjugate proceeds through a series of steps. In step I, an exogenously added conjugate binds to an extracellular membrane receptor via the ligand moiety. In step II, the plasma membrane surrounding the receptor–conjugate complex invaginates to form an intracellular vesicle called an endosome (step III). Membrane-associated proton pumps then acidify the lumen of the endosome, thereby promoting conformational changes within the receptor's ligand binding region. This process permits the conjugate to dissociate from the receptor, and this dissociation continues during the formation of multivesicular bodies (MVB; step IV). Even though the mechanisms still remain unresolved, a variety of vesicle–vesicle or vesicle–organelle fusion events occur that permit the endocytosed ligand–macromolecule conjugates that have undergone endocytosis to shuttle throughout the cell. Thus, conjugates may be sent to the Golgi apparatus (step V), deposited directly into the cytosol (step VI), or simply delivered to the lysosomes for degradation (step VII). In addition, transport vesicles may bud from the Golgi (step VIII) and recycle internalized receptors or conjugates back to the plasma membrane for another round of endocytosis (step IX). Alternatively, the budding vesicles may be directed to the lysosomes for destruction (step X) or they may even travel in retrograde direction to the endoplasmic reticulum (ER; step XI).

Hormone Receptors

Peptide hormones control a wide variety of physiological processes including cell division, pigment formation, appetite sensation, lactation, blood pressure, cell differentiation, and glucose metabolism. Many hormone receptors are relatively ubiquitous among the body's various tissues (e.g., insulin), whereas others exhibit a very limited distribution (e.g., Steel factor and erythropoietin). Because of the mitogenic effects of some hormones, a variety of cancers are thought to be benefited by up-regulating the levels of their cognate receptors (*11*). As a consequence, much interest has accrued in exploiting growth hormone receptors as a means of targeting toxins to tumors.

"Hormonotoxins" have proven to be suitable for ablation of many cancer cells in vitro. For example, epidermal growth factor (EGF) coupled via a disulfide bond to either ricin A (*12*) or diphtheria toxin fragment A (DTA) (*11*) was shown to be extremely toxic to a variety of cultured cancer cells. Similarly, a fusion construct of transforming growth factor-α and the binding-incompetent *Pseudomonas* exotoxin A fragment (PE) was lethal only to cells expressing EGF receptors (*13*). In an analogous scheme, the binding domain of the exotoxin was genetically replaced with interleukin-6, and the resulting conjugate was lethal to myeloma, hepatoma, and various carcinoma cell types, including cells with as few as 200 receptors per cell (*10*).

Other examples of hormonotoxin cell targeting include

- thyrotropin-releasing hormone–DTA toxicity to GH3 cells (*14*);
- human chorionic gonadotropin–ricin A chain (*15*, *16*) or DTA conjugate toxicity (*17*) to Leydig cells;
- ovine luteinizing hormone–gelonin toxicity to Leydig cells (*18*);
- interleukin-2–DTA conjugate toxicity to a variety of interleukin-2 receptor-positive cell types (*19*); and
- insulin–DTA conjugate toxicity to Swiss 3T3 cells (*20*).

Interestingly, the insulin–DTA conjugate also was used to select DTA-resistant variants of the Swiss 3T3 fibroblasts (*21*), a result suggesting a similar selection for resistant cells could conceivably occur in vivo. Obviously, hormone–toxin conjugates are also toxic to nontransformed receptor-expressing cells, and so their clinical future is still uncertain.

Although many of the hormone-assisted delivery systems have been designed to deposit toxic polypeptides into target cells for cell destruction, some hormones also were effective for delivering nonlethal, functionally active proteins for other purposes. For example, insulin facilitated delivery of functionally active cholesteryl esterase to the lysosomes of skin cells from a patient suffering from cholesteryl ester storage disease. Here, the insulin–esterase conjugate reduced levels of cholesterol esters by 50% after 2 h with a concomitant increase in free cholesterol (*22*). Similarly, an insulin-1,4-glucosidase conjugate targeted to muscle cells, hepatocytes, and lymphocytes delivered sufficient enzyme to be considered as a possible treatment for patients with type 2 glycogenosis (*23*).

Transferrin Receptors

All actively metabolizing cells require iron that is taken up as a transferrin–iron complex (i.e., holotransferrin) via receptor-mediated endocytosis (*24*). Following internalization and acidification of the endosomes containing transferrin and its receptor, transferrin releases its iron, and together with its receptor apotransferrin is recycled to the cell surface where it can undergo another round of delivery (*25*). Like other receptor-mediated pathways, the transferrin pathway was capable of mediating the specific uptake and functional deposition of proteins such as ricin A chain (*26*) and ribonuclease A (*27*) into the cytosol of cultured cells. Furthermore, antitransferrin receptor monoclonal antibodies conjugated to ricin A chain were cytotoxic to receptor-bearing cells (*28*). Perhaps the most novel contribution of transferrin-mediated delivery technology comes from "transferrinfection" studies, where the metalloprotein has been used to introduce DNA into receptor-bearing cells via electrostatic complexation with poly-L-lysine–transferrin conjugates (*29*, *30*). Importantly, such studies have revealed that simple transferrin-mediated DNA internalization alone does not guarantee efficient access of the attached gene to the cytosol. Thus, endosome disrupting agents, like adenovirus particles, may also be needed to maximize gene delivery efficiency (*vide infra*) (*31*).

IgG to Cell Surface Antigens

Conjugation of cell-surface reactive antibodies to ribosome-inactivating enzymes produces conjugates well suited for killing selected populations of cells. These so-called "immunotoxins" have been the focus of many reviews (*32–34*) and continue to receive widespread attention. For example, ricin A chain conjugated via a disulfide bond to a tumor-specific monoclonal antibody was recently found to kill sarcoma 791 T cells (*35*). In contrast to hormone–toxin conjugates, this immunotoxin entered cells almost exclusively through smooth pits and at a slow rate. Thus, it is conceivable that immunotoxins follow the uptake pathways of their antigen receptors, which in many cases may only enter cells via bulk membrane flow. Aside from ricin A chain, immunotoxins have been successfully prepared using other ribosome-inactivating proteins like bryodin (*36*), gelonin (*37*), saporin (*38*), and momordin (*39*). Obviously, the success of this approach depends on the tissue selectivity of the antibody and the ability of the immunotoxin conjugate to extravasate and reach the target cell surface.

Asialoglycoprotein Receptors

The asialoglycoprotein receptor of the liver functions to remove galactose-terminal (asialo-) glycoproteins from the sinusoidal circulation by internalizing the bound protein via endocytosis. Following uptake, the asialoglycoprotein dissociates from its receptor in response to intra-endosomal acidic pH, and the receptor is recycled back to the cell surface while the asialoglycoprotein travels to the lysosomes for destruction (*40*). Like most receptor pathways, this pathway is also capable of mediating the uptake of both protein and DNA into target cells. For example, covalent attachment of DTA to asialoorosomucoid or asialofetuin yielded conjugates that specifically killed rat hepatocytes (*41*, *42*). Furthermore, analogous to the aforementioned transferrin conjugates, asialoglycoprotein–poly-L-lysine complexes were useful for DNA delivery into both receptor-bearing liver cells and the liver-derived HepG2 cell line (*43*, *44*). Obviously, a limitation of this uptake strategy is that most of the peptide/protein will follow the asialoglycoprotein to the lysosomes where it will be digested. Why a small fraction of the internalized asialoglycoprotein conjugate escapes this unwanted fate is currently unknown (*43*, *44*).

Vitamins

Cells internalize some water-soluble vitamins by receptor-mediated endocytosis (*45*; *46*; Leamon, Yang, Lee, and Low, unpublished results). However, unlike the ligands discussed previously, vitamins are not primarily delivered to lysosomes for destruction (*47*). Consequently, vitamin uptake pathways may prove to be very useful for macromolecular delivery as long as the desired macromolecule can be covalently attached to the vitamin at a site that does not interfere with receptor recognition. This process has been possible, in fact, for many vitamins (unpublished results). For illustration purposes, the example of folate-mediated macromolecule delivery will be summarized.

Folic acid enters cells by two routes:

- via a transporter or carrier with a Michaelis constant (K_m) for folate of ~5×10^{-6} M and
- via a receptor that binds folate with a binding affinity (K_d) of ~5×10^{-10} M (*45*, *48–50*).

Importantly, the first route is present on virtually all cells, whereas the second route is expressed at high levels mainly on cancer cells (*51–63*). Those cancers reported to overexpress the folate receptor include cancers of the ovary, mammary gland, colon, lung, prostate, nose, throat, and brain (*51–63*). Further, leukemias such as chronic myelogenous leukemia may also overexpress the folate binding protein.

We found that molecules of virtually any size can be targeted to and delivered into the cytoplasm of tumor cells by folate receptor-mediated endocytosis, if the molecules are attached to the γ-carboxylate of folate.

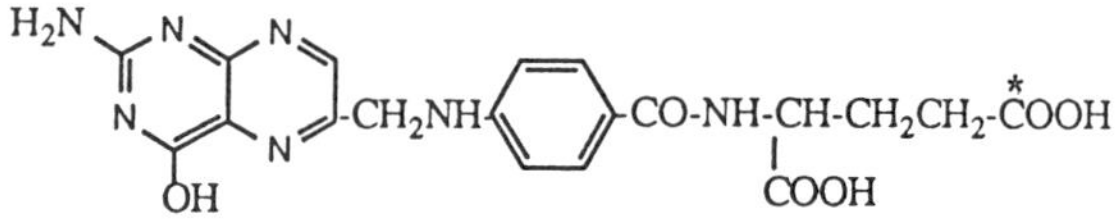

Examples of exogenous components delivered nondestructively into cancer cells include enzymes, antibodies, dextrans, hormones, plasmids, viruses, antisense DNA, and 150-nm liposomes (*47*, *63–71*). In all cases, uptake has been shown to occur via the folate receptor because

- nonfolate-labeled components do not enter cells;
- uptake by cells is competitively and quantitatively inhibited by high concentrations of free folate;
- uptake is also blocked by an antibody to the human folate receptor;
- folate conjugates exhibit the same affinity for cells as free folate; and
- selective cleavage of the folate receptor from the cancer cell surface (via a phospholipase C specific to glucose phosphate isomerase) obliterates folate conjugate uptake (*63*).

By using folates labeled with 15-nm colloidal gold particles, we have shown that folate conjugates first bind to cancer cell surfaces in invaginations termed caveolae (*47*), consistent with the itinerary of free folate (*72*, *73*). Thereafter, the conjugates internalize via folate-receptor-mediated endocytosis and rapidly enter a multivesicular sorting compartment. After moving on to an unidentified membrane-surrounded region where they are intensely concentrated, some of the conjugates are finally released into the cytoplasm in unmodified form. Few if any conjugates ever enter lysosomes for destruction (*47*), an observation consistent with the fact that vitamins are brought into cells primarily for consumption and not for destruction.

The electron microscopy data showing little or no deposition of folate conjugates in lysosomes (*47*) is confirmed by activity studies demonstrating that folate conjugates remain functionally operational following release into a receptor-bearing cell. Thus, folate conjugates of ribosome-inactivating proteins (e.g., momordin, pseudomonas exotoxin catalytic domain, and diphtheria toxic catalytic domain) kill cancer cells with 50% inhibition concentration (IC_{50}) values between 10^{-9} and 10^{-12} M (*64, 67, 68*), a result indicating that they are catalytically competent following folate-mediated uptake. Folate conjugates of enzymes such as horseradish peroxidase are also fully functional following internalization, and radiolabeled proteins like folate-^{125}I-serum albumin retain their parent molecular weights for hours to days following internalization by cells (*65*). Because folate-linked plasmids are expressed following endocytosis, and because folate-conjugated antisense oligonucleotides lead to enhanced suppression of endogenous genes (*71*), we conclude that folate uptake pathways do not involve a degradation pathway. The cell biology of this pathway was characterized by ourselves (*47*) and others (*72, 73*).

Overexpression of the folate receptor on many cancer cells suggested that folate might be exploited to deliver toxins specifically to tumor cells in the presence of normal cells. To test this possibility, a number of co-cultures of tumor and normal cells in the same culture flasks were established, and at ~50% confluence the co-cultures were treated with folate conjugates of momordin or the catalytic domain of *Pseudomonas* exotoxin (i.e., membrane-impermeant ribosome inactivating proteins). In the absence of folate conjugation, the cancer cells invariably dominated the culture flasks, eventually choking out the normal cells (*67*). However, upon addition of the conjugates, the cancer cells were quantitatively killed, while the normal cells rapidly filled the flask to confluence. Because similarly selective cancer cell killing was observed for other folate–toxin conjugates (*67, 70*), it is conceivable that folate might be exploited in vivo for the selective elimination of cancer cells.

Methods of Conjugation and Use of Dissolvable Linkers

There are many ways to covalently attach a targeting ligand to a protein for cellular delivery (for reviews *see* refs. 9, 23, and 74). We will focus on linkers that remain stable under normal physiological conditions, yet hydrolyze or disjoin after exposure to the reducing or acidic intracellular milieus.

Certain of the toxins used in the previously described studies (namely ricin A and DTA) occur in nature as heterodimers. Each toxin catalytic subunit is connected to its binding subunit through a disulfide bond that requires intracellular reduction, probably in the Golgi cells (*75*), before cytosolic release. In the design of ligand–toxin conjugates, it seemed obvious to attach the surrogate targeting ligand to the toxic subunit at the same site as that of the original binding subunit, and with the same chemistry (disulfide bonds). Interestingly, the majority of the aforementioned ligand–toxin conjugates were prepared by using agents that supplied such an interchain disulfide bond (*76–79*). This disulfide linkage typically, but not exclusively, was generated using a heterobifunctional cross-linker that tethered a primary amino group from the first molecule to a free sulfhydryl group on the second molecule (*80*). Whereas one could conceivably form the ligand–protein tether with a nonreducible thioether linkage (*9*), such linkages consistently were observed to deliver lower quantities of functionally active enzymes to the cytosol than analogous disulfide linkages. Presumably by allowing for reductive separation of the toxin from its targeting ligand, the disulfide cross-link permits more efficient escape from an endosomal compartment (*68, 75, 81*).

Novel cross-linking reagents also were developed that are relatively stable at extracellular pH but undergo acid-catalyzed hydrolysis within the endosomes. These reagents are based on ortho ester, acetal, and ketal functionalities (*82*). For example, Neville et al. (*83*) reported the potency of DTA–immunotoxin conjugates dramatically increases when noncleavable linkers are replaced with acid-labile ones. Further, Welhoner et al. (*84*) measured an increased quantity of free cross-linked antigen-binding fragment, $F(ab')_2$, in K562 cells when using an acid-labile linker in a transferrin–$F(ab')_2$ conjugate. In contrast, McIntyre et al. (*76*) found that a structurally unrelated acid-sensitive linker, (4-(iodoacetamido)-1-cyclohexene-1,2-dicarboxylic acid), did not improve the cytosolic delivery of the protein toxin, gelonin. However, this cyclohexene-based linker should be further tested in other conjugates and cell types before it is considered ineffective, because the general principle seems to be emerging that dissolvable/dissociable cross-links generate more potent toxin–ligand conjugates than their nonreleasable counterparts.

Encapsulation of Peptides and Proteins for Bulk Delivery

Liposome-Mediated Peptide Delivery

Liposomes are microscopic particles composed of one or more phospholipid bilayer membranes. These structures can carry water-soluble drugs in their aqueous compartments and lipid-soluble drugs within their bilayers (*85*). The advantage of using liposomes over direct targeting of a pharmacological agent lies simply in the fact that liposomes can encapsulate and deliver many thousands of drug molecules simultaneously. The disadvantage is that extravasation and migration through the interstitial spaces to target cells can be very slow. The application of liposomes for drug delivery has been the focus of many reviews and books (*85–92*). Therefore, this section will summarize only a few of the liposome-based techniques used for delivering peptides into cells.

Nontargetable Liposomes

Nontargetable liposomes can interact with cells by adsorption, lipid exchange, and pinocytosis (*90*). During liposome–cell adsorption, the contents of the liposome are not released directly into the cell. Instead, the contents slowly leak into the extracellular fluid where a portion may cross the cell membrane by nonspecific processes or mechanisms. The overall process is inefficient, unless the liposome is fortuitously engulfed and internalized by pinocytosis. Although methods recently were developed to encapsulate native proteins into liposomes (*93–95*), the lack of cell-specific targeting and problems with efficient intracellular release have limited development of these delivery vehicles.

Ligand-Targeted Liposomes

When efficient intracellular delivery of liposomal contents is observed, it is generally accepted that the entry is dependent on endocytosis and not direct fusion of the liposome with the plasma membrane (*96*). Nevertheless, liposome endocytosis alone does not guarantee deposition of the contents within the cytoplasm (*97*), because escape from endosomal compartments can be slow and even endocytosed liposomes can be returned to the cell surface before unloading their contents into the cytoplasm (*85*). Thus, each ligand-targeting strategy must be independently evaluated for its ability to intracellularly deliver an effective dose of any pharmaceutical agent.

Several classes of ligands such as sugars, lectins, peptide hormones, haptens, antibodies, vitamins, and other proteins were used successfully to target liposomes to cells (*69, 85, 98*). In one such study, a dye encapsulated in folate–PEG–liposome was employed to document that fluorescent dyes could be specifically targeted to cancer cells in coculture with normal cells (i.e., only cancer cells become labeled). In the same experiment, folate-free liposomes transferred no dye to any of the cultured cells (*69*). In a second study, doxorubicin was substituted for the dye in the folate-labeled liposomes, and selective killing of the cancer cells was demonstrated (*70*). Similarly, antisense oligonucleotides to the EGF receptor have been delivered via folate–PEG–liposomes specifically into cancer cells, and expression of the EGF receptor and cell growth were essentially eliminated (*71*). In another series of studies, asialofetuin-tethered liposomes were employed to deliver interferon-γ to HepG2 cells infected with hepatitis B virus, and viral proliferation was inhibited (*99*, *100*). Importantly, in these and related studies, deletion of the targeting ligand completely abrogated the biological response. Thus, compared to nontargeted liposomes, ligand-tethered liposomes may be preferred when intracellular delivery is essential. As will be discussed later, pH-sensitive liposomes can further enhance delivery efficiency.

Virosomes

Virosomes are essentially artificial viral envelopes composed of similar lipid and protein compositions as viral membranes (*101*, *102*). The viral protein provides both a docking and fusion element that allows for attachment and unloading of virosome contents via membrane fusion with the target cell. Coupled with a targeting moiety, virosomes can serve as delivery vehicles for drugs, peptides, toxins, and gene constructs (*101*). For example, ^{125}I-lysozyme recently was delivered into HepG2 cells by encapsulation into virosomes containing only the Sendai virus fusion protein (*103*). Although cell docking was exclusively at the asialoglycoprotein receptor, lysosomal deposition was largely avoided because virosome–cell fusion was complete at the plasma membrane and virosome contents were delivered before virosome internalization.

Cationic Liposomes

Cationic liposomes were used for delivering electrostatically complexed DNA molecules into cells (*104*, *105*). The mechanism of action and examples of use are beyond the scope of this review. However, there is at least one example of protein delivery into cells performed by using this technique. Transcription factors prebound to plasmid DNA at a glucocorticoid response element could be complexed with the cationic lipid, DOTMA, and transported into the cytosol of target cells (*106*). Although the extent of association between transcription factor and lipid was not determined, this result does suggest that other polyanionic proteins may be similarly deliverable if they can electrostatically associate with cationic liposomes. However, as with other cationic liposome complexes, cell toxicity and nonspecificity will have to be overcome before they can achieve widespread clinical use.

Escape from Endocytic Elements

Random Leak

Many of the aforementioned receptor-mediated delivery systems did not include a functional vesicle/endosome disruption element in their construction. Nevertheless, the data indicate that at least a fraction of the internalized proteins do manage to breach the endosomal membrane and enter the cytosol *(21, 36, 37, 41, 42, 64, 76, 77)*. For example, macromolecules delivered by using EGF, insulin, or asialoglycoprotein ligands would be expected to be either deposited into lysosomes for destruction or returned to the cell surface. Yet, the expression of some cytosolic activity demonstrates that escape into the cytoplasm does occur even when no defined pathway exists to promote it.

Shortly after their scission from the plasma membrane, newly formed endosomes tend to swell and fuse with one another to form large tubulovesicular elements and multivesicular bodies (*6*). Conceptually, the interendosomal fusion event may not be perfect. Indeed, if transient lesions in the bilayer form during vesicle fusion, solutes and macromolecules contained within these compartments could diffuse into the cytosol. Obviously, conjugates that remain bound (by virtue of their ligand) to their membrane receptors would not be able to escape during such transient processes, perhaps accounting for the enhanced delivery of peptides/proteins connected to their ligands via dissolvable/hydrolyzable linkers.

Facilitated Escape

The idea of enhancing the rate and quantity of macromolecular escape from endocytic elements has been the focus of many recent investigations. An obvious solution to this problem is to include a component within the molecular conjugate design that functions by penetrating or disrupting the endocytic bilayer. Following is a brief summary of some of the current methods used to breach this endocytic barrier.

By nature's design, certain plant, viral, and bacterial protein toxins contain intrinsic mechanisms for destabilizing or penetrating lipid bilayers. These so-called translocation domains, or fusogenic peptides, are induced to unfold and insert into membranes in vivo when exposed to the low pH milieu found in endocytic compartments (*107*, *108*). The conformational change has been best described as a partial unfolding event in which tertiary structure is disrupted, but in which secondary structure largely remains intact. Such proteins are said to adopt a molten globule conformation during the insertion process (*109*).

There are many strategies for employing translocation or "membrane-destabilization" elements for the enhancement of conjugate escape from endocytic compartments. For instance, the molecule intended for delivery, while impermeable to intact cells, may contain an endocytic translocation domain designed to facilitate cytosolic entry. Perhaps the best example of this process is in construction of ligand–toxin conjugates by using recombinant forms of *Pseudomonas* exotoxin (PE). PE is found naturally as a three-domain polypeptide (*110*). Domain I mediates the toxin's binding to the cell surface, and to improve targeting specificity it is usually replaced with a more specific ligand such as a hormone or cytokine. Domain II is responsible for translocation of the protein from an endoplasmic reticulumlike compartment into the cytosol, whereas domain III mediates the adenosine 5′-diphosphate ribosylation of elongation factor 2 causing the inhibition of protein synthesis and cell death (*111*, *112*). Exploiting this modular construction, a potent immunotoxin was formed by replacing domain I of PE with a tumor-associated glycoprotein-specific monoclonal antibody. This conjugate inhibited the growth of a variety of human colon, pancreatic, and cervical cancers as well as xenograft tumors in nude mice (*113*). Similar conjugates of PE prepared using other epitope-specific monoclonal antibodies were equally effective both in vitro and in vivo (*112*, *114–117*). Alternatively, PE domain I was replaced with the vitamin folate for targeted killing of cultured and xenographic folate receptor-bearing tumor cells in vitro and in vivo (*68*, unpublished observations). Although these examples have all used the PE translocation domain to promote escape of the natural toxic subunit from the endosome, there is also evidence that the translocation domain can function independently in the assisted cytosolic delivery of heterologous protein constructs (*118*).

An alternative to the use of bacterial toxins and derivatives for endocytic bilayer disruption has been the use of replication-incompetent viral particles. In this case, the endosome-entrapped therapeutic protein is thought to escape along with the viral nucleocapsid as the virus promotes its own entry into the cell (*31*, *119*). Interestingly, delivery efficiency can be enhanced even further if the adenovirus particle is coupled (covalently or noncovalently) to the molecular conjugate before delivery (*31*, *120*). In this configuration, however, the viral moiety functions both as an endosome lysis agent and as an alternative ligand domain that can influence cell specificity (*121*). Still, Michael et al. (*121*) recently demonstrated that the virus cell binding activity can be blocked without affecting its endosome disruption capability, thereby increasing the targeting specificity of the molecular conjugate.

A third strategy for promoting endocytic membrane destabilization is to incorporate fusogenic peptides into the conjugate's design. Following receptor-mediated endocytosis of influenza virus, the low pH of the endosome triggers a conformational change in the hemagglutinin viral coat protein forcing exposure of a once-buried hydrophobic fusogenic peptide that can insert into the endosomal bilayer (*107*). This process is believed to potentiate the fusion between the viral and endosomal membranes with resultant release of the viral nucleocapsid into the cytosol (*108*). Conceptually, naked fusion peptides should be capable of disrupting the endosomal bilayer in a similar manner. Indeed, when added simultaneously to cultured cells, various synthetic forms of influenza virus hemagglutinin fusion peptide dramatically enhanced the cytosolic delivery of molecules such as fluorescent dextrans and DNA–polylysine complexes (*120, 122*).

Importantly, there are nonimmunogenic alternatives to the use of viral proteins for membrane destabilization. For example, Parente et al. (*123*) synthesized the water-soluble 30-mer peptide of sequence $(EALA)_n$ and found that it selectively partitions into lipid bilayers under mildly acidic conditions. This partitioning derives from the peptide's transition from a random coil at pH 7.5 to an amphipathic alpha helix at pH 5, and it can lead to destabilization of the lipid bilayer and to membrane fusion. Incorporation of these peptides into targeted protein constructs could conceivably facilitate endosome release much like viral proteins do.

pH-Sensitive Liposomes

pH-sensitive liposomes are commonly composed of mixtures of unsaturated phosphatidylethanolamine and a weakly acidic amphiphile (*95, 124*). Because of the tendency of dioleylphosphatidylethanolamine (DOPE) to undergo a bilayer-to-hexagonal phase transition when adjacent lipids do not structurally complement its cone-shaped geometry, the pH-sensitive liposome rapidly destabilizes and becomes fusion competent when the adjacent acidic amphiphile is titrated to its neutral form in the low pH of the endosomes (*95*). For example, when hen egg lysozyme was encapsulated into pH-sensitive (1:1 palmitoylhomocysteine/DOPE) or pH-insensitive (1:1 dioleylphosphatidylserine/DOPE) liposomes, pH-sensitive liposomes effectively emptied the peptide contents into the cytosol, whereas the pH-insensitive liposomes delivered the peptide to the lysosomes (*125*). Likewise, pH-sensitive liposomes loaded with ovalbumin sensitized mouse thymoma cells to lysis by major histocompatibility complex (MHC), class I restricted, cytotoxic T lymphocytes, suggesting that the peptide readily gained access to the cytosol (*95*). More recently, ovalbumin loaded into pH-sensitive liposomes via a dehydration/rehydration technique generated a class I MHC response in vivo (*93*). Finally, in the field of cytotoxin delivery, DTA encapsulated in pH-sensitive liposomes comprised of DOPE and cholesterol hemisuccinate was orders of magnitude more lethal to RAW 264.7 cells (a macrophagelike cell line) than was DTA in pH-insensitive (DOPE and dioleylphosphatidylcholine) liposomes (*126*). Clearly, design of an endosome escape mechanism is as important to protein delivery as endocytosis by the target cell.

Problems and Prospects

Despite the excellent progress made toward development of numerous protein delivery strategies, there are still many obstacles that limit realization of the full potential of proteinaceous drugs. These obstacles and their possible solutions are listed in Table 2.1. Even though it is beyond the scope of this review to address each obstacle individually, it should be mentioned that many of the obstacles can be solved with relatively minor manipulations. For example, use of folate as a targeting ligand avoids the problems associated with protein size (No. 2, Table 2.1), lysosome delivery (No. 4), immunogenicity of the ligand (No. 5), ligand affinity (No. 8), release of soluble receptors (No. 9), nonspecificity (No. 10), competition with endogenous ligands (No. 11), chemical inhomogeneity (No. 13), and toxic side effects (No. 14). Careful choice of other specific ligands for related purposes likely can achieve similar objectives. Thus, with continued research and judicious choice of ligand and protein drug, the intracellular targeting of peptides/proteins may soon achieve clinical prominence.

TABLE 2.1. Obstacles to Development of Ligand-Targeted Peptide/Protein Drugs for Treatment of Intracellular Defects

Obstacle	*Solution*
1. Protein is proteolytically degraded.	1. a. Engineer protease-resistant peptide. b. Protect protease-sensitive sites.
2. Protein is too large to extravasate.	2. a. Develop smaller protein drug. b. Coadminister vascular permeabilization drug.
3. Protein is not sufficiently water soluble.	3. a. Engineer a more soluble protein. b. Derivatize protein with solubilizing groups.
4. Targeting ligand delivers protein to lysosomes.	4. a. Employ a targeting ligand that avoids lysosomes. b. Attach an endosome escape element.
5. Protein or ligand is immunogenic.	5. a. Engineer epitope-deleted protein. b. Derivatize protein with polyethylene glycol (PEG) or related protectant.
6. Protein is rapidly filtered by kidneys.	6. a. Cross-link protein to oligomer. b. Derivatize with PEG.
7. Number of ligand receptors on target cell is too low.	7. a. Target other receptors on same cell. b. Encapsulate multiple protein copies in targeted liposome or virosome.
8. Receptor affinity for coupled ligand is too weak.	8. Use multivalent ligand so that protein can dock simultaneously with several receptors.
9. Target cells release soluble receptors into circulation.	9. a. Add excess ligand–protein conjugate. b. Presaturate serum receptors with ligand multimers that cannot extravasate and compete for cell surface receptors.
10. Ligand receptors are found in tissues not to be targeted.	10. a. Use a localized injection of targeted protein. b. Identify more specific ligand.
11. Endogenous ligands compete with protein-conjugated ligand for target cell receptor.	11. Use multivalent ligand so that protein can dock simultaneously with several receptors, exponentially raising cell surface affinity.
12. Protein is unstable during storage.	12. a. Engineer more stable protein. b. Improve formulation.
13. Protein–ligand conjugate is not chemically homogeneous.	13. a. Employ more specific chemical cross-linking agents. b. Engineer the ligand into the polypeptide chain of the protein.
14. Targeting ligand exhibits expected but unwanted biological effects.	14. a. Convert ligand from an agonist to an antagonist b. Select an alternative ligand.

References

1. Burton, P. S.; Conradi, R. A.; Hilgers, A. R. *Adv. Drug Delivery Rev.* **1991**, *7*, 365–386.
2. Stahl, P.; Schwartz, A. L. *J. Clin. Invest.* **1986**, *77*, 657–662.
3. Smythe, E.; Warren, G. *Eur. J. Biochem.* **1991**, *202*, 689–699.
4. Low, P. S.; Chandra, S. *Annu. Rev. Plant Physiol. Plant Mol. Biol.* **1994**, *45*, 609–631.
5. Marsh, M.; Helenius, A. *J. Mol. Biol.* **1980**, *142*, 439–454.
6. Pastan, I.; Willingham, M. C. *The Pathway of Endocytosis;* Plenum: New York, 1985.
7. Wileman, T.; Harding, C.; Stahl, P. *Biochem. J.* **1985**, *232*, 1–14.
8. (a) Willingham, M. C.; Pastan, I. *Cell* **1980**, *21*, 67–77. (b) Basu, S. K. *Biochem. Pharmacol.* **1990**, *40*, 1941–1946.
9. Brinkley, M. *Bioconjugate Chem.* **1992**, *3*, 2–13.
10. Siegall, C. B.; FitzGerald, D. J.; Pastan, I. *J. Biol. Chem.* **1990**, *265*, 16318–16323.
11. Shaw, J. P.; Akiyoshi, D. E.; Arrigo, D. A.; Rhoad, A. E.; Sullivan, B.; Thomas, J.; Genbauffe, F. S.; Bacha, P.; Nichols, J. C. *J. Biol. Chem.* **1991**, *266*, 21118–21124.
12. Cawley, D. B.; Herschman, H. R.; Gilliland, D. G.; Collier, R. J. *Cell* **1980**, *22*, 563–570.
13. Theuer, C. P.; FitzGerald, D.; Pastan, I. *J. Biol. Chem.* **1992**, *267*, 16872–16877.
14. Bacha, P.; Murphy, J. R.; Reichlin, S. *J. Biol. Chem.* **1983**, *258*, 1565–1570.
15. Oeltmann, T. N.; Heath, E. C. *J. Biol. Chem.* **1979**, *254*, 1028–1032.
16. Sakai, A.; Sakakibara, R.; Ishiguro, M. *J. Biochem. (Tokyo)* **1989**, *105*, 275–280.
17. Oeltmann, T. N. *Biochem. Biophys. Res. Commun.* **1985**, *133*, 430–435.
18. Singh, V.; Sairam, M. R.; Bhargavi, G. N.; Akhras, R. G. *J. Biol. Chem.* **1989**, *264*, 3089–3095.
19. Bacha, P.; Williams, D. P.; Waters, C.; Williams, J. M.; Murphy, J. R.; Strom, T. B. *J. Exp. Med.* **1988**, *167*, 612–622.
20. Miskimins, W. K.; Shimizu, N. *Biochem. Biophys. Res. Commun.* **1979**, *91*, 143–151.
21. Miskimins, W. K.; Shimizu, N. *Proc. Natl. Acad. Sci. U.S.A.* **1981**, *78*, 445–449.
22. Poznansky, M. J.; Hutchinson, S. K.; Davis, P. J. *FASEB J.* **1989**, *3*, 152–156.
23. Poznansky, M. J.; Singh, R.; Singh, B. *Science (Washington, D.C.)* **1984**, *223*, 1304–1306.
24. Wagner, E.; Zenke, M.; Cotten, M.; Beug, H.; Birnstiel, M. L. *Proc. Natl. Acad. Sci. U.S.A.* **1990**, *87*, 3410–3414.
25. Karin, M.; Mintz, B. *J. Biol. Chem.* **1981**, *256*, 3245–3252.
26. Raso, V.; Basala, M. *J. Biol. Chem.* **1984**, *259*, 1143–1149.
27. Rybak, S. M.; Saxena, S. K.; Ackerman, E. J.; Youle, R. J. *J. Biol. Chem.* **1991**, *266*, 1202–21207.
28. Martell, L. A.; Agrawal, A.; Ross, D. A.; Muraszko, K. M. *Cancer Res.* **1993**, *53*, 1348–1353.
29. Cotten, M.; Langle-Rouault, F.; Kirlappos, H.; Wagner, E.; Mechtler, K.; Zenke, M.; Beug, H.; Birnstiel, M. L. *Proc. Natl. Acad. Sci. U.S.A.* **1990**, *87*, 4033–4037.
30. Zenke, M.; Steinlein, P.; Wagner, E.; Cotten, M.; Beug, H.; Birnstiel, M. L. *Proc. Natl. Acad. Sci. U.S.A.* **1990**, *87*, 3655–3659.
31. Curiel, D. T.; Agarwal, S.; Wagner, E.; Cotten, M. *Proc. Natl. Acad. Sci. U.S.A.* **1991**, *88*, 8850–8854.
32. Frankel, A. E. *J. Biol. Response Modif.* **1985**, *4*, 437–446.
33. Olsnes, S.; Sandvig, K.; Petersen, O. W.; van Deurs, B. *Immunol. Today* **1989**, *10*, 291–295.
34. Vitetta, E. S.; Fulton, R. J.; May, R. D.; Till, M.; Uhr, J. W. *Science (Washington, D.C.)* **1987**, *238*, 1098–1104.
35. Byers, V. S.; Pawluczyk, I. A.; Hooi, D. W.; Price, M. R.; Carroll, S.; Embleton, M. J.; Garnett, M. C.; Berry, N.; Robins, R. A.; Baldwin, R. W. *Cancer Res.* **1991**, *51*, 1990–1995.
36. Stirpe, F.; Wawrzynczak, E. J.; Brown, A. F.; Knyba, R. E.; Watson, G. J.; Barbieri, L.; Thorpe, P. E. *Eur. J. Cancer* **1988**, *58*, 558–561.
37. Thorpe, P. E.; Brown, A. F.; Ross, W. J.; Cumber, A. J.; Detre, S. I.; Edwards, D. C.; Davies, A. S.; Stirpe, F. *Eur. J. Biochem.* **1981**, *116*, 447–454.
38. Thorpe, P. E.; Brown, A. F.; Bremner, J. G.; Foxwell, B. J.; Stirpe, F. *J. Natl. Cancer Inst.* **1985**, *75*, 151–159.
39. Dinota, A.; Barbieri, L.; Gobbi, M.; Tazzari, P. L.; Rizzi, S.; Bontadini, A.; Bolognesi, A.; Tura, S.; Stirpe, F. *Br. J. Cancer* **1989**, *60*, 315–319.
40. Fallon, R. J.; Schartz, A. L. *Adv. Drug Delivery Rev.* **1989**, *4*, 49–63.
41. Cawley, D. B.; Simpson, D. L.; Herschman, H. R. *Proc. Natl. Acad. Sci. U.S.A.* **1981**, *78*, 3383–3387.

42. Simpson, D. L.; Cawley, D. B.; Herschman, H. R. *Cell* **1982,** *29,* 469–473.
43. Wu, G. Y.; Wu, C. H. *Biochemistry* **1988,** *27,* 887–892.
44. Wu, G. Y.; Wu, C. H. *Adv. Drug Delivery Rev.* **1993,** *12,* 159–167.
45. Kamen, B. A.; Capdevila, A. *Proc. Natl. Acad. Sci. U.S.A.* **1986,** *83,* 5983–5987.
46. Vesely, D. L.; Kemp, S. F.; Elders, M. J. *Biochem. Biophys. Res. Commun.* **1987,** *143,* 913–916.
47. Turek, J. J.; Leamon, C. P.; Low, P. S. *J. Cell Sci.* **1993,** *106,* 423–430.
48. Anthony, A. C. *Blood* **1992,** *79,* 2807–2820.
49. Henderson, G. B. *Annu. Rev. Nutr.* **1990,** *10,* 319–335.
50. Rothberg, K. G.; Ying, Y.; Kolhouse, J. F.; Kamen, B. A.; Anderson, R. W. *J. Cell Biol.* **1990,** *110,* 637–649.
51. Campbell, I. G.; Jones, T. A.; Foulkes, W. D.; Trowsdale, J. *Cancer Res.* **1991,** *51,* 5329–5338.
52. Coney, L. R.; Tomassetti, A.; Carayannopoulos, L.; Frasca, V.; Kamen, B. A.; Colnaghi, M. I.; Zurawski, V. R. *Cancer Res.* **1991,** *51,* 6125–6132.
53. Weitman, S. D.; Weinberg, A. G.; Coney, L. R.; Zurawski, V. R.; Jennings, D. S.; Kamen, B. A. *Cancer Res.* **1992,** *52,* 6708–6711.
54. Garin-Chesa, P.; Campbell, I.; Saigo, P. E.; Lewis, Jr., J. L.; Old, L. J.; Rettig, W. J. *Am. J. Pathol.* **1993,** *142,* 557–567.
55. Holm, J.; Hansen, S. I.; Sondergaard, K.; Hoier-Madsen, M. In *Chemistry and Biology of Pteridines and Folates;* Ayling, J. E., Ed.; Plenum: New York, 1993; pp 757–760.
56. Brigle, K. E.; Seither, R. L.; Westin, E. H.; Goldman, I. D. *J. Biol. Chem.* **1994,** *269,* 4267–4272.
57. Brigle, K. E.; Spinella, M. J.; Westin, E. H.; Goldman, I. D. *Biochem. Pharmacol.* **1994,** *47,* 337–345.
58. Jansen, G.; Westerhof, G. R.; Kathmann, I.; Rademaker, B. C.; Rijksen, G.; Schornagel, J. H. *Cancer Res.* **1989,** *49,* 2455–2459.
59. Vincent, M. L.; Russell, R. M.; Sasak, V. *Hum. Nutr. Clin. Nutr.* **1985,** *39C,* 355–360.
60. Mendelbaum-Shavit, F. In *Chemistry and Biology of Pteridines and Folates;* Ayling, J. E., Ed.; Plenum: New York, 1993; pp 787–790.
61. Ross, J. F.; Chaudhuri, P. K.; Ratnam, M. *Cancer* **1994,** *73,* 2432–2443.
62. Mantovani, L. T.; Miotti, S.; Menard, S.; Canevari, S.; Raspagliesi, F.; Bottini, C.; Bottero, F.; Colnaghi, M. I. *Eur. J. Cancer* **1994,** *30A,* 363–369.
63. Leamon, C. P.; Low, P. S. *Biochem. J.* **1993,** *291,* 855–860.
64. Leamon, C. P.; Low, P. S. *J. Biol. Chem.* **1992,** *267,* 24966–24971.
65. Leamon, C. P.; Low, P. S. *Proc. Natl. Acad. Sci. U.S.A.* **1991,** *88,* 5572–5576.
66. Lee, R. J.; Wang, S.; Low, P. S. *Biochim. Biophys. Acta* **1996,** in press.
67. Leamon, C. P.; Low, P. S. *J. Drug Targeting* **1994,** *2,* 101–112.
68. Leamon, C. P.; Pastan, I.; Low, P. S. *J. Biol. Chem.* **1993,** *268,* 24847–24854.
69. Lee, R. J.; Low, P. S. *J. Biol. Chem.* **1994,** *269,* 3198–3204.
70. Lee, R. J.; Low, P. S. *Biochim. Biophys. Acta* **1995,** *1233,* 134–144.
71. Wang, S.; Lee, R. J.; Cauchon, G.; Gorenstein, D. G.; Low, P. S. *Proc. Natl. Acad. Sci. U.S.A.* **1995,** *92,* 3318–3322.
72. Rothberg, K. G.; Heuser, J. E.; Donzell, W. C.; Ying, Y.-S., Glenney, J. R.; Anderson, R. W. *Cell* **1992,** *68,* 673–682.
73. Kamen, B. A.; Wang, M.-T.; Streckfuss, A. J.; Peryea, X.; Anderson, R. W. *J. Biol. Chem.* **1988,** *263,* 13602–13609.
74. Ji, T. H. *Methods Enzymol.* **1983,** *91,* 580–609.
75. Feener, E. P.; Shen, W. C.; Ryser, H. P. *J. Biol. Chem.* **1990,** *265,* 18780–18785.
76. Chang, T. M.; Kullberg, D. W. *J. Biol. Chem.* **1982,** *257,* 12563–12572.
77. McIntyre, G. D.; Scott, C. F.; Ritz, J.; Blattler, W. A.; Lambert, J. M. *Bioconjugate Chem.* **1994,** *5,* 88–97.
78. Raso, V.; Basala, M. *Study of the Transferrin Receptor Using a Cytotoxic Human Transferrin–Ricin A Chain Conjugate;* Plenum: New York, 1984; pp 73–86.
79. Roth, R. A.; Madddux, B. *J. Cell. Physiol.* **1983,** *115,* 151–158.
80. Carlsson, J.; Drevin, H.; Axen, R. *Biochem. J.* **1978,** *173,* 723–737.
81. Vitetta, E. S.; Uhr, J. W. *Cell* **1985,** *41,* 653–654.
82. Srinivasachar, K.; Neville, D. M. *Biochemistry* **1989,** *28,* 2501–2509.
83. Neville, D. M.; Srinivasachar, K.; Stone, R.; Scharff, J. *J. Biol. Chem.* **1989,** *264,* 14653–14661.
84. Welhoner, H. H.; Neville, D. M.; Srinivasachar, K.; Erdmann, G. *J. Biol. Chem.* **1991,** *266,* 4309–4314.
85. Weinstein, J. N. *Cancer Treat. Rep.* **1984,** *68,* 127–135.

86. Betageri, G. V.; Jenkins, S. A.; Parsons, D. L. *Liposome Drug Delivery Systems;* Technomic: Lancaster, PA, 1993.
87. Gregoriadis, G. *Liposome Technology: Entrapment of Drugs and Other Materials*; CRC Press: London, 1993.
88. Lasic, D. D. *Liposomes: From Physics to Applications;* Elsevier: New York, 1993.
89. New, R. C. *Liposomes: a Practical Approach;* IRL Press: New York, 1990.
90. Ostro, M. J.; Cullis, P. R. *Am. J. Hosp. Pharm.* **1989,** *46,* 1576–1587.
91. Ranade, V. V. *J. Clin. Pharmacol.* **1989,** *29,* 685–694.
92. Sullivan, S. M.; Conner, J.; Huang, L. *Med. Res. Rev.* **1986,** *6,* 171–195.
93. Collins, D. S.; Findlay, K.; Harding, C. V. *J. Immunol.* **1992,** *148,* 3336–3341.
94. Meyer, J.; Whitcomb, L.; Collins, D. *Biochem. Biophys. Res. Commun.* **1994,** *199,* 433–438.
95. Zhou, F.; Rouse, B. T.; Huang, L. *J. Immunol. Methods* **1991,** *145,* 143–152.
96. Straubinger, R. M.; Honk, K.; Friend, D. S.; Papahadjopoulos, D. *Cell* **1983,** *32,* 1069–1079.
97. Weinstein, J. N.; Blumenthal, R.; Sharrow, S. O. *Biochim. Biophys. Acta* **1978,** *509,* 272–288.
98. Leserman, L.; Machy, P. *Ligand Targeting of Liposomes;* Marcel Decker: New York, 1987; pp 157–194.
99. Ishihara, H.; Hara, T.; Aramaki, Y.; Tsuchiya, S.; Hosoi, K. *Pharm. Res.* **1990,** *7,* 542–546.
100. Ishihara, H.; Hayashi, Y.; Hara, T.; Aramaki, Y.; Tsuchiya, S.; Koike, K. *Biochem. Biophys. Res. Commun.* **1991,** *174,* 839–845.
101. Chander, R.; Schreier, H. *Life Sci.* **1992,** *50,* 481–489.
102. Stecenko, A. A.; Walsh, E. E.; Schreier, H. *Pharm. Pharmacol. Lett.* **1992,** *1,* 127–129.
103. Bagai, S.; Sarkar, D. P. *J. Biol. Chem.* **1994,** *269,* 1966–1972.
104. Behr, J. P. *Bioconjugate Chem.* **1994,** *5,* 382–389.
105. Felgner, P. L.; Ringold, G. M. *Nature (London)* **1989,** *337,* 387–388.
106. Debs, R. J.; Freedman, L. P.; Edmunds, S.; Gaensler, K. L.; Duzgunes, N.; Yamamoto, K. R. *J. Biol. Chem.* **1990,** *265,* 10189–10192.
107. Lear, J. D.; DeGrado, W. F. *J. Biol. Chem.* **1987,** *262,* 6500–6505.
108. London, E.; Ulbrandt, N. D.; Tortorella, D.; Jiang, J. X.; Abrams, F. S. *Insights into Membrane Protein Folding and Translocation from the Behavior of Bacterial Toxins: Models for Membrane Translocation;* The Rockefeller University: New York, 1993.
109. London, E. *Mol. Microbiol.* **1992,** *6,* 3277–3282.
110. FitzGerald, D.; Pastan, I. *Semin. Cell Biol.* **1991,** *2,* 31–37.
111. Hwang, J.; FitzGerald, D. J.; Adhya, S.; Pastan, I. *Cell* **1987,** *48,* 129–136.
112. Pastan, I.; Chaudhary, V.; FitzGerald, D. J. *Annu. Rev. Biochem.* **1992,** *61,* 331–354.
113. Debinski, W.; Karlsson, B.; Lindholm, L.; Siegall, C. B.; Willingham, M. C.; FitzGerald, D.; Pastan, I. *J. Clin. Invest.* **1992,** *90,* 405–411.
114. Batra, J. K.; FitzGerald, D. J.; Chaudhary, V. K.; Pastan, I. *Mol. Cell. Biol.* **1991,** *11,* 2200–2205.
115. Batra, J. K.; Jinno, Y.; Chaudhary, V. K.; Kondo, T.; Willingham, M. C.; FitzGerald, D. J.; Pastan, I. *Proc. Natl. Acad. Sci. U.S.A.* **1989,** *86,* 8545–8549.
116. Kondo, T.; FitzGerald, D.; Chaudhary, V. K.; Adhya, S.; Pastan, I. *J. Biol. Chem.* **1988,** *263,* 9470–9475.
117. Pai, L. H.; Batra, J. K.; FitzGerald, D. J.; Willingham, M. C.; Pastan, I. *Proc. Natl. Acad. Sci. U.S.A.* **1991,** *88,* 3358–3362.
118. Prior, T. I.; FitzGerald, D. J.; Pastan, I. *Cell* **1991,** *64,* 1017–1023.
119. Carrasco, L. *FEBS Lett.* **1994,** *350,* 151–154.
120. Wagner, E.; Plank, C.; Zatloukal, K.; Cotten, M.; Birnsteil, M. L. *Proc. Natl. Acad. Sci. U.S.A.* **1992,** *89,* 7934–7938.
121. Michael, S. I.; Huang, C. H.; Romer, M. U.; Wagner, E.; Hu, P. C.; Curiel, D. T. *J. Biol. Chem.* **1993,** *268,* 6866–6869.
122. Plank, C.; Oberhauser, B.; Mechtler, K.; Koch, C.; Wagner, E. *J. Biol. Chem.* **1994,** *269,* 12918–12924.
123. Parente, R. A.; Nadasdi, L.; Subbarao, N. K.; Szoka, F. C. *Biochemistry* **1990,** *29,* 8713–8719.
124. Chu, C. J.; Szoka, F. C. *J. Liposome Res.* **1994,** *4,* 361–395.
125. Harding, C. V.; Collins, D. S.; Slot, J. W.; Geuze, H. J.; Unanue, E. R. *Cell* **1991,** *64,* 393–401.
126. Chu, C. J.; Dijkstra, J.; Lai, M. Z.; Hong, K.; Szoka, F. C. *Pharm. Res.* **1990,** *7,* 824–834.

3

DNA Delivery Vectors for Somatic Cell Gene Therapy

Joel M. Kupfer

The potential benefits of gene therapy can be realized only with the development of techniques for efficient transfer and control of expression of recombinant DNA at the target sites. Effective gene delivery requires efficient internalization of the gene and subsequent escape from lysosomal degradation. pH-sensitive liposomes, cationic liposomes, and recombinant viruses have been developed and applied as gene delivery vectors. This chapter describes characteristics of liposome- and recombinant virus-mediated DNA deliveries into the cells.

Somatic cell gene therapy is a medical intervention designed to treat human disease by correcting genetic deficiencies or by producing a therapeutic protein. In some ways gene therapy is analogous to organ transplantation. Whereas organ transplantation focuses on correcting a deficiency by replacing the organ's entire metabolic function, gene therapy targets specific genetic deficiencies or pathophysiologic pathways. Because the host cell uses its own regulatory machinery to control production and secretion of recombinant protein, the targeted organ becomes an "in situ factory" for the release of a therapeutic agent (i.e., drug).

The first published report of mammalian gene transfer appeared in 1980. In that study methotrexate-susceptible murine bone marrow cells were transformed into drug-resistant cells by introducing the gene for dihydrofolate reductase, a protein capable of metabolizing methotrexate (*1*). In 1984, investigators were successful in partially correcting growth retardation in mutant dwarf mouse (growth hormone deficient) by incorporating recombinant rat-growth hormone DNA into the germ line of these mice (*2*). Shortly thereafter, methods for producing replication-defective recombinant retrovirus vectors were described. These vectors were shown to be capable of efficiently transferring recombinant DNA to susceptible cells. By the end of 1994 more than 300 patients have been enrolled in 100 active or pending clinical gene therapy trials (*3*).

Somatic cell gene therapy has the potential for treating a spectrum of human diseases: from hereditary single gene disorders to complex disor-

Copyright © 1997 American Chemical Society

ders such as cardiovascular disease, diabetes, hypertension, and malignancy. For example, 25% of the cases of severe combined immunodeficiency (SCID) are due to adenosine deaminase (ADA) deficiency, an enzyme that catalyzes the conversion of adenosine and deoxyadenosine to inosine and deoxyinosine, respectively. In the absence of this enzyme T and B cells are dysfunctional and cause a syndrome associated with increased susceptibility to infections. Isolation, cloning, and sequencing of the normal human ADA gene opened the possibility to correct this disorder by transferring the normal ADA gene to affected individuals and restoring normal ADA metabolism (*4*). Clinical studies evaluating the benefits of recombinant gene therapy for ADA deficiency are in progress.

Another example is the treatment of cystic fibrosis (CF). This disease is characterized by production of highly viscous mucus by the respiratory epithelium causing pulmonary obstruction and chronic infections. CF remains the most common lethal autosomal recessive disease in the United States and has no specific cure. The underlying etiology of CF was elucidated after it was discovered that CF patients lack a specific protein responsible for controlling electrolyte transport into and out of epithelial cells. This protein, known as the cystic fibrosis transmembrane conductance regulator protein (CFTR), is abnormal in patients with CF. The cloning and sequencing of the normal CFTR gene opened the possibility of using gene therapy to correct this genetic deficiency by introducing recombinant CFTR DNA into the respiratory epithelium of affected patients. Clinical trials are now underway to evaluate the safety and efficacy of gene therapy in CF patients (*5*).

A third example is the treatment of familial hypercholesterolemia. This is an autosomal dominant disease characterized by diminished capacity to metabolize low-density lipoprotein (LDL), extremely elevated serum cholesterol levels, accelerated atherosclerosis, and premature coronary artery disease. The genetic basis for this disease lies within the gene encoding for low-density lipoprotein receptors (LDL-R) resulting in diminished or absent expression of these receptors. Transfer of LDL-R DNA to livers of Watannabe rabbits (a strain of hypercholesterolemic rabbits) lowers serum LDL levels by correcting the receptor deficiency (*6*). Employing retrovirus-mediated gene delivery to autologous reimplanted hepatocytes of affected individuals has been shown to be safe and result in long-term expression of recombinant LDL-Rs (*7*). These three examples illustrate how gene therapy can be used to correct molecular-genetic defects. Unfortunately, many common diseases such as hypertension, diabetes, and atherosclerosis are genetically and environmentally complex. The exact role gene therapy will play in these cases remains unknown.

A separate but related and equally important challenge is the development of techniques for efficiently transferring and controlling expression of recombinant DNA at target sites. The process of altering phenotypic behavior with exogenous recombinant DNA can be viewed as consisting of five consecutive stages (Figure 3.1). These phases are

1. delivery of DNA to the target site,
2. cellular penetration and internalization,
3. nuclear translocation,
4. transcription and translation, and
5. protein processing.

Unfortunately, the physiochemical properties of DNA (large size and negative charge) and the protective barrier of the cell membrane limit uptake and nuclear translocation of native extracellular DNA. Extracellular DNA is rapidly degraded by acid proteases and deoxyribonuclease present in lysosomes. Thus, most normal mammalian cells are resistant to transformation by foreign DNA.

Because cellular penetration and internalization is so critical for accomplishing DNA transfer, methods have been developed to overcome these natural barriers. The earliest methods used microprecipitates of DNA to facilitate absorption to the

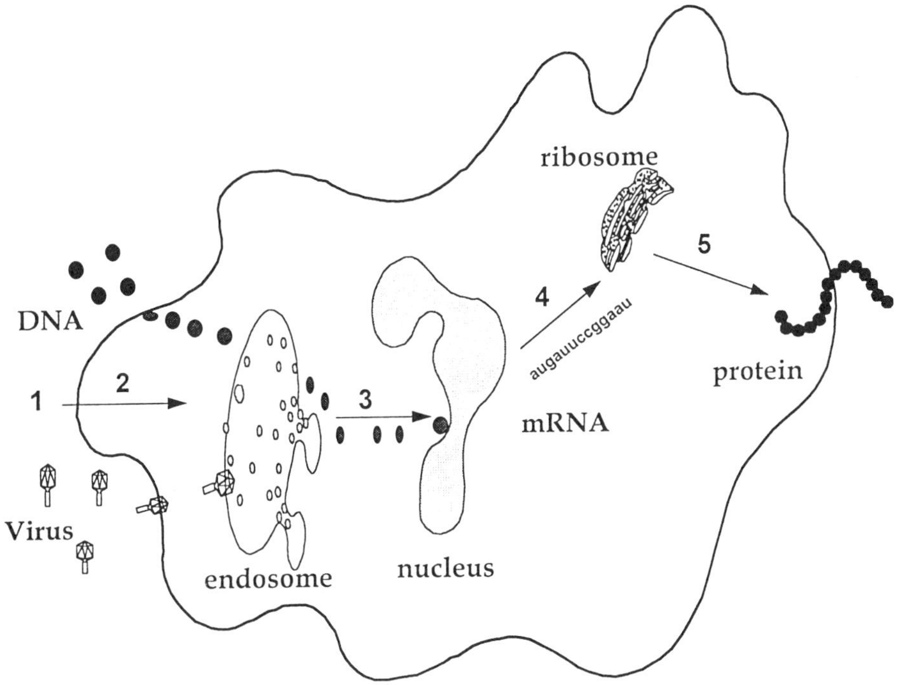

FIGURE 3.1. Steps in gene transfer. Gene transfer and gene therapy involve several key steps. Step 1, DNA must be delivered to the target site. This process can be accomplished by direct iv injection or by incubating DNA with cultures of target cells. The vehicle that delivers DNA to the target site is known as the DNA delivery substrate or vector, and it can be a liposome or a virus. Step 2, once the substrate comes in contact with the target it must successfully penetrate the outer plasma membrane and be internalized into low-pH endosomal complexes. If the substrate has no mechanism for disrupting the endosomal membrane most of the internalized DNA will be degraded. On the other hand if the substrate possesses fusogenic properties the endosomal disruption occurs and the DNA is released into the cytoplasm. Step 3, cytoplasmic DNA is transported to the nucleus and decoded to messenger RNA. Step 4, mRNA is then transported back to the cytoplasm to ribosomal sites where protein synthesis takes place. Step 5, recombinant protein must be processed and inserted into its site of action or secreted.

cell surface (*8*, *9*). Coprecipitation of DNA with calcium phosphate results in the formation of tight nuclease-resistant complexes. Studies have shown that even though much of the precipitated DNA is internalized only a small proportion is transported to the nucleus (*10*). Consequently, the efficiency of gene transfer by these methods is inadequate for therapeutic applications (*11*). In addition, DNA delivered by microprecipitates cannot be given systemically. Finally, DNA delivered by this route remains in an extrachromosomal or episomal state and is inefficiently inserted into the host genome (random insertion at extremely low efficiency, <0.001%, does take places). Episomal DNA, unless specifically designed, will not undergo replication, is not passed on to all daughter cells, and is eventually degraded. Therefore, this form of gene transfer results in transient recombinant gene expression. Two terms commonly used to measure the adequacy of gene transfer are transfection efficiency and transformation frequency. Transfection efficiency refers to the percentage of cells manifesting recombinant gene expression within 24–48 h after incubation with recombinant DNA. Transformation frequency refers to the percentage of cells displaying long-term recombinant gene expression (over 2 weeks) and is used as an estimate of insertional frequency.

The optimal DNA delivery vector would be one that accomplishes both high transfection and transformation efficiency. To accomplish these goals methods for improving cellular penetration and nuclear translocation have been developed. These methods are divided into two broad categories: liposome-mediated DNA delivery and recombinant viruses (retrovirus [RNA viruses] and adenovirus [DNA virus]). Both classes of delivery systems are approved for clinical use.

Methods for Introducing DNA into Cells

Liposome-Mediated DNA Delivery

Interest in using liposomes to deliver DNA evolved from the premise that their phospholipid composition would facilitate delivery of macromolecules by promoting membrane fusion. Moreover, they are noninfectious and biodegradable and can incorporate ligands for cell-specific targeting. Liposomes are classified according to charge (negative, cationic, or neutral), phospholipid composition, pH-sensitivity, and structure: unilamellar or multilamellar. Earliest liposome preparations were large unilamellar (LUV) or multilamellar negative-charged vesicles. Preparation is time consuming and employs organic solvents. Encapsidation efficiency (the proportion of entrapped aqueous-phase DNA) is variable and dependent on lipid composition and the method of preparation (*12*). LUV composed of cholesterol and phosphotidylserine (chol-PS) prepared by reversed-phase evaporation can encapsulate almost 50% of the DNA contained in an aqueous buffer (*13*). However, entrapment efficiency declines rapidly as DNA size increases (*14*). An additional drawback of liposomes is their inherent instability in aqueous solutions and membrane-induced leakage. Liposomes lacking cholesterol or PS were demonstrated to leak a significant amount of their contents when they come in contact with biological membranes (*15, 16*).

Despite these limitations the development of LUVs represented an important advancement in the field of DNA transfer. The earliest studies found that encapsulating infectious poliovirus particles into LUVs provided resistance against neutralizing antiserum. Furthermore, encapsulated poliovirus was capable of infecting poliovirus-resistant cells presumably by bypassing the need for receptor-mediated internalization and delivering virus directly to the cytoplasm (*17*). Subsequently, investigators demonstrated the capacity of LUVs to directly transfer poliovirus RNA or DNA to susceptible mammalian cells (*18, 19*). The efficiency of LUV-mediated viral DNA transfer (measured as the number of viral progeny or plaques per microgram of DNA) is dependent on the amount of encapsulated DNA, phospholipid composition of the LUV, and the culture conditions (*20*). Lysosomotrophic agents such as chloroquine enhance liposome-mediated infectivity presumably by inhibiting lysosome-mediated DNA degradation.

LUV–DNA complexes also have been used to transfer eukaryotic DNA to mammalian cells. Transfection and transformation frequencies are summarized in Table 3.1. In general, LUV–DNA complexes give mediate transfection efficiency compared with Ca^{2+} phosphate precipitation (10% vs. ≤2%, respectively). However, transfection efficiency varies considerably with cell type and culture conditions. Cells that are difficult to maintain in culture or have active lysosomal degradation tend to be more resistant. The frequency of stable transformation is also highly variable and depends on target-cell type and culture conditions (*21–23*). Typical transformation rates are 0.1–1% and are only marginally higher than microprecipitation techniques.

An important property of any clinically useful DNA delivery substrate is its compatibility with in vivo applications. The plasma clearance of LUVs following intravenous injection is strongly influenced by phospholipid composition and LUV size.

TABLE 3.1. Summary of Liposome-Mediated Gene Transfer

Liposome Type	*Cell Type*	*Transfection Efficiency (%)*	*Transformation Frequency (%)*	*Reference*
LUVs	C6 glial	10	NP	35
	Mouse LtK⁻	10	0.02	21
	Suspension FM3AtK⁻	NP	2	23
	LtK⁻ monolayer	NP	2	23
	Human HL60	NP	0.01	23
	CV-1	NP	0.001	20
	CV-1	NP	0.1	22
pH-Immunoliposomes	C6 glial	48	NP	22
	LtK⁻	48	2	33
	HeLa	<2	NP	44
RSVE	CV-1	20	NP	38
RSVE	LtK⁻	NP	0.08–0.1	37
HVJ-Liposomes	LtK⁻	95–100	1	42
	CHO	NP	0.2	42
	HeLa	NP	0.09	42
	LLCMK2	18	NP	41
	Rat VSMCs	20	NP	40
Cationic	Multiple cell types	25–30	0.001	43
	HeLa	5–20	NP	44
	Rat VSMCs	5	NP	40
	Myoblasts	8–15	NP	47
	Adult hepatocytes	10	NP	45

NOTE: NP, not reported; RSVE, reconstituted Sendai virus vesicles; HVJ-liposomes, human Sendai virus liposomes; CV-1, monkey cells; LtK, mouse cells; CHO, chinese hamster ovary cells; HeLa, human embryonic cells; VSMC, vascular smooth muscle cells.

LUVs high in negatively charged PS or sphingomyelin (SM) are associated with less leakage and deliver greater amounts of encapsulated material. Studies conducted in lipoprotein-deficient mice suggest that high-density lipoproteins (HDL) present in plasma are responsible for inducing leakage by extracting neutral phospholipid molecules and destabilizing the liposome bilayer (*24*). Negative-charged phospholipids appear to resist HDL-mediated extraction. Plasma clearance (circulation time) is also affected by liposome size. Liposomes exceeding 0.2 μm reduced circulation time owing to avid hepatic and splenic entrapment (*25*).

Following intravenous injection LUV-entrapped agents rapidly accumulate in the reticuloendothelial cells of the liver, and smaller quantities accumulate in spleen and bone (*26*, *27*). Kupffer cells appear to be the primary hepatocellular target, and lesser amounts of DNA localize to endothelial cells and hepatocytes. In one study intravenous injection of LUV–DNA complexes was used to deliver preproinsulin DNA to the liver of laboratory rats. Sequential measurement of blood glucose levels revealed the appearance of marginal and transient hypoglycemia (*26*). Even though analysis of subcellular fractions of hepatic tissue indicated the presence of functional preproinsulin DNA in hepatic endosomes, only 14% of the input DNA was estimated to localize to the nucleus, thus explaining the poor physiologic response to this treatment (*28*, *29*). These data highlight the requirement for efficient endosomal release and nuclear translocation. In summary, major limitations of traditional LUVs are the following:

- low transformation efficiency,
- absence of mechanisms for escaping lysosomal degradation and poor nuclear translocation,
- transient recombinant gene expression, and
- nontargeted delivery.

To overcome the limitations of traditional LUVs and increase the amount of cytoplasmic delivered DNA, new liposomes were designed that contained special elements capable of inducing disruption of the lysosomal membrane. This property was accomplished by incorporating acid-induced fusogenic elements into the phospholipid bilayer. On exposure to the low pH environment within the endosome these elements fuse with the endosomal membrane causing disruption and release of entrapped material into the cytoplasm. In one specific strategy dioleylphosphatidylethanolamine (DOPE), a synthetic phospholipid, was incorporated into the lipid bilayer in association with small amounts of fatty acid derivatives, such as palmitoyl-immunoglobulin (pIgG). Incorporation of pIgG imparts target-specific binding characteristics to the liposome by targeting antigen expressing cells. At neutral pH, DOPE is a stable element. However, at low pH this synthetic phospholipid undergoes a phase change and becomes fusogenic. Binding studies indicate that the affinity of these immunoliposomes for cells expressing the cross-reactive antigen is at least three orders of magnitude greater than corresponding nontargeted liposomes (*30*). Moreover, uptake of fluorescent macromolecules is specific for antigen-expressing cells and inhibited by lysosomotropic agents indicating internalization into endocytic vesicles. In comparison with traditional LUVs, pH-sensitive immunoliposomes have high target-cell specificity and deliver more intracellular macromolecules (*31*, *32*). Thus, by designing the phospholipid bilayer to contain acid-induced fusogenic components endosomal disruption and targeted delivery can be integrated into the same delivery vehicle.

pH-sensitive immunoliposomes are also capable of mediating highly efficient and cell-specific DNA delivery. Short-term transfection efficiency is significantly higher compared with LUVs (48% vs. 10%, *see* Table 3.1), whereas stable transformation is comparable (*33*). One problem with first generation pH-sensitive immunoliposomes is poor stability and substantial leakage on binding to the target cell. This problem was overcome by separating the phospholipid stabilizer from the targeting molecule (*34*). By using this design and different cell-specific targeting antibody, transfection efficiencies in C6 glial cells of 42% were obtained, compared with 10% for bare LUVs (*35*). Even though pH-sensitive immunoliposomes are efficient DNA vectors, they are highly toxic and difficult to prepare. Moreover, large quantities of DNA are required because the efficiency of encapsulating DNA is only 17–20%.

A second method for improving cytoplasmic delivery is to capitalize on the fusogenic properties of human Sendai virus (HVJ). Specific glycoproteins present in the HVJ viral envelope become fusogenic when exposed to weakly acidic environments (*36*). Several methods for incorporating HVJ glycoproteins into liposomal delivery vesicles have been developed. In one method, viral envelope glycoproteins and phospholipids are solubilized in non-ionic detergent and then isolated by centrifugation. DNA added to the solubilized envelope solution is entrapped within reformed vesicles when the detergent is extracted (*37*). Studies demonstrate that reconstituted Sendai virus envelopes (RSVE) are highly fusogenic and efficient DNA delivery substrates. Reported transfection efficiencies range between 3 and 20%, whereas the frequency of stable transformation is 0.08–0.8% (*38*).

In another method intact HVJ particles are conjugated directly to the liposome by including gangliosides (GS) within the phospholipid bilayer to act as receptors for HVJ particles. The presence of GS elements stabilizes and increases the affinity between HVJ particles and liposomes at neutral pH (*39*). DNA delivery by HVJ liposomes is extremely efficient in mouse and rodent cell lines. Transfection efficiencies of 90% and corresponding stable transformation rates of 0.1–1% have been reported. By contrast transfection efficiency drops to 15–20% for primary cells, such as monkey kidney cells and rat aortic smooth muscle cells, and a corresponding drop occurs in stable transformation (*40–42*).

All of the liposome preparations described are cumbersome to synthesize and require the use of highly toxic organic solvents. Recent developments in the synthesis of cationic lipids capable of spontaneously interacting with DNA in physiologic buffered solutions have greatly simplified liposome-mediated DNA delivery. The first such agents used a synthetic cationic lipid (DOTMA) in combination with equal amounts (wt/wt) of phosphatidylethanolamine (PtdEtn) to form stable liposomes (DOTMA/PtdEtn) (*43*). Association between plasmid DNA and DOTMA liposomes is nearly 100%. Moreover, these vesicles fuse to plasma membranes with extremely high efficiency. A large variety of cationic liposomes are now commercially available.

Many studies have evaluated the DNA delivery capacity of cationic liposomes. Transfection efficiencies range from 8 to 20% for hepatocytes, endothelial cells, glial cells, and myoblasts (*11, 44–47*). Moreover, these levels of transfection are achieved with less toxicity and shorter incubation times and require one-fifth of the DNA compared with other liposome-based delivery methods. In direct comparison to other transfection techniques, including pH-sensitive liposomes, cationic liposomes yield greater transfection efficiencies and higher recombinant gene expression (*44*). It has been proposed that charge neutralization and DOTMA-mediated collapse of DNA into small condensed structures that are more efficiently encapsulated and membrane-fusion competent account for the improved transfection efficiency (*48*). However, as with other liposome preparations the frequency of stable transformation is still ≤1%.

Following intravenous injection cationic liposome–DNA complexes localize to lung, spleen, heart, and liver cells (*49*). Furthermore, deleterious toxic effects on cardiac and hepatic function have not been observed (*50*). Because of their favorable organ distribution and low toxicity, cationic liposomes have been used to deliver DNA to lung epithelium, liver, spleen, and metastatic tumor nodules (*51–53*). In a recent clinical trail, direct injection of DNA–liposome complexes into metastatic melanoma nodules resulted in site-specific recombinant protein expression and a beneficial therapeutic outcome in one patient (*54*). Currently there are about 12 approved clinical trails employing cationic liposome–DNA complexes. Most of these studies are designed to stimulate cytotoxic T-cell responses to solid tumors by locally injecting the HLA-B27 gene directly into tumor masses. In addition, several trials are evaluating intratracheal administration of liposome–DNA complexes carrying normal CFTR sequences to patients with cystic fibrosis.

In summary, significant advancements in the design of liposome–DNA complexes have been made. Detailed analyses of lung, muscle, cardiac, and hepatic tissues show the absence of chromosomal integration. Thus, most of the currently available data suggests that liposome–DNA complexes display similar toxic profiles to conventional drugs. Unlike virus-mediated gene delivery, liposome–DNA complexes pose no significant biosafety hazard to the patient. In addition, cationic liposome preparations are relatively simple to use and widely available through numerous commercial sources. In laboratory practice liposomes are more efficient and easier to use than calcium or *O*-(diethylaminoethyl)dextran microprecipitates. Because transfection efficiency is variable, cell-type-dependent experimental conditions need to be empirically determined. Potential disadvantages of liposome–DNA complexes are low in vivo efficiency (because of limited contact time), the need for large amounts of DNA, and absence of a mechanism for efficient chromosomal integration.

Recombinant Retrovirus-Mediated DNA Delivery

Retroviridae comprise a large family of enveloped single-stranded RNA viruses. Intact retroviral particles average 70–100 nm in diameter and have an outer lipid envelope coated by viral glycoproteins.

The internal nucleocapsid is composed of retrovirus-encoded structural proteins and houses the RNA genome along with virus-encoded protease, integrase, and reverse transcriptase (RT). Traditionally, these viruses are grouped according to their pathogenicity and host range. Murine retroviruses are classified as *ecotropic* if they infect only mouse cells, *xenotropic* if they infect most mammalian cells except murine cells, and as *amphotropic* if they can infect both murine and nonmurine cells (*55*). Recombinant retroviruses used in human gene therapy are exclusively of murine origin. As part of their life cycle these viruses integrate their DNA into the host genome. This process leads to permanent transformation of the host cell and passage of the viral genome to all daughter cells. However, the ability to integrate and permanently alter the host genome also carries the risk of producing cancer or infectious disease causing mutations. Methods for impairing replication without inhibiting cellular penetration and integration had to be developed before these viruses could safely be used in clinical practice.

The ability to efficiently penetrate the cell membrane is an essential property of all viruses. Retroviruses are internalized by forming stable complexes with specific high affinity receptors on the cell membrane. A close correlation exists between infectivity and receptor density. Recently, cloning and sequencing of some of these receptors revealed them to be amino acid transport molecules, indicating that these viruses use host-cell nutritional pathways to gain entry to the cell (*56*). Virus-encoded envelope glycoprotein plays an important role in determining host range and successful internalization. The subunit glycoprotein (estimated weight of 70 kilodaltons [kDa]) functions as the primary receptor interacting component and antigenic determinant. Mutational analysis of subunit suggests that the N-terminal region influences host-cell range, a feature that has implications for designing cell-specific vectors. A second smaller transmembrane spanning protein (TM) is processed from a larger precursor molecule having an estimated size of 15 kDa. This larger protein is cleaved by a virus-specific protease into a smaller 16 amino acid R-fragment and the TM protein with a truncated molecular weight of 12 kDa (p12). Both p12 and R play a role in membrane fusion by mechanisms not yet fully elucidated (*57*). Stable receptor–viral complexes can be found in clathrin pits and endosomal vesicles. Once inside the endosome, acid-induced fusogenic properties of the viral envelope cause endosomal disruption and release of the nucleocapsid into the cytoplasm.

At the molecular genomic level, retroviruses contain 2 copies of identical strands of RNA (diploid number) averaging 7–10 kb in size. Chromosomal integration and generation of viral progeny requires converting the RNA template to a DNA provirus capable of chromosomal insertion. To accomplish this process, viral reverse transcriptase copies the RNA template into double-stranded DNA, a process that takes place in the cytoplasm. Viral DNA in association with capsid proteins and viral integrase is then transported to the nucleus. Even though the mechanism of integration is too detailed to review here, extensive research shows that more than one copy of viral DNA can integrate randomly at different sites and that specific sequences within the target DNA are not required (*58*). This research is relevant to gene therapy, because random integration potentially can cause activation of cellular oncogenes or inactivation of tumor-suppresser genes. Furthermore, recent experimental data indicate that insertion of multiple copies may result in early suppression of recombinant gene expression (*59*).

Once integrated proviral DNA state has been established the life cycle of the retrovirus is geared toward viral replication and generation of viral progeny. This process is illustrated in Figure 3.2. Proviral DNA contains four transcriptionally active genes called *gag*, *pro*, *pol*, and *env* (*pro* gene is not indicated in the figure). *Gag* gene products encode for at least 3 distinct peptides located in the viral capsid. *Pro* and *pol* gene products encode for viral proteases and integrase, and *env* gene products encode for envelope proteins subunit and TM. These genes are always arranged in the same order: *gag* being the most 5′, followed by *pro*,

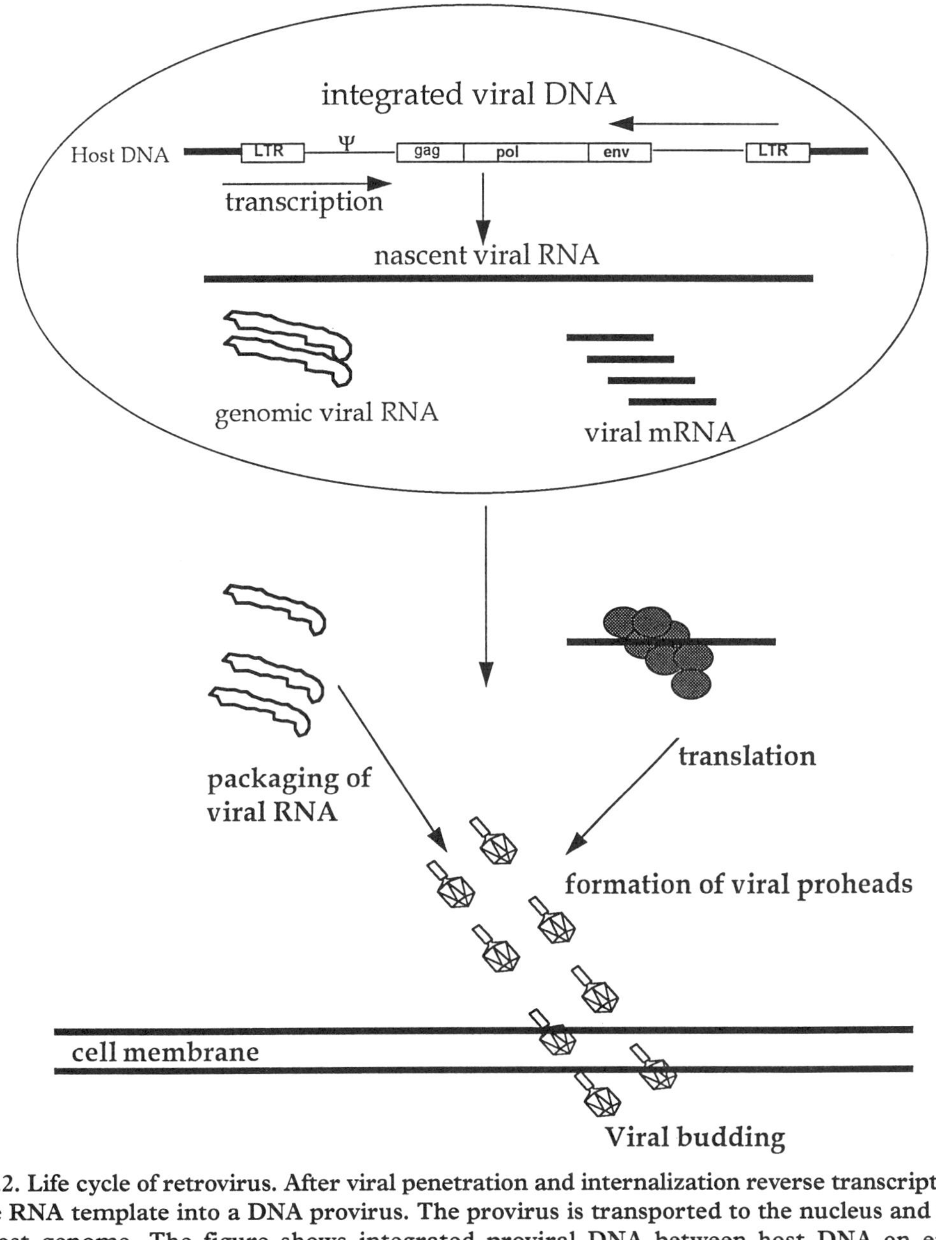

FIGURE 3.2. Life cycle of retrovirus. After viral penetration and internalization reverse transcriptase is used to copy the RNA template into a DNA provirus. The provirus is transported to the nucleus and integrated into the host genome. The figure shows integrated proviral DNA between host DNA on either side. Functionally the provirus can be divided into transcriptional regulatory units (LTR), an untranslated packaging domain (Ψ), and three translated viral genes: *gag, pol,* and *env.* Transcription is initiated from the LTR and can occur bidirectionally. Nascent viral RNA is synthesized and spliced into a genomic viral RNA (subsequently to be packaged) and viral mRNA, which is used to direct the synthesis of viral structural proteins. Newly synthesized genomic viral RNA is packaged into viral proheads. New virions are released by budding through the cell membrane during which time viral glycoproteins are inserted into the capsid. In addition a portion of the host membrane is also kept on the viral surface.

pol, and *env* at the 3′ end. Other functionally important but nontranslated regions include a packaging domain (Ψ) and flanking sequences (LTRs) that regulate transcriptional activity. Once the proviral state has been established, host-cell mechanisms are used to transcribe, translate, and process viral proteins. It has been estimated that nearly 10% of the host-cell synthetic capacity may be turned over to viral replication.

The life cycle of the virus continues with the synthesis of viral proteins and RNA. Through poorly understood mechanisms viral capsid proteins, integrase, protease, and viral RNA are transported to a site within the cytoplasm where viral assembly takes place. The capsid structure is assembled first into empty viral proheads lacking genomic RNA (*60*). Packaging of viral RNA is highly efficient, a characteristic suggesting that DNA sequences within the viral genome aid in discriminating between host cell and viral RNA. Deletion analysis of the viral genome indicated that the Ψ region located in the untranslated region just upstream to the *gag* domain must be present for efficient packaging of viral RNA. Proviral DNA lacking this region are inefficiently packaged although viral proteins are synthesized. In summary, several important features of this process should be recognized:

- Infection and gene transfer are dependent on specific receptors and escape from lysosomal degradation.
- Integration requires host-cell proliferation and viral integrase.
- Expression of heterologous DNA is dependent on integration.

Before retroviruses could be used as gene delivery substrates, a method for inserting recombinant foreign DNA into infectious viral particles had to be developed. Figure 3.3 illustrates the steps used to accomplish this goal. Initially a competent (wild-type) virus was used to rescue recombinant DNA into viral particles (*61, 62*). As shown in the figure the first step involved making a mutant replication-defective viral genome in which a recombinant drug-selectable marker gene was inserted in place of *gag, pol,* and *env* sequences, while leaving intact other elements such as LTRs and Ψ (pLpL). Deletion of these important structural genes renders the DNA replication deficient but retains the ability to transform drug-susceptible cells into drug-resistant cells by inserting into the host genome. Cells incorporating pLpL into their genomes will survive long-term exposure to selecting agents (in this case tissue culture medium containing 30 μM hypoxanthine, 1.0 μM amethopterin, and 20 μM thymidine [HAT-medium]). The surviving resistant colonies then can be selectively grown in culture. The drug-resistant marker gene (in this case hypoxanthine phosphoribosyltransferase [HPRT]) can be rescued into infectious viral progeny by adding a replication competent murine leukemic virus (MuLV) helper virus to provide a source of *gag, pol,* and *env* proteins. Rescue of recombinant DNA occurs because the helper virus supplies the necessary packaging and structural information to generate new virions. However, under these conditions two populations of viruses are generated: replication-competent helper virus and replication-defective recombinant virus. Furthermore, the titer of recombinant virus (number of colony [CFU] or plaque forming units [PFU] per milliliter of tissue culture medium) is usually 10 to 100 times lower than the competent virus. Thus, propagation of retroviral vectors with a helper virus contaminates the viral supernatants with large numbers of competent virus.

It quickly became the focus of several research groups to develop a system of packaging recombinant retrovirus vectors with expanded host range and free of contaminating wild-type viruses. This process is illustrated in Figure 3.4. The first step was to generate a cell line capable of constitutively producing essential viral packaging proteins, *gag, pol,* and *env.* This step was accomplished by making a mutant retrovirus genome lacking the critical Ψ region (pMOV-Ψ^-) (*63*). Viral DNA missing this region is incapable of being packaged into newly formed viral proheads. Analysis of pMOV-Ψ^- transformed 3 T3 murine cells revealed reverse

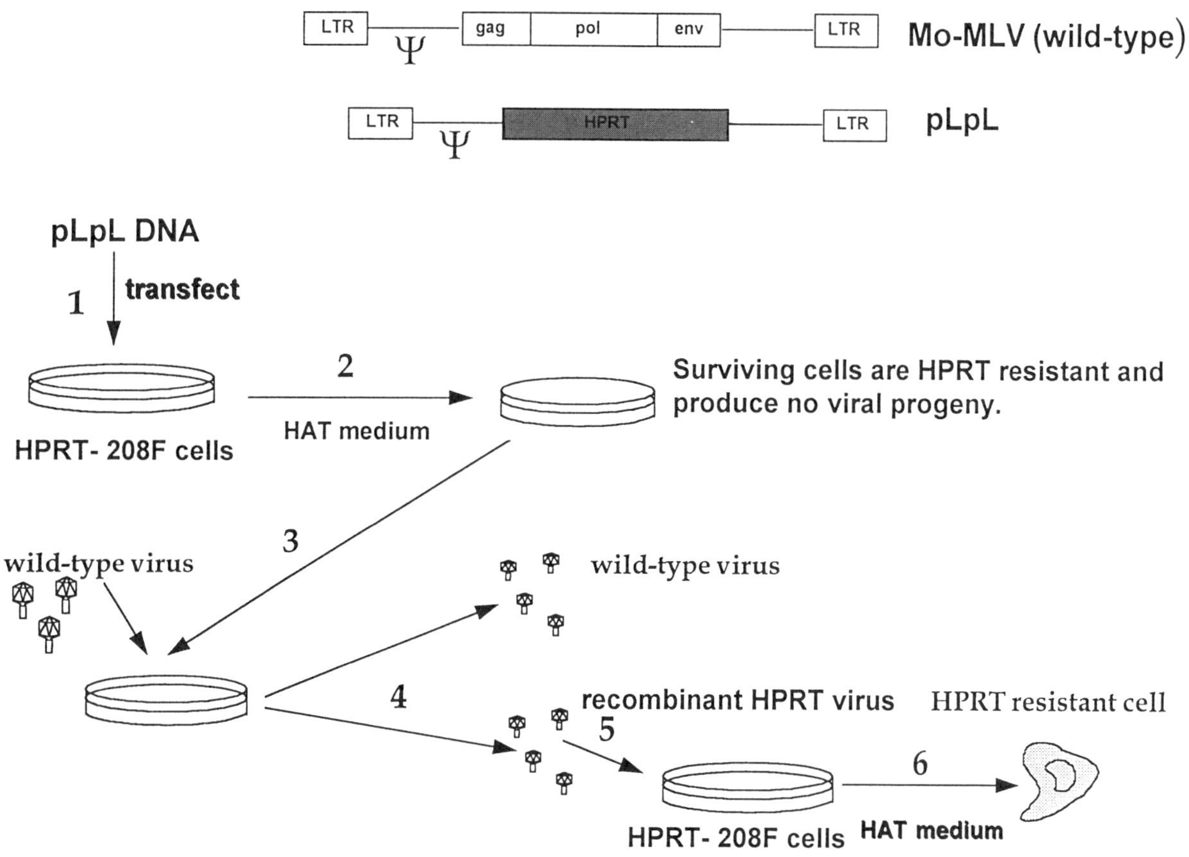

FIGURE 3.3. Rescue of recombinant DNA using helper virus. Wild-type viral *gag, pol,* and *env* sequences are removed and substituted by recombinant DNA encoding for human hypoxanthine phosphoribosyltransferase (HPRT). The new DNA (pLpL) is transfected into HPRT-susceptible cells. Application of growth medium containing 30 μM hypoxanthine, 1.0 μM amethopterin, and 20 μM thymidine [HAT-medium] selects for cells that have been transformed by pLpL and are therefore able to metabolize these substrates. Analysis of DNA from HAT-resistant cells indicates that pLpL had integrated into the genome as indicated in Figure 3.2. Note that while the packaging domain is intact, new virus particles are not formed because the structural genes *gag, pol,* and *env* are missing. By taking the HAT-resistant cells and exposing them to wild-type virus (Mo-MLV) the *gag, pol,* and *env* functions can be reconstituted. As a result new virions are produced, some of which contain the HPRT gene. Collection of medium conditioned by HAT-resistant cells producing new virus particles is able to confer HAT resistance to fresh plates of HPRT-208 F cells indicating the presence of retrovirus containing HPRT DNA.

transcriptase activity without the production of viral progeny, indicating pMOV-Ψ^- was indeed replication defective. To demonstrate that Ψ sequences could be supplied independently, a second mutant DNA containing the missing Ψ sequences and a drug-selectable marker gene (*E. coli* xanthine–guanine phosphoribosyltransferase, XGPRT) was constructed (pMSVgpt) and co-transfected with pMOV-Ψ^- into 3 T3 cells. Rescue of XGPRT DNA into viral progeny capable of transmitting drug resistance was determined by treating fresh plates of 3 T3 cells with culture

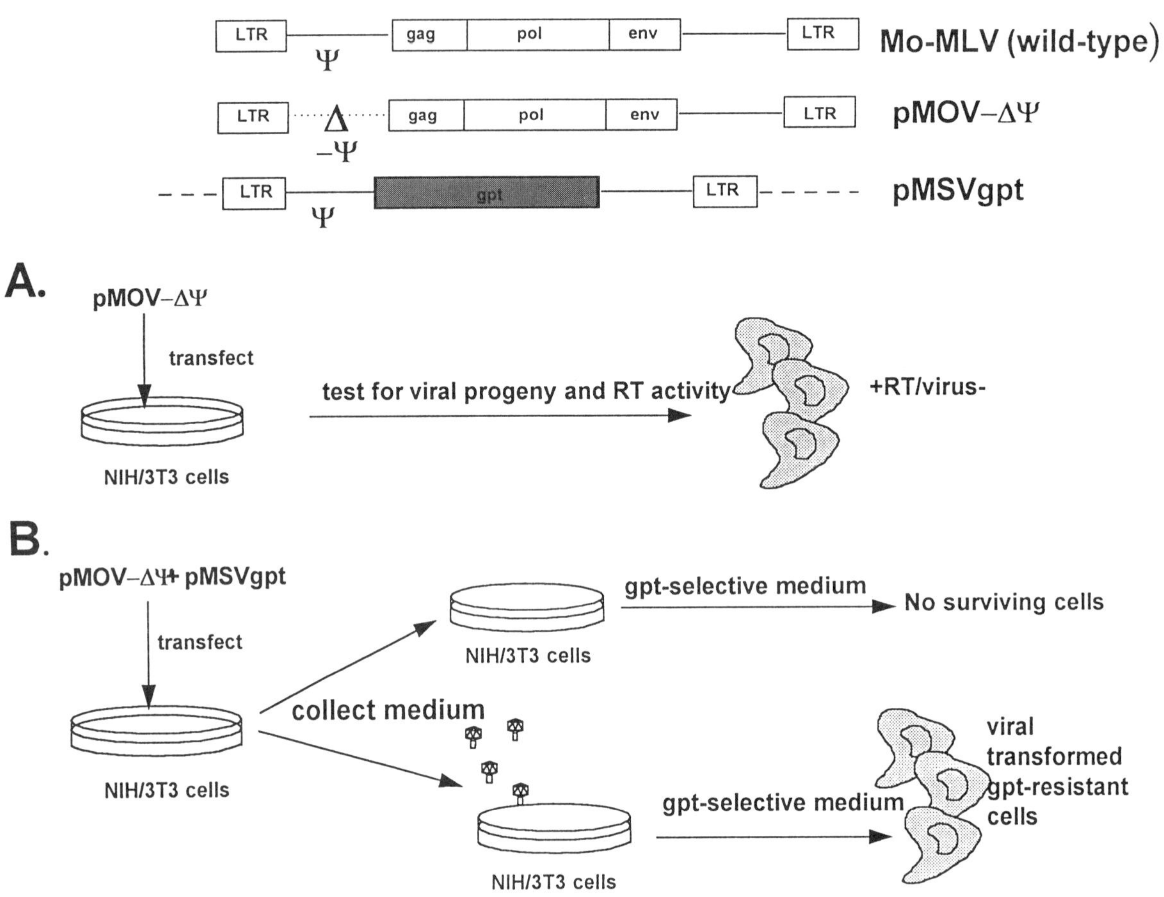

FIGURE 3.4. Helper-virus free packaging cell lines are necessary to avoid contamination with wild-type pathogenic virus. In this strategy the packaging domain, Ψ, is removed leaving the three structural genes, *gag, pol,* and *env,* intact. When NIH/3T3 cells are transfected by pMOV-ΔΨ transformed cells can be recognized by the production of reverse transcriptase (RT). However, because Ψ is missing new virions are not formed. To rescue recombinant DNA into retrovirus particles the missing Ψ domain must be supplied. This is accomplished by taking wild-type virus DNA and substituting the *gag, pol,* and *env* regions with a drug-resistance gene leaving Ψ intact (pMSVgpt). When NIH/3T3 cells are cotransfected with both DNAs (pMSVgpt + pMOV-ΔΨ) some cells will begin to produce recombinant virus capable of transferring GPT resistance to fresh cells.

medium collected from the cotransfected cells. The appearance of surviving cell in the presence of selective medium indicated the presence of viral particles capable of transmitting drug resistance to fresh plates of cells. Further analysis revealed the presence of wild-type virus, suggesting that a recombination event had taken place.

On the theory that recombination was random and infrequent these investigators reasoned that some of the original pMOV-Ψ^- transformed cells had not experienced recombination. To identify these cells pMOV-Ψ^- and a different drug-selectable DNA were cotransfected into fresh 3 T3 cells. After applying selective medium, surviving colonies were isolated and analyzed for the presence of viral proteins, such as reverse transcriptase, *gag*, *pol*, and *env*. Clones that were positive for these viral products were further shown by genomic analysis to have integrated the Ψ-deleted provirus. Supernatant from one of these clones (Ψ-2) was free of infectious particles showing recombination had not occurred. After transfecting these cells with pMSVgpt (a source of Ψ sequences) viral progeny capable of transferring XGPRT resistance were produced at a titer of 10^4 CFU/mL. This time contamination with helper viruses was not detected. These elegant experiments established the first helper-free packaging cell line (Ψ-2 cells) capable of producing defective recombinant ecotropic retrovirus free of wild-type viruses.

Extending the host range was achieved by substituting envelope sequence from the amphotropic virus 407 A for the original ecotropic *env* sequences, present in pMOV-$Ψ^-$ (*64*). By using the same strategy as described previously, a new cell line (Ψ-AM) producing moderate titers of amphotropic virus capable of infecting human and primate cells was identified. Several important features should be emphasized. First, the parent plasmid pMOV was derived from ecotropic murine sarcoma virus in which a single deletion (removal of Ψ) had been made. Thus, only a single recombination event would be necessary for emergence of wild-type virus. Given the high frequency of retrovirus recombinations and rearrangements these cells began to produce competent viruses after multiple passages.

Increasing the number of mutations directly affects the number of recombination events necessary to generate competent viruses (Figure 3.5). Currently approved U.S. Food and Drug Administration packaging lines are generated from proviral DNA containing at least three deletions or mutations. In the PA317 cell line the right-sided LTR is replaced by SV40 poly(A) signal sequences, the left-sided LTR is truncated, and the Ψ region is deleted (*65*). These modifications resulted in cell lines capable of producing high viral titers in the absence of contamination. Lastly, by mutating the remaining *gag* sequences in the rescuing vector the risk of contamination has been virtually eliminated.

Many factors determine the transduction efficiency of retrovirus-mediated DNA delivery. These factors relate to the density and type of retroviral receptors present on the target cell, the mitotic rate of the target cell, the method of transduction (in vivo vs. in vitro), and the type and titer of retrovirus (*66–68*). Slowly proliferating cells having few receptors are poor targets for retroviral transduction. A commonly determined parameter is the multiplicity of infection (MOI) and relates viral titer to transduction efficiency. The MOI is calculated by determining the transduction efficiency at a given level of input virions per cell. For example, 50 mL of viral supernatant containing 1×10^6 CFU/mL would be required to transduce 5×10^6 cells at an MOI of 10. Therefore, the need for high MOIs reflects poor transduction efficiency.

The method of viral transduction is also an important determinant of DNA delivery efficiency. In vitro transduction is the most efficient because high MOIs and prolonged contact time between vector and target cell can be maintained. For example, hematopoietic stem cells can be

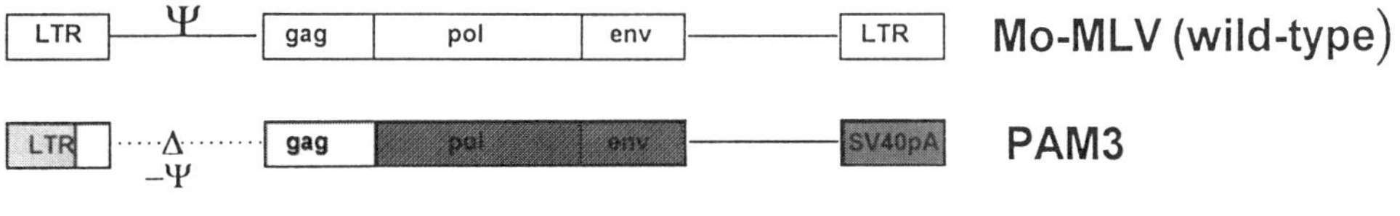

FIGURE 3.5. Genomic structure of a commonly used packaging vector. PAM3 packaging DNA is shown with respect to wild-type DNA. The downstream LTR has been replaced by a heterologous SV40 polyadenylation sequence, the upstream LTR has been truncated, Ψ is deleted, and the *gag* gene has been mutated.

transduced by cocultivating bone marrow harvests with retroviral producer cells, followed by long-term culturing and periodic replenishing of vector producing cells. Progenitor stem cells present in marrow extracts are the preferred targets, because these cells give rise to all the blood elements: neutrophils, lymphocytes, platelets, and red blood cells. Recombinant DNA that integrates is passed down and permanently marks all cells deriving from that progenitor.

Initial experiments with murine (*61*, *69*), canine (*70*), and human bone marrow (*71*) were disappointing because of inefficient and often variable transduction. Depending on the stem cell lineage (granulocyte-macrophage progenitors or erythroid CFUs) efficiencies varied from 0.3 to 25%. Recently, significantly higher transduction efficiencies (40–70%) were reported by using high titer producer cells (10^7–10^8 CFU/mL) coupled with procedures for inducing expression of retrovirus receptors and long-term maintenance of progenitor cells in culture (*67*, *72–76*). Reinfusion of bone marrow transduced in this way results in long-term detection of recombinant DNA in all blood formed elements. By marking progenitor stem cells with a reporter gene the cellular origin of leukemic relapses was demonstrated to arise from residual leukemic cells and not new mutations, a result implying that more intense chemotherapy to purge these cells could result in improved leukemic-free survival times (*77*). In this regard the nontherapeutic application of gene transfer was used to assist in furthering our understanding of this disease.

The observation that retroviral infectivity can be blocked by anti-gp70 antibodies suggested that host range specificity could be redirected to specific populations of cells. Current strategies for accomplishing this process include modifying the viral envelope with specific ligands or using recombinant techniques to produce a chimeric viral particle with altered host range susceptibility. The modification approach has been used to redirect infection of ecotropic MuLV to human cells, normally resistant to this virus, by using a streptavidin bridge to couple biotinylated anti-gp70 antibody to biotinylated antihistocompatibility complexes (MHC I and II) (*78*). In this configuration anti-gp70 antibodies block natural infectivity by forming stable complexes with viral envelope gp70, whereas the MHC complexes are used to reconstitute uptake to antigen expressing cells. By using this method, human-derived HeLa cells that are normally resistant to ectropic retrovirus can be transduced with an efficiency of 0.4% at MOIs of 0.1–1. Higher transduction efficiencies can be achieved if anti-MHC class II antibodies are used as the targeting ligand and the cells are treated with interferon to upregulate the density of class II surface receptors. Other redirecting ligands such as biotinylated-EGF (epidermal growth factor) have also been used (*79*). In another technique ecotropic murine virus is modified chemically with sodium cyanoborohydride in the presence of lactose, a sugar taken up through liver specific asialoglycoprotein receptors (*80*). Previously unsusceptible HepG2 (human liver cells) can be infected with an efficiency of 35% by these modified viruses.

Theoretically the gp70 protein or other envelope constituents can be replaced by heterologous proteins. This process was suggested after it was observed that pseudotype RSV virions (Rous sarcoma virus) containing the RSV genome encapsidated by vesicular stomatitis virus (VSV) glycoproteins could be generated by superinfecting chronically infected RSV cells with VSV (*81*). Further experiments demonstrated that the host range of murine leukemia virus (Mo-MuLV) could be extended by substituting the wild-type *env* gene with sequences encoding for avian or Gibbon ape leukemia virus envelope protein. Such chimeric particles can be produced in high titer and can infect a wide range of cells including human cells. Pseudotyped gibbon envelope viruses have been shown to accomplish high transduction levels to human bone marrow progenitor cells and hepatocytes (*82*). Moreover, the envelope protein need not be of retroviral origin. For example, influenza virus hemagglutinin (HA) sequences can be substituted for RSV *env* sequences to produce pseudotype virions containing RSV RNA encapsidated by HA envelope proteins (*83*). Chimeric RSV genes containing the entire HA coding sequence fused in frame to the RSV signal sequence (sHA) or replaced by the RSV *env* transmembrane and cytoplasmic domains (sHAa) can be synthesized. In sHAa, a chimeric protein is synthesized in which the extracellular ligand domain of HA is anchored to the viral envelope by wild-type RSV transmembrane and cytoplasmic domains. sHA and sHAa chimeric virions can infect both avian and human cells with comparable efficiency. In another method pseudotyped Mo-MuLV derived retroviral vectors containing the G glycoprotein of vesicular stomatitis virus were used to achieve gene transfer efficiencies of nearly 60% in rat hepatocytes (*84*).

Adenovirus-Mediated DNA Delivery

Another class of viruses that have found their way into clinical gene therapy are adenoviruses (Ad). These viruses are nonenveloped isometric double-stranded DNA viruses, having a diameter of 70–90 nm and a molecular weight of 170×10^6. The viral genome consists of a single linear double-stranded DNA molecule containing about 35,000 base pairs divided into 100 map units (mu). About 33 human and nearly 24 mammalian (nonhuman) serotypes have been identified. Human Ad are highly trophic and capable of infecting a wide variety of different cell types.

Several properties of Ad make them very attractive as gene therapy vectors. First, human Ad is not associated with malignant transformation in human. Second, penetration and replication is extremely efficient and does not require active host-cell proliferation. Therefore, nonproliferating target cells such as hepatocytes, neural tissue, and vascular cells can be efficiently transfected by Ad vectors. This characteristic contrasts to the limited tropism of retroviruses and the requirement for ongoing cellular proliferation to achieve integration. Third, Ad are easily propagated and crude viral supernatants can be concentrated to high titer (in the range of 10^{12} particles/mL) by cesium chloride banding. However, Ad have no mechanism for inserting their DNA into the host-cell chromosome.

Another property of Ad is its ability to escape the endosomal compartment and enter the cytoplasm with high efficiency. Studies have demonstrated that nearly 80% of input virus is internalized into receptosomes within minutes of target cell contact (*85*). The low pH of the receptosome alters the configuration of the outer capsid penton proteins resulting in rupture of the endocytic vesicle and release of large numbers of intact viral particles into the cytoplasm. Virus particles, which have now partially shed their outer protein coat, are transported through nuclear pores into the nucleus. It has been estimated that nearly 40% of input virus particles enter the nucleus. Compared with retroviruses, adenoviruses have simple life cycles (*86*). First, reverse transcription from viral RNA to proviral DNA is not required. Second, adenovirus DNA remains episomal. Therefore, Ad replication is a transient phenomenon culminating in lysis and death of the host cell with release of a large number of newly formed infectious particles. Consequently, adenovirus-mediated gene transfer and expression of recombinant protein is also transient.

The replicative cycle of adenovirus can be divided into an immediate early transcriptional phase (E) and a late replicative phase (L), corresponding to early, early intermediate, and late phase transcriptional units (Figure 3.6). Early phase viral genes are further subdivided into E1A, E1B, E2a, E2b, E3, and E4 regions. Kinetic analysis indicates that viral RNA abundance increases rapidly after infection and in parallel with a 75-kDa DNA-binding protein. Temperature-sensitive mutations within the 75-kDa protein gene suggest that this protein is required for efficient viral DNA synthesis (*87*). Additional studies on the replicative properties of E1-deleted virus indicate that at high multiplicities of infection the kinetics of viral production are delayed but proceed at normal, albeit lower, levels. These findings suggest that the

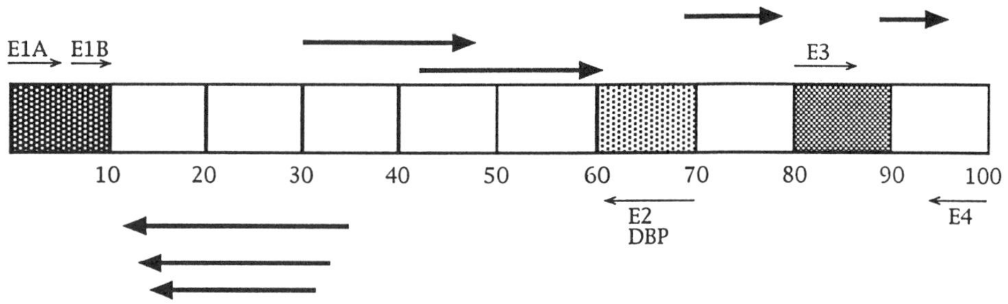

FIGURE 3.6. Structure of the adenovirus genome. Shown above is the genomic map of Ad-5. It consists of 36,000 base pairs of double-stranded linear DNA divided into 100 map units (mu). Approximate locations of early genes, E1, E2, E3, and E4 are shown as thin arrows. Note that transcription is bi-directional and the 72-kDa DNA-binding protein is transcribed from the E2 region. Intermediate and late gene elements are depicted as thick arrows. The filled boxes represent transcriptional elements that can be deleted or mutated for the purpose of making recombinant adenovirus. Most Ad-vectors are E1A/E1 B deleted. Foreign DNA and heterologous promoter elements can be inserted into this region. These vectors are replication deficient. To increase the size of recombinant DNA that can be packaged additional deletions in the E3 region can be made. However, E3-delete vectors are replication competent and therefore must be combined with E1-deletions. Temperature-sensitive mutations within the E2 region have been incorporated into the design of second-generation vectors. At the nonpermissive temperature, E2 region transcription is impaired and the amount of DBP synthesized is significantly inhibited. Use of these vectors appears to be associated with reduced immunogenic reactions as described in the text.

E1 region is responsible for inhibiting host-cell mechanisms for repressing viral transcription. Thus, the most likely function of immediate early phase (E1) transcription is to promote viral RNA synthesis from E2 and E3 regions. For the most part viral transcription proceeds sequentially from E1 to E4, and then on to late transcription (L1). Packaging of adenovirus takes place within the cytoplasm and new virions are released through lysis of the host cell. Adenovirus particles are nonenveloped and do not contain any host-cell proteins in their outer coat.

About 105% of viral genome can be packaged efficiently into virion particles. Thus, nearly 2 kb of extra DNA can be inserted without comprising viral packaging. However, such vectors are replicative competent. For human gene therapy applications, methods for encapsidating and propagating replication-defective virions had to be developed. Because deletion of the E1 region impairs viral propagation, a method for providing E1 function independent of the viral genome had to be developed. This adaptation was done by transfecting human embryonic kidney cells with E1 region DNA and selecting transformed cells (labeled 293 cells) that were constitutively expressing E1 proteins (*88*). E1-deleted adenovirus can be propagated efficiently in these cells, indicating that E1 protein function can be supplied in trans—similar to retroviral packaging cells described previously. By deleting the E1 region the amount of heterologous DNA that can be packaged is increased to nearly 5 kb. Additional deletions in the E3 region further increase vector capacity to nearly 7 kb. However, deletions within the E3 region do not affect viral replication. Therefore, all clinically approved vectors are E1-deleted, although additional E3 deletions may be included, depending on the size of recombinant DNA requiring packaging.

Recombinant replication-defective adenovirus can be generated by in vitro ligation of recombinant DNA to the full-length viral genome followed by transfection into permissive 293 cells or by rescuing foreign genes into the viral genome through a process known as homologous recombination

(*89*). In our laboratory, homologous in vivo recombination routinely is used to rescue foreign genes into Ad5 genome. In this process foreign DNA is inserted into a bacterial plasmid containing a small portion of the left side of the viral genome. The resulting new plasmid is then cotransfected with a larger plasmid containing an overlapping portion of the left-side sequences and the remaining right-handed sequences into subconfluent monolayers of 293 cells (liposomes are used). With the onset of a cytopathic effect, conditioned medium is collected and viral DNA is analyzed by polymerase chain reaction (PCR) for the presence of inserted DNA (*90*). Conditioned medium testing positive by PCR is analyzed further by plaque assays on 293 cells. Several plaques are isolated, retested for the presence of inserted DNA by PCR and restriction digestion, and finally purified so that a pure population of recombinant virus is obtained.

Many in vitro and in vivo studies demonstrate the high gene transfer efficiency of E1-deleted adenovirus vectors. Depending on the MOI, transfection efficiencies above 70% are obtainable in hepatic cells, lung epithelium, vascular endothelium, vascular smooth muscle, neural cells, cardiac and skeletal myocytes, and synovial sites (*91–94*). In our hands transfection efficiency to human vascular smooth muscle cells is nearly 100% when MOIs of >50 are used (Figure 3.7). Following intravenous injection adenovirus primarily localizes to the liver. By using this technique, cholesterol clearance in mice was accelerated rapidly by Ad5 LDL-R vectors (*95*). About 90% of the liver cells are transfected and express LDL-R. However, the duration of recombinant gene expression is short lived.

The greatest limitations to adenovirus-mediated gene transfer are immunogenic responses to viral proteins and the transient nature of recombinant gene expression. Several reasons for the limited duration of recombinant gene expression have been postulated. First, adenovirus-delivered DNA remains episomal and susceptible to cellular degradation. Second, intravenous injection of E1-deleted vectors stimulates production of neutralizing circulating antibodies (*96*). These antibodies are serotype specific and severely impair subsequent gene transfer from identical serotyped vectors. Third, E1-deleted vectors stimulate a significant cell-mediated immune reaction. Delivery of E1-deleted vectors to lung epithelium produces a pneumonia like illness in laboratory rats (*97*) and results in flu-like illness and the development of diffuse but self-limiting patchy infiltrates on chest roentgenograms in humans (*5*). These data suggest that immunomediated reactions play an important role in limiting the duration of Ad-mediated recombinant gene expression. Further support for this hypothesis comes from experiments demonstrating that gene transfer to bronchial xenografts and immunosuppression significantly prolong the duration of recombinant gene expression (*98*).

On the basis of these data it has been postulated that background viral transcription stimulates cytotoxic T-cell responses aimed against infected cells. These cells are destroyed and replaced by nontransduced cells. As a result the pool of recombinant expressing cells is reduced rapidly. Presumptive evidence in support of this theory has been shown by using ad vectors that have temperature-sensitive mutations in the E2 region (*99*). These vectors mediated significantly prolonged recombinant expression in the absence of immunosuppression. Detailed molecular analysis revealed that at the nonpermissive temperature (body temp) transcription of the 75-kDa DNA binding protein is significantly impaired and background viral transcription is reduced. Thus, less viral antigens are available for stimulating T-cell responses.

Applications

An important issue is the ability to use gene therapy to demonstrate relevant physiologic and therapeutic endpoints. In this respect studies in animal models have been encouraging and exciting. In one approach, transduction of intrahepatic tumors in rats with a suicide gene (retrovirus-mediated

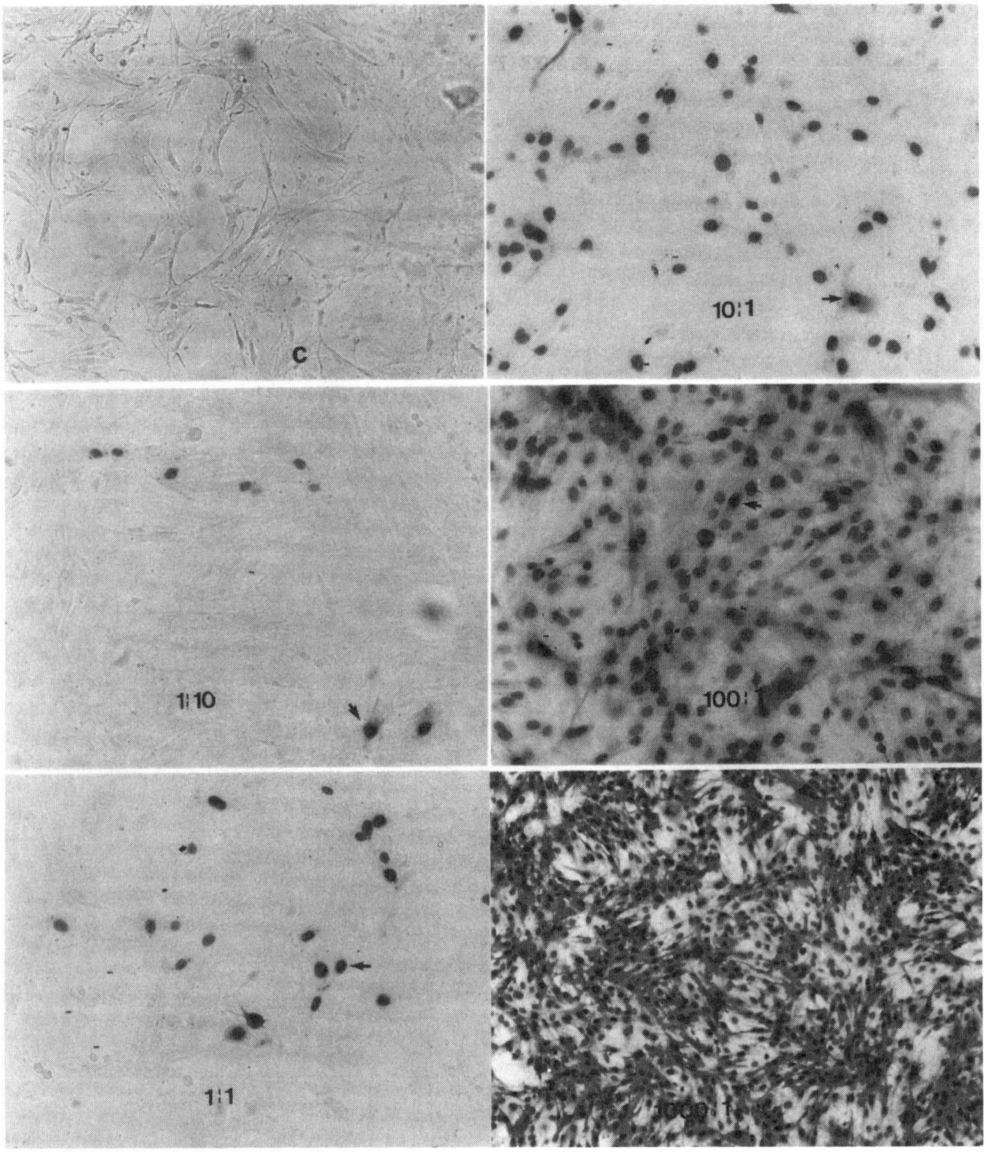

FIGURE 3.7. Example of high-efficiency gene transfer to vascular smooth muscle cells. Human vascular smooth muscle cells obtained from the aorta of patients undergoing coronary bypass were grown in tissue culture and incubated with recombinant E1-deleted adenovirus containing the *E. coli* β-galactosidase gene (AdLacZ) linked to a nuclear translocation signal peptide. Cells were incubated for 1 h at the indicated multiplicity of infections (infectious units per cell). Twenty-four hours later the medium was washed out, and the cells were fixed with glutaldehyde (0.05%) and then incubated for 4 h with chromogenic substrate, X-gal (5-bromo-4-chloro-3-indolyl-β-D-galactopyranoside). Cells expressing recombinant β-galactosidase can be identified by their blue nuclei. C, control SMCs transduced with E1-deleted adenovirus not containing the LacZ element, note the absence of blue nuclei. As the MOI is increased from 1:10 to 100:1 the percentage of blue staining cells increases. At an MOI of 100:1 all the cells are expressing β-galactosidase.

delivery) was demonstrated to reduce tumor mass by 60-fold compared with controls (*100*). In this example, the herpes simplex virus thymidine kinase gene (HSVTK) was used to confer sensitivity to nucleoside analogs such as ganciclovir (GA). Cells expressing the HSVTK gene are able to phosphorylate these analogs into triphosphates that can then be incorporated into DNA during cell division causing chain termination and premature cell death. What is interesting about this study was the lack of correlation between transduction efficiency and regression of tumor mass. It has now been established that the efficacy was partly due to a "bystander effect" in which phosphorylated GA could be transferred to neighboring cells, even though these cells had not incorporated the HSVTK gene into their genome. Similarly, adenovirus-mediated delivery of a suicide gene has been used to induce regression of experimental gliomas in mice and to inhibit arterial balloon-injury induced neointimal hyperplasia (*101*, *102*). In these studies transfection efficiency and degree of tumor regression or inhibition of neointimal hyperplasia were not correlated, results reflecting the bystander effect.

In other strategies, bystander effects do not contribute to physiologic outcome and a relationship between transduction efficiency and treatment efficacy exists. However, correction of metabolic defects rarely requires 100% replacement. For example, hemophilic B patients require only 1–5% of active factor IX to prevent bleeding. Thus, injection of as little as 1×10^9 PFU of recombinant adenovirus containing factor IX into the tail vein of mice leads to sufficient production of hepatic factor IX to reach plasma levels that would be therapeutic for hemophilia B patients (*103*).

Extrapolation of these exciting results to human outcomes has been more difficult to demonstrate. For example, children who have undergone gene therapy for SCID still require administration of recombinant ADA. Therefore, the efficacy of gene therapy as monotherapy in this case has not been established. Studies are underway to determine if these children can be safely weaned off ADA or can tolerate lower doses without infectious complications. Another application of human gene therapy has been in the treatment of familial hypercholesterolemia (*7*). In this pilot study partial hepatectomies were performed on affected patients. The hepatocytes were isolated and transduced in tissue culture (ex vivo) with recombinant retrovirus containing normal low-density lipoprotein receptor (LDL-R) DNA. Hepatocytes were then reinfused through a catheter placed in the portal vein (autologous transplantation). Unfortunately, the extent of LDL reduction and its duration where clinically not significant. One of the reasons for this poor result may be due to low transduction efficiency of human hepatocytes by retroviruses. Analysis of transduced hepatocytes revealed long-term expression of functional LDL-Rs, indicating that the limitations of this initial trail could be overcome by developing methods for improving hepatocyte transduction frequency.

Summary and Future Directions

This chapter focused on the development and application of liposomes and recombinant retroviruses as DNA delivery vectors. Specifically, important developments in pH-sensitive liposomes, cationic liposomes, and recombinant viruses were discussed. We also explored the development of methods to achieve cell-specific DNA transfer by incorporating ligands into the design of the vector. Two fundamental points should be evident. First, effective DNA delivery requires efficient internalization and subsequent escape from lysosomal degradation. If these conditions are not satisfied internalized DNA is rapidly degraded and results in ineffective expression of recombinant protein. Second, the nature of the vector determines the way in which delivered DNA is processed by the cell. For retroviral delivery, provirus integration is an important corollary resulting in permanent mutation of the host genome. Even though currently approved retroviruses are murine derived and have never been associated with malignant transformation in humans, the high frequency of errors that can occur during reverse transcription, replication, or RNA synthesis could lead to mutations and recombinations in the viral genome. Whether a truly "safe" vector exists will only be determined once significant numbers of patients have been treated and followed for many years.

In little more than 14 years the field of gene transfer has moved from a laboratory science to clinical applications. Much remains to be done, such as the design and testing of better ways of regulating and directing expression of recombinant gene products. New vectors, such as recombinant adenovirus, recombinant adenoassociated virus, papillomavirus, and herpesvirus are being designed and tested. Equally important is the development of better in vivo delivery systems to target. Continued research into the molecular and genetic origins of human disease is most important. Only when we fully understand these components can the full weight of molecular medicine be brought to bear.

References

1. Cline, M. J.; Stang, H.; Mercola, K.; Morse, L.; Rupricht, R.; Browne, J.; Slaser, W. *Nature (London)* **1980,** *284,* 422–425.
2. Hammer, R. E.; Palmiter, R. D.; Brinster, R. L. *Nature (London)* **1984,** *311,* 65–67.
3. Anderson, W. F. *Hum. Gene Ther.* **1994,** *5,* 1431–1432.
4. Blaese, R. M.; Culver, K. W.; Chang, L.; Anderson, W. F.; Mullen, C.; Nienhuis, A.; Carter, C.; Dunbar, C.; Leitman, S.; Berger, M. *Hum. Gene Ther.* **1993,** *4,* 521–527.
5. Crystal, R. G.; McElvaney, N. G.; Rosenfeld, M. A.; Chu, C. S.; Mastrangeli, A.; Hay, J. G.; Brody, S. L.; Jaffe, H. A.; Eissa, N. T.; Danel, C. *Nat. Genet.* **1994,** *8,* 42–51.
6. Chowdhury, J. R.; Grossman, M.; Gupta, S.; Chowdhury, N. R.; Baker, J. R.; Wilson, J. M. *Science (Washington, D.C.)* **1991,** *254,* 1802–1805.
7. Grossman, M.; Raper, S. E.; Kozarsky, K.; Stein, E. A.; Engelhardt, J. F.; Muller, D.; Lupien, P. J.; Wilson, J. M. *Nat. Genet.* **1994,** *6,* 335–341.
8. Chen, C.; Okayama; H. *Mol. Cell Biol.* **1987,** *7,* 2745–2752.
9. Sussman, D. J.; Milman, G. *Mol. Cell Biol.* **1984,** *4,* 1641–1643.
10. Loyter, A.; Scangos, G. A.; Ruddle, F. H. *Proc. Natl. Acad. Sci. U.S.A.* **1982,** *79,* 422–426.
11. Ray, J.; Gage, F. H. *BioTechniques* **1992,** *13,* 598–603.
12. Mannino, R. J.; Fogerite-Gould, S. *BioTechniques* **1988,** *6,* 682–689.
13. Szoka, F.; Paphadjopoulous, D. *Annu. Rev. Biophys. Bioeng.* **1980,** *9,* 467–508.
14. Mannino, R. J.; Allebach, E. S.; Strohl, W. A. *FEBS Lett.* **1979,** *101,* 229–232.
15. Mayhew, E.; Rustum, Y. M.; Szoka, F.; Papahadjopoulos, D. *Cancer Treat. Rep.* **1979,** *63,* 1923–1928.
16. Kirby, C.; Gregoriadis, G. *Life Sci.* **1980,** *27,* 2223–2230.
17. Wilson, T.; Papahadjopoulos, D.; Taber, R. *Proc. Natl. Acad. Sci. U.S.A.* **1977,** *74,* 3471–3478.
18. Wilson, T.; Papahadjopoulos, D.; Taber, T. *Cell* **1977,** *17,* 77–84.
19. Papahadjopoulos, D.; Wilson, T.; Taber, T. *In Vitro* **1980,** *16,* 49–54
20. Fraley, R.; Straubinger, R. M.; Rule, G.; Springer, E. L.; Papahadjopoulos, D. *Biochemistry* **1981,** *20,* 6978–6987.
21. Schaefer-Ridder, M.; Wang, Y.; Hofschneider, P. H. *Science (Washington, D.C.)* **1982,** *215,* 166–168.
22. Rizzo, W. B.; Schulman, J. D.; Mukherjee, A. B. *J. Gen. Virol.* **1983,** *64,* 911–919.
23. Itani, T.; Ariga, H.; Yamaguchi, N.; Tadakuma, T.; Yasuda, T. *Gene* **1987,** *56,* 267–276.
24. Senior, J.; Gregoriadis, G.; Mitropoulos, K. A. *Biochim. Biophys. Acta* **1983,** *760,* 111–118.
25. Gregoriadis, G.; Senior, J.; Wolff, B.; Kirby, C. *Ann. N.Y. Acad. Sci.* **1985,** *446,* 319–340.
26. Nicolau, C.; Le Pape, A.; Soriano, P.; Fargette, F.; Juhel, M. F. *Proc. Natl. Acad. Sci. U.S.A.* **1983,** *80,* 1068–1072.
27. Eichler, H. G.; Senior, J.; Stadler, A.; Ganer, S.; Pfundner, P.; Gregoriadis, G. *Eur. J. Clin. Pharmacol.* **1988,** *34,* 475–479.
28. Cudd, A.; Nicolau, C. *Biochim. Biophys. Acta* **1985,** *845*(3), 477–491.
29. Nandi, P. K.; Legrand, A.; Nicolau, C. *J. Biol. Chem.* **1986,** *261*(35), 16722–16726.
30. Huang, A.; Kennel, S. J.; Huang, L. *J. Biol. Chem.* **1982,** *258,* 14034–14040.

31. Conner, J.; Huang, L. *J. Cell Biol.* **1985,** *101,* 582–588.
32. Ho, R. J. Y.; Rouse, B. T.; Huang, L. *Biochemistry* **1986,** *25,* 5500–5506.
33. Wang, C. Y.; Huang, L. *Biochemistry* **1989,** *28,* 9508–9514.
34. Pinnaduwage, P.; Huang, L. *Biochemistry* **1992,** *31,* 2850–2855.
35. Holmberg, E. G.; Reuer, Q. R.; Geisert, E. E.; Owens, J. L. *Biochem. Biophys. Res. Commun.* **1994,** *201,* 888–893.
36. Novick, S.; Hoekstra, D. *Proc. Natl. Acad. Sci. U.S.A.* **1988,** *85*(20), 7433–7437.
37. Vainstein, A.; Razin, A.; Graessmann, A.; Loyter, A. *Methods Enzymol.* **1983,** *101,* 492–512.
38. Loyter, A.; Vainstein, A.; Graessmann, M.; Graessmann, A. *Exp. Cell Res.* **1983,** *143,* 415–425.
39. Klappe, K.; Wilschut, J.; Nir, S.; Hoekstra, D. *Biochemistry* **1986,** *25,* 8252–8260.
40. Morishita, R.; Gibbons, G. H.; Kaneda, Y.; Ogihara, T.; Dzau, V. J. *Hypertension (Dallas)* **1993,** *21,* 894–899.
41. Kato, K.; Nakanishi, M.; Kaneda, Y.; Uchida, T.; Okada, Y. *J. Biol. Chem.* **1991,** *266*(6), 3361–3364.
42. Kaneda, Y.; Uchida, T.; Kim, J.; Ishiura, M.; Okada, Y. *Exp. Cell Res.* **1987,** *173,* 56–69.
43. Felgner, P.; Gadek, T. R.; Holm, M.; Roman, R.; Chan, H. W.; Wenz, M.; Northrop, J. P.; Ringold, G. M.; Danielsen, M. *Proc. Natl. Acad. Sci. U.S.A.* **1987,** *84,* 7413–7417.
44. Legrendre, J. Y.; Szoka, F. C. *Pharmacol. Res.* **1992,** *9,* 1235–1242.
45. Jarnigan, W. R.; Debs, R. J.; Wang, S. S.; Bissell, D. M. *Nucleic Acids Res.* **1992,** *20,* 4205–4211.
46. Brigham, K. L.; Meyrick, B.; Christman, B.; Berry, L. C.; King. G. *Am. J. Resp. Cell Mol. Biol.* **1989,** *1,* 95–100.
47. Albert, N.; Tremblay, J. P. *Transplant. Proc.* **1992,** *24,* 2784–2786.
48. Gershon, H.; Ghirlando, R.; Guttman, S. B.; Minsky, A. *Biochemistry* **1993,** *32,* 7143–7151.
49. Zhu, N.; Liggitt, D.; Debs, R. *Science (Washington, D.C.)* **1993,** *261,* 209–211.
50. Nabel, E. G.; Gordon, D.; Yang, Z. Y.; Xu, L.; San, H.; Plautz, G. E.; Wu, B. Y.; Gao, W.; Huang, L.; Nabel, G. J. *Hum. Gene Ther.* **1992,** *3,* 649–656.
51. Yoshimura, K.; Rosenfeld, M. A.; Nakamura, H.; Scherer, E. M.; Pavirani, A.; Lecoq, J.-P.; Crystal, R. G. *Nucleic Acids Res.* **1992,** *20,* 3233–3240.
52. Leibiger, B.; Leibiger, I.; Sarrach, D.; Zuhlke, H. *Biochem. Biophys. Res. Commun.* **1991,** *174,* 1223–1231.
53. Plautz, G. E.; Yang, Z.; Wu, B.; Gao, X.; Huang, L.; Nabel, G. J. *Proc. Natl. Acad. Sci. U.S.A.* **1993,** *90,* 4645–4649.
54. Nabel, G. J.; Nabel, E. G.; Yang, Z. Y.; Fox, B. A.; Plautz, G. E.; Gao, X.; Huang, L.; Shu, S.; Gordon, D.; Chang, A. E. *Proc. Natl. Acad. Sci. U.S.A.* **1993,** *90,* 11307–11311.
55. Coffin, J. M. In *Virology,* 2nd ed.; Fields, B. N., Ed.; Raven Press: New York, 1990; pp 1437–1497.
56. Wang, H.; Kavanaugh, M. P.; North, R. A.; Kabat, D. *Nature (London)* **1991,** *352,* 729–731.
57. Ragheb, J. A.; Anderson, W. F. *J. Virol.* **1994,** *68,* 3220–3231.
58. Swain, A.; Coffin, J. M. *Science (Washington, D.C.)* **1992,** *255,* 841–845.
59. Berg, P. E.; Sheffery, M.; King, R. S.; Gong, Y.; Anderson, W. F. *Exp. Cell Res.* **1987,** *168,* 376–388.
60. Hunter, E. *Semin. Virol.* **1994,** *5,* 71–83.
61. Joyner, A.; Keller, G.; Phillips, R. A.; Bernstein, A. *Nature (London)* **1983,** *305,* 556–558.
62. Miller, A. D.; Jolly, D. J.; Friedmann, T.; Verma, I. M. *Proc. Natl. Acad. Sci. U.S.A.* **1983,** *80,* 4709–4713.
63. Mann, R.; Mulligan, R. C.; Baltimore, D. *Cell* **1983,** *33,* 153–159.
64. Cone, R. D.; Mulligan, R. C. *Proc. Natl. Acad. Sci. U.S.A.* **1984,** *81,* 6349–6353.
65. Miller, A. D.; Buttimore, C. *Mol. Cell. Biol.* **1986,** *6,* 2895–2902.
66. Wu, J. Y.; Robinson, D.; Kung, H. J.; Hatzoglou, M. *J. Virol.* **1994,** *68,* 1615–1623.
67. Crooks G. M.; Kohn, D. B. *Blood* **1993,** *82,* 3290–3297.
68. Springett, G. M.; Moen, R. C.; Anderson, S.; Blaese, R. M.; Anderson, W. F. *J. Virol.* **1989,** *63,* 3865–3869.
69. Williams, D. A.; Lemischka, I. R.; Nathan, D. G.; Mulligan, R. C. *Nature (London)* **1984,** *310,* 476–480.
70. Stead, R. B.; Kwok, W. W.; Storb, R.; Miller, A. D. *Blood* **1988,** *71,* 742–747.
71. Hock, R. A.; Miller, A. D. *Nature (London)* **1986,** *320,* 275–277.
72. Bodine, D. M.; McDonagh, K. T.; Brandt, S. J.; Ney, P. A.; Agricola, B.; Byrne, E.; Nienhuis, A. W. *Proc. Natl. Acad. Sci. U.S.A.* **1990,** *87,* 3738–3742.

73. Bodine, D. M.; Karlsson, S.; Nienhuis, A. W. *Proc. Natl. Acad. Sci. U.S.A.* **1989,** *86,* 8897–8901.
74. Schuening, F. G.; Strob, R.; Stead, R. B.; Goehle, S.; Nash, R.; Miller, A. D. *Blood* **1989,** *74,* 152–155.
75. Sorrentino, B. P.; Brandt, S. J.; Bodine, D.; Gottesman, M.; Pastan, I.; Cline, A.; Nienhuis, A. W. *Science (Washington, D.C.)* **1992,** *257,* 99–102.
76. Xu, L.; Stahl, S. K.; Dave, H. P.; Schiffmann, R.; Correll, P. H.; Kessler, S.; Karlsson, S. *Exp. Hematol. (Charlottesville, Va.)* **1994,** *22,* 223–230.
77. Brenner, M. K.; Rill, D. R.; Moen, R. C.; Krance, R. A.; Heslop, H. E.; Mirro, J., Jr.; Anderson, W. F.; Ihle, J. N. *Ann. N.Y. Acad. Sci.* **1994,** *716,* 204–214.
78. Roux, P.; Jeanteur, P.; Piechaczk, M. *Proc. Natl. Acad. Sci. U.S.A.* **1989,** *86,* 9079–9083.
79. Julan-Estienne, M.; Roux, P.; Carillo, S.; Jeanteur, P; Piechaczyk, M. *J. Gen. Virol.* **1992,** *73,* 3251–3255.
80. Neda, H.; Wu, C. H.; Wu, G. Y. *J. Biol. Chem.* **1991,** *266,* 14143–14146.
81. Weiss, R. A.; Boettiger, D.; Murphy, H. M. *J. Virol.* **1977,** *76,* 808–825.
82. von Kalle, C.; Kiem, H. P.; Goehle, S.; Darovsky, B.; Heimfeld, S.; Torok-Storb, B.; Storb, R.; Schuening, F. G. *Blood* **1994,** *84,* 2890–2897.
83. Dong, J.; Roth, M. G.; Hunter, E. *J. Virol.* **1992,** *66,* 7374–7382.
84. Yess, J. K; Miyanohara, A.; Laporte, P.; Bouic, K.; Burns, J. C.; Friedmann, T. *Proc. Natl. Acad. Sci. U.S.A.* **1994,** *91,* 9564–9568.
85. Greber, U. F.; Willets, M.; Webster, P.; Helenius, A. *Cell* **1993,** *75,* 477–486.
86. Horwitz, M. S. In *Virology,* 2nd ed.; Fields, B. N., Ed.; Raven Press: New York, 1990; pp 1679–1722.
87. Brough, D. E.; Groguett, G.; Horwitz, M. S.; Klessig, D. F. *Virology* **1993,** *196,* 269–281.
88. Graham, F. L.; Smiley, J.; Russell, W. C.; Nairn, R. *J. Gen. Virol.* **1977,** *36,* 59–74.
89. Graham, F. L.; Prevec, L. *Methods Mol. Biol. (Totowa, N.J.)* **1991,** *7,* 109–127.
90. Zhang, W. W.; Fang, X.; Branch, C. D.; Mazur, W.; French, B. A.; Roth, J. A. *BioTechniques* **1993,** *15,* 868–872.
91. Lemarchand, P.; Jaffe, A. H.; Danel, C.; Cid, M. C.; Kleinman, H. K.; Perricaudet-Stratford, L. D.; Perricaudet, L.; Pavirani, A.; Lecocq, J. P. *Proc. Natl. Acad. Sci. U.S.A.* **1992,** *89,* 6482–6486.
92. Mastrangeli, A.; Danel, C.; Rosenfeld, M. A.; Perricaudet-Stratford, L.; Perricaudet, M.; Pavirani, A.; Lecocq, J. P.; Crystal, R. G. *J. Clin. Invest.* **1993,** *91,* 225–234.
93. Quantin, B.; Perricaudet, L. D.; Tajbakish, S.; Mandel, J. L. *Proc. Natl. Acad. Sci. U.S.A.* **1992,** *89,* 2581–2584.
94. Roessler, B. J.; Allen, E. D.; Wilson, J. M.; Hartman, J. W.; Davidons, B. L. *J. Clin. Invest.* **1993,** *92,* 1085–1092.
95. Herz, J.; Gerard, R. D. *Proc. Natl. Acad. Sci. U.S.A.* **1993,** *90,* 2812–2816.
96. Zabner, J.; Petersen, D. M.; Puga, A. P.; Graham, S. M.; Couture, L. A.; Keyes, L. D.; Lukason, M. J.; St. George, J. A.; Gregory, R. J.; Smith, A. E.; Welsh, M. J. *Nat. Genet.* **1994,** *6,* 75–83.
97. Ginsberg, H. S.; Horswood, R. L.; Chanock, R. M.; Prince, G. A. *Proc. Natl. Acad. Sci. U.S.A.* **1990,** *87,* 6191–6195.
98. Engelhardt, J. F.; Yang, Y.; Stratford-Perricaudet, L. D.; Allen, E. D.; Kozarsky, K.; Perricaudet, M.; Yankaskas, J. R.; Wilson, J. M. *Nat. Genet.* **1993,** *4,* 27–34.
99. Yang, Y.; Nunes, F. A.; Berencsi, K.; Gonczol, E.; Engelhardt, J. F.; Wilson, J. M. *Nat. Genet.* **1994,** *7,* 362–369.
100. Caruso, M.; Panis, Y.; Gagandeep, S.; Houssin, D.; Salzmann, J.; Klatzman, D. *Proc. Natl. Acad. Sci. U.S.A.* **1993,** *90,* 7024–7028.
101. Chen, S. H.; Shine, H. D.; Goodman, J. C.; Grossman, R. G.; Woo, S. L. C. *Proc. Natl. Acad. Sci. U.S.A.* **1994,** *91,* 3054–3057.
102. Ohno, T.; Gordon, D.; San, H.; Pompili, V. J.; Imperiale, M. J.; Nabel, G. J.; Nabel, E. G. *Science (Washington, D.C.)* **1994,** *265,* 781–785.
103. Smith, T. G. A.; Mehaffey, M. G.; Kayda, D. B.; Saunders, J. M.; Yei, S.; Trapnell, B. C.; McClelland, A.; Kaleko, M. *Nat. Genet.* **1993,** *5,* 397–402.

4
Targetable Polymeric Drugs

Kazunori Kataoka

A widespread consensus exists that efficacy of drug targeting is mainly owing to the properties of drug carriers. This chapter focuses on the promising features of water-soluble polymers as carriers used for drug targeting. Considerable discussion is made on the physicochemical properties as well as biological properties of targetable polymeric drugs, and earlier works done in the field are examined. Advantageous features of polymeric drugs for targeting therapy are summarized for modulated biodistribution (passive targeting) and modulated cellular uptake and retention (active targeting). Critical issues in polymeric drugs are also discussed in this chapter, emphasizing the importance of tailoring novel polymers to overcome present problems in targeting therapy.

D*rug targeting* is defined as a concept of delivering an adequate amount of drug to the target site in the body compartment at an appropriate time. More than 2000 years ago, the Greek philosopher Hippocrates described an idea of drug targeting in his literary work. Then, in the beginning of the 20th century, P. Ehrlich embodied this idea in his concept of *magic bullets* (*1*), consisting of haptophore (binding component to the target) and toxophore (cytotoxic part). As a haptophore, he proposed the use of an antibody. Nevertheless, it took another 70 years for drug targeting to become an active area of research, arousing the interest of many people.

When we look back on the history of drug targeting, several epoch-making discoveries are noticeable in the period of the 1960s to 1970s. On one hand, Ehrlich's concept of using an antibody as haptophore was experimentally proven to work in the treatment of several tumors inoculated into experimental animals. Toxophores used in these systems were radiotherapeutic agents (radioactive iodine) (*2*, *3*), cytotoxic drugs (e.g., chlorambucil) (*4*), and diphtheria toxin (*5*).

On the other hand, it was discovered that drug uptake by cells could be modulated by conjugating the drug to appropriate polymeric carriers. Polymers follow a different cellular uptake mechanism than low-molecular-weight analogs, and they are introduced into the cell neither by permeation through the plasma membrane nor by means of transport proteins. Polymers are internalized into the cellular compartment by a process called endocytosis (*6*). Endocytosis begins with enclosure of the polymer by a part of the plasma membrane to form an intracellular vesicle.

Copyright © 1997 American Chemical Society

The endocytic vesicles with ingested substances (polymers) are transferred to organelles called endosomes, whence they are eventually carried to lysosomes. In the acidic milieu of lysosomes (pH ~5), substances suffer digestion by lysosomal enzymes (e.g., acid hydrolases). Drugs conjugated to polymeric carriers by a bond cleavable by lysosomal enzymes should then be liberated from the carrier in the lysosomal compartment, followed by the penetration of the drug into the cytoplasm through lysosomal membranes. Compounds that are taken up selectively into the lysosomes are called lysosomotropic. An impermeable drug can be sent into cells through the lysosomal pathway by conjugating the drug to a polymeric carrier with lysosomotropic character (a Trojan horse strategy). This concept of the lysosomotropic chemotherapy was established by DeDuve in 1974 (*7*). In this approach, an ideal conjugate is stable and pharmacologically inactive in the circulation but, after cellular internalization through endocytosis, becomes active by hydrolysis in the lysosomes. Finally, in 1976, a great breakthrough in hybridoma technology was achieved by Kohler and Milstein to obtain monoclonal antibodies, powerful haptophore components of magic bullets (*8*).

Water-soluble polymers were first applied in the biomedical field as plasma expanders. The following list summarizes major water-soluble polymers that were investigated as plasma expanders:

- poly(glutamic acid),
- poly[N^5-(2-hydroxyethyl)-L-glutamine] (PHEG),
- β-poly(2-hydroxyethyl aspartamide),
- β-poly(aspartyl hydrazide),
- dextran,
- poly(vinyl alcohol),
- polyvinylpyrrolidone, and
- poly[*N*-(2-hydroxypropyl)methacrylamide].

In a rational manner these candidate plasma expanders became used as starting components to design polymer–drug conjugates (polymeric drugs). A sophisticated model of polymeric drugs was made by Ringsdorf (*9*) in the mid 70s (Figure 4.1). In this model, he married the concepts of magic bullets and lysosomotropic drugs to design polymeric drugs consisting of components with different functionalities such as solubilizer, pharmacon, and transport systems (homing device and nonspecific resorption enhancer). Since then, this model has been widely accepted as the guideline for the design of various kinds of targetable polymeric drugs.

Another important aspect of drug targeting is the modulation of distribution and disposition of drugs by binding with appropriate carrier systems. After the administration of polymeric drugs via the intravenous (iv) route, their transport in the body compartment is principally governed by diffusion and convection. Even if we install effective homing moieties in the conjugate, these moieties are of no use unless the conjugates reach the vicinity of the target cell to ensure close contact.

Let us take an example of solid tumors. Tumor mass is located outside of the vasculature. Therefore, after the process of extravasation, polymeric drugs meet the tumor mass. How is it possible for the polymeric drugs to penetrate into the interstitial tissue of the tumor from the blood compartment? To obtain an answer to this question, we should have a proper understanding of the unique feature of the tumor vasculatures.

It is now well accepted that, because of the activation of the kinin-generating cascade and the secretion of vascular permeability factor, blood capillaries at tumorous tissues develop in a considerably high density with enhanced permeability due to the loose interendothelial junctions. This process leads to an enhanced passive transport of macromolecular substances, such as proteins and polymeric drugs, across the blood vessel into the interstitial space of the tumorous tissues. On the other hand, the development of a lymphatic drainage system is insufficient in tumorous tissues, resulting in a poor tissue drainage of macromolecular substances. Consequently, macromolecular substances may exhibit a considerable accumulation in the tumor due to the synergistic effect of the increased vascular permeability and the decreased tissue drainage.

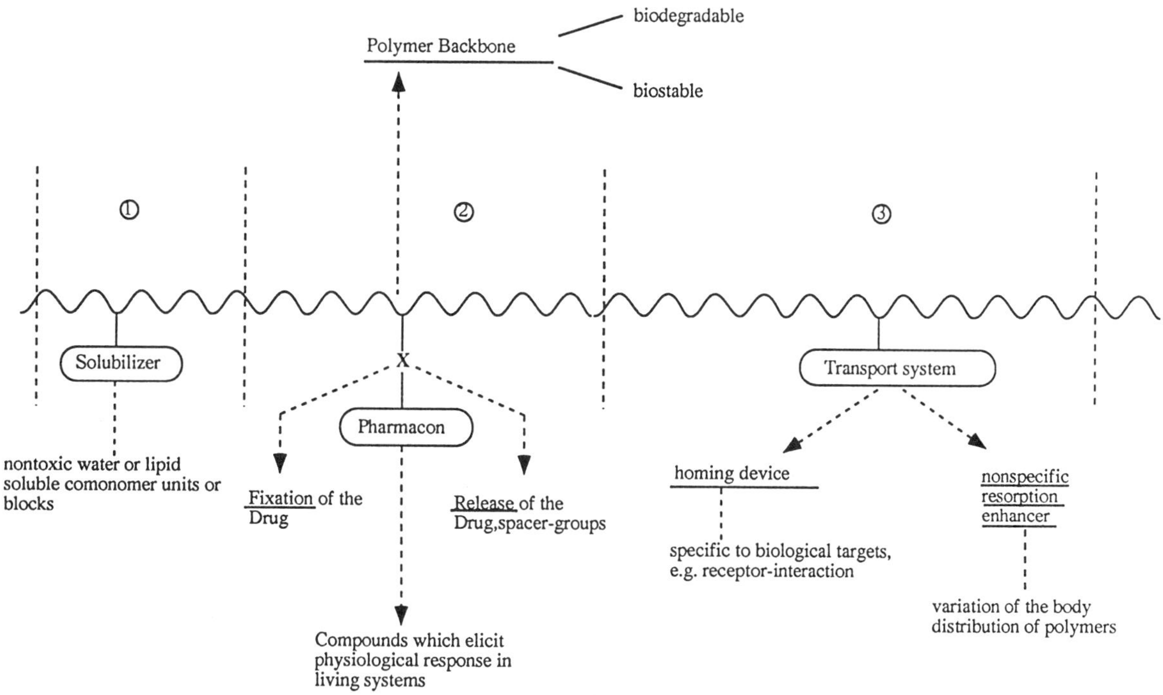

FIGURE 4.1. Model for targetable polymeric drugs.

This effect has been systematically studied by the Maeda group at Kumamoto University, Japan, and is termed "enhanced permeability and retention" (EPR) effect (*10*, *11*). It has now become one of the major guiding principles in drug targeting using polymeric carries. Macromolecular accumulation due to the enhanced permeability of blood vessels is also expected to take place at the site of inflammatory reactions, as a result of several causes, including microbial infections (*11*).

Obviously, the EPR effect is a great advantage of using polymeric drugs for targeting. Low-molecular-weight analogs readily suffer from glomerular excretion from the bloodstream, whereas polymeric drugs are expected to achieve a prolonged half-life in the bloodstream because of decreased glomerular excretion. Obviously, extended circulation is a requisite for sufficient EPR effect. However, an optimum molecular weight exists in terms of achieving the most efficient accumulation by the EPR effect. This characteristic is due to a decreased rate of permeation of high-molecular-weight compounds through inter-endothelial junctions of the blood vessels. Furthermore, macromolecular carriers in the bloodstream may be recognized by a group of scavenger cells, namely the reticuloendothelial system (RES). These cells are located at such organs as liver, spleen, and lung. RES recognition is particularly serious for colloidal and vesicular carrier systems, including latices and liposomes. Thus, it is essential to develop appropriate carrier systems that achieve long circulation in the bloodstream avoiding glomerular excretion and RES recognition, allowing for a significant EPR effect.

Research on the targetable polymeric drugs using synthetic polymers as well as natural macromolecules (including antibodies) as carriers grew in the period of the mid 70s to 80s. Recently, several clinical successes have been reported, which certainly provide a firm basis for this area of research. The field of drug targeting is of a multidisciplinary nature, and thus progress in the past several decades resulted from the concomitant achievements in related fields, including development of novel routes in polymer synthesis, establishment of pharmacokinetics, finding of novel cellular markers, clarification of intracellular trafficking, and progress in hybridoma and gene technologies.

The major facet of this chapter is to overview the progress in targetable polymeric drugs from the standpoint of polymer chemistry. Stimulating reviews on this topic from various aspects are already available (*12–16*). Readers are encouraged to access these review articles to gain more insights into this area of research.

Advantageous Features of Polymeric Drugs

Modified Biodistribution

A major purpose of conjugating drugs to polymer carrier is to modulate the disposition of drugs in the body, a process that allows high therapeutic efficacy with low adverse side effects. In this section, the biodistribution of polymeric drugs will be featured, and these drugs will be compared with their low-molecular-weight analogs.

Let us consider the distribution pathways of drugs administered into the body compartments through different routes. Drug dosages are separated into two major categories: local administration and systemic administration. Systemic administration involves the following (*17*):

- transdermal administration,
- transmucosal administration,
- peroral administration,
- intramuscular injection,
- intraperitoneal (ip) injection,
- iv injection, and
- intra-arterial (ia) injection.

The movement pathway of drugs in the body alters with the administration route. Yet, in this section, a major focus will be given to the fate of drugs administered by the iv route. To increase the targeting efficiency of drugs through the iv route, it is essential to decrease nonspecific removal of drugs by such organs as kidney (urinary excretion) and liver. Furthermore, elimination through RES, which is located in the liver and the spleen, is quite significant for polymer and vesicular carrier systems. When the target site is located outside of the capillaries (e.g., solid tumors), the efficacy of extravasation of carrier systems also crucially influences drug accumulation at target sites.

An obvious effect expected for drug conjugation to polymeric carriers is the extended blood half-life due to decreased kidney excretion. This effect is of course not always the case because biodistribution is influenced by the chemical structure of the polymers. The threshold molecular weight for glomerular excretion is approximately 40,000–70,000 in the case of neutral water-soluble polymers. The molecular weight of serum albumin, the most abundant plasma protein, is 66,500.

Detailed data on the threshold molecular weight for glomerular excretion were reported for several polymeric plasma expanders. For example, Seymour et al. (*18*) carried out extensive research on the glomerular excretion of poly(*N*-(2-hydroxypropyl)methacrylamide) (PHPMA) polymers with varying molecular weights, which were prepared by the radical polymerization of HPMA followed by fractionation using gel filtration chromatography (*18*). The threshold molecular weight (M_w) of PHPMA was determined to be approximately 45,000.

As expected, the threshold M_w for glomerular excretion varies with the conformation of the polymer chain in solution. Dextran has lower flexibility than neutral and water-soluble vinyl polymers and has a rather high threshold M_w of ~70,000 in terms of its glomerular excretion (*19*). The value of renal clearance for dextran with M_w of 70,000 is about 1/100 of low-molecular-weight analogs. Of interest, introduction of a small amount of anionic groups (e.g., carboxylate) into dextran molecules resulted in a further decrease in the renal clearance by 25% of neutral dextran, a result indicating that a slight modification in the chemical structure of macromolecules leads to a considerable change in their disposition in blood compartments when their molecular weight is critical in terms of the glomerular filtration (*19*). Thus, significant prolongation of plasma half-life may be achieved by an appropriate modification of polymer to decrease the rate of renal clearance.

Glomerular filtration is also significant for proteins with a relatively small molecular weight. Generally, proteins with M_w lower than 40,000 show a very short plasma half-life because of rapid renal clearance (*11*). This problem becomes serious when we intend to use peptides and enzymes of relatively low molecular weight for therapeutic purposes. A promising approach to solve this problem is to increase the apparent molecular weight of protein molecules by conjugation with water-soluble polymers. Urinary excretion of soybean trypsin inhibitor (M_w ~20,000) can be significantly reduced by the conjugation with water-soluble polymers including poly(ethylene glycol) (PEG) and dextran (*20*).

However, prolonged circulation is not always obtained for drugs and proteins through the conjugation with polymers. As will be described in detail in a later section, *Undesirable Change in Biodistribution Due to Drug Binding*, an increase in hydrophobicity or introduction of cationic charges in the carrier significantly alters the disposition characteristics of conjugates accumulating specifically in the liver, resulting in rapid clearance from the bloodstream. In terms of prolonged circulation, conjugation to a hydrophilic polymer with neutral or slightly anionic character generally gives good results. On the other hand, conjugates showing high accumulation in the liver can be designed by introducing moieties (cationic groups, galactose, etc.) having high affinity for liver cells.

Among the diverse applications of targetable polymeric drugs, the treatment of solid tumors is one in which intensive research has been devoted. As described in the preceding section, macromolecular compounds with considerable half-lives in the bloodstream tend to accumulate into tumor sites by the EPR effect (*21*). Reasons for the hyperpermeability of tumor capillaries are given in detail in another review (*11*). Briefly, tumor cells are known to produce a protein called vascular permeability factor to stimulate endothelial cell growth and to enhance the permeability of vasculature, allowing the transport of nutrients that are essential for tumor cell growth. Consequently, this increased permeability results in the accumulation of therapeutic compounds with a relatively high molecular weight, and these compounds usually tend not to extravasate into the interstitial space of the tissue.

Quite a number of examples of polymeric conjugates show significant accumulation in solid tumors through the EPR effect. The group of Kopeček and Duncan has worked systematically on polymeric drugs based on the neutral and water-soluble polymer PHPMA (*15*, *16*). They showed that the conjugates of daunorubicin (Dau) or doxorubicin (Dox) with PHPMA linked by a peptide spacer tended to accumulate into the tumor tissue due to the EPR effect (*22*, *23*). Conjugation of Dau to PHPMA resulted in a 15-fold increase in plasma half-life compared to free Dau. Concurrently, Dau levels in the tumor tissue over the first 24 h increased at least fourfold compared to free Dau.

Takakura et al. (*24*) observed a similar effect in the case of mitomycin C (MMC) conjugated to carboxymethyldextran (CM-Dex) (M_w = 70,000), a dextran derivative with a slightly anionic character due to the presence of carboxylate groups. In this conjugate, MMC is bound at the aziridinyl nitrogen atom to a spacer moiety by an amide linkage (**1**). Because of the ring strain of the three-membered ring, this amide linkage is susceptible to hydrolysis. In sharp contrast to a simple amide linkage, which is quite stable at physiological conditions, the amide linkage between MMC and the spacer slowly cleaves and results in a half-life of 36 h to achieve sustained release of MMC at the tumor site. The MMC–CM-Dex conjugate could achieve a superior anticancer effect compared to free MMC for sarcoma 180 subcutaneously implanted to mice by iv treatment. From the measurement of tissue and

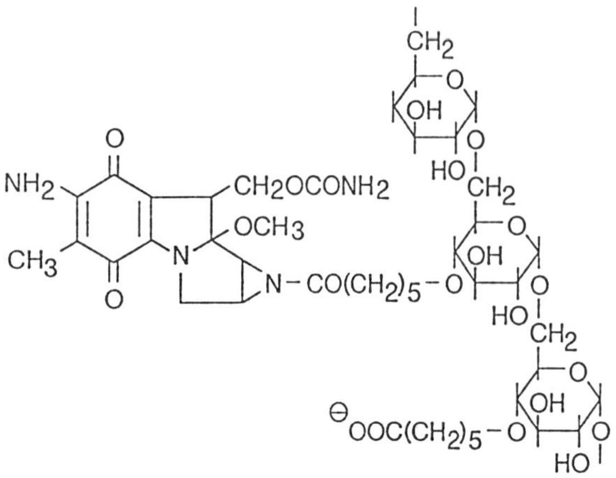

1

organ clearance rates (uptake rates) of the conjugates, a low clearance of conjugate excretion at liver, spleen, and kidney was concluded to be more essential for a significant EPR effect than a high clearance of the conjugate into the tumor site.

The liver is one of the major organs responsible for the carrier removal from the blood compartment. Liver capillaries are so-called sinusoidal capillaries characterized by an absence of basement membranes and a sieve-like structure through which macromolecules and particles with diameter of several tens of nanometers can pass. Thus, in the liver, substances can traffic between the blood compartment and the interstitial space. Furthermore, the liver has a large surface area and possesses the cells of the RES including Kupffer cells. These cells routinely take up foreign substances, particularly vesicles and particles. Thus, in terms of both structure and function, the liver works as the filter in the blood compartment. Consequently, polymeric drugs should escape recognition by the liver to achieve prolonged circulation in the bloodstream.

Among the common water-soluble polymers, PEG is probably the most effective for the purpose of eliminating liver uptake of polymeric conjugates. Nontoxicity as well as high-flexibility and a high degree of hydration are unique biological and physical features of PEG, and it is well-accepted that the high flexibility of PEG chains attached to surfaces contributes to protein-resistant properties (properties leading to minimal protein adsorption) of the surfaces through steric exclusion mechanism (*25*). Indeed, many reports confirm the extended plasma half-life as well as the decreased liver uptake of enzymes conjugated with PEG (*26–28*). For the purpose of protein modification, a variety of PEG derivatives with activated end-groups (e.g., activated esters) are available (*29*).

A similar effect of PEG on the elimination of liver uptake was observed for colloidal carrier systems. In sharp contrast to a prompt liver uptake of polystyrene nanoparticles after iv administration, a significant reduction in the liver uptake for the nanoparticles precoated with a nonionic polymeric surfactant (Pluronic), poly(ethylene oxide)/poly(propylene oxide)/poly(ethylene oxide) block copolymer, was observed (*30*). A brush-like layer of PEG chains of the Pluronic on the nanoparticle surface may work as a repulsive barrier against the contact with cells and proteins, allowing an extended circulation of the nanoparticles in the bloodstream. Complement activation was also reduced by the treatment of nanoparticle surfaces with Pluronic (*31*).

Block copolymers consisting of PEO and other hydrophobic polymer segments are amphiphilic and are known to associate intermolecularly in aqueous milieu by forming micelles with a hydrophobic core surrounded by a corona of hydrophilic PEO chains (*32, 33*) (Figure 4.2). The hydrophobic core serves as a microreservoir of drugs. Drugs can be loaded into the core by physical entrapment or by covalent linkage to the side-chain functional groups of the core-forming segment directly or by means of a spacer (*34, 35*).

We found that the conjugation of Dox to side-chain carboxyl groups of PEG/poly(α,β-aspartic acid) block copolymers (PEG-PAsp) (**2**) led to the spontaneous formation of micelles in aqueous milieu (*34*). The obtained micelle is essentially monodispersive with a diameter of about 40 nm, and it is so stable that it can retain its structure even in the presence of serum proteins (*36*).

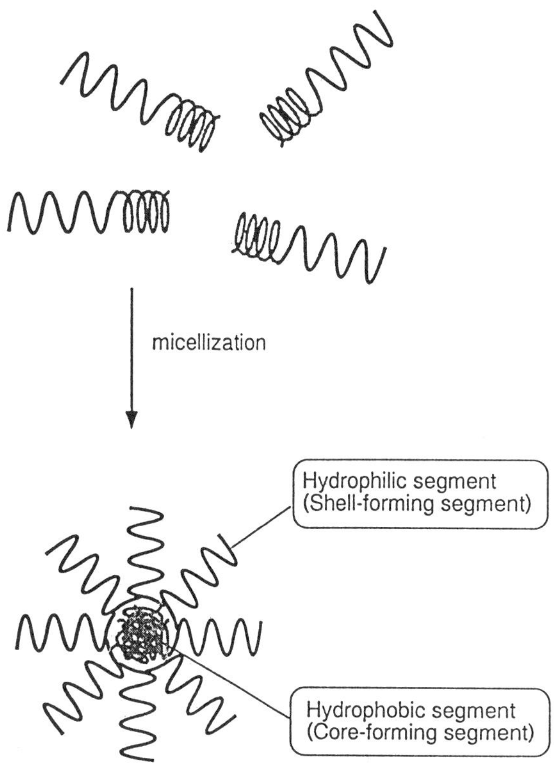

FIGURE 4.2. Multimolecular micellization of amphiphilic block copolymers.

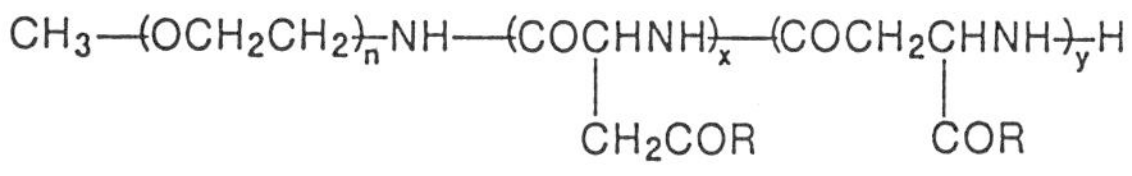

2

R = OH or

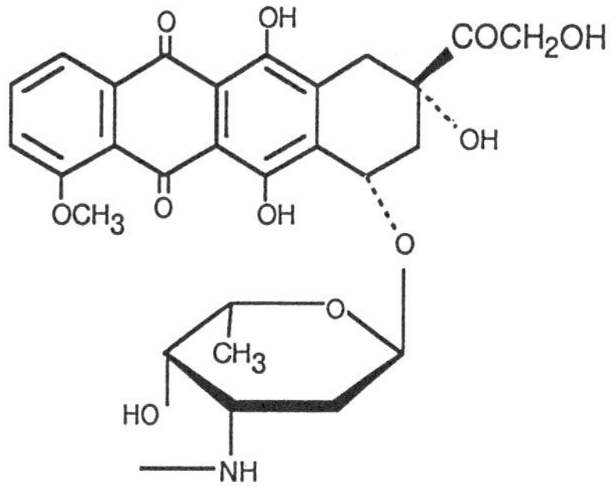

Although the molecular weight of the block copolymer itself (M_w ~14,000) is lower than the threshold molecular weight for glomerular filtration, the apparent molecular weight of the micelle exceeds one million. Thus, this micelle-forming Dox conjugate is expected to achieve a prolonged circulation in the blood compartment as long as it retains the micelle structure with PEG palisades. Indeed, the Dox conjugate with PEG/P(Asp) was revealed to have a considerably long half-life in the bloodstream of mice after iv injection and low accumulation to RES-related organs including liver and spleen. Ten percent of the initial dose of the conjugate still remained in the blood compartment 24 h after the iv injection (*37*). This result suggests that the micelle formed from the conjugate should be stable even in the blood compartment. Indeed, gel filtration chromatography verified the presence of the micelle in the serum sample collected from the mice.

PEG/P(Asp)–Dox conjugate showed a remarkable accumulation to the solid tumor (colon 26) implanted subcutaneously in mice by iv administration (*38*). Taking into account a prolonged plasma half-life of the conjugate, this characteristic tumor accumulation of the conjugate might be caused by the EPR effect. Even compounds like polymeric micelles with a significant molecular weight (>10^6) can extravasate and localize in the tumor site through passive targeting when they keep a considerable stability in the blood compartment. Further, the micelle-forming conjugate showed a 50-fold higher accumulation to the tumor compared with normal muscles in terms of the ratio of %dose/g organ (*38*). On the other hand, the tumor/muscle ratio of free Dox was only 1.5. Dox causes life-threatening cardiac toxicity. The micelle-forming conjugates demonstrate an order of magnitude decrease in heart/tumor ratio in comparison to free Dox, a result indicating a low incidence of cardiac toxicity of conjugated Dox compared to free Dox. As a consequence of these relevant features of the micelle-forming conjugates, higher in vivo anticancer activity, including complete regression, against a number of solid tumors compared to free Dox was exhibited for the conjugates (*39*).

The passive localization of a conjugate through the EPR effect is also an important process for delivery of bioactive agents into a particular cell population through receptor-mediated endocytosis by a conjugate with homing moieties (active targeting). A prerequisite for effective active targeting is to keep a considerable concentration of the conjugate in the vicinity of the target cell by the EPR effect. Indeed, PEG-modified $F(ab')_2$ fragment of the monoclonal antibody (A7) against glycoprotein molecules expressed on human colon cancer cells achieved enhanced localization to SW1116 (human colon cancer) tumor implanted into nude mice compared to the unmodified $F(ab')_2$ fragment because of its long-circulating property in the blood compartment (*40*, *41*).

In summary, effective targeting of polymeric drugs into solid tumor through systemic administration can be achieved through the design of long-circulating (stealth) carriers, allowing modulation of the pharmacokinetics of cytotoxic agents and considerable accumulation at tumor sites. Followed by accumulation, cytotoxic action takes place at the tumor region through several mechanisms, including sustained release of cytotoxic agents in the vicinity of cancer cells and intracellular liberation of cytotoxic agents from the conjugate following endocytosis (lysosomotropic system). An obvious advantage of the intracellular liberation system is that systemic toxicity of the drug is preventable because the drug is in its inactive form (prodrug) while combined to the polymer, and it is only released in the lysosomes through the cleavage of the lysosome-cleavable spacer between the drug and the polymer. Further, a local drug concentration is controllable for a definite time period by controlling the rate of enzymatic degradation of the spacers. As will be discussed in a later section, there is a possibility of overcoming multidrug resistance, a serious problem in cancer treatment, by changing the cellular-uptake route of the cytotoxic agent from membrane penetration to endocytosis through the conjugation with appropriate polymeric carriers.

Modified Cellular Uptake and Retention (Active Targeting)

When a receptor-mediated endocytosis mechanism is in place, we can deliver a drug into a particular cell population in the target region. For this purpose, a pilot moiety having a specific affinity toward a cellular receptor should be introduced into the conjugate. Polymers without a specific affinity to plasma membrane are taken up by cells through a process called "fluid-phase endocytosis" in which polymers are engulfed into endosomes concomitantly with liquid components. On the other hand, when polymers show an affinity for a plasma membrane and adsorb on it, they are internalized into the cell by "adsorptive endocytosis", which generally has a 10-fold faster uptake rate compared with fluid-phase endocytosis.

Duncan et al. (*42*) studied intensively the endocytic uptake of HPMA copolymers bearing tyrosineamide (Tyr-NH_2) residues (**3**) by rat visceral yolk sacs. A steep increase in the uptake rate was observed when 10 mol% of the copolymer side groups were substituted with Tyr-NH_2. This result was considered to be due to an increase in the contribution of nonspecific adsorptive endocytosis with an increase in hydrophobicity of the copolymer chain owing to the increased amount of the hydrophobic aromatic residue, tyrosine. Amounts

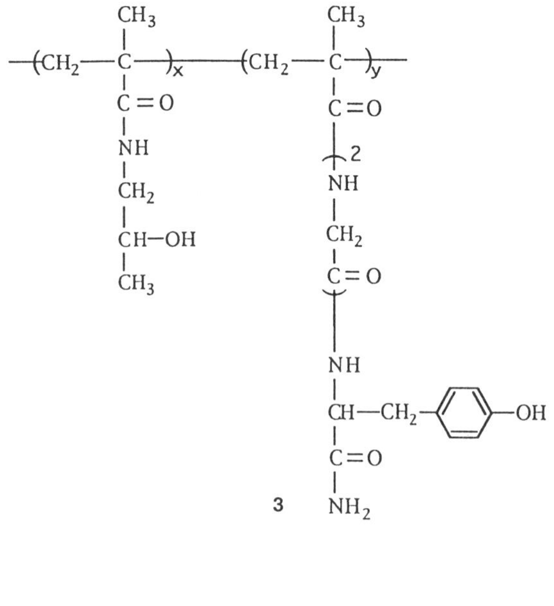

3

of Tyr-NH_2 in the copolymer were increased up to a maximum of about 20 mol% above which the polymers were insufficiently water soluble. In this range of substitution, a good correlation was observed between amounts of Tyr-NH_2 in the copolymer and the rate of cellular uptake.

Nonspecific adsorptive endocytosis is significant for the cellular uptake of cationic polymers because of the negatively charged features of cellular plasma membranes. Nishida et al. (*43*) obtained interesting results on the hepatic disposition of charged dextran by using a rat in vivo model. Endocytic indices (μL per h per 10^8 cells) for liver parenchymal cells in this in vivo disposition experiment increased as follows: carboxymethyldextran (CM-Dex, anionic), 0.9; dextran (Dex, neutral), 3.9; and diethylaminoethyldextran (DEAE-Dex, cationic), 15.2. Values obtained for CM-Dex and Dex were comparable with that for poly(vinyl pyrrolidone) (3.11 μL per h per 10^8 cells) (*44*), which is known to be taken up by cells through fluid-phase endocytosis. On the other hand, the endocytic index for DEAE-Dex was definitely higher than the value expected for the fluid-phase endocytic mechanism, a result indicating that nonspecific adsorptive endocytosis through electrostatic interaction plays a substantial role.

In the case of endocytosis through receptors expressed on the surface of cellular plasma membranes (receptor-mediated endocytosis), the uptake rate reaches even 1,000-fold higher value compared with that observed for fluid-phase endocytosis. Receptor-mediated endocytosis is quite common for many kinds of cell species. For example, low-density lipoprotein (LDL), a natural vehicle for cholesterol, is known to be taken up by liver parenchymal cells through LDL receptors expressed on these cellular surfaces. As a matter of fact, receptor-mediated endocytosis is not only efficient but also highly specific. Thus, it can be used as a route to deliver a drug into a particular cell population located in a particular site in the body (cellular-specific targeting).

The asialoglycoprotein receptor of liver parenchymal cells has been most intensively studied from the standpoint of drug delivery to liver. The asialoglycoprotein receptor specifically recognizes exposed or terminal galactose residues on macromolecules (*45*, *46*). Through this receptor, liver cells effectively internalize asialoglycoproteins (*46*) and chemically glycosylated proteins (neoglycoproteins) (*47*). Thus, these proteins have been widely used as carrier macromolecules for liver targeting (*48*, *49*). For example, an increased uptake into hepatoma cells of Dau covalently attached to galactosylated albumin was reported (*49*). The liver is the most suitable organ for targeting therapy using glycoconjugate primarily due to the presence of the asialoglycoprotein receptor. Furthermore, as mentioned, the liver has a large surface area with a vasculature of sinusoidal structure. The sieve plates of the liver endothelial lining are reported to have a diameter of about 100 nm (*30*). Eventually, even substances with high molecular weights can easily extravasate from the blood compartment to the interstitial space where they meet the parenchymal cell layer.

The asialoglycoprotein receptor of liver parenchymal cells has a remarkably high affinity toward clustered galactose moieties. Thus, oligosaccharides with triantennary galactose residues have a significantly stronger affinity for the receptor than monoantennary oligosaccharide (*50*). High affinity binding is also expected for synthetic water-soluble polymers having many galactose residues at the side chain. Indeed, there are many observations that polymers with pendant galactose groups promptly accumulate into the liver after iv injection (*51–55*).

The application of this type of polymer conjugate with a cytotoxic agent in the treatment of hepatoma was reported (*56*). One major advantage of using glycoconjugated synthetic polymer over neoglycoproteins is a decreased immunogenicity. However, a neoplastic transformation of liver cells is often accompanied with the diminishment of galactose receptors, a process making galactose-mediated targeting ineffective (*57*). Further, even if galactose receptors are expressed on hepatoma surfaces, hepatoma targeting by glycoconjugates is not exclusive because normal liver parenchymal cells also express galactose receptors.

Thus, an efficacy of hepatoma-specific targeting by glycoconjugates is still controversial, although liver accumulation can be achieved undoubtedly with high efficiency.

Many kinds of sugar receptor proteins (animal lectins) are expressed on mammalian cell surface (*58*). Major constituent cells of liver are parenchymal cells (PC) occupying 92.5% of liver proteins (*59*). Other component cells of the liver are generally called nonparenchymal cells (NPC), which consist of endothelial cells (3.3% of liver protein (*59*)) and Kupffer cells (2.5% of liver protein (*59*)). Both endothelial cells and Kupffer cells are known to have mannose receptors. Thus, distribution of the conjugate in the liver can be modulated by changing the structure of the sugar moieties in the conjugate. Let us take CM-Dex as an example. Galactosylated CM-Dex is preferentially taken up by the PC, whereas mannosylated CM-Dex is accumulated in the NPC rather than the PC (*53*). Some groups of sugars including glucose are recognized by both galactose and mannose receptors (*60*), indicating that recognition is not exclusive in terms of sugar structures.

Several drugs other than anticancer reagents were conjugated to carriers and delivered into the liver via sugar receptors. Examples of such drugs are antiviral drugs against hepatitis B and other viral infections (*61–65*). Differential tissue macrophage (MO) allows the replication of viruses, including the herpes simplex virus (HSV) and the human immunodeficiency virus (HIV) (*64*). Thus, it is obviously a promising approach to deliver antiviral drugs into MO through the mannose receptor. Midoux et al. (*64*) prepared the conjugate shown in structure **4** from poly(L-lysine). In this conjugate, mannosyl groups (55/molecule) and 9-(2-phosphonylmethoxyethyl)adenine (PMEA) (20/molecule), an anti-HSV reagent, were covalently attached to the main chain through a glycylglycine spacer. Cationic amino groups of poly(L-lysine) were quantitatively masked by the reaction with excess amounts of γ-gluconolactone and acetic anhydride. This conjugate showed a higher anti-HSV activity compared with PMEA itself in the assay using cultured cells infected with HSV (*64*).

Targeting via galactose receptors recently has received growing attention in the field of gene delivery (*65, 66*). Wu and Wu (*67*) designed a delivery system for genes based on the complexation of DNA, which has an anionic character from water-soluble polycations such as poly(L-lysine) (*67*). They coupled asialoorosomucoid (ASOR) to poly(L-lysine) forming an ASOR–poly(L-lysine) conjugate. Then, the plasmid was complexed to the conjugate through electrostatic interactions. The plasmid containing a gene for chloramphenicol acetyltransferase (CAT), a bacterial enzyme that catalyzes acetylation of chloramphenicol, was used as a model gene. Because mammalian cells lack this gene, appearance of CAT activity in target cells can be used as a marker for gene transformation. Both in vitro (*67*) and in vivo (*68*) studies revealed that the CAT gene was expressed in target cells through the treatment with this ASOR–poly(L-lysine)/DNA conjugate. In vivo liver uptake was then determined by using ^{32}P-plasmid DNA, and it was verified that 85% of the dose accumulated into the liver within 10 min for the conjugate, whereas there was only 17% accumulation for nonconjugated plasmid.

Wu and Wu (*69*) applied this ASOR–poly(L-lysine) to complex antisense oligonucleotides. A 21-mer oligodeoxynucleotide complementary to the polyadenylation signal for human HBV was complexed with ASOR–poly(L-lysine), and this complex was applied to HBV-infected cells in culture. Consequently, an 80% reduction of viral antigen expression on the cell surface was achieved by this treatment (*69*).

Other than ASOR used in the Wu and Wu study, several ligands capable of binding with asialoglycoprotein receptor are now under investigation by different research groups. They involve poly(L-lysine) with tetra-antennary galactose ligands (*70*) (**5**) and a system using human serum albumin as an intermediate carrier (*71*).

There is no doubt that adsorptive- and receptor-mediated endocytoses are quite promising routes for the delivery of bioactive substances specifically into target cells. To achieve cellular-specific targeting, we should keep in mind that, as

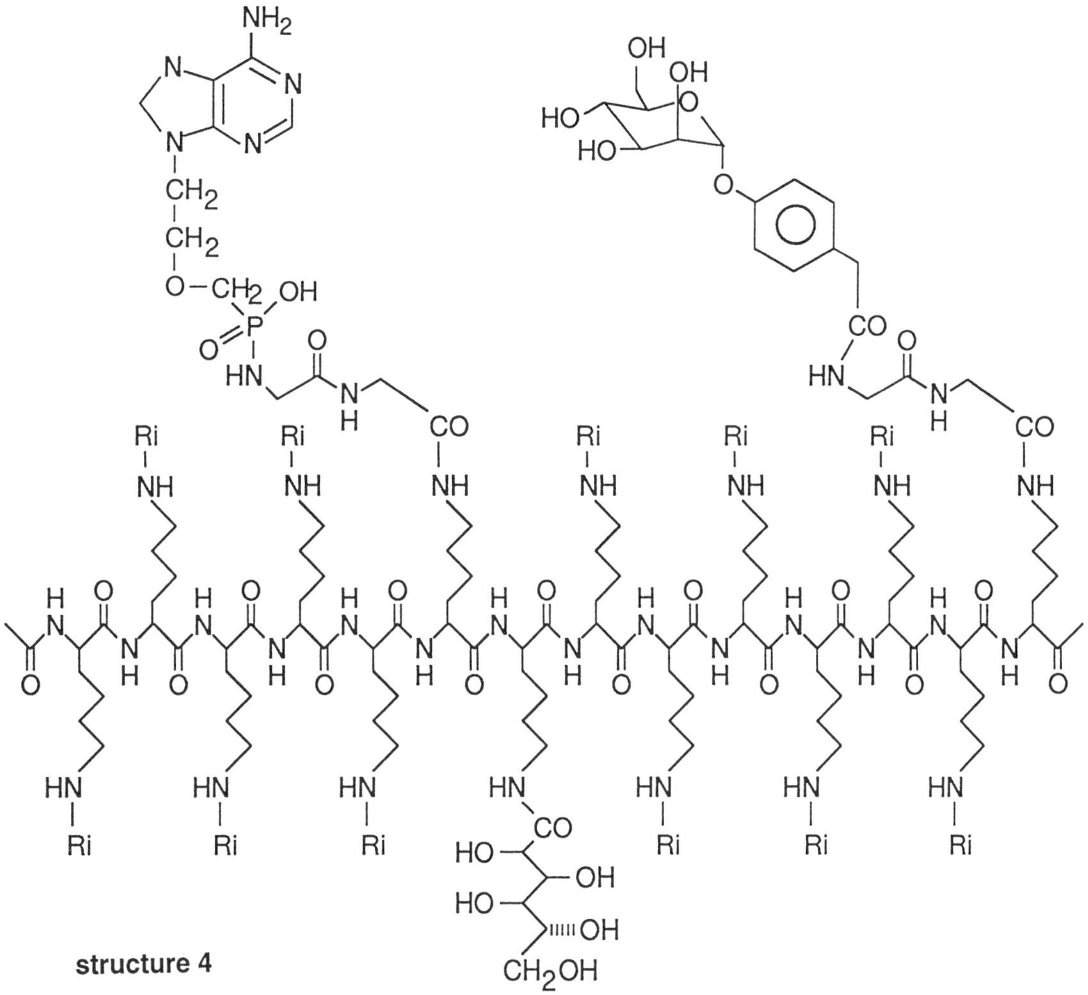

structure 4

a prerequisite, the conjugate itself should have minimal nonspecific interactions with body components (stealth property), allowing a sufficient EPR effect. Another issue of concern is the intracellular fate of the conjugate after endocytosis.

As mentioned, conjugate that underwent endocytosis is transported intracellularly to the acidic (pH ~5) compartment of endosome, and successively to the lysosome, where enzymatic degradation of the conjugate takes place. In the concept of lysosomotropic polymer–drug conjugates, the drug should be cleaved from the carrier in the lysosomal compartment, followed by penetration into the cytoplasm through the lysosomal membrane. Thus, mode and efficacy of drug cleavage at the lysosome are important determining factors for the lysosomotropic type of conjugates. A promising approach for the selective cleavage of a drug from the conjugate is to bind the drug to a polymer by a spacer group that is cleavable by proteases existing in the lysosomal compartment.

Pioneering work was done by Kopeček et al. (*72, 73*) in the late 1970s on synthetic water-soluble polymers having enzyme-cleavable oligopeptide side chains. *p*-Nitroanilide as a model drug

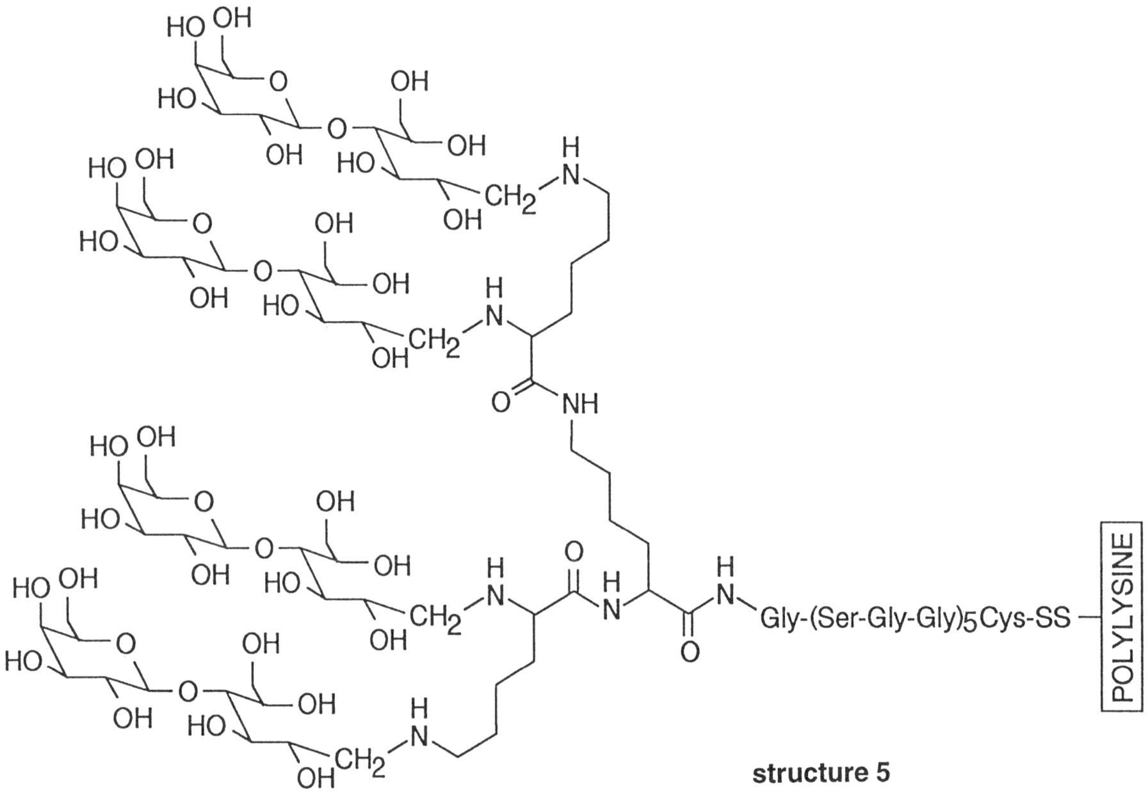

structure 5

was introduced in the side chain of PHPMA through oligopeptide spacers with varying lengths and compositions, and its release from the polymer catalyzed by chymotrypsin was investigated (*73*). The length of the spacer group had a crucial effect on the susceptibility to chymotrypsin attack: steric hindrance by the polymer was predominant for the conjugate with a dipeptide spacer, whereas the substrate specificity became clear by increasing the length of the spacer from dipeptide to tri- or tetrapeptides. When the susceptibility toward chymotrypsin attack was estimated by the ratio of hydrolytic rate constant (k_{cat}) and Michaelis constant (K_M) (k_{cat}/K_M), an approximately 10-fold increase in the susceptibility was observed for the conjugate with –Ala–Gly–Val–Phe– spacer (k_{cat}/K_M = 14,200) compared with that with –Gly–Gly–Phe–Phe– spacer (k_{cat}/K_M =1,520) (*73*). The k_{cat}/K_M ratio is only 32 for the conjugate with the dipeptide (–Gly–Phe) spacer, a result indicating that the rate of cleavage of the drug from the conjugate is controllable in a wide range by varying the length and composition of oligopeptide spacers. Later, a collaborative research team of Kopeček and Duncan (*74, 75*) demonstrated the intralysosomal digestion of PHPMA–oligopeptide–*p*-nitroaniline by using a rat yolk sac in organ culture as an experimental system, proposing the feasibility of the PHPMA–oligopeptide spacer system as a lysosomotropic carrier.

In 1982, Trouet et al. (*76*) performed in vitro and in vivo studies to show the effectiveness of the polymer–cytotoxic drug (Dau) conjugate with an oligopeptide spacer in cancer chemotherapy. Dau was conjugated to succinylated serum albumin (carrier) by means of an oligopeptide spacer arm varying from one to four amino acids (**6**). Because leucyl–Dau is cleavable by lysosomal enzymes, they chose –Leu–Ala–Leu– and –Ala–Leu–Ala–

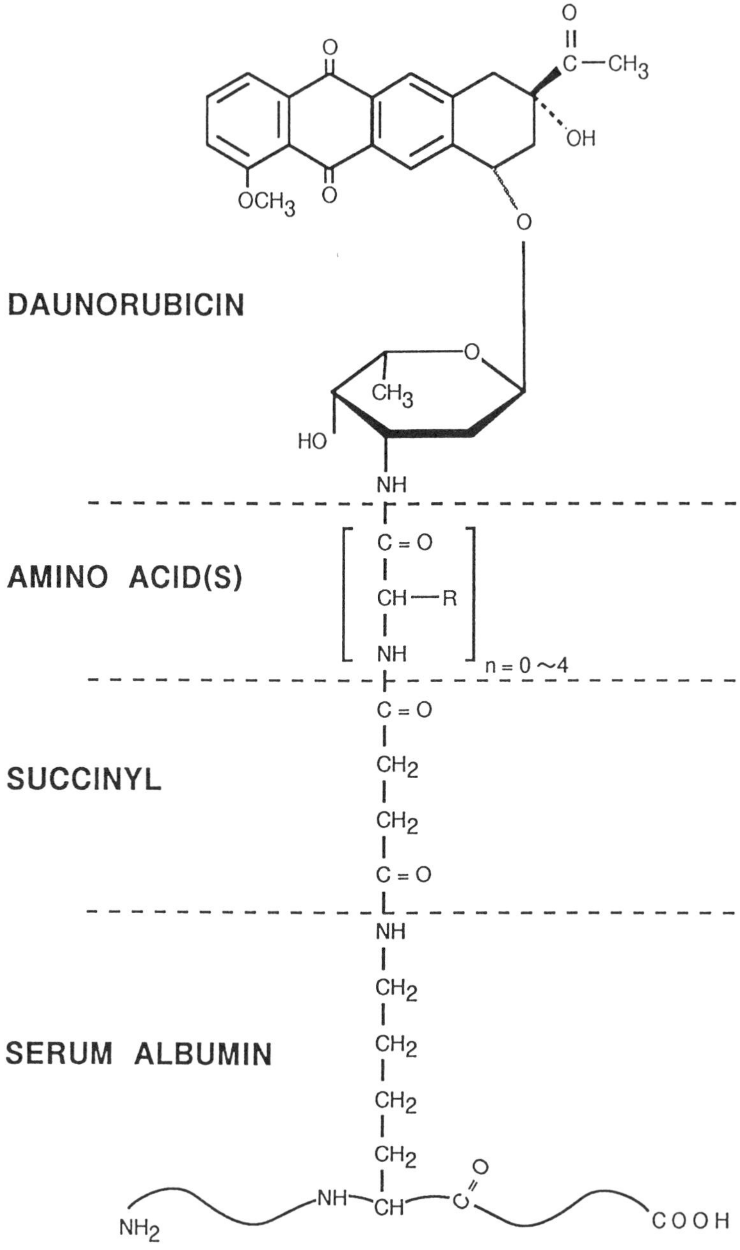
O
C—CH3
OH
O
OCH3
O
O
DAUNORUBICIN
O
CH3
HO
NH
AMINO ACID(S)
C = O
CH—R
NH
n = 0 ~4
SUCCINYL
C = O
CH2
CH2
C = O
SERUM ALBUMIN
NH
CH2
CH2
CH2
CH2
NH2
NH
CH
O
C
COOH

structure 6

Leu– sequences for the spacer cleavable by lysosomal enzymes. These conjugates with tri- and tetrapeptide spacers showed more remarkable therapeutic effects than free Dau against L1210 leukemia developed in the murine ip cavity by ip injection, allowing a high percentage of long-term survivors. Because the conjugate without spacer arms was completely inactive, this remarkable chemotherapeutic efficiency was concluded to be due to the effective intralysosomal release of Dau from the conjugate through enzymatic cleavage. A superior therapeutic index of the conjugate compared with free Dau by ip injection was explained as follows (*76*):

- prolonged retention of Dau moieties in the ip cavity due to the conjugation to macromolecular carrier (depot effect) and
- enhanced uptake of the conjugate into L1210 cells due to the succinylation of albumin (contribution of adsorptive endocytosis).

From a clinical viewpoint, lysosomotropic targeting via the iv route has a wide applicability, as reviewed elsewhere (*77–80*). An essential requirement for the lysosomotropic conjugate dosed through iv injection is that the linkage between drug and conjugate is stable in the bloodstream, whereas it is susceptible to the attack of lysosomal enzymes. In this context, a combination of amino acids in oligopeptide spacer plays a substantial role. In the case of Dox or Dau conjugated to HPMA copolymers via different peptidyl spacers, the rate of drug liberation from the conjugates in the presence of a mixture of lysosomal enzymes from liver (tritosomes) was remarkably affected by the spacer structure and was in the order of –Gly–Gly < –Gly–Phe–Gly– < –Gly–Leu–Gly– < –Gly–Phe–Leu–Gly– < –Gly–Leu–Phe–Gly– (*81*). All of these conjugates with different spacer structures were stable in human serum with less than 1% drug release for 24-h incubation. The in vivo therapeutic efficiency was not in the same order with the drug release rate in vitro. When tested in vivo against the ip form of L1210 leukemia by ip injection, the conjugate with –Gly–Phe–Leu–Gly– spacer was more cytotoxic than that with –Gly–Leu–Phe–Gly– spacer, the sequence that is most susceptible to lysosomal enzymes. Even the conjugates with tripeptide spacers showed a better therapeutic index than those with tetrapeptide spacers. This result indicates that a slowly maintained liberation of the cytotoxic agent in the lysosomal compartment seems to have a considerable therapeutic potential for a maintained antitumor activity (*81*). Thus, sustained liberation of the cytotoxic agent from the retentive conjugate at the tumor site may be required to achieve a high tumoricidal index. Recently, a detailed report on the preclinical trial of PHPMA–peptidyl spacer–Dox (or –Dau) was published (*82*), in which drug liberation in the lysosome of liver parenchymal cells was clearly evidenced.

Another promising approach for the design of a lysosomotropic polymer–drug conjugate is to bind the drug moiety to the polymer backbone via acid-labile linkage. The lysosomal (or endosomal) compartment has a relatively acidic pH (~5.5), and thus the drug linked through an acid-labile bond should easily be liberated from the polymer carrier, penetrating through the lysosomal membrane into the cytoplasm. Hurwitz et al. (*83*) prepared the acid-sensitive conjugates shown in structure **7** through the reaction of Dau with a polymeric hydrazide based on dextran and poly(glutamic acid). These polymers linked with Dau through a hydrazone bond cleave under slightly acidic conditions. Compared with free Dau, the conjugates were equally or more effective against ip Yac lymphoma by iv injection (*83*).

Drug conjugation via a hydrazone linkage was also applied to the preparation of Dox-immunoconjugate, where Dox was linked to a monoclonal antibody via a hydrazone linkage (*84–86*). Complete regressions and cures of xenografted human lung, breast, and colon carcinomas growing subcutaneously in athymic mice were achieved by iv injection of hydrazone-type immunoconjugate of Dox with a monoclonal antibody (BR 96) against a tumor-associated antigen. This antigen is closely related to Lewis Y (Le^{y}), which is abundantly expressed (>200,000 molecules per cell) on human carcinoma cell lines (*87*). A similar ap-

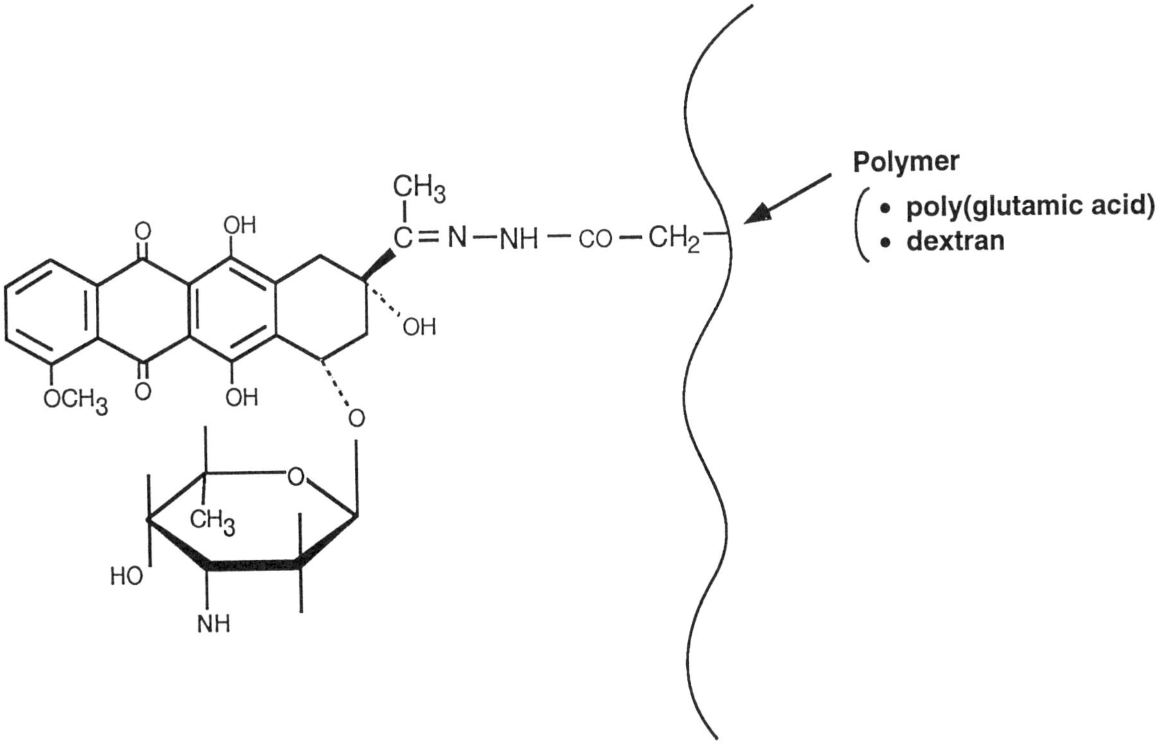

structure 7

proach was reported by the use of a *cis*-aconityl moiety (*88*) (Scheme 4.1) or maleamyl moiety (*89*) as acid-labile linkage between Dox and polymeric carriers to affect intracellular drug release after pinocytosis.

A promising feature of lysosomotropic drug targeting is indicated in the treatment of tumors with multidrug resistance (MDR). Poly(L-lysine), a cationic polyamino acid, is known to be internalized into cells through nonspecific adsorptive endocytosis due to the strong binding onto cellular surfaces via electrostatic interaction. Shen and Ryser (*90*) demonstrated that the conjugation of methotrexate (MTX) with poly(L-lysine) led to a marked increase in the cytotoxicity against Chinese hamster ovarian (CHO) cell lines that were resistant to MTX. This result suggests that MDR tumor might be successfully overcome by the use of a polymer–drug conjugate that is taken up by the tumor cell by a different pathway compared with free drug.

A consensus now exists that a plasma membrane glycoprotein called P-glycoprotein (Pgp) plays a key role in the extrusion of xenobiotics, including some anticancer drugs, to cause MDR. Indeed, overexpression of Pgp was observed on MDR cell lines. Of interest, Pgp recognizes xenobiotics in the plasma membrane, yet it cannot recognize xenobiotics in the cytoplasm. Thus, the MDR problem may be overcome when anticancer drugs are successfully transported into the cytoplasm via a pathway other than the plasma membrane penetration. Recently, Bennis et al. (*91*) succeeded in obtaining an enhanced cytotoxicity of Dox against Dox-resistant cell line by the use of nanospheres (200–300 nm in diameter) made

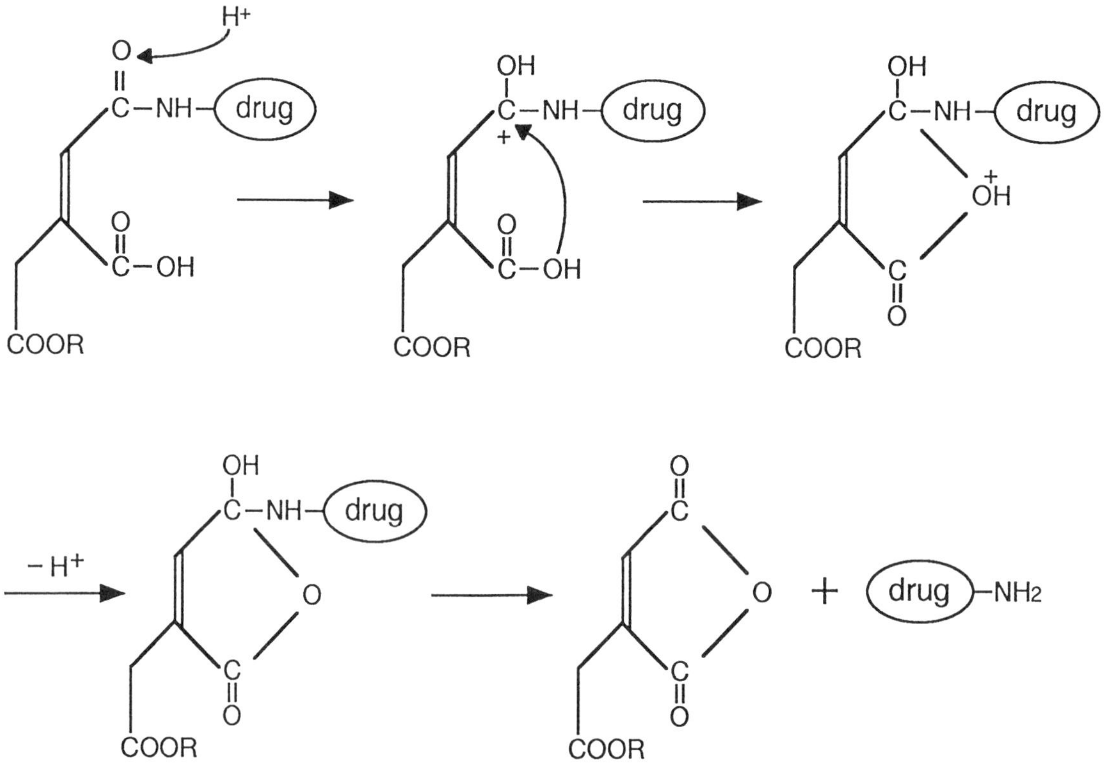

SCHEME 4.1. Mechanism of intramolecular hydrolysis of *cis*-aconityl spacer in acidic milieu.

from poly(isohexylcyanoacrylate) as carriers for Dox. In this way, Dox is considered to be transported into the cytoplasm via the endocytosis route, bypassing Pgp existing on the plasma membrane.

Critical Issues in Polymeric Drugs

Immunogenicity

Immunogenicity of polymeric carriers is a matter of great concern in the development of polymeric drug conjugates. Poly(amino acid) derivatives have been widely used as carrier macromolecules, yet their potential to become an antigen should be taken into account (*92*). In general, amino acid homopolymers with a flexible nature, such as poly(glutamic acid), show no or only weak immunogenicity. Nevertheless, binding of a drug moiety to such a carrier often causes a problem of antibody induction against the conjugate because sequences including drug-linked residues are recognized as epitopes when the drug molecule can act as a hapten. For example, Dox acts as a hapten when it is directly linked to the side chain COOH group of poly(glutamic acid) (*93*) by means of amide linkage. Of interest, a conjugate of Dox to poly(glutamic acid) via a degradable peptidyl spacer was nonimmunogenic, a result suggesting that the nature of the spacer groups has a crucial influence on the immunogenicity of the conjugates.

Dextran, the most common carbohydrate carrier, is nonimmunogenic itself. However, it is cross-reactive with antibodies against bacterial polysaccharides that are widely distributed in the normal human population (*94*). This issue of cross-reactivity with bacterial polysaccharides may also be a problem of using various carbohydrates as drug carriers even if they are not immunogenic. One promising approach to solve this problem is a modification of dextran with PEG (*94*). PEG is a typical example for a synthetic neutral polymer with high flexibility and hydrophilicity, which is generally nonimmunogenic and often reduces the biorecognition of immunogenic substances by antibodies when it is chemically conjugated to these substances (*95*). This approach of chemical conjugation seems to be the most practical one to reduce a possible immunogenicity of polymer–drug conjugates.

Chronic Nonspecific Accumulation

As mentioned before, glomerular filtration plays a substantial role in the clearance of macromolecular substances from the blood compartment. Thus, a more effective EPR effect is expected for the polymeric carrier with higher molecular weight. On the other hand, an increase in the molecular weight of the carrier may cause an adverse effect of chronic accumulation toxicity. To overcome this hurdle, design criteria for polymeric carriers necessitate a carrier that has sufficiently high molecular weight to avoid renal clearance for a certain period of time, followed by a programmed decay into components with lower molecular weight to achieve smooth excretion by the kidney. This type of carrier with programmed degradation properties may be designed by the following approaches:

- preparation of a biodegradable polymer with controlled decay profile. An important prerequisite of this approach is to ensure controlled biodegradability even after the binding of drug moieties. Further, the degradation rate of the conjugates should be balanced with their disposition rate into the target site. The polymer should not be immunogenic. The following are major polymers that were studied as biodegradable carrier systems: poly(β-malic acid) (*96–99*), poly(α-malic acid) (*100*), poly(L-lysine citramide) (*101*), polydepsipeptides (e.g., poly[(glycolic acid)-*alt*-(L-aspartic acid)] (*102*)), poly[N^5-2-(hydroxyethyl)-L-glutamine] (PHEG) (*103–105*), and hyaluronic acid derivatives (*106*, *107*).
- extension of the polymer chain having a moderate molecular weight (M_w ~10,000) by a biodegradable linker. In this way, the polymer is excreted by the kidney after decaying into the fractions through the cleavage of the biodegradable linker. Kopeček et al. (*72*, *108*) prepared such a polymer through cross-linking of PHPMA with moderate molecular weight (M_w ~20,000) at a level below the gel point by the use of oligopeptide cross-linkers. The polymer was confirmed to be stable in plasma, and it liberated a lower molecular weight polymer chain only after the incubation with lysosomal enzymes. This concept was further applied to the extension of PEG chains with biodegradable linkers (*109*).
- increasing apparent molecular weight of the carrier by an intermacromolecular association through a noncovalent force (supramolecular assembly). The typical example of this type of carrier is based on the multimolecular micellization of amphiphilic block copolymers (*110*, *111*). Block copolymer micelles generally have association numbers of approximately 200 and are hardly excreted by the kidney. On the other hand, constituent polymer chains are expected to suffer renal clearance when their molecular weight is designed to be less than the threshold of glomerular filtration. Thus, block copolymer micelles with regulated dissociation properties should be quite promising as carriers, achieving both prolonged circulation in the bloodstream and controlled clearance from the body. The core-forming segment of the block copolymer should have strong cohesive forces to keep the micelle structure for a certain period of time after introduction into the body compartment.

The polymeric micelles have a higher thermodynamical stability compared with common surfactant micelles. Details on the design of block copolymer micelles with programmed decaying properties in the body compartment are given in several review papers cited in references 34, 35, 112, and 113. As a closely related carrier to block copolymer micelles, colloidal particles have been investigated by many research groups (*see* Chapter 5 of this book).

Undesirable Change in Biodistribution Due to Drug Binding

A most serious problem accompanying the conjugate preparation by covalently binding a drug as a pendant group to a soluble polymer is an untoward change in the biodistribution of the conjugate due to changes in physicochemical properties (i.e., hydrophobicity, charge, etc.). Flanking drug moieties along random-coiled polymer chains always have interactions with substances in the surroundings and result in a considerable variation in biodistribution. This interaction is rather significant when a hydrophobic moiety is introduced into a polymer as a pendant group. For example, when a copolymer of HPMA and *N*-(2[4-hydroxyphenyl]ethyl)acrylamide (HPEA) (HPMA:HPEA = 97.6:2.4, M_w = 98,000) was derivatized with only 2–4 mol% of *N*-(2-[(cholest-5-en-3-yl(3β)-oxycarbonyl)oxypropyl]methacrylamide residues (cholesteryl residues), the derivatized copolymers formed aggregates in the solution and revealed a significantly different biodistribution compared with the underivatized compound (*114*). Twenty four hours after the iv administration to rat, only 17% of the injected dose of cholesterol-derivatized copolymer remained in the blood, and 22% was present in the liver. On the other hand, underivatized copolymer showed only 2% liver uptake, and 50% of the injected dose still remained in the bloodstream.

In the case of dextran modification, only 0.7% introduction of fluorescein isothiocyanate (FITC) into dextran resulted in a faster internalization of the conjugate by nonparenchymal cells in the liver (*115*). FITC modification of dextran, leading to an incorporation of negatively charged carboxyl groups, was assumed to promote the interaction with nonparenchymal cells through scavenger receptors that are known to bind anionic substances in the blood compartment. However, an opposing effect was observed by Nishikawa et al. (*53*): slightly anionic dextran (CM-Dex), prepared by carboxymethylation, showed a decreased uptake into the liver and resulted in a prolonged circulation in the bloodstream.

One of the promising approaches to avoid a modified biodistribution of the conjugate due to flanking drug moieties is to segregate the drug moieties from the outer environment by caging them into the compartment. Some amphiphilic polyelectrolytes undergo coil-to-globule transition corresponding to the dissociation degree. In the globular state, only dissociative groups located outside of the globule undergo the dissociation, and the rest of the dissociation groups are buried in the hydrophobic core region of the globule. Vert and Huguet (*116*) used this hydrophobic region of the globular state of amphiphilic polyelectrolyte as microreservoir for a hydrophobic drug (estradiol) by constructing a pH-sensitive drug-releasing vehicle in which all the drug molecules in the globule are released at once at the critical pH of globule-to-coil cooperative transition.

The most typical example for polymer association to form a microcontainer segregated from the outer environment is the micellization of amphiphilic macromolecules. It is well known that the introduction of appropriate hydrophobic groups into a water-soluble polymer often leads to the formation of multimolecular associates (*32–35, 112–114, 117, 118*). From a standpoint of using these associates as a vehicle for drug targeting, exceptional stability, even in the plasma, is needed to ensure the effective delivery of the drug into the target site.

As schematically shown in Figure 4.2, AB block copolymers composed of hydrophilic (A chain) and hydrophobic (B chain) segments form a quite stable micelle with a core–shell structure. Block copolymer micelles, in general, have a lower critical micelle concentration compared with surfactant micelles because of their relatively large cohesive force and low combinatorial entropy (*34, 35*). The micelle with a solid core has a particularly remarkable stability. Furthermore, the dissociation rate of the constituent polymer chains (unimers) from the micelle is extremely small compared with surfactant micelles (*34, 35*). This slow dissociation is a great practical advantage in using polymer micelle systems as drug carriers because in this way the drug in the micelle container can reach the target before the micelle dissociation, followed by a slow liberation of the drug from the micelle through a programmable dissociation. Drugs can be loaded into the micelle core either by covalent linkage (*112, 119–121*) or by a simple physical entrapment (*122–125*), which is quite effective for hydrophobic drugs. Detailed reviews on block copolymer micelles as long-circulating vehicle systems are available (*34, 35, 113*).

Cytoplasmic Delivery

At the present moment, most of the efforts in targeting therapy are focused on the treatment of solid tumors by delivering cytotoxic agents using an appropriate carrier. Cytotoxic agents may penetrate through the plasma membrane of malignant cells after their release from the carrier at the site of tumors. As mentioned, another path is by means of the lysosomes in which drugs detach from the carrier, followed by penetration into the cytoplasm through the lysosomal membrane. Further, recent progress in cellular and molecular biology clarifies substantial roles of peptides and nucleotides in various diseases and accelerates the development of delivery systems for these novel compounds. Because peptides and nucleotides, including antisense oligonucleotides and plasmids, are rather unstable in a biological environment and have difficulties penetrating plasma membranes, a novel concept is required to effectively deliver these compounds into cytoplasm, and further, into the nucleus. In this context, the vehicle system should have the functionality of controlling its intracellular trafficking.

A kind of malignant cells is known to have a receptor for folate, called folate-binding protein (FBP), on the plasma membrane (*126*). FBP mediates the transport of folate into the cytoplasm after the process of endocytosis. Of interest, the folate–protein conjugate is able to reach the cytoplasm without any degradation following entry into the cells (*127–130*). This result is quite promising for designing a conjugate that enters the cytoplasm bypassing transport into lysosomes. There is also an approach to install fusion proteins of influenza virus on the carrier to disrupt endosome membrane, allowing plasmid DNA to enter the cytoplasm (*70*). In this way, intracellular targeting of various biologically active compounds can be achieved by the use of appropriate polymeric carrier systems.

Conclusion

This review chapter was specially focused on the treatment of solid tumors by targetable polymeric drugs. Nevertheless, a wide variety of diseases exist where effective treatment is strongly awaited. For example, treatment of congenital diseases related to gene deficiency has received growing interest. Some of these topics are described in detail in other chapters of this book.

Targetable polymeric drugs offer, as described comprehensively in this chapter, special advantages in their availability to the modulation of the cellular uptake of drugs as well as to the regulation of biodistribution and disposition. Further, intracellular trafficking of drugs might be modulated through the binding with appropriate polymeric carriers. Nevertheless, the future success of these approaches is indeed dependent on novel carrier systems with excellent biocompatibility and high selectivity. Thus, a more intensive approach from the field of chemistry for tailoring novel carriers for drug targeting is strongly encouraged.

Acknowledgments

I thank Glen S. Kwon, University of Alberta; Sandrine Cammas, Tokyo Women's Medical College; and Carmen Scholz, Science University of Tokyo, for useful discussions and for the critical reading of the manuscript.

References

1. Ehrlich, P. *Collected Studies on Immunity;* John Wiley: New York, 1906; Vol. 2.8.
2. Ghose, T.; Cerini, M.; Carter, M.; Nairn, R. C. *Br. Med. J.* **1967,** *1,* 90.
3. Bale, W. F.; Spar, J. L.; Goodland, R. L. *Cancer Res.* **1960,** *20,* 1488.
4. Ghose, T.; Norvell, S. T.; Guclu, A.; Cameron, D.; Bodurtha, A.; MacDonald, A. S. *Br. Med. J.* **1972,** *3,* 495.
5. Moolten, F. L.; Cooperband, R. S. *Science (Washington, D.C.)* **1970,** *169,* 68.
6. Alberts, B.; Bray, D.; Lewis, J.; Raff, M.; Roberts, K.; Watson, J. D. In *Molecular Biology of the Cell,* 2nd ed.; Garland: New York, 1989; pp 323–340.
7. DeDuve, C.; DeBarsy, T.; Poole, B.; Trouet, A.; Tulkens, P.; van Hoof, F. *Biochem. Pharmacol.* **1974,** *23,* 2495.
8. Kohler, G.; Howe, S. C.; Milstein, C. *Eur. J. Immunol.* **1976,** *6,* 292.
9. Ringsdorf, H. *J. Polym. Sci. Symp.* **1975,** *51,* 135.
10. Matsumura, Y.; Maeda, H. *Cancer Res.* **1986,** *46,* 6387.
11. Maeda, H.; Seymour, L. W.; Miyamoto, Y. *Bioconjugate Chem.* **1992,** *3,* 351.
12. Hoes, C. J. T.; Feijen, J. In *Drug Carrier Systems;* Roerdink, F. H. D.; Kroon, A. M., Eds.; John Wiley: New York, 1989; pp 57–109.
13. Vert, M. *CRC Crit. Rev. Ther. Drug Carrier Syst.* **1985,** *2,* 291.
14. Sezaki, H.; Takakura, Y.; Hashida, M. *Adv. Drug Delivery Rev.* **1989,** *3,* 247.
15. Duncan, R.; Kopeček, J. *Adv. Polym. Sci.* **1984,** *57,* 51.
16. Duncan, R. *Anti-Cancer Drugs* **1992,** *3,* 175.
17. Hashida, M.; Nishikawa, M.; Yamashita, F.; Takakura, Y. *Drug Dev. Ind. Pharm.* **1994,** *20,* 581.
18. Seymour, L. W.; Duncan, R.; Strohalm, J.; Kopeček, J. *J. Biomed. Mater. Res.* **1987,** *21,* 1341.
19. Takakura, Y.; Fujita, T.; Hashida, M.; Sezaki, H. *Pharm. Res.* **1990,** *7,* 339.
20. Takakura, Y.; Fujita, T.; Hashida, M.; Maeda, H.; Sezaki, H. *J. Pharm. Sci.* **1989,** *78,* 219.
21. Maeda, H. *J. Controlled Release* **1992,** *19,* 315.
22. Cassidy, J.; Duncan, R.; Morrison, G. J.; Strohalm, J.; Plocova, D.; Kopeček, J.; Kaye, S. B. *Biochem. Pharmcol.* **1989,** *38,* 875.
23. Seymour, L. W.; Ulbrich, K.; Strohalm, J.; Kopeček, J.; Duncan, R. *Biochem. Pharmacol.* **1990,** *39,* 1125.
24. Takakura, Y.; Takagi, A.; Hashida, M.; Sezaki, H. *Pharm. Res.* **1987,** *4,* 293.
25. Jeon, S. I.; Lee, J. H.; Andrade, J. D.; de Gennes, P. G. *J. Colloid Interface Sci.* **1991,** *142,* 149.
26. Abuchowski, A.; McCoy, J. R.; Palczuk, N. C.; van Es, T.; Davis, F. F. *J. Biol. Chem.* **1977,** *252,* 3582.
27. Kamisaki, Y.; Wada, H.; Yagura, T.; Matsushima, A.; Inada, Y. *J. Pharm. Exp. Ther.* **1981,** *216,* 410.
28. Fuertges, F.; Abuchowski, A. In *Advances in Drug Delivery Systems;* Anderson, J. M.; Kim, S. W.; Kuntson, K., Eds.; Elsevier: Amsterdam, Netherlands, 1990; Vol. 4, pp 139–148.
29. Zalipsky, S.; Lee, C. In *Poly(ethylene glycol) Chemistry;* Harris, J. M., Ed.; Plenum: New York, 1992; pp 347–370.
30. Davis, S. S.; Illum, L. In *Site-Specific Drug Delivery;* Tomlinson, E.; Davis, S. S., Eds.; John Wiley: Chichester, England, 1986; pp 93–110.
31. Kreuter, J. In *Colloidal Drug Delivery Systems;* Kreuter, J., Ed.; Dekker: New York, 1994; pp 219–342.
32. Riess, G.; Hurtrez, G.; Bahadur, P. In *Encyclopedia of Polymer Science and Engineering,* 2nd ed.; Wiley-Interscience: New York, 1985; Vol. 2, pp 324–434.
33. Xu, R.; Winnik, M. A.; Hallett, F. R.; Riess, G.; Croucher, M. D. *Macromolecules* **1991,** *24,* 87.
34. Kataoka, K.; Kwon, G. S.; Yokoyama, M.; Okano, T.; Sakurai, Y. *J. Controlled Release* **1993,** *24,* 119.
35. Kataoka, K. *J. Macromol. Sci. Pure Appl. Chem.* **1994,** *A31,* 1759.
36. Yokoyama, M.; Okano, T.; Sakurai, Y.; Kataoka, K. *J. Controlled Release* **1994,** *32,* 269.
37. Kwon, G. S.; Yokoyama, M.; Okano, T.; Sakurai, Y.; Kataoka, K. *Pharm. Res.* **1993,** *10,* 970.
38. Kwon, G.; Suwa, S.; Yokoyama, M.; Okano, T.; Sakurai, Y.; Kataoka, K. *J. Controlled Release* **1994,** *29,* 17.

39. Yokoyama, M.; Okano, T.; Sakurai, Y.; Ekimoto, H.; Shibazaki, C.; Kataoka, K. *Cancer Res.* **1991,** *51,* 3229.
40. Kitamura, K.; Takahashi, T.; Takashima, K.; Yamaguchi, T.; Noguchi, A.; Tsurumi, H.; Toyokuni, T.; Hakomori, S. *Biochem. Biophys. Res. Commun.* **1990,** *171,* 1387.
41. Kitamura, K.; Takahashi, T.; Yamaguchi, T.; Noguchi, A.; Noguchi, A.; Takashima, K.; Tsurumi, H.; Inagake, M.; Toyokuni, T.; Hakomori, S. *Cancer Res.* **1991,** *51,* 4310.
42. Duncan, R.; Cable, H. C.; Rejmanová, P.; Kopeček, J.; Lloyd, J. B. *Biochim. Biophys. Acta* **1984,** *799,* 1.
43. Nishida, K.; Mihara, K.; Takino, T.; Nakane, S.; Takakura, Y.; Hashida, M.; Sezaki, H. *Pharm. Res.* **1991,** *8,* 437.
44. Munniksma, J.; Noteborn, M.; Kooistra, T.; Stienstra, S.; Bouma, J. M. W.; Gruber, J.; Brower, A.; Praaning-Van Dalen, D. P.; Knook, D. L. *Biochem. J.* **1980,** *192,* 613.
45. Ashwell, G.; Morell, A. G. *Adv. Enzymol.* **1974,** *41,* 99.
46. Ashwell, G.; Harford, J. *Annu. Rev. Biochem.* **1982,** *51,* 531.
47. Lee, R. T.; Lee, Y. C. *Biochemistry* **1980,** *19,* 156.
48. Bodmer, J. L.; Dean, R. T. *Methods Enzymol.* **1985,** *112,* 298.
49. Meijer, D. K. F.; Molema, G.; Jansen, R. W.; Moolenaar, F. In *Trends in Drug Research;* Timmerman, H., Ed.; Pharmacochemistry Library Vol. 13; Elsevier: Amsterdam, Netherlands, 1990; pp 303–332.
50. Lee, Y. C.; Townsend, R. R.; Hardy, M. R.; Lönngren, J.; Arnarp, J.; Haraldsson, M.; Lönn, H. *J. Biol. Chem.* **1983,** *258,* 199.
51. Duncan, R.; Kopeček, J.; Rejmanová, P.; Lloyd, J. B. *Biochim. Biophys. Acta* **1983,** *755,* 518.
52. Pimm, M. V.; Perkins, A. C.; Duncan, R.; Ulbrich, K. *J. Drug Targeting* **1993,** *1,* 125.
53. Nishikawa, M.; Kamijo, A.; Fujita, T.; Takakura, Y.; Sezaki, H.; Hashida, M. *Pharm. Res.* **1993,** *10,* 1253.
54. Vansteenkiste, S.; Schacht, E.; Seymour, L.; Duncan, R. In *Polymeric Delivery Systems: Properties and Applications;* El-Nokaly, M. A.; Piatt, D. M.; Charpentier, B. A., Eds.; ACS Symposium Series 520; American Chemical Society: Washington, DC, 1993; pp 362–370.
55. Goto, M.; Yura, H.; Chang, C.-W.; Kobayashi, A.; Shinoda, T.; Maeda, A.; Kojima, S.; Kobayashi, K.; Akaike, T. *J. Controlled Release* **1994,** *28,* 223.
56. O'Hare, K. B.; Hume, I. C.; Scarlett, L.; Chytry, V.; Kopecková, P.; Kopeček, J.; Duncan, R. *Hepatology (New York)* **1989,** *10,* 207.
57. Virgolini, I.; Muller, C.; Klepetko, W. *Br. J. Cancer* **1990,** *61,* 937.
58. Monsigny, M.; Kieda, C.; Roche, A.-C. *Biol. Cell* **1983,** *47,* 95.
59. Van Bergel, T. J. C.; Kruijt, J. K.; Harkes, L.; Nagelkerke, J. F.; Spanjer, H.; Kempen, H.-J. M. In *Site-Specific Drug Delivery;* Tomlinson, E.; Davis, S. S., Eds.; John Wiley: Chichester, England, 1986; pp 49–68.
60. Nishikawa, M.; Ohtsubo, Y.; Ohno, J.; Fujita, T.; Koyama, Y.; Yamashita, F.; Hashida, M.; Sezaki, H. *Int. J. Pharm.* **1992,** *85,* 75.
61. Fiume, L.; Busi, C.; Mattioli, A.; Balboni, P. G.; Barbanti-Brodano, G.; Wieland, Th. In *Targeting of Drugs;* Gregoriadis, G.; Senior, J.; Trouet, A., Eds.; NATO ASI Series A47; Plenum: New York, 1982; pp 1–17.
62. Roche, A. C.; Midoux, P.; Pimpaneau, V.; Negre, E.; Mayer, R.; Monsigny, M. *Res. Virol.* **1990,** *141,* 243.
63. Groman, E. V.; Ensiquez, P. M.; Jung, C.; Josephson, L. *Bioconjugate Chem.* **1994,** *5,* 547.
64. Midoux, P.; Negre, E.; Roche, A. C.; Mayer, R.; Monsigny, M.; Balzarini, J.; De Clercq, E.; Mayer, E.; Ghaffar, A.; Gangemi, J. D. *Biochem. Biophys. Res. Commun.* **1990,** *167,* 1044.
65. Strauss, M. *Gene Ther.* **1994,** *1,* 156.
66. Michael, S. I.; Curiel, D. *Gene Ther.* **1994,** *1,* 223.
67. Wu, G. Y.; Wu, C. H. *J. Biol. Chem.* **1987,** *262,* 4429.
68. Wu, G. Y.; Wu, C. H. *J. Biol. Chem.* **1988,** *263,* 14621.
69. Wu, G. Y.; Wu, C. H. *J. Biol. Chem.* **1992,** *267,* 12436.
70. Plank, C.; Zatloukal, K.; Cotten, M.; Mechtler, K.; Wagner, E. *Bioconjugate Chem.* **1992,** *3,* 533.
71. Merwin, J. R.; Noell, G. S.; Thomas, W. L.; Chiou, H. C.; DeRome, M. E.; McKee, T. D.; Spitalny, G. L.; Findeis, M. A. *Bioconjugate Chem.* **1994,** *5,* 612.
72. Kopeček, J. *Makromol. Chem.* **1977,** *178,* 2169.
73. Kopeček, J.; Rejmanová, P.; Chytry, V. *Makromol. Chem.* **1981,** *182,* 799.
74. Duncan, R.; Lloyd, J. B.; Kopeček, J. *Biochem. Biophys. Res. Commun.* **1980,** *94,* 284.
75. Duncan, R.; Rejmanová, P.; Kopeček, J.; Lloyd, J. B. *Biochem. Biophys. Acta* **1981,** *678,* 143.

76. Trouet, A.; Masquelier, M.; Banrain, R.; Deprez-DeCampeneere, D. *Proc. Natl. Acad. Sci. U.S.A.* **1982,** *79,* 626.
77. Duncan, R. *Anti-Cancer Drugs* **1992,** *3,* 175.
78. Kopeček, J.; Duncan, R. *J. Controlled Release* **1987,** *6,* 315.
79. Kopeček, J. *J. Controlled Release* **1990,** *11,* 279.
80. Ulbrich, K. *J. Bioact. Compat. Polym.* **1991,** *6,* 348.
81. Šubr, V.; Strohalm, J.; Ulbrich, K.; Duncan, R.; Hume, I. C. *J. Controlled Release* **1992,** *18,* 123.
82. Duncan, R.; Seymour, L. W.; O'Hare, K. B.; Flanagan, P. A.; Spreafico, F.; Grandi, M.; Ripamonti, M.; Farao, M.; Suarato, A. *J. Controlled Release* **1992,** *19,* 331.
83. Hurwitz, E.; Wilchek, M.; Pitha, J. *J. Appl. Biochem.* **1980,** *2,* 25.
84. Mueller, B. M.; Wrasidlo, W. A.; Reisfield, R. A. *Bioconjugate Chem.* **1990,** *1,* 325.
85. Kaneko, T.; Willner, D.; Monkovic, I.; Knipe, J. O.; Braslawsky, G. R.; Greenfield, R. S.; Vyas, D. M. *Bioconjugate Chem.* **1991,** *2,* 133.
86. Willner, D.; Trail, P. A.; Hofstead, J.; King, H. D.; Lasch, S. J.; Braslawsky, G. R.; Greenfield, R. S.; Kaneko, T.; Firestone, R. A. *Bioconjugate Chem.* **1993,** *4,* 521.
87. Trail, P. A.; Lasch, S. J.; Henderson, A. J.; Hofstead, S.; Casazza, A. M.; Firestone, R. A.; Hellström, I.; Hellström, K. E. *Science (Washington, D.C.)* **1993,** *261,* 212.
88. Shen, W.-C.; Ryser, J.-P. *Biochem. Biophys. Res. Commun.* **1981,** *102,* 1048.
89. Hoes, C. J. T.; Boon, P. J.; Kaspersen, F.; Bos, E. S.; Feijen, J. *Makromol. Chem. Macromol. Symp.* **1993,** *70/71,* 119.
90. Shen, W.-C.; Ryser, J.-P. *Mol. Pharmacol.* **1979,** *16,* 614.
91. Bennis, S.; Chapey, C.; Couvreur, P.; Robert, J. *Eur. J. Cancer* **1994,** *30A,* 89.
92. Vermeersch, H.; Remon, J. P. *J. Controlled Release* **1994,** *32,* 225.
93. Hoes, C. J. T.; Grootoonk, J.; Duncan, R.; Hume, I. C.; Bhakoo, M.; Bouma, J. M. W.; Feijen, J. *J. Controlled Release* **1993,** *23,* 37.
94. Chiu, H.-C.; Konák, C.; Kopecková, P.; Kopeček, J. *J. Bioact. Compat. Polym.* **1994,** *9,* 388.
95. Fuertges, F.; Abuchowski, A. *J. Controlled Release* **1990,** *1–3,* 139.
96. Guerin, P.; Vert, M.; Braud, C.; Lenz, R. W. *Polym. Bull.* **1983,** *14,* 187.
97. Braud, C.; Bunel, C.; Vert, M. *Polym. Bull.* **1985,** *13,* 293.
98. Fournie, P.; Domurado, D.; Guerin, P.; Braud, C.; Vert, M. *J. Bioact. Compat. Polym.* **1990,** *5,* 381.
99. Fournie, P.; Domurado, D.; Guerin, P.; Braud, C.; Vert, M.; Pontikis, R. *J. Bioact. Compat. Polym.* **1992,** *7,* 113.
100. Ouchi, T.; Kobayashi, H.; Hirai, K.; Ohya, Y. In *Polymeric Delivery Systems: Properties and Applications;* El-Nokaly, M. A.; Piatt, D. M.; Charpentier, B. A., Eds.; ACS Symposium Series 520; American Chemical Society: Washington, DC, 1993; pp 382–394.
101. Boustta, M.; Huguet, J.; Vert, M. *Makromol. Chem. Macromol. Symp.* **1991,** *47,* 345.
102. Ouchi, T.; Shiratani, M.; Jinno, M.; Hirao, M.; Ohya, Y. *Makromol. Chem. Rapid Commun.* **1993,** *14,* 825.
103. Hoes, C. J. T.; Grootoonk, J.; Feijen, J.; Boon, P. J.; Kaspersen, F.; Loeffen, P.; Schlachter, I.; Winters, M.; Bos, E. S. *J. Controlled Release* **1992,** *19,* 59.
104. De Marre, A.; Seymour, L. W.; Schacht, E. *J. Controlled Release* **1994,** *31,* 89.
105. Bayley, D.; Sancho, M.-R.; Brown, J.; Brookman, L.; Petrak, K.; Goddard, P.; Steward, A. *J. Bioact. Compat. Polym.* **1993,** *8,* 51.
106. Pouyani, T.; Prestwich, D. *Bioconjugate Chem.* **1994,** *5,* 339.
107. Wada, T.; Chirachanchai, S.; Izawa, N.; Inaki, Y.; Takemoto, K. *J. Bioact. Compat. Polym.* **1994,** *9,* 429.
108. Cartlidge, S. A.; Duncan, R.; Lloyd, J.; Kopecková-Rejmanová, P.; Kopeček, J. *J. Controlled Release* **1987,** *4,* 265.
109. Ulbrich, K.; Strohalm, J.; Kopeček, J. *Makromol. Chem.* **1986,** *187,* 1131.
110. Tuzar, Z.; Kratochvil, P. *Adv. Colloid Interface Sci.* **1976,** *6,* 201.
111. Riess, G.; Hurtrez, G.; Bahadur, P. In *Encyclopedia of Polymer Science and Engineering,* 2nd ed.; Wiley Interscience: New York, 1985; Vol. 2, pp 324–434.
112. Bader, H.; Ringsdorf, H.; Schmidt, B. *Angew. Makromol. Chem.* **1984,** *123/124,* 457.
113. Kwon, G. S.; Kataoka, K. *Adv. Drug Delivery Rev.* **1995,** *16,* 295.
114. Ambler, L. E.; Brookman, L.; Brown, J.; Goddard, P.; Petrak, K. *J. Bioact. Compat. Polym.* **1992,** *7,* 223.

115. Vansteenkiste, S.; Schacht, E.; Duncan, R.; Seymour, L.; Pawluczyk, I.; Baldwin, R. *J. Controlled Release* **1991**, *16*, 91.
116. Vert, M.; Huguet, J. *J. Controlled Release* **1987**, *6*, 159.
117. Ulbrich, K.; Konak, C.; Tuzar, Z.; Kopeček, J. *Makromol. Chem.* **1987**, *188*, 1261.
118. Nukui, M.; Hoes, K.; van den Berg, H.; Feijen, J. *Makromol. Chem.* **1991**, *192*, 2925.
119. Yokoyama, M.; Miyauchi, M.; Yamada, N.; Okano, T.; Kataoka, K.; Inoue, S. *J. Controlled Release* **1990**, *11*, 269.
120. Yokoyama, M.; Kwon, G. S.; Okano, T.; Sakurai, Y.; Seto, T.; Kataoka, K. *Bioconjugate Chem.* **1992**, *3*, 295.
121. Yokoyama, M.; Okano, T.; Sakurai, Y.; Kataoka, K. *J. Controlled Release* **1994**, *32*, 269.
122. Kwon, G. S.; Naito, M.; Yokoyama, M.; Okano, T.; Sakurai, Y.; Kataoka, K. *Langmuir* **1993**, *9*, 945.
123. Kwon, G. S.; Naito, M.; Yokoyama, M.; Okano, T.; Sakurai, Y.; Kataoka, K. *Pharm. Res.* **1995**, *12*, 192.
124. Kabanov, A. V.; Batrakova, E. V.; Melik-Nubarov, N. S.; Fedoseev, N. A.; Dorodnich, T. Y.; Alakhov, V. Y.; Chekhonin, V. P.; Nazarova, I. R.; Kabanov, V. A. *J. Controlled Release* **1992**, *22*, 141.
125. Rolland, A.; O'Mullane, J.; Goddard, P.; Brookman, L.; Petrak, K. *J. Appl. Polym. Sci.* **1992**, *44*, 1195.
126. Campbell, I. G.; Jones, T. A.; Foulkes, W. D.; Trowsdale, J. *Cancer Res.* **1991**, *51*, 5329.
127. Leamon, C. P.; Low, P. S. *Proc. Natl. Acad. Sci. U.S.A.* **1991**, *88*, 5572.
128. Leamon, C. P.; Low, P. S. *J. Drug Targeting* **1994**, *2*, 101.
129. Leamon, C. P.; Pastan, I.; Low, P. S. *J. Biol. Chem.* **1993**, *268*, 24847.
130. Leamon, C. P.; Low, P. S. *Biochem. J.* **1993**, *291*, 855.

5

Particle Engineering of Biodegradable Colloids for Site-Specific Drug Delivery

P. D. Scholes, A. G. A. Coombes, M. C. Davies, L. Illum, and S. S. Davis

Significant advances are being made in the design, formulation, and surface modification of biodegradable colloids, which improve the prospects for site-specific drug delivery following intravenous administration. This chapter first defines the areas of application for targeted drug delivery and describes various carrier systems including macromolecular drug carriers, micellar systems, liposomes, and microparticles. The biological barriers to targeted drug delivery by colloidal particles, principally phagocytosis and opsonization, are identified. The physicochemical factors influencing circulation times and the fate of microparticulate systems in vivo are presented and discussed. The concept of colloid surface modification is considered in depth because it dominates strategies aimed at achieving passive and active targeting. To fully exploit the successes achieved to date with nondegradable particles in avoiding phagocytosis, biodegradable materials that adequately mimic or improve these characteristics must be used. The chapter concludes with a description of the synthesis, biodegradation behavior, and microsphere formulation techniques applicable to lactide polymers and copolymers. These factors profoundly influence drug incorporation strategies, the potential for particle size reduction, and surface modification, which will ultimately decide the success or failure of a site-specific delivery system.

Targeting of drugs to specific tissue sites would undeniably be of great benefit in the therapy of many disease states. Mills and Davis (*1*) suggested that to derive the maximum benefit from a drug it should be delivered to the target site at a rate and concentration that permits its optimal therapeutic activity whilst reducing undesirable side effects to a minimum. This rationale is particularly relevant in the case of cancer chemotherapy. Reduction of systemic toxicity, while retaining or increasing the tumoricidal effect, ideally requires exposure of the tumor to the drug and limited drug contact with normal tissue (*2*, *3*). Considerable interest and applications for targeted systems also apply in the field of gene therapy (*4*).

A detailed understanding of the characteristics of the target tissue, the drug structure and action, and the delivery system is vital for achieving maxi-

Copyright © 1997 American Chemical Society

mum therapeutic benefit (*2*). The essential characteristics of an ideal targeting system have been defined by Mills and Davis (*1*) and Davis et al. (*5*) as follows:

- compatibility with the body in terms of toxicity, biodegradability, and antigenicity;
- protection of the drug until it reaches its site of action;
- maintenance of the drug-carrier integrity until the target is reached;
- avoidance of interaction with normal cells;
- an ability to traverse intervening membranes;
- target recognition and association;
- controlled drug release to achieve the desired therapeutic effect; and
- carrier elimination from the body following drug release.

Several definitions of drug targeting have been proposed. Poste and Kirsch (*6*) pointed out that targeting could occur at three different levels: organ, cellular, and subcellular. Artursson (*7*) described targeting as active or passive depending on whether or not natural physiological processes were exploited, whereas Gupta (*2*) used the definition of absolute or partial targeting, reflecting either exclusive or preferential drug delivery. Davis and Illum (*8*) emphasized the need to consider the nature of the drug at an early stage of the design process of targeted delivery systems, because its physicochemical characteristics, dose, and required release pattern can largely dictate design feasibility.

Drug Targeting Systems

Macromolecular Drug Carriers

In the search for the elusive "magic bullet" (*9*), considerable attention has been focused on the use of soluble, macromolecular drug carriers (*10–12*) such as *N*-(2-hydroxypropyl)methacrylamide (HPMA). These systems possess the inherent advantage of solubility and hence a greater potential for accessing the target tissue by transport across the capillary endothelium. The uptake of soluble macromolecules by cells occurs by pinocytosis and not phagocytosis and provides the main basis of their targeting potential (*13*). The use of an Adriamycin–dextran conjugate, for example, was more efficacious than the free drug alone (*14*). Passive targeting of mitomycin C–(anionic) dextran conjugates to tumor sites in mice was accomplished (*15*). Furthermore, the use of natural macromolecules such as glycoproteins affords the possibility of interaction with receptors on cellular subsets (*16, 17*). Recent studies demonstrated the capacity for extending the circulatory half-life of a model molecule attached to polysialic acids as the carrier. These conjugates are potentially advantageous in terms of being both biodegradable and nonimmunogenic (*18*). However, disadvantages include the low carrier capacity and the possibility that the chemical coupling procedure results in loss of activity.

Antibody-based drug delivery systems have received widespread attention over the past decade, particularly for cancer chemotherapy (*2, 19, 20*). Successful delivery to brain parenchyma in the rat was reported for methotrexate coupled to an antibody specific for capillary endothelia (*21*). However, major concerns over the immunogenicity and activity of many antibody–drug conjugates (*2, 7* have driven research into alternative strategies such as indirect coupling of the drug to the antibody and the use of antibody fragments (*2*). The use of immunotoxins to deliver highly toxic molecules (for example, ricin in Hodgkin's disease) also was investigated (*22*). In addition, antibodies have been proposed as a novel delivery system for enzymes, which could subsequently activate a nontoxic prodrug locally at the target site (*23, 24*).

Micellar Drug Carriers

Micellar drug carriers are formed from amphiphilic block copolymers composed of hydrophilic and hydrophobic segments. The hydrophobic segments form the inner core of the micelle and are surrounded by an outer shell consisting of the hydrophilic copolymer segments. The use of block copolymers as drug carriers recently was reviewed by Yokoyama (*25*). Micelles are commonly of the order of 50 nm, which compares with the dimension of viruses, and thus may be able to penetrate the sinusoidal and fenestrated capillaries that have pores approximately 100 nm in size.

Micellar carriers, however, are generally considered to be poor delivery systems for parenteral administration. Micellar complexes are in dynamic equilibrium with free molecules in solution and continuously break down and reform (*26*) and are unstable on dilution. This process results in disassociation of drug and carrier and low circulation times. Drug loading in micelles is also usually low. Investigations of the solubility of several steroids including testosterone in polyoxyethylated cetyl alcohols (*27*) revealed that only 2–9 steroid molecules were associated with each micelle formed from long-chain poly(oxyethylene) surfactants at 25 °C, representing a maximum of 3% of the micellar weight.

Despite their disadvantages, interest continues to be shown in micelles, or *self-assembling, supramolecular complexes,* as microcontainers for drug targeting. A micelle-forming polymeric anticancer drug was synthesized by covalently bonding Adriamycin and poly(ethylene glycol)–poly(aspartic acid) block copolymer (*14*). The micellar delivery system exhibited excellent in vivo anticancer activity in mice and lower toxic side effects compared to the free drug.

The use of micelles for targeting brain tissue after intravenous (iv) administration also was described. The neuroleptic action of haloperidol injected into mice in micellar solutions of nonionic, poly(oxyethylene)–poly(oxypropylene)–poly(oxyethylene) (POE–POP–POE) block copolymer surfactant (Pluronic P-85) in water increased relative to aqueous haloperidol solutions (*28*). Molecules of the drug were solubilized in the inner hydrophobic POP core of the micelles. The outer hydrophilic shell of such micelles is formed by nontoxic and nonimmunogenic POE blocks. Targeting of such microcontainers to specific cells has been attempted by conjugating the Pluronic molecules with antibodies against a target-specific antigen or with protein ligands selectively interacting with target cell receptors (*29*). These same authors also reported the ability of the low molecular weight compound adenosine 5′-triphosphate, solubilized in Pluronic micelles, to penetrate an intact cell in vitro.

Certain amphiphilic block copolymers synthesized from poly(lactic acid) and poly(ethylene glycol) (PLA–PEG) are also capable of forming micelles of approximately 20 nm in size (*30*). The carrying capacity of testosterone in PLA–PEG micelles was characteristically low (0.3% w/v). Approximately 20% of an administered dose of PLA–PEG could be retained in the bloodstream of rats for 3 h if the concentration of PLA–PEG was high enough to retain the micelle form on dilution rather than the disassociated unimer form.

Nonbiodegradable, amphipathic, ABA-type block copolymers based on poly(oxyethylene–isoprene–oxyethylene) also form micelles in aqueous dispersions (*31*). Stable nanospheres of approximately 100 nm in size were subsequently prepared from these micelles by cross-linking the chains in the hydrophobic (polyisoprene) core. Nanosphere samples remained in circulation in experimental animals with a half-life in excess of 50 h.

The use of natural, endogenous transport vehicles such as lipoproteins (*32, 33*), erythrocytes (*34*), and lymphocytes (*35*) also was proposed to avoid possible immunogenic responses.

Colloidal Carriers

Colloidal carriers such as liposomes and microparticles have attracted much research interest for site-specific drug delivery (*36–39*). Their use has been proposed for purposes of tissue targeting, protection of encapsulated drugs, and amplification of therapeutic effects in the field of drug delivery (*40, 41*).

Liposomes

Liposomes, or phospholipid vesicles, are formed by equilibration of natural phospholipids such as lecithin with excess water or aqueous salt solution. Multilamellar liposomes consist of concentric lipid bilayers, which can solubilize hydrophobic drugs, alternating with aqueous compartments, which can entrap hydrophilic drugs (*42*). Sonication of the multilamellar liposomes formed initially can give rise to unilamellar liposomes. Unmodified liposomes, essentially lipid-based microparticles, are rapidly cleared from the circulation and accumulate in the phagocytic cells of the reticuloendothelial system (RES), especially the Kupffer cells of the liver and splenic macrophages. The use of liposomes has traditionally suffered from disadvantages of poor stability in the presence of serum and poor drug loading capacity compared with other colloidal carriers (*43*). However, during the last 5 years these problems have largely been solved to give sophisticated drug delivery and targeting systems (*44*).

Nonionic surfactant vesicles (niosomes) have been proposed as synthetic analogs of liposomes: investigations indicated an increase in the circulation time and tumor uptake of entrapped doxorubicin (*45*). In related investigations the use of lipid emulsions containing dexamethasone and prostaglandins were described (*14*) for delivery to vascular sites.

Microparticles

A wide range of particulate materials of both natural and synthetic origin have been investigated for targeted drug delivery, and some of the most prominent applications are listed in Tables 5.1 and 5.2. The composition and in vivo behavior of colloidal particles receiving most attention for iv administration are discussed in detail subsequently.

Barriers to Tissue Targeting by Colloidal Drug Carriers

Following iv injection, prospective microparticulate delivery systems will be confronted with two major obstacles that can greatly limit their utility: the body's phagocytic defense systems and the means of migrating from the vasculature into target tissue sites.

TABLE 5.1. Examples of Compartmental Delivery with Colloidal Carriers

Route	*Colloid*	*Disease*	*Ref.*
Intraarticular	Microspheres	Rheumatoid arthritis	285
Intraperitoneal (IP)	Microspheres	Tumor cells, IP	286
Intraocular	Microspheres	Glaucoma	287, 288
	Microspheres	Vitreoretinopathy	289
Intramedullary	Microspheres	Osteomyelitis	219
Pulmonary	Liposomes	Asthma, infection	290

TABLE 5.2. Potential Therapeutic Applications of IV Administered PACA Microparticles in Passive Delivery

Drug	*Site*	*Purpose*	*Ref.*
Primaquine	Kupffer cells	Treats leishmaniasis	291
Ampicillin	Liver/spleen	Treats *Salmonella*/*Listeria* intracellular infections	174
MTP-Chol[a]	Macrophages	Antimetastatic activity	292
Mitoxantrone	RES	Decreases side effects	293
Mitomycin C	RES	Decreases toxicity and mutagenicity	294
Doxorubicin	—	Decreases cardiac toxicity	295

[a]MTP-Chol is muramyl dipeptide-L-alanyl-cholesterol.

Extravasation and Problems of Tumor Targeting

To be effective, site-specific drug delivery systems will often be required to leave the vasculature to interact with extravascular sites. Extravasation of macromolecules can occur by diffusion, convection, and transcytosis, processes under strict anatomical and physiological control (*46*). Vascular capillaries generally consist of a continuous endothelium and an uninterrupted basement membrane. Morphological differences arise where gaps are present in the endothelium (fenestrations). In some tissues (such as the liver, spleen, and bone marrow) the absence of the basement membrane gives rise to the presence of sinusoids and as such offers opportunities for delivery of microparticles to the underlying parenchyma (*40, 46, 47*).

An increased potential exists for extravasation as a result of the increased vascular permeability that attends certain pathological conditions, such as inflammation, ischemia, or hypertensive vascular lesions, and which also occurs in tumor endothelia. In reality, the often poor extravasation encountered in tumors may be explained by high interstitial pressure coupled with a decreased vascular pressure (*46*). Further difficulties to be faced in tumor targeting include poor blood supply (*40, 48*), low level of biochemical differentiation between host and tumor cells (*49*), and heterogeneity of neoplastic cells (*2*).

Phagocytosis and Opsonization

Phagocytosis is the defense mechanism of the body that clears invading pathogens, unwanted cells, and small particles (*50, 51*). Most bacteria are nonpathogenic because of their spontaneous phagocytosis: pathogenic organisms are those that can avoid ingestion and breakdown. Phagocytic cells primarily consist of circulating polymorphonuclear leukocytes and mononuclear phagocytes. These phagocytes originate in the bone marrow but become fixed in certain tissue sites, primarily the liver, spleen, and bone marrow, to form the mononuclear phagocytic system or RES (*51, 52*). Particulate uptake or ingestion proceeds by endocytosis following adhesion to the phagocyte or, alternatively, in the apparent absence of an adhesion step by pinocytosis, whereby small particles are internalized by endocytosis (*51*). Once internalized, the endosome or phagosome so formed will fuse with lysosomes, which will expose the carrier to highly active enzyme systems (*53*).

In the biological environment following iv administration, microspheres or other drug carriers will first interact with plasma proteins. The adsorption of these plasma proteins is known as opsonization. The amount and type of protein bound to the surfaces will be influenced by the protein, its concentration, the charges of the two phases, and the hydrophilicity/hydrophobicity of the substrate (*54, 55*). Davis and Illum (*40*) suggested that serum albumin will most likely adsorb first, followed by a later rearrangement and displacement by proteins of lower concentrations but with higher affinity for the particle surface. The result of these competing adsorption processes will govern subsequent interactions of the carrier with other surfaces, such as the cells of the RES (*54*).

The process of opsonization mediates the adhesion stage in the phagocytic process. Several biological entities are known to either promote (opsonins) or retard (dysopsonins) this event (*56, 57*). The serum proteins associated with the opsonization process are known to include tuftsin (*58*), fibronectin (*59, 60*), C-reactive protein (*56*), and most importantly immunoglobulin (Ig) G and complement C3 b (*56, 61, 62*). Phagocytic cells are known to possess receptors for the Fc region of immunoglobulin and the C3 b fragment of complement (*62*).

The influence of a range of serum proteins on the phagocytic uptake of microparticles has been studied by several authors (*63–66*). Coating particles with gamma-globulin, fibronectin, tuftsin, or gelatin increased phagocytic uptake (*64*). Albumin, in contrast, had the opposite effect (*63, 64*).

A number of endogenous and exogenous substances are also known to act as dysopsonins and inhibit phagocytic ingestion (*57*). Adsorption of

IgA for example is known to reduce the phagocytosis of several bacteria (*67*).

Evidence for tissue-specific opsonins and dysopsonins has arisen from the biological behavior of liposomes that have either a low or high cholesterol content (*43, 57*). In vitro and in the presence of serum, Kupffer cells avidly take up cholesterol-poor but not cholesterol-rich liposomes, whereas for splenic macrophages the reverse is true (*68*). Evidence suggested that opsonins specific for hepatic and splenic phagocytic cells were present in the serum, and these cells had different affinities for the two liposome types (*68, 69*). Further studies with liposomes consisting of sphingomyelin or saturated phospholipids showed an inhibited uptake by Kupffer cells, which was attributed to the attraction of serum dysopsonins (*70*). Thus, the surface properties of the liposomes appeared to be responsible for regulating the selection and adsorption of serum components that control uptake and hence the biodistribution of the carrier.

In addition, structure–activity relationships conducted with ganglioside GM1 incorporated into liposomes emphasized the importance of molecular structure for the functional ability to prolong liposome circulation. Dysopsonin recognition and binding were hypothesized to be the most likely explanation for its activity, and not absolute prevention of opsonization (*71*).

Specific receptor-mediated interactions can also be responsible for phagocytic uptake. Artursson et al. (*72*) found, for example, that carbohydrate receptors were responsible for the removal of starch and mannan microspheres by the macrophages of the liver and spleen.

Physicochemical Factors Influencing Biodistribution of Particulates

Following the iv administration of small latex particles to rabbits, typically 90% of the dose is cleared within a matter of a few minutes, predominantly by the liver, and to a lesser extent by the spleen (*73*). Lenaerts et al. (*74*) studied the intrahepatic distribution of poly(alkyl cyanoacrylate) (PACA) microparticles and found that they were located in the Kupffer cells rather than endothelial or parenchymal tissue. An enhanced uptake of particulate material by endothelial cells, due to specific antibody Fc receptors, was reported (*75*) in the case of rats having a depleted population of Kupffer cells.

The highly efficient extraction of particles from the bloodstream by the phagocytic cells of the RES, particularly the Kupffer cells of the liver, would appear to limit severely the prospect of drug delivery to cells other than these phagocytic cells (*76*). Many targeting systems, judged to be highly promising on the basis of an array of in vitro tests, have proved disappointing in vivo (*77*). However, because the process of phagocytosis and hence the biodistribution of iv administered colloids is critically dependent on their interaction with the biological environment, the physicochemical properties of the carrier can play a pivotal role in influencing this process (*47, 51*).

Particle Size

Particle size has been shown to be of primary importance in determining the biodistribution of the carrier. Administered particles of several micrometers in diameter for example, become entrapped within the lung capillaries (*78, 79*). Similarly, smaller, submicrometer systems are normally rapidly phagocytosed by the Kupffer cells of the liver (*73, 80*). Lenaerts et al. (*74*) found no difference in the in vivo sequestration by the RES of PACA microspheres ranging from 0.08 to 0.215 μm in the rat, although Pratten and Lloyd (*81*) showed that in the case of submicrometer particulates, uptake increased by rat peritoneal macrophages with an increased particle size. Tabata and Ikada (*64*) reported that in vitro phagocytosis of polystyrene and polyacrolein microspheres by mouse peritoneal macrophages was maximal when the particles were between 1.0 and 2.0 μm in size. When the particle size is reduced below 100 nm a concentration of the carrier can occur in the bone marrow for both radiopharmaceuticals (*82*) and liposomes (*83*).

Surface Charge

The role of surface charge in influencing protein (fibrinogen) adsorption and thus the surface nature of polyamide microcapsules was illustrated by Kondo (*63*). However, evidence for the influence of surface charge on particulate uptake has appeared confusing and often contradictory. This confusion may reflect the fact that changes in surface charge are also likely to alter other surface properties, such as hydrophobicity, which also influence the opsonization process (*84*). Wilkins and Myers (*85*) found that a negatively charged colloid was directed to the liver, whereas a positive charge caused an initial uptake in the lungs, followed by a secondary redistribution toward the spleen. Interestingly, in rat serum both types of colloid had the same zeta potential (*85*). Gregoriadis and Neerunjun (*86*) and Juliano and Stamp (*87*) found that the rate of hepatic removal of liposomes was greatest when they were negatively charged rather than of positive or neutral charge. More recently however, Park et al. (*88*) suggested that while some negatively charged liposomes can increase liver and spleen uptake, others could actually avoid uptake leading to prolonged circulation time. Davis et al. (*89*) reported no effect on the biodistribution of emulsions on reversal of the charge. Similarly, Tabata and Ikada (*64*) showed no differences between negatively or positively charged particles if they had the same absolute value. The same authors also demonstrated that for cellulose microspheres, the extent of phagocytosis increased with increasing negative zeta potential values (*65*). Similar results were obtained by Holden et al. (*90*) for polystyrene microspheres, although Illum et al. (*91*) showed no direct correlation between uptake and charge.

Surface Hydrophobicity

The importance of surface hydrophobicity in the pathogenicity of bacteria was clearly illustrated by Van Oss (*92*) and Absolom (*93*). If the contact angles (a measure of surface hydrophobicity) were less than those for neutrophils then phagocytosis was avoided, whereas a more hydrophobic surface led to sequestration (*93*). In addition to affecting the extent of ingestion of nonopsonized bacteria, the surface hydrophobicity will also strongly influence the degree of nonspecific IgG adsorption and complement activation, which promote phagocytic uptake (*93*). Van Oss (*51*) found a direct relationship between increased surface hydrophobicity of bacteria and enhanced phagocytic uptake in serum. The adsorption of dysopsonic IgA, however, increased the surface hydrophilicity and hence decreased phagocytosis (*51*).

The tendency toward increased phagocytosis with increased surface hydrophobicity is also evident in the case of synthetic polymeric microparticles. Tabata and Ikada (*64, 65*) showed a greater uptake of hydrophobic compared to hydrophilic microspheres. Extensive investigations have since reinforced the apparent link between the in vitro surface hydrophobicity and a diminished ability to avoid phagocytic sequestration for a range of surface-modified polystyrene and poly(methyl methacrylate) particles (*94, 95*).

Approaches to Targeted Drug Delivery

Local Administration of Drug Delivery Systems

The local (regional or compartmental) administration of drug-loaded microparticles to the affected tissue site offers a direct method of drug targeting and restricted drug delivery (e.g., injections of albumin microspheres in joints). Several types of particulate systems investigated for compartmental drug delivery are listed in Table 5.1. This approach to targeting is, however, of limited use in treating disseminated disease states. In cases where the target tissue is inaccessible or when frequent dosing is necessary, the iv route is more appropriate.

Passive Targeting: Exploitation of Natural Deposition Processes

The natural biological fate of particulate carriers was used for passive delivery to specific organ and tissue sites. Bradfield (*62*) summarized the potential of a macrophage delivery system for imaging, assessment of RES function, slow release of active agents, and drug and enzyme delivery.

One of the first examples of passive delivery was for the treatment of the intracellular infection leishmaniasis in the Kupffer cells (*96*). A 700-fold increase in drug efficacy was reported by administering antimony compounds entrapped in liposomes. Other intracellular diseases residing in macrophages and hence potentially treatable by passive targeting include brucellosis, listeriosis, salmonellosis, and mycobacterium infections (*97*). Considerable attention has also focused on the delivery of immunomodulators such as cytokines and muramyl di- and tripeptides to phagocytic cells (*66, 98, 99*). Macrophage activation can lead to increased metabolic and digestive processes and hence could be particularly advantageous in the treatment of neoplastic disorders (*51, 66*). The use of liposomal amphotericin B has attracted great interest for reducing the renal toxicity of the drug in the treatment of systemic fungal infections in patients with cancer, acquired immunodeficiency syndrome, and other immunodeficiencies (*100*).

Similar studies also were conducted using microparticulate carrier systems. The use of PACA microspheres containing a variety of therapeutic agents recently attracted considerable attention (Table 5.2). Schafer et al. (*101*) also indicated the in vitro potential of using PACA nanoparticles for the treatment of macrophages infected with human immunodeficiency virus.

Microparticulate drug delivery systems were investigated extensively for the treatment of tumors by the technique of chemoembolization (*102*). This technique involves the administration of particles, loaded with an antitumor drug, into the artery supplying the affected organ. The microparticles become lodged in the smaller blood vessels, hence providing a localized depot for drug release. The use of ethylcellulose microspheres loaded with mitomycin C for example, was described by Kato et al. (*103*) and Goldberg et al. (*104*) for the treatment of colorectal liver metastases. The beneficial effects reported were an increased drug half-life, a decrease in systemic exposure to the drug and hence side effects, and a decrease in tumor size (*103, 104*). The prior administration of angiotensin II in the therapy schedule recently was suggested in an attempt to increase the relative blood flow to the tumor tissue (*105, 106*). A novel targeting system recently was described by Madison et al. (*107*), whereby a latex nanosphere delivery system was sequestered by neuronal subpopulations following local injection into the brains of mice.

Potential for Active Targeting of Drug Delivery Systems

The concept of active targeting supposes interference in the way the body would normally respond to the targeting system (*108*). The use of external influences to direct carriers to localized areas has been investigated, such as applying an external magnetic field to direct magnetite-loaded albumin microspheres (*109*) or heating the target tissue to release drugs from heat-sensitive liposomes (*110*). Efforts also were directed toward developing pH-sensitive liposomes that destabilize in acidic microenvironments (*111*).

In an alternative approach aimed at modifying the response of the RES to administered microparticles, Illum et al. (*112*) suppressed RE activity by administering either dextran sulfate or a predose of polystyrene microspheres. Liver uptake of a subsequent dose of latex particles decreased from 90 to 30% and was accompanied by a corresponding increase in the lung levels from 5 to 60%. Similar observations were reported with the use of liposomes by Souhami et al. (*113*) and Patel et al. (*114*). Clearly, such blockade of the RES can have little justification in clinical practice, and hence recent attention has focused on the modification of the surface properties of the carrier for achieving site-specific drug delivery. The potential exists for using various surface-attached targeting moieties to improve the interaction between colloidal particles and specific tissue. Colloidal gold coated with lactosylated bovine serum albumin,

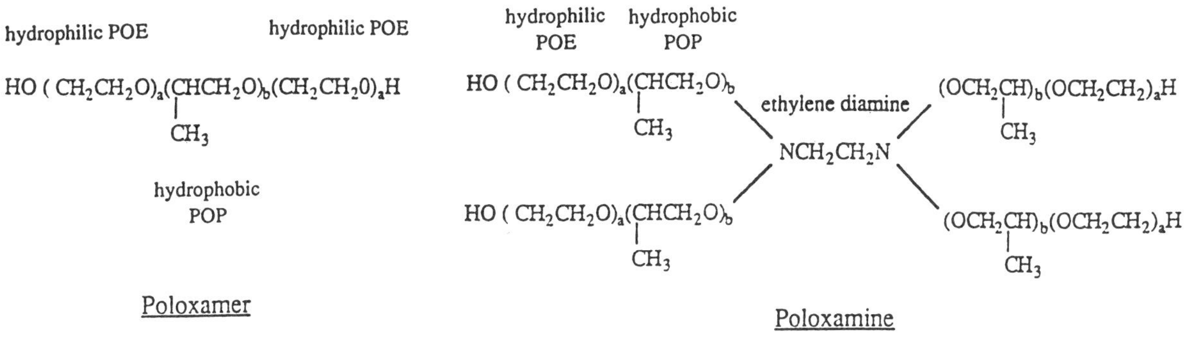

CHART 5.1. Structure of poloxamer and poloxamine POE–POP block copolymers.

for example, interacts with galactose ligands on rat hepatocytes (*115*), whereas corchorusin D bearing liposomes are captured largely by the nonparenchymal cells of the liver (*116*).

Targeting by Surface Modification of the Carrier

In nature, various cells and organisms have the capability of avoiding sequestration by the RES. As outlined previously, this property is believed to be related to the surface nature of the organisms whether it be the oligosaccharide units associated with blood cells or with the enveloping hydrophilic polysaccharide capsules of pathogenic streptococci bacteria (*40*). Bacterial antiphagocytic mechanisms were discussed in detail by Van Oss (*51*), who identified three principal types:

- expression of specific surface groups, for example protein A in *Staphylococcus aureus*, which binds preferentially to the Fc fragment of antibodies;
- presence of a surface membrane not susceptible to phagocytic digestion (as in mycobacteria); and
- formation of a hydrophilic surface.

Recently streptococcal M protein was shown to inhibit complement by binding the serum control protein, factor H (*117*). Consequently, attempts were made to exploit these principles and experimental findings to develop effective delivery systems capable of avoiding RES sequestration.

The use of microparticles for targeted drug delivery has been the subject of several recent texts and reviews (*8, 36, 49, 118*). Areas of research may be conveniently divided into nondegradable model systems and biodegradable carriers.

Nondegradable Systems: Adsorption of Block Copolymers

Coating of polystyrene microspheres by POE–POP block copolymers reduced dramatically their sequestration by the liver (*73, 119, 120*). In some cases their biodistribution was modified further such that a large proportion of the administered dose reached the bone marrow. (*121–124*). The chemical structure and composition of several POE–POP copolymers are shown in Chart 5.1 and Table 5.3, respectively. Anchoring of the copolymers to the particle surface occurs by attractive hydrophobic interactions between the polystyrene and the POP segments. This anchoring results in extension of the hydrophilic POE chains into the suspension medium. The resulting modification of the surface properties of the microparticles is believed to be responsible for the ensuing dramatic changes in their biological behavior. For example, coating of 60-nm polystyrene microspheres with poloxamer 338 caused a reduction in Kupffer cell uptake from 90 to 46%

TABLE 5.3. Composition of POE–POP Copolymers

Copolymer	Molecular Blocks (average values in moles) POE	POP	Average M_w of Polymer (Da)
Poloxamer 188	75	30	8,400
Poloxamer 238	97	39	11,400
Poloxamer 338	128	54	14,600
Poloxamer 407	98	67	12,600
Poloxamine 904	15	17	6,700
Poloxamine 908	122	22	25,000

of the injected dose (*73*). Coating with poloxamine 908 had a more pronounced effect, decreasing the amount removed to less than 25% (*119*).

If polystyrene microspheres are coated with poloxamine 908 and are larger than about 250–300 nm in size, then around 50% of the injected dose is captured by a filtration mechanism in the red pulp of the spleen following administration to rats (*122*). However, possible species-to-species differences have been identified, complicating the issue of colloidal drug delivery, because similar results are not obtained in the rabbit. Targeted delivery to the spleen could be advantageous for the delivery of drugs, such as hemoxygenase inhibitors, and both imaging and therapeutic radiopharmaceuticals (*122*).

The challenge of bone-marrow-directed delivery using microparticles has been comprehensively reviewed by Moghimi et al. (*125*). The applications and prospects were listed as diagnostic imaging, treatment of infections, proliferative diseases, and storage disorders and also protection or stimulation of the marrow hematopoietic functions. If polystyrene nanoparticles (less than 150 nm in diameter) were coated with poloxamer 407, redirection to the endothelial cells of the rabbit bone marrow occurred (*123, 124*). This targeting effect appears to be highly dependent on the molecular weight profile of the coating polymer (*124*).

Other investigators studied the effects on particle uptake of poly(ethylene oxide) groups grafted onto the surface of polystyrene microspheres (*126*). In vitro cellular interaction studies indicated that the approach adopted was not as effective as coating with poloxamer 238 (Table 5.3) in terms of reducing Kupffer cell uptake. However, a combination of the two surface modifications had a synergistic effect.

The ability to extend the blood circulation times of poly(methyl methacrylate) nanoparticles by adsorption of nonionic POE–POP block copolymers particles also was confirmed by several groups of workers. Leu et al. (*127*) (using poloxamer 188) and Troster et al. (*128*) (using poloxamer 338 and poloxamine 908) reported a decreased liver uptake and a prolonged circulation time of the coated particles. In contrast, coating of PACA microspheres with poloxamer 338 and poloxamine 908 was not reported to affect their in vivo behavior, probably due to desorption of the coating in vivo (*129*).

An alternative approach to surface modification of microparticles with PEG chains was reported by Artursson et al. (*130*). The circulatory half-life of acrylic microspheres could be increased by coating the particles with an albumin–PEG conjugate.

Biodegradable Carriers

Surface Modification of Liposomes

Ganglioside GM1

The incorporation of carbohydrate sialic acid residues in the form of ganglioside GM1 in the surface of liposomes greatly extends their circulation times (*131*). The activity of these so-called Stealth liposomes was discussed by Gabizon and Papahadjopoulos (*132*), who concluded that the retardation of uptake was due to the presence of the hydrophilic carbohydrate and a sterically hindered negatively charged group at the liposomal surface. Prolonged liposomal circulation times enhanced localization of drugs in murine colon carcinoma (*133*) and murine lymphosarcoma and in mammary adenocarcinoma (*134*).

Liu et al. (*135*) also reported that over 50% of the injected dose of GM1-containing liposomes can accumulate in the spleen, providing they are over 300 nm in size. This response was attributed to gradual filtration of the liposomes having prolonged circulation times. Exploitation of the targeting specificity of liposomes to the spleen, to confer adjuvant effects, was reported by Liu et al. (*136*) for obtaining a humoral response from an entrapped antigen.

Poly(ethylene glycol)
As predicted from the work of Illum and Davis, modification of liposomal surfaces by the incorporation of PEG–phospholipid derivatives prolonged circulation times and reduced phagocytic removal (*137, 138*). These effects were reported to be greater than those observed with GM1 (*137, 139*). Senior et al. (*140*) showed that the surface PEG groups acted as a barrier to the adsorption of proteins (opsonins) responsible for phagocytic removal. Blume and Cevc (*141*) indicated that an important factor in achieving these effects is the surface density of the PEG groups. The surface hydrophilicity is of lesser importance provided that a minimum value is exceeded. A detailed investigation of liposomal compositions concluded that for PEG incorporation with a molecular weight in the range of 1000 to 5000 Da, the prolonged circulation times (of up to 35% of the dose after 24 h) were not affected by wide variations in lipid compositions (*142*). PEG liposomes also accumulate in the spleen (*143*), as noted for GM1 containing liposomes.

Other Formulations

Liposomes modified with an uronic acid derivative, palmityl-D-glucuronide, were capable of avoiding RES uptake (*144*). Furthermore, following administration to tumor-bearing mice, liposomal tumor levels were 3–4 times higher than the control, and the liver:tumor ratio was only 2:1 (*145*).

The incorporation of phosphatidylinositol (PI) at the liposome surface also retarded in vivo clearance (*146*). When administered to rats with lung infections, PI-liposomes were collected in the infected tissue in amounts 10 times greater than control liposomes (*147*). Again, this effect would appear to be dependent on prolonging the circulation time of the liposomes.

Immunoliposomes

The targeting of liposomes coupled with monoclonal antibodies was more efficacious if the liposomes exhibit prolonged circulation times (*148*). The effect of the presence of GM1 and 2000- and 5000-Da molecular weight PEG on the efficacy of immunoliposomes was studied by Mori et al. (*149*). Only the larger PEG molecule caused a reduction in the target binding, which was attributed to the presence of a steric barrier. Immunoliposomes containing GM1 and a monoclonal antibody specific for pulmonary endothelia locate effectively in the lung at around 50% of the injected dose (*150, 151*). Furthermore, when doxorubicin was loaded into PEG immunoliposomes specific for lung cancer cells, the delivery system was capable of eradicating the tumors in mice (*152*).

Surface Modification and Biodistribution

The explanation of the effects of surface modification of particles by POE–POP copolymers and PEG ligands on carrier biodistribution is considered to reside in two main hypotheses involving steric stabilization and dysopsonin attraction.

Steric Stabilization and Surface Hydrophilicity
Steric stabilization of colloidal particles by adsorbed or grafted polymer chains arises from the formation of an enthalpic and entropic energy barrier to particle–particle interactions (*40, 153*). Napper (*153*) indicated that the coating polymers that provide the best stability are amphipathic block or graft copolymers. The adsorption of POE–POP block copolymers to particle surfaces

results in the protrusion of hydrophilic POE chains into the aqueous medium, thereby creating a steric barrier to the opsonization and phagocytic processes. A similar argument was proposed for the resistance to phagocytosis of liposomes surface-modified with PEG (*43*).

The in vitro characteristics of polystyrene lattices coated with a range of block copolymers was extensively studied (*91*, *94*) to quantify the changes occurring at colloid surfaces and hence to provide an interpretation and prediction of biological behavior. The coating-layer thicknesses, zeta potentials, surface hydrophobicity, and critical flocculation temperatures were investigated by Illum et al. (*91*), who concluded that for effective steric stabilization of the colloidal particles the dimensions of the stabilizing chains had to exceed the van der Waals range of attraction between particles (10 nm for a 60-nm particle) and perhaps between a particle and a macrophage. Similarly, Rudt and Muller (*154*) reported that particle uptake by human granulocytes could be reduced by increasing the POE chain length, an effect also attributed to changes in the adsorbed layer thickness and surface hydrophobicity.

The influence of the steric barrier on the opsonization of polystyrene microspheres was studied by Norman et al. (*155*, *156*). A reduction in the adsorption of human serum albumin following polymer coating was observed, an effect that increased with increasing POE content. An extensive study carried out with several proteins, however, did not reveal any significant differences in the protein adsorption patterns on microspheres coated with different POE–POP copolymers, and thus did not provide possible explanations for their different biological behavior (*156*). However, these experiments confirmed the general ability of POE-treated surfaces to minimize protein adsorption on biomedical materials (*157–159*).

Dysopsonin Theory

The effect of opsonins and dysopsonins in mediating phagocytosis and hence the biodistribution of particulates was discussed previously. Further evidence of dysopsonic activity in relation to synthetic polymeric microparticles was provided by Muir et al. (*120*). In vitro experiments showed that in the presence of serum, the uptake of both uncoated and poloxamine-908-coated polystyrene microspheres by isolated rat Kupffer cells was reduced: a serum dysopsonic factor was responsible for this behavior as well as the hydrophilic coating. In addition, the uptake observed for poloxamer-407-coated polystyrene latex by bone marrow endothelial cells would indicate that the steric barrier was overcome by some recognition process, possibly mediated by a bone-marrow-specific opsonin (*123*).

Biodegradable Particulate Delivery Systems

To develop and fully exploit the successes described previously in delivering nondegradable particles to specific target sites, it will be necessary to use biodegradable microparticles that adequately mimic the physicochemical properties, size, and surface nature of successful nondegradable systems. Consequently, increasing attention is now turning toward the potential applications of targeted nanoparticulate systems for drug delivery that are formulated from biodegradable polymers.

Biodegradable Polymers

An increasing requirement for the modulated delivery of both conventionally and biotechnology generated drugs of a high molecular weight and short half-life has generated considerable interest in the development of biodegradable polymers and their formulation into drug delivery systems (*160–163*). The use of biodegradable polymers confers the inherent advantage of alleviating the need for surgical removal of the delivery system at a later date. However, despite extensive investigations of drug delivery systems and vaccines, comparatively little work has involved the formulation of carriers from these materials capable of achieving site-specific delivery.

Biodegradable polymers used in drug delivery research may be broadly classified as of natural or synthetic origin. The majority of investigations into the use of natural polymers as drug delivery systems have concentrated on the use of proteins (e.g., collagen, gelatin, and albumin) and polysaccharides (e.g., starch, dextran, inulin, cellulose, and hyaluronic acid). Bogdansky (*164*) explained the advantages of these materials as being natural products of living organisms, readily available, relatively inexpensive, and capable of a multitude of chemical modifications. Many natural polymers have been formulated into microparticle form for drug delivery and have been the subject of several review articles (Table 5.4). A major drawback, however, concerns the possible immune response to these materials.

Synthetic Polymers

The synthetic polymers listed in Table 5.4 all were investigated for formulation of controlled drug delivery systems. They may be synthesized with specific properties to suit particular applications. The chemical structures are presented in Table 5.5 and Charts 5.2 and 5.3.

Polycaprolactone (PCL), for example, is an aliphatic polyester having a relatively slow degradation rate, which may be increased by copolymerization (*165*). PCL has a high permeability to many drugs, which consequently provides a rapid release characteristic from PCL devices (*166*, *167*).

TABLE 5.4. Natural and Synthetic Biodegradable Polymers Used in the Manufacture of Drug Delivery Systems

Type	*Material*	*Ref.*
Natural	Albumin	296, 297
	Gelatin	298
	Starch	299
Synthetic	Lactide polymers	300, 301
	(PLA, PGA, PLGA)	177, 302
	Polycaprolactone	303
	Polyanhydrides	169
	Poly(ortho ester)s	168
	Polyphosphazenes	170
	Poly(alkylcyanoacrylate)s	49
	Poly(malic acid) copolymers	172, 173

TABLE 5.5. Structure of Synthetic Biodegradable Polymers

Polymer	*Structure*
Aliphatic polyesters	
Poly(glycolide)	$-[-O-CH_2-CO-]-$
Poly(L-lactide)	$-[-O-CHMe-CO-]-$
Poly(ε-caprolactone)	$-[-O-(CH_2)_5-CO-]-$
Polyanhydrides	
Poly[bis(*p*-carboxyphenoxy)propane anhydride]	$-[-CO-C_6H_5-O-(CH_2)_3-O-C_6H_5-COO-]-$
Poly(carboxyphenoxyacetic acid)	$-[-CO-C_6H_5-O-(CH_2)-COO-]-$
Poly(carboxyphenoxyvaleric acid)	$-[-CO-C_6H_5-O-(CH_2)_4-COO-]-$
Polyphosphazenes	
Aryloxyphosphazene polymer	$-[-N{=}P(OC_5H_5)_2-]-$
Amino acid ester system	$-[-N{=}P(NHCH_2COOEt)_2-]-$
Poly(ortho ester)s	R | -[-O-C-O-]- | O-

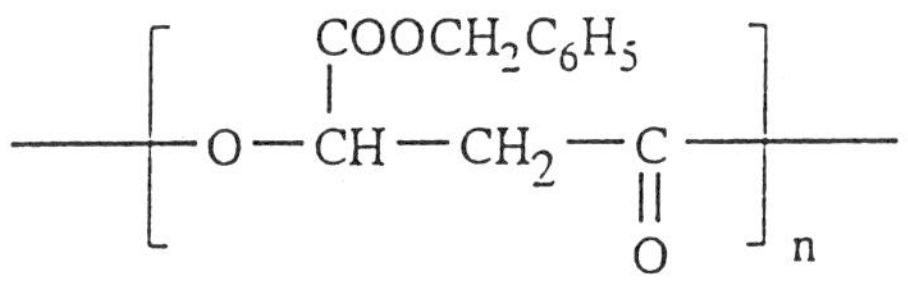

Poly(benzyl β-malate) or PMLABe100

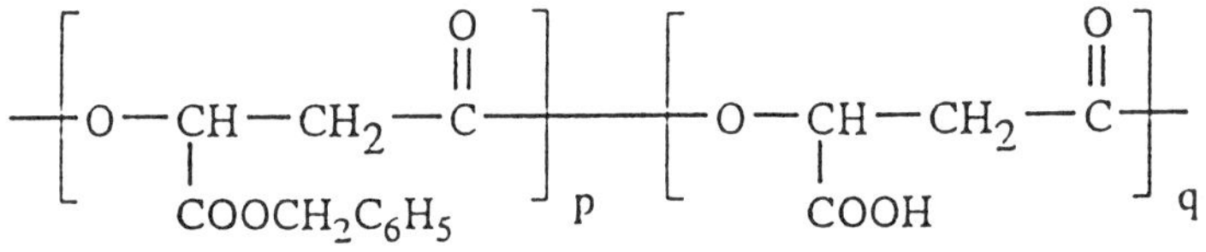

Poly(benzyl β-malate - co -β-malic acid) or PMLABe(100-x) Hx

CHART 5.2. Structure of poly(malic acid) derivatives.

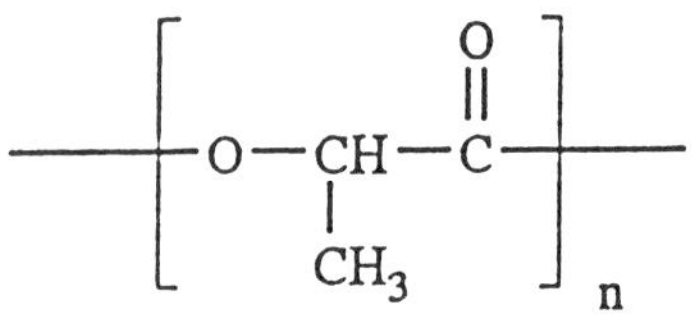

Poly(lactic acid) or PLA

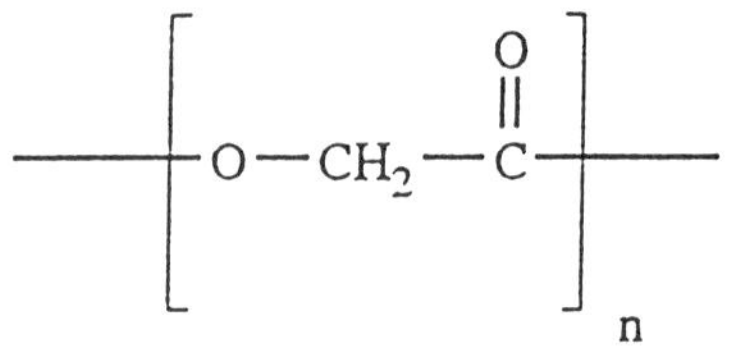

Poly(glycolic acid) or PGA

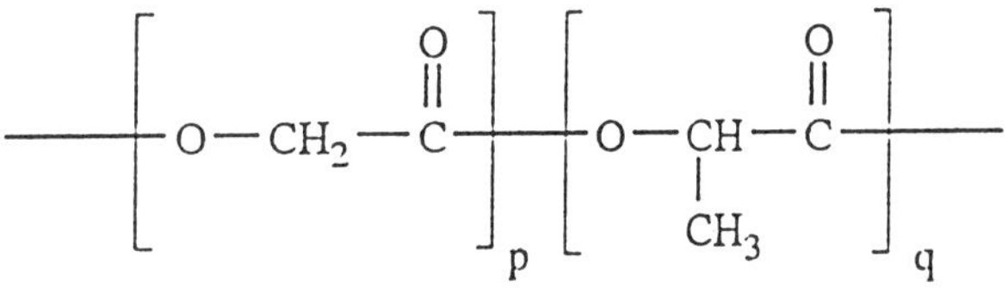

Poly(lactic acid-*co*-glycolic acid)

CHART 5.3. Structure of poly(lactic acid), poly(glycolic acid), and poly(lactic acid-*co*-glycolic acid).

Poly(ortho ester)s have the advantage over some polymeric materials of undergoing controlled degradation. This process is important for polypeptides that have a molecular weight too large to be released by a diffusion mechanism (*168*).

To maximize control over the process of drug release from polymeric matrices, surface degradation is desirable: that is, the rate of polymer hydrolysis at the surface of the device should be greater than the rate of water penetration into the bulk (*162*). Consequently, polyanhydride polymers and copolymers have attracted considerable interest for the fabrication of drug delivery systems because of the highly labile anhydride linkages in the polymer structure (*162, 169*).

Polyphosphazenes are a relatively new group of biodegradable polymers. Their degradation rates are controlled by side-group modification (*170, 171*). Also, reactive drug molecules can be linked to the polymer backbone.

Poly(malic acid) (PMLA) and its derivatives form a relatively new range of biodegradable polymers that have received attention recently for formulation of nanoparticulate drug delivery systems (*172, 173*). Poly(malic acid-*co*-benzyl malolactonate), the fully benzylated product, is nondegradable (Chart 5.2). The fully carboxylated polymer PMLA H_{100} is water soluble and degrades in a matter of days, but the hydrophilic/hydrophobic balance of PMLA derivatives may be tailored for control of degradation rate and solubility. Poly(alkyl cyanoacrylate) carriers have attracted considerable interest in the field of drug delivery (*74*, *174*), but reservations still exist with respect to the unresolved issue of toxicity (*175*, *176*).

Poly(lactic acid), Poly(glycolic acid), and Their Copolymers

The most widely investigated and advanced synthetic polymers in terms of the available toxicological and clinical data are the linear aliphatic polyesters based on the hydroxyacids lactic acid and glycolic acid (Chart 5.3). Poly(lactic acid) (PLA), poly(glycolic acid) (PGA), and poly(lactide-*co*-glycolide) (PLGA) display important advantages of biocompatibility, predictability of biodegradation kinetics, ease of fabrication, and regulatory approval (*177*). Importantly, these polymers can be purchased commercially. They are used clinically as absorbable sutures (*178*) and have been investigated for bone plates (*179*), implant materials (*180*), bone graft substitutes (*181*), and nerve graft substitutes (*182*). Yolles et al. (*183*) were among the first to use PLA for parenteral drug delivery. In recent years, these polymers have been applied widely in microsphere/microcapsule formulation (*177*, *184*) and are already used clinically for drug delivery in the form of implants (Parlodel SA, Sandoz; Zoladex, Zeneca; and Prostap, Lederle).

The synthesis, degradation behavior, and microparticle formulation techniques applicable to the poly(α-hydroxy acid)s are discussed subsequently because these factors profoundly influence drug incorporation strategies and the potential for microparticle size control and surface modification, which will ultimately decide the success or failure of a site-specific delivery system.

Synthesis

PLA and PGA are produced by ring-opening melt condensation of the cyclic dimers lactide and glycolide (*177*, *185*). Polymerization is usually conducted over 2–6 h at 175 °C in the melt in the presence of an organic catalyst such as stannous chloride or stannous octoate. Controlled polymerization requires a low monomer acidity and low humidity. The use of catalysts, however, may be undesirable, and hence several recent reports have detailed the synthesis of low molecular weight polymers via direct polycondensation in the absence of catalysts (*186*, *187*). Because of an asymmetric β-carbon of lactic acid, D and L stereoisomers are formed and hence polymers exist in the D, L, or racemic DL forms.

Properties

A broad spectrum of performance characteristics is required to span the wide range of biomedical applications of lactide polymers. The physical properties of PLA, PGA, and PLGA polymers can be fairly well defined at the synthesis stage by attention to the monomer stereochemistry, comonomer ratio, polymer chain linearity, and the polymer molecular weight (*177*). PLGA, for example, is reported as having suitable characteristics such as strength, flexibility, and hydrophobicity for use in drug delivery applications (*188*, *189*).

Polymer crystallinity is known to be an important determinant of polymer degradation (*177*), which influences the release profile of encapsulated proteins and peptides. Racemic DL-PLA is amorphous and has a lower softening point than either of the crystalline stereoisomers, D-PLA or L-PLA. Polyglycolide is also a crystalline solid with a high melting point (*185*). The copolymers of L-lactide and glycolide are less crystalline than either of the two homopolymers. Crystallinity decreases with increased copolymer content (*185*), and copolymers consisting of L-PLA and 25–70% of glycolic acid are amorphous. If DL-PLA is used for copolymer synthesis instead of L-PLA, the amorphous range extends from 0–70% because DL-PLA is intrinsically amorphous (*185*).

PGA is almost insoluble in common solvents, but when randomly copolymerized with 30–50% polylactide, the physical properties become more amenable to processing (*161*). Polylactide and PLGA copolymers with less than 50% glycolide are soluble in a range of both halogenated and nonhalogenated solvents (*177*) and are thus amenable to microparticle preparation by emulsification and precipitation techniques.

The rate of water uptake (hydration) will be influential in determining the rate of polymer degradation and hence drug release from targeted devices. The hydrophobic nature of the polymers also varies with the structural composition. Using water uptake into polymer films as an indication of polymer hydrophilicity, Gilding and Reed (*185*) showed that PLA was more hydrophobic than PGA because of the methyl side group in the chain structure. By increasing the glycolic acid content to 70%, the water level (hydrophilicity) increased. The block or random structure of the copolymer also is important in this process (*190*).

The physical and behavioral properties of PLA and PLGA copolymers may be further modified by two processes. The first process is by copolymerization with other polymers such as polycaprolactone (PCL) (*165, 166, 191*), aromatic hydroxyacids (*192*), poloxamers (*165*), or POE (*193–195*). A novel polymeric material of lactic acid linked to a tripeptide sequence was described by Yoshida et al. (*196*) for the manufacture of a microsphere drug delivery system. The second approach involves the formation of polymeric blends of PLGA with, for example, poly(vinyl alcohol) (PVA) (*197*) or poloxamers (*198, 199*).

Biocompatibility

The good biocompatibility of PLA implants was first noted by Kulkarni et al. (*179*). In vivo evaluation of PLA and PLGA copolymer delivery systems, in microsphere form, subsequently was evaluated by several groups of workers. In general, tissue responses were mild with no abnormal reactions, and these result explain the rapid clinical acceptance of lactide polymers (*200–202*).

The observations of Visscher et al. (*203*) are typical in that intramuscular administration of PLGA microcapsules of various particle sizes to rats evokes a sharply localized acute inflammatory reaction followed by minimal connective tissue, macrophage, and foreign body cell responses. These responses decline in parallel with microsphere degradation. Menei et al. (*204*) studied the brain tissue reaction to PLGA microspheres following implantation in the rat. A slight proliferation in astrocytes and an inflammatory reaction was temporarily observed, but with no adverse immunological or toxicological effects. A dose-dependent effect was observed by Julienne et al. (*205*) following iv injection of PLGA particles in mice. At a level of 665 mg/kg, microspheres were well tolerated; whereas at 1000 mg/kg, 20% of the mice died from cyanosis. No adverse effects on ocular tissues were observed following the vitreous injection of PLA microparticles in rabbits. Smith and Hunneyball (*206*) however, did report in vitro cell death of mouse peritoneal macrophages following incubation with high concentrations of PLA microcapsules. No differences in tissue response were observed between PLA and PLGA microcapsules (*207*), although interestingly, following intraperitoneal injection of PLA microparticles in mice, a greater inflammatory response was observed if L-PLA was used rather than DL-PLA (*208*).

Biodegradation

Biodegradation of the aliphatic polyesters (PLA, PGA, and PLGA) proceeds by homogenous or bulk degradation due to hydrolytic deesterification to their constituent monomers (*177*). The suggestion of enzymatic involvement in the degradation process remains controversial, although Holland et al. (*209*) suggested a possible involvement in the later stages of degradation. Lactic and glycolic acids are bioresorbable metabolites that are subsequently taken up into the endogenous pool of the body and eliminated through the Krebs cycle, primarily as carbon dioxide and in urine (*177, 210*).

The rate of microparticle degradation is critically dependent on the compositional and structural properties of the polymer and can greatly influence the rate of release of encapsulated drugs. In the case of semicrystalline copolymers hydrolytic attack starts in the amorphous regions, whereas crystalline areas degrade at a slower rate

(*211*). The extraction of short chain segments or the crystallization of more structurally regular chain segments produced by degradation can have the net result of increasing polymer crystallinity during the degradation process (*212*). Amorphous DL-PLA degrades faster than semicrystalline L-PLA. The degradation rate of PLA may also be increased by copolymerization with PGA (*185*). Miller et al. (*211*) showed that the shortest degradation half-life was exhibited by a 50:50 PLGA copolymer, whereas the longest corresponded with the L-PLA homopolymer.

Degradation rates decrease with an increase in polymer molecular weight (*213*). Consequently, Bodmeier et al. (*214*) suggested varying PLA microsphere degradation times by blending high and low molecular weight fragments. Other influential factors include the repeat structure of the copolymer (block versus random arrangement of comonomers) (*190*), the pH and ionic strength of the external medium (*215*), and the nature of the drug incorporated into the delivery device (*216, 217*).

Sterilization

Gamma irradiation has generally been employed for sterilization of resorbable polymer devices (*204, 218, 219*) although the dramatic decrease in polymer molecular weight on treatment is a well-known concern (*202*). Tsai et al. (*220*) investigated steam and ethylene oxide sterilization techniques for PLA microcapsules loaded with mitomycin C and found them unsuitable due to moisture penetration and interaction with the drug, respectively. The use of ethylene oxide did not affect the physical characteristics of drug-free PLA microspheres (*221, 222*).

Production Techniques for Biodegradable Polyester Microspheres

The preparation of microspheres and microcapsules from PLA, PGA, PLGA, and related polyesters for use as drug delivery devices has been the subject of several recent reviews (*184, 223, 224*). The technology currently available for preparation of drug-loaded nanoparticles from natural and synthetic polymers recently was reviewed by Allemann et al. (*225*). Various aspects of targeting, sustained drug release, avoidance of the RES, and oral and ocular administration were also discussed. Microparticle production techniques are summarized subsequently. The selection of formulation method will be influenced by such factors as the particle size required, the nature of any drug to be incorporated, and the required surface characteristics for site-specific delivery. In certain cases each stage of a manufacturing technique can offer opportunities for surface modification, as exemplified below, for the emulsification–solvent evaporation approach and the nanoprecipitation method.

Emulsification–Solvent Evaporation

The emulsification–solvent evaporation technique for preparation of microspheres was first described by Beck et al. (*226*) for the delivery of contraceptive steroids. It is arguably the most widely used manufacturing technique for biodegradable microspheres and has recently been reviewed by Tice and Gilley (*227*), Arshady (*228*), and Watts et al. (*229*). Microsphere formation consists essentially of three stages:

- droplet formation,
- droplet stabilization, and
- microsphere hardening.

First, a dispersed phase containing the polymer is emulsified in an immiscible continuous phase containing a stabilizing agent. The second stage involves the diffusion of the solvent from the emulsion droplet into the continuous phase and its subsequent evaporation. Simultaneous inward diffusion of the nonsolvent into the droplet causes polymer precipitation, microsphere formation, and hardening. Depending on the nature of the two phases, the process may be termed an oil-in-water (o/w) or water-in-oil (w/o) method. The o/w process has been the most widely used for production of PLGA microparticles and is discussed in more detail.

Oil-in-Water Emulsification–Solvent Evaporation: Organic Phase

Selection of an appropriate organic solvent for the disperse phase will be influenced by several requirements such as low toxicity, immiscibility with the continuous phase, a lower boiling point than the continuous phase, and drug solubility (*229*). Bodmeier and McGinity (*230*) examined a large range of solvents and their effects on drug loading and microsphere properties. They concluded that water-immiscible solvents with a high water solubility resulted in a higher rate of precipitation of the polymer and hence better drug entrapment. Dichloromethane is the most widely used organic solvent, whereas other reports have documented the use of chloroform (*205, 231*) or dichloromethane mixtures with water-miscible solvents (*222, 232*). A dichloromethane–acetone mixture recently was used in the o/w solvent evaporation method by Niwa et al. (*233*) and Coombes et al. (*234*) to prepare PLGA microspheres below 1 μm in size.

Surfactant Stabilizers

The solvent evaporation process requires the use of a surfactant to stabilize the dispersed-phase droplets formed during emulsification and inhibit coalescence. Surfactants are amphipathic in nature and hence will align themselves at the droplet surface, so promoting stability by lowering the free energy at the interface between the two phases. Furthermore, the creation of a charge or steric barrier at the droplet surface confers resistance to coalescence and microsphere flocculation. Interestingly, Bodmeier and McGinity (*235*) reported that by employing a continuous phase of pH 12, a satisfactory yield of PLA microspheres was obtained without an emulsifying agent. Droplet stabilization was considered to result from the presence of negative charges due to the polymer carboxyl end groups. Surfactants employed in the o/w process tend to be hydrophilic in nature, and examples are listed in Table 5.6. Of these surfactants, PVA is by far the most widely used and would appear to be the most effective for formation of particles below 1 μm in diameter.

Emulsification

The mixing conditions (equipment geometry, speed, and time) primarily influence the size of the dispersed-phase droplets and hence the final microsphere size (*223, 228*). The types of emulsification system used for microparticle production have included low or high speed mechanical stirring (*205*), sonication (*186*), and microfluidization (*236*). The emulsification forces generated by these techniques include high shear, turbulence, and cavitation (*237*). The use of low speed stirrers (typically below 2000 rpm) tends to result in the formation of particles above 20 μm in size. By increasing the emulsification speed there is both a decrease in the particle size and a narrowing of the

TABLE 5.6. Surfactant Stabilizers Used for Microsphere Production by the Emulsification–Solvent Evaporation Technique

Method	*Surfactant*	*Ref.*
O/W	PVA	226, 238, 243, 304
	Gelatin	305, 306, 307
	Polysorbate 80	206, 239
	Sodium alginate	305
	Sodium oleate	201, 308
	Methyl cellulose	238, 309
	Albumin	241
	Sodium dodecyl sulfate	243
	Sodium lauryl sulfate	200
	Polysorbate 20	252
	Pluronic-F68/Polysorbate 20	240
W/O, O/O	Span 40	310
	Span 65	220
	Span 85	308
	Soybean lecithin	311, 312

NOTE: O/W is oil-in-water; W/O is water-in-oil; and O/O is oil-in-oil.

size distribution (*238*). The design of the mixing vessel also was reported to influence particle size: smaller particles may be prepared by introducing baffles into the apparatus (*223, 228, 239*). The preparation of submicrometer particles, however, seems to be favored by a two-stage emulsification process having a high energy component (*240–243*).

Solvent Evaporation and Microsphere Recovery

Following emulsification, the removal of remnant solvent and complete microsphere hardening is usually accomplished by gentle agitation of the suspension, typically by magnetic stirring at room temperature and pressure. The speed of this process can be altered by changing the temperature or pressure. However, rapid evaporation of solvent can affect the particle morphology, polymer crystallinity, and drug crystallinity (*227, 244, 245*). A potential problem associated with solvent evaporation involves the formation of drug crystals on the microsphere surface (*238*). The authors suggested, however, that this problem could be prevented by interrupting the evaporation process, removing the surfactant phase, and resuspending the microparticles in emulsifier-free media.

Incomplete removal of solvent from the microspheres, particularly if halogenated, may be of toxicological concern. Spenlehauer et al. (*246*) and Gangrade and Price (*247*) have both found residual amounts of dichloromethane in microspheres, the actual quantity being related to the evaporation temperature and the microsphere properties such as porosity.

After evaporation of the solvent, the final stage of the emulsification–solvent evaporation process is the isolation of the microsphere from the dispersed phase containing free surfactant (and drug). This process has generally been achieved by centrifugation or filtration, and it is usually followed by a further cleaning process in which particles are washed several times with distilled water.

Water-in-Oil, Oil-in-Oil, and Multiple Emulsion Techniques

Whereas an o/w system is ideal for encapsulating lipophilic drugs, it is unsuitable for hydrophilic substances because of partition in the aqueous phase. For such substances w/o or o/o emulsion systems have been proposed. A dispersed phase of a water-miscible solvent such as acetonitrile or dimethylformamide containing the drug and polymer is emulsified in an oily continuous phase. The stabilizer used is consequently more lipophilic in nature than those employed in the o/w process (Table 5.6). However, this approach has the major disadvantages of product cleaning, oil-phase recovery, and microparticle agglomeration (*248*).

The manufacture of microspheres containing a hydrophilic peptide, leuprolide acetate, by a double-emulsion technique was first described by Ogawa et al. (*249*). The first stage of this process involved the formation of a primary w/o emulsion of an aqueous drug in an oil phase of PLGA or PLA in dichloromethane. The primary emulsion is subsequently dispersed in an external aqueous phase containing PVA to give a w/o/w double emulsion. Several other groups have since used this process to encapsulate water-soluble agents (*189, 250, 251*).

Solvent Extraction

A solvent extraction method was described for the preparation of PLA and PLGA microspheres by Pavanetto et al. (*252*) and Cowsar et al. (*253*). The process is similar in nature to the o/w emulsification technique except that following emulsification the preformed microspheres are poured into a large volume of nonsolvent (diluent) to extract the remaining organic solvent (less than 30 min), and thus cause rapid hardening of the particles. Pavanetto et al. (*252*) reported that microspheres formed by this method were more regular in shape, smaller, and had a narrower size distribution and higher porosity than those formed by the traditional o/w emulsification approach.

Phase Separation

The process of coacervation/phase separation recently was reviewed by Arshady (*254*). It consists essentially of suspending the drug in an organic solution of polymer and initiating phase separation by the addition of a second organic solvent. The encapsulation of a luteinizing hormone releasing hormone (LHRH) analog, nafarelin acetate, into PLGA microspheres was reported by Saunders et al. (*255*). The peptide was dissolved in water and suspended in a dichloromethane phase containing PLGA, producing a w/o emulsion. A nonsolvent was then added to precipitate the polymer around the droplet. Hardening of the microspheres and solvent extraction are then completed by the addition of a larger volume of nonsolvent. The phase-separation technique has since been used to entrap other drugs and LHRH analogues (*256–258*).

Nanoprecipitation

A simple method for preparing PLA and PLGA microspheres and microcapsules below 200 nm in size was described by Fessi et al. (*259*). This procedure involves dissolving the polymer, drug, and (for microcapsule formation) a phospholipid mixture in an organic acetone phase that is then slowly added (with gentle stirring only) to an aqueous solution of poloxamer 188. The miscibility of the two phases results in spontaneous and rapid particulate formation. A variety of drugs were encapsulated using this method (*260*).

Reversible Salting-Out

Ibrahim et al. (*261*) and Allemann et al. (*262*) recently described a salting out procedure to prepare submicrometer microspheres from a range of polymers including PLA. An aqueous phase saturated with electrolyte and containing PVA is added to an organic phase of PLA in acetone to form an o/w emulsion: the addition of excess water then dilutes the saturated salt solution and induces miscibility of the acetone with the aqueous phase, resulting in microsphere formation.

Spray-Drying

The preparation of PLA microspheres by spray drying was described by Bodmeier and Chen (*263*), Wang et al. (*213*), and Pavanetto et al. (*252*). A solution of the polymer in dichloromethane, with or without a suspended drug, is passed through the nozzle of a spray-drier causing solvent evaporation and microsphere hardening. Microspheres of between 5 and 10 μm can be produced by this approach.

Interfacial Deposition

A novel method of producing PLA microcapsules was described by Makino et al. (*215*) in which an o/w emulsion was first formed by using *n*-hexane and 1.5% poloxamer 188. A solution of PLA in dichloromethane was then added dropwise to the primary emulsion causing the PLA to precipitate at the surface of the *n*-hexane droplets. Microcapsules of approximately 1.5 μm diameter were produced in this way.

Melt Processing

Wichert and Rohdewald (*264*) described a melting method to prepare PLA that alleviates the need for organic solvents. The polymer and drug were melted together, cooled, then emulsified in a hot aqueous surfactant solution. The microsphere isolated by centrifugation or spray-drying had diameters of around 10 μm.

Current Applications of Polyester Microspheres

Embolic Material

Particles fabricated from poly(vinyl alcohol) sponge or Ivalon have been widely used as embolic materials to treat vascular and tumoral lesions (*265, 266*). However, the material suffers from the major disadvantages of having an irregular shape and a poorly defined size range (*222*), to the extent where the use of Ivalon has been associated with fatalities (*267*). Polylactide microspheres recently were investigated as an alternative material because of the more reproducible preparation techniques and an improved particle size distribution. Grandfils et al. (*222*) reported the controlled preparation of microspheres in the optimal tumor embolization range of 100–160 µm by using the solvent evaporation method. These particles were applied successfully in humans to reduce tumor vascularization and also to prevent hemorrhage after surgery by preoperative devascularization (*221, 222*).

Vaccine Adjuvants

The need for repeated administration of vaccines against tetanus and diphtheria to maintain satisfactory antibody levels is expected to be eliminated by combining microparticle delivery systems with differing degradation rates and hence antigen release times. The use of polymeric microsphere carriers for improving the immunogenicity of antigens was demonstrated by several groups (*41, 268*). O'Hagan et al. (*268*) have shown enhanced and prolonged IgG responses in mice following the entrapment of a model antigen (ovalbumin) into biodegradable PLGA microparticles. The biodegradable carrier may act as a reservoir for the antigen, allowing its slow release on polymer degradation and hence sustaining the immune response (*269*). In vivo studies have been conducted using PLA and PLGA microspheres with encapsulated diphtheria toxoid (*270*), tetanus toxoid (*271*), and staphylococcal enterotoxin B toxoid (*272*).

Drug Delivery

Drug delivery applications to date can be placed in two main categories. First, PLGA microspheres have been employed for regional tumor therapy by the technique of chemoembolization as described previously. Verrijk et al. (*3*) reported that in the treatment of hepatic tumors, PLGA microspheres loaded with cisplatin had a comparable antitumor efficacy to that of free drug but reduced side effects.

Most interest has been generated in the use of PLA, PGA, and PLGA microspheres as controlled release rate devices for a variety of conventional and peptide drugs (Table 5.7) (*177, 184*).

TABLE 5.7. PLA, PGA, and PLGA Microspheres Used for Controlled Drug Delivery In Vivo

Microsphere	*Drug*	*Route*	*Species*	*Ref.*
PLA, PGA	Norethisterone	IM	Humans	313
PLGA	Norethisterone	IM	Humans	218
PLA	Testosterone	IM	Rats	200
PLA	Cisplatin	IM	Rabbits	314, 315
PLA	Cisplatin	IP	Mice	316
PLA	5-Fluorouracil	IO	Rabbits	289
PLA	Doxorubicin	IP	Humans	314, 315
PLA	Bromocriptine	IM	Humans	317
PLGA	Dopamine	Brain	Rats	318
PLGA	Ampicillin	Bone	Rabbits	219
PLA	Insulin	SC	Rats	319
PLGA	LHRH antagonist	IM	Rats	258
PLGA	LHRH agonist	SC	Rats	320
PLA	Neurotensin analogue	SC	Rats	232
PLGA	Somatostatin analogue	IM	Rabbits	250

NOTE: IM is intramuscular; IP is intraperitoneal; IO is intraocular; and SC is subcutaneous.

Their fabrication by the widely used o/w solvent evaporation process, however, is often problematical because most drugs will tend to be water soluble, resulting in poor levels of drug incorporation due to partitioning into the aqueous phase. This situation may be improved by devices such as presaturating the aqueous phase with drug (*246*), altering the aqueous phase pH (*239*), using lipophilic prodrugs (*273*), adding 3 M saline (*244*), using more miscible solvents to increase polymer precipitation (*230*), and also producing multiple w/o/w emulsions (*249*). The continued development of such techniques will be highly important for delivery of new peptides and proteins emanating from advances in biotechnology.

Site-Specific Delivery of Biodegradable Microparticles Following Intravenous Administration

Relatively few investigations of PLA or PLGA particles as iv targeting systems have been reported. Bazile et al. (*241*) indicated that PLA microspheres coated with albumin were incapable of avoiding Kupffer cell sequestration: approximately 90% of the administered dose was recovered from the liver. Similarly, PLGA microspheres were found predominantly in the liver and spleen following iv injection (*274*).

Preparation of Submicrometer Particles

One accepted strategy for avoiding RES uptake and targeting nanoparticles to sites other than the RES requires colloidal systems less than 50–60 nm to enable transendothelial passage via fenestrations (50–60 nm) or sinusoids (<100 nm) (*275*). Small size liposomes are capable of reaching hepatocytes by penetrating the endothelial barrier through the fenestrations (*276*). Seijo et al. (*277*) attempted to exploit the size control of drug targeting by preparing nanoparticles less than 50 nm from isobutyl and isohexyl cyanoacrylate by using an emulsion polymerization method. These nanoparticles were loaded with hydrophilic (ampicillin) and hydrophobic drugs (dexamethasone).

Several studies in recent years have described the preparation of biodegradable microspheres from PLA, PLGA, and other polyesters in the nanometer size range by using the techniques outlined previously. These procedures are listed in Table 5.8. Most success has been achieved using

TABLE 5.8. Preparation of Submicrometer Microspheres from Biodegradable, Synthetic Polyesters

Polymer	*Method*	*Size (nm)*	*Ref.*
PLA, PLGA	O/W, sonication mix with PVA	200	236
PLA	O/W, sonication mix with gelatin	500–700	231
PLA	O/W	450	321
PLGA, PCL	O/W, high-speed mix with PVA	200–300	205
PLGA	O/W, mechanical, homogenization mix with PVA	100–200	242
PLGA	O/W, mechanical–sonication, mix with PVA, SDS	250	243
PLA, PCL	O/W, mechanical–microfluidizer, Tween 20/F68	200	240
PLA	O/W, mechanical–microfluidizer mix with albumin	100	241
PLGA	O/W, high-speed mix with PVA	200	233
PLA	O/W, sonication mix with PVA	<1000	186
PLGA	O/W, high-speed mix with PVA and sonication	92	284
PLA	Reversible salting out	200	262
PLA, PLGA	Interfacial polymer deposition	100	259

NOTE: O/W is oil-in-water and solvent evaporation. SDS is sodium dodecylsulfate, PLA is poly(lactide); PGA is poly(glycolide); PLGA is poly(lactide-*co*-glycolide); and PCL is polycaprolactone.

the o/w emulsification–solvent evaporation method. However, the lower size restraint of approximately 100 nm imposed on PLG nanoparticles prepared by the methods listed in Table 5.8 is expected to restrict transendothelial transport.

Surface Modification by Adsorption of Block Copolymers

POE–POP Copolymers

The approach to targeting by surface modification has not yet produced biological results comparable with those of polystyrene latex when applied to a variety of colloidal systems including liposomes (*278*, *279*), poly(methyl methacrylate) (*95*, *128*), and poly(alkyl cyanoacrylate) particles (*175*). These findings serve to underline the fact emerging from the earlier discussion that the mechanisms governing particulate identification and sequestration by the cells of the RES are many and complex.

To date, very few reports have been published detailing the biological fate of PLGA particulates coated with the POE–POP–POE copolymers responsible for reducing the uptake of polystyrene carriers by the RES. Muller and Wallis (*243*) studied the in vitro physicochemical properties of PLGA microspheres manufactured in the presence of poloxamers, but they concluded that on the basis of several numerical values identified for the avoidance of RES uptake (*94*), the biodegradable particles would not be successful in vivo.

The design principles established for avoiding RES uptake by using polystyrene nanoparticles with adsorbed POE–POP block copolymers recently were applied successfully to biodegradable PLGA systems (*280*). Nanoparticles were first prepared by nanoprecipitation methods as previously described. As expected, coating with various copolymers (Poloxamer 407, Poloxamine 908, and Poloxamine 407; *see* Table 5.3 for copolymer composition) resulted in differences in adsorbed layer thickness and surface charge corresponding directly with the chain length of the POE segment within the copolymer. Coating with Poloxamer 407 resulted in the most favorable biodistribution profile in rats, characterized by dramatically reduced liver uptake relative to uncoated particles and high blood circulation times (Figure 5.1). Around 44% of the injected dose of nanoparticles remained in the circulation after 3 h. Further work is planned to elucidate the relationship between the chemical and configurational structure of the POE–POP coating copolymer and bone-marrow targeting.

PLA–PEG Copolymers

Biodegradable block copolymers produced from PLA and POE have attracted interest for a variety of biomedical applications. Zhu et al. (*194*) prepared super microcapsules of 40–100 nm diameter for drug delivery by phase separation of poly(ethylene oxide)–polylactide star-type copolymers.

PLGA nanoparticles (140 nm) prepared by the nanoprecipitation technique have been coated with water-soluble, PLA–PEG diblock copolymers to increase surface hydrophilicity and protein resistance (*173*). The importance of the interaction between the particle substrate and the hydrophobic segment of the coating polymer, for extending the blood circulation time of delivery systems, was one of the points highlighted by the study. Coating of poly(butyl 2-cyanoacrylate) nanoparticles with POE–POP copolymers (Poloxamer 338 and Poloxamine 908) was ineffective in preventing RES uptake (*281*). Polystyrene nanoparticles coated with PLA–PEG copolymers exhibited high blood circulation levels initially on administration to rats, but the organ distribution after 3 h was similar to uncoated nanoparticles. In contrast, coating of PLGA nanoparticles with PLA–PEG copolymers resulted in marked improvements in their biodistribution, and approximately 40% of the administered dose was retained in the circulation after 3 h (Figure 5.2).

The hydrophobic PLA segment of the PLA–PEG copolymer was thought to provide a more stable anchorage with the structurally similar PLGA substrate compared with polystyrene. Consequently the hydrophilic coating was more resistant to desorption and binding with plasma components. In addition, a similar biodistribution pattern was observed for PLGA nanoparticles

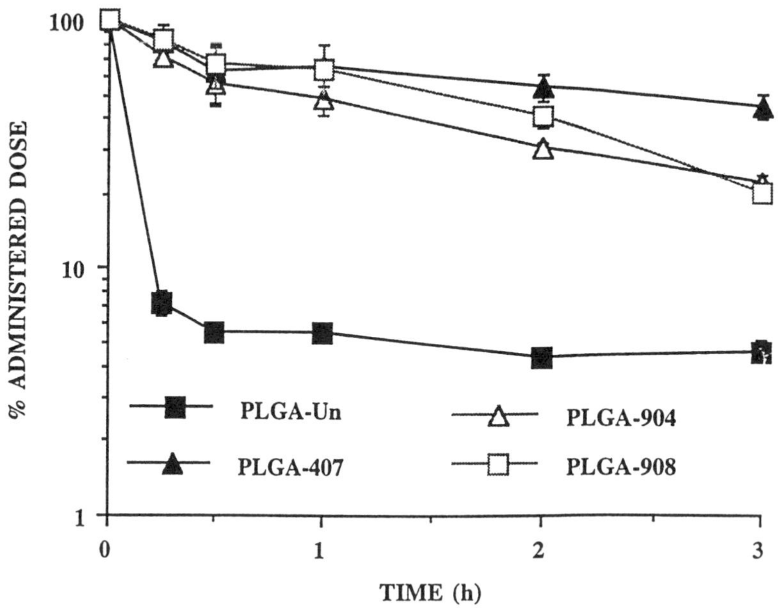

FIGURE 5.1. Blood clearance profiles for PLGA nanoparticles coated with POE–POP copolymers in the rat model.

coated with PLA–PEG copolymers having either 2000 or 5000 MW PEG chains and hence different coating thicknesses. This finding lends support to the views of Blume and Cevc (*141*) that the surface hydrophilicity of particulates is of lesser importance for avoiding RES sequestration once a minimum value is reached and that the surface density of PEG groups is then more influential in the process of phagocytosis.

Production and In Situ Surface Modification of PLGA Nanoparticles

The possibility of desorption or displacement of adsorbed coatings by blood components is recognized as a disadvantage of colloidal particles modified by adsorption and could lead to their eventual accumulation in the liver (*31*). Watrous-Peltier et al. (*282*) observed significant loss of POE–POP-coated polystyrene microspheres from the blood pool approximately 6 h after injection, a result that was believed to result from desorption of the coating with time. In addition, the hydrophilic polymer coating on polystyrene microspheres minimizes phagocytosis by splenic macrophages following filtration and interaction with other splenic cells unless the coating is gradually lost (*122, 278*). Such concerns explain the interest in surface modification of colloidal particles by grafted hydrophilic POE chains (*126, 280*).

PLGA nanoparticles surface modified by PEG chains may be prepared directly from PLGA–PEG copolymers, obviating the need for postadsorption and providing inherently more stable surface layers. Gref et al. (*283*) prepared biodegradable nanoparticles (90–150 nm) containing approximately 10% w/w PEG from amphiphilic, diblock copolymers based on PLGA (or PCL) and PEG by an emulsification–solvent evaporation technique, which resulted in a PEG surface layer. Increased blood circulation times in mice and reduced liver accumulation were measured. In addition, up to 45% by weight of the drugs, lidocaine or prednisolone, could be incorporated in the nanoparticles. Blood circulation times increased as the molecular weight of the PEG component

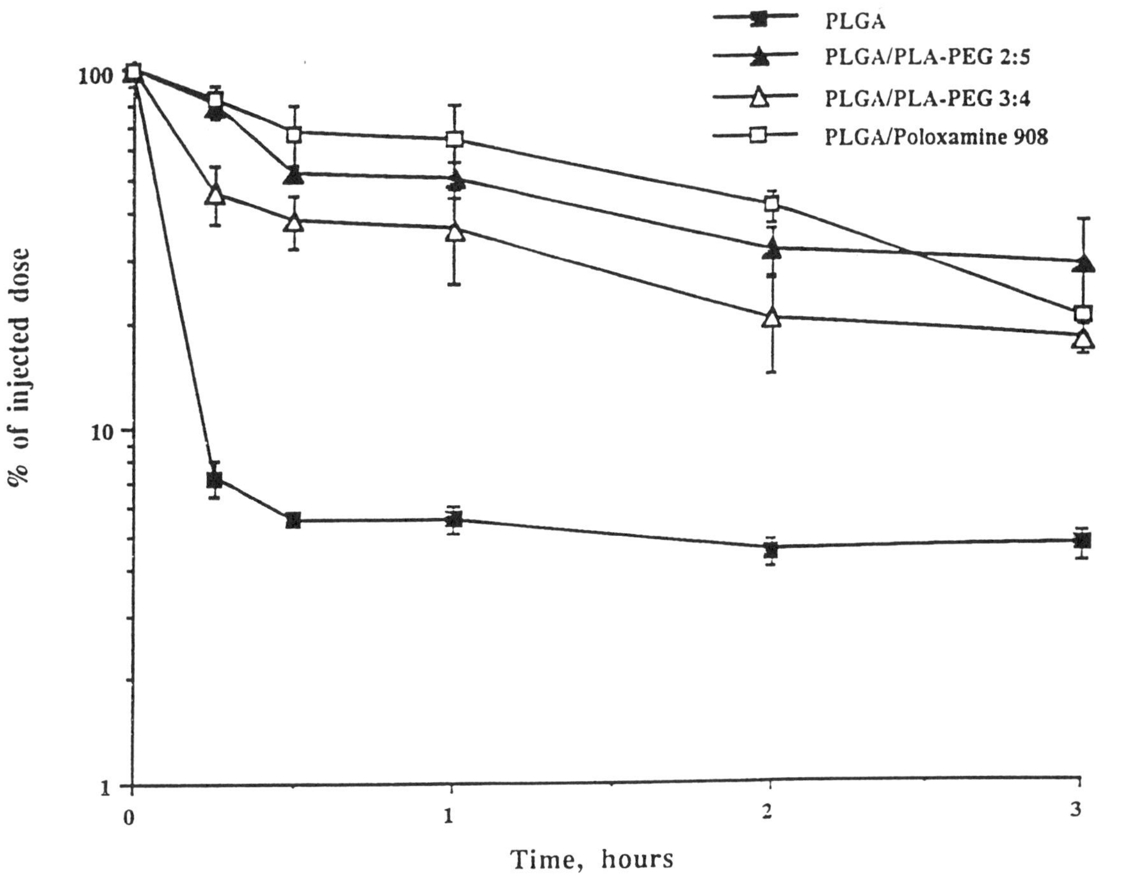

FIGURE 5.2. Blood clearance profiles for PLGA nanoparticles coated with PLA–PEG copolymers in the rat model.

increased, and this result was explained in terms of the corresponding increase in thickness of the hydrophilic PEG surface layer. The accumulation of 20-kDa PEG modified nanospheres in the liver was less than 30% after 5 h, whereas unmodified PLGA nanoparticles were cleared rapidly by the liver and spleen in less than 15 min. Opportunities for endocytosis and specific targeting of the PLGA–PEG nanoparticles were envisaged by surface attachment of proteins such as transferrin and antibodies, respectively.

An alternative approach to surface modification of PLGA microparticles by PEG chains was developed by Coombes et al. (*234*). It exploits the finding that the stabilizers used in the emulsification–solvent evaporation technique are highly resistant to removal by washing and dialysis. Thus, opportunities were presented for simultaneous manufacture and surface modification of microspheres by some of the POE–POP copolymers previously applied as coatings on nondegradable nanoparticles. Improved binding of stabilizer molecules with the carrier surface was anticipated, possibly leading to improved resistance to desorption and displacement by blood components. Surface analysis revealed the presence of POE–POP copolymer at the microparticle surface after extended periods of dialysis. The smallest nanoparticles produced in the study had a mean size of 322 nm, and these nanoparticles are potentially suitable for targeting the spleen (*122, 278*). The preparation of POE–POP modified nanoparticles of around 160 nm by using the emulsification–solvent evaporation technique subsequently was described by Scholes (*284*).

We are currently investigating surface modification of PLGA nanoparticles (500–800 nm) by hydrophilic species such as polysaccharides during production by the emulsification–solvent evaporation technique. This approach involves the conjugation of hydrophilic species to PEG chains, which then function as particle stabilizers. The PEG component of the conjugate provides the anchor to the particle surface, whereas the pendant hydrophilic segment provides steric stabilization and possibly a targeting function or alternatively acts as a spacer for attachment of a separate targeting ligand.

Conclusions

Extensive investigations over a period approaching 30 years of the interaction of particulate materials with the RES have provided several well-defined strategies and design guidelines for avoiding phagocytosis and formulating site-specific drug delivery systems. Advances in nanoparticle technology and techniques of surface modification allied to fundamental investigations of RES function have resulted in the production of biodegradable particles that display an increase in their iv half-life. This result markedly improves the prospects for targeted drug delivery to sites of inflammation and tumors. The present goal would appear to lie in effectively combining targeting agents, such as those applied to macromolecular carriers and liposomes, with camouflage techniques to provide circulating colloidal delivery systems with a homing ability and directionality.

References

1. Mills, S. N.; Davis, S. S. In *Polymers in Controlled Drug Delivery;* Illum, L.; Davis, S. S., Eds.; IOP Publishing: Bristol, U.K., 1987; pp 1–14.
2. Gupta, P. K. *J. Pharm. Sci.* **1992,** *79,* 949–962.
3. Verrijk, R.; Smolders, I. J. H.; Bosnie, N.; Begg, A. C. *Cancer Res.* **1992,** *52,* 6653–6656.
4. Stewart, M. J.; Plautz, G. E.; Del Buono, L.; Yang, Z. Y.; Xu, L.; Gao, X.; Huang, L.; Nabel, E. G.; Nabel, G. J. *Hum. Gene Ther.* **1992,** *3,* 267–275.
5. Davis, S. S.; Hunneyball, I. M.; Illum, L.; Ratcliffe, J. H.; Smith, A.; Wilson, C. G. *Drugs Exp. Clin. Res.* **1985,** *11,* 633–640.
6. Poste, G.; Kirsch, R. *Biotechnology* **1983,** *1,* 869–878.
7. Artursson, P. In *Polymers in Controlled Drug Delivery;* Illum, L.; Davis, S. S., Eds.; IOP Publishing: Bristol, U.K., 1987; pp 15–24.
8. Davis, S. S.; Illum, L. In *Drug Carrier Systems;* Roedink, F. H. D.; Kroon, A. M., Eds.; Wiley: New York, 1989; pp 131–153.
9. Ehrich, P. *Collected Studies on Immunity;* Wiley and Sons: New York, 1906; p 442.
10. Schacht, E. In *Polymers in Controlled Drug Delivery;* Illum, L.; Davis, S. S., Eds.; IOP Publishing: Bristol, U.K., 1987; pp 131–151.
11. Kopeček, J.; Duncan, R. In *Polymers in Controlled Drug Delivery;* Illum, L.; Davis, S. S., Eds.; IOP Publishing: Bristol, U.K., 1987; pp 152–170.
12. Seymour, L. W. *Crit. Rev. Ther. Drug Carrier Syst.* **1992,** *9,* 135–187.
13. Kopeček, J. *J. Controlled Release* **1990,** *11,* 279–290.
14. Yokoyama, K.; Ueda, Y.; Kikukawa, A.; Yamanouchi, K. In *Targeting of Drugs 2: Optimization Strategies;* Gregoriadis, G.; Allison, A. C.; Poste, G., Eds.; Plenum: New York, 1990; pp 1–20.
15. Takakura, Y.; Takagi, A.; Hashida, M.; Sezaki, H. *Pharm. Res.* **1987,** *4,* 293–300.
16. Chakraborty, P.; Bhaduri, A. N.; Das, P. K. *J. Protozool.* **1990,** *37,* 358–364.
17. Kaneo, Y.; Tanaka, T.; Iguchi, S. *Chem. Pharm. Bull.* **1991,** *39,* 999–1003.
18. Gregoriadis, G.; McCormack, B.; Wang, Z.; Lifely, R. *FEBS Lett.* **1993,** *315,* 271–276.
19. Courtenay-Luck, N. S.; Epenetos, A. A. *Curr. Opin. Immunol.* **1990,** *2,* 880–883.
20. Buchsbaum, D. J.; Lawrence, T. S. *Antibody Immunoconjugates Radiopharm.* **1991,** *4,* 245–272.
21. Friden, P. M.; Walus, L. R.; Musso, G. F.; Taylor, M. A.; Malfroy, B.; Starzyk, R. M. *Proc. Natl. Acad. Sci. U.S.A.* **1991,** *88,* 4771–4775.
22. Engert, A.; Thorpe, P. In *Targeting of Drugs 2: Optimization Strategies;* Gregoriadis, G.; Allison, A. C.;

Poste, G., Eds.; Plenum: New York, 1990; pp 39–48.

23. Bagshawe, K. D. *Br. J. Cancer* **1989,** *60,* 275–281.
24. Roffler, S. R.; Wang, S.-M.; Chern, J.-W.; Teh, M.-Y.; Tung, E. *Biochem. Pharmacol.* **1991,** *42,* 2062–2065.
25. Yokoyama, M. *Crit. Rev. Ther. Drug Carrier Syst.* **1992,** *9,* 213–248.
26. Florence, A. T.; Attwood, D. *Physicochemical Principles of Pharmacy,* 2nd ed.; Macmillan: Basingstoke, U.K., 1988.
27. Barry, B. W.; El Eini, D. I. D. *J. Pharm. Pharmacol.* **1976,** *28,* 210–218.
28. Kabanov, A. V.; Chekhonin, V. P.; Alakhov, V. Y.; Batrakova, E. V.; Lebedev, A. S.; Melik-Nubarov, N. S.; Arzakov, S. A.; Levashov, A. V.; Morozov, G. V.; Severin, E. S.; Kabanov, V. A. *FEBS Lett.* **1989,** *258,* 343–345.
29. Kabanov, A. V.; Batrakova, E. V.; Melik-Nubarov, N. S.; Fedoseev, N. A.; Dorodnich, T.; Alakhov, V.; Chekhonin, V. P.; Nazarova, I. R.; Kabanov, V. A. *J. Controlled Release* **1992,** *22,* 141–158.
30. Hagan, S. A.; Coombes, A. G. A.; Garnett, M. C.; Dunn, S. E.; Davies, M. C.; Illum, L.; Davis, S. S.; Harding, S. E.; Purkiss, S.; Gellert, P. *Langmuir* **1996,** *12,* 2153–2161.
31. Rolland, A.; O'Mullane, J.; Goddard, P.; Brookman, L.; Petrak, K. *J. Appl. Polym. Sci.* **1992,** *44,* 1195–1203.
32. Van Berkel, T. J. C.; De Smidt, P. C.; Van Dijk, M. C. M.; Ziere, G. J.; Bijsterbosch, M. K. *Biochem. Soc. Trans.* **1990,** *18,* 748–750.
33. Bijsterbosch, M. K.; Van Berkel, T. J. C. *Mol. Pharmacol.* **1991,** *41,* 404–411.
34. Tonetti, M.; Polvani, C.; Zocchi, E.; Guida, L.; Benatti, U.; Biassoni, P.; Romei, F.; Guglielmi, A.; Aschele, C.; Sobrero, A.; DeFlora, A. *Eur. J. Cancer* **1991,** *27,* 947–948.
35. Rosenberg, S. A.; Aebersold, P.; Cornetta, K.; Kasid, A.; Morgan, R. A.; Moen, R.; Karson, E. M.; Lotze, M. T.; Yang, J. C.; Topalian, S. L.; Merino, M. J.; Culver, K.; Miller, D.; Blaese, M.; Anderson, W. F. *N. Engl. J. Med.* **1990,** *323,* 570–578.
36. *Microspheres and Drug Therapy, Pharmaceutical, Immunological and Medical Aspects;* Davis, S. S.; Illum, L.; McVie, J. G.; Tomlinson, E., Eds.; Elsevier: Amsterdam, Netherlands, 1984.
37. Gregoriadis, G. *Liposomes as Drug Carriers, Recent Trends and Progress;* Wiley and Sons: New York, 1988.
38. Speiser, P. P. *Methods Find. Exp. Clin. Pharmacol.* **1991,** *13,* 337–342.
39. Moghimi, S. M.; Illum, L.; Davis, S. S. In *Advances in Molecular and Cellular Biology;* Zetter, B. R., Ed.; JAI Press: London, 1994; Vol. 9.
40. Davis, S. S.; Illum, L. *Biomaterials* **1988,** *9,* 111–115.
41. Kreuter, J. *J. Microencapsulation* **1988,** *5,* 115–127.
42. Gregoriadis, G. *Trends Biotechnol.* **1995,** *13,* 527–537.
43. Woodle, M. C.; Lasic, D. D. *Biochim. Biophys. Acta* **1992,** *1113,* 171–199.
44. Lasic, D. *Am. Sci.* **1992,** *80,* 20–31.
45. Florence, A. T.; Cable, C.; Cassidy, J.; Kaye, S. B. In *Targeting of Drugs 2: Optimization Strategies;* Gregoriadis, G.; Allison, A. C.; Poste, G., Eds.; Plenum: New York, 1990; pp 117–126.
46. Tomlinson, E. In *Targeting of Drugs 2: Optimization Strategies;* Gregoriadis, G.; Allison, A. C.; Poste, G., Eds.; Plenum: New York, 1990; pp 1–20.
47. Davis, S. S.; Illum, L. In *Targeting of Drugs, Anatomical and Physiological Considerations;* Gregoriadis, G.; Poste, G., Eds.; Plenum: New York, 1988; pp 177–187.
48. Jain, R. K. *Int. J. Radiat. Biol.* **1991,** *60,* 85–100.
49. Douglas, S. J.; Davis, S. S.; Illum, L. *Crit. Rev. Ther. Drug Carrier Syst.* **1987,** *3,* 233–261.
50. Horwitz, M. A. *Rev. Infect. Dis.* **1982,** *4,* 104–123.
51. Van Oss, C. *Methods Enzymol.* **1986,** *132,* 3–15.
52. Saba, T. M. *Arch. Intern. Med.* **1970,** *126,* 1031–1052.
53. Smythe, E.; Warren, G. *Eur. J. Biochem.* **1991,** *202,* 689–699.
54. Norde, W. In *Microspheres and Drug Therapy, Pharmaceutical, Immunological and Medical Aspects;* Davis, S. S.; Illum, L.; McVie, J. G.; Tomlinson, E., Eds.; Elsevier: Amsterdam, Netherlands, 1984; pp 39–60.
55. Norde, W. *Adv. Colloid Interface Sci.* **1986,** *25,* 267–340.
56. Absolom, D. R. *Methods Enzymol.* **1986,** *132,* 281–318.
57. Patel, H. M. *Crit. Rev. Ther. Drug Carrier Syst.* **1992,** *9,* 39–90.
58. Najjar, V. A.; Konopinska, D.; Lee, J. *Methods Enzymol.* **1986,** *132,* 318–334.
59. Blumenstock, F. A.; Saba, T. M.; Cardarelli, P.; Dayton, C. H. *Methods Enzymol.* **1986,** *132,* 334–349.
60. Cardarelli, P. M.; Blumenstock, F. A.; McKeown-Longo, P. J.; Saba, T. M.; Mazurkiewicz, J. E.; Dias, J. A. *J. Leukocyte Biol.* **1990,** *48,* 426–437.

61. Bjornson, A. B.; Michael, J. G. *J. Infect. Dis.* **1973,** *128,* 182–186.
62. Bradfield, J. W. B. In *Microspheres and Drug Therapy, Pharmaceutical, Immunological and Medical Aspects;* Davis, S. S.; Illum, L.; McVie, J. G.; Tomlinson, E., Eds.; Elsevier: Amsterdam, Netherlands, 1984; pp 25–38.
63. Kondo, T. In *Microspheres and Drug Therapy, Pharmaceutical, Immunological and Medical Aspects;* Davis, S. S.; Illum, L.; McVie, J. G.; Tomlinson, E., Eds.; Elsevier: Amsterdam, Netherlands, 1984; pp 61–72.
64. Tabata, Y.; Ikada, Y. *Biomaterials* **1988,** *9,* 356–362.
65. Tabata, Y.; Ikada, Y. *J. Colloid Interface Sci.* **1989,** *127,* 132–140.
66. Tabata, Y.; Ikada, Y. *Pharm. Res.* **1989,** *6,* 296–301.
67. Magnusson, K. E.; Stendahl, O.; Stjernstrom, I.; Edebo, L. *Immunology* **1979,** *36,* 439–447.
68. Moghimi, S. M.; Patel, H. M. *FEBS Lett.* **1988,** *233,* 143–147.
69. Moghimi, S. M.; Patel, H. M. *Biochim. Biophys. Acta* **1989,** *984,* 379–383.
70. Moghimi, S. M.; Patel, H. M. *Biochim. Biophys. Acta* **1989,** *984,* 384–387.
71. Park, Y. S.; Huang, L. *Biochim. Biophys. Acta* **1993,** *1166,* 105–114.
72. Artursson, P.; Johanssen, D.; Sjoholm, I. *Biomaterials* **1988,** *9,* 241–246.
73. Illum, L.; Davis, S. S. *FEBS Lett.* **1984,** *167,* 79–82.
74. Lenaerts, V.; Nagelkerke, J. F.; Van Berkel, T. J. C.; Couvreur, P.; Grislain, L.; Roland, M.; Speiser, P. *J. Pharm. Sci.* **1984,** *73,* 980–982.
75. Bogers, W. M. J. M.; Stad, R.-K.; Janssen, D. J.; Van Rooijen, N.; Van Es, L. A.; Daha, M. R. *Clin. Exp. Immunol.* **1991,** *86,* 328–333.
76. Patel, H. M.; Moghimi, S. M. In *Targeting of Drugs 2: Optimization Strategies;* Gregoriadis, G.; Allison, A. C.; Poste, G., Eds.; Plenum: New York, 1990; pp 87–94.
77. Illum, L.; Jones, P. D. E.; Baldwin, R. W.; Davis, S. S. *J. Pharmacol. Exp. Ther.* **1984,** *230,* 733–736.
78. Kanke, M.; Morlier, E.; Geissler, R.; Powell, D.; Kaplan, A.; DeLuca, P. P. *J. Parenter. Sci. Technol.* **1986,** *40,* 114–118.
79. Illum, L.; Davis, S. S.; Wilson, C. G.; Frier, M.; Hardy, J. G.; Thomas, N. W. *Int. J. Pharm.* **1982,** *12,* 135–146.
80. Poste, G. *Biol. Cell* **1983,** *47,* 19–38.
81. Pratten, M. K.; Lloyd, J. B. *Biochim. Biophys. Acta* **1986,** *881,* 307–313.
82. Desai, A. G.; Thakur, M. L. *Semin. Nucl. Med.* **1985,** *3,* 229–239.
83. Senior, J.; Crawley, J. C. W.; Gregoriadis, G. *Biochim. Biophys. Acta* **1985,** *839,* 1–8.
84. Muller, R. H.; Davis, S. S.; Illum, L.; Mak, E. In *Targeting of Drugs with Synthetic Systems;* Gregoriadis, G.; Senior, J.; Poste, G., Eds.; Plenum: New York, 1986; pp 239–263.
85. Wilkins, D. J.; Myers, P. A. *Br. J. Exp. Pathol.* **1966,** *47,* 568–576.
86. Gregoriadis, G.; Neerunjun, D. E. *Eur. J. Biochem.* **1974,** *47,* 179–185.
87. Juliano, R. L.; Stamp, D. *Biochem. Biophys. Res. Commun.* **1975,** *63,* 651–658.
88. Park, Y. S.; Maruyama, K.; Huang, L. *Biochim. Biophys. Acta* **1992,** *1108,* 257–260.
89. Davis, S. S.; Illum, L.; Washington, C.; Harper, G. *Int. J. Pharm.* **1992,** *82,* 99–105.
90. Holden, C. A.; Olliff, C. J.; Lloyd, A. W.; Paul, F.; Taylor, P. *J. Pharm. Pharmacol.* **1992,** *44,* 46.
91. Illum, L.; Jacobsen, L. O.; Muller, R. H.; Mak, E.; Davis, S. S. *Biomaterials* **1987,** *8,* 113–117.
92. Van Oss, C. J. *Annu. Rev. Microbiol.* **1978,** *32,* 19–39.
93. Absolom, D. R. *Methods Enzymol.* **1986,** *132,* 16–94.
94. Muller, R. H.; Heinemann, S.; Blunk, T.; Rudt, S. In *Laser Light Scattering in Biochemistry;* Harding, S. E.; Sattelle, D. B.; Bloomfield, V. A., Eds.; Royal Society of Chemistry: Cambridge, U.K., 1992; pp 425–440.
95. Muller, R. H.; Wallis, K. H.; Troster, S. D.; Kreuter, J. *J. Controlled Release* **1992,** *20,* 237–246.
96. Alving, C. R.; Steck, E. A.; Chapman, W. L.; Waits, V. B.; Hendrick, L. D.; Swartz, G. M.; Hanson, W. L. *Proc. Natl. Acad. Sci. U.S.A.* **1978,** *75,* 2959–2963.
97. Nassander, U. K.; Storm, G.; Peeters, P. A. M.; Crommelin, D. J. A. In *Biodegradable Polymers as Drug Delivery Systems;* Chasin, M.; Langer, R., Eds.; Marcel Dekker: New York, 1990; pp 261–337.
98. Fidler, I. J.; Schroit, A. J. *J. Immunol.* **1984,** *133,* 515–518.
99. Tabata, Y.; Ikada, Y. *Crit. Rev. Ther. Drug Carrier Syst.* **1990,** *7,* 121–148.

100. Lopez-Berestein, G.; Bodey, G. P.; Fainstein, V.; Keating, M.; Frankel, L. S.; Zeluff, B. *Arch. Intern. Med.* **1989,** *149,* 2533–2536.
101. Schafer, V.; Von Brisen, H.; Andreesen, R.; Steffan, A.-M.; Royer, C.; Troster, S.; Kreuter, J.; Rubsamen-Waigmann, H. *Pharm. Res.* **1992,** *9,* 541–546.
102. Kerr, D. J.; Kaye, S. B. *Crit. Rev. Ther. Drug Carrier Syst.* **1991,** *8,* 19–37.
103. Kato, T.; Nemoto, R.; Mori, H.; Takahashi, M.; Tamakawa, Y.; Harada, M. *JAMA J. Am. Med. Assoc.* **1981,** *245,* 1123–1127.
104. Goldberg, J. A.; Kerr, D. J.; Blackie, R.; Whately, T. L.; Pettit, L.; Kato, T.; McArdle, C. S. *Cancer* **1991,** *67,* 952–955.
105. Goldberg, J. A.; Murray, T.; Kerr, D. J.; Willmott, N.; Bessent, R. G.; McKillop, J. H.; McArdle, C. S. *Br. J. Cancer* **1991,** *63,* 308–310.
106. Willmott, N.; Goldberg, J.; Anderson, J.; Bessent, R.; McKillop, J.; McArdle, C. S. *Int. J. Radiat. Biol.* **1991,** *60,* 195–199.
107. Madison, R.; Macklis, J. D.; Thies, C. *Brain Res.* **1990,** *522,* 90–98.
108. Illum, L. Doctor of Science Thesis, Royal Danish School of Pharmacy, Copenhagen, 1987.
109. Widder, K. J.; Senyei, A. E.; Scarpelli, D. G. *Proc. Soc. Exp. Biol. Med.* **1978,** *158,* 141–146.
110. Weinstein, J. N.; Leserman, L. D. *Pharm. Ther.* **1984,** *24,* 207–228.
111. Yatvin, M. B.; Krentz, W.; Horwitz, B. A.; Shinitzky, M. *Science (Washington, D.C.)* **1980,** *210,* 1253–1254.
112. Illum, L.; Thomas, N.; Davis, S. S. *J. Pharm. Sci.* **1986,** *75,* 16–22.
113. Souhami, R. L.; Patel, H. M.; Ryman, B. E. *Biochim. Biophys. Acta* **1981,** *674,* 254–371.
114. Patel, K. R.; Li, M. P.; Baldeschwieler, J. D. *Proc. Natl. Acad. Sci. U.S.A.* **1983,** *80,* 6518–6522.
115. Dini, L.; Lentini, A.; Mantile, G.; Massimi, M.; Devirgiliis, L. C. *Biol. Cell* **1992,** *74,* 217–224.
116. Medda, S.; Das, N.; Bachhawat, B. K.; Mahato, S. B.; Basu, M. K. *Biotechnol. Appl. Biochem.* **1990,** *12,* 537–543.
117. Horstmann, R. D.; Sievertsen, H. J.; Knobloch, J.; Fischetti, V. A. *Proc. Natl. Acad. Sci. U.S.A.* **1988,** *85,* 1657–1661.
118. Davis, S. S.; Illum, L.; Moghimi, S. M.; Davies, M. C.; Porter, C. J. H.; Muir, I. S.; Brindley, A.; Christy, N. M.; Norman, M. E.; Williams, P.; Dunn, S. E. *J. Controlled Release* **1993,** *24,* 157–163.
119. Illum, L.; Davis, S. S.; Muller, R. H.; Mak, E.; West, P. *Life Sci.* **1987,** *40,* 367–374.
120. Muir, I. S.; Moghimi, S. M.; Illum, L.; Davis, S. S.; Davies, M. C. *Biochem. Soc. Trans.* **1991,** *19,* 329.
121. Illum, L.; Davis, S. S. *Life Sci.* **1987,** *40,* 1553–1560.
122. Moghimi, S. M.; Porter, C. J. H.; Muir, I. S.; Illum, L.; Davis, S. S. *Biochem. Biophys. Res. Commun.* **1991,** *177,* 861–866.
123. Porter, C. J. H.; Moghimi, S. M.; Illum, L.; Davis, S. S. *FEBS Lett.* **1992,** *305,* 62–66.
124. Porter, C. J. H.; Moghimi, S. M.; Davis, S. S.; Illum, L. *Int. J. Pharm.* **1992,** *83,* 273–276.
125. Moghimi, S. M.; Illum, L.; Davis, S. S. *Crit. Rev. Ther. Drug Carrier Syst.* **1990,** *7,* 187–209.
126. Harper, G. R.; Davies, M. C.; Davis, S. S.; Tadros, T. H. F.; Taylor, D. C.; Irving, M. P.; Waters, J. A. *Biomaterials* **1991,** *12,* 695–700.
127. Leu, D.; Manthey, B.; Kreuter, J.; Speiser, P.; DeLuca, P. P. *J. Pharm. Sci.* **1984,** *73,* 1433–1437.
128. Troster, S. D.; Muller, U.; Kreuter, J. *Int. J. Pharm.* **1990,** *61,* 85–100.
129. Douglas, S. J.; Davis, S. S.; Illum, L. *Int. J. Pharm.* **1986,** *34,* 145–152.
130. Artursson, P.; Laakso, T.; Edman, P. *J. Pharm. Sci.* **1983,** *72,* 1415–1420.
131. Allen, T. M.; Chonn, A. *FEBS Lett.* **1987,** *223,* 42–46.
132. Gabizon, A.; Papahadjopoulos, D. *Biochim. Biophys. Acta* **1992,** *1103,* 94–100.
133. Mayhew, E. G.; Lasic, D.; Babbar, S.; Martin, F. J. *Int. J. Cancer* **1992,** *51,* 302–309.
134. Forssen, E. A.; Coulter, D. M.; Proffitt, R. T. *Cancer Res.* **1992,** *52,* 3255–3261.
135. Liu, D.; Mori, A.; Huang, L. *Biochim. Biophys. Acta* **1991,** *1066,* 159–165.
136. Liu, D.; Wada, A.; Huang, L. *Immunol. Lett.* **1992,** *23,* 177–181.
137. Klibanov, A. L.; Maruyama, K.; Torchilin, V. P.; Huang, L. *FEBS Lett.* **1990,** *268,* 235–237.
138. Blume, G.; Cevc, G. *Biochim. Biophys. Acta* **1990,** *1029,* 91–97.
139. Maruyama, K.; Yuda, T.; Okamoto, A.; Ishikura, C.; Kojima, S.; Iwatsuru, M. *Chem. Pharm. Bull.* **1991,** *39,* 1620–1622.

140. Senior, J.; Delgardo, C.; Fisher, D.; Tilcock, C.; Gregoriadis, G. *Biochim. Biophys. Acta* **1991**, *1062*, 77–82.
141. Blume, G.; Cevc, G. *Biochim. Biophys. Acta* **1993**, *1146*, 157–168.
142. Woodle, M. C.; Matthay, K. K.; Newman, M. S.; Hidayat, J. E.; Collins, L. R.; Redemann, C.; Martin, F. J.; Papahadjopoulos, D. *Biochim. Biophys. Acta* **1992**, *1105*, 193–200.
143. Litzinger, D. C.; Huang, L. *Biochim. Biophys. Acta* **1992**, *1127*, 249–254.
144. Namba, Y.; Oku, N.; Ito, F.; Sakakibara, T.; Okada, S. *Life Sci.* **1992**, *50*, 1773–1779.
145. Oku, N.; Namba, Y.; Okada, S. *Biochim. Biophys. Acta* **1992**, *1126*, 255–260.
146. Gabizon, A.; Papahadjopoulos, D. *Proc. Natl. Acad. Sci. U.S.A.* **1988**, *85*, 6949–6953.
147. Bakker-Woudenberg, I. A. J. M.; Lokerese, A. F.; Kate, M. T.; Storm, G. *Biochim. Biophys. Acta* **1992**, *1138*, 318–326.
148. Maruyama, K.; Kennel, S. J.; Huang, L. *Proc. Natl. Acad. Sci. U.S.A.* **1990**, *87*, 5744–5748.
149. Mori, A.; Klibanov, A. L.; Torchilin, V. P.; Huang, L. *FEBS Lett.* **1991**, *284*, 263–266.
150. Litzinger, D. C.; Huang, L. *Biochim. Biophys. Acta* **1992**, *1104*, 179–187.
151. Hughes, B. J.; Kennel, S.; Lee, R.; Huang, L. *Cancer Res.* **1989**, *49*, 6214–6220.
152. Ahmad, I.; Longenecker, M.; Samuel, J.; Allen, T. M. *Cancer Res.* **1993**, *53*, 1484–1488.
153. Napper, D. H. *J. Colloid Interface Sci.* **1977**, *58*, 390–407.
154. Rudt, S.; Muller, R. H. *J. Controlled Release* **1993**, *25*, 51–59.
155. Norman, M. E.; Williams, P.; Illum, L. *Biomaterials* **1992**, *13*, 841–849.
156. Norman, M. E.; Williams, P.; Illum, L. *Biomaterials* **1992**, *14*, 193–202.
157. Lee, J.; Martic, P. A.; Tan, J. S. *J. Colloid Interface Sci.* **1989**, *131*, 252–259.
158. Jeon, S. I.; Andrade, J. D. *J. Colloid Interface Sci.* **1991**, *142*, 159–165.
159. Jeon, S. I.; Lee, J. H.; Andarade, J. D.; DeGennes, P. G. *J. Colloid Interface Sci.* **1991**, *142*, 149–157.
160. Wood, D. A. *Int. J. Pharm.* **1980**, *7*, 1–18.
161. Baker, R. *Controlled Release of Biologically Active Agents;* John Wiley and Sons: New York, 1987.
162. Chasin, M.; Domb, A.; Ron, E.; Mathiowitz, E.; Leong, K.; Laurencin, C.; Brem, H.; Grossman, S.; Langer, R. In *Biodegradable Polymers as Drug Delivery Systems;* Chasin, M.; Langer, R., Eds.; Marcel Dekker: New York, 1990; pp 43–70.
163. Smith, K. L.; Schimpf, M. E.; Thompson, K. E. *Adv. Drug Delivery Rev.* **1990**, *4*, 343–357.
164. Bogdansky, S. In *Biodegradable Polymers as Drug Delivery Systems;* Chasin, M.; Langer, R., Eds.; Marcel Dekker: New York, 1990; pp 231–260.
165. Sawhney, A. S.; Hubbell, J. A. *J. Biomed. Mater. Res.* **1990**, *24*, 1397–1411.
166. Pitt, C. G.; Jeffcoat, A. R.; Zweidinger, R. A.; Scindler, A. *J. Biomed. Mater. Res.* **1979**, *13*, 497–507.
167. Guzman, M.; Molpeceres, J.; Garcia, F.; Aberturas, M. R.; Rodriguez, M. *J. Pharm. Sci.* **1993**, *82*, 498–502.
168. Heller, J.; Sparer, R. V.; Zentner, G. M. In *Biodegradable Polymers as Drug Delivery Systems;* Chasin, M.; Langer, R., Eds.; Marcel Dekker: New York, 1990; pp 121–162.
169. Tamada, J.; Langer, R. *J. Biomed. Sci. (Basel)* **1992**, *3*, 315–353.
170. Allcock, H. R. In *Biodegradable Polymers as Drug Delivery Systems;* Chasin, M.; Langer, R., Eds.; Marcel Dekker: New York, 1990; pp 163–194.
171. Crommen, J. H. L.; Schacht, E. H.; Mense, E. H. G. *Biomaterials* **1992**, *13*, 511–520.
172. Vert, M. In *Polymers in Controlled Drug Delivery;* Illum, L.; Davis, S. S., Eds.; IOP Publishing: Bristol, U.K., 1987.
173. Stolnick, S.; Davies, M. C.; Illum, L.; Davis, S. S.; Boustta, M.; Vert, M. *J. Controlled Release* **1994**, *30*, 57–67.
174. Couvreur, P.; Fattal, E.; Alphandary, H.; Puisieux, F.; Andremont, A. *J. Controlled Release* **1992**, *19*, 259–268.
175. Muller, R. H.; Lherm, C.; Herbort, J.; Blunk, T.; Couvreur, P. *Int. J. Pharm.* **1992**, *84*, 1–11.
176. Lherm, C.; Muller, R. H.; Puisieux, F.; Couvreur, P. *Int. J. Pharm.* **1992**, *84*, 13–22.
177. Lewis, D. H. In *Biodegradable Polymers as Drug Delivery Systems;* Chasin, M.; Langer, R., Eds.; Marcel Dekker: New York, 1990; pp 1–42.
178. Frazza, E. J.; Schmitt, E. E. *Biomed. Mater. Symp.* **1971**, *1*, 43.
179. Kulkarni, R. K.; Pani, K. C.; Neuman, C.; Leonard, F. *Arch. Surg. (Chicago)* **1966**, *93*, 839–843.
180. Wise, D. L.; Fellmann, T. D.; Sanderson, J. E.; Wentworth, R. L. In *Drug Carriers in Biology and Medicine;* Gregoriadis, G., Ed.; Academic: New York, 1979; pp 237–270.

181. Coombes, A. G. A.; Heckman, J. D. *Biomaterials* **1992,** *13,* 297–307.
182. Hoppen, H. L.; Leenslag, J. W.; Pennings, A. J.; Van der Lei, B.; Robinson, P. H. *Biomaterials* **1990,** *11,* 286–290.
183. Yolles, S.; Eldridge, J. E.; Woodland, J. H. R. *Polym. News* **1971,** *1,* 9–14.
184. Jalil, R.; Nixon, J. R. *J. Microencapsulation* **1990,** *7,* 297–325.
185. Gilding, D. K.; Reed, A. M. *Polymer* **1979,** *20,* 1459–1464.
186. Yoshikawa, H.; Nakao, Y.; Takada, K.; Muranishi, S.; Wada, R.; Tabata, Y.; Hyon, S.-H.; Ikada, Y. *Chem. Pharm. Bull.* **1989,** *37,* 802–804.
187. Asano, M.; Fukuzaki, H.; Yoshida, M.; Kumakura, M.; Mashimo, T.; Yuasa, H.; Imai, K.; Yamanaka, H.; Kawaharada, U.; Suzuki, K. *Int. J. Pharm.* **1991,** *67,* 67–77.
188. Kitchell, J. P.; Wise, D. L. In *Methods in Enzymology: Drug and Enzyme Targeting;* Widder, K. J.; Green, R., Eds.; Academic: Orlando, FL, 1985; Vol. 112, pp 436–448.
189. Cohen, S.; Yoshioka, T.; Lucarelli, M.; Hwang, L. H.; Langer, R. *Pharm. Res.* **1991,** *8,* 713–720.
190. Dunn, R. L.; English, J. P.; Strobel, J. D.; Cowsar, D. R.; Tice, T. R. In *Polymers in Medicine III;* Migliaresi, M., Ed.; Elsevier: Amsterdam, Netherlands, 1988; pp 149–160.
191. Gu, Z.-W.; Ye, W.-P.; Yang, J.-Y.; Li, Y.-X.; Chen, X.-L.; Zhong, G.-W.; Feng, X.-D. *J. Controlled Release* **1992,** *22,* 3–14.
192. Imasaka, K.; Yoshida, M.; Fukuzaki, H.; Asano, M.; Kumakura, M.; Mashimo, T.; Yamanaka, H.; Nagai, T. *Int. J. Pharm.* **1992,** *81,* 31–38.
193. Cohn, D.; Younes, H. *Biomaterials* **1989,** *10,* 466–474.
194. Zhu, K. J.; Bihai, S.; Shilin, Y. *J. Polym. Sci. Polym. Chem. Ed.* **1989,** *27,* 2151–2159.
195. Deng, X. M.; Xiong, C. D.; Cheng, L. M.; Xu, R. P. *J. Polym. Sci. Polym. Lett. Ed.* **1990,** *28,* 411–416.
196. Yoshida, M.; Asano, M.; Kumakura, M.; Katakai, R.; Mashimo, T.; Yuasa, H.; Imai, K.; Yamanaka, H. *Colloid Polym. Sci.* **1990,** *268,* 726–730.
197. Pitt, C. G.; Cha, Y.; Shah, S. S.; Zhu, K. J. *J. Controlled Release* **1992,** *19,* 189–200.
198. Park, T. G.; Cohen, S.; Langer, R. *Pharm. Res.* **1992,** *9,* 37–39.
199. Park, T. G.; Cohen, S.; Langer, R. *Macromolecules* **1992,** *25,* 116–122.
200. Gurny, R.; Peppas, N. A.; Harrington, D. D.; Banker, G. S. *Drug Dev. Ind. Pharm.* **1981,** *7,* 1–25.
201. Fong, J. W.; Nazareno, J. P.; Pearson, J. E.; Maulding, H. V. *J. Controlled Release* **1986,** *3,* 119–130.
202. Spenlehauer, G.; Vert, M.; Benoit, J. P.; Boddaert, A. *Biomaterials* **1989,** *10,* 557–563.
203. Visscher, G. E.; Pearson, J. E.; Fong, J. W.; Argentieri, G. J.; Robison, R. L.; Maulding, H. V. *J. Biomed. Mater. Res.* **1988,** *22,* 733–746.
204. Menei, P.; Daniel, V.; Montero-Menei, C.; Brouillard, M.; Pouplard-Bathelaix, A.; Benoit, J.-P. *Biomaterials* **1993,** *14,* 470–478.
205. Julienne, M. C.; Foussard, F.; Benoit, J. P. *Proc. Int. Symp. Controlled Release Bioact. Mater.* **1989,** *16,* 77–78.
206. Smith, A.; Hunneyball, I. M. *Int. J. Pharm.* **1986,** *30,* 215–220.
207. Maulding, H. V. *J. Controlled Release* **1987,** *6,* 167–176.
208. Van Sliedregt, A.; Knook, M.; Hesseling, S. C.; Koerten, H. K.; de Groot, K.; Van Blitterswijk, C. A. *Biomaterials* **1992,** *13,* 819–824.
209. Holland, S. J.; Tighe, B. J.; Gould, P. L. *J. Controlled Release* **1986,** *4,* 155–171.
210. Vert, M. *Angew. Makromol. Chem.* **1989,** *166,* 155–168.
211. Miller, R. A.; Brady, J. M.; Cutright, D. E. *J. Biomed. Mater. Res.* **1977,** *11,* 711–719.
212. Vert, M.; Li, S.; Garreau, H. *J. Controlled Release* **1991,** *16,* 15–26.
213. Wang, H. T.; Palmer, H.; Linhardt, R. J.; Flanagan, D. R.; Schmitt, E. *Biomaterials* **1990,** *11,* 679–685.
214. Bodmeier, R.; Oh, K. H.; Chen, H. *Int. J. Pharm.* **1989,** *51,* 1–8.
215. Makino, K.; Arakawa, M.; Kondo, T. *Chem. Pharm. Bull.* **1985,** *33,* 1195–1201.
216. Maulding, H. V.; Tice, T. R.; Cowsar, D. R.; Fong, J. W.; Pearson, J. E.; Nazareno, J. P. *J. Controlled Release* **1986,** *3,* 103–117.
217. Cha, Y.; Pitt, C. G. *J. Controlled Release* **1989,** *8,* 259–265.
218. Beck, L. R.; Ramos, R. A.; Flowers, C. E.; Lopez, G. Z.; Lewis, D. H.; Cowsar, D. R. *Am. J. Obstet. Gynecol.* **1981,** *140,* 799–806.
219. Jacob, E.; Setterstrom, J. A.; Bach, D. E.; Heath, J. R.; McNiesh, L. M.; Cierny, G. *Clin. Orthop. Relat. Res.* **1991,** *267,* 237–244.

220. Tsai, D. C.; Howard, S. A.; Hogan, T. F.; Malanga, C. J.; Kandzari, S. J.; Ma, J. K. H. *J. Microencapsulation* **1986,** *3,* 181–193.
221. Flandroy, P.; Grandfils, C.; Collignon, J.; Thibaut, A.; Nihant, N.; Barbette, S.; Jerome, R.; Teyssie, P. H. *Neuroradiology* **1990,** *32,* 311–315.
222. Grandfils, C.; Flandroy, P.; Nihant, N.; Barbette, S.; Jerome, R.; Teyssie, P. H.; Thibaut, A. *J. Biomed. Mater. Res.* **1992,** *26,* 467–479.
223. Arshady, R. *J. Controlled Release* **1991,** *17,* 1–22.
224. Conti, B.; Pavanetto, F.; Genta, I. *J. Microencapsulation* **1992,** *9,* 153–166.
225. Allemann, E.; Gurny, R.; Doelker, E. *Eur. J. Pharm. Biopharm.* **1993,** *39,* 173–191.
226. Beck, L. R.; Cowsar, D. R.; Lewis, D. H.; Cosgrove, R. J.; Riddle, C. T.; Lowry, S. L.; Epperly, T. *Fertil. Steril.* **1979,** *31,* 545–551.
227. Tice, T. R.; Gilley, R. M. *J. Controlled Release* **1985,** *2,* 343–352.
228. Arshady, R. *Polym. Eng. Sci.* **1990,** *30,* 915–924.
229. Watts, P. J.; Davies, M. C.; Melia, C. D. *Crit. Rev. Ther. Drug Carrier Syst.* **1990,** *7,* 235–259.
230. Bodmeier, R.; McGinity, J. W. *Int. J. Pharm.* **1988,** *43,* 179–186.
231. Krause, H.-J.; Schwartz, A.; Rohdewald, P. *Int. J. Pharm.* **1985,** *27,* 145–155.
232. Yamakawa, I.; Tsushima, Y.; Machida, R.; Watanabe, S. *J. Pharm. Sci.* **1992,** *81,* 899–903.
233. Niwa, T.; Takeuchi, H.; Hino, T.; Kunou, N.; Kawashima, Y. *J. Controlled Release* **1993,** *25,* 89–98.
234. Coombes, A. G. A.; Scholes, P. D.; Davies, M. C.; Illum, L.; Davis, S. S. *Biomaterials* **1994,** *15,* 673–680.
235. Bodmeier, R.; McGinity, J. W. *J. Microencapsulation* **1987,** *4,* 289–297.
236. Koosha, F. Ph.D. Thesis, University of Nottingham, U.K., 1989.
237. Silvestri, S.; Ganguly, N.; Tabibi, E. *Pharm. Res.* **1992,** *9,* 1347–1350.
238. Benita, S.; Benoit, J. P.; Puisieux, F.; Thies, C. *J. Pharm. Sci.* **1984,** *73,* 1721–1724.
239. Bodmeier, R.; McGinity, J. W. *J. Microencapsulation* **1987,** *4,* 279–288.
240. Coffin, M. D.; McGinity, J. W. *Pharm. Res.* **1992,** *9,* 200–205.
241. Bazile, D. V.; Ropert, C.; Huve, P.; Verrecchia, T.; Marland, M.; Frydman, A.; Veillard, M.; Spenlehauer, G. *Biomaterials* **1992,** *13,* 1093–1102.
242. Julienne, M. C.; Alonso, M. J.; Gomez Amoza, J. L.; Benoit, J. P. *Drug Dev. Ind. Pharm.* **1992,** *18,* 1063–1077.
243. Muller, R. H.; Wallis, K. H. *Int. J. Pharm.* **1993,** *89,* 25–31.
244. Kishida, A.; Dressman, J. B.; Yoshioka, S.; Aso, Y.; Takeda, Y. *J. Controlled Release* **1990,** *13,* 83–89.
245. Izumikawa, S.; Yoshioka, S.; Aso, Y.; Takeda, Y. *J. Controlled Release* **1991,** *15,* 133–140.
246. Spenlehauer, G.; Veillard, M.; Benoit, J. P. *J. Pharm. Sci.* **1986,** *75,* 750–755.
247. Gangrade, N.; Price, J. C. *J. Pharm. Sci.* **1992,** *81,* 201–202.
248. Bodmeier, R.; Chen, H.; Tyle, P.; Jarosz, P. *J. Controlled Release* **1991,** *15,* 65–77.
249. Ogawa, Y.; Yamamoto, M.; Okada, H.; Yashiki, T.; Shimamoto, T. *Chem. Pharm. Bull.* **1988,** *36,* 1095–1103.
250. Bodmer, D.; Kissel, T.; Traechslin, E. *J. Controlled Release* **1992,** *21,* 129–138.
251. Jeffery, H.; Davis, S. S.; O'Hagan, D. T. *Pharm. Res.* **1993,** *10,* 362–368.
252. Pavanetto, F.; Conti, B.; Genta, I.; Giunchedi, P. *Int. J. Pharm.* **1992,** *84,* 151–159.
253. Cowsar, D. R.; Tice, T. R.; Gilley, R. M.; English, J. P. In *Methods in Enzymology: Drug and Enzyme Targeting;* Widder, K. J.; Green, R., Eds.; Academic: Orlando, FL, 1985; Vol. 112, pp 101–116.
254. Arshady, R. *Polym. Eng. Sci.* **1990,** *30,* 905–914.
255. Saunders, L. M.; Kent, J. S.; McRae, G. I.; Vickery, B. H.; Tice, T. R.; Lewis, D. H. *J. Pharm. Sci.* **1984,** *73,* 1294–1297.
256. Vidmar, V.; Smolcic-Bulbalo, A.; Jalsenjak, I. *J. Microencapsulation* **1984,** *1,* 131–136.
257. Ruiz, R. M.; Tisser, B.; Benoit, J. P. *Int. J. Pharm.* **1989,** *49,* 69–77.
258. Korkut, E.; Bokser, L.; Comaru-Schally, A. M.; Groot, K.; Schally, A. V. *Proc. Natl. Acad. Sci. U.S.A.* **1991,** *88,* 844–848.
259. Fessi, H.; Puisieux, F.; Devissaguet, J. P. European Patent Application 0274961A1, 1988.
260. Fessi, H.; Puisieux, F.; Devissaguet, J. P.; Ammoury, N.; Benita, S. *Int. J. Pharm.* **1989,** *55,* 1–4.
261. Ibrahim, H.; Bindschaedler, C.; Doelker, E.; Buri, P.; Gurny, R. *Int. J. Pharm.* **1992,** *87,* 239–246.
262. Allemann, E.; Gurny, R.; Doelker, E. *Int. J. Pharm.* **1992,** *87,* 247–253.

263. Bodmeier, R.; Chen, H. *J. Pharm. Pharmacol.* **1988,** *40,* 754–757.
264. Wichert, B.; Rohdewald, P. *J. Controlled Release* **1990,** *14,* 269–283.
265. Tadavarthy, S. M.; Moller, J. H.; Amplatz, K. *AJR Am. J. Roentgenol.* **1975,** *125,* 609–616.
266. DuCret, R. P.; Adkins, M. C.; Hunter, D. W.; Yedlicka, J. W.; Engeler, C. M.; Castaneda-Zuniga, W. R.; Amplatz, K.; Sirr, S. A.; Boudreau, R. J.; Kuni, C. C.; White, J. G. *Radiology (Easton, Pa.)* **1990,** *177,* 571–575.
267. Repa, I.; Moradian, G. P.; Dehner, L. P.; Tadavarthy, S. M.; Hunter, D. W.; Castaneda-Zuniga, W. R.; Wright, G. B.; Katkov, H.; Johnson, P.; Chrenka, B.; Amplatz, K. *Radiology (Easton, PA)* **1989,** *170,* 395–399.
268. O'Hagan, D. T.; Rahman, D.; McGee, J. P.; Jeffery, H.; Davies, M. C.; Williams, P.; Davis, S. S.; Challacombe, S. J. *Immunology* **1991,** *73,* 239–242.
269. Eldridge, J. H.; Staas, J. K.; Tice, T. R.; Gilley, R. M. *Res. Immunol.* **1992,** *143,* 557–563.
270. Singh, M.; Singh, A.; Talwar, G. P. *Pharm. Res.* **1991,** *8,* 958–961.
271. Almeida, A. J.; Alpar, H. O.; Brown, M. R. W. *J. Pharm. Pharmacol.* **1993,** *45,* 198–203.
272. Eldridge, J. H.; Staas, J. K.; Meulbroek, J. A.; Tice, T. R.; Gilley, R. M. *Infect. Immun.* **1991,** *59,* 2978–2986.
273. Seki, T.; Kawaguchi, T.; Endoh, H.; Ishikawa, K.; Juni, K.; Nakano, M. *J. Pharm. Sci.* **1990,** *79,* 985–987.
274. Kanke, M.; Morlier, E.; Geissler, R.; Powell, D.; Kaplan, A.; DeLuca, P. P. *J. Parenter. Sci. Technol.* **1986,** *40,* 114–118.
275. Bundgaard, M. *Annu. Rev. Physiol.* **1980,** *42,* 325–336.
276. Spanjer, H. H.; van Galen, M.; Roerdink, J. F.; Regis, J.; Scherphof, G. I. *Biochim. Biophys. Acta* **1984,** *770,* 195–202.
277. Seijo, B.; Fattal, E.; Roblot-Treupel, L.; Couvreur, P. *Int. J. Pharm.* **1990,** *62,* 1–7.
278. Moghimi, S. M.; Porter, C. J. H.; Illum, L.; Davis, S. S. *Int. J. Pharm.* **1991,** *68,* 121–126.
279. Woodle, M. C.; Newman, M. S.; Martin, F. J. *Int. J. Pharm.* **1992,** *88,* 327–334.
280. Dunn, S. E.; Coombes, A. G. A.; Garnett, M. C.; Davies, M. C.; Davis, S. S.; Illum, L. *J. Controlled Release,* in press.
281. Douglas, S. J.; Davis, S. S.; Illum, L. *Int. J. Pharm.* **1986,** *34,* 145–152.
282. Watrous-Peltier, N.; Uhl, J.; Steel, V.; Brophy, L.; Merisko-Liversidge, E. *Pharm. Res.* **1992,** *9,* 1177–1183.
283. Gref, R.; Minamitake, Y.; Peracchia, M. T.; Trubetskoy, V.; Torchilin, V.; Langer, R. *Science (Washington, D.C.)* **1994,** *263,* 1600–1603.
284. Scholes, P. D.; Coombes, A. G. A.; Illum, L.; Davis, S. S.; Watts, J. F.; Ustariz, C.; Vert, M.; Davies, M. C. submitted for publication in *J. Controlled Release.*
285. Ratcliffe, J. H.; Hunneyball, I. M.; Smith, A.; Wilson, C. G.; Davis, S. S. *J. Pharm. Pharmacol.* **1984,** *36,* 431–436.
286. Cuvier, C.; Roblot-Treupel, L.; Millot, J. M.; Lizard, G.; Chevillard, S.; Manfait, M.; Couvreur, P.; Poupon, M. F. *Biochem. Pharmacol.* **1992,** 44, 509–517.
287. Losa, C.; Alonso, M. J.; Vila, J. L.; Orallo, F.; Martinez, J.; Saavedra, J. A.; Pastor, J. C. *J. Ocul. Pharmacol.* **1992,** *8,* 191–198.
288. Losa, C.; Marchal-Heussler, L.; Orallo, F.; Vila Jato, J. L.; Alonso, M. J. *Pharm. Res.* **1993,** *10,* 80–87.
289. Moritera, T.; Ogura, Y.; Honda, Y.; Wada, R.; Hyon, S.-H.; Ikada, Y. *Invest. Opthalmol. Visual Sci.* **1991,** *32,* 1785–1789.
290. Taylor, K. M. G.; Newton, J. M. *Thorax* **1992,** *47,* 257–259.
291. Gaspar, R.; Opperdoes, F. R.; Preat, V.; Roland, M. *Ann. Trop. Med. Parasitol.* **1992,** *86,* 41–49.
292. Yu, W.-P.; Barratt, G. M.; Devissaguet, J.-P. H.; Puisieux, F. *Int. J. Immunopharmacol.* **1991,** *13,* 167–173.
293. Beck, P.; Kreuter, J.; Reszka, R.; Fichtner, I. *J. Microencapsulation* **1993,** *10,* 101–114.
294. Blagoeva, P. M.; Balansky, R. M.; Mircheva, T. J.; Simeonova, M. I. *Mutat. Res.* **1992,** *268,* 77–82.
295. Kattan, J.; Droz, J.-P.; Couvreur, P.; Marino, J.-P.; Boutan-Laroze, A.; Rougier, P.; Brault, P.; Vranckx, H.; Grognet, J.-M.; Morge, X.; Sancho-Garnier, H. *Invest. New Drugs* **1992,** *10,* 191–199.
296. Tomlinson, E.; Burger, J. J. In *Polymers in Controlled Drug Delivery;* Illum, L.; Davis, S. S., Eds.; IOP Publishing: Bristol, U.K., 1987; pp 25–48.
297. Arshady, R. *J. Controlled Release* **1990,** *14,* 111–131.
298. Oppenheim, R. C. In *Polymers in Controlled Drug Delivery;* Illum, L.; Davis, S. S., Eds.; IOP Publishing: Bristol, U.K., 1987; pp 73–86.

299. Edman, P.; Artursson, P.; Laakso, T.; Sjoholm, I. In *Polymers in Controlled Drug Delivery;* Illum, L.; Davis, S. S., Eds.; IOP Publishing: Bristol, U.K., 1987; pp 87–98.
300. Juni, K.; Nakano, M. *Crit. Rev. Ther. Drug Carrier Syst.* **1987**, *3,* 209–231.
301. Juni, K.; Nakano, M. In *Polymers in Controlled Drug Delivery;* Illum, L.; Davis, S. S., Eds.; IOP Publishing: Bristol, U.K., 1987; pp 49–59.
302. Jalil, R. *Drug Dev. Ind. Pharm.* **1990**, *16,* 2353–2367.
303. Pitt, C. G. In *Biodegradable Polymers as Drug Delivery Systems;* Chasin, M.; Langer, R., Eds.; Marcel Dekker: New York, 1990; pp 71–120.
304. Jeffery, H.; Davis, S. S.; O'Hagan, D. T. *Int. J. Pharm.* **1991**, *77,* 169–175.
305. Wakiyama, N.; Juni, K.; Nakano, M. *Chem. Pharm. Bull.* **1981**, *29,* 3363–3368.
306. Juni, K.; Ogata, J.; Nakano, M.; Ichihara, T.; Mori, K.; Akagi, M. *Chem. Pharm. Bull.* **1985**, *33,* 313–318.
307. Wichert, B.; Rohdewald, P. *J. Microencapsulation* **1993**, *10,* 195–207.
308. Wang, H. T.; Schmitt, E.; Flanagan, D. R.; Linhardt, R. J. *J. Controlled Release* **1991**, *17,* 23–32.
309. Rosilio, V.; Benoit, J. P.; Deyme, M.; Thies, C.; Madelmont, G. *J. Biomed. Mater. Res.* **1991**, *25,* 667–682.
310. Jalil, R.; Nixon, J. R. *J. Microencapsulation* **1989**, *6,* 473–484.
311. Wada, R.; Tabata, Y.; Hyon, S.-H.; Ikada, Y. *Bull. Inst. Chem. Res. Kyoto Univ.* **1988**, *66,* 241–250.
312. Sakakura, C.; Takahashi, T.; Hagiwara, A.; Itoh, M.; Sasabe, T.; Lee, M.; Shobayashi, S. *J. Controlled Release* **1992**, *22,* 69–74.
313. Rivera, R.; Alvarado, G.; Aldaba, C. F. S.; Hernandez, A. *J. Steroid Biochem.* **1987**, *27,* 4–6.
314. Ike, O.; Shimizu, Y.; Hitomi, S.; Wada, R.; Ikada, Y. *Chest* **1991**, *99,* 911–915.
315. Ike, O.; Shimizu, Y.; Wada, R.; Hyon, S.-H.; Ikada, Y. *Biomaterials* **1991**, *13,* 230–234.
316. Hagiwara, A.; Takahashi, T.; Sasabe, T.; Ito, M.; Lee, M.; Sakakura, C.; Shobayashi, S.; Muranishi, S.; Tashima, S. *Anti-Cancer Drugs* **1992**, *3,* 237–244.
317. Svanberg, L.; Lancranjan, I.; Arvidsson, T.; Andersch, B. *Acta Obstet. Gynecol. Scand.* **1987**, *66,* 61–62.
318. McRae, M.; Hjorth, S.; Dahlstrom, A.; Dillon, L.; Mason, D.; Tice, T. *Mol. Chem. Neuropathol.* **1992**, *16,* 123–141.
319. Kwong, A. K.; Chou, S.; Sun, A. M.; Sefton, M. V.; Goosen, M. F. A. *J. Controlled Release* **1986**, *4,* 47–62.
320. Okada, H.; Heya, T.; Ogawa, Y.; Toguchi, H.; Shimamoto, T. *Pharm. Res.* **1991**, *8,* 584–587.
321. Gurny, R.; Peppas, N. A.; Harrington, D. D.; Banker, G. S. *Drug Dev. Ind. Pharm.* **1981**, *7,* 1–25.

6
Site-Specific Drug Delivery in the Gastrointestinal Tract

Randall J. Mrsny

Anatomical, biochemical, and physiological barriers dictate the stability, absorption, and mucosal metabolism of a drug at various sites along the gastrointestinal tract. In this chapter, the design of site-specific delivery schemes and issues related to drug targeting are discussed. Direct application, chemical conjugates, polymeric carriers, and specially designed devices all show promise as delivery-specific methods. Targeting to cell-surface receptors, transporters, bacterial adhesins, and lectins or with microparticulates provides exquisite selectivity in mucosal association and putative drug uptake. Finally, cellular components used in such targeting approaches may be altered in some pathological conditions, providing the possibility for specific opportunities of targeted drug delivery.

Issues of drug delivery have come to the forefront of the pharmaceutical industry. Today, the industry employs many new and exciting technologies that have revolutionized methodologies associated with the identification, characterization, evaluation, synthesis, purification, and stabilization of new drug candidates. Many of these new drug candidates are still considered to be somewhat novel therapeutics: proteins, peptides, and even oligonucleotides. Not surprisingly, there has been a dramatic increase in both the number and rate at which new potential drug entities are being moved into clinical trials. Typically, one of the initial screening criteria for a new drug is that of oral bioavailability, and many potentially exciting drugs are passed over because of poor or variable absorption profiles. Could some of these poorly absorbed drugs be given in a way that might improve their oral bioavailability? Site-specific delivery in the gastrointestinal (GI) tract may provide a solution for some of these otherwise poorly absorbed drug candidates. Although it is initially more labor intensive to identify the limitations associated with the oral delivery of a specific drug candidate, site-specific delivery or targeting to selected tissues or cells of the GI tract can ultimately provide several putative benefits: improved drug stability, a more desirable blood profile, and improved patient compliance due to reductions in dosing regimens.

Copyright © 1997 American Chemical Society

Site-specific delivery in the GI tract can be used to obtain any of a number of possible outcomes. The appropriate set of desired outcomes (and thus an appropriate delivery scheme) can be identified by the answers to a short series of questions. Is there a particular target tissue or cell type where delivery might be optimal? If the answer is yes, then taking advantage of some common biochemical or morphological feature in these target sites should be considered. Would it be advantageous to target a discrete region of the GI tract? If so, then systems that maintain the drug at that site or deliver it to that site selectively by using a temporal or physiological stimulus could be considered. Once a drug entity is delivered in a site-specific fashion, the outcome of that delivery must also be anticipated. Was the delivery intended to be local or systemic? If it is intended for systemic delivery, would it matter if the drug is absorbed primarily into the blood or the lymph? Once systemic delivery has been accomplished, it may also be possible to target a drug to specific organs or tissues, but this subsequent step is beyond the scope of this chapter. Instead, this chapter will focus on ways in which site-specific drug delivery in the GI tract might be accomplished and thus provide examples by which targeted delivery to the GI tract can be achieved. A number of these approaches, those already described in the literature as well as some potentially novel methods to target in the GI tract, will be discussed. A review related to some of these topics recently was published (*1*).

The GI Tract

The GI tract is defined as beginning at the opening of the mouth and ending at the opening of the anus. Essentially all of the defined segments of the GI tract have been investigated as potential sites for targeted drug delivery. The usual portals of entry to the GI tract are those that define its limits: the mouth and the anus. Drugs released from depots in the nose or eye ultimately drain into the oropharynx (back of the throat) and thus provide a potential delivery to the GI tract by the act of swallowing. Since the specific aspects of ocular and nasal drug delivery are beyond the scope of this chapter, site-specific targeted delivery to the mouth, esophagus, stomach, duodenum, jejunum, ileum, colon, and rectum will be the only areas covered. When targeting the delivery of a drug to these regions of the GI tract, there are a number of anatomical, physiological, and biochemical issues that must be considered. Some of these issues are outlined below.

Anatomical Considerations

The GI tract is a continuous tube that, during development, differentiates into separate but contiguous structures (Figure 6.1). Such differential development results in regions with unique roles in the sequential processes of nutrient assimilation: mastication, insalivation, deglutition, digestion, absorption, and elimination of foodstuffs.

The mouth cavity is the site of mastication and insalivation. It has a surface area of approximately 100 cm^2 and is composed of a stratified squamous epithelium with varied regional degrees of cornification. Most regions have an extremely thin cornified layer that allows for a water permeability of about 10 times greater than that of human skin (*2*). The residence time in the mouth is brief, typically seconds to minutes. Solutions or solids held in the mouth are diluted by saliva (derived from the parotid, submandibular, and sublingual glands), which is produced at a rate of about 20 mL/h. Extensive lymphatic tissues are present at the posterior of the mouth cavity, a region known as the isthmus faucium, and this region is covered by a thinner epithelium than that present in the rest of the mouth.

The esophagus is a muscular tube with a surface area of about 200 cm^2 through which foodstuffs pass rapidly: a 10-mL bolus travels from mouth to stomach in about 15 s (*3*). Swallowing produces a 2–3-cm shortening of the esophagus from its normal 23–25-cm length (*4, 5*), which would also act to reduce this residence time and

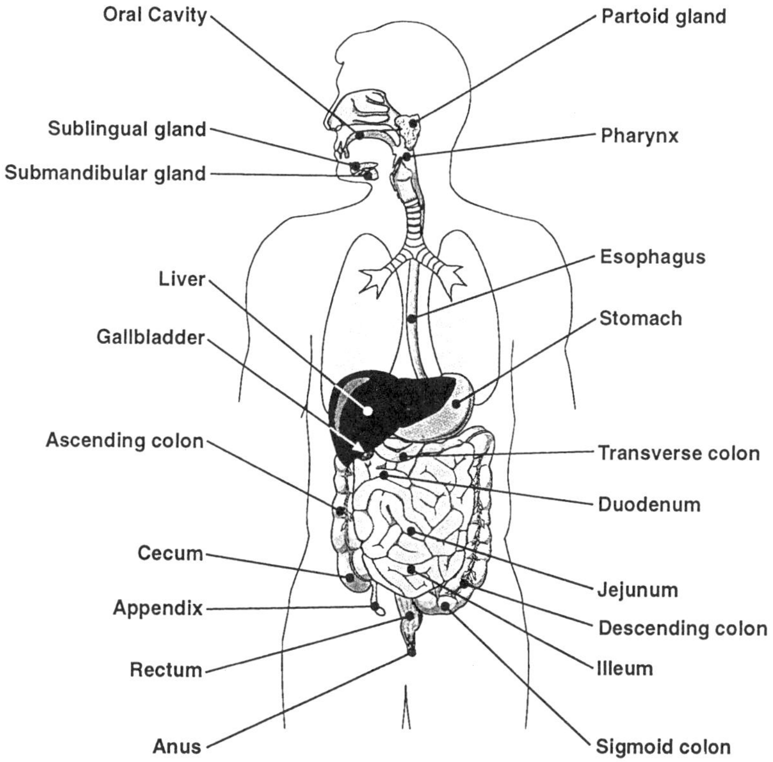

GI tract segment	*Approx. surface area*	*Approx. segment length*	*Approx. residence time*	*Approx. pH of segment*	*Principal catabolic activities*
Oral cavity	100 cm^2	–	Seconds to minutes	6.5	Polysaccharidases
Esophagus	200 cm^2	23-25 cm	seconds		–
Stomach	3.5 m^2 (variable)	0.25 m (variable)	90 minutes (variable)	1-2	Proteases; lipases
Duodenum	1.9 m^2	0.35 m	30-40 minutes	4-5.5	Polysaccharidases; oligosaccharidases proteases; peptidases; lipases
Jejunum	184 m^2	2.8 m	1.5-2.0 hours	5.5-7.0	Oligosaccharidases; peptidases; lipases
Ileum	276 m^2	4.2 m	5-7 hours	7.0-7.5	Oligosaccharidases; peptidases; lipases
Colon and Rectum	1.3 m^2	1.5 m	1-60 hours (35-36 hours on average)	7.0-7.5	Broad spectrum of bacterial enzymes

FIGURE 6.1. Anatomical relations and physiological parameters of the GI tract related to drug delivery.

effective surface area slightly. A nonkeratinizing, stratified squamous epithelium is present along the entire length of this organ, and intercellular glycoconjugate material (probably glycoproteins) in the upper layers of the mucosa provides the major barrier to drug diffusion (*6*). Mucosal and submucosal glands located at various regions along the esophagus continually secrete mucus onto the surface of this epithelium.

On exiting the esophagus, foodstuffs enter the stomach and experience an environment of low pH (~1–2) and enzymatic activities. The mucosal surface of the stomach is a simple columnar epithelium containing millions of cardiac and gastric glands. These glands secrete enzymes, notably pepsinogens, and 0.1 N hydrochloric acid in response to a variety of hormones including histamine, gastrin, and acetylcholine. The surface area of the stomach, ~3.5 m^2 in the adult male, varies somewhat because it can extend significantly to accommodate large meals. Residence time in the stomach also varies: an average residence time for human subjects has been described of approximately 90 min (*7*), although this time can be extended by the use of certain dosage forms (*8*). Further, stomach contents appear to have a significant impact on this residence time: fatty meals typically have the longest residence time, protein meals leave a little sooner, and meals containing a high proportion of carbohydrate usually have the shortest gastric residence time. Opening of a sphincter at the distal portion of the stomach controls the transit of chyme (partially digested foodstuffs) into the small intestine.

The duodenum is the most proximal segment of the small intestine. It is one foot long, has an approximate surface area of 1.9 m^2 and a residence time of ~30–40 min. The two subsequent small intestinal segments, the jejunum and ileum, have approximate lengths of 8 feet and 12 feet, surface areas of 184 m^2 and 276 m^2, and residence times of approximately 1.5–2.0 h and 5–7 h, respectively (*7*). Residence times of the three segments of the small intestine are fairly consistent despite variations in meal contents that might affect initial gastric residence times. Functional differences of the three small intestinal regions can be briefly summarized as follows:

- The duodenum contains extensive glands (Brunner's), which secrete neutral or alkaline mucus and few goblet cells (which secrete acidic glycoproteins).
- The jejunum contains decreased Brunner's glands and increased goblet cell number.
- The ileum has many goblet cells and lymphoid tissue localized into structures known as Peyer's patches.

The colon (or large intestine) and rectum represent the last two regions of the GI tract. They have a combined surface area of ~1.3 m^2 and an extremely variable residence time commonly ranging between 1 and 60 h with an average of 35–36 h (*9*, *10*). Also, there does not seem to be any differences in the transit times of solids and liquids through the colon (*11*), although bioactive molecules can be used to dramatically delay the transit time through this region of the GI tract (*12*). A striking change in the mucosa of the GI tract occurs at the ileocecal junction (the point of transition from small intestine to colon). Although the size and extent of villi steadily decline from the duodenum to the jejunum to the ileum, colonic and rectal mucosa contain no villi at all and are composed only of crypts. This alteration in the mucosal surface results in a reduction in absorptive surface area for the colon and rectum compared with small intestinal segments.

Physiological Considerations

Hydrophilic components absorbed from the latter part of the stomach, the small intestine, the colon, and the proximal rectum enter the portal venous system, which delivers blood directly to the liver. This important mechanism protects the body by allowing for detoxification of noxious agents before their systemic distribution in the body. Molecules delivered to the liver are subjected to (and typically degraded by) a wide range of destructive (protective) enzymatic activities, an event referred to in a generic fashion as first-pass metabolism. Hydrophilic molecules absorbed rostral to the midstomach and caudal to the latter third of the rectum do not enter into the portal venous system, but instead enter into the systemic circulation directly. Hydrophobic molecules absorbed all along the GI tract typically enter the lymphatics and are spared from the initial fate of drugs subjected to first-pass metabolism.

Several fairly rapid pH changes occur along the GI tract (*13*). The first occurs on entrance to the stomach: a drop to pH 1.0–2.5 due to hydrochloric acid produced by the gastric mucosa. From the stomach to the duodenum there is a significant increase in pH to ~5.5 as a result of the acidic chyme mixing with bicarbonate based (pH 8.5) pancreatic juices. During subsequent small intestinal transit the luminal pH bathing the digesting foodstuffs gradually climbs and is ~7.5 at the termination of the ileum. At the ileocecal junction, the pH appears to fall sharply (~5.5–6.0) and then climbs back to neutrality during transit through the colon, as free fatty acids are absorbed. More specifically, the pH of the ascending colon appears to be about 7.0; the transverse to sigmoid colon is close to 7.4 and the pH of the rectum is closer to 7.1 (*14*). Potential application of these pH changes will be discussed later.

Mucus secretion as well as water and ion transport are controlled by a variety of agents throughout the GI tract through direct epithelial actions or through nervous control (for review, *see* ref. 15). Acetylcholine, vasoactive intestinal polypeptide, secretin, gastric inhibitory peptide, and neurotensin induce secretory responses, whereas neuropeptide Y and somatostatin have the opposite effect on epithelial cells. Calcitonin gene-related peptide, cholecystokinin, and galanin also act to regulate epithelial ion transport. Stimulation of secretory events also can be induced through activation of local neural receptors by acetylcholine, 5-hydroxytryptamine, and substance P. Enkephalins (acting through delta opioid receptors) and norepinephrine (acting through alpha-2 adrenergic receptors) can stimulate absorption events in mucosal epithelia. Some of these agents are released from endocrine cell types associated with the GI tract and can produce both autocrine as well as paracrine events. Alterations in mucus secretion or the balance between the absorption and secretion of fluid and electrolytes can result in local mucosal trauma as well as constipation or diarrhea, and thus drug delivery can be significantly affected.

Biochemical Considerations

Digestive tube, a term commonly used to describe the GI tract, automatically (and accurately) paints a grim picture for the delivery of drugs that can be attacked by resident catabolic enzymes. Carbohydrate catabolism begins in the oral cavity by enzymes secreted from salivary glands. Additional enzymatic activities (from pancreatic secretions) are introduced in the duodenum. The small polysaccharides produced from complex carbohydrates as a result of these luminal enzymatic activities are degraded further by enzymes associated with either the cell surface or glycocalyx (proteins and glycoproteins coating the cell surface) along the remainder of the small intestine. Monosaccharides and disaccharides are absorbed into intestinal cells through Na^+-dependent sugar transport proteins present in the apical membrane of epithelial cells of the intestine (enterocytes) and colon (colonocytes).

Peptide and protein degradation is not significant in the oral cavity. It begins in earnest in the stomach and is essentially complete by the ileocecal junction (Figure 6.1). A broad spectrum of proteases and peptidases, from pancreatic secretions, are introduced into the duodenum. Along the remainder of the small intestine, proteins are broken down into peptides, which act as substrates for additional peptidase activities associated with the organized microvilli (brush-border) present at the apical surface enterocytes. These enzyme activities wax and wane at various segments of the GI tract (*16–18*). Peyer's patches may also have lower peptidase activity (or at least a different spectrum of expressed peptidase activities) than adjacent jejunal epithelium (*19*). Some protein and peptide hydrolysis can also occur as a result of bacterial enzymatic activities. This process is typically not extensive in the small intestine because of the low bacterial load, but it can become significant in the colon from the large numbers of resident bacteria. Final assimilation of amino acids (as well as some small peptides) occurs through Na^+-dependent transporters present in the apical membrane of intestinal cells. Evidence exists that some absorptive processes for small peptides also exist in enterocytes, but this transport is H^+-dependent and appears to be under the regulatory control of protein kinase C (*20*).

Lipid metabolism begins in the stomach; about 10–30% of fat digestion occurs by the action of acid lipases (*21*). The resulting fatty acids help to emulsify other fats (*22*) and facilitate emptying of fats from the stomach into the duodenum, where bile salts from the liver (via the gall bladder) can complete the emulsification process (*23*). Fat in the stomach also stimulates the release of pancreatic lipase into the duodenum, resulting in the hydrolysis of emulsified fats into glycerol and long-chain fatty acids. Absorption of hydrolyzed fats occurs primarily in the duodenum and upper jejunum while bile salts are resorbed through specific transporters located in the distal ileum (*24*). Lipid absorption can also occur through a passive mechanism of monomer diffusion (*25*) as well as via uptake of surfactant-like particles that contain triacylglycerol (*26*). Ultimately, bile salts used in the emulsification process of fats in the small intestine are returned to the gall bladder through a mechanism involving recollection in the liver.

Site Specificity Through Delivery

The most direct and perhaps simplest method of site-specific delivery along the GI tract is that of direct drug application. Although a number of formulation issues must be satisfied for such a delivery to be successful, it is relatively straightforward to deliver a mouthwash to the oral cavity or an enema to the rectum and lower colon. A mouthwash could be held in the oral cavity for up to several minutes without significant discomfort. A formulation could then be swallowed, or ejected (if one wishes to reduce the potential of drug delivery to GI tract segments distal to the oropharynx). Such a delivery of transforming growth factor β3 (TGF-β3), for example, may be beneficial for oral mucositis (*27*). Alternately, 0.1% gelatin solutions containing either alpha, beta, or gamma interferon (IFN) have been shown to produce a biological response after oral administration (*28*). Clearly, retention of a formulation at the surface oral mucosa through the use of a buccal patch would act to improve the systemic delivery characteristics of most drugs and will be discussed later. Oral delivery of TGF-β3 and IFN, however, appears to provide examples where local site-specific topical delivery is effective, presumably because these molecules can access the mucosal epithelial cells or lymphocytes of the isthmus faucium to activate specific cell-surface receptors. Therefore, a desired clinical endpoint for targeted delivery to the oral mucosa need not require systemic delivery.

Enema formulations have been shown to spread and coat the entire descending and sigmoid colonic segments as well as the rectum (*29*), a result making rectal administration for site-specific delivery of drugs a viable drug delivery route. Suppository deliveries can provide essentially the same site-specific delivery advantages as an enema delivery. A quick scan of the *Physician's Desk Reference* suggests at least a half-dozen drugs are currently being marketed as suppositories for indications including migraines, nausea, fever, inflammation, and pain relief. Unfortunately, patients can sometimes eliminate suppositories before their complete dissolution (resulting in variability of the administered dose). There are definite advantages to rectal delivery, however, including bypass of first-pass metabolism and reduced mucosal metabolism in some cases.

Site-specific targeting of drugs through delivery can also be accomplished by chemical conjugation as well as by using polymeric matrices and devices. Application of such delivery schemes can result in site-specific delivery along the GI tract. Further, such approaches can take advantage of unique environments present in the GI tract that occur as a result of normal physiological events or pathological states to increase oral bioavailability or to focus the delivery of a drug. Some of these delivery routes and schemes are presented in an idealized fashion in Figure 6.2. Drug conjugation might offer improved drug stability or improved uptake characteristics as well as allow for polymer association. Drug or drug conjugates could also be delivered without the use of a device or polymeric carrier. Release of drug or drug conjugate from polymers or devices as well as conversion of drug conjugates back to active drug might be accomplished through a number of mechanisms, including catalysis by an enzymatic activity or by an environmental (e.g., pH) change. Epithelial transport involving paracellular or transcellular pathways, such as endocytosis, carrier-mediated uptake, and lipophilic diffusion, may involve the transit of drug molecules in native or conjugated form. Not all drugs, however, would benefit from such a site-specific delivery approach. Indeed, drugs such as nonsteroidal anti-inflammatory drugs can produce extensive mucosal damage when delivered in a focused or targeted fashion (*30*).

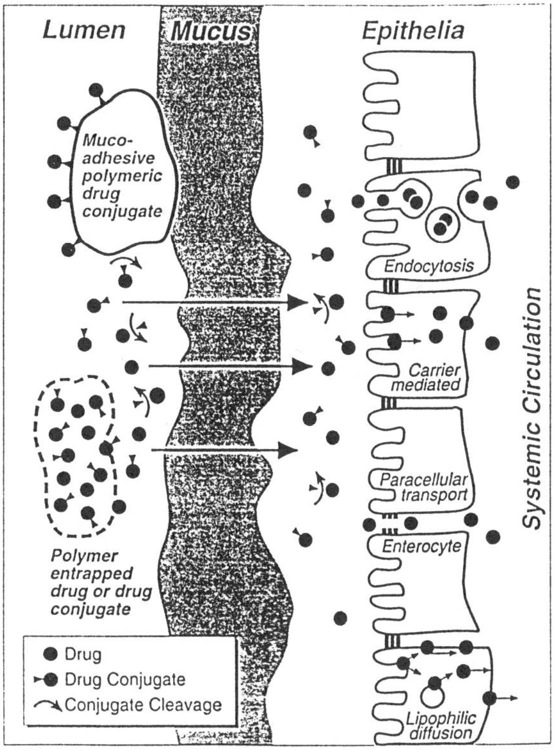

FIGURE 6.2. Some methods of site-specific drug delivery in the GI tract. Drug molecules or conjugated drug molecules can be incorporated into a device (not shown) or polymeric matrix (shown). These platforms might or might not be able to adhere to either mucus (shown) or the epithelial surface (not shown).

Chemical Conjugates

Several examples of chemical conjugates appear to successfully augment or modify the delivery of a drug. For example, conjugation of misoprostol, a prostaglandin analog, with the polymer hydroxypropyl methylcellulose appears to offer a sustained gastric delivery (*31*, *32*). Prodrugs of 5-fluorouracil that shift the absorption of this drug to more distal regions of the small intestine have also been generated (*33*). 5-Aminosalicylic acid has been conjugated in such a way so that it provides targeted delivery to the small intestine following hydrolysis by brush-border enzyme activities (*34*). In an opposite fashion, glycerolipidic peptides have been synthesized in an effort to evade metabolism by brush-border enzymes of the small intestine (*35*). Resident bacterial enzymes can also be used to produce a site-specific delivery. Glycoside prodrugs of steroids (*36*, *37*) are examples of chemical conjugates that are converted by colon-specific bacterial enzymes to produce active drug. Such approaches can have tremendous benefit in reducing the occurrence of unwanted side effects (*38*).

Successful use of drug conjugates for site-specific delivery requires a unique environment in the GI tract where drug release or prodrug conversion can occur. Bacterial enzymes present in the colon or endogenous enzyme activities that are differentially expressed along the GI tract have been exploited for this purpose. Nonenzymatic activators can also be used to activate drug conjugates—an acid-stable conjugate that is activated and absorbed as it leaves the stomach and enters the more alkaline pH of the duodenum. This approach to site-specific delivery is hampered, however, by individual variability in gastric acidity and from differences in gastric acidity associated with fed versus fasted states (*39*).

Polymeric Carriers

Although a number of factors can be used to control the release of a drug from a polymeric matrix (*40*), polymeric carrier–drug complexes are typically designed to respond to some activating event and provide a sustained, high concentration of drug to facilitate absorption. Some hydrogel materials, for example, swell or shrink in response to temperature (*41*, *42*); others can respond to pH (*42*, *43*), thereby opening the possibility for either specific delivery to sites of increased temperature (e.g., at the site of an acute inflammation) or shifts in pH (e.g., at the gastroduodenal or ileocecal junctures or at a site where metabolic events may produce a slightly acidic or alkaline environment). Erosion-controlled devices coated with materials such as pH- or temperature-sensitive polymers could also be used to initiate a slow release or extended release of drug along a segment of the GI tract distal to the site of polymer activation (*44*, *45*). Enzymatic disruption of a polymeric carrier such as pectin (*46*) or a polymer cross-linked through azo bonds (*44*, *47*) by colonic bacterial enzymes can provide site-specific delivery to the distal GI tract. Even more complex polymeric systems that can be activated to release drug in the presence of a specific substrate have been designed (*48*). Such systems could be used for site-specific delivery to regions of the GI tract with unique enzymatic activities and thus unique enzyme products. Alternately, a specific substrate for one of these enzymes could be coformulated into the polymeric carrier as a means of initiating a site-specific drug delivery.

Adhesive polymers, such as polycarbophil, can adhere strongly to the gastric mucosa in rats (*49*), but similar application of this bioadhesive in humans is not as effective because of a more rapid gastric mucus production that acts to slough the polymer–drug mixture and reduce gastric residence time (J. Robinson, personal communication). The mucus coating of the stomach is significantly thicker than distal segments of the GI tract, and the stomach is more responsive to the mucus-secreting actions of cholinergic agents than other segments (*50*). On the basis of differences in mucus turnover, sensitivity to mucus secretagogues, and adherence properties, the colon and cecum therefore may be more applicable for mucoadhesive drug delivery than the stomach or jejunum (*50*). Activation of the enterogastric reflex might be one approach to slowing stomach emptying and enhancing gastric retention. This reflex acts to keep the duodenum from becoming overloaded by inhibiting the release of gastric contents. Alternately, counteracting the actions of gastrin (which stimulates stomach emptying) with other intestinal hormones such as cholecystokinin, secretin, or gastrin inhibiting peptide might result in improved gastric retention. Coordinated delivery of these hormones and stimulation of the enterogastric reflex are not trivial problems, however.

Although most polymeric drug carriers are designed with putative toxicity issues in mind, the metabolic fate of polymers and their breakdown products must be assessed. For example, some colon-specific delivery bioadhesive devices (*51*) might be broken down by bacterial fermentation to produce products such as butyrate and fiber that have been shown to attenuate hypertrophic and hyperplastic events in the colon (*52*). While the application of adhesives (e.g., chitosan derivatives and polycarbophil) to retain a drug on the mucosal surface and promote drug uptake appears effective (*53*), the concern of adhesive biocompatibility must still be addressed (*54*). Further, some of the commercially available adhesives being investigated are synthetic polymers with unknown routes of degradation and toxicity. Although it is not likely that significant amounts of these very large molecular weight adhesives would be absorbed, it would be of interest to have a clear picture of the toxicology of these polymers and any of their potential break-down products, because low levels of large-molecular-weight compounds can still be absorbed by mucosal surfaces (*55*).

Devices

Delivery to the oral cavity might be achieved by a device, similar to a lozenge, that the patient would retain in the vestibule of the mouth. Several natural factors, however, limit the success of such a simple, straightforward approach. The oral mucosa has rather poor permeability characteristics, thus limiting the absorption of many drug molecules. Further, dilution by saliva would reduce the effectiveness of such a delivery scheme. Additionally, the patient might accidentally swallow or inhale the device or intentionally eject it prematurely. This last point is a common problem associated with using a device. Although devices are engineered to precisely deliver drug over a set time frame and at a set rate (e.g., Alzet devices), the difficulty of retaining the device at a particular segment of the GI tract and the extensive variability of transit through the different GI tract segments can negate this precision.

To overcome such concerns for buccal delivery, adhesive devices have been studied (for review, *see* ref. 56), and approaches that use a polymeric drug reservoir retained at the buccal mucosa via a bioadhesive have shown promise (*57*). Polysulfone microporous hollow fibers have also been used as a platform to deliver peptides in the buccal cavity (*58*). Hydroxyethyl methacrylate hydrogel has similarly been used to deliver diclofenac sodium (*59*) or low-molecular-weight heparin across buccal mucosa (*60*). Several hours of application are required before steady-state blood level can be achieved. Also, blood levels are maintained for some time after the patch is removed, a result suggesting a continued delivery of drug from a reservoir that develops in the mucosa.

An interesting device has been described for site-specific drug delivery to the esophagus. Since the residence time at this surface is extremely short, a complex of magnetic ferrite granules suspended in hydroxypropyl cellulose and Carbopol 934 was designed to be loaded with drug and held in place with a magnet (*61*). Whereas esophageal delivery is complicated by extremely short retention time, gastric retention is not as rapid but highly variable. For this reason, a number of devices designed for prolonged residency in the stomach have been investigated for device size, composition, and density (*62*, *63*). Since stomach emptying is stimulated by nerve impulses in response to distension and stomach gastrin (released in response to certain types of food such as partially digested proteins and ethanol), device retention in the stomach can be very different in either the fasted or fed state. Furthermore, variability of device residence in the stomach can be affected by the type of food and quantity of liquid residing in the stomach.

In general, residence time (and drug stability) at particular sites along the GI tract must be considered when contemplating a site-specific delivery in the GI tract. Although knowledge of GI tract transit kinetics are useful in assessment of potential site-specific delivery to various intestinal segments (*64*), the inherent variability associated with GI transit can be great and therefore difficult to overcome (*7*). Methods to deliver drugs to the small intestine and colon (*65*) and methods to evaluate the in vivo transit of various dosage forms (*8*) have been described. Some polymers with mucoadhesive properties, such as polystyrene lattices (*66*) and polycarbophil and chitosan derivatives (*53*), can be used as a device or depot to deliver drugs in a site-specific fashion. Polymers such as poly(bis(carboxylatophenoxy)phosphazene) form hydrogels in the presence of divalent cations (*67*), resulting in structures that might also have application in targeted drug delivery to the GI tract.

Site Specificity Through Targeting

Both the oral and rectal routes of delivery have been used to help facilitate site-specific delivery, but site-specific delivery to other mucosal surfaces of the GI tract may require some sort of targeting approach. Following oral or rectal delivery, some drugs are self-targeting—being absorbed at only certain locations along the GI tract as a result of physicochemical properties of the drug itself (*68*). Some devices may also be self-targeting to some extent. Microspheres of poly(lactide-*co*-glycolide) appear to be taken up preferentially into Peyer's patches (areas of concentrated lymphoid tissue) in the intestine (*69*). Most drugs or drug carriers, however, are not directed by some unique property and therefore must be targeted in some fashion to achieve a site-specific delivery within the GI tract. Several approaches for such targeting that take advantage of components associated with the surface of cells lining the mucosal surface of the GI tract have been described. For each of these approaches, however, there are still a number of unanswered questions related to efficiency of delivery and uptake.

Bacterial Adhesins and Lectins

Adhesion of devices to various regions of the GI tract appears to facilitate drug absorption when compared with application of solution formulations (*60*). A number of bacterial species adhere to mucosal cell surfaces at discrete regions of the GI tract. For example, *E. coli* F-18 association with the colonic mucosa appears to specifically occur through receptors that can be blocked by D-mannose (*70*). Several bacterial adherence factors have also been investigated for targeting antigens to the gut mucosa for mucosal immunization (*71*). The physiological role of these adherence factors may be related to pathogenic states (*72*), and although the mucosal barrier of the GI tract can typically resist invasion by adhering bacteria, the impact of additional adherence factors is unknown.

The presence of resident bacteria would obviously compete with drug–polymers or drug conjugates targeted to bacterial binding sites associated with a specific region of the GI tract. Because the bacterial flora of the intestine can vary from individual to individual and change dramatically as a result of systemic antibiotic therapy, the results obtained from such an approach could have significant variability due to changes in the number of binding sites for targeted delivery. It is also possible that, by flooding a region with a specific adhesin-containing drug delivery system, bacteria that would normally use that targeting adhesin could be displaced, thus changing the bacterial flora normally present within a discrete segment of the GI tract. Such an event could produce a spectrum of effects that might not be readily apparent or anticipated.

A number of lectins have been identified that bind selectively to unique sugar residues or complexes. The glycocalyx associated with much of the mucosal surface of the GI tract therefore can be targeted by lectins that bind selectively to specific sugars (*73*). One could envision several possible applications for lectin binding in targeted drug delivery. Lectins might be applied to deliver an agent that produces the final drug species rather than deliver the drug itself. For example, a lectin–enzyme conjugate could be targeted to a particular region of the GI tract. Ingestion of a poorly absorbed prodrug would be converted selectively at this site. A somewhat similar approach has been applied using immunoliposomes conjugated with an enzyme to activate a prodrug in a site-specific fashion (*74*). Antibody-directed enzyme prodrug therapy (ADEPT) might also provide a method of GI tract targeting (*75*). Regions of the GI tract associated with the immune system also might be targeted with a lectin. M cells (present in the gut-associated lymphoid tissues of the gut) of the rabbit cecum are labeled selectively by lectins specific for fucose or *N*-acetylgalactosamine when compared with adjacent tissue (*76*). Also, some evidence suggests that lectin-binding properties of lymphoid tissues from various regions of GI tract may also differ (*77*), and thus potentially increase the specificity of targeted delivery of antigen or immunomodulators.

The ultimate fate of lectins is of concern if one wishes to use them to target the delivery of drugs in the GI tract. To date, studies have failed to find any significant epithelial uptake and no marked impact on epithelial integrity for a limited number of lectins (*73*). Although not exhaustive, these findings are not surprising considering the large amount of plant lectins the GI tract is exposed to every day from a normal diet. One must, however, consider the potential variability among individuals of expressed lectin-binding sites and the observation that some lectins, such as phytohemagglutinin, can produce dramatic morphological modification to the small intestinal microvilli (*78*). The significance of these hypertrophic effects is still unclear.

Cell-Surface Receptors and Transporters

Some of the most elegant delivery systems described to date are those employed by pathogens that take advantage of a resident cell-surface receptor or transporter as part of their infective process. One such example is the vaccinia virus, which binds to epidermal growth factor (EGF) receptors in the gut (*79*). EGF is a protein hormone that plays an important role in the maturation and stabilization of the intestinal mucosa. By using this receptor in one of the stages of its life cycle, vaccinia has demonstrated the potential of using such a mucosal surface receptor as an avenue of targeted drug delivery. A number of other receptors, present at the mucosal surface at various locations along the GI tract, could also act as targets for delivery. Other receptor systems may be less specific or more ubiquitous and thus require other targeting approaches to deliver drug conjugates or mimetics to selected regions of the GI tract, as discussed earlier. The poly(D-lysine) carrier is an example of a nonselective transport system (*80*).

Several transcellular pathways of the GI tract are specific for a basolateral to apical movement of a particular solute. For example, pathways for both the polymeric IgA receptor (*81*) and polycations such as polylysine (*82*) are unidirectional, appear to involve a trimeric G protein function, and are brefeldin A-sensitive. Such transcytosis pathways typically use an apical endosomal structure, which is distinct from that used in recycling internalized membrane proteins and is not accessible to apically internalized fluid-phase components. Site-specific delivery into this pathway from the luminal cell surface (in regions of the GI tract where such a pathway is active) may provide a viable transcellular delivery route for drugs including macromolecules. Such an approach is important, because the apical uptake pathway employs a mechanism to limit the nonselective uptake of luminal solutes; overcoming this mechanism may be very difficult.

Nutrient transporters have been used as viable routes for uptake of certain drugs. Since these transporters are expressed at different levels at various segments of the GI tract, such uptake can produce a targeted delivery (*83*). Amino acid transporters are one prominent class that can be further divided into a series of systems (*84*); there may also be crossover between the route of amino acid uptake and an unexpected endocytotic uptake route. For example, the cationic amino acid transporter (termed the System y^+ transporter) is also a receptor for the murine ecotropic retrovirus (*85, 86*). The apical membrane of enteroendocrine cells of the small intestine appears to be the primary site for a neutral and basic amino acid transporter (*87*). Targeting to this transporter, primarily in the proximal regions of the small intestine, might provide not only a region-specific delivery but also a cell-specific delivery.

There are several factors to consider when targeting to amino acid transporters for drug transport. These transporters frequently require the cotransport of an ion (Na^+, H^+, or Cl^- are common); their expression can be modulated by diet (*88*) and some are hormone-dependent (*89*). Amino acid carriers selective for taurine (*90*), β-alanine (*91*), or imino acids (*92*) might be useful in the uptake of prodrug (e.g., chemical conjugates) from the jejunum and ileum. Although these carriers are commonly described as being Cl^--dependent, a requirement for this halide has not been clearly established in the human intestine (*93, 94*).

Two peptide/H^+ cotransport systems recently were described (*95, 96*). Apparently, these transport systems can be exploited for the delivery of β-lactam antibiotics (*95, 97*), bestatin (*98, 99*), angiotensin-converting enzyme inhibitors (*95*), and possibly for a spectrum of peptidomimetic compounds (*100*). Small peptide delivery has also been demonstrated through conjugation to bile acids and absorption through the bile acid transporters present in the apical membrane of distal ileal enterocytes (*101*). H^+-antiport systems, normally involved in the uptake of organic cations (*102*), might also be useful systems for the uptake of drugs or drug conjugates that can mimic this class of molecules. Several angiotensin-converting enzyme inhibitors also appear to be absorbed from the intestinal lumen through a sodium-dependent carrier-mediated process (*103*).

Vitamin transporters may be useful for selective drug targeting to the small intestine (*104*). Ascorbic acid is absorbed in the proximal ileum via an active transport process requiring Na^+ cotransport. Biotin, nicotinic acid, and thiamin are actively absorbed in the jejunum and also through Na^+-coupled processes. Whereas some of these transporters, such as the ascorbic acid transporter, can have high maximal absorption rates, others specific for biotin, thiamin, and riboflavin can be limited by down-regulation of the transporter in the presence of excessive vitamin (*104*). Although dietary deprivation of a vitamin could be used to enhance the uptake of vitamin look-alike or drug–vitamin conjugates through these transport systems, the practicality of such an approach must be questioned. Other factors associated with vitamin uptake and metabolism may further complicate the approach of targeting these transporters. The cyanocobalamin (vitamin B_{12}) uptake system has been studied sufficiently to demonstrate some of these problems.

Vitamin B_{12} combines with intrinsic factor (IF), a 45-kDa glycoprotein secreted from gastric parietal cells, and it is absorbed as a complex in the distal ileum through a limited number of IF–B_{12} receptor complexes on the surface of enterocytes in the distal ileum. During transcytosis, conversion of the cobalamin to deoxyadenosylcobalamin may occur in mitochondria before release from the basolateral membrane and into the portal circulation where it associates with a globulin protein referred to as transcobalamin II. This complex is rapidly cleared from the portal blood by the liver through a receptor-mediated endocytosis. A further complication is that cobalamin absorption declines in the elderly (*105*).

The use of transcellular transport routes results initially in the topical or local delivery of the drug moiety. If the drug has its biological actions at the site of absorption, then a successful delivery only requires the drug to cross the apical membrane of the enterocyte. However, for a systemic delivery, the drug must also cross the basolateral membrane. In some instances, there appears to be a defined pathway across the basolateral membrane (*106*), but in general no obvious or specialized route across the basolateral membrane exists for most absorbed nutrients. Since nutrients reach the systemic circulation, one must assume that the basolateral membrane is a less imposing barrier to transport across than the apical membrane. Differences in the physical properties of the apical and basolateral membrane bilayers of polarized epithelial cells may account for at least some of these transport properties (*107*).

Cellular Versus Systemic Absorption Issues

Site-specific targeting in the GI tract can result in a variety of pharmacokinetic and pharmacodynamic outcomes that reflect the rate of drug absorption and the subsequent metabolic fate of the drug during and after absorption. Drug absorption, which occurs primarily through the paracellular route, results in a systemic delivery that has several possible outcomes. If the drug is hydrophilic, it will usually be taken up into the blood capillary bed and will enter the circulation. If the site-specific delivery has occurred in either the oral cavity, the esophagus, the proximal part of the stomach, or the distal segments of the rectum, then the drug will bypass the portal venous system (which is fed by the capillary bed of more distal segments of the GI tract) and enter the systemic circulation directly. Schemes for site-specific targeting of drugs in the GI tract must avoid the portal venous system if significant first-pass liver metabolism is anticipated for a particular drug entity. If the transported drug acts directly on the absorbing epithelial cell or on a population of cells present in the submucosal region, concerns about first-pass metabolism in the liver would be lessened. Indeed, first-pass metabolism can be used to lessen the potential systemic actions of a drug intended to work locally at the intestinal mucosa. The local action of budesonide for the treatment of inflammatory bowel disease is one example of using hepatic clearance to advantage (*108*).

The uptake rate of a molecule from the lumen of the GI tract and its ultimate systemic delivery are dependent on many factors, some of which are discussed in this chapter. One important factor involves the potential for P-glycoprotein present in the apical membrane of the intestinal epithelium to pump absorbed small molecules back out and into the lumen of the GI tract, because this protein is expressed on the apical surfaces of jejunal and colonic epithelia (*109*). P-glycoprotein functions as an adenosine 5′-triphosphate-dependent transporter of small drug molecules out of cells, conferring a multidrug resistance (*110*, *111*). Information about substrate specificity of the multidrug-resistance pump (*112*) and methods of slowing its activity by calcium channels blockers (*113*) is growing. Expression of P-glycoprotein-dependent multidrug resistance is a frequent occurrence in human colorectal tumors (*114*) and may limit the efficacy of anticancer drugs delivered to the colon in a site-specific fashion. The P-glycoprotein, however, is not always a detractor to intestinal drug absorption. Cyclosporin A uptake by the small intestine through P-glycoprotein channels appears to be significant (*115*). A similar mechanism may also apply to uptake of celiprolol, a cardioselective β-adrenoreceptor antagonist (*116*).

Disease States

Disease states can affect the application of site-specific targeting schemes. The adherent layer of mucus on the colonic mucosa is important to normal health of the gut and is modified in disease states (*117*). Targeting to components of these structures may be inefficient in certain diseases. Acute inflammation of the GI tract can affect the potential of devices or polymeric carriers to successfully deliver to targeted regions of the GI tract. Colitis is associated with an exaggerated postcibal satiety signal. The resulting suppression of eating can affect stomach emptying (*118*) and thus disrupt the timing of systems delivered to various regions of the GI tract. Disease states can affect meal intake and meal composition: Both affect GI tract transit times. For example, fat intake might be restricted in some diseases. Since intestinal fat digestion appears to be responsible for induction of additional gastric acid secretion and gastrin release (*119*), this effect could result in an increased acidity of the stomach that might in turn affect the timing of an acid-activated device or reduce the stability of an acid-labile formulation destined for site-specific delivery.

The mucosal barrier of the GI tract typically, in the absence of frank ulcerations or gross pathologies, can resist significant uptake of macromolecules present in the lumen of the GI tract. Several disease states, including scleroderma modification of the oral mucosa, disturbance of the lower esophageal sphincter caused by achalasia, gastritis, improper hydrochloric acid production in the stomach, pancreatic insufficiency, enteritis, intestinal infections, inflammatory bowel disease, Crohn's disease, and diarrhea (*39*), can alter drug absorption from the GI tract and greatly modify attempts at site-specific delivery. HIV infection also appears to affect the absorptive and secretory function of enterocytes (*120*), thereby putatively impacting drug delivery. Alterations in affected mucosal epithelia might provide an opportunity for targeted delivery of a drug species. Certainly, alterations in local blood flow could affect the potential delivery of a targeted drug. Some pathologies reduce mesenteric blood flow, whereas others show an apparent increase. In the case of peritonitis, a decreased blood flow may account for impaired intestinal absorption of amino acids (*121*). Certain drugs, such as chlorpromazine, can reduce blood flow in subepithelial capillaries and reduce the intestinal absorption of other drugs (*122*). Other pathologies increase epithelial permeability and possibly increase blood flow. In such cases a potential role for endogenous mast cell activation has been suggested (*123*).

An interesting approach for targeting non-small cell lung cancers may also have appeal for targeting in the GI tract. An adenovirus polylysine mediated gene transfer method has been used to direct the local expression of a chimeric toxin gene in cancer cells that have an increased transcriptional activity (*124*). A similar approach might be employed along the GI tract where expression of an augmented or unique cell component is associated with a specific cell type or segment of the GI tract involving a pathological state. Initially, colorectal cancer (CRC) might seem a promising arena for such a targeting approach. Successful site-specific delivery to CRC cells, however, will be compromised by the variability present in the environmental and genetic stimuli involved (for review, *see* ref. 125). These factors reduce the possibility of a truly universal targeting approach. A similar targeting approach might take advantage of the presence of cell-surface components that are not typically present at the luminal or apical surface of cells lining the GI tract. For example, Na^+/K^+–ATPase is a transport complex typically expressed only in the basolateral plasma membrane. However, in response to ischemia, a disruption of the actin cytoskeleton leads to a loss of polarization of epithelial cells and the redistribution of the Na^+/K^+–ATPase to the apical membrane (for review, *see* ref. 126). Therefore, targeting of the Na^+/K^+–ATPase or other plasma membrane components that redistribute to the apical plasma membrane might provide a means of site-specific delivery to ischemic tissues in the GI tract.

Summary and Conclusions

Successful site-specific drug delivery to the GI tract has been achieved in a limited number of instances. The majority of these successes have employed either a device of some kind or a drug (or drug conjugate) with unique transport properties. Devices that physically deliver a drug payload to a specific site of the GI tract can use some parameter such as time, pH, local enzymatic activity, or physical retention to facilitate targeted delivery. Alternately, some specific cell- or tissue-associated marker or property may provide a target or avenue for site-specific delivery. The use of a device might increase the success of this approach; however, one can still proceed without such an additional component.

Although it is difficult to say what the future will hold for site-specific delivery to the GI tract, there is good cause for an optimistic outlook. Clearly, advances in our understanding of the basic biology of the these mucosal surfaces will provide new avenues for targeting and transporting drug entities. Improvements in current technologies as well as the identification of new ones should also lead to greater application of site-specific delivery schemes.

Acknowledgments

Suggestions made by Tue Nguyen and thoughtful discussions with Joseph Robinson are gratefully acknowledged. I also thank Allison Bruce for the figure graphics.

References

1. Wilding, I. R.; Davis, S. S.; O'Hagan, D. T. *Pharmacol. Ther.* **1994,** *62,* 97–124.
2. Squier, C. A.; Cox, P.; Wertz, P. W. *J. Invest. Dermatol.* **1991,** *96,* 123–126.
3. Russell, C. H.; Hill, L. D.; Holmes, E. R.; Hull, D. A.; Gannon, R.; Pope, C. E. *Gastroenterology* **1981,** *80,* 887–892.
4. Edmundowicz, S. A.; Clouse, R. E. *Am. J. Physiol.* **1991,** *260,* G512–G516.
5. DeNardi, F. G.; Riddell, R. H. *Am. J. Surg. Pathol. (Raven, N.Y.)* **1991,** *15*(3), 296–309.
6. Orlando, R. C.; Lacy, E. R.; Tobey, N. A.; Cowart, K. *Gastroenterology* **1992,** *102,* 910–923.

7. Coupe, A. J.; Davis, S. S.; Wilding, I. R. *Pharmacol. Res.* **1991,** *8,* 360–364.
8. Digenis, G. A.; Sandefer, E. *Crit. Rev. Ther. Drug Carrier Syst.* **1991,** *7,* 309–345.
9. Adkin, D. A.; Davis, S. S.; Sparrow, R. A.; Wilding, I. R. *J. Controlled Release* **1993,** *23,* 147–156.
10. Moës, A. J. *Crit. Rev. Ther. Drug Carrier Syst.* **1993,** *10,* 143–195.
11. Proano, M.; Camilleri, M.; Phillips, S. F.; Thomforde, G. M.; Brown, M. L.; Tucker, R. L. *Am. J. Physiol.* **1991,** *260,* G13–G16.
12. Madsen, J. L. *Dig. Dis. Sci.* **1990,** *35,* 1500–1504.
13. Evans, D. F.; Pye, G.; Bramley, R.; Clark, A. G.; Dyson, T. J.; Hardcastle, J. D. *Gut* **1988,** *29,* 1035–1041.
14. McDougall, C. J.; Wong, R.; Scudera, P.; Lesser, M.; DeCossse, J. *Dig. Dis. Sci.* **1993,** *38,* 542–545.
15. Mrsny, R. J. *Oral Colon-Specific Drug Delivery;* Friend, D. R., Ed.; CRC Press: Boca Raton, FL, 1992; pp 45–84.
16. Bai, J. F. *Life Sci.* **1993,** *53,* 1193–1201.
17. Bai, J. F. *Pharm. Res.* **1994,** *11,* 897–900.
18. Langguth, P.; Breves, G.; Stöckli, A.; Merkle, H. P.; Wolffram, S. *Pharmacol. Res.* **1994,** *11,* 1640–1645.
19. Haseto, S.; Ouchi, H.; Isoda, T.; Mizuma, T.; Hayashi, M.; Awazu, S. *Pharmacol. Res.* **1994,** *11,* 361–364.
20. Brandsch, M.; Miyamoto, Y.; Ganapathy, V.; Leibach, F. H. *Biochem. J.* **1994,** *299,* 253–260.
21. DeNigris, S. J.; Hamosh, M.; Kasbekar, D. K.; Fink, C. S.; Lee, T. C.; Hamosh, P. *Biochim. Biophys. Acta* **1985,** *836,* 67–72.
22. Armand, M.; Borel, P.; Dubois, C.; Senft, M.; Peyrot, J.; Salducci, J.; Lafont, H.; Lairon, D. *Am. J. Physiol.* **1994,** *266,* G372–G381.
23. Cortot, A.; Phillips, S. F.; Malagelada, J. R. *Gastroenterology* **1981,** *80,* 922–927.
24. Carey, M. C. In *The Liver: Biology and Pathobiology,* 2nd ed.; Arias, I. M.; Jacoby, W. B.; Popper, H.; Schachter, D.; Shafritz, D. A., Eds.; Raven Press: New York, 1988; pp 429–465.
25. Schulthess, G.; Lipka, G.; Compassi, S.; Boffelli, D.; Weber, F. E.; Paltauf, F.; Hauser, H. *Biochemistry* **1994,** *33,* 4500–4508.
26. Mahmood, A.; Yamagishi, F.; Eliakim, R.; DeSchryver-Kecskemeti, K.; Gramlich, T. L.; Alpers, D. H. *J. Clin. Invest.* **1994,** *93,* 70–80.
27. Sonis, S. T.; Lindquist, L.; Van Vugt, A.; Stewart, A. A.; Stam, K.; Qu, G.-Y.; Iwata, K. K.; Haley, J. D. *Cancer Res.* **1994,** *54,* 1135–1138.
28. Fleischmann, W. R., Jr.; Fields, E. E.; Wang, J.-L.; Hughes, T. K.; Stanton, G. J. *Proc. Soc. Exp. Biol. Med.* **1991,** *197,* 424–430.
29. Chapman, N. J.; Brown, M. L.; Phillips, S. F.; Tremaine, W. J.; Schroeder, K. W.; Deewanjee, M. K.; Zinsmeister, A. R. *Mayo Clin. Proc.* **1992,** *67,* 245–248.
30. Hollander, D. *Am. J. Med.* **1994,** *96,* 274–281.
31. Tremont, S. J.; Collins, P. W.; Perkins, W. E.; Fenton, R. L.; Forster, D.; McGrath, M. P.; Wagner, G. M.; Gasiecki, A. F.; Bianchi, R. G.; Casler, J. J.; Ponte, C. M.; Stolzenbach, J. C.; Jones, P. H.; Gard, J. K.; Wise, W. B. *J. Med. Chem.* **1993,** *36,* 3087–3097.
32. Perkins, W. E.; Bianchi, R. G.; Tremont, S. J.; Collins, P. W.; Casler, J. J.; Fenton, R. L.; Wagner, G. M.; McGrath, M. P.; Stolzenbach, J. C.; Kowalski, D. L.; Gasieski, A. F.; Forster, D.; Jones, P. H. *J. Pharmacol. Exp. Ther.* **1994,** *269,* 151–156.
33. Lee, V. L.; Yamamoto, A.; Buur, A.; Bundgaard, H. *Proc. Int. Symp. Controlled Release Bioact. Mater.* **1989,** *16,* 56.
34. Pellicciari, R.; Garzon-Aburbeh, A.; Natalini, B.; Marinozzi, M. *J. Med. Chem.* **1993,** *36,* 4201–4207.
35. Delie, F.; Couvreur, P.; Nisato, D.; Michel, J.-B.; Puisieux, F.; Letourneux, Y. *Pharm. Res.* **1994,** *11,* 1082–1087.
36. Haeberlin, B.; Rubas, W.; Nolen, H. W., III; Friend, D. R. *Pharm. Res.* **1993,** *10,* 1553–1562.
37. Kimura, T.; Yamaguchi, T.; Usuki, K.; Kurosaki, Y.; Nakayama, T.; Fujiwara, Y.; Matsuda, Y.; Unno, K.; Suzuki, T. *J. Controlled Release* **1994,** *30,* 125–135.
38. McLeod, A. D.; Fedorak, R. N.; Friend, D. R.; Tozer, T. N.; Cui, N. *Gastroenterology* **1994,** *106,* 405–413.
39. Dressman, J. B.; Bass, P.; Ritschel, W. A.; Friend, D. R.; Rubinstein, A.; Ziv, E. *J. Pharm. Sci.* **1993,** *82,* 857–872.
40. Langer, R. *Acc. Chem. Res.* **1993,** *26,* 537–542.
41. Hoffman, A. S. *J. Controlled Release* **1987,** *6,* 297–305.
42. Kim, Y.-H.; Bae, Y. H.; Kim, S. W. *J. Controlled Release* **1994,** *28,* 143–152.
43. Dong, L.-C.; Hoffman, A. S. *J. Controlled Release* **1991,** *15,* 141–152.
44. Brondsted, H.; Kopeček, J. *Pharm. Res.* **1992,** *9,* 1540–1545.
45. Kopeček, J.; Kopečková, P.; Brondsted, H.; Řihová, B.; Rathi, R.; Yeh, P.-Y.; Ikesue, K. *J. Controlled Release* **1992,** *19,* 121–130.

46. Ashford, M.; Fell, J.; Attwood, D.; Sharma, H.; Woodhead, P. *J. Controlled Release* **1994,** *30,* 225–232.
47. Saffran, M.; Kumar, G. S.; Savariar, C.; Burnham, J. C.; Williams, F.; Neckers, D. C. *Science (Washington, D.C.)* **1986,** *233,* 1081–1084.
48. Heller, J. *Crit. Rev. Ther. Drug Carrier Syst.* **1993,** *10,* 253–305.
49. Ch'ng, H. S.; Park, H.; Kelly, P.; Robinson, J. R. *J. Pharm. Sci.* **1985,** *74,* 399–405.
50. Rubinstein, A.; Tirosh, B. *Pharm. Res.* **1994,** *11,* 794–799.
51. Kopečková, P.; Rathi, R.; Takada, B.; Říhová, B.; Berenson, M. M.; Kopeček, J. *J. Controlled Release* **1994,** *28,* 211–222.
52. Bugaut, M.; Benéjac, M. *Annu. Rev. Nutr.* **1993,** *13,* 217–241.
53. Luessen, H. L.; Lehr, C.-M.; Rentel, C.-O.; Noach, A. B. J.; de Boer, A. G.; Verhoef, J. C.; Junginger, H. E. *J. Controlled Release* **1994,** *29,* 329–338.
54. Guo, J.-H. *J. Controlled Release* **1994,** *28,* 272–273.
55. Hoogstraate, A. J.; Cullander, C.; Nagelkerke, J. F.; Senel, S.; Verhoef, J. C.; Junginger, H. E.; Boddé, H. E. *Pharm. Res.* **1994,** *11,* 83–89.
56. De Vries, M. E.; Boddé, H. E.; Verhoef, J. C.; Junginger, H. E. *Crit. Rev. Ther. Drug Carrier Syst.* **1991,** *8,* 271–303.
57. Merkle, H. P.; Wolany, G. M. In *Biological Barriers to Protein Delivery;* Audus, K. L. Raub, T. J., Eds.; Pharmaceutical Biotechnology; Plenum: New York, 1993; Vol. 4, pp 131–160.
58. Burnside, B. A.; Keith, A. C.; Snipes, W. *Proc. Int. Symp. Controlled Release Bioact. Mater.* **1989,** *16,* 93–94.
59. Cassidy, J.; Berner, B.; Chan, K.; John, V.; Toon, S.; Holt, B.; Rowland, M. *Proc. Int. Symp. Controlled Release Bioact. Mater.* **1989,** *16,* 91–92.
60. Ebert, C. D.; Heiber, S. J.; Dave, S. C.; Kim, S. W.; Mix, D. *J. Controlled Release* **1994,** *28,* 37–44.
61. Ito, R.; Machida, Y.; Sannan, T.; Nagai, T. *Int. J. Pharm.* **1990,** *61,* 109–117.
62. Bechgaard, H.; Ladefoged, K. *J. Pharm. Pharmacol.* **1978,** *30,* 690–692.
63. Bechgaard, H. *Acta Pharm. Technol.* **1982,** *28,* 149–157.
64. Tozer, T. N.; Friend, D. R.; McLeod, A. D. *S.T.P. Pharma Sci.* **1995,** *5,* 5–12.
65. Christensen, F. N.; Davis, S. S.; Hardy, J. G.; Taylor, M. J.; Whalley, D. R.; Wilson, C. G. *J. Pharm. Pharmacol.* **1985,** *37,* 91–95.
66. Durrer, C.; Irache, J. M.; Puisieux, F.; Duchêne, D.; Ponchel, G. *Pharmacol. Res.* **1994,** *11,* 680–683.
67. Andrianov, A. K.; Cohen, B.; Visscher, K. B.; Payne, L. G.; Allcock, H. R.; Langer, R. *J. Controlled Release* **1993,** *27,* 69–77.
68. Chan, K. H.; Buch, A.; Glazer, R. D.; John, V. A.; Barr, W. H. *Pharmacol. Res.* **1994,** *11,* 432–437.
69. Morris, W.; Steinhoff, M. C.; Russell, P. K. *Vaccine* **1994,** *12,* 5–11.
70. Wadolkowski, E. A.; Laux, D. C.; Cohen, P. S. *Infect. Immun.* **1988,** *56,* 1035–1043.
71. Walker, R. I. *Vaccine* **1994,** *5,* 387–400.
72. Boedeker, E. C. *Gastroenterology* **1994,** *106,* 255–257.
73. Lehr, C.-M.; Lee, V. L. *Pharm. Res.* **1993,** *10,* 1796–1799.
74. Vingerhoeds, M. H.; Haisma, H. J.; van Muijen, M.; van de Rijt, R. J.; Crommelin, D. A.; Storm, G. *FEBS Lett.* **1993,** *336,* 485–490.
75. Bagshawe, K. D. *Clin. Pharmacokinet.* **1994,** *27,* 368–376.
76. Gebert, A.; Hach, G. *Gastroenterology* **1993,** *105,* 1350–1361.
77. Jepson, M. A.; Clark, M. A.; Simmons, N. L.; Hirst, B. H. *Histochemistry* **1993,** *100,* 441–447.
78. Hagen, S. J.; Trier, J. S.; Dambrauskas, R. *Gastroenterology* **1994,** *106,* 73–84.
79. Epstein, D. A.; Marsh, Y. V.; Schreiber, A. B.; Newman, S. R.; Todaro, G. J.; Nestor, J. J., Jr. *Nature (London)* **1985,** *318,* 663–665.
80. Taub, M. E.; Wan, J.; Shen, W.-C. *Pharm. Res.* **1994,** *11,* 1250–1256.
81. Barroso, M.; Sztul, E. S. *J. Cell Biol.* **1994,** *124,* 83–100.
82. Taub, M. E.; Shen, W.-C. *J. Cell Sci.* **1993,** *106,* 1313–1321.
83. Saier, M. H., Jr.; Daniels, G. A.; Boerner, P.; Lin, J. *J. Membr. Biol.* **1988,** *104,* 1–20.
84. Kilberg, M. S.; Stevens, B. R.; Novak, D. A. *Annu. Rev. Nutr.* **1993,** *13,* 137–165.
85. Wang, H.; Kavanaugh, M. P.; North, R. A.; Kabat, D. *Nature (London)* **1991,** *352,* 729–731.
86. Kim, J. W.; Closs, E. I.; Albritton, L. M.; Cunningham, J. M. *Nature (London)* **1991,** *352,* 725–728.
87. Pickel, V. M.; Nirenberg, M. J.; Chan, J.; Mosckovitz, R.; Udenfriend, S.; Tate, S. S. *Proc. Natl. Acad. Sci. U.S.A.* **1993,** *90,* 7779–7783.
88. Salloum, R. M.; Souba, W. W.; Fernandez, A.; Stevens, B. R. *J. Surg. Res.* **1990,** *48,* 635–638.

89. Nakanishi, M.; Kagawa, Y.; Narita, Y.; Hirata, H. *J. Biol. Chem.* **1994**, *269*, 9325–9329.
90. Miyamoto, Y.; Tiruppathi, C.; Ganapathy, V.; Leibach, F. H. *Am. J. Physiol.* **1989**, *257*, G65–G72.
91. Munck, L. K.; Munck, B. G. *Am. J. Physiol.* **1992**, *262*, G609–G615.
92. Munck, L. K. *Am. J. Physiol.* **1993**, *265*, G979–G986.
93. Munck, L. K.; Munck, B. G. *Am. J. Physiol.* **1994**, *266*, R997–R1007.
94. Munck, B. G.; Munck, L. K.; Rasmussen, S. N.; Polache, A. *Am. J. Physiol.* **1994**, *266*, R1154–R1161.
95. Dantzig, A. H.; Hoskins, J.; Tabas, L.; Bright, S.; Shepard, R. L.; Jenkins, I. L.; Duckworth, D. C.; Sportsman, J. R.; Mackensen, D.; Rosteck, P. R., Jr.; Skatrud, P. L. *Science (Washington, D.C.)* **1994**, *264*, 430–433.
96. Fei,Y.-J.; Kanai, Y.; Nussberger, S.; Ganapathy, V.; Leibach, F. H.; Romero, M. F.; Singh, S. K.; Boron, W. F.; Hediger, M. A. *Nature (London)* **1994**, *368*, 563–566.
97. Inui, K.; Okano, T.; Maegawa, H.; Kato, M.; Takano, M.; Hori, R. *J. Pharmacol. Exp. Ther.* **1988**, *247*, 235–241.
98. Inui, K.; Tomita, Y.; Katsura, T.; Okano, T.; Takano, M.; Hori, R. *J. Pharmacol. Exp. Ther.* **1992**, *260*, 482–486.
99. Takano, M.; Tomita, Y.; Katsura, T.; Yasuhara, M.; Inui, K.-I.; Hori, R. *Biochem. Pharmacol.* **1994**, *47*, 1089–1090.
100. Stewart, B. H.; Chan, O. H.; Lu, R. H.; Reyner, E. L.; Schmid, H. L.; Hamilton, H. W.; Steinbaugh, B. A.; Taylor, M. D. *Pharm. Res.* **1995**, *12*, 693–699.
101. Kramer, W.; Wess, G.; Neckermann, G.; Schubert, G.; Fink, J.; Girbig, F.; Gutjahr, U.; Kowalewski, S.; Baringhaus, K.-H.; Böger, G.; Enhsen, A.; Falk, E.; Friedrich, M.; Glombik, H.; Hoffman, A.; Pittius, C.; Urmann, M. *J. Biol. Chem.* **1994**, *269*, 10621–10627.
102. Iseki, K.; Sugawara, M.; Saitoh, N.; Miyazaki, K. *Biochim. Biophys. Acta* **1993**, *1152*, 9–14.
103. Zhou, X. H.; Po, A. W. *Biochem. Pharmacol.* **1994**, *47*, 1121–1126.
104. Said, H. M.; Mohammadkhani, R. *Gastroenterology* **1993**, *105*, 1294–1298.
105. Scarlett, J. D.; Read, H.; O'Dea, K. *Am. J. Hematol.* **1992**, *39*, 79–83.
106. Saito, H.; Inui, K.-I. *Am. J. Physiol.* **1993**, *265*, G289–G294.
107. Lande, M. B.; Priver, N. A.; Zeidel, M. L. *Am. J. Physiol.* **1994**, *267*, C367–C374.
108. Greenberg, G. R.; Feagan, B. G.; Martin, F.; Sutherland, L. R.; Thomson, A. B.; Williams, C. N.; Nilsson, L. G.; Persson, T. *N. Engl. J. Med.* **1994**, *331*, 836–841.
109. Thiebaut, F.; Tsuruo, T.; Hamada, H.; Gottesman, M. M.; Pastan, I.; Willingham, M. C. *Proc. Natl. Acad. Sci. U.S.A.* **1987**, *84*, 7735–7738.
110. Kartner, N.; Ling, V. *Sci. Am.* **1989**, *260*, 44–51.
111. Doige, C. A.; Yu, X.; Sharom, F. J. *Biochim. Biophys. Acta* **1992**, *1109*, 149–160.
112. Beck, W. T.; Qian, X.-D. *Biochem. Pharmacol.* **1992**, *43*, 89–93.
113. Jaffrézou, J.-P.; Herbert, J.-M.; Levade, T.; Gau, M.-N.; Chatelain, P.; Laurent, G. *J. Biol. Chem.* **1991**, *266*, 19858–19864.
114. Kramer, R.; Weber, T. K.; Morse, B.; Arceci, R.; Staniunas, R.; Steele, G., Jr.; Summerhayes, I. C. *Br. J. Cancer* **1993**, *67*, 959–968.
115. Augustijns, P. F.; Bradshaw, T. P.; Gan, L.-S. L.; Hendren, R. W.; Thakker, D. R. *Biochem. Biophys. Res. Commun.* **1993**, *197*, 360–365.
116. Karlsson, J.; Kuo, S.-M.; Ziemniak, J.; Artursson, P. *Br. J. Pharmacol.* **1993**, *110*, 1009–1016.
117. Pullan, R. D.; Thomas, G. O.; Rhodes, M.; Newcombe, R. G.; Williams, G. T.; Allen, A.; Rhodes, J. *Gut* **1994**, *35*, 353–359.
118. McHugh, K.; Castonguay, T. W.; Collins, S. M.; Weingarten, H. P. *Am. J. Physiol.* **1993**, *265*, R1001–R1005.
119. Shiratori, K.; Watanabe, S.-I.; Takeuchi, T. *Dig. Dis. Sci.* **1993**, *38*, 2267–2272.
120. Asmuth, D. M.; Hammer, S. M.; Wanke, C. A. *J. Controlled Release* **1993**, *23*, 147–156.
121. Gardiner, K. R.; Gardiner, R. E.; Barbul, A. *Br. J. Surg.* **1994**, *81*, 361–364.
122. Chung, Y. H.; Nishigaki, R.; Iga, T.; Hanano, M. *Pharmacobiodyn (Pharm. Soc. Jpn.)* **1983**, *6*, 829–835.
123. Kanwar, S.; Wallace, J. L.; Befus, D.; Kubes, P. *Am. J. Physiol.* **1994**, *266*, G222–G229.
124. Smith, M. J.; Rousculp, M. D.; Goldsmith, K. T.; Curiel, D. T.; Garver, R. I., Jr. *Hum. Gene Ther.* **1994**, *5*, 29–35.
125. Finlay, G. J. *Mutat. Res.* **1993**, *290*, 3–12.
126. Leiser, J.; Molitoris, B. A. *Biochim. Biophys. Acta* **1993**, *1225*, 1–13.

Self-Regulated Drug Delivery

7

Feedback-Controlled Drug Delivery

Jorge Heller

Feedback-controlled drug delivery systems are devices that are implanted in the body and are capable of releasing drugs in response to the concentration of a specific external molecule. A significant portion of this research deals with the development of delivery devices capable of releasing insulin in response to glucose blood plasma levels. Two types of devices are covered. In one type, drug release is continuously varied as the concentration of the external stimulus varies. In the other type, appearance of a specific external molecule triggers drug delivery from a normally passive device and once triggered, drug release proceeds by preprogrammed release kinetics. Major effort deals with development of devices that release naltrexone in response to external morphine in the treatment of narcotic addiction.

The optimal way to administer drugs is in a manner that precisely matches drug delivery and physiological need. To do so, drug delivery rate control has to be coupled with the physiological need by means of some feedback mechanism. In this chapter, I review means by which such a coupling can be accomplished. In doing so, I will focus entirely on the interaction of specific chemical moieties with appropriate delivery device components. The use of biosensors, environmental stimuli such as temperature or pH, and encapsulation of living cells are covered in other chapters of this book.

In covering feedback-controlled drug delivery systems, I will classify these into modulated devices and triggered devices (*1*). In a modulated device, rate of drug delivery can change continuously in response to the concentration of a specific external moiety, whereas in a triggered device no drug release takes place until it is activated by a specific external moiety. Once activated, drug release proceeds at a preprogrammed rate.

Modulated Devices

The most extensively investigated devices that can alter rate of drug release in response to changes in concentration of a specific external moiety are devices that release insulin in response to changes in glucose concentration. This extensive activity is driven by the need for better control of diabetes because current treatment by periodic insulin injection provides only a poor approximation of

Copyright © 1997 American Chemical Society

normal insulin output (*2*). This need is especially true for the approximately 10% insulin-dependent diabetics whose blood glucose concentration varies widely despite conventional insulin therapy. Poor control of blood glucose levels can result in a variety of degenerative conditions such as neuropathy and increased vascular disease including blindness. The various approaches used in the development of modulated devices can be classified into complex-controlled devices and enzyme–substrate controlled devices.

Complex-Controlled Devices

The basic principle of this approach is the formation of a reversible complex that can dissociate when exposed to free glucose. The dissociation occurs in proportion to glucose concentration and is thus used to control release of insulin.

Competitive Desorption

Lectin-Glycosylated Insulin-Controlled Devices. The first chemical approach for achieving insulin delivery that is proportional to the concentration of external glucose was published by Brownlee and Cerami (*3, 4*). This approach is based on the preparation of insulin derivatives with chemically attached oligosaccharides, which are complementary to the binding sites of the lectin concanavalin A (Con A). Brownlee and Cerami found that when a sugar–insulin conjugate is complexed with Con A, it can be displaced from the complex by free glucose. Because the displacement occurs in direct proportion to the concentration of glucose, this method can form the basis of a delivery device where the insulin conjugate is displaced from Con A in direct proportion to blood glucose levels. The proportionality of external glucose concentration and amount of displaced insulin with covalently attached mannotriose is shown in Figure 7.1 (*4*).

This method has been under intensive investigation by Kim and co-workers for a number of years, and a series of glycosylated insulin derivatives have been synthesized and characterized (*5*).

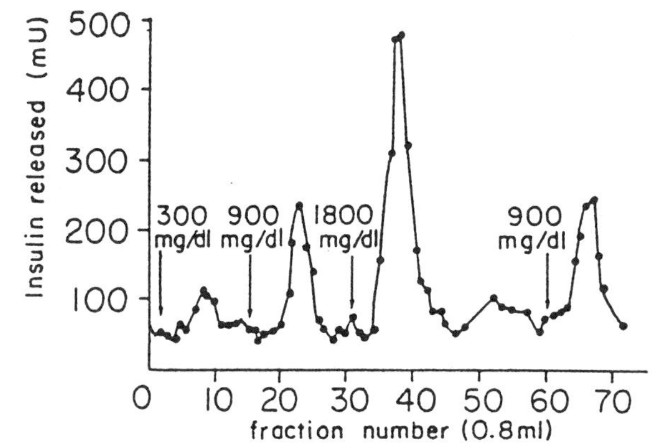

FIGURE 7.1. Glucose displacement of mannotriose–insulin complexed to Con A. Glucose displacement characteristics were evaluated by using 2.0-mL pulses of phosphate-buffered saline containing varying concentrations of glucose. These concentrations are indicated in the figure by labeling arrows. (Reproduced with permission from reference 4. Copyright 1983 American Diabetes Association.)

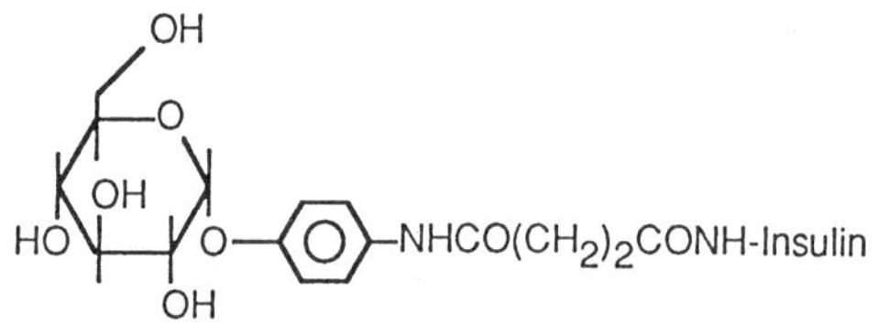

1

In vitro glucose exchange diffusion studies with various glycosylated insulins bound to Con A have shown that *p*-succinyl amidophenyl glucopyranoside (SAPG) insulin bound to Con A had an affinity for Con A 80 times higher than that of glucose, whereas *p*-succinyl amidophenyl mannopyranoside (SAPM) bound to Con A had an affinity for Con A 400 times higher than that of glucose. A large difference in binding affinities is necessary to prevent displacement of glycosylated insulins at any glucose concentration that could lead to life-threatening hypoglycemia. On the basis of these studies, SAPG insulin (**1**) was selected for further study. SAPG was characterized in vivo and had no undesirable immunological effects (*6–8*). Further, it retained full biological activity and was metabolized at rates identical to those of native insulin.

Because Con A is a toxic material, its release from the device must be prevented. This prevention has been achieved by surrounding the device with a porous membrane that has pores large enough to permit free passage of glucose and SAPM-insulin, but which excludes the much larger Con A molecules. A schematic representation of such a device is shown in Figure 7.2 (*9*).

Initial in vivo work used a poly(hydroxyethyl methacrylate) porous membrane fabricated into a pouch and implanted intraperitoneally in pancreatectomized dogs. However, a poly(hydroxyethyl methacrylate) membrane did not have the required mechanical strength to ensure that no rupture would take place and was replaced with regenerated cellulose tubing to produce a device about 1.8 cm in diameter and about 5.7 cm in length. In actual use, glucose diffuses through the membrane into the device at rates proportional to its plasma concentration, and once inside the device competes with the glycosylated insulin for the saccharide binding sites on Con A. Glycosylated insulin is then displaced and diffuses out of the device. Results of an in vivo study are shown in Figure 7.3 (*9*).

Even though the device functions extremely well—diabetic dogs with an implanted device have blood glucose profiles identical to those of normal dogs—the safety of the device was still of concern. Therefore, safety was further enhanced by covalently binding Con A to Sepharose beads, which were then enclosed in a macroporous membrane (*10*). The in vitro behavior of such a device is shown in Figure 7.4 (*10*).

Because the delayed "off" response of this device could lead to hypoglycemia, microcapsules and microspheres were investigated (*10, 11*). These devices have large surface areas that should allow rapid diffusion of both glucose and glycosylated insulin so that response time should be greatly reduced. Hydrophilic microcapsules were prepared by using an interfacial polycondensation reaction that enclosed SAPG-insulin and Con A in a thin nylon membrane formed from 1,6-hexanediamine and sebacoyl chloride. The behavior of microspheres is shown in Figure 7.5 (*11*). Stepwise glucose challenges result in rapid "on" and "off" response. During the "on" time, there is a gradual decline in the amount of SAPG-insulin released. When glucose concentration is decreased from 500 to 50 mg/dL, the lag-time is less than 10 min.

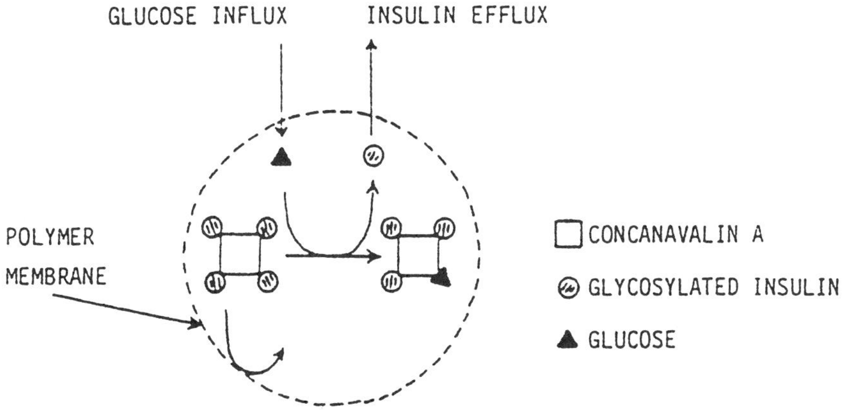

FIGURE 7.2. Schematic representation of self-regulated insulin delivery system. As the glucose level increases, the influx of glucose into the pouch increases, displacing glycosylated insulin (G-insulin) from the Con A substrate (□). Increased displaced G-insulin results in efflux of G-insulin from the pouch into the body. (Reproduced with permission from reference 10. Copyright 1990 Elsevier Science.)

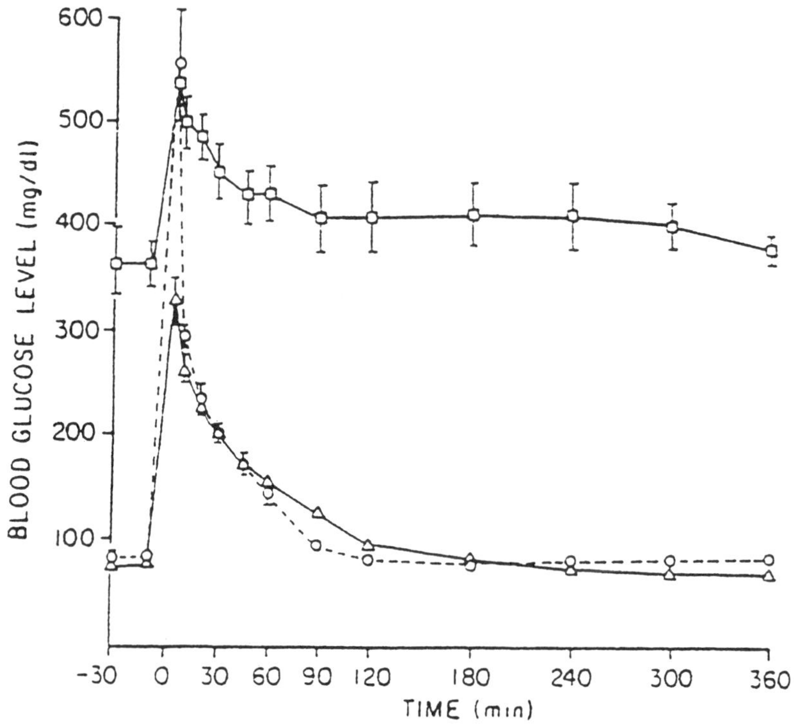

FIGURE 7.3. Peripheral blood glucose profiles of dog-administered bolus dextrose (500 mg/kg) during an intravenous glucose tolerance test. Normal dogs (○) had an intact pancreas, diabetic dogs (□) had undergone a total pancreatectomy, and implant dogs (△) had been intraperitoneally implanted with a cellulose pouch containing Con A–G-insulin complex. Blood glucose levels at $t = -30$ min show the overnight fasting levels 30 min before bolus injection. (Reproduced with permission from reference 9. Copyright 1985 Elsevier Science.)

However, release of the potentially immunogenic Con A still remained a concern. For this reason, the possibility of using cross-linked Con A was explored. Several groups have reported that Con A can be cross-linked and still retain its saccharide-binding properties (*12–14*). The use of cross-linked Con A would minimize the possibility of leakage from the pouch, and the use of porous microspheres would result in rapid exchange of bound SAPG–insulin. Preparation of cross-linked Con A microspheres was achieved by first blocking the sugar-binding sites of Con A with α-D-methylmannopyranoside and then reacting with glutaraldehyde. The sugar-binding sites were then deblocked by repeated washing with cold acetate buffer (pH 4.5, 4 °C). In a final step, the Schiff base linkages were reduced with sodium cyanoborohydride.

Figure 7.6 shows release of SAPG–insulin from the cross-linked, porous microspheres. The data show a rapid exchange of bound SAPG–insulin with glucose and a short response time. However, there is little difference in the amount of SAPG–insulin released from poly(vinylidene difluoride) pouches having surface areas of 16.7 cm^2 and 31.8 cm^2. This result may be due to sedimentation of the microspheres in the pouch. In actual use, the microspheres would be placed in an implanted pouch and periodically replenished after removal of the expended microspheres. This recent development represents a significant advance in the development of an implantable self-

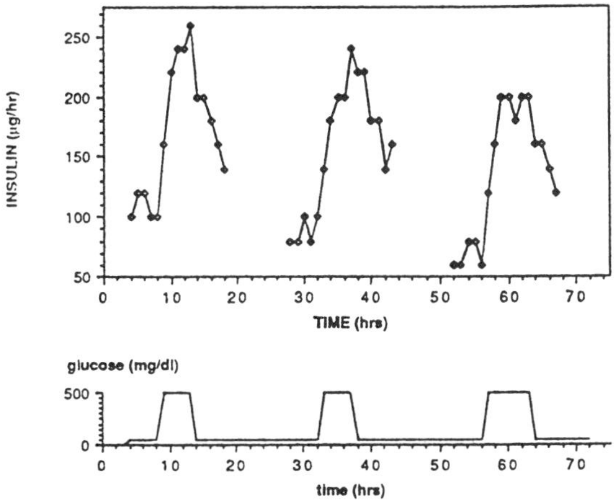

FIGURE 7.4. A 3-day, 2 glucose challenge cycle release profile for SAPG–insulin from a 0.22-mm pore size, 31 cm^2 Durapore pouch. (Reproduced with permission from reference 10. Copyright 1990 Elsevier Science.)

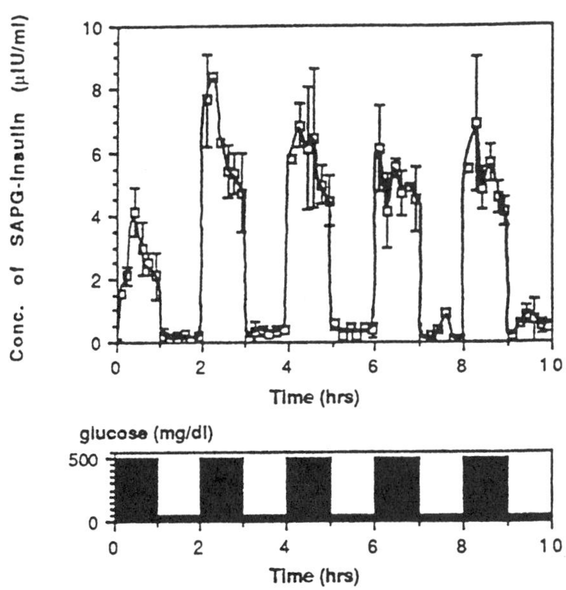

FIGURE 7.5. Release of SAPG–insulin in response to stepwise glucose challenge. (Reproduced with permission from reference 11. Copyright 1990 Elsevier Science.)

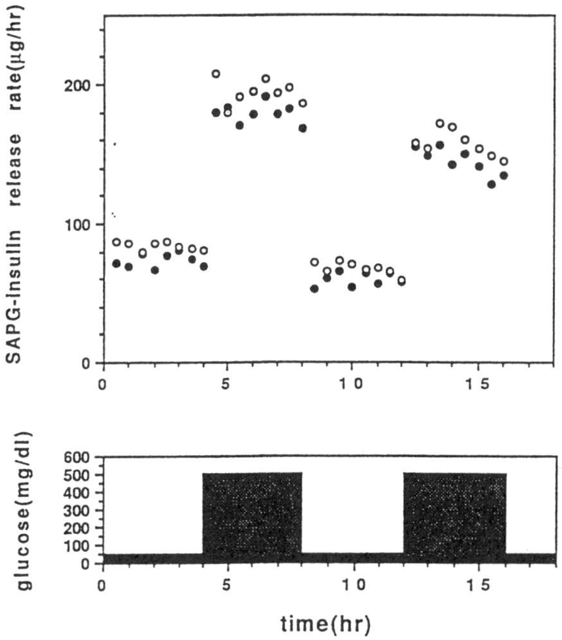

FIGURE 7.6. Release rate of SAPG–insulin in response to stepwise challenges of glucose from (●) 16.7 cm^2 and (○) 31.8 cm^2 polyvinylidene fluoride pouches (0.1-µm pore size) containing 6 mg of SAPG-insulin/100 mg of Con A microspheres. (Reproduced with permission from reference 14. Copyright 1992 American Pharmaceutical Association.)

modulated insulin delivery system, and in vivo studies using pancreatectomized dogs are in the planning stages.

Insulin-Controlled Devices Using Hydroxylated Insulin and Boronic Acid. Because phenylboronic acid can reversibly bind dihydroxy compounds when the two hydroxyl groups are in a coplanar configuration (*15–17*), polymers containing phenylboronic acid can form complexes with poly(vinyl alcohol), which can be dissociated with glucose and can thus form the basis of insulin delivery systems. The use of this approach will be discussed later.

The same principle can also be used to displace insulin by glucose from a polymer contain-

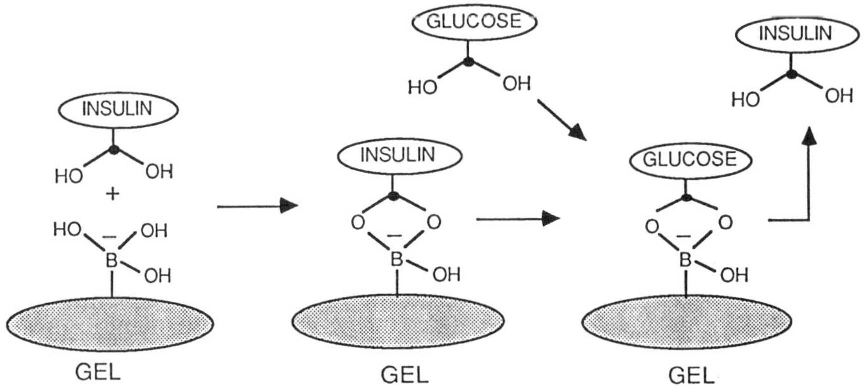

SCHEME 7.1. Schematic of release mechanism of glycosylated insulin from poly[*m*-methacrylamidophenylboronic acid-*co*-*N,N*-dimethylaminopropylacrylamide-*co*-acrylamide-*co*-*N,N′*-methylenebis(acrylamide)] gel beads. (Reproduced with permission from reference 18. Copyright 1994 Butterworth-Heinemann.)

ing phenylboronic acid groups (*18*). However, to do so requires placing two hydroxyl groups on insulin. This approach is shown in Scheme 7.1.

Hydroxylation of insulin has been accomplished by protecting one amino group with *p*-methoxybenzyloxycarbonylazide and then attaching gluconic acid to the unreacted amino groups by using *N*-hydroxysuccinimide coupling. The structure of the hydroxylated insulin is shown as structure **2**.

Gels containing phenylboronic acid moieties were prepared by copolymerizing *m*-methacrylamidophenyl boronic acid with acrylamide and *N,N′*-methylenebis(acrylamide). The structure of the gel is shown as structure **3**.

Release of gluconic acid–insulin conjugate (G-Ins) from the gel as a function of glucose pulses is shown in Figure 7.7. Clearly, the device is functional, and because typical diabetics show blood glucose concentration in excess of 200 mg/dL, it is also very sensitive. However, considerable additional work will need to be carried out to control the amount of G-Ins so that hypoglycemia is avoided.

Sol–Gel Controlled Devices

In this approach, hydrogels are prepared by synthesizing the polymer shown in structure **4** and forming cross-links by interaction between the chemically bound glucose and lectin (*19, 20*). Cross-linking takes place because at physiologic

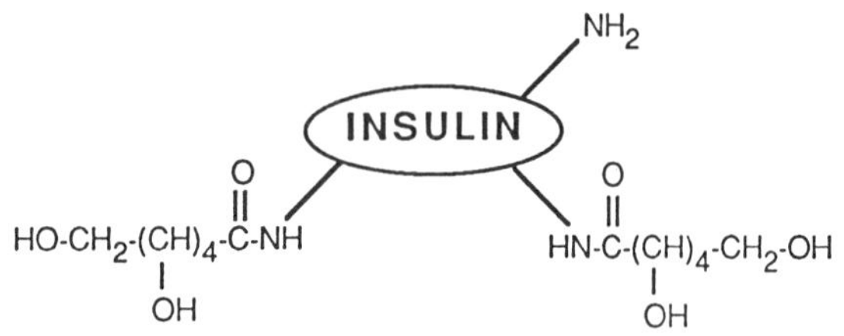

2

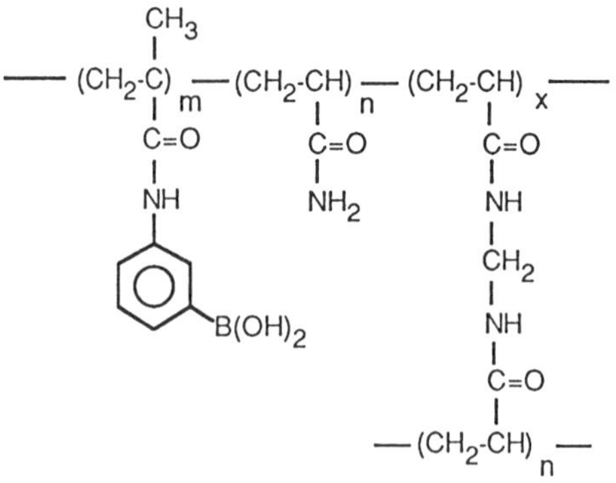

3

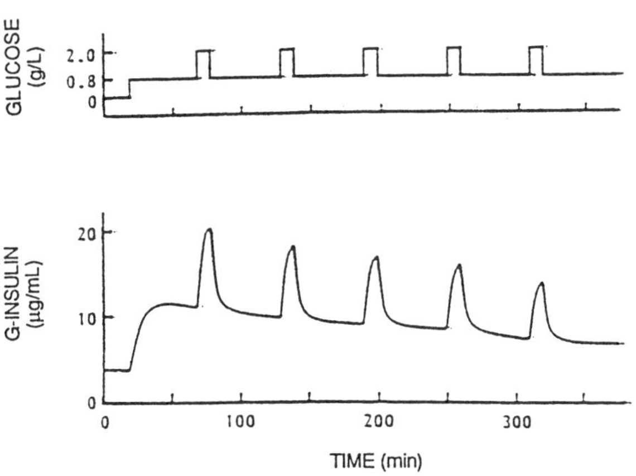

FIGURE 7.7. Release of G-insulin from PBA gel by concentration changes of glucose (80–2.0 g/L). Column volume, 0.314 cm^2 (5 mm diameter × 16 mm); boron content, 4.95 × 10^{-6} mol; loaded G-insulin 1.35 × 10^{-6} mol = 8.2 mg (after 5.19 × 10^{-6} mol of G-insulin loaded, rinsing was carried out with glucose-containing buffer solution at 2.0 g/L, six times, for 10 min); flow rate, 0.2 mL/min; eluant, pH 8.5, 50 mM *N*-(2-hydroxyethyl)piperazine-*N'*-2-ethanesulfonic acid, 10 mM $MgCl_2$. (Reproduced with permission from reference 18. Copyright 1994 Butterworth-Heinemann.)

pH, Con A exists as a tetramer and each subunit has a glucose binding site.

Because the cross-links are not covalent bonds, they can dissociate by competition between free glucose and covalently attached glucose in proportion to free glucose concentration. Because diffusion of insulin through the dissociated sol would be much faster than that through the gel, this approach provides a novel means of controlling release of insulin in response to blood sugar levels.

A preliminary account of a similar approach was recently described (*21*). In this approach, a gel is formed by complexing Con A with polysucrose or dextran. A thin layer of the gel is then held between two cellulose filters of pore size 0.1 µm and used as a rate-limiting membrane in a diffusion cell. Preliminary data using tartrazine as a model drug show that the system is indeed glucose-sensitive.

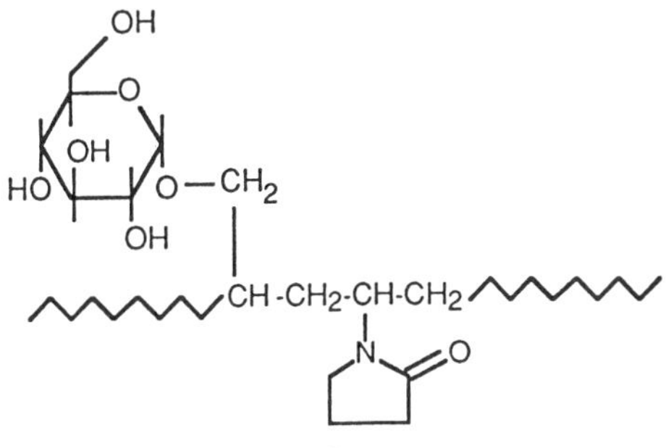

4

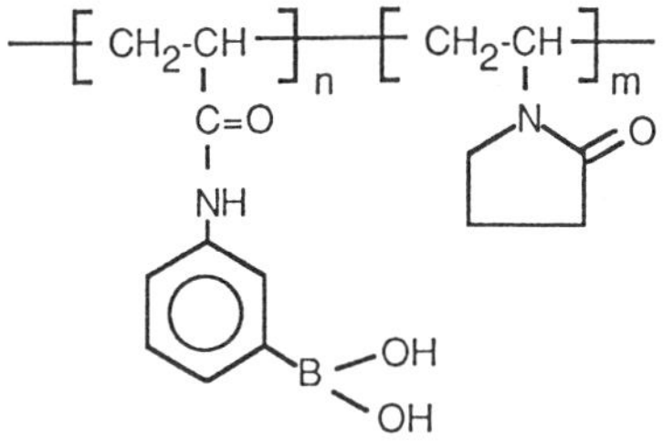

5

Polymeric Boronic Acid Controlled Devices

As already described, polymers containing a phenylboronic acid moiety can form reversible covalent complexes with molecules that have diol units, provided that that the two hydroxyl groups are held in a coplanar configuration. By using this principle, hydrogels have been formed by first copolymerizing *N*-vinyl-2-pyrrolidone and *m*-acrylamidophenylboronic acid (NVP-*co*-PBA) (**5**) and then complexing this water-soluble polymer with poly(vinyl alcohol) (*22, 23*).

This gel can be converted to a sol by the addition of a given amount of glucose, which competes with poly(vinyl alcohol) for the boronic acid moieties. Thus, by using the competitive binding of glucose with PBA and PVA, a self-modulated insulin delivery system has been proposed (Figure 7.8).

Enzyme-Controlled Devices

In this approach, the rate-control mechanism is an enzyme–substrate reaction that results in a pH change and a polymer system that can respond to

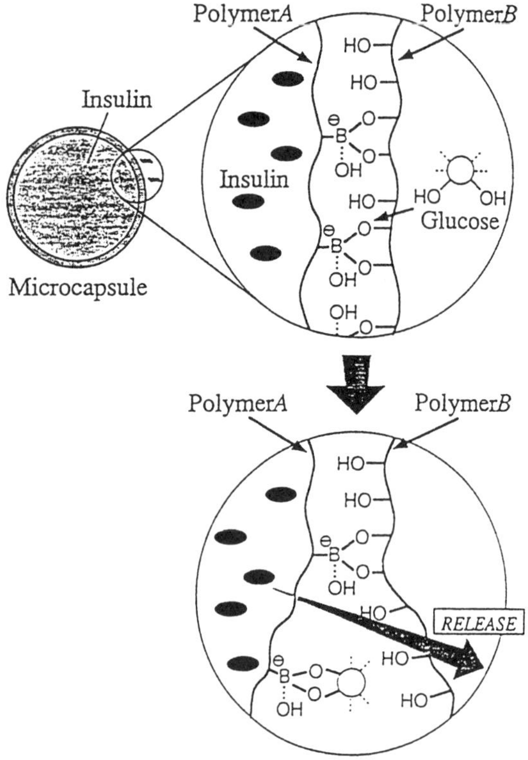

FIGURE 7.8. Concept of glucose-sensitive insulin release system using PVA–poly(NVP-*co*-PBA) complex system (polymer capsule type). (Reproduced with permission from reference 23. Copyright 1992 Elsevier Science.)

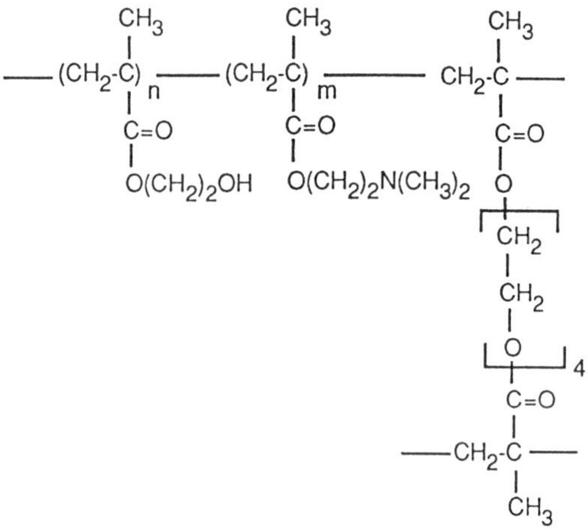

6

that change. Three approaches have been investigated. One approach uses membranes that change permeability in response to a decrease in pH. Another approach uses acid-sensitive polymers that change erosion rates with a change in pH. Yet another approach uses pH-induced changes in solubility.

pH-Sensitive Membranes

Membranes Containing Tertiary Amine Groups. In this approach, membranes are constructed from polymers that contain tertiary amine groups. Such membranes swell reversibly in acid because of the protonation of the amine groups and consequent repulsion of the charged groups. The first report using this approach for the construction of self-modulated insulin delivery devices was by Horbett, Kost, and Ratner (*24*), who immobilized glucose oxidase in a cross-linked polymer prepared from *N,N*-dimethylaminoethyl methacrylate, hydroxyethyl methacrylate, and tetraethylene glycol dimethacrylate (**6**).

In the absence of external glucose the amine groups are unprotonated and the porosity of the membrane is such that insulin molecules are unable to diffuse through the membrane. However, when glucose diffuses into the membrane, it is oxidized by glucose oxidase to gluconic acid, which protonates the amine groups and the porosity of the membrane increases due to the electrostatic repulsion between the protonated amine groups. The increased porosity then allows insulin

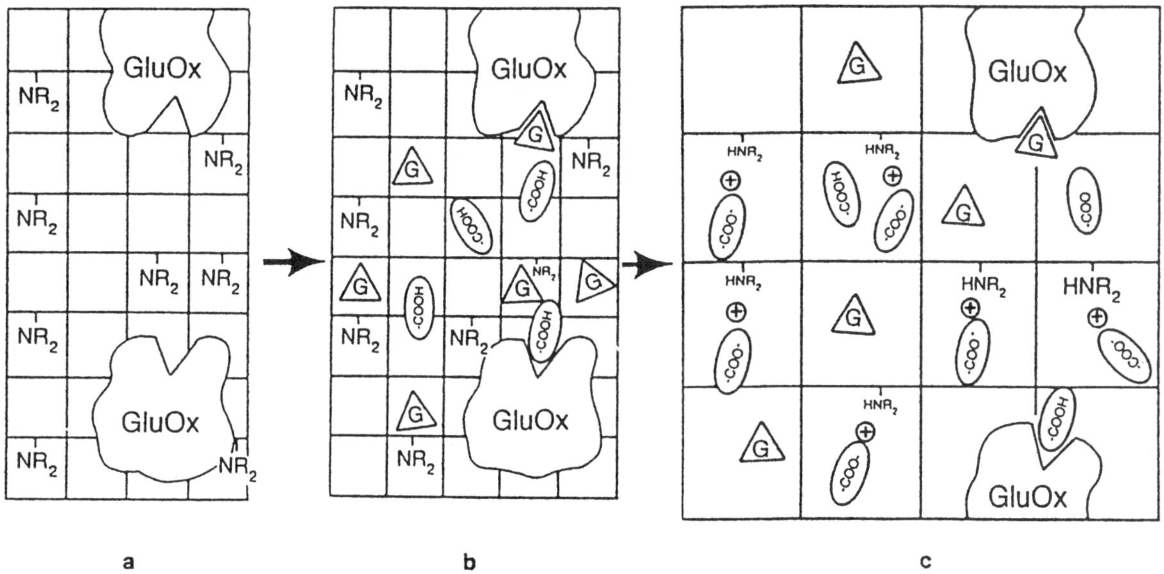

FIGURE 7.9. Schematic representation of the proposed mechanism of action of glucose-sensitive membrane. (a) In the absence of glucose, at physiologic pH, few of the amine groups are protonated. (b) In the presence of glucose, the glucose oxidase produces gluconic acid that can (c) protonate the amine groups. The fixed positive charge on the polymeric network led to electrostatic repulsion and membrane swelling. (Reproduced with permission from reference 25. Copyright 1984 Plenum.)

to diffuse through the membrane. This process is shown in Figure 7.9 (*25*).

Membranes were prepared by mixing the monomers, glucose oxidase, and an ethylene glycol–water solvent mixture and pouring between two glass plates separated by shims. The assembly was frozen at –70 °C and irradiated in a ^{60}Co source with a dose of 0.25 Mrad (*26, 27*). Under these conditions, enzyme activity is preserved. Provided that the solvent mixture contains more than 50% water, polymer precipitation takes place and leads to the formation of membranes having micropores in the micrometer range (*28, 29*). Such membranes have an insulin permeability in the 10^{-7} cm^2/s range, and it has been calculated that a device having a therapeutically useful insulin delivery rate could be constructed with a 0.1-mm thick membrane having a diameter of about 10 mm (*30*).

Insulin transport studies through such membranes containing immobilized glucose oxidase before and after adding enough glucose to the "downstream" side of the diffusion cell to bring the concentration momentarily to 400 mg% are shown in Table 7.1 (*30*). These data show that these macroporous gels are 8–12 times more permeable to insulin at pH 4.0 than at pH 7.4, and gels containing glucose oxidase are 2.4–5.5 more permeable to insulin when 400 mg% glucose is introduced into an initially glucose-free environment. Theoretical models for the design of glucose-sensitive insulin delivery systems were recently published (*31, 32*).

A similar approach was published by Ishihara and co-workers (*33*). In this system, glucose oxidase was immobilized in a cross-linked polyacrylamide membrane that was sandwiched with a responsive membrane, in this case a linear copolymer of *N,N*-diethylaminoethyl methacrylate and 2-hydroxypropyl methacrylate. This system swells in response to glucose concentration. However, because the uncontrolled release of large amounts

TABLE 7.1 Insulin Transport Studies Through Insulin-Permeable Membranes Containing Immobilized Glucose Oxidase

Membrane	*pH*	*Glucose (mg%)*	$P \times 10^{-7}$ *(cm²/s)*
A	7.4	0	0.34
A	4.0	0	3.73
B	7.4	0	0.79
B	7.4	400	4.32

NOTE: Membrane components (volume fraction) were as follows: for membrane A, hydroxyethyl methacrylate, 0.198; *N,N*-dimethylaminoethyl methacrylate, 0.0495; and H_2O, 0.752; for membrane B, hydroxyethyl methacrylate, 0.198; *N,N*-dimethylaminoethyl methacrylate, 0.0496; H_2O, 0.451; and ethylene glycol, 0.031. *P* is permeability of membrane.

SOURCE: Adapted from reference 30.

of insulin is lethal, a clinically useful device must have adequate mechanical properties to minimize the possibility of membrane rupture, and this requires the right balance of polymer hydrophilicity and hydrophobicity. For this reason, a copolymer prepared from 2-hydroxyethyl acrylate (hydrophilic monomer), 4-trimethylsilylstyrene (hydrophobic monomer), and *N,N*-diethylaminoethyl methacrylate (pH-sensitive monomer) was investigated (*34*). Devices were prepared by encapsulating insulin and glucose oxidase in a copolymer containing these three monomers in 0.6, 0.2, and 0.2 mole ratios, respectively. The structure of the terpolymer is shown in structure 7.

This membrane undergoes a change in water content from 30% to 53% and a change in permeability from 5.5×10^{-9} cm²/s to 2.3×10^{-7} cm²/s between the narrow pH range of 6.1–6.3. Further, the membrane, even in the highly hydrated state at pH 6.1, was not permeable to glucose oxidase.

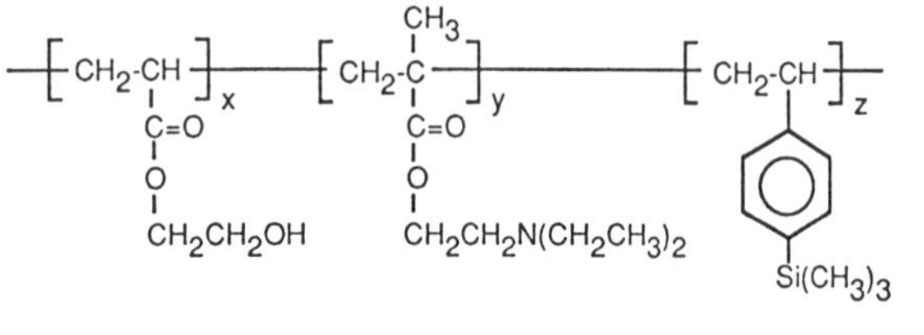

7

Release of insulin from the capsules as a function of external glucose is shown in Figure 7.10.

Early work by Ishihara and co-workers (*35*) published in 1983, the same year Horbett and co-workers published their approach using membranes with tertiary amine groups (*24*), described a glucose-sensitive membrane that uses hydrogen peroxide generated from the glucose–glucose oxidase reaction. The structure of the hydrogen peroxide sensitive copolymer is shown in Scheme 7.2.

In this system, glucose oxidase was immobilized in a cross-linked polyacrylamide membrane that was sandwiched with the polymer shown in Scheme 7.2. Oxidation of the dihydronicotinamide portion of the polymer by hydrogen peroxide increased polymer hydrophilicity and consequently increased insulin permeability. However, even though this system was able to increase insulin flux in response to an increase of glucose concentration, the increase was not reversible and this approach is clearly not therapeutically useful unless suitably modified to make it readily reversible.

Ishihara and co-workers (*36, 37*) also described a urea-sensitive polymer. Even though

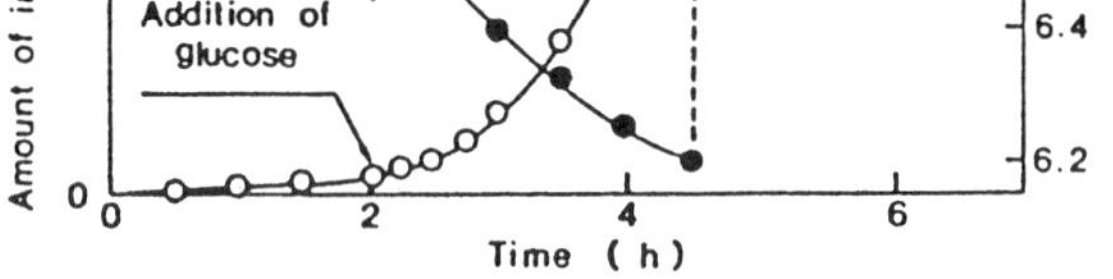

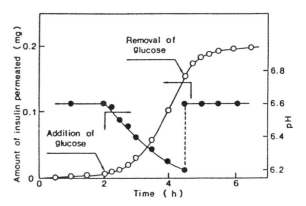

FIGURE 7.10. Effect of the addition and removal of glucose on insulin permeation through the HEA–DEA–TMS (HEA is 2-hydroxyethyl acrylate, DEA is *N,N*-diethylaminoethyl methacrylate, and TMS is 4-trimethylsilylstyrene) copolymer membrane and on the pH of the feed side solution at 30 °C. (Reproduced with permission from reference 34. Copyright 1986 John Wiley & Sons.)

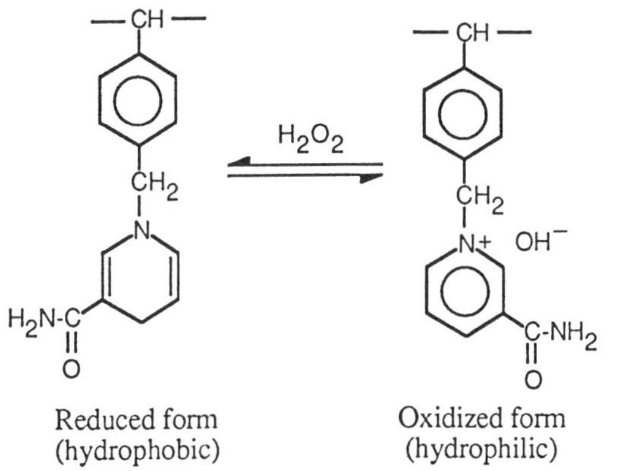

SCHEME 7.2. Oxidation of pendant dihydronicotinamide groups to nicotinamide groups by hydrogen peroxide. (Reproduced with permission from reference 35. Copyright 1983 Huethig and Verlag.)

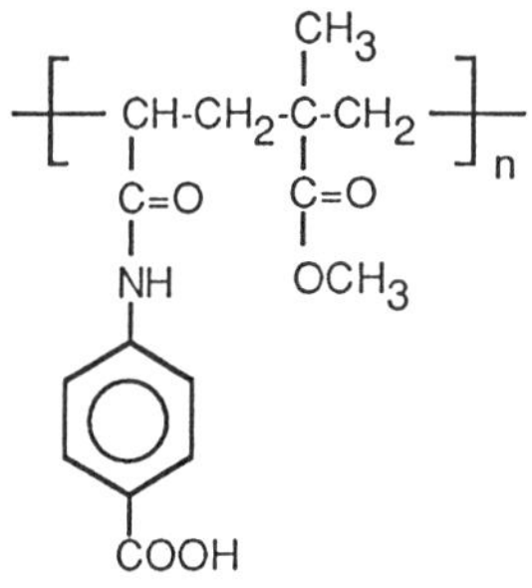

8

this system has no obvious therapeutic relevance, it is of interest because unlike glucose oxidase, which generates acidic products, urease generates basic products and thus requires a polymer that can respond to pH increases. One such polymer is a copolymer of 4-carboxyacrylanilide and methyl methacrylate (**8**).

Urea-responsive behavior is achieved by immobilizing urease in a cross-linked polyacrylamide membrane, which is then sandwiched with the 4-carboxyacrylanilide/methyl methacrylate copolymer shown in structure **8**. Interaction between urea and urease generates ammonium hydroxide and ammonium bicarbonate, and the resulting pH increase causes ionization of the carboxyl groups with consequent increase in permeability of the copolymer. An earlier example of a urea-sensitive system using a bioerodible polymer was published by Heller and Trescony (*38*).

Membranes Containing Grafted Carboxylic Acids. In this approach, first described by Iwata and Matsuda (*39*), a porous poly(vinylidene fluoride) film having pore sizes of 0.22 μm was first pretreated with air plasma and then acrylamide graft polymerized on the treated surface. A portion of the polyacrylamide was hydrolyzed to poly(acrylic acid), and another portion was converted to poly(vinyl amine) by a Hofmann rearrangement. In the pH region of 5–7, poly(acrylic acid) is ionized and chains are extended because of the repulsion of the negative charges. The extended chains thus effectively close the pores of the membrane. When the pH is lowered, the chains are not ionized and collapse, thus effectively opening the pores. An all poly(acrylic acid) grafted membrane became permeable in the pH region of 2–4, which is too low for a glucose oxidase–glucose modulated system. However, this pH range could be shifted to higher values by adjusting the molar ratio of poly(acrylic acid) to poly(vinyl amine). This process is shown in Figure 7.11.

Imanishi and co-workers (*40*) adapted this approach to a self-regulated insulin delivery system by first grafting poly(acrylic acid) on regenerated cellulose having 0.2-μm pore sizes by using ceric ion or plasma-mediated graft polymerization and then immobilizing glucose oxidase on the grafted membrane by coupling with 1-ethyl-3-(3-dimethylaminopropyl) carbodiimide. In the presence of glucose, an increase in insulin permeation was noted, but the increase was less than twofold.

In an attempt to produce a system that would not release insulin in the absence of glucose and that would have increased sensitivity, another approach was investigated (*41*). In this complex approach, insulin was immobilized on a poly(methyl methacrylate) membrane that was grafted with poly(acrylic acid) by using plasma-mediated polymerization. The grafted poly(acrylic acid) chains were then used to immobilize glucose dehydrogenase (GDH), the enzyme co-factors nicotin-

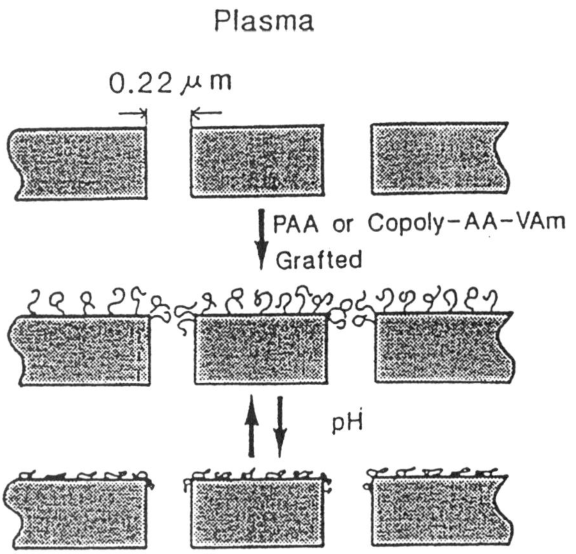

FIGURE 7.11. Schematic representation of grafting method and mechanism of sensitivity to environment of membrane. (Reproduced with permission from reference 39. Copyright 1988 Elsevier Science.)

amide adenine dinucleotide (NAD) and flavin adenine dinucleotide (FAD), and ethylene diamine by using 1-ethyl-3-(3-dimethylaminopropyl) carbodiimide. The free amino groups of ethylene diamine were then used to immobilize insulin, also by using 1-ethyl-3-(3-dimethylaminopropyl) carbodiimide. The functionality of this grafted membrane is based on the oxidation of glucose with generation of electrons. The enzyme co-factors NAD and FAD then act as electron mediators and reduce the disulfide bond that tethers insulin to the membrane with consequent release of insulin. In initial studies, in the absence of glucose, no insulin was released, but a large amount of insulin was released in response to a 9.8 mM concentration of glucose. Only a negligible amount of insulin was released in response to 0.98 mM concentration of glucose. In view of the high amount of insulin released, hypoglycemia would be a serious problem.

pH-Sensitive Bioerodible Polymers

In this approach, an enzyme–substrate reaction produces a pH change that is used to modulate the erosion rate of a hydrolytically labile, pH-sensitive polymer containing a dispersed therapeutic agent.

Although not therapeutically relevant, an early version of this approach used a pH-sensitive polymer in combination with the enzyme urease (*38*). In this system, urease acts on urea to produce NH_4HCO_3 and NH_4OH. This process leads to a pH increase that requires a polymer capable of increasing erosion rate with small increases in pH. A polymer with such characteristics is a partially esterified copolymer of methyl vinyl ether and maleic anhydride that was previously shown to undergo surface erosion with an erosion rate that is extraordinarily pH-dependent (*42*). This pH sensitivity is shown in Figure 7.12. As with the polymer described by Ishihara on a similar system, already discussed, the polymer dissolves by ionization of a carboxylic acid group (Scheme 7.3).

A schematic representation of the experimental device is shown in Figure 7.13. The pH-sensitive polymer shown in Scheme 7.3 containing dispersed hydrocortisone is surrounded with urease immobilized in a hydrogel prepared by cross-link-

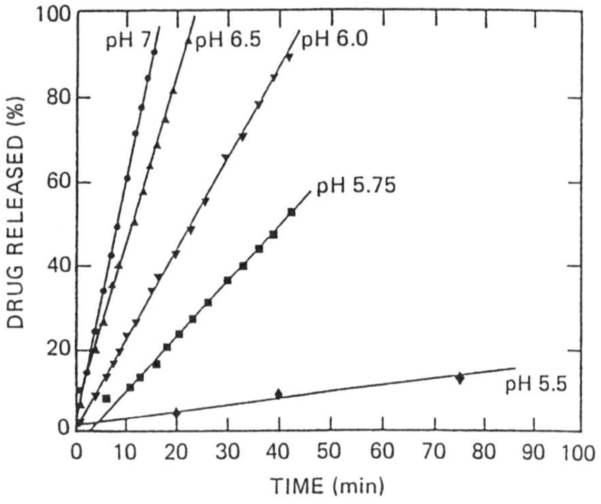

FIGURE 7.12. Effect of pH of erosion medium on rate of erosion of half-ester of methyl vinyl ether–maleic anhydride copolymer. (Reproduced with permission from reference 42. Copyright 1978 John Wiley & Sons.)

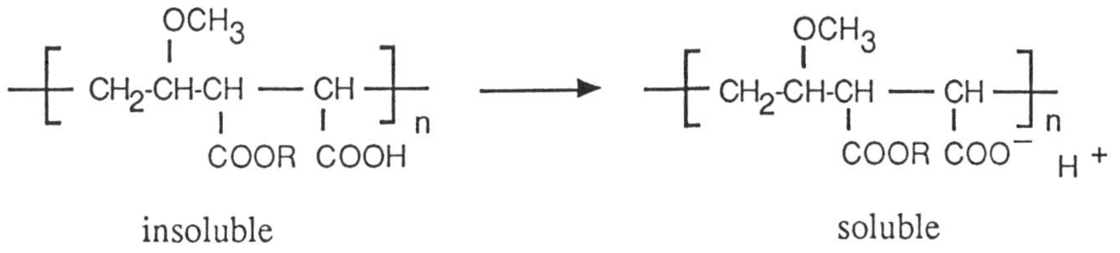

SCHEME 7.3. Solubilization of a partially esterified copolymer of methyl vinyl ether and maleic anhydride by ionization of pendant carboxylic acid functions. (Reproduced with permission from reference 42. Copyright 1978 John Wiley and Sons.)

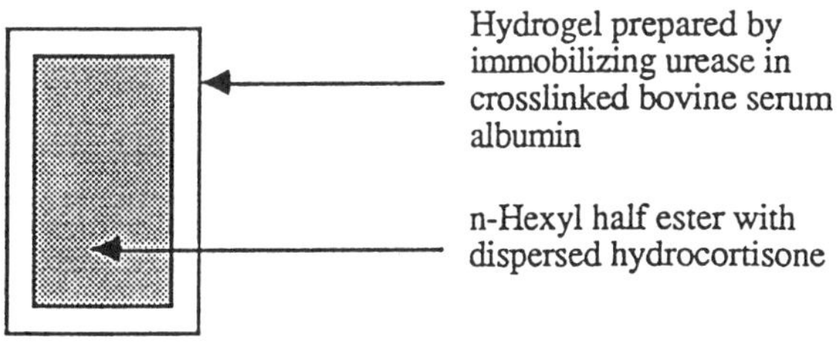

FIGURE 7.13. Schematic representation of urea-sensitive drug delivery system.

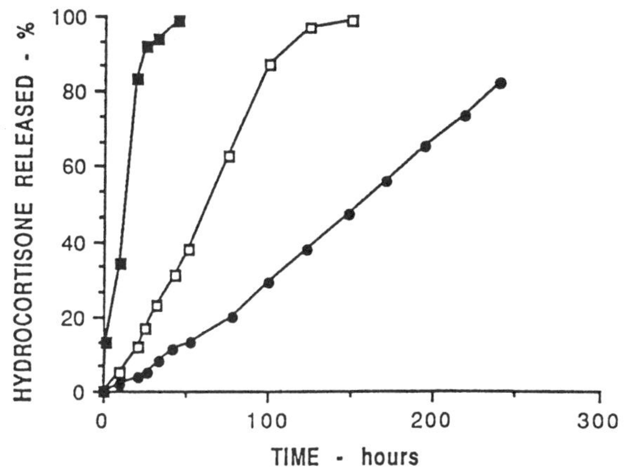

FIGURE 7.14. Hydrocortisone release rate at 35 °C from an *n*-hexyl half ester of a copolymer of methyl vinyl ether and maleic anhydride at pH 6.25 in the absence and presence of external urea: (●) no urea, (■) 0.1 M urea, (□) 0.01 M urea. (Reproduced with permission from reference 38. Copyright 1979 American Pharmaceutical Association.)

ing a mixture of urease and bovine serum albumin with glutaraldehyde. When urea diffuses into the hydrogel, its interaction with the enzyme leads to a pH increase that modulates erosion of the pH-sensitive polymer with concomitant changes in the release of hydrocortisone.

Results of that study are shown in Figure 7.14. Clearly, the device is functional and this study establishes the feasibility of this concept and also represents the earliest published demonstration of a self-regulated drug delivery system (*38*).

The same approach, using the reaction between glucose and glucose oxidase, has been used in the development of a system that releases insulin in response to blood glucose concentration (*43*). Because this reaction leads to a pH decrease, a polymer that undergoes an increased erosion rate with decreasing pH is required.

In extensive past work, poly(ortho ester)s underwent an increased erosion rate with decreasing pH (*44*). However, usefulness in a self-regulated insulin delivery device requires a polymer that is capable of undergoing a greatly increased erosion rate with only a very modest pH decrease. Because conventional poly(ortho ester)s do not have the required pH sensitivity, their structure was modified by incorporating tertiary amine groups into the polymer backbone (*43*) (Scheme 7.4).

When insulin was physically dispersed into this polymer and the polymer–insulin disk-shaped devices subjected to well-defined pH pulses, results shown in Figure 7.15 were obtained. Clearly, control over insulin release is excellent and it is especially gratifying to note that the response of the polymer to a decrease in pH is virtually instantaneous and that insulin release stops as soon as the

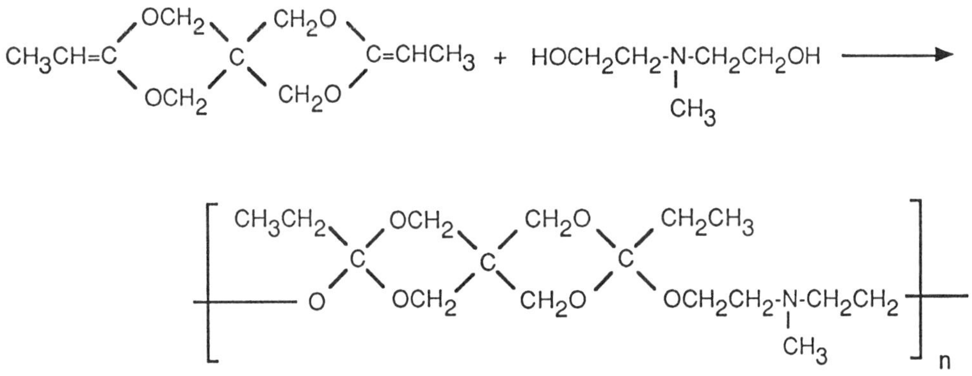

SCHEME 7.4. Preparation of poly(ortho ester)s containing tertiary amine groups in polymer backbone. (Reproduced with permission from reference 43. Copyright 1990 Elsevier Science.)

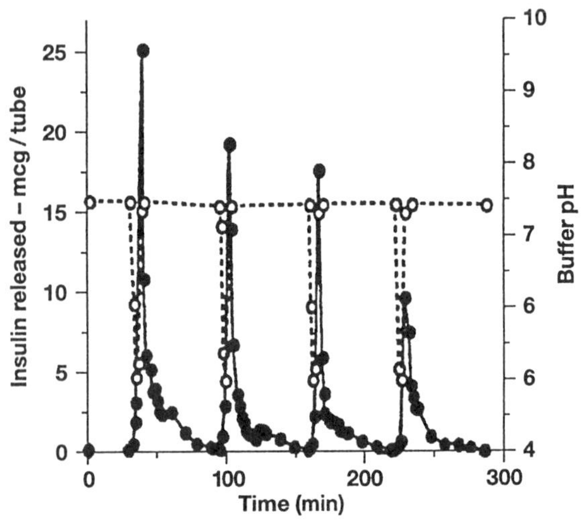

FIGURE 7.15. Release of insulin from a linear polymer prepared from 3,9-bis(ethylidene)-2,4,8,10-tetraoxaspiro[5.5]undecane and *N*-methyldiethanolamine as a function of external pH variations between pH 7.4 and 5.0 at 37 °C: disks, 3 × 0.55 mm; insulin loading, 10 wt%; nonrecycling media was continuously perfused at a flow rate of 2 mL/min; total efflux was collected at 1–10 min intervals; (○) buffer pH; (●) insulin release. (Reproduced with permission from reference 43. Copyright 1990 Elsevier Science.)

pH increases. This characteristic is essential if serious problems associated with hypoglycemia are to be avoided. The apparent descending release with repeated stimulation is due to gradual depletion of insulin from the device. Figure 7.16 shows the response of the device to pulses of decreasing pH in the same flow system. These studies also show that release of insulin is a function of the pH change, a highly desirable property for an eventual therapeutically useful device.

Because such a device requires that insulin, a large molecule, be able to traverse the hydrogel containing immobilized glucose oxidase, it is essential that the hydrogel be macroporous. Such hydrogels were prepared by first derivatizing insulin with acrylic anhydride and then covalently incorporating the derivatized insulin into a macroporous hydrogel prepared from hydroxyethyl methacrylate cross-linked with ethylene glycol dimethacrylate (*45*).

Although this system has excellent performance with well-defined pulses of phosphate buffer of well-defined pH values, the system was not functional when coupled to a glucose–glucose oxidase reaction. The difficulty was traced to the fact that the amine-containing polymer shown in Scheme 7.4 undergoes general acid catalysis where the catalyzing species is not a hydronium ion but rather the specific buffer molecule used.

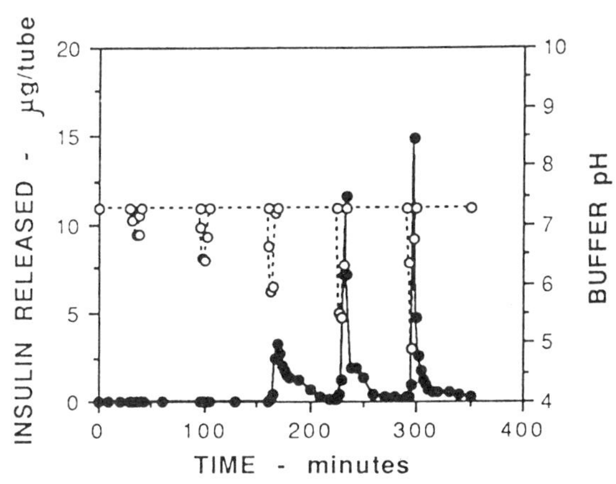

FIGURE 7.16. Release of insulin from a linear polymer prepared from 3,9-bis(ethylidene)-2,4,8,10-tetraoxaspiro[5.5]undecane and *N*-methyldiethanolamine as a function of external pulses of decreasing pH at 37 °C: disks, 3 × 0.55 mm; insulin loading 10 wt%; nonrecycling media was continuously perfused at a flow rate of 2 mL/min; total efflux was collected at 1–10 min intervals; (○) buffer pH; (●) insulin release. (Reproduced with permission from reference 43. Copyright 1990 Elsevier Science.)

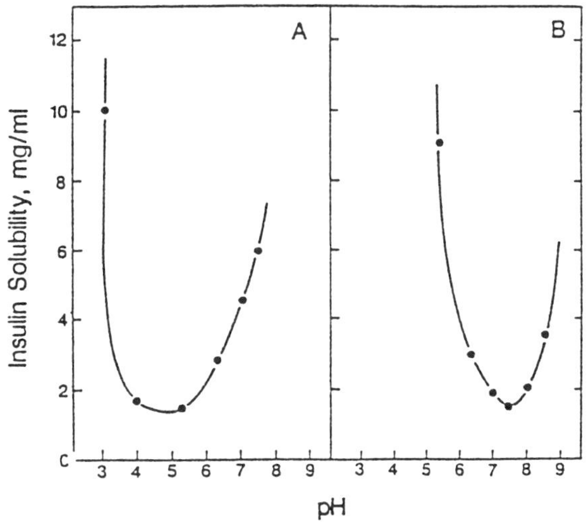

FIGURE 7.17. Solubility dependence on pH for regular (A) and trilysyl (B) insulin. Each solution was prepared by adding insulin above its saturation level to phosphate-buffered saline at different pH values. The solutions were centrifuged, and the concentration of insulin in each supernatant was measured spectrophotometrically. (Reproduced with permission from reference 46. Copyright 1988 National Academy of Sciences.)

Thus, further development of this system will require the development of a bioerodible polymer that not only has adequate pH sensitivity but that also undergoes specific ion catalysis (*45*).

Solubility-Controlled Devices

The solubility dependence of regular insulin can be shifted to higher pH values by using trilysyl insulin (Figure 7.17) (*46*). Because trilysyl insulin exhibits a large change in solubility between pH 7 and 5, the interaction between glucose and glucose oxidase will significantly change the rate of dissolution and allow self-regulated release. Devices were prepared by first immobilizing glucose oxidase onto Sepharose beads by using the cyanogen bromide method and then incorporating insulin and the immobilized glucose oxidase into an ethylene–vinyl acetate copolymer. This procedure produces channels within the polymer containing the enzyme and insulin, and preliminary data show that when the microenvironment within the channels is modified by the interaction between the enzyme and glucose, reversibly enhanced insulin release is observed.

Triggered Devices

In this section, we will only cover triggering occurring as a consequence of interaction with specific external molecules. Triggering that takes place by changes in external pH or temperature is covered in other chapters of this book.

A major driving force for the development of triggered drug delivery devices has been the desire to develop a more effective means of treating narcotic addiction. A major goal has been the development of an implantable device that would be normally passive, but on exposure to morphine, a metabolite of heroin, it would release the narcotic antagonist naltrexone.

Unfortunately, development of triggered drug delivery devices has not yet received the attention it deserves, and aside from applications to narcotic addiction, only two other applications have been considered. These applications are contraception and the release of chelating agents.

Hapten–Antibody Interactions

A device where activation is based on hapten–antibody interactions has been described by Pitt and co-workers (*47, 48*) (Scheme 7.5). In this device, a polymer fabricated into a monolithic or reservoir device containing the drug is first derivatized by covalent attachment of a hapten and then exposed to the hapten antibody, which results in a device that is surface coated with antibodies. If the monolithic device is constructed from a flexible polyester that has been shown to undergo surface attack by endogenous esterases (*49*), then the antibodies covering the surface will block access of the esterases to the polymer and prevent enzyme-induced polymer erosion and concomitant release of the incorporated drug. If the device contains the drug in a reservoir, then the antibodies covering the surface will impede diffusional release. The device will be activated when an external free hapten competes for the antibody attached to the polymeric device with consequent removal and activation.

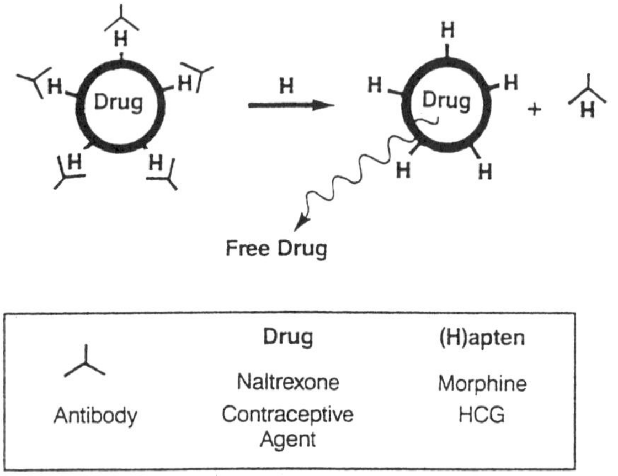

SCHEME 7.5. Triggered release based on reversible antibody binding to haptens attached to the surface of permeable or biodegradable polymers. (Reproduced with permission from reference 48. Copyright 1985 Elsevier Science.)

The development of such devices has not been completed, but two examples are under consideration (*48*). In one device, a morphine antibody is attached to the polymer, and activation by free morphine will release the narcotic antagonist naltrexone. Such a device would be useful in the treatment of heroin addiction. In another device, the β-subunit of human chorionic gonadotropin (HCG) is attached to the polymer and activation occurs by the appearance of HCG in the circulatory system. Because this appearance is the first biochemical indication of pregnancy, such a device would be useful in contraception.

Reversibly Inactivated Enzymes

A device designed to release naltrexone in response to external morphine has been under development by Heller and co-workers for a number of years (*50–56*). Activation is based on the reversible inactivation of enzymes achieved by the covalent attachment of a hapten close to the active site of the enzyme and a subsequent complexing of the enzyme–hapten conjugate with the hapten antibody (*57*).

Because the antibodies are large molecules, access of the substrate to the enzyme's active site is sterically inhibited, thus effectively rendering the enzyme inactive. Triggering of drug release is initiated by the appearance of morphine in the tissue surrounding the device, diffusion into the device, and dissociation of the enzyme–hapten–antibody complex rendering the enzyme active. The activated enzyme then removes a protective coating allowing release of naltrexone from a rate-limiting, bioerodible polymer. The reversible inactivation of an enzyme is shown in Figure 7.18.

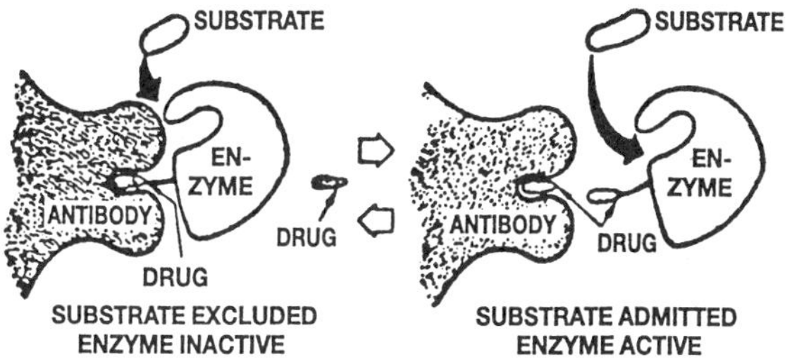

FIGURE 7.18. Reversible enzyme inactivation by hapten–antibody interactions. (Reproduced with permission from reference 57. Copyright 1973 American Association for Clinical Chemistry.)

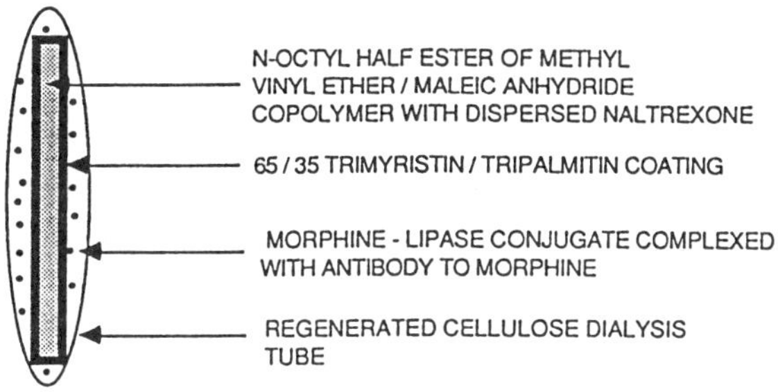

FIGURE 7.19. Schematic representation of triggered naltrexone delivery device.

The device currently under development is shown schematically in Figure 7.19. The device contains a rod-shaped core about 0.4 × 3.0 cm prepared from the *n*-octyl half ester of methyl vinyl ether and maleic anhydride copolymer containing physically dispersed naltrexone at a concentration of about 30 wt%. The bioerodible polymer used for the core material was that shown in Scheme 7.3, where R = *n*-octyl. This polymer previously was shown to undergo excellent surface erosion (*42*), and it releases incorporated naltrexone loaded at 30 wt% for the desired 2 weeks. At that erosion rate and loading, naltrexone is released at a rate of 5 mg/day, which has been shown to be sufficient to maintain effective blockage of opiate receptors (*58*). An additional advantage of this polymer is that it solubilizes without backbone cleavage so that the soluble polymer remains within the semipermeable bag, thus simplifying toxicological testing.

Solubilization of the naltrexone-containing core is prevented by an enzyme-degradable 65/35 trimyristin/tripalmitin coating that ensures core stability in the absence of external morphine. The coating is applied to the polymer at a 5:1 core–coating ratio making the final dimensions about 0.7 × 3.4 cm. The core is enclosed within a molecular porous regenerated cellulose dialysis tubing closed securely at each end with a double knot, and the knotted portions are covered with Silastic grade silicon tubing. The dialysis tubing is freely permeable to morphine and naltrexone but completely excludes the antigenic high molecular weight morphine antibody, lipase–morphine conjugate, and the bioerodible polymer.

The cellulose bag also contains the enzyme lipase that was reversibly inactivated by the covalent attachment of morphine and subsequent complexing with an antibody to morphine (*59*). The inactive lipase is capable of reactivation by free morphine, which can displace the lipase–morphine conjugate from the antibody. The active enzyme can then rapidly destroy the integrity of the protective coating by enzymatic hydrolysis of the triglyceride. The contents of the bag are made isoosmotic with poly(*N*-vinyl pyrrolidone).

The most difficult and the most challenging aspect in the development of a triggered drug delivery device is the ability to reversibly and completely inactivate an enzyme and yet be able to rapidly disassociate the complex with concentrations of the trigger molecule as low as 10^{-8} to 10^{-9} M. To achieve this sensitivity, lipase was conjugated with several morphine analogs and complexed with affinity chromatography purified polyclonal antimorphine antibodies. Purification of polyclonal antimorphine antibodies by affinity chromatography allows a decrease in the amount of antibody required to inhibit the morphine conjugated lipase, and changing the tether between the lipase and morphine affects the reactivation of the lipase conjugate. Although this work has not been completed, it is expected that the correct combination of morphine tether and affinity purified antibody will yield a device with the required sensitivity.

Recently, a study was completed that determined the amount of morphine injected intravenously into rabbits that penetrates an implanted placebo device (*60*). That study has shown that when a dose of 150 mg/kg is injected, the measured concentration in the device is only about 10-fold less than that measured in rabbit blood. Thus, the concentration of morphine in a device implanted in a typical heroin-addicted patient is estimated to be about 10^{-7} to 10^{-8} M. At this point, this concentration is lower than that necessary to reactivate the currently available complex. However, recent studies have shown that reaching such a sensitivity is possible.

Chelation-Enhanced Hydrolysis

A triggered system that can deliver a chelating agent on demand has been described by Pitt and co-workers (*48*). The objective of such a device would be a more effective method of administration of drugs that function by chelation, because release of the chelating agent would only occur in the presence of the target metal, would be proportional to the metal concentration, and would be specific for the metal.

The method is based on the well-known fact that certain metal ions under the right conditions dramatically accelerate the rate of hydrolysis of carboxylic esters, phosphate esters, and amides (*61*). The enhancement is believed to occur via complexation of the metal to the ester carbonyl group (*62*). The principle of this method is shown in Scheme 7.6.

Summary and Conclusions

In the early days of drug delivery research, the most sought-after delivery system was one that could deliver a drug by perfect zero-order kinetics. However, as drug delivery methodologies became more sophisticated, another goal gradually emerged. In this goal, the drug delivery system would not necessarily deliver the drug by zero-order kinetics, but instead would deliver the drug as needed, much in the same way as the body does. Thus, the field of self-modulated drug delivery has emerged.

Even though self-modulated devices can be constructed by using sensors coupled to an appropriately responsive delivery module, this approach is cumbersome and necessitates a pump connected to a stimulus responsive catheter. These devices are best reserved for life-threatening situations. For this reason, research attempting the development of self-modulated devices that use purely chemical or physical methods is clearly preferable. However, this research is complex and to this day remains almost exclusively focused on finding better ways to treat diabetes, a high priority, but certainly not the only high priority objective.

Research on triggered drug delivery devices has focused largely in finding a better way to rehabilitate individuals that have developed an addiction to opiates, principally heroin. Again, this

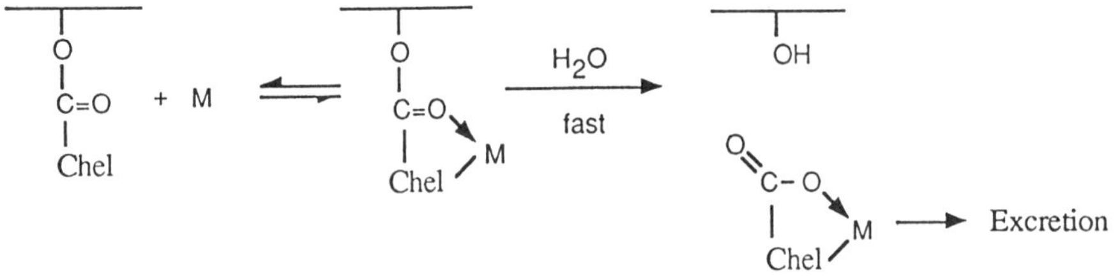

SCHEME 7.6. Concept of a self-regulated delivery system for metal chelators, based on metal promoted hydrolysis of carboxylic acids. (Reproduced with permission from reference 49. Copyright 1984 Elsevier Science.)

objective is certainly a high priority, but not the only high priority objective.

Because development of such complex systems with their attendant toxicological problems is many years removed from commercialization, this type of research has been largely confined to academic environments. The necessary focus of that research is channeled to areas where funding is available, which dictates the current emphasis. This result is unfortunate because more broadly focused research cannot be realized until at least a portion of the immense resources available to commercial institutions can be focused on the development of what is clearly the next generation of delivery devices.

References

1. Heller, J. *Crit. Rev. Ther. Drug Carrier Syst.* **1993,** *10,* 253–305.
2. Cahill, G. F.; Etzwiller, D. D.; Freinkel, N. *Diabetes* **1976,** *25,* 137–138.
3. Brownlee, M.; Cerami, A. *Science (Washington, D.C.)* **1979,** *206,* 1190–1191.
4. Brownlee, M.; Cerami, A. *Diabetes* **1983,** *32,* 499–504.
5. Jeong, S. Y.; Kim, S. W.; Eenink, M. D.; Feijen, J. *J. Controlled Release* **1984,** *1,* 57–66.
6 Sato, S.; Jeong, S. Y.; McRea, J. C.; Kim, S. W. *J. Controlled Release* **1984,** *1,* 67–77.
7. Seminoff, L. A.; Olsen, G. B.; Kim, S. W. *Int. J. Pharm.* **1989,** *54,* 241–249.
8. Seminoff, L. A.; Gleeson, J. M.; Zheng, J.; Olsen, G. B.; Holmberg, D.; Mohammad, S. F.; Wilson, D.; Kim, S. W. *Int. J. Pharm.* **1989,** *54,* 251–257.
9. Jeong, S. Y.; Kim, S. W.; Holmberg, D. L.; McRea, J. C. *J. Controlled Release* **1985,** *2,* 143–152.
10. Kim, S. W.; Pai, C. M.; Makino, K.; Seminoff, L. A.; Holmberg, D. L.; Gleeson, J. M.; Wilson, D. E.; Mack, E. J. *J. Controlled Release* **1990,** *11,* 193–201.
11. Makino, K.; Mack, E. J.; Okano, T.; Kim, S. W. *J. Controlled Release* **1990,** *12,* 235–239.
12. Donneley, E. H.; Goldstein, I. *Biochem. J.* **1970,** *118,* 679–680.
13. Kamra, A.; Gupta, N. M. *Biochem. Int.* **1988,** *16,* 679–687.
14. Pai, C. M.; Bae, Y. H.; Mack, E. J.; Wilson, D. E.; Kim, S. W. *J. Pharm. Sci.* **1992,** *81,* 532–536.
15. Boeseken, J. *Adv. Carbohydr. Chem.* **1949,** *47,* 189–210.
16. Aronoff, S.; Chen, T.; Cheveldayoff, M. *Carbohydr. Res.* **1979,** *40,* 299–309.
17. Foster, A. B. *Adv. Carbohydr. Chem.* **1957,** *12,* 81–115.
18. Shiino, D.; Murata, Y.; Kataoka, K.; Koyama, Y.; Yokoyama, M.; Okano, T.; Sakurai, Y. *Biomaterials* **1994,** *15,* 121–128.
19. Lee, S. J.; Park, K. *Polym. Prepr. (Am. Chem. Soc. Div. Polym. Chem.)* **1994,** *35*(2), 391–392.
20. Lee, S. J.; Park, K. *Proc. Int. Symp. Controlled Release Bioact. Mater.* **1994,** *21,* 93–94.
21. Taylor, M. J.; Tanna, S.; Cockshott, S.; Vaitha, R. *Abstracts of Papers,* Third European Symposium on Controlled Drug Delivery, Noordwijkaan Zee, Netherlands, 1994; pp 240–245.
22. Kataoka, K.; Koyama, Y.; Okano, T.; Sakurai, Y. *Makromol. Chem. Rapid Commun.* **1991,** *12,* 227–233.
23. Kitano, S.; Koyama, Y.; Kataoka, K.; Okano, T.; Sakurai, Y. *J. Controlled Release* **1992,** *19,* 162–170.
24. Horbett, T. A.; Kost, J.; Ratner, B. D. *Polym. Prepr. (Am. Chem. Soc. Div. Polym. Chem.)* **1983,** *24*(1), 34–35.
25. Horbett, T. A.; Kost, J.; Ratner, B. D. In *Polymers as Biomaterials;* Shalaby, S. W.; Hoffman, A. S.; Horbett, T. A.; Ratner, B. D., Eds.; Plenum: New York, 1984; pp 193–207.
26. Kost, J.; Horbett, T. A.; Ratner, B. D.; Singh, M. *J. Biomed. Mater. Res.* **1985,** *19,* 1117–1133.
27. Kaetsu, I.; Kumakura, M.; Yoshida, M. *Biotechnol. Bioeng.* **1979,** *21,* 847–861.
28. Ronel, S. H.; D'Andrea, M. J.; Hashiguchi, H.; Klomp, G. F.; Dobelle, W. H. *J. Biomed. Mater. Res.* **1983,** *17,* 855–864.
29. Refojo, M. F.; Yasuda, H. *J. Appl. Polym. Sci.* **1965,** *9,* 2425–2435.
30. Albin, G.; Horbett, T. A.; Ratner, B. D. *J. Controlled Release* **1985,** *2,* 153–164.
31. Albin, G. W.; Horbett, T. A.; Miller, S. R.; Ricker, N. L. *J. Controlled Release* **1987,** *6,* 267–291.
32. Klumb, L. A.; Horbett, T. A. *J. Controlled Release* **1992,** *18,* 59–80.
33. Ishihara, K.; Kobayashi, M.; Ishimaru, N.; Shinohara, I. *Polym. J.* **1984,** *16,* 625–631.
34. Ishihara, K.; Matsui, K. *J. Polym. Sci. Polym. Lett. Ed.* **1986,** *24,* 413–417.
35. Ishihara, K.; Kobayashi, M.; Shinohara, I. *Makromol. Chem. Rapid Commun.* **1983,** *4,* 327–331.

36. Ishihara, K.; Muramoto, N.; Fujii, H.; Shinohara, I. *J. Polym. Sci. Polym. Chem. Ed.* **1985,** *23,* 2841–2850.
37. Ishihara, K.; Muramoto, N.; Fujii, H.; Shinohara, I. *J. Polym. Sci. Polym. Lett. Ed.* **1985,** *23,* 531–535.
38. Heller, J.; Trescony, P. V. *J. Pharm. Sci.* **1979,** *68,* 919–921.
39. Iwata, H.; Matsuda, T. *J. Membr. Sci.* **1988,** *38,* 185–199.
40. Ito, Y.; Casolaro, M.; Kono, K.; Imanishi, Y. *J. Controlled Release* **1989,** *10,* 195–203.
41. Chung, D.-K.; Ito, Y.; Imanishi, Y. *J. Controlled Release* **1992,** *18,* 45–54.
42. Heller, J.; Baker, R. W.; Gale, R. M.; Rodin, J. O. *J. Appl. Polym. Sci.* **1978,** *22,* 1991–2009.
43. Heller, J.; Chang, A. C.; Rodd, G.; Grodsky, G. M. *J. Controlled Release* **1990,** *13,* 295–302.
44. Heller, J.; Penhale, D. H.; Helwing, R. F. *Contracept. Delivery Syst.* **1983,** *4,* 43–53.
45. Heller, J.; Franson, N. M., SRI International, personal communication, 1990.
46. Fishel-Ghodsian, F.; Brown, L.; Mathiowitz, E.; Langer, R. *Proc. Natl. Acad. Sci. U.S.A.* **1988,** *85,* 2403–2406.
47. Pitt, C. G. *Pharm. Int.* **1986,** 88–91.
48. Pitt, C. G.; Gu, Z.-W.; Hendren, R. W.; Thompson, J.; Wani, M. C. *J. Controlled Release* **1985,** *2,* 363–374.
49. Pitt, C. G.; Hendren, R. W.; Schindler, A.; Woodward, S. C. *J. Controlled Release* **1984,** *1,* 3–14.
50. Heller, J.; Pangburn, S. H.; Penhale, D. W. H. In *Controlled Release Technology: Pharmaceutical Applications;* Lee, P. I.; Good, W. R., Eds.; ACS Symposium Series 348; American Chemical Society: Washington, DC, 1987; pp 172–187.
51. Heller, J.; Penhale, D. H.; Pangburn, S. H. In *Proceedings of 3rd International Conference on Polymers in Medicine, Biomedical and Pharmaceutical Applications;* Nicolais, H.; Migliaresi, C.; Chielini, E., Eds.; Elsevier: Amsterdam, Netherlands, 1988; pp 175–188.
52. Heller, J.; Pangburn, S. H.; Roskos, K. V. In *Topics in Pharmaceutical Sciences;* Breimer, D. D.; Crommelin, D. A.; Midha, K. K., Eds.; Amsterdam Medical Press: Noordwijk, Netherlands, 1989; pp 39–49.
53. Heller, J. In *Pulsed and Self-Regulated Drug Delivery;* Kost, J., Ed.; CRC Press: Boca Raton, FL, 1990; pp 93–108.
54. Roskos, K. V.; Tefft, J. A.; Fritzinger, B. K.; Heller, J. *J. Controlled Release* **1992,** *19,* 145–160.
55. Tefft, J. A.; Roskos, K. V.; Heller, J. *J. Biomed. Mater. Res.* **1992,** *26,* 713–724.
56. Roskos, K. V.; Tefft, J. A.; Heller, J. *Clin. Mater.* **1993,** *13,* 109–119.
57. Schneider, R. S.; Lindquist, P.; Wong, E. T.; Rubenstein, K. E.; Ullman, E. F. *Clin. Chem. (Washington, D.C.)* **1973,** *9,* 821–825.
58. Chiang, C. N.; Hollister, L. E.; Gillespie, H. K.; Foltz, R. L. *Drug Alcohol Depend.* **1985,** *16,* 1–8.
59. Nakayama, G. R.; Roskos, K. V.; Fritzinger, B. K.; Heller, J. *J. Biomed Mater. Res.* **1995,** *29,* 1389–1396.
60. Roskos, K. V.; Fritzinger, B. K.; Tefft, J. A.; Nakayama, G. R.; Heller, J. *Biomaterials* **1995,** *16,* 1235–1239.
61. Buckingham, D. A. In *Biological Aspects of Inorganic Chemistry;* Addison, A. W.; Cullen, W. R.; Dolphin, D.; James, B. R., Eds.; Wiley Interscience: New York, 1976; pp 140–141.
62. Alexander, M. D.; Bush, D. H. *J. Am. Chem. Soc.* **1966,** *88,* 1130–1138.

8
Stimuli-Sensitive Drug Delivery

You Han Bae

Environmental factors such as chemical concentrations and physical signals influence the physical and chemical properties (swelling, solubility, configuration or conformational change, redox states, and crystalline/amorphous transition) of some materials. A drug delivery system composed of these responding materials will sense environmental changes and result in stimuli-sensitive drug delivery. Such a system will find potential applications to many clinical situations that require drug delivery at specific times, in a pulsed manner, or depending on metabolite concentration. Most technologies related to this purpose are still in an infant stage, and there are many obstacles to overcome for clinical trials.

A variety of concepts have been introduced for controlled drug delivery with the use of polymeric materials as drug carriers. Nearly all of the systems developed in the early phase of research release drugs with a decreasing rate or at a constant rate. More recent activities are related to targeting drugs to specific body sites or releasing drugs when needed. Typical examples include insulin delivery based on the body's glucose concentration for insulin-dependent diabetic patients; cell, tissue, or organ targeting for maximum drug effect and minimum side effects; and time-programmed drug delivery in harmony with physiological cycles (time-dependent physiological activities, pharmacokinetics, and pharmacodynamics). One approach to such systems is to use the changes of physiological signals in the body, such as temperature and pH, or externally applied stimuli to trigger drug release from drug delivery devices. Figure 8.1 presents the concept of an ideal stimuli-sensitive drug delivery system for self-regulation. The system containing drugs (drug reservoir) could change its properties (sensor) in response to environmental stimulus (signal), and then it could affect drug release (actor) from the drug reservoir.

Most technologies related to this purpose are still in an infant stage. Also, many obstacles must be overcome for clinical trials, such as system sensitivity to signal changes in a physiological range, biodegradability, biocompatibility, monitoring dose size after implant, and the external generation of signals. However, the potential for the development of a practical system is growing as all the related technologies advance.

In the design of a stimulus-sensitive system, the stimulus may be either chemical or physical signals. Chemical signals, such as pH, metabolites, and ionic factors, will alter the molecular interac-

Copyright © 1997 American Chemical Society

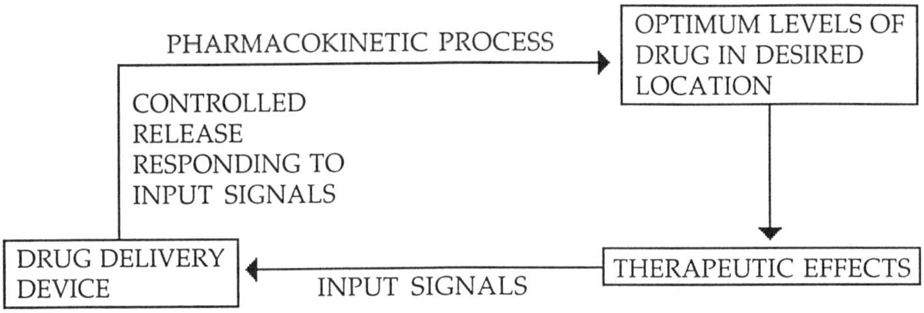

FIGURE 8.1. Schematic illustration of an ideal stimuli-sensitive drug delivery system.

tions between polymer chains or between polymer chain and solutes present in a system. The physical stimuli, such as temperature or electrical potential, may provide various energy sources for molecular motions and altering molecular interactions. Mechanistically, these various stimuli will influence drug delivery systems in relation to the constituent materials of the systems (e.g., polymers, lipids, and liquid crystals) and the changes of properties or structures of a given material (e.g., swelling, solubility, and conformation of a polymer matrix or polymer chain).

This chapter will summarize modulated solute release under the influence of an applied stimuli regardless of physiological relevancy, because the solutes can be applied in the agricultural and industrial fields. The chapter will also summarize the potential physiological and pharmaceutical applications. The chapter is closely related to Chapters 7 and 24; if a stimulus is altered by the effect of released drugs, a system is considered as feed-back controlled delivery (or closed-loop delivery), and intelligent materials are an important class of the means for stimuli-sensitive or feed-back controlled delivery systems.

Chemical Stimuli

pH

The environmental change in pH affects the degree of ionization of a polyelectrolyte, which in turn influences the solubility or conformation of the polyelectrolyte (*1–6*). When the polyelectrolyte is cross-linked, the gel swelling is a function of pH as well as other environmental factors, such as ionic strength and counter ions in solution (*7*). The pH sensitivity of polyelectrolyte gel membranes is influenced by the nature of ionizable groups, cross-linking density, polymer composition, or the hydrophobicity of the backbone polymer. The effect of functional groups on the pH sensitivity is illustrated by the examples of weakly acidic (*8, 9*) and basic polymers (*10, 11*) that show opposite swelling changes in response to the same pH difference.

The effect of cross-linking density on the permeability of solutes through pH-sensitive polymers was studied by Weiss et al. (*12*) with weakly acidic poly(methacrylic acid). The membrane permeability to solutes decreased as cross-linking density increased. However, this tendency was not significant at a higher pH region. The dependence of permeability on the pH of solution was more significant for high molecular weight solutes than small solutes, probably due to the screening effect of the network mesh.

The hydrophobicity of pH-sensitive polymers was controlled by copolymerizing hydrophobic monomers such as alkyl methacrylate with a pH-sensitive monomer (*13*). The dependence of permeability on pH increased as the hydrophobicity of the polymer network increased, although the overall permeability decreased with increased polymer hydrophobicity.

The conformational transitions such as helix to random coil for polypeptides (*14–16*) and compact coil to random coil for synthetic polymers (*17*, *18*) were used for the pH-sensitive drug delivery system. The pH-dependent permeabilities of water-soluble solutes such as sugars to synthetic polypeptide membranes were attributed to the conformational change of the grafted polypeptides from rigid α-helical structure at low pH to random coil at high pH, resulting in higher swelling at high pH (*19*, *20*).

pH-sensitive release of solutes using structural transition of synthetic polymers from tightly coiled chain was also demonstrated (*21*). The pH-sensitive copolymers of maleic acids were coated on the surface of polystyrene microcapsules, and the permeability of *n*-propyl alcohol was controlled by the structural changes of the coated copolymer. Responsive release of drugs to pH from microcapsules (*21–25*) or liposomes (*26*, *27*) usually uses this conformational change of the constituting polymer.

The pH-sensitive polymers were used for enteric coating materials (*28*), site-specific targeting (*29*), or tumor-specific delivery (*27*). When they are used as enteric coating materials, acid labile drugs can be protected in the stomach. Also, foul tastes can be masked by coatings. Brøndsted and Kopeček (*30*) investigated the feasibility of targeting drugs to the colon, where enzymatic proteolysis is relatively low, by using hydrogels that contain both acidic comonomers and enzymatically degradable azoaromatic cross-links. At low pH solute release from a polymer matrix or a polymer-coated device was minimal, whereas the release was enhanced by increased swelling and polymer degradation in the colon.

The pH-sensitive polymers were used as tumor-specific delivery systems, because it was reported that the pH around tumor cells is lower than that of normal cells. This difference in pH is due to the large amount of neuramic acid derivatives on the surface of tumor cells or active metabolic function of tumor cells (*31*, *32*).

Other Chemicals

Permeation of NaCl from the inner aqueous phase of a porous nylon capsule coated with sodium dodecyl phosphate bilayer membrane was investigated (*33*). The calcium ion (Ca^{2+}) altered the bilayer structure or capsule membrane through the chelation of the phosphate head groups and resulted in increased permeability of solute. The permeability then could be decreased to the original rate by washing with ethylenediaminetetraacetic acid disodium salt aqueous solution.

Ion effects on the solute permeability have been related to the phase transition of the polymeric structure (*34*). Morphological changes of collagen membrane from a crystalline to amorphous state in the presence of Ca^{2+} occurred, followed by increased solute permeability.

Poly(ethylene oxide) (PEO) is well known to form complexes with poly acid, mediated by hydrogen bonding. Poly(methacrylic acid) membrane was used in ultrafiltration to investigate water and protein permeation. In this system, mechanochemical expansion and contraction of the pores, regulated by the complexation of PEO with poly(methacrylic acid), caused changes in the permeability to solutes (*35*).

Charge-transfer complex formation between electron acceptor and electron donor system was used to modulate controlled release of amino compounds (*36*). A poly(2-hydroxyethyl methacrylate)-*co*-methacryloyl-*b*-hydroxylethyl-3,5-dinitrobenzoate) device showed an increased release rate of methyl orange in the presence of triethyl amine (TEA). This change in release rate was explained through the influence of the swelling of the polymer membrane induced from the formation of a charge-transfer complex between the dinitrophenyl group in the polymer and TEA.

Physical Stimuli

Temperature

The body temperature often deviates from the normal temperature, kept constant near 37 °C, by the physiological presence of pathogens or pyrogens. This temperature change may be a useful stimulus that can modulate the delivery of therapeutic drugs for diseases accompanying fever. In addition, externally controlled temperature can also be used to modulate drug release.

Polymer

The water swelling of most hydrogels is influenced by temperature in a different degree in terms of sensitivity and dependency: increase (positive thermosensitivity) or decrease (negative thermosensitivity) of swelling with increasing temperature. The negative thermosensitivity is the characteristic of the gel composed by a polymer having a lower critical solution temperature (LCST) in an aqueous solution. The LCST is defined as a critical temperature at which a polymer solution undergoes phase transition from a soluble to an insoluble state when the temperature is raised. Many hydrophilic/hydrophobic (HPL/HPB) balanced water-soluble polymers exhibit LCSTs, and the HPL/HPB balance is from either the monomeric structure of homopolymers or the polymer composition of copolymers. Thus, the LCST and swelling transition of the hydrogels may be altered by copolymerization. For instance, the swelling transition temperature and transition pattern of poly(*N*-isopropylacrylamide) (poly(NiPAAm)), one of the most pronounced temperature-sensitive polymers with an LCST of around 32 °C, were modified by copolymerization with acrylamide, acrylic acid, and alkylmethacrylates (*37–39*). However, when this polymer was modified by grafting (*40*) or interpenetrating second network (*39*, *41–43*), the swelling transition temperature remained almost constant.

Figure 8.2 shows how various stimuli can influence drug release via temperature-induced

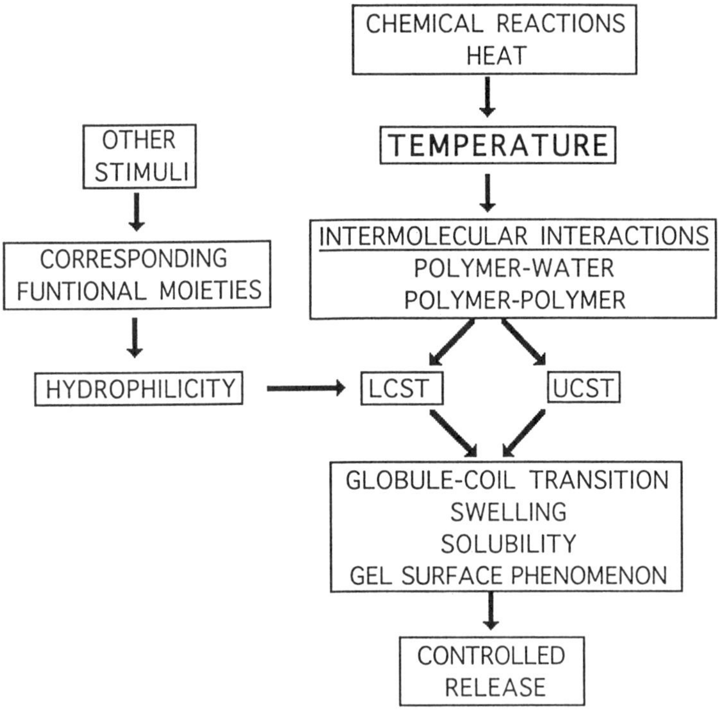

FIGURE 8.2. Temperature and other stimuli that influence a polymeric drug delivery system by means of temperature-induced polymer property changes. LCST is lower critical solution temperature, and UCST is upper critical solution temperature.

polymer property changes. Environmental temperature affects the molecular interactions among polymer chains or between polymer and solvent molecules. When such interactions occur in an aqueous system, LCST or UCST (upper critical solution temperature) are often accompanied. In particular, as the temperature fluctuates around the critical temperature, polymer properties such as degree of swelling or solubility are strongly influenced by temperature, resulting in temperature-controlled drug release. An interesting possibility is that when the copolymer contains a moiety that is sensitive in terms of hydrophilicity to other stimuli, these stimuli can shift the LCST of the polymers. This characteristic may lead to a polymeric system responding to multiple signals.

Solute permeability to temperature-sensitive polymer membranes was proven to be affected by the degree of swelling change by temperature rather than thermal effect (*44*). Also, temperature modulation below or across the LCST is of importance in solute permeation control (*45*).

Poly(NiPAAm) and its copolymers have shown two patterns of temperature-modulated drug release: bulk squeezing and surface regulation (Figure 8.3). Hoffman et al. (*37*) focused on cross-linked poly(NiPAAm) to study the role of these gels as drug carriers. Water-soluble solutes, such as vitamin B_{12} and methylene blue, were loaded into these gels below their LCSTs. The release behavior of solutes above the LCSTs of the gels was investigated by varying cross-linking density and synthetic conditions. In addition, equilibrium water swelling properties of these gels in drug solution at low temperature were reported. When the swollen gels in drug solution at low temperature were transferred to release media kept at 50 °C, the initial rapid release was followed by a slow release rate. This phenomenon was interpreted as the *squeezing* effect accompanying gel deswelling, which causes outflux of dissolved drug with water flow in addition to diffusional flux. On the other hand, the copolymers of NiPAAm with more hydrophobic comonomers such as *n*-butylmethacrylate showed dense skin formation during deswelling process when the temperature increased

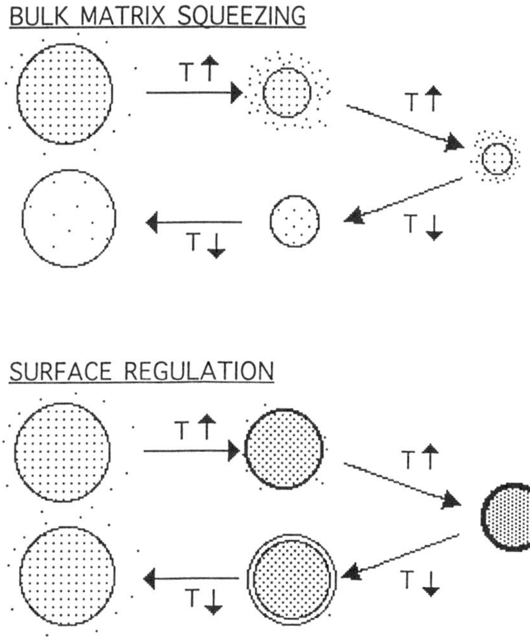

FIGURE 8.3. Two patterns of temperature-modulated drug release from NiPAAm copolymer gels.

past LCST values (*42*, *43*, *46*, *47*). This dense skin was able to block the release of loaded drugs from the matrix or diffusion through the membrane, resulting in the on–off control of solute release. The repeated cycle of constant release of loaded drugs during the *on* stage was explained by the redistribution of drug concentration inside the matrix during the *off* stage. Similar results were obtained from the interpenetrating polymer networks (IPNs) composing poly(NiPAAm) and poly(tetramethylene oxide) (*48*) or poly(NiPAAm) and poly(ethylene oxide–dimethyl siloxane–ethylene oxide) (*42*, *43*). The detailed release behaviors after temperature changes for on or off release was influenced by the applied temperatures, and this result demonstrated that the off stage accompanies small surface squeezing phenomenon even for a short period. Also, the lag time was observed when the temperature decreased for drug release (*48*). The larger temperature fluctuation caused less squeezing in the off stage and more immediate start of drug release in the on stage.

Katono et al. (*49*) reported that the opposite drug release pattern was obtained by using an IPN structure of poly(acrylamide-*co*-butylmethacrylate) and poly(acrylamide). The similar surface phenomenon was observed for on–off modulation. The gel swelling and drug release were explained by temperature-dependent hydrogen bonding interaction between two polymer networks that becomes weaker with increasing temperature, resulting in positive thermosensitivity. Structure **1** is a hydrogen bonding complex of poly(acrylic acid) with polyacrylamide.

Okahata et al. (*50*) reported temperature-modulated solute release from nylon capsule membrane with asymmetric pores. In the case of grafted poly(NiPAAm), the permeability of naphthalene-disulfonate through the capsule decreased with increasing temperature. Around the polymer cloud point (C_p) of the poly(NiPAAm), temperature changes induced larger permeability changes. Poly(NiPAAm) grafted on the capsule surface in water at temperature below C_p, thereby permitting permeants to pass smoothly through the porous capsule membrane. In contrast, at temperature above C_p the compact, insoluble, grafted polymers cover the porous capsule membrane. The permeation was significantly reduced depending on the molecular size of the permeants. Also, different polymers (poly(*N*,*N*-alkyl substituted acrylamide)) with different C_p values were grafted and demonstrated the same behavior as poly(NiPAAm).

More elegant control of temperature-induced on–off release of water-soluble drugs was demonstrated by combining thermosensitive hydrogels and a housing with a diffusion channel (*51*). When a hydrogel consists of NiPAAm and a more hydrophilic comonomer such as acrylamide in a drug solution, the gel shrinks in a rigid housing as the temperature increases. The drug solution then is squeezed out in the housing as described previously (bulk squeezing), and the squeezed drug diffuses through a diffusion channel like small holes on the housing. When the temperature is lowered, the gel reabsorbs the drug solution in the housing, resulting in minimal drug diffusion. The triggering temperature for drug release can be adjusted by keeping the gel in the drug solution at the desired setting temperature until equilibrated and charged into the housing. If the temperature increases above this setting temperature, the gel deswells; below the setting temperature, the additional swelling is restricted by the rigid housing. Thus, any temperature below the LCST of the gel can be set as a triggering temperature. This principle can be applied to hydrogels whose swelling changes in response to signals. Figure 8.4 demonstrates a result from such a system comprising polypropylene capsule with a pinhole and slightly cross-linked poly(NiPAAm-*co*-acrylic acid) equilibrated in an acetaminophen solution at 36 °C before loading into the capsule.

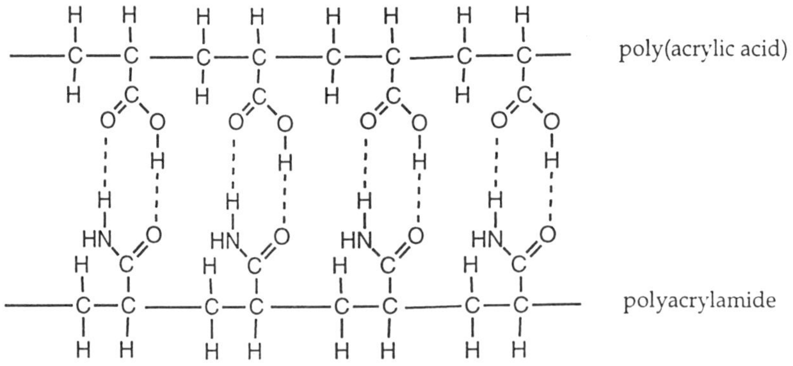

1

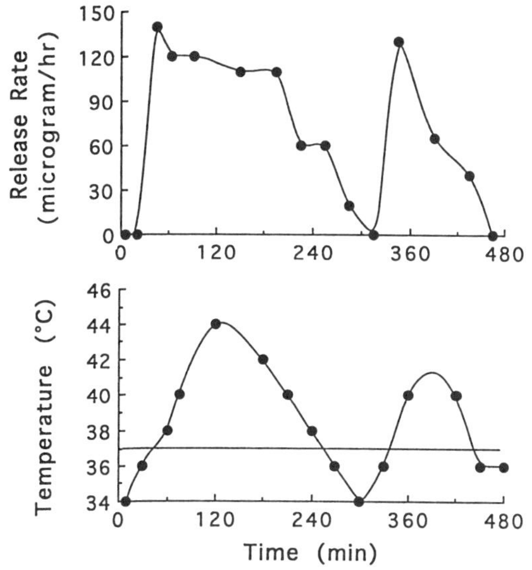

FIGURE 8.4. Temperature-modulated acetaminophen release (top) from a system comprised of polypropylene capsule with a pinhole and slightly cross-linked poly(NiPAAm-*co*-acrylic acid) equilibrated in an acetaminophen solution at 36 °C before loading into the capsule when the temperature was fluctuated as a function of time (bottom).

Lipid Bilayer

Tirrel (*52*) reported the vesicle-to-micelle transition of phosphatidylcholine in the presence of poly(2-ethylacrylic acid) in the solution or with a 2-ethylacrylic acid copolymer that associated into the lipid bilayer by a hydrophobic anchoring moiety. The transitions were very sensitive to environmental pH and temperature. pH effects arise from the ionization of polyelectrolytes. Under a fixed pH (protonated state of the polyelectrolyte), the membrane reorganization was a function of temperature. Even though the entrapped drug release from the vesicle was not performed with temperature, the concept can be used for temperature-modulated release system.

Instead of the transition of vesicle to micelle, another example of temperature effects on solute permeation through lipid bilayer was reported by Okahata et al. (*53*). Nylon capsules loaded with NaCl were transferred to a dodecane/dialkyldimethyl ammonium bromide solution kept at 60 °C for 5–10 minutes. This process introduced the dialkyl surfactant onto the lumen of the capsule membrane pores. The existence of an amphiphilic bilayer on the capsule membrane was confirmed by differential scanning calorimetry, and the permeability of NaCl through the capsule was measured from 5 °C to 60 °C. Permeability changes induced by temperature were observed at the phase-transition temperature (T_c), in contrast to the case of the uncoated capsules. The permeability of NaCl was increased at temperature above T_c and decreased below T_c under the influence of changes in the conformation of the bilayer chains in response to temperature.

Liquid Crystal

Shinkai et al. (*54*) reported alkali metal cation permeation through composite membranes consisting of polycarbonate/liquid crystal, *N*-(4-ethoxybenzylidene)-4′-butylaniline (EBBA)/amphiphilic crown ethers at different temperatures. The on–off control of permeability to potassium ions through one of these membranes was explained by EBBA fluidity changes in the membrane. A self-organized, continuous phase in the composite membrane was reversibly formed as the temperature changed above or below the crystal–nematic liquid crystal transition temperature, T_{kn} (21 °C), of EBBA.

A site-to-site jump mechanism for ionic diffusion was proposed to explain the different permeabilities observed above and below T_{kn}. More specifically, metal complexes could diffuse through the fluid crystal membrane phase of the composite membrane via the site-to-site jump mechanism at 40 °C (above T_{kn}) because of the liquid crystal organization at this temperature. At 20 °C, which is below the T_{kn}, the thermal molecular motion of EBBA was frozen, and diffusion of the crown ether–metal complexes across such crystalline membrane was observed to be negligible.

Furthermore, the effect of temperature fluctuation (10 °C → 40 °C → 10 °C → 40 °C) on permeability changes of metal ions was investigated by using a composite membrane. Results demonstrated the permeation of potassium ions at 10 °C was initially negligible. When the temperature was increased from 10 °C to 40 °C, increased permeability of ions was observed. Observed lag times in response to temperature changes were much larger for changes from 10 °C to 40 °C than from 40 °C to 10 °C. In the case of decreasing temperature, the slow response was apparently due to slow reorganization of the liquid crystal phase in the gel phase of the polymer matrix. When the temperature was changed from 10 °C to 40 °C during a second cycle, no lag time was observed. This result is different from the initial temperature change and suggests that crown ether–ion complexes already formed in the membrane from the previous cycle start to release ions immediately with increasing temperature.

Electric Field

Electric field or current can affect physicochemical properties or redox states of polymeric systems via various mechanisms (Figure 8.5). Electric stimulus-sensitive drug release will be associated with these property changes.

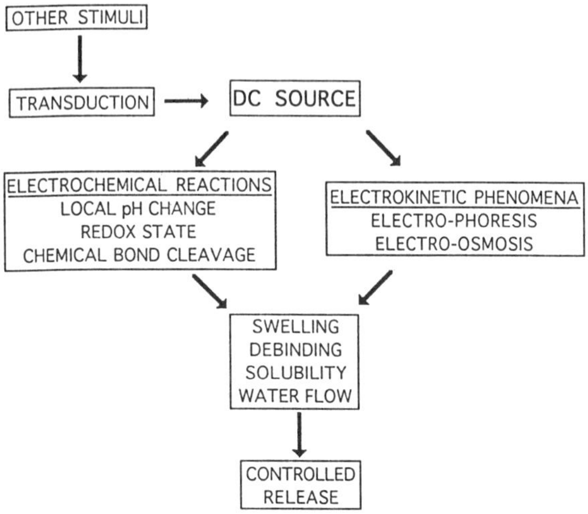

FIGURE 8.5. Electric field or current effects on drug release by means of changes of physicochemical properties or redox states of polymeric system.

Polyelectrolyte

Solute permeation through polyelectrolyte gels under an applied electric field have been controlled by several electrokinetic phenomena, such as electrodiffusion processes, electroosmotic or electrophoretic augmentation of solute flux, and electrostatic partitioning of charged solutes into the charged gels (*55–61*). When an electric field is applied across a polyelectrolyte membrane that supports a concentration gradient of mobile ions, the field will affect the intramembrane concentration profiles of these ions by electrodiffusion. This phenomenon is governed by a balance between the competing processes of ion migration by an electric field, diffusion due to a concentration gradient, and convection if there is fluid flow within the membrane. Changes in intramembrane ionic strength or pH can modulate the ionic double-layer repulsion forces between the charged polymer chains in deformable, charged membranes. Thus, an increase in intermolecular and interfibrillar forces may increase swelling. An electric field can provide the means for fast, localized control of intramembrane ionic strength and pH. This process affects permeability changes for neutral solutes or proteins through the polyelectrolyte membrane, such as poly(methacrylic acid) and collagen.

Electrical-release modulation of pilocarpine, glucose, and insulin from the polyelectrolyte gels of poly(acrylic acid), poly(methacrylic acid), and poly(*N*,*N*-dimethylaminopropyl acrylamide) was obtained (*62*). This modulation was explained by the gel swelling/deswelling control by applied electric current. The small solute release was enhanced during deswelling by a squeezing mechanism, whereas large molecule release was affected by the degree of swelling resulting in more release during swelling process. Similar pulsatile-release control of hydrocortisone using calcium alginate/poly(acrylic acid) composite (semi-IPN structure) as a gel matrix was demonstrated (*63*).

Ion Exchange

The release of an ionically bound drug to an electrolyte gel having strong acid groups, such as sulfonate groups, with varying intensity of applied electric stimulation in distilled–deionized water showed a complete on–off release profile (*64*). The magnitude of the release rate was regulated by the electric current. This release pattern was explained by ion exchange between a positively charged solute and hydrogen ions produced at the anode by water electrolysis. The released, positively charged solute migrates to the cathode and diffuses out of the membrane. The release of solute is enhanced by the squeezing effect at the anode side and by electroosmosis.

Polymer Complexes

Polyethyloxazoline and either poly(methacrylic acid) (PMAA) or poly(acrylic acid) form complexes via intermolecular hydrogen bonding. This complex was formed below pH 5 and dissociated above pH 5.4. This result was attributed to the deionization and ionization of the carboxylic groups of PMAA with pH changes. The discrepancy between the precipitation and dissolution pH values may result from a change in pK_a for the PMAA carboxylic groups before and after complex formation or cooperative interaction in complex formation.

During the application of an electric current, the solid matrix surface facing the cathode began to dissolve (*65*). Because the cathode produces hydroxyl ions (electrolysis), the local pH near the cathode increased and hydrogen bonding between the two polymers was disrupted. As a result, the polymer complex disintegrated into two water-soluble polymers. The rate of polymer weight loss was constant until 80% of initial disk weight was lost. The linearity of weight loss versus time implies that this process occurs through erosion of the cathode-facing surface. By applying a step function of electric current to the insulin-loaded matrix, insulin was released in an on–off manner until 70% of loaded insulin was released.

Heparin, a bioactive polyanion, was complexed with polyallylamine. At neutral pH, the positively charged amine groups complexed with $-COO^-$ (from iduronic acid and glucuronic acid units) or $-SO_3^-$ (from sulfoiduronic acid and 2,6-disulfoglucosamine units) in heparin. This complex was formed from pH 3 to pH 10, and the resulting complex dissociated below pH 2 (deionization of carboxylic and sulfate groups) and above pH 11 (deionization of amine groups). During the application of electric current, the solid matrix dissolved from the cathode-facing surface because of an increase of pH (*66*). This phenomenon was similar to that of the previous hydrogen-bonding complex system, except that dissociation at a given electric current was slower for this system than for the previous complex. The release pattern of heparin showed a complete on–off profile in response to the applied electric field. The release rate was dependent on the intensity of the applied electric current but was not linearly proportional to the applied electric current.

Redox Reaction

To mimic neurotransmitter release from presynaptic terminal in a neuron junction by an electric potential stimulation, an experimental approach was conducted to release dopamine (*67*, *68*), glutamic acid (*69*), or γ-aminobutyric acid (*69*) from polymer coatings on an electrode surface such as glassy carbon. The underlying concept is presented in Scheme 8.1, which shows a modified polystyrene having an isonicotinamide unit, an electron acceptor, and neurotransmitter units attached via cathodically cleavable amide bonds. In pH 7 solution, at a potential more negative than 0.9 V (standard calomel electrode, SCE), cathodic current causes cleavage of amide linkage and release of the neurotransmitters. However, at pH 7, the reduction process is limited to the adjacent layer to the electrode surface, and propagation of the reduction to the additional polymer layer is slow, resulting in limited cleavage of neurotransmitters (only about 10% of coupled chemicals was released).

Conductive Polymer

For a more effective approach for easy preparation and more sufficient release of incorporated ions, the use of a conductive polymer film was introduced. Many of the conductive polymers switch

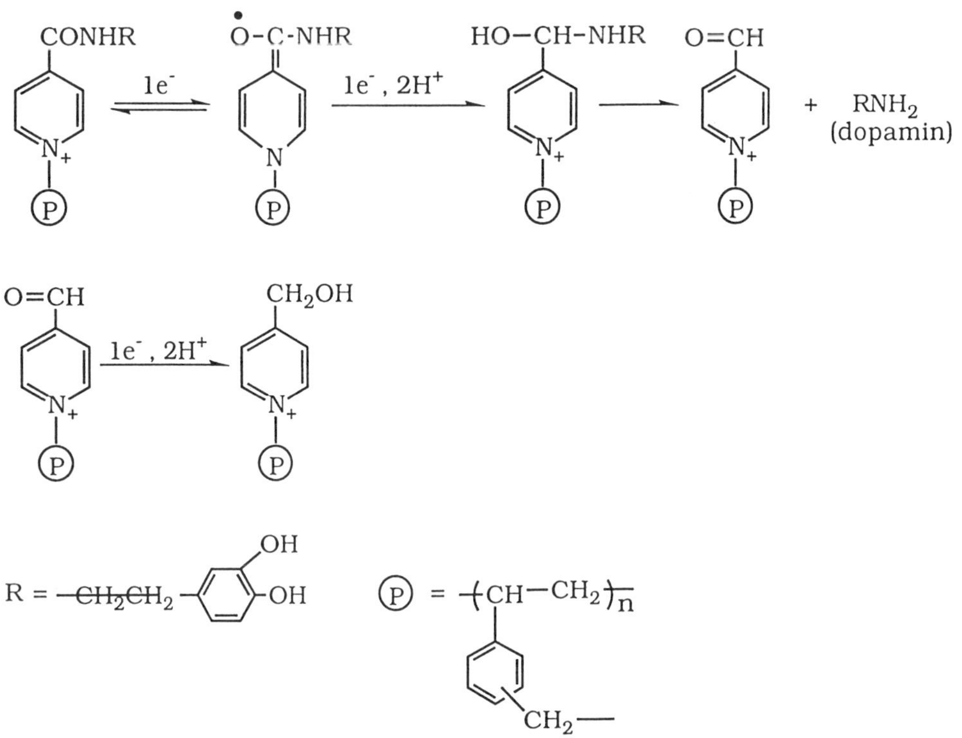

SCHEME 8.1. Electric-current-induced dopamine release from modified polystyrene having isonicotinamide unit, an electron acceptor, and neurotransmitter units attached by means of cathodically cleavable amide bond.

from a conductive form doped with anions to a nonconductive form releasing anions on reduction. In this way, anions such as ferrocyanide and salicylate can be loaded reversibly and released by electrochemical control (*70–72*). The conductive polymers were obtained by electrochemical oxidation of pyrrole, its derivatives, or 3-methoxythiophene.

For cation delivery systems, a conductive composite film composed of oxidized poly(*N*-methylpyrrole) (PMP^+) and polymeric anions, such as poly(styrene sulfonate) (PSS^-) or Nafion, was prepared (*73*, *74*). In aqueous cationic solution, when the composite was reduced at more negative potential than 0.2 V (SCE), the conductive polymer became un-ionized (PMP). To neutralize the entrapped polymeric anion in the composite film, cationic small molecules such as dopamine•HBr, protonated procaine, 1-benzyl nicotinamide chloride, and methyl viologen were incorporated. This binding/debinding controlled by electric potential is presented in Scheme 8.2. Similar conductive membranes have been applied for ion gates for on–off permeation control (*75–79*).

Photoirradiation

Photosensitive compounds such as azobenzene, stilbene, spiropyran, and rhodopsin undergo a conformational or configurational change on photoirradiation (Scheme 8.3). Photoresponsive polymers can be prepared by incorporation of these compounds to a polymeric backbone (**2, 3**). These polymers were used for photochemical control of the permeation of various solutes, such as metal salts (*80–83*), proteins (*84*), amino acids (*85*), and other solutes (*86*). Novel experiments were per-

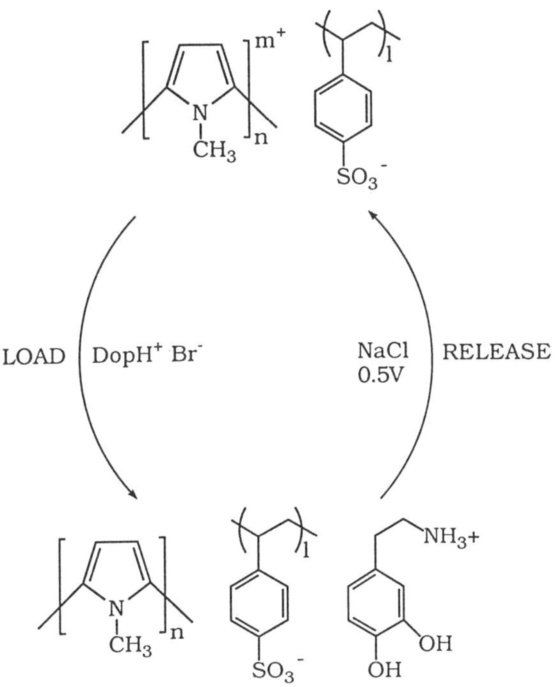

SCHEME 8.2. Cationic drug loading and release from polypyrrole doped with polyanion.

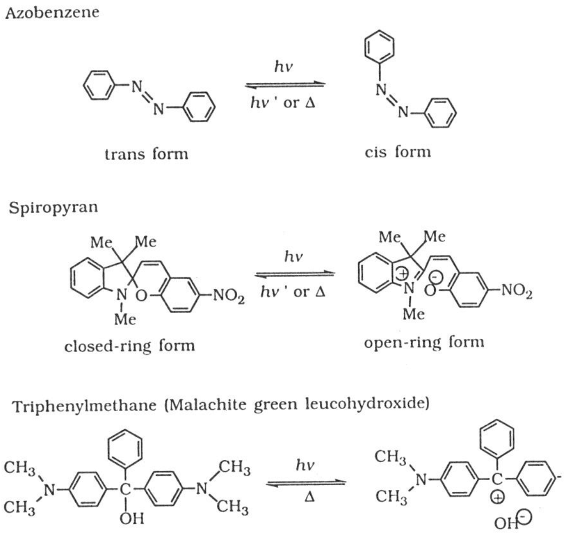

SCHEME 8.3. Photoinduced structural changes of photochromic compounds.

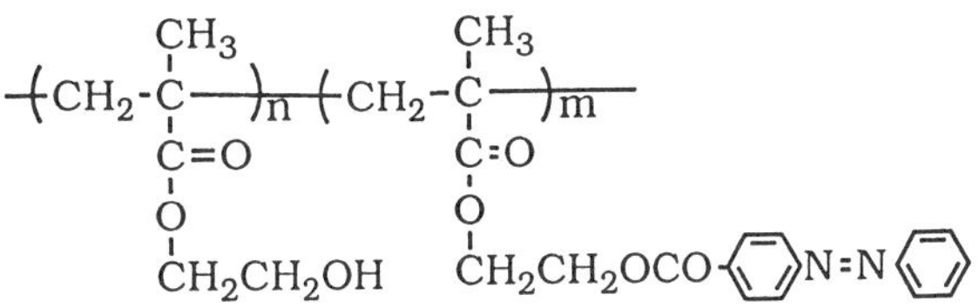

2

HEMA-AEMA copolymer

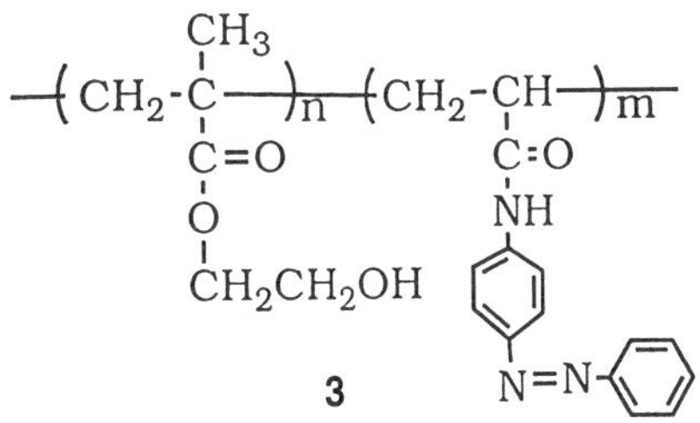

3

HEMA-PAAn copolymer

formed for the active transports of metal ions and amino acids across membrane-containing photochromes by using visible or UV light as energy sources.

One approach used azobenzene-bridged crown (AZO-CR) ether in a ternary composite membrane (polymer/liquid crystal/AZO-CR) (*1*). The cis–trans isomers of AZO-CR have different affinity to metal ions dependent on the ion size. The cis isomer, under UV irradiation, has an affinity for K^+ ions. These complexes can move to an opposite side of the membrane, as a result of a concentration gradient, and release K^+ ions under visible light (trans form of AZO-CR). The trans isomers diffuse back to the original side as a result of the chemical potential gradient of trans form and are converted to cis form by UV irradiation. As a result, the permeation of K^+ ion can be facilitated by an on–off pattern by light.

A similar experiment was carried out using a different polymer matrix composite (*81*). The same mechanism was applied to realize active transport of ions and amino acids using spiropyran derivatives (*82*, *85*). These compounds undergo ring opening (a charged form) by UV irradiation and reversibly return to the closed structure (a neutral form) by visible light or elevated temperature. When in the open form, alkali univalent cations or amino acids complexed with the compounds. The reversibility of the system led to dissociation of the complex enabling active transport. The spiropyran derivatives were incorporated into a liquid membrane and lipid bilayer as carrier in the experiments.

An on–off modulation for protein permeation by UV irradiation was also achieved (*84*). Cross-linked random copolymers of HEMA with a monomer containing azobenzene groups in the side chain slightly changed swelling levels by cis–trans isomerization of the azobenzene group. The cis isomer, under UV irradiation, caused less swelling due to dipole moment change across the azo bond of *cis*-azobenzene moiety. These interactions desolvated the hydroxy group (*87*), and the resulting small changes of swelling were effective in controlling the permeabilities of large molecules in an on–off manner.

Magnetic Fields

When a poly(ethylene-*co*-vinyl acetate) matrix bearing magnetic beads (1.4-mm diameter) and bovine serum albumin (BSA) was exposed to an oscillating magnetic field, the release rate of BSA increased about twofold when compared with that in the absence of the field (*88*). This modulation effect was improved up to eightfold at different experimental conditions (*89*).

As the dispersed BSA was released, a relatively large channel of about 100–200-mm diameter (corresponding to the incorporated BSA particles) was developed, and BSA was released continuously from the polymer matrix (*90*). The proposed mechanism for enhanced release when using embedded magnets was magnet movement inside the matrix in response to the field strength change. This phenomenon causes strain to the matrix, compression and expansion of the preformed channels, and alteration of the integrity of the matrix. This oscillating strain may not induce net convective motion in either direction, but it is able to enhance diffusive mass transfer. Mathematical modeling based on this concept was able to approximate the enhanced release rate as a function of field strength (*91*). Indeed, a gap between surrounding matrix and magnets, about 100-mm wide, was observed by scanning electron microscopy after repeated exposure to an oscillating magnetic field (*89*). The magnet movement was assumed to be restricted by the matrix rigidity, and an experiment was conducted with various vinyl acetate contents (*92*). With higher vinyl acetate content, the polymer matrix was more water absorptive and had a lower Young's modulus. The oscillating field-induced modulation increased with increasing vinyl acetate content, and the effect was greatly improved when the content reached 50%. The modulation was observed by in vivo experiments with implant into healthy (*93*) and diabetic rats (*94*).

Similar approaches where a hydrogel system containing dispersed strontium ferrite microparticles (1 mm) and dispersed drug (bovine insulin powder) demonstrated up to 50-fold increase of release rate on exposure to the oscillating magnetic field (*95*). The hydrogel bead was fabricated by dropping sodium alginate solution into calcium solution followed by surface complexing with polycations such as poly(L-lysine) and poly(ethylene imine). When not complexed, immediate response was observed; however, with complexed hydrogel, a lag time for enhanced release was found. The degree of modulation was a function of molecular weight of alginate, complexing density, as well as applied magnetic field frequency (*96*). Because of the low passive diffusivity of insulin through the complexed surface, a different mechanism would exist from the previous system, but the local force exerted by the magnet particles to polymer networks may be responsible for the modulation effect.

Ultrasound

The first example of the use of acoustic energy for controlled release was reported by Okahata and

Noguchi (*97*). When ultrasonic wave (20 kHz, 10–15 W/cm^2) was applied to large porous nylon capsules coated with lipid bilayer, the ion permeability through the lipid phase was reversibly controlled. The permeability ratio of NaCl with-to-without ultrasonic irradiation ranged from 1.7 to 6.2 depending on the gel–fluid-like transition temperature (T_c) of the coated layer. When the cell used for permeability test was thermostated at 25 °C, maximum permeability enhancement was observed with lipids having T_c of 25–35 °C. The ion permeability was not enhanced with ultrasonic irradiation lower than 7 W, and above 20 W power the coating bilayers were readily damaged. The observed results are partially explained because the state of lipid was switched from a rigid gel state to a fluid-like liquid crystalline state by the irradiation above a critical power level.

External ultrasound irradiation has also been applied to enhance solutes released from polymeric devices. By using an ultrasound frequency of 75 kHz, the release rate of *p*-nitroaniline (10% loading) from a biodegradable poly[bis(*p*-carboxyphenoxy)methane] disk compression molded with the solute at a high temperature was repeatedly regulated by ultrasound on–off. The release rate was closely related to the degradation rate of the polymer, but the ratio of release rates after-to-before irradiation (10–20) was higher than that of degradation rate ratios (1.5–5) (*98*). Similar results were obtained with copolymer of bis(*p*-carboxyphenoxy)propane and sebacic acid, polylactide, and polyglycolide devices (*99*).

The direct effects of ultrasonic application on the surrounding liquid around the devices are an increased temperature (*99–101*) and reduced boundary effect (*99*). Although these two factors contribute to the enhanced release rates to some extent, they cannot explain the augmented release rate by ultrasound. Recent studies (*102*) on the mechanism for polymer erosion induced by ultrasound (therapeutic range, 1 MHz, <2 W/cm^2) and solute release revealed that ultrasound enhances permeation of water into the degradable polymer matrices. This permeation exposes labile linkage for hydrolysis and mechanical shear stress caused by the micro liquid jet produced by cavitation phenomenon. These mechanisms were supported by the observation of more rapid decrease of molecular weight of the degradable polymers sampled from the device surface when exposed to ultrasound, the morphological change (surface disintegration) by mechanical shear stress, and other experimental approaches such as gas content dependence of ultrasonic effect (*103*). The ultrasound-enhanced solute permeability could be supported by the greatly enhanced solute release from or permeation through nondegradable polymers such as poly(ethylene-*co*-vinyl alcohol) (*100*, *101*) and poly(ethylene-*co*-vinyl acetate) (*99*), although the mechanism is not fully understood yet. By considering the reversibility with on–off ultrasound application to nondegradable polymers, the damage of constituting polymers by ultrasound seems not to be responsible for the enhanced permeation. In vivo studies with implanted degradable (*104*) or nondegradable matrices (*100*) have proven the augmented release of incorporated solutes (5-fluorouracil, insulin, or *p*-aminohippuric acid) from the devices.

Combined Stimuli

As illustrated in Figure 8.2, a polymeric system that is sensitive to several stimuli can be obtained by copolymerizing temperature-sensitive components with other comonomers that are sensitive to other stimuli such as pH. The mutual effects between two components, temperature effect on pH sensitivity and pH effect on temperature sensitivity, were well investigated (*105*). The system used in Figure 8.4 was also sensitive to pH with sharp on–off release of insulin around pH 5 (*51*). Similar polymers were used for protein drug loading and release by modulating pH and temperature (*106*). Another potential example for a drug delivery system sensitive to temperature and UV was demonstrated in Figure 8.6, where a chromophore in the NiPAAm copolymer converts the irradiated light (488 nm) to heat, increasing the

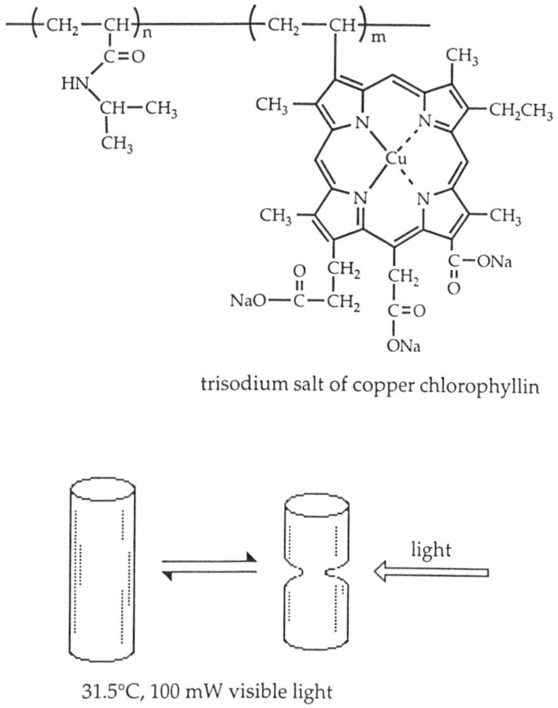

FIGURE 8.6. Temperature- and UV-sensitive system for potential drug delivery application where a chromophore in the NiPAAm copolymer converts the irradiated UV light (488 nm) to heat, causing gel shrinkage.

local temperature of the gel, which in turn causes local gel shrinkage (*107*).

Concluding Remarks

Although a variety of approaches to stimuli-sensitive drug delivery systems using chemical and physical signals have been reported, few systems have yet challenged clinical trial. For human implant applications, biodegradable systems with simpler, safer, more predictable, and more reproducible features should be developed in the future. However, currently accumulating technologies will find more practical applications to nonimplantable drug delivery and to other areas that require modulated chemical delivery.

References

1. Gliozzi, A.; Vittoria, V.; Ciferri, A. *J. Membr. Biol.* **1972,** *8,* 149–162.
2. Grignon, J.; Scallan, A. M. *J. Appl. Polym. Sci.* **1980,** *25,* 2829–2843.
3. Firestone, B. A.; Siegel, R. A. *Polymer* **1988,** *29,* 204–208.
4. Drummond, W. R.; Knight, M. L.; Brannon, M. L.; Peppas, N. A. *J. Controlled Release* **1988,** *7,* 181–183.
5. Brannon-Peppas, L.; Peppas, N. A. *Biomaterials* **1990,** *11,* 635–644.
6. Gerheke, S. H.; Cussler, E. L. *Chem. Eng. Sci.* **1989,** *44,* 559–566.
7. Brøndsted, H.; Kopeček, J. In *Polyelectrolyte Gels: Properties, Preparation, and Applications;* Harland, R. S.; Prud'homme, R. K., Eds.; ACS Symposium Series 480; American Chemical Society: Washington, DC, 1992; pp 285–304.
8. Touitou, E.; Rubinstein, A. *Int. J. Pharm.* **1986,** *30,* 95–99.
9. Kou, J. H.; Amidon, G. L.; Lee, P. I. *Pharmacol Res.* **1988,** *5,* 592–597.
10. Siegel, R. A.; Falamarzian, M.; Firestone, B. A.; Moxley, B. C. *J. Controlled Release* **1988,** *8,* 179–182.
11. Siegel, R. A.; Firestone, B. A. *Macromolecules* **1988,** *21,* 3254–3259.
12. Weiss, A. M.; Grodzinsky, A. J.; Yarmush, M. L. *AIChE Symp. Ser.* **1986,** *82,* 85–98.
13. Kirstein, D.; Brasselmann, H.; Vacik, J.; Kopeček, J. *Biotechnol. Bioeng.* **1985,** *27,* 1382–1384.
14. Nagasawa, M.; Holtzer, A. *J. Am. Chem. Soc.* **1964,** *86,* 538–543.
15. Kono, K.; Kimura, S.; Imanishi, Y. *J. Membr. Sci.* **1991,** *58,* 1–9.
16. Pefferkorn, E.; Schmitt, A.; Varoqui, R. *Biopolymers* **1982,** *21,* 1451–1463.
17. Jager, J.; Engberts, J. N. *Eur. Polym. J.* **1987,** *23,* 579–584.
18. Hassan, R. M.; El-Shatoury, S. A.; Mousa, M. A.; Hassan, A. *Eur. Polym. J.* **1988,** *24,* 1173–1175.
19. Chung, D. W.; Higuchi, S.; Maeda, M.; Inoue, S. *J. Am. Chem. Soc.* **1986,** *108,* 5823–5826.
20. Higuchi, S.; Mozawa, T.; Maeda, M.; Inoue, S. *Macromolecules* **1986,** *19,* 2263–2267.
21. Kokufuta, E.; Shimizu, N.; Nakamura, I. *Biotechnol. Bioeng.* **1988,** *32,* 289–294.
22. Bala, K.; Vasudevan, P. *J. Pharm. Sci.* **1982,** *71,* 960–962.

23. Okahata, Y.; Seki, T. *J. Am. Chem. Soc.* **1984,** *106,* 8065–8070.
24. Kokufuta, E.; Sodeyama, T.; Katano, T. *J. Chem. Soc. Chem. Commun.* **1986,** 641–642.
25. Okahata, Y.; Noguchi, H.; Seki, T. *Macromolecules* **1987,** *20,* 15–21.
26. Yatvin, M. B.; Kreutz, W.; Horwitz, B. A.; Shinitzky, M. *Science (Washington, D.C.)* **1980,** *210,* 1253–1255.
27. Kitano, K.; Wolf, H.; Ise, N. *Macromolecules* **1990,** *23,* 1958–1961.
28. Friend, D. *Adv. Drug Delivery Rev.* **1991,** *7,* 149–201.
29. Pradny, M.; Kopeček, J. *Makromol. Chem.* **1990,** *191,* 1887–1897.
30. Brøndsted, H.; Kopeček, J. *Biomaterials* **1991,** *12,* 584–592.
31. Kahler, H.; Robertson, W. B. *J. Natl. Cancer Inst.* **1943,** *3,* 495–501.
32. Gullino, P. M.; Grantham, F. H.; Smith, S. H.; Haggerty, A. C. *J. Natl. Cancer Inst.* **1965,** *34,* 857–869.
33. Okahata, Y.; Lim, H.; Nakamura, G. *Chem. Lett.* **1983,** 755–758.
34. Bartolini, A.; Gliozzi, A.; Richardson, I. W. *J. Membr. Biol.* **1973,** *13,* 283–298.
35. Osada, Y.; Takeuchi, Y. *J. Polym Sci. Polym. Lett. Ed.* **1983,** *15,* 279–284.
36. Ishihara, K.; Muramoto, N.; Shinohara, I. *J. Appl. Polym. Sci.* **1984,** *29,* 211–217.
37. Hoffman, A. S.; Afrassiabi, A.; Dong, L. C. *J. Controlled Release* **1986,** *4,* 213–222.
38. Gutowska, A.; Bae, Y. H.; Kim, S. W. *J. Controlled Release* **1992,** *22,* 95–104.
39. Bae, Y. H.; Okano, T.; Kim, S. W. *Pharm. Res.* **1991,** *8,* 531–537.
40. Chen, J. P.; Yang, H. J.; Hoffman, A. S. *Biomaterials* **1990,** *11,* 625–630.
41. Gutowska, A.; Bae, Y. H.; Jacobs, H.; Feijen, J.; Kim, S. W. *Macromolecules* **1994,** *27,* 4167–4175.
42. Mukae, K.; Bae, Y. H.; Okano, T.; Kim, S. W. *Polym. J.* **1990,** *22,* 206–217.
43. Mukae, K.; Bae, Y. H.; Okano, T.; Kim, S. W. *Polym. J.* **1990,** *22,* 250–265.
44. Bae, Y. H.; Okano, T.; Kim, S. W. *J. Controlled Release* **1989,** *9,* 271–279.
45. Feil, H.; Bae, Y. H.; Feijen, J.; Kim, S. W. *J. Membr. Sci.* **1991,** *64,* 283–294.
46. Bae, Y. H.; Okano, T.; Hsu, R.; Kim, S. W. *Macromol. Chem. Rapid Commun.* **1987,** *8,* 481–485.
47. Okano, T.; Bae, Y. H.; Jacobs, H.; Kim, S. W. *J. Controlled Release* **1990,** *11,* 255–265.
48. Bae, Y. H.; Okano, T.; Kim, S. W. *Pharm. Res.* **1991,** *8,* 624–628.
49. Katono, H.; Maruyama, A.; Sanui, K.; Ogata, N.; Okano, T.; Sakurai, Y. *J. Controlled Release* **1991,** *16,* 215–228.
50. Okahata, Y.; Noguchi, H.; Seki, T. *Macromolecules* **1986,** *19,* 493–494.
51. Bae, Y. H.; Valuev, L. I.; Kim, S. W. U.S. Patent 5,226,902, 1993.
52. Tirrel, D. A. *J. Controlled Release* **1987,** *6,* 15–21.
53. Okahata, Y.; Lim, H.; Nakamura, G.; Hachiya, S. *J. Am. Chem. Soc.* **1983,** *104,* 4855–4859.
54. Shinkai, S.; Nakamura, S.; Ohara, K.; Tachiki, S.; Manabe, O.; Kajiyama, T. *Macromolecules* **1987,** *20,* 97–103.
55. Weiss, A. M.; Grodzinsky, A. J.; Yarmush, M. L. *AIChE Symp. Ser.* **1986,** *82,* 85–98.
56. Nussbaum, J. H.; Grodzinsky, A. H. *J. Membr. Sci.* **1981,** *8,* 193–219.
57. Eisenberg, S. R.; Grodzinsky, A. J. *J. Membr. Sci.* **1984,** *19,* 173–194.
58. Eisenberg, S. R. *J. Biomech. Eng.* **1987,** *109,* 79–89.
59. Grimshaw, P. E.; Grodzinsky, A. J.; Yarmush, M. L.; Yarmush, D. M. *Chem. Eng. Sci.* **1989,** *44,* 827–840.
60. Grimshaw, P. E.; Grodzinsky, A. J.; Yarmush, M. L.; Yarmush, D. M. *Chem. Eng. Sci.* **1990,** *45,* 2917–2929.
61. Grimshaw, P. E.; Nussbaum, J. H.; Grodzinsky, A. J.; Yarmush, M. L. *J. Chem. Phys.* **1990,** *93,* 4462–4472.
62. Sawahata, K.; Hara, M.; Yasunaga, H.; Osada, Y. *J. Controlled Release* **1990,** *14,* 253–262.
63. Yuk, S. H.; Cho, S. H.; Lee, H. B. *Pharm. Res.* **1992,** *9,* 955–957.
64. Kwon, I. C.; Bae, Y. H.; Okano, T.; Kim, S. W. *J. Controlled Release* **1991,** *17,* 149–156.
65. Kwon, I. C.; Bae, Y. H.; Kim, S. W. *Nature (London)* **1991,** *354,* 291–293.
66. Kwon, I. C.; Bae, Y. H.; Kim, S. W. *J. Controlled Release* **1994,** *30,* 155–159.
67. Lau, A. K.; Miller, L. L. *J. Am. Chem. Soc.* **1983,** *105,* 5217–5277.
68. Miller, L. L.; Lau, A. K.; Miller, E. K. *J. Am. Chem. Soc.* **1982,** *104,* 5242–5244.
69. Lau, A. K.; Miller, L. L.; Zinger, B. *J. Am. Chem. Soc.* **1983,** *105,* 5278–5284.

70. Chang, A. C.; Miller, L. L. *J. Electroanal. Chem.* **1988,** *247,* 173–184.
71. Miller, L. L.; Zinger, B.; Zhou, Q.-X. *J. Am. Chem. Soc.* **1987,** *109,* 2267–2272.
72. Zinger, B.; Miller, L. L. *J. Am. Chem. Soc.* **1984,** *106,* 6861–6863.
73. Miller, L. L.; Zhou, Q.-X. *Macromolecules* **1987,** *20,* 1594–1597.
74. Zhou, Q.-X.; Miller, L. L.; Valentine, J. R. *J. Electroanal. Chem.* **1989,** *261,* 147–164.
75. Mirmohseni, A.; Price, W. E.; Wallace, G. G. *Polym. Gels Networks* **1993,** *1,* 61–77.
76. Zhao, H.; Price, W. E.; Wallace, G. G. *J. Electroanal. Chem.* **1992,** *334,* 111–120.
77. Burgmayer, P.; Murray, R. W. *J. Am. Chem. Soc.* **1982,** *104,* 6139–6140.
78. Burgmayer, P.; Murray, R. W. *J. Electroanal. Chem.* **1983,** *147,* 339–344.
79. Burgmayer, P.; Murray, R. W. *J. Phys. Chem.* **1984,** *88,* 2515–2521.
80. Kumano, A.; Niwa, O.; Kajiyama, T.; Takayamagi, M.; Kano, K.; Shinkai, S. *Chem. Lett.* **1983,** 731–734.
81. Anzai, J.; Ueno, A.; Sasaki, H.; Shimokawa, K.; Osa, T. *Makromol. Chem. Rapid Commun.* **1983,** *4,* 731–734.
82. Shimidazu, T.; Yoshikawa, M. *J. Membr. Sci.* **1983,** *13,* 1–13.
83. Okahata, Y.; Lim, H.; Hachiya, S. *Makromol. Chem. Rapid Commun.* **1983,** *4,* 303–306.
84. Ishihara, K.; Shinohara, I. *J. Polym. Sci. Polym. Lett. Ed.* **1984,** *22,* 515–518.
85. Sunamoto, J.; Iwamoto, K.; Mohri, Y.; Koninato, T. *J. Am. Chem. Soc.* **1982,** *104,* 5504–5506.
86. Ishihara, K.; Hamada, N.; Kato, S.; Shinohara, I. *J. Polym. Sci. Polym. Chem. Ed.* **1984,** *22,* 881–884.
87. Ishihara, K. Ph.D. Thesis, Waseda University, Japan, 1984.
88. Hsieh, D. T.; Langer, R.; Folkman, J. *Proc. Natl. Acad. Sci. U.S.A.* **1981,** *78,* 1863–1867.
89. Edelman, E. R.; Kost, J.; Bobeck, H.; Langer, R. *J. Biomed. Mater. Res.* **1985,** *19,* 67–83.
90. Langer, L.; Rhine, W.; Hsieh, D. T.; Folkman, J. *J. Membr. Sci.* **1980,** *7,* 333–350.
91. McCarthy, M.; Soong, D.; Edelman, E. R. *J. Controlled Release* **1984,** *1,* 143–147.
92. Kost, J.; Niecker, R.; Kunica, E.; Langer, R. *J. Biomed. Mater. Res.* **1985,** *19,* 935–940.
93. Edelman, E. R.; Brown, L.; Taylor, J.; Langer, R. *J. Biomed. Mater. Res.* **1987,** *21,* 339–353.
94. Kost, J.; Wolfrum, J.; Langer, R. *J. Biomed. Mater. Res.* **1987,** *21,* 1367–1373.
95. Saslawski, O.; Weingarten, C.; Benoit, J. P.; Couvreur, P. *Life Sci.* **1988,** *42,* 1521–1528.
96. Saslawski, O.; Couvreur, P.; Peppas, N. *Proc. Int. Symp. Controlled Release Bioact. Mater.* **1988,** *15,* 26–27.
97. Okahata, Y.; Noguchi, H. *Chem. Lett.* **1983,** 1517–1520.
98. Kost, J.; Leong, K. W.; Langer, R. *Proc. Int. Symp. Controlled Release Bioact. Mater.* **1984,** *11,* 84–85.
99. Kost, J.; Leong, K.; Langer, R. *Proc. Natl. Acad. Sci. U.S.A.* **1989,** *86,* 7663–7666.
100. Miyazaki, S.; Yokouchi, C.; Takada, M. *J. Pharm. Pharmacol.* **1988,** *40,* 716–717.
101. Miyazaki, S.; Hou, W. M.; Takada, M. *Chem. Pharm. Bull.* **1985,** *33,* 428–431.
102. Liu, L.-S.; Kost, J.; D'Emanuele, A.; Langer, R. *Macromolecules* **1992,** *25,* 123–128.
103. D'Emanuele, A.; Kost, J.; Domb, A.; Langer, R. *Macromolecules* **1992,** *25,* 511–515.
104. Kost, J.; Leong, K.; Langer, R. *Proc. Int. Symp. Controlled Release Bioact. Mater.* **1987,** *14,* 186–187.
105. Feil, H.; Bae, Y. H.; Fein, J.; Kim, S. W. *Macromolecules* **1992,** *25,* 5528–5530.
106. Kim, Y. H.; Bae, Y. H.; Kim, S. W. *J. Controlled Release* **1994,** *28,* 143–152.
107. Suzuki, A.; Tanaka, T. *Nature (London)* **1990,** *346,* 345–346.

9

Sensocompatibility

Design Considerations for Biosensor-Based Drug Delivery Systems

*A. Adam Sharkawy, Michael R. Neuman, and William M. Reichert**

For an implanted sensor to function successfully, many issues resulting from the mutual interaction between the sensor and the host response must be addressed. This chapter examines the challenges presented by encapsulation, the ultimate host response to an implanted sensor. Further, it explores a set of conceptual solutions and ways in which they may be incorporated into sensor design to overcome those challenges.

Closed-Loop Insulin Delivery

Closed-loop control allows implanted devices to perform independently of external feedback. This process usually requires an implanted sensor to continuously determine the status of the physiological variable being controlled. Because these sensors are relatively inaccessible when implanted, they must be capable of flawless function for months at a time to be useful. The most widely successful example of a closed-loop implanted device is the rate-responsive cardiac pacemaker. The sensors needed to close the loop must function as long as the pacemaker does, which currently is expected to last at least 10 years. In this case, the paced heart responds in much the same way as the unpaced heart would under normal physiological situations. For example, exercising muscles generate more heat because of increased metabolic activity, and they cause the blood returning from these muscles to be at a slightly higher temperature than if the muscles were less active. A thermal sensor coupled to the control system can sense this temperature increase and induce the pacemaker to increase the heart rate.

The insulin delivery system based on a glucose sensor, or the artificial pancreas, is the current "Holy Grail" of closed-loop drug delivery

*Corresponding author

Copyright © 1997 American Chemical Society

systems. A summary of a clinical survey by the European Community's Concerted Action on Chemical Sensors reveals that "Glucose easily tops the list of analytes that clinicians would like to measure in vivo with a sensor" (*1*).

Although several groups claim to be on the verge of developing a functional device, none of them works reliably for prolonged periods. The problem is that a sensor for serum glucose is required, and this sensor must be in contact with either blood or well-perfused tissue. Unlike a thermal sensor, the analyte, glucose, must be transported from the blood to the detection region of the sensor for a measurement to be made. Therefore, both the detection region and the analyte transport must remain unaffected by the interaction between the sensor and the adjacent tissue. If affected, it must be in a way that can be compensated through recalibration. Thus, for the glucose sensor, a clear understanding of the mechanisms of this interaction is important to the success of the system. A similar argument could be made for the cardiovascular sensors of chemical constituents or for blood pressure, as well as sensors for other prosthetic and drug delivery devices. For the purpose of illustration, however, the glucose sensor of the artificial pancreas provides an example of the more general problems facing biosensor-based feedback control of drug delivery.

Ideally, drug delivery systems would be controlled by the plasma levels of the chemical substance they are releasing. Indeed, much of the human body's maintenance of chemical homeostasis is based on the classic negative-feedback loop. Consider the natural regulation of blood glucose levels. As the concentration of glucose rises, increasing metabolic activity of the pancreatic islet β cells stimulates the release of insulin. Circulating insulin in turn stimulates the oxidation and storage of existing glucose molecules while decreasing the synthesis of new glucose molecules, resulting in a global decrease of blood glucose concentration (*2*). Because of the high sensitivity and low response time of pancreatic β cells as blood glucose levels decrease, insulin release is immediately depressed, and a swing in the opposite direction is prevented. Although the individual activities involved in blood glucose regulation are much more complex, the overall process is a straightforward negative-feedback loop.

Pancreatic β cells are destroyed and degenerated in patients suffering from diabetes mellitus. The resulting inability to synthesize and release insulin inhibits the assimilation of nutrients and can ultimately lead to death. In the past 60–70 years, insulin replacement therapy has allowed patients to survive this disease (*3*). However, unlike other hormone replacement therapies, the timing and dosages of insulin vary greatly across individuals. The critical consequences of too much or too little insulin and the dynamic nature of its regulation require the patient to frequently test glucose levels to estimate insulin dosages. Therefore, insulin replacement therapy comprises two processes: determination of blood glucose levels and administration of insulin.

Studies have suggested that frequent monitoring of blood glucose levels coupled with the appropriate administration of insulin may significantly reduce the complications associated with diabetes (*4*). Self monitoring by direct blood testing has been the most common way to determine plasma glucose levels. Patients may determine their glucose levels by testing their urine. However, with the lag time between blood and urine glucose levels, this method does not accurately reflect the insulin dosage required at any given time. On the other hand, direct blood testing performed several times a day can be tedious and rather painstaking. Moreover, some patients undergo significant fluctuations in glucose levels between their readings and thus require constant monitoring. The realization of an artificial pancreas would be the ultimate solution.

Unlike in vitro detection by blood sampling, which only provides information at *discrete* times, to create an artificial pancreas, plasma glucose levels must be *continuously* detected; insulin must be released accordingly, in a manner similar to the function of pancreatic islet β cells. Recently, there has been a great deal of progress in developing insulin pumps to continuously release insulin in response to hyperglycemia. These pumps, comprising the delivery component of closed-loop insulin delivery systems, are either extracorporeal or implantable. Extracorporeal pumps have been successfully miniaturized such that they can easily be worn on a belt or shoulder harness. Whether they are peristaltic or syringe type pumps, they are usually microprocessor controlled to deliver variable rates of insulin (*5*). The cannulas through which the insulin is pumped can either be implanted subcutaneously or intravenously. Because of the exceeding complications of intravascular placement due to thrombosis at the cannula tip, subcutaneous injection is less troublesome. However, insulin diffusion through the subcutaneous tissue into the vasculature can vary among patients and with site of implantation. Therefore, administration of insulin through subcutaneous injection is not as controllable as direct intravascular injection. The second type of insulin pump is the implantable type. The components of implantable pumps are housed within a hermetically sealed container to protect them against biological responses. Such pumps are usually implanted in the abdominal subcutaneous tissue and deliver insulin intraperitoneally. Implantable pumps require further miniaturization before they are ready for clinical use.

Closed-loop insulin delivery systems are currently limited by the lack of a means to continuously detect plasma glucose levels (*1*). Certainly, an implantable glucose sensor would be an ideal complement to the insulin pump for the creation of such a system. As the front end of the closed-loop delivery system, the output of a glucose sensor can easily be digitized and fed into the microprocessor of an insulin pump. In Figure 9.1, a schematic of a closed-loop insulin delivery system is shown. On the basis of preprogrammed calibration, the microprocessor can control the pump's release of insulin to maintain a blood glucose concentration within normal limits.

Although sensors in general have been developed based on calorimetry and piezoelectric activity, optical and electrochemical sensors are by far the most common. Currently, these two technologies hold the most promise for the development of commercialized glucose sensors. One is based on near-IR spectroscopy in which the sensor detects changes in optical properties of indicator molecules in response to varying glucose concentrations (*6*). This method has the great advantage of

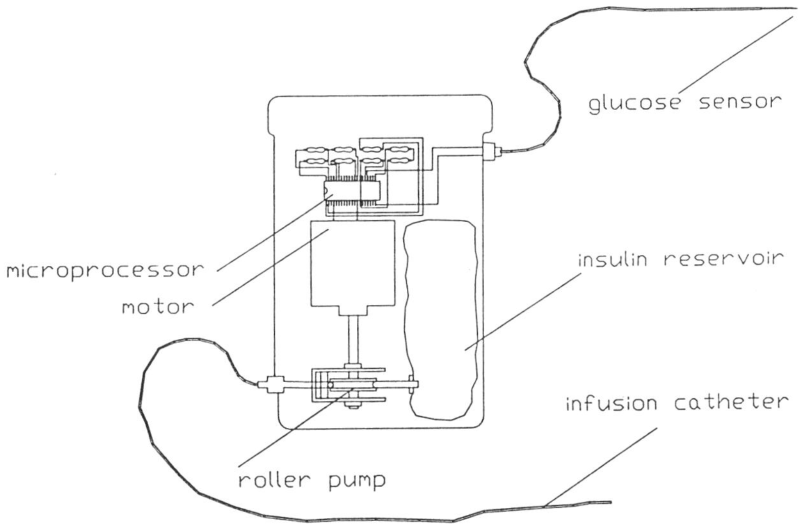

FIGURE 9.1. Schematic diagram of a closed-loop insulin delivery system.

being noninvasive. However, this system attempts to detect a fairly small glucose signal within much larger signals from proteins, lipids, and scattered light. Other issues such as signal variability and system miniaturization remain to be addressed before near-IR spectroscopy based systems are feasible for commercial use. Most research has focused on advancing the amperometric glucose sensor that detects the rate of an enzyme-catalyzed electrochemical reaction that is proportional to glucose concentration (*5*).

Most amperometric glucose sensors are governed by the following reaction:

$$\text{glucose} + O_2 \xrightarrow{\text{glucose oxidase}} \text{gluconic acid} + H_2O_2 \quad (1)$$

Because this sensor depends on the action of a biological component, in this case catalysis by glucose oxidase, it is a *biosensor*. By coupling this reaction to a standard electrochemical reaction, the concentration of glucose can be determined by the rate of O_2 consumption or H_2O_2 production. This determination is accomplished by immobilizing the glucose oxidase enzyme at the working electrode and polarizing it at a suitable voltage with respect to a reference electrode (Figure 9.2). As glucose is oxidized to gluconic acid, hydrogen peroxide will be oxidized at the anode:

$$H_2O_2 \xrightarrow{+700\ \text{mV}} O_2 + 2H^+ + 2e^- \quad (2)$$

The free electrons produced from the oxidation of peroxide flow to the cathode where they reduce oxygen:

$$O_2 + 4e^- + 4H^+ \xrightarrow{-700\ \text{mV}} 2H_2O \quad (3)$$

To decrease the sensitivity of such sensors to background oxygen tension, the transfer of electrons between the enzyme and the electrode is often mediated by a redox couple such as ferrocene (*7*). By replacing oxygen as the electron acceptor, ferrocene can decrease sensor dependency on oxygen. In either case, because the current flowing between the two electrodes will be proportional to the amount of glucose being oxidized, the relative concentration of glucose can be determined.

Sensocompatibility

According to current literature, biosensors as a class of devices are currently incapable of remain-

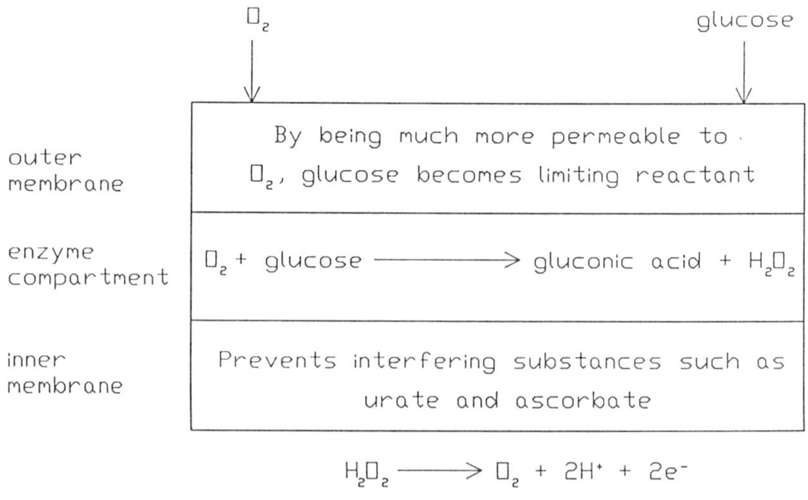

FIGURE 9.2. Typical amperometric glucose sensor.

ing functional during long-term implantation. This fact limits the clinical use of biosensors to in vitro testing of samples extracted from body fluids. However, sensors such as the amperometric glucose sensor described previously must be capable of long-term implantation before closed-loop in vivo drug delivery is realizable. Some of the hurdles that must be overcome before sensors are successfully implanted can be classified under the following categories:

- corrosion of leads and contacts due to tissue salinity,
- water absorption shorting electrical connections due to tissue moisture,
- biodegradation and biofouling of key sensor components, and
- host isolation or extrusion of the sensor.

The first three of these were reviewed (*8–11*). Furthermore, a set of tests to assess sensor function in vivo has also been reviewed (*12*).

Solving the implantation problem extends far beyond minimizing the harm that the sensor imparts on the body: that is, *biocompatibility*. One must also consider the reverse condition of minimizing the harm that the body will impart on a sensor. For this purpose, we have coined the term *sensocompatibility* as the minimization of in vivo sensor degradation. The objective of the following sections is to review the last of the hurdles to sensocompatibility, the process of host isolation of the sensor, and possible approaches to overcoming it.

Many researchers have claimed success with implanted sensors for short periods of time (*13–15*). In a study conducted by Johnson et al. (*16*), glucose sensors were implanted subcutaneously in the abdominal tissue of humans for 72 h. They obtained a strong correlation between subcutaneous glucose levels measured by their biosensor as compared with measurements of plasma glucose concentrations from direct blood samples. Moreover, they reported minor inflammatory reactions. A few years before that study, Ertefai and Gough (*17*) implanted glucose sensors inside skin folds made in the back of male rats. They reported no formation of an avascularized capsule, no indication of inflammation, and consequently good permeability of glucose and oxygen to the implanted sensor. Moatti-Sirat et al. (*18–20*) obtained accurate glucose measurements with a sensor implanted for over 10 days. They also observed little inflammation and a well-vascularized capsule around their implanted sensor. In all of the studies, sensors were implanted for short times not exceeding a few weeks. However, it is generally recognized that long-term implantation will invoke the formation of dense avascular capsules that may limit the sensor's access to the tissue analytes of interest and render the devices useless.

If an implanted sensor can resist short circuiting and corrosion, which can be caused by tissue moisture and salinity, and if it can survive the inflammatory biochemical assault, it must overcome the challenge posed by the ultimate host response, encapsulation. Woodward (*21*) directly posed this challenge when he stated that the collage of fibroblasts, collagen, and giant cells provides adherent, impermeable, avascular barriers when it encapsulates implants. Pfeiffer (*22*) also implicated encapsulation as a source of glucose sensor output degradation when implanted in both animals and humans. On the basis of these hypotheses, we investigated (*23*) the diffusion properties of capsules that formed around several types of materials implanted in rats as compared with those of normal rat dorsal subcutaneous tissue. In those findings, effective diffusion coefficients of low molecular weight analytes through the tissue encapsulating implants were as low as one-half those through surrounding subcutaneous tissue. Finite difference simulation showed that such avascular capsules could retard the steady-state response time by nearly threefold and decrease the magnitude of the transient response by an order of magnitude. Such findings reconfirm that the tissue that encapsulates implanted sensors can impose a transport barrier to plasma analytes.

Overcoming host isolation caused by the transport properties of encapsulating tissue may be key to the successful function of long-term implanted sensors. Unfortunately, very little literature focuses on host isolation of implanted sensors. On the basis of a review of research in related fields, this chapter will discuss a few exploratory ideas for overcoming this ultimate hurdle to sensocompatibility.

Definition of the Problem

Here, the elimination or reduction of the body's isolation of an implanted sensor is viewed as a design problem. The solution of a design problem depends very sensitively on how the problem is defined: Slight differences in the definition of the problem may lead to drastically varying solutions. Before discussing possible schemes of overcoming host isolation of an implanted sensor, it is necessary to identify the nature of the problem from which the design objective can be defined.

The few reports that describe methods of extending the life of an implanted sensor usually take a prophylactic approach; they identify a synthetic material that benignly minimizes host capsule formation. By doing so, however, the possible paths to overcoming host isolation have been severely limited. Specifically, the *underlying* problem caused by host encapsulation is not the formation of the capsule itself, but that the analyte levels at the sensor do not correlate to those in the tissue of interest. Defining the problem more generally as a means of providing a closer correlation between sensor and tissue glucose levels opens a broader range of solutions discussed subsequently.

Brief Review of Foreign-Body Response

Like any other artificial object, the implantation of a sensor into the subcutaneous tissue will evoke a host-tissue response (*24–27*). The overall response can be considered as a combination of two processes: wound healing due to the surgical excision of the subcutaneous tissue and the foreign-body response due to the synthetic nature of the implant. Because the two are highly coupled, the overall response to an implanted sensor will be overviewed. Even though it does not delineate the high degree of interdependency between the cells and chemical factors involved, the most simple depiction of the tissue response to an implant is a chronological sequence of events. These events can be divided into two phases: *inflammation* (Figure 9.3) and *repair* (Figure 9.4). The intensity and length of each phase is highly dependent on the chemical and physical properties of the implant as well as the general immunological reactiveness of the host.

The surgical procedure alone causes direct injury to the microvascular bed that perfuses tissue. Cell damage, complement pathway activation, and blood coagulation initiate a series of events that result in the release of a myriad of chemical mediators such as histamine, leukotrienes, prostaglandins, and thromboxanes. Although each mediator has its own specific effect, the combination of the mediators usually leads to vasodilation and increased permeability of vessel walls. By increasing local blood flow, vasodilation causes the redness and heat often associated with inflammation. Increased permeability leads to the extravasation of blood macromolecules that would normally be constrained to the vasculature. The resulting edema causes the local swelling and pain accompanying the first phase, inflammation.

Polymorphonuclear leukocytes, or *neutrophils*, are the first cells to arrive to the scene of injury. Neutrophils begin the phagocytosis of foreign particles and debris. However, another primary role of these cells is to undergo lysis and release many substances stored within their granules. These substances include proteolytic and hydrolytic enzymes that attempt to lyse and break down the foreign body. Because of their short lifetime and inability to reproduce, continued presence of these cells indicates that the foreign implant is either leaching toxic chemical additives or has somehow caused a bacterial infection. The chemical mediators are directly responsible for the influx of neutrophils to the injured tissue by two mechanisms. First, they increase the permeability of the sur-

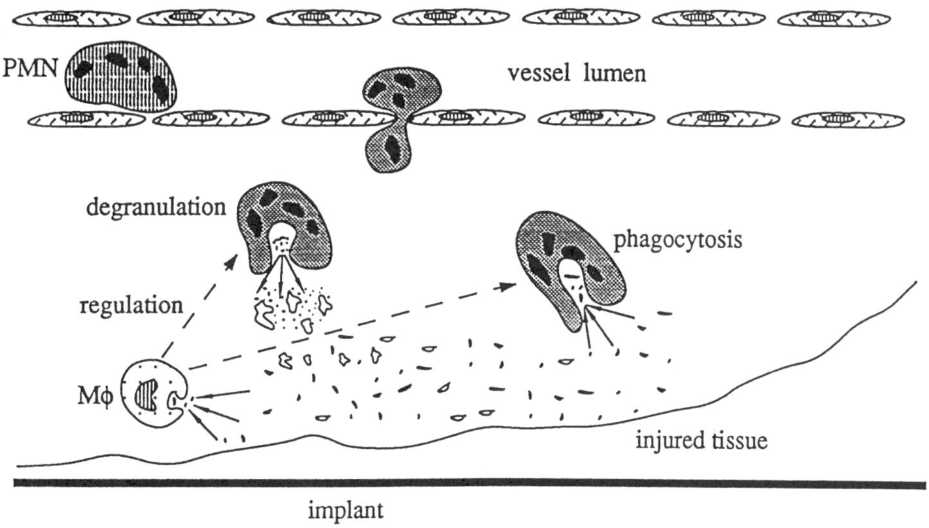

FIGURE 9.3. Extravasation and phagocytosis by neutrophils (PMN), which degranulate enzymes to break down debris and dead cells, and macrophages (Mϕ), which control the intensity and duration of inflammation by intricate and complex signaling.

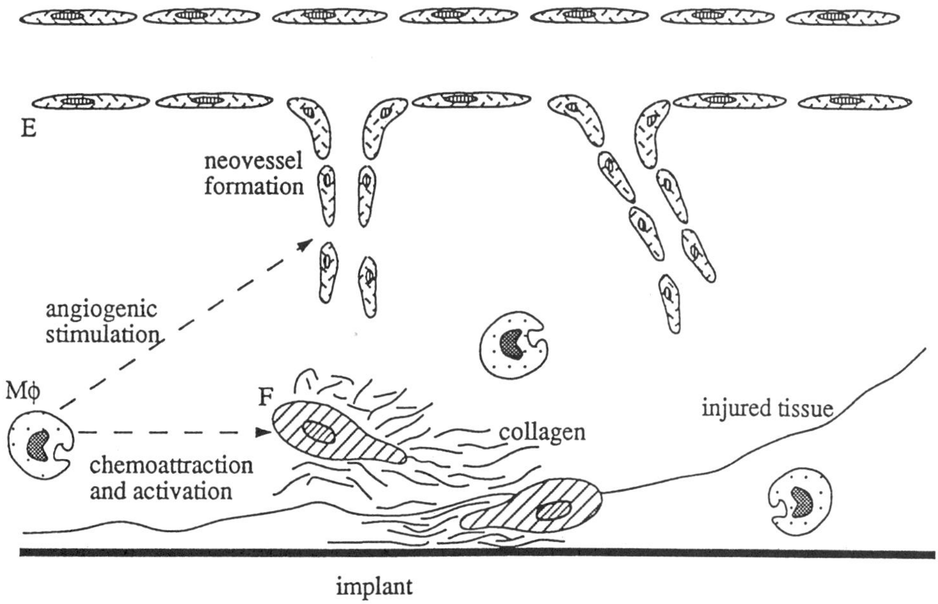

FIGURE 9.4. Migration and proliferation of endothelial cells (E) creating new vessels by budding off of existing ones. Fibroblasts (F) are signaled and activated to synthesize and secrete collagen, proteoglycans, and other extracellular matrix components to repair injured tissue and isolate nondegradable foreign bodies.

rounding vessel walls to allow leukocytes to migrate into the extravascular space. Second, they serve as chemoattractants to such cells as neutrophils to guide them to the site of injury.

Ordinarily, with minor tissue damage due to a surgical incision, neutrophil pervasion to the extracellular matrix would cease after a few days. However, the duration of this inflammatory phase is highly dependent on the nature of the implanted object. If the implant continues to cause damage to the surrounding tissue, either physically by abrasion or chemically by leaching toxic substances, the acute phase of inflammation will prolong.

In either case, the next leukocyte arrives at the site of injury, the *macrophage*. As opposed to the initial neutrophil assault, macrophages accumulate more slowly. As will be discussed, they perhaps play the most central role in cellular response to tissue damage and implanted objects. Circulating monocytes transform into activated macrophages in the extravascular space. In addition to their phagocytic capabilities, macrophages mediate many of the processes associated with inflammation and wound healing through the myriad chemical factors they release. As a matter of fact, directly or indirectly, the macrophage affects virtually every aspect of the response to tissue damage and foreign bodies.

Among its mediatory roles, the macrophage initiates the formation of granulation tissue as the beginning of the second phase, repair. To prime the injured area for new tissue growth, the macrophage releases enzymes such as elastase, collagenase, and plasminogen activator that degrade damaged tissue. More importantly, it releases factors that mediate the release of such enzymes by other cells such as the fibroblast, and this process will be discussed later.

During this phase, *granulation tissue*, the scaffold for repair, is laid down. Granulation tissue formation can be considered to consist of two primary processes: the formation of new connective tissue and its neovascularization. Although both of these processes are carried out by different effector cells, they are heavily mediated by the macrophage. If the inflammatory phase does not prolong, granulation tissue begins to surround the implant within several days. The vascularity of the granulation tissue allows for rapid transport of oxygen and metabolic substances required by all the cells involved in the tissue response.

Just as the inflammatory phase was affected by the chemical nature of the implant, the duration and characteristics of the repair phase are highly dependent on the implant's physical surface properties. When the implant, or foreign body in general, is not successfully cleared or continues to cause trauma to the host, acute inflammation develops into granulomatous inflammation. It is important to introduce one of the most prominent cells of granulomatous inflammation, the multinucleated *giant cell*, which comprises fused macrophages. These cells are given their name because they are relatively large in comparison to the other inflammation cells, which usually range from 5 to 20 μm in size. The giant cell can be as large as 200 μm in size and contain over 100 nuclei. Although one function of giant cells is to continue to ingest the foreign implant, this process may not be their primary role. In general, a very smoothly surfaced implant (e.g., stainless steel, poly(tetrafluoroethylene) (PTFE), or silicon) elicits much less granulomatous inflammation with an almost complete absence of giant cells. However, when the implant is roughly surfaced or fibrous (e.g., felts or velours), giant cells are present in abundance and strongly attach themselves to the implant. This result supports the notion that the primary role of the giant cell is actually to create a protective barrier between irritating rough surfaces of an implant and the host tissue. If by this time the implant is not biodegraded and absorbed by the continuous enzyme secretion and ingestion started by the neutrophils and continued by the macrophages and giant cells, the body takes a new approach. Instead of attempting to break it down, the host attempts to completely isolate the foreign implant by *encapsulating* it.

Encapsulation is a specialized form of the generation of reparative tissue in the healing of subcutaneous wounds and is mediated by the *fibroblasts*.

Although the majority of their activity begins to take place in repair, fibroblasts become present in the inflammatory phase. These cells originate in perivascular connective tissue and increase in number during the chronic inflammatory phases. As the name implies, fibroblasts deposit collagen fibers, the primary constituent of the connective tissue that ultimately will be formed to encapsulate the implant. Additionally, if any damage was done to the vasculature by implanting a foreign body, the fibroblasts are followed to the wound area by endothelial cells. These cells migrate and undergo mitosis to create new vascular pathways (capillaries) within the area being encapsulated. This process establishes the necessary blood flow in the granulation tissue to provide the required metabolic compounds for fibroblasts to form collagen.

Possible Approaches To Minimize Host Isolation

One can argue that the intensity of the inflammatory phases is a reflection of biocompatibility, how the body reacts to the sensor. However, it is really the final phase, host encapsulation, that dictates how the sensor reacts to the body. Therefore, it is this phase that has greatest bearing on sensocompatibility.

From the outset, there seems to be two general approaches to overcoming the effects of host encapsulation and isolation of an implanted sensor. The first is an attempt to prevent or minimize the host's encapsulation of the foreign body. Even though there are many ways to regulate the level of acute and chronic inflammation in response to a foreign body, there is no evidence in the literature of any way to avoid the formation of connective tissue around the implant. One line of research in this area would most likely aim to ascertain methods of regulating fibroblast, macrophage, and giant cell activity at the sensor face in efforts to decrease the thickness of the host capsule. A second approach is to accept the formation of the host capsule but attempt to increase the vascularity of its connective tissue such that the analyte concentrations within the granulation tissue would closely represent that of plasma. A third possibility that will not be discussed here is to calibrate measured levels to actual levels by using a transport limited model.

Common to all approaches is the need to anchor the body of the sensor to the surrounding tissue. Studies have shown that preventing relative motion between the sensor and tissue is critical to minimizing the inflammatory response. For example, in a study conducted by Walton and Brown (*28*), when a porous PTFE specimen was implanted in a rabbit skin flap scaffold, strong fibrous ingrowth was observed. This ingrowth allowed anchoring of the implant to the tissue and resulted in a light, acellular inflammatory response. However, in the second part of their study, grafted skin from the rabbit was placed between the implant and scaffold. An acute inflammatory response was observed. Histological comparisons between the first and second parts of the test showed that the second case displayed a separation between the implant and the surrounding tissue. Walton concluded that the observed acute inflammation in the second part of the test was a result of poor cellular adhesion between the tissue and the implant due to the added skin graft.

Minimization of Capsule Thickness and Density

Although little seems to be known about the mechanism of its actions, perfluorosulfonic acid, trademarked Nafion, is showing promise as a material that elicits little acute or chronic inflammation and invokes only a thin layer of connective tissue encapsulation. In a study by Turner et al. (*29*), small cannulas made of Nafion were implanted for several days, explanted, and observed. The fibrous capsule in each of the implants was rather uniform and thin and ranged from 10 to 100 μm. Moreover, they reported that the connective tissue surrounding the Nafion implants was well vascularized. More recently, Nafion-coated glucose sensors were first tested in vivo by Moussy et al. (*30*). Although the response time slightly increased and sensitivity was reduced, the subcutaneously implanted sensors responded to changes in blood glucose levels. At two weeks, the sensor failed from degradation of the electrode, showing that its ability to survive encapsulation may have outlived the electrodes. Nafion also lends itself nicely to glucose sensor applications because of its higher O_2 permeability with respect to that of glucose. Moreover, because Nafion is negatively charged, it excludes negatively charged substances such as ascorbate and urate, which often interfere with amperometric measurement (*31*).

Physical Regulation

Another thought may be to coat the tip of the sensor with a material of low fibroblast adhesivity such that connective tissue deposition is inhibited in the sensing region. Materials such as Bioglass (SiO_2 doped with CaO, Na_2O, and P_2O_5) inhibit fibroblast growth.

In a study by Matsuda et al. (*32*), flasks were divided into several groups, one of which contained normal silica glass plates, whereas another contained Bioglass plates. Fibroblast cells derived from connective tissue were added to flasks of each group. Cell counts revealed that the Bioglass flasks had less than one-fifth the cell population of the other two groups. In addition, the pH of the suspension medium decreased from 7.5 at the time that cells were added to the flasks to 7.09 after 7 days for the silica glass plate group. In contrast, the Bioglass group pH actually increased to 7.61. Because acidic pH levels are indicative of cell metabolism and growth, Bioglass seemed to inhibit fibroblast proliferation.

Poor adhesivity to Bioglass is thought to be the mechanism that inhibits fibroblast growth. It is well established that for fibroblast cells to undergo mitosis, they must achieve good adhesion to a substrate surface. Therefore, if it is possible to coat the sensor surface with a material that is porous enough to allow entry of certain analytes yet small enough to prevent cell entry while retaining the property of poor adhesivity, fibroblasts may be unable to achieve anchorage-dependent growth.

On the other hand, it is important to note that in vivo, inhibiting adhesion to an implant can result in a competing effect. Campbell and von Recum (*33*) studied the effects of various material surface parameters on soft tissue response. More specifically, they studied the effect of material pore size on capsule thickness. Each implanted specimen was made of the same material (a grade of PTFE), only varying in pore size. Implants with a mean pore size smaller than 1 μm resulted in a thick granulous capsule (Figure 9.5a). Pores between 1 μm and 2 μm resulted in acellular, fibrous capsules that were significantly thinner (Figure 9.5b). When pore sizes were further increased, a chronic inflammatory response was again evoked, and a thick, granulous capsule with a layer of giant cells surrounding the implant resulted (Figure 9.5c). From these results, Campbell and von Recum hypothesized that surfaces that elicited a chronic inflammatory response promoted the formation of a layer of giant cells that prevented fibroblasts from adhering well to the implant surface. They argued that this result led to inadequate anchoring of the implant resulting in a granulous and consequently thick capsule. Pore sizes smaller than 1 μm presented little surface texture onto which fibroblasts could grab and hold, whereas pore sizes much greater than 3 μm appeared as macroscale surface irregularities to the surrounding tissue. Both of these pore sizes triggered chronic inflammation. By contrast, surfaces that allowed stronger fibroblast adhesion eliminated micromotion between the implant and the surrounding tissue. In such cases, the capsule, composed mainly of fibrous tissue, was void of any macrophages and giant cells. In addition, the capsule was relatively thin and uniform. Table 9.1 summarizes the responses of the different pore size implants tested.

To resolve such competing effects of fibroblast adhesion, it may be useful to study the tissue response of implanted Bioglass substrates as compared to a control group of standard silica glass, which usually promotes good fibroblast adhesion. Particularly interesting would be a study attempting to differentially control the thickness of a cap-a

b

c

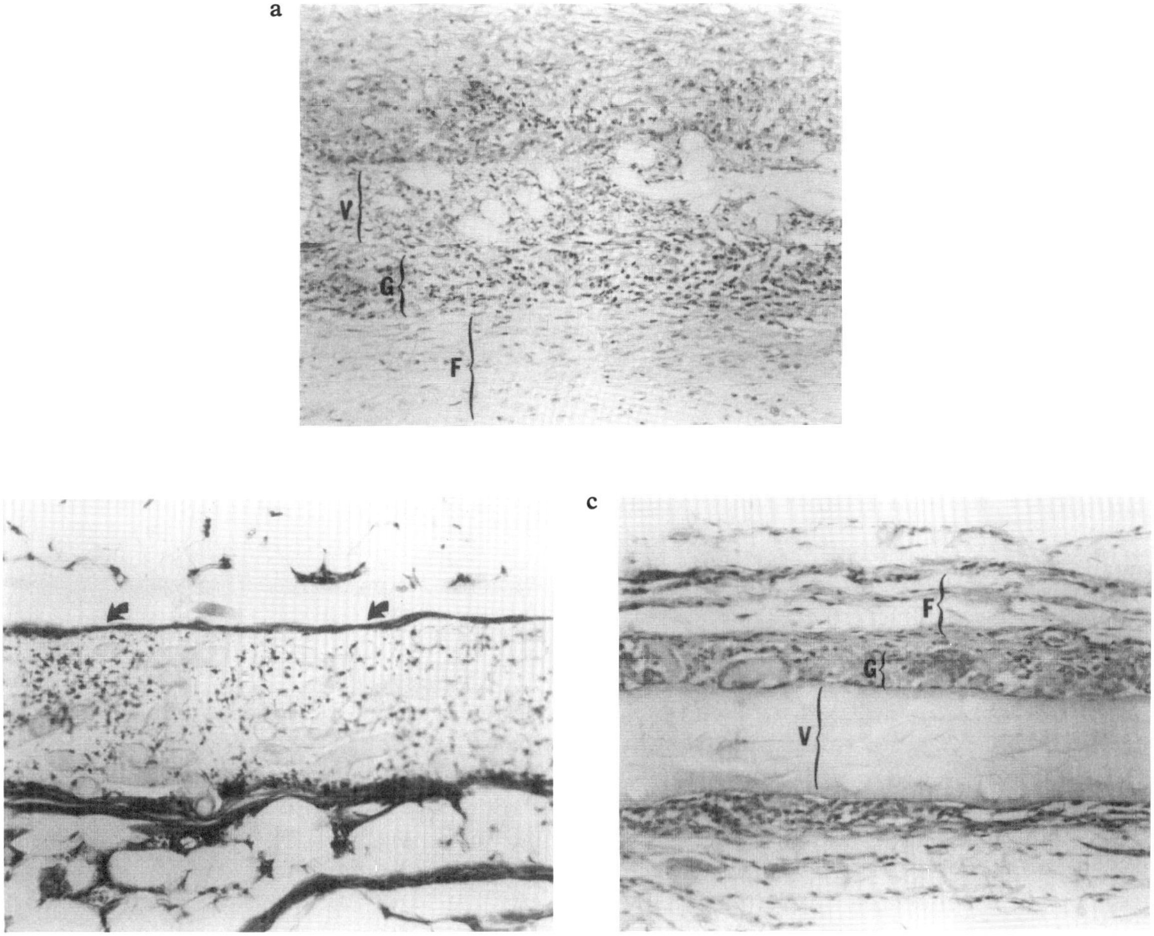

FIGURE 9.5. a, Histological section of tissue around 0.42-μm pore implant; b, tissue around 1.42-μm pore implant; and c, tissue around 3.33-μm pore implant. V indicates the implant and F and G denote the fibrous capsule and giant cells, respectively. (Reproduced with permission from reference 33. Copyright 1989 Taylor & Francis.)

sule such that it is thick around the body of the sensor and thin around the sensing element. If successful, perhaps the body of a sensor could be constructed from a material that promotes fibroblast adhesion to anchor the device and eliminate the motion that can elicit a chronic inflammatory response, whereas the tip could be coated with a porous Bioglass tip that inhibits fibroblast adhesion yet allows the passage of analytes.

Chemical Regulation

Along similar lines, consider the concept of releasing a chemical agent around the sensor tip, which may reduce the thickness of the connective tissue deposited around the sensing element. Agents that inhibit tissue growth have been identified by tumor regression studies. For example, the effects of some interferons (glycoprotein antiviral agents produced mostly by lymphocytes, fibroblasts, and

TABLE 9.1. Characteristic Histological Responses to Various (Mean) Pore Sizes of (Hydrophilic) Versapor Filter Material After 12-Week Implantation in Dogs

Histological Parameters	*Mean Pore Diameter (μm)*				
	0.42	*1.42*	*1.83*	*3.33*	*3–14*
Tissue capsule					
thickness (μm)	210	5–25	5–30	115–350	120
quality	Granulous	Fibrous	Fibrous	Granulous	Granulous
Surface contact with					
macrophase and giant cells	Yes	No	No	Yes	Yes
fibroblasts	No	Yes	Yes	No	No
collagen	No	Yes	Yes	No	No
Surface anchorage	No	Yes	yes	No	No
Capsular contraction	Yes	No	No	Yes	Yes
Histocompatibility rating	Poor	Optimal	Optimal	Fair	Poor
Manufacturer's identification	V–200	V–1200	V–3000	V–5000	V–10,000

SOURCE: Reproduced with permission from reference 33. Copyright 1989 Taylor & Francis.

macrophages) were studied on wound healing. They were shown to markedly reduce fibroblast activity and hence the rate of formation of connective tissue around an incited lesion (*34*). However, they also seem to inhibit vascularization of connective tissue and may cause damage to the vascular endothelium (*35*).

Topographical Regulation

The structure and density of tissue encapsulating implants can be influenced by the topography of the implant. Capsules forming around nonporous polyvinyl alcohol material (without an open interconnecting pore structure) were composed of much more tightly packed tissue than those forming around identical material with open pores of 60 and 350 μm sizes (*23*). Because fibrovascular ingrowth occurred in the porous implants, the capsules that formed around them were almost an extension of the surrounding subcutaneous tissue. This result was further evidenced by the fact that the effective diffusion coefficients of tissue encapsulating the porous implants was similar to those of subcutaneous tissue and double those of the tissue encapsulating the nonporous implants. Such findings imply that capsules may be more permeable if the implant allows integration of surrounding tissue such as the case of a porous implant.

Maximizing Analyte Transport Through the Capsule

Even if the thickness of connective tissue can be decreased, a dense, collagenous capsule could certainly decrease sensor access to the measured analyte. Another approach would be to increase the sensor's access to the analyte by maximizing analyte transport through changes in the capsule properties. One way this can be achieved is to increase the rate of diffusion through the capsule by decreasing the density of the connective tissue encapsulating the sensor. Alternatively, the extent to which the capsule is perfused could be increased by inducing capillary growth through the capsule. The induction of capillary growth may be more feasible in the short term because in the past 10–20 years, *angiogenesis*, the growth of new capillaries, has been gaining much attention in tumor growth research (*36*). If angiogenesis can be induced within the connective tissue formation encapsulating an implanted sensor, it may allow the analyte levels at the surface of the sensing mechanism to be similar to that of the surrounding tissue. Increasing the vascularity of the inevitable capsule to facilitate analyte transport to the sensor may be the most plausible solution to the host isolation problem.

However, in considering the diffusion barrier presented by the fibrous capsule, we must be concerned not only with the diffusion properties of the capsule and its vascularity, but also with consumption-driven flux of analyte into the sensor. Some sensors such as amperometric oxygen sensors consume analyte, whereas others such as glass pH electrodes have only a minuscule flux of hydrogen ions into the sensor. If capsules having the same perfusion and diffusion properties surround each of these sensors, the effect of the barrier will be quite different. There will be little effect on the pH sensor due to the low analyte flux into the sensor. On the other hand, the oxygen sensor may yield a lower reading than it would in the absence of the capsule due to the greater barrier, which decreases flux of oxygen into the sensor. If the capsules surrounding each sensor were histologically the same, the diffusion barrier to oxygen would also be greater than that for hydrogen ions because hydrogen ions diffuse more easily through tissue than do oxygen molecules. This characteristic would even further accentuate the differences between the sensor readings and actual tissue values for oxygen molecules and hydrogen ions.

Normally, angiogenesis in the adult only occurs during ovulation, menstruation, inflammation, and tissue repair. Factors regulating angiogenesis in tissue repair are of most interest here. Because the vast majority of this work has focused on the identification of angiogenesis factors, few studies have investigated *how* angiogenesis occurs and gained a clear understanding of the mechanisms that control it (*37*). Perhaps the current literature investigating angiogenic mechanisms can be divided into one of two approaches: chemical and physical.

Chemically Induced Angiogenesis

One possible approach to increasing the vascularity of the capsule formed around foreign implants is to leach angiogenic substances into the surrounding tissue. By far the most studied angiogenic substances have been *growth factors*, mostly polypeptides released to induce general or specific cell growth usually named after their source of derivation. Only recently, through advances in purification techniques, have many factors been identified. As was pointed out by Whalen and Zetter (*38*) 10 years ago, few factors had been identified; now, it is quite easy to lose track of them all.

Certain angiogenic factors seem to have direct effects on endothelial cell growth. For example, basic fibroblast growth factor (bFGF) (*39*) and transforming growth factor (TGF-a) (*40*) directly promote endothelial cell growth and proliferation. Other factors such as transforming growth factor (TGF-b) only have angiogenic effects in vivo, a characteristic indicating that they indirectly stimulate the release of other angiogenic factors from cells such as macrophages or platelets (*41*).

Although little has been reported on 1-butyryl glycerol in comparison with other angiogenic factors, it deserves special mention. Of all angiogenic factors, this one may seem especially attractive for sensor-related testing because it can easily be synthesized. Whereas most angiogenic factors are polypeptides that must be purified from fibroblast or macrophage cells, Dobson et al. (*42*) reported the lipid, 1-butyryl glycerol, also known as monobutyrin, to be strongly angiogenic. Before their study, it had been determined that 3T3 adipocytes produce an angiogenic factor that is highly coupled to the differentiation of adipose tissue cells in a developing embryo. After identifying the factor as 1-butyryl glycerol ($HOCH_2CH[OH]CH_2O_2C$-C_3H_7), they chemically synthesized the compound and studied its effects in a chick CAM (chorioallantoic membrane) assay (Figure 9.6). Although there have been several methods of quantifying the extent of neovascularization, a positive response in the CAM assay is defined as a visually clear increase in neovessel formation. The chemically synthesized monobutyrin was angiogenic at amounts as low as 20 pg. When studied in vitro, monobutyrin was shown to have almost no mitogenic properties, having little or no effect on endothelial cell proliferation; rather, it greatly increased endothelial cell motility. It was also shown that its effects were specific to endothelial

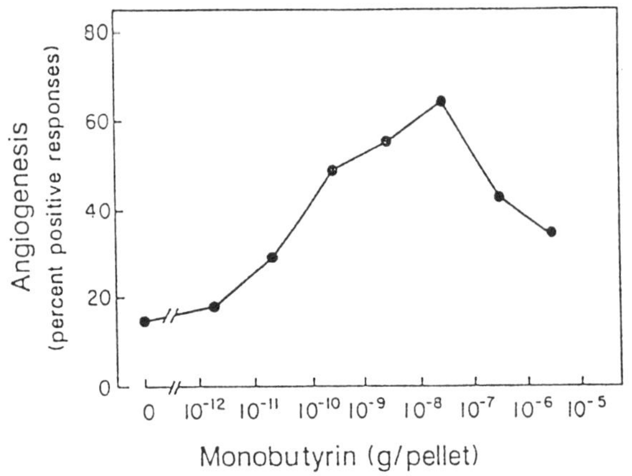

FIGURE 9.6. Angiogenic activity of synthetic monobutyrin. (Reproduced with permission from reference 42. Copyright 1990 Cell Press.)

cells because for other cell types such as fibroblast and smooth muscle, cell motility remained unaffected by the lipid.

Because most of the compounds that induce angiogenesis also stimulate fibroblast growth and activity, it is possible that they promote thickening of the capsule as well, thereby resulting in competing effects. The overall consequence of each growth factor with respect to sensor glucose access is difficult to speculate. For example, basic fibroblast growth factor (bFGF) stimulates extracellular matrix deposition while inhibiting collagen fibril formation. In contrast, TGF-β promotes the opposite effects (*39*). Both are angiogenic. Studies need to be performed that assess the contribution of each effect on sensocompatibility.

Physically Induced Angiogenesis

Although some investigators have shown that there is increased angiogenesis in tissue surrounding electrodes that provide electrical stimuli to tissue (*43, 44*), the mechanism responsible for this process is not understood. We can speculate as to the possible causes of increased tissue perfusion around such electrodes. First, electrochemical reactions at the electrode surface often accompany electrical stimulation of tissue. These reactions result in waste products at elevated concentrations in the vicinity of the electrodes, and this increase may stimulate angiogenesis in an effort to provide a method to remove the offending substance. Consider an alternative mechanism in which the electrode is a cathode. Because the usual electrochemical reaction consumes oxygen, if the electrode surface area is great enough, sufficient quantities of oxygen will be consumed to make the surrounding tissue hypoxic. Reduced oxygen tension has been known to promote neovascularization. Other chemical factors might play a role as well. For example, slight dissolution of the electrode material or its insulator might produce gradients of these trace elements around the electrode that could promote the formation of new capillaries.

If these mechanisms could be better understood, it might be possible to use them to promote angiogenesis over the sensitive portion of an implanted chemical sensor. For example, one could incorporate electrochemical electrodes at the sensitive surface of an implanted glucose sensor. By continuously passing a current between these electrodes early in the life of the implant, the formation of new capillaries in the fibrous capsule would be encouraged as the capsule develops. This process, in turn, would reduce the diffusion barrier associated with the capsule. It would be difficult to know which of the three mechanisms described (if any) was responsible for the angiogenesis, but one could certainly design studies to determine if vascularization of the surrounding tissue was indeed enhanced.

Topographically Induced Angiogenesis

Another means of physically stimulating angiogenesis is by varying the topography of the implanted sensor. Although variations in implant topography may not affect fibroblast adhesion to the extent of greatly reducing the thickness of the host capsule, topography certainly may be used to orient endothelial cells and guide their migration, possibly enhancing vascularization of the capsule. Curtis and Clark (*45*) assembled a rather comprehensive review of topographical effects on cell adhesion and migration. Even though they do not directly discuss endothelial cell migration, many inferences can be made from their collection of works.

In summary of the Curtis and Clark review, ridges and grooves have definitely been shown to align cells. Even though the width of a ridge or groove and the lateral spacing between them have some bearing on alignment, the parameter with the greatest effect is the depth of the groove. Deeper grooves increase the alignment and hence motion guidance of a cell. This result is probably somewhat related to the effects of a single vertical plane; even though gravitational effects on cell migration may be neglected, the probability of a cell ascending or descending a vertical plane was inversely proportional to the plane height. A groove can be considered as two opposing vertical planes. Therefore, one would expect increasing groove depth to discourage lateral movement across the grooves and encourage longitudinal movement along the grooves. As for the effects of tubes and cylinders, basically concave and convex curvature, orientation is inversely proportional to the radius of curvature. The effects are not equal however, because cells seem to prefer aligning themselves on the concave rather than convex curvature. The general effect of topography was summarized best by Dunn and Heath (*46*), who stated that the probability of a cell traversing certain topography was inversely related to the amount of cytoskeletal disruption required to do so. Dunn's theory would predict that a cell is more likely to traverse a structure with gradual curvature rather than one with a sudden or small radius of curvature. Moreover, a cell would be more likely to migrate across an obtuse rather than an acute angle. These predictions seem consistent with observed findings.

Now consider material texture, topographical effects on the microscale. Walton and Brown (*28*) studied the vascular integration of an "isolated" scaffold for surgical reconstruction applications. A small slab of porous PTFE was placed adjacent to a vascular pedicle in a rabbit's ear. The slab and pedicle were isolated from surrounding tissue by sealing them in silicone sheets. Neovascularization from the adjacent pedicle was observed. At six weeks after implantation, the PTFE specimen were almost completely vascularized. Walton's findings suggest that the porosity alone may promote endothelial cell penetration and vascular ingrowth. Attempts to immunoisolate implanted cells by containing them within a partially permeable membrane found that pore size and pore density were critical factors in the promotion of vascularity of the tissue-membrane interface (*47*). Brauker et al. (*47*) found that PTFE membrane material with pores larger than 0.8 μm and pore densities over 70% allowed the entry of tissue host cells, promoting good vascularization adjacent to the membrane. Lesser values for those two factors increased the likelihood of poor vascularization and a more typical foreign body response. In a related study, hydrogel sponges were polymerized from poly(2-hydroxylethyl methacrylate) in varying percentages of water (*48*). Those sponges polymerized in 70% water had a smaller pore size than those polymerized in 80% and 90% water. Moreover, the 80% and 90% groups had interconnecting pores. As a result, greater and deeper vascularization occurred in those groups with a higher water content.

Picha and Levy (*49*) considered the role of porosity in promoting neovascularization from a different perspective. In their study, disc-shaped segments of nitrocellulose of various porosities were drilled radially with a single hole and used as healing conduits for an arteriovenous (AV) shunt between the femoral artery and vein of rats. Lexan sheets sealed the two ends of each cylinders and solid Lexan discs served as a control. Among other things, they observed the patency of their AV anastomosis with the varying porosity cylindrical membranes. With the AV anastomosis positioned in the center of each disc, the discs were implanted and then harvested for observation. For the control group, that patency was 0%. For the pore sizes of 0.025, 1.2, and 8.0 μm, the vessel patency was 10%, 96%, and 60%, respectively. They hypothesized that exudate from the post-anastomosis traumatized vessels contained such compounds as superoxide ions, kininogens, and prostaglandins that are involved in the inflammatory response. The removal of these compounds may be necessary for vessel patency. In effect, this study asserts that porosity around the sensor element is not only crucial for endothelial cell migration and vascular ingrowth, it may also be important to maintain the health of newly formed vessels by allowing an outlet for waste and other products that may be harmful to the endothelial cells composing such vessels.

Only recently has there been an organized effort in identifying factors that promote or inhibit angiogenesis. Research in this field is still attempting to more clearly define the cause and effect relationship between physical or chemical factors and the growth of new capillaries. It may be quite some time before angiogenesis can actually be regulated. Research that aims to quantify and model angiogenic increase in tissue analyte levels, however, is not only important to increase capsule vascularization. Because foreign bodies exposed to blood trigger an even more complex host response, subcutaneous implantation is generally recognized to be less challenging. Therefore, such studies would be useful in strengthening the correlation between subcutaneous tissue and blood analyte levels.

Sensor Design Considerations

Several alternatives to the biosensor-based artificial pancreas are currently being pursued. One example is the near-IR spectroscopy sensor mentioned previously. The obvious advantage of such a system is that the sensor detects glucose noninvasively. This process would obviate issues of biocompatibility and sensocompatibility. On the opposing extreme is the development of a fully implantable *bioartificial pancreas* membrane permeable to glucose and insulin but impermeable to immune cells and secreted antibodies. Being a fully contained system, without the need for any ex vivo components, this solution is ideal and most closely resembles the natural pancreas. Even though the bioartificial pancreas may be the ultimate solution to diabetes, it may be some time yet before it is developed for clinical application. Each of the three concepts have their own niche. Even though the implantable sensor-based system is likely to be replaced with some of these more ambitious concepts in the future, it probably remains the most feasible at this time.

Many challenges remain in the face of an implantable glucose sensor, the most prominent are those pertaining to sensocompatibility. On the basis of findings of general foreign body response and angiogenesis literature, a few concepts are proposed (Figure 9.7). These configurations suggest directions for further research and act as bases for solutions to the current sensocompatibility problems.

On the basis of the earlier discussion of chemically induced angiogenesis, one approach would be to vascularize the capsule and tissue surrounding an implanted sensor. To that end, consider the coaxial catheter illustrated in Figure 9.7a. The inner lumen contains a standard glucose sensor, similar to those discussed earlier. The outer lumen serves as a conduit for the passage of angiogenic factors that are then slowly released through the small port holes near the sensor tip.

Further, Figure 9.7b depicts an example of how topography may be used to improve the vascularity of tissue near a sensor. In this case, consider the combination of chemical induction of angiogenesis with physical guidance of capillaries to maximize the sensor's access to the analyte. For example, an outer ring that has been etched with grooves pointing radially inward toward the sensing mechanism was added to the sensor configuration in Figure 9.7a. The concept here is that when angiogenic factors are released, the grooves guide endothelial cells toward the angiogenic factor and the middle of the sensor, thereby "magnifying" the tissue's access to glucose or any other analyte of interest near the sensing element. Obviously, such a configuration would most certainly introduce greater complexity of manufacture and may be difficult to implement. However, with advancing technology in material processing techniques such as photolithography and ion-beam etching, the topography of an implant is more readily controllable.

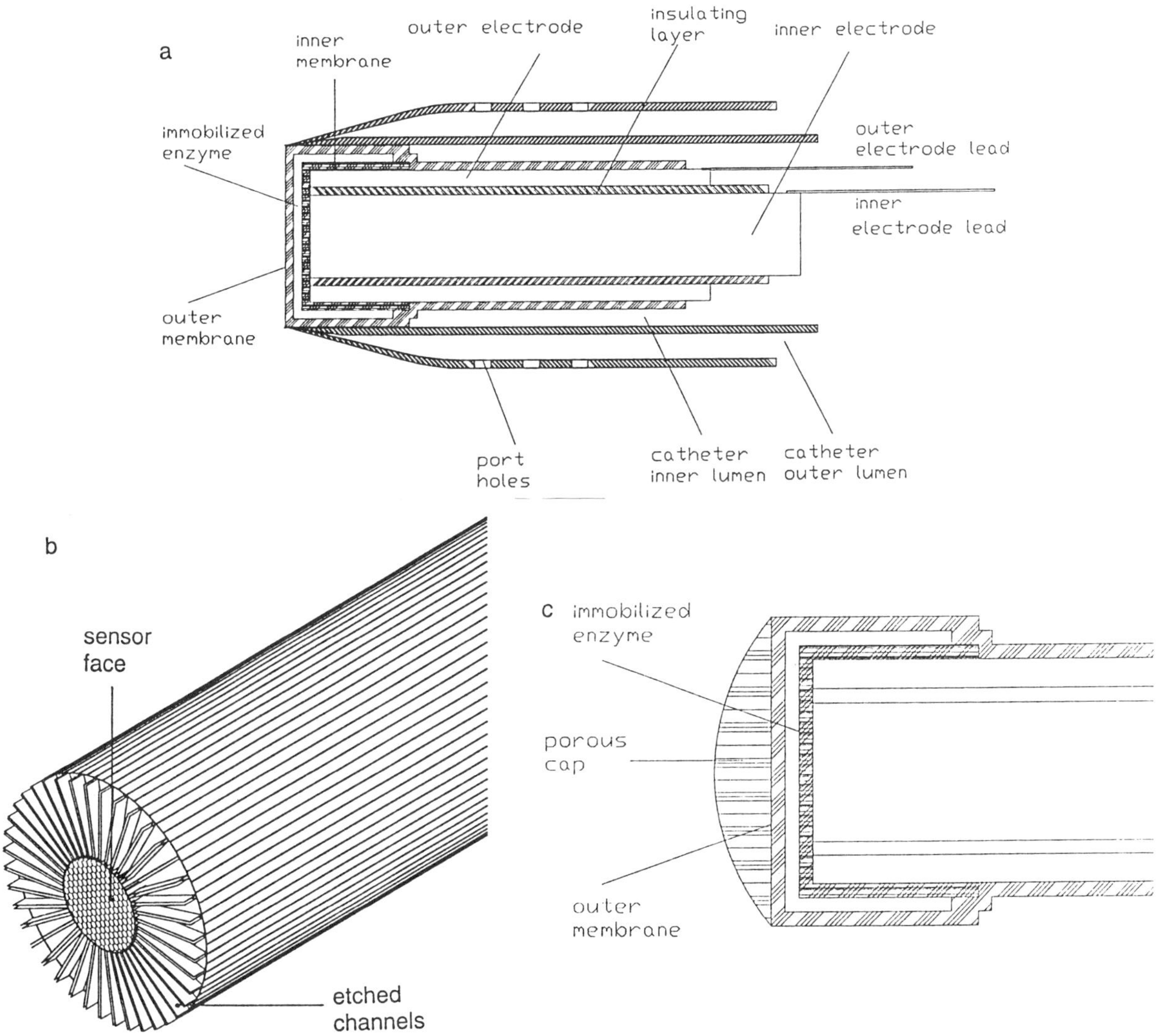

FIGURE 9.7. a, Glucose sensor that releases angiogenic factors at the sensor tip to promote capillary growth near the sensing element; b, microgrooved sensor tip to guide capillary growth; and c, textured or porous sensor tip to minimize thickness and increase the vascularity of the capsule near the sensing element.

Coating the tip of the sensor with a porous material that would promote vascular ingrowth as shown in Figure 9.7c may be more feasible. On the basis of the findings of implanted porous substrates and our own unpublished data, the capsule at the tip of such a sensor should be well vascularized. To avoid chronic inflammation and giant cell encapsulation, the material used to coat the tip must be chemically inert. Porous PTFE used in Walton's study and polyether polyurethane are candidate materials.

The concepts presented here, which range from the straightforward to the purely speculative, all aim to suggest possible directions for future research in sensocompatibility. Two concepts that seem particularly promising are minimizing the thickness of the capsule by coating the sensor face with a textured or porous cap and inducing a well-vascularized capsule through the promotion of angiogenesis.

However, a textured surface probably will never completely eliminate the formation of a connective tissue capsule around the sensor; likewise, sensor configurations that induce angiogenesis will never result in predictable capsule vascularity. Therefore, the solution to the encapsulation problem will undoubtedly incorporate some sort of recalibration to each patient's blood glucose level after the transient stages of the foreign body response. This result reconfirms the way the design problem was originally defined in the beginning of the chapter. It is unrealistic and unnecessary to attempt eliminating the capsule altogether, because the key to successful implantation is not necessarily to equate analyte concentrations around the sensor to those in the subcutaneous tissue or plasma but rather to dynamically *stabilize* the correlation between the two levels.

References

1. Pickup, J. C. *Diabetes Care* **1993,** *16,* 535–539.
2. Genuth, S. M. In *Physiology;* Berne, R. M.; Levy, M. N., Eds.; Mosby Year Book: St. Louis, MO, 1993; pp 895–948.
3. Reach, G.; Wilson, G. S. *Anal. Chem.* **1992,** *64,* 381A–386A.
4. Diabetes Control and Complications Trial Research Group. *N. Engl. J. Med.* **1993,** *329,* 977–986.
5. Brunetti, P.; Benedetti, M.; Calabrese, G.; Reboldi, G. P. *Int. J. Artif. Organs* **1991,** *14,* 216–226.
6. Zeller, H.; Novak, P.; Landgraf, R. *Int. J. Artif. Organs* **1990,** *12,* 129–135.
7. Fisher, U. *Diabetic Medicine (New York)* **1991,** *8,* 309–321.
8. Reichert, W. M.; Saavedra, S. S. In *Medical and Dental Materials;* Williams, D. F., Ed.; VCH Publishers: New York, 1992; pp 303–343.
9. Regnault, W. F.; Piccolo, G. L. *J. Biomed. Mater. Res.* **1987,** *21,* 163–180.
10. Ko, W. H. In *Implantable Sensors for Closed Loop Prosthetic Systems;* Ko, W. H., Ed.; Futura Publishing: New York, 1985; pp 89–197.
11. Donaldson, P. E. *IEEE Trans. Biomed. Eng.* **1976,** *BME-23,* 281–285.
12. Sharkawy, A. A.; Reichert, W. M. In *Handbook of Biomaterials Evaluation;* von Recum, A. F., Ed.; Macmillan Publishing: New York, in press.
13. Shichiri, M.; Kawomori, R.; Hakui, N.; Yamasaki, Y.; Abe, H. *Diabetes* **1984,** *33,* 1200–1202.
14. Rebrin, K.; Fischer, U.; von Dorsche, H.; von Woetke, T.; Abel, P.; Brunstein, E. *J. Biomed. Eng.* **1992,** *14,* 33–40.
15. Bobbioni-Harsch, E.; Rohner-Jeanrenaud, F.; Koudelka, M.; de Rooij, N.; Jeanrenaud, B. *J. Biomed. Eng.* **1993,** *15,* 457–463.
16. Johnson, K. W.; Mastrototaro, J. J.; Howey, D. C.; Brunelle, R. L.; Burden-Brady, P. L.; Bryan, N. A.; Andrew, C. G.; Rowe, H. M.; Allen, D. J.; Noffke, B. W. *Biosens. Bioelectron.* **1992,** *7,* 709–714.
17. Ertefai, S.; Gough, D. A. *J. Biomed. Eng.* **1989,** *11,* 362–368.
18. Moatti-Sirat, D.; Capron, F.; Poitout, V.; Reach, G.; Bindra, D. S.; Zhang, Y.; Wilson, G. S.; Thevenot, D. R. *Diabetologia* **1992,** *35,* 224–230.
19. Wilson, G. S.; Zhang, Y.; Reach, G.; Moatti-Sirat, D.; Poitout, V.; Thevenot, D. R.; Lemonnier, F.; Klein, J. C. *Clin. Chem. (Washington, D.C.)* **1992,** *38,* 1613–1617.
20. Poitout, V.; Moatti-Sirat, D.; Reach, G.; Zhang, Y.; Wilson, G. S.; Lemonnier, F.; Klein, J. C. *Diabetologia* **1993,** *36,* 658–663.
21. Woodward, S. C. *Diabetes Care* **1982,** *5,* 278–281.
22. Pfeiffer, E. F. *Horm. Metab. Res. Suppl. Ser.* **1990,** *24,* 154–164.
23. Sharkawy, A. A.; Klitzman, B.; Truskey, G. A.; Reichert, W. M. *J. Biomed. Mater. Res.* in press.
24. Clark, R. F. In *The Molecular and Cellular Biology of Wound Repair;* Clark, R. F.; Henson, P. M., Eds.; Plenum: New York, 1988; pp 1–34.
25. Irvin, T. T. *Wound Healing—Principles and Practices;* Chapman and Hall: London, 1981; pp 2212–2220.
26. Anderson, J. M. In *Inflammation and the Foreign Body Response;* Klitzman, B., Ed.; J. B. Lippincott: Philadelphia, PA, 1994; pp 147–160.
27. Woodward, S. C.; Salthouse, T. N. In *Handbook of Biomaterials Evaluation;* von Recum, A. F., Ed.; MacMillan: New York, 1986; pp 351–412.
28. Walton, R. L.; Brown, R. E. *Ann. Plast. Surg. (Boston, Mass.)* **1993,** *30,* 105–110.

29. Turner, R. B.; Harrison, D. J. *Biomaterials* **1991,** *12,* 361–368.
30. Moussy, F.; Harrison, D. J.; O'Brein, D. W.; Rajotte, R. V. *Anal. Chem.* **1993,** *65,* 2072–2077.
31. Harrison, D. J.; Turner, R. F.; Baltes, H. P. *Anal. Chem.* **1988,** *60,* 2002–2007.
32. Matsuda, T.; Yamauchi, K.; Ito, G. *J. Biomed. Mater. Res.* **1987,** *21,* 499–507.
33. Campbell, C. E.; von Recum, A. F. *J. Invest. Surg. (New York)* **1989,** *2,* 51–74.
34. Stout, A. J.; Gresser, I.; Thompson, W. D. *Int. J. Exp. Pathol.* **1993,** *74,* 69–85.
35. Dvorak, H. F.; Gresser, I. *J. Natl. Cancer Inst.* **1989,** *81,* 497–502.
36. Gross, J. L.; Herblin, W. F.; Horlick, R.; Brem, S. S. In *Angiogenesis: Key Principles—Science, Technology, Medicine;* Steiner, R.; Weisz, P. B.; Langer, R., Eds.; Birkhäuser Verlag: Basel, Switzerland, 1992; pp 421–427.
37. Konerding, M. A.; von Ackern, C.; Steinberg, F.; Streffer, C. In *Angiogenesis: Key Principles—Science, Technology, Medicine;* Steiner, R.; Weisz, P. B.; Langer, R., Eds.; Birkhäuser Verlag: Basel, Switzerland, 1992; pp 40–57.
38. Whalen, G. F.; Zetter, B. R. In *Wound Healing;* Cohen, I. K.; Diegelmann, R. F.; Lindblad, W. J., Eds.; Saunders: Philadelphia, PA, 1992; pp 1145–1149.
39. Pierce, G. F.; Tarpley, J. E.; Yangihara, D.; Mustoe, T. A.; Fox, G. M.; Thomason, A. *Am. J. Pathol.* **1992,** *140,* 1375–1388.
40. Mahadevan, V.; Hart, I. R.; Lewis, G. P. *Cancer Res.* **1989,** *49,* 415–419.
41. Liebovich, S. J.; Wiseman, D. M. *Growth Factors and Other Aspects of Wound Healing: Biological and Clinical Implications;* Liss: New York, 1988; pp 131–145.
42. Dobson, D. E.; Kambe, A.; Block, E.; Dion, T.; Lu, H.; Castellot, J. J., Jr.; Spiegelman, B. M. *Cell* **1990,** *61,* 223–230.
43. Huddlicka, O.; Tyler, K. R. *J. Physiol. (London)* **1984,** *353,* 435–445.
44. Huddlicka, O.; Price, S. *Eur. J. Physiol.* **1986,** *417,* 67–72.
45. Curtis, A. S.; Clark, P. *Crit. Rev. Biocompat.* **1990,** *5,* 343–362.
46. Dunn, G. A.; Heath, J. P. *Exp. Cell Res.* **1976,** *101,* 1.
47. Brauker, J.; Martinson, L. A.; Hill, R. S.; Young, S. K.; Carr-Brendel, V. E.; Johnson, R. C. *Transplant. Proc.* **1992,** *24,* 2924.
48. Chirila, T. V. *Biomaterials* **1993,** *14,* 26–38.
49. Picha, G. J.; Levy, D. *Plast. Reconstr. Surg.* **1991,** *87,* 509–517.

Delivery of Peptide and Protein Drugs

10
Formulation of Proteins and Peptides

Steven L. Nail

Key formulation and processing issues involved in preparation of pharmaceutically acceptable, injectable protein products are reviewed in this chapter. They include preformulation studies, the influence of route of administration, chemical and physical stability issues, common excipients in injectable protein formulations, freeze-drying of proteins, and processing effects, such as filtration and shear denaturation.

Development of formulations of proteins and peptides has become an increasingly important aspect of pharmaceutical science and technology in recent years because more new products are biologicals—either from recombinant DNA technology, hybridoma technology, or traditional biological products such as polyclonal antibodies. From 1993 through 1995, a total of 42 new biological products were approved by the Center for Biologics Evaluation and Research branch of the U.S. Food and Drug Administration (*1*). This number compares with a total of six new biological product approvals each year from 1986 through 1988 (*1*). Important new biological products approved in the past two years include dornase alfa (Pulmozyme) for treatment of cystic fibrosis; interferon beta-1b (Betaseron) for multiple sclerosis; a recombinant blood clotting factor (Kogenate) for treatment of hemophilia A; pegaspargase (Oncaspar) for treatment of acute lymphocytic leukemia; and several human immune gamma globulin products for such indications as prevention of Rh isoimmunization and prevention of infections in bone marrow transplant patients and in children with human immunodeficiency virus. Vaccine development has figured prominently in new biological product development. The list of new product approvals includes vaccines for hepatitis A; varicella virus; and various combinations of *Hemophilus* influenza, tetanus, diphtheria, and pertussis (*1*). Sales of peptide drugs, which vary in number of amino acids from three (thyrotropin-releasing hormone) to 32 (calcitonin), are currently in excess of $1 billion worldwide (*2*).

The development of a pharmaceutically acceptable dosage form of a protein or peptide is generally more complicated and requires more

Copyright © 1997 American Chemical Society

development time than traditional drug compounds, primarily because of the greater structural complexity of proteins. From the standpoint of analytical development, protein drugs are often a challenge with respect to implementation of assays that are predictive of biological activity and indicative of stability of the product. Bioassays often are not highly reproducible and may have relative standard deviations of 20% or more. Successful formulation and process development requires an awareness that the integrity of a protein depends on proper folding of the protein. The stability of this folded state involves the interplay of a large number of both stabilizing and destabilizing interactions, and the net stabilizing interaction often amounts to only a few hydrogen bonds, hydrophobic patches, or ion pairs (*3*). The free energy of stabilization of the biologically active folded state in aqueous solution is, on average, only about 50 kJ/mole (*3, 4*). Because disruption of proper folding does not require a large input of energy, seemingly subtle aspects of manufacturing such as the stirring rate in a mixing vessel can have a substantial impact on the critical attributes of the final product. In addition, these quality attributes must be maintained during storage of the product for at least 18 months despite numerous possible routes of both physical and chemical degradation of the proteins. As a result, development of a safe, effective, and reliable protein formulation requires an in-depth understanding of the properties of the protein, particularly its susceptibility to either chemical or physical instability, development of suitable stability indicating assays, design of a dosage form appropriate to the route of administration (generally either an aqueous solution or a freeze-dried powder for an injectable product), development of the formulation including both the protein and other components (excipients) intended to enhance stability or other pharmaceutical properties, and selection of processing methods for manufacture of the final dosage forms that accomplish their purpose without adversely affecting the final product.

This brief review will highlight the issues that must be addressed by the formulation scientist responsible for product development of protein pharmaceuticals intended to be administered by injection. References to detailed reviews in specific subject areas are included.

Preformulation Studies

Dose and Solubility

The dosage of protein drugs varies enormously, from microgram or submicrogram doses for proteins such as interferons to several grams for gamma globulins. The dose, the solubility of the protein, and the volume of solution to be administered determine the concentration of protein in the formulation. The concentration, in turn, affects the need for formulation components that inhibit adsorption to surfaces such as materials used in manufacture, the container/closure system, or the devices used to administer the drug. For high concentrations of protein, saturation of such surfaces may result in an immeasurably small decrease in protein concentration in solution, whereas the same quantity of protein adsorbed from a solution at a low concentration of protein could result in removal of essentially all of the protein from the formulation.

The aqueous solubility of proteins also varies over several orders of magnitude–from virtually insoluble to several hundred milligrams per milliliter—and is an important consideration in preparing an aqueous solution of a protein where the required dose must be administered in a volume of solution appropriate to the intended route of administration. Experimentally, measuring the solubility of a protein in water can be more complicated than measuring the solubility of a conventional, low-molecular-weight drug substance for several reasons. First, the solubility of a protein is generally more affected by experimental parameters such as the type of buffer, buffer concentration, and ionic strength than traditional drug compounds. Even though parameters such as pH and temperature affect the solubility of many low-molecular-weight drugs, the relationship among pH, temperature, and solubility is often more complex for protein solutes. As a result, more experimental data are needed to obtain a reasonably complete picture of the solubility behavior of a protein. Second, the equilibration of a protein, which is determined by using the classical technique of suspending excess solid in the dissolution medium and measuring concentration in the supernatant, may take weeks instead of hours or days. Attempts to accelerate equilibration, such as vigorous agitation of the system, must be approached with caution because of the risk of inducing interfacial denaturation of the protein. Third, two distinct phases—solid protein and aqueous protein solution—may not form. Instead, measurements may be complicated by the presence of gels, nonequilibrium suspensions, and separation into protein-rich and protein-poor phases (*5*). Specialized techniques for minimizing equilibration time have been reported, particularly a simple column technique where the approach to equilibration is measured from two directions using both "undersaturated" and "oversaturated" buffers (*6*). A method of circumventing the problem of multiple phase formation is to induce precipitation of the protein by addition of another solute such as polyethylene glycol (PEG). Plots of the logarithm of protein solubility versus PEG concentration are often linear, and extrapolation to no PEG concentration allows an estimate of the aqueous solubility of the protein alone (*5*).

Preformulation studies should include examination of the influence of temperature, ionic strength, and pH on protein solubility. In general, protein solubility increases with temperature up to the temperature where unfolding starts. Solubility measurements should include, at a minimum, room temperature and 4 °C. Increases in solubility with increased ionic strength are common at low ionic strength, but solubility generally decreases with increased ionic strength (salting out) at higher ionic strength. Solubility is generally minimal in the region of the pI of the protein (the pH at which the net charge on the protein is zero), and it increases on both the acidic and basic side of the pI. Again, exceptions are common, and the possibility of more than one minimum in the pH/solubility profile should not be overlooked.

Protein solubility studies should also include the influence of other anticipated components of the formulation, such as stabilizing solutes, surfactants, bulking agents (for lyophilized dosage forms), and preservatives. Direct binding of other solutes to the protein, as well as preferential hydration and effects of ionic strength, can all influence solubility.

Routes of Administration

Protein drugs are usually administered by injection: usually either intravenous (iv) (either as a bolus dose or as an infusion), intramuscular, or subcutaneous, although other routes are possible. The intended route of administration determines the maximum volume that can be administered, as shown below:

Route	*Volume Administered*
Intravenous	
Bolus	About 10 mL
Infusion	No upper limit, as long as rate of fluid administration does not exceed the rate of fluid elimination
Intramuscular	
Deltoid muscle	About 2 mL
Lateral thigh	About 5 mL
Gluteal muscle	About 5 mL
Subcutaneous	About 2 mL

In addition to retention of the labeled activity of the drug substance, sterility, freedom from extraneous particulate matter, and freedom from pyrogens are critical quality attributes of all injectable dosage forms. However, for certain critical routes of administration, such as into the brain cavity, spinal column, or eye, other critical quality attributes apply. For these routes, physiological pH, the same pH as normal plasma (7.4), is critical, as well as isotonicity, the same osmotic pressure as normal physiological fluids. For these critical routes of administration, the presence of a preservative in the formulation is not acceptable. Damage to sensitive tissues can result if these additional criteria are not observed. Obviously, the formulation composition is always dictated by the most critical route of intended use.

For iv, subcutaneous, or intramuscular routes of administration, physiological pH and isotonicity are desirable properties of the formulation to minimize pain or tissue irritation on injection. However, other factors may dictate against either physiological pH or isotonicity of the formulation if acceptable physical or chemical stability cannot be obtained under these conditions. Many injectable products administered by one of these routes of administration are formulated at a wide range of pH and osmolarity (*7*).

Relevant Routes of Degradation

The stability characteristics of proteins are broadly classified as chemical stability, which refers to breaking and reforming covalent bonds, and physical stability, which refers to changes in secondary, tertiary, or quaternary structure of the protein that lead to loss of biological activity or loss of the protein from solution due to precipitation or adsorption to surfaces. Chemical and physical stability of proteins has been a subject of enormous interest over the past few years, and several excellent reviews have been published. The reader is referred to reviews by Manning et al. (*8*), Cleland et al. (*9*), Arakawa et al. (*10*), and Wang and Hanson (*11*) and to the book edited by Ahern and Manning (*12*). Key aspects of chemical and physical instability of proteins are briefly highlighted below.

Chemical stability refers to reactions such as deamidation, oxidation, and disulfide exchange. The peptide bond that forms the backbone of proteins is much more resistant to hydrolysis than, for example, an ester, and the peptide bond is generally considered stable within the pH range of 6 to 8 unless hydrolysis is assisted by a side group (9) or catalyzed by a protease.

Deamidation is the hydrolysis of the side chain amide of an asparagine (Asn) or a glutamine (Gln) residue to form a carboxylic acid. Numerous pharmaceutically relevant proteins have been shown to undergo deamidation in solution, including insulin (*13, 14*), interleukin 1-β (*15*), monoclonal antibodies (*16*), adrenocorticotropic hormone (*17*), luteinizing-hormone-releasing hormone (LHRH) antagonist peptide (*18*), and thyrotropin-releasing hormone (*19*). Even though most of these studies focused on solution stability, deamidation in the freeze-dried solid state also was demonstrated (*20, 21*).

Oxidation of methionine, tryptophan, tyrosine, and cysteine residues can contribute to protein instability, both in solution and in the solid state. From a formulation perspective, oxidation occurring in the presence of air under the conditions of formulation and processing is of the most concern. Methionine readily oxidizes under mild conditions to form methionine sulfoxide. This reaction causes loss of biological activity for several peptide hormones (*8*). Cysteine undergoes oxidation in the presence of air to form cystine disulfide in a reaction catalyzed by metal ions (*22*). This reaction contributes to both chemical and physical instability of protein formulations, because denaturation can cause exposure of cysteine residues that are not accessible for disulfide formation in the native state. Intermolecular disulfide bond formation following unfolding can lead to covalent aggregate formation and precipitation (*15*). The presence of trace metals as a contaminant in pharmaceutical formulations of human serum albumin has been shown to correlate with oxidation of tryptophan and thiol groups (*23*). A comprehensive summary of oxidation of pharmaceutically relevant proteins is included in the review by Cleland, Powell, and Shire (*9*).

Proteolysis refers to hydrolysis of a peptide bond that is catalyzed by the protein itself or by a contaminating protease. A common example of self-proteolysis is the intramolecular catalysis by the carboxyl group of an asparagine residue in acidic solution (*8*).

Racemization refers to the formation of a mixture of D- and L-enantiomers, one of which may be biologically inactive, via either a base-catalyzed or acid-catalyzed reaction involving a carbanion intermediate. Electron-withdrawing groups such as asparagine, tyrosine, serine, and phenylalanine all promote racemization (*11*). Racemization has been reported as one of the degradation pathways of luteinizing-hormone-releasing hormone (*24*).

β-Elimination usually refers to the destruction of disulfide bonds by means of β-elimination from a cystine residue. β-Elimination is accelerated under alkaline conditions and has been reported as a route of degradation for human macrophage colony stimulating factor (*25*) and lysozyme (*26*). Other amino acids can undergo β-elimination under alkaline conditions, such as serine, threonine, phenylalanine, and lysine (*8*).

The Maillard reaction, often referred to as nonenzymatic browning, is the reaction between free amino groups of the protein, such as lysine residues, and the carbonyl groups of reducing sugars. This reaction results in the formation of a Schiff base, which undergoes further degradation to produce colored products. This reaction is very common in intermediate-moisture foods, and it is most frequently observed in protein formulations as a result of reaction between the protein and a sugar such as dextrose or sucrose, either in solution or in the solid state.

Physical stability refers to processes such as adsorption to surfaces, aggregation, and precipitation. These processes usually occur without breaking or reforming covalent bonds, but rather by disruption of the native folded conformation of the protein. One of the main forces that drive protein folding is reduction of exposure of hydrophobic groups to an aqueous environment. Loss of this tertiary structure, called denaturation, is frequently followed by one or more of the previously mentioned processes. In the area of physical stability, development of protein pharmaceuticals differs most dramatically from the development of conventional low-molecular-weight drug entities, because traditional drugs are not as prone to physical stability problems.

Adsorption to surfaces is of greatest concern when the protein concentration in solution is low, because even a small amount of protein adsorbed to a surface can result in significant loss of activity. Adsorption to a surface may, or may not, be accompanied by a conformational change. Given that the amino acids composing proteins vary greatly in their hydrophilicity, all proteins are, at least to some extent, amphipathic and therefore surface active. Characterization of a protein should include examination of the tendency of the protein to adsorb to materials such as glass, plastic tubing, filters used for sterilization of the formulation as well as in-line filters used in iv administration sets, rubber closures, and stainless steel. Adsorption studies are usually done by placing a solution of the protein of interest at a known concentration appropriate for the formulation in a container with minimum tendency to adsorb the protein. The solution is then exposed to the surface of interest under either flowing (for components such as filters or plastic tubing) or static conditions (for components such as rubber closures or plastic iv bags). Any depletion in protein is measured by a suitable assay such as absorption at 278 nm.

No single physical–chemical property of a protein, such as pI or molecular weight, seems to predict the tendency to adsorb to surfaces. However, the driving force seems to be the tendency of hydrophobic regions, which are normally buried in the interior of the protein in its native folded state, to make contact with hydrophobic surfaces. Adsorption behavior is usually quantitated by adsorption isotherms, where the amount of protein adsorbed per unit area of surface is plotted against the protein concentration in solution. The quantity of protein adsorbed is usually measured by depletion of protein from solution, so it is important to use enough sorbent surface to deplete the concentration significantly. Alternatively, the quantity of protein adsorbed to a surface can be measured directly by using radiolabeled protein. Typical isotherms are illustrated in Figure 10.1a for high-affinity adsorption, which is characteristic of conformationally flexible polymers such as albumin. Such isotherms are characterized by an initially very sharp increase in the quantity of protein adsorbed, followed by a continuous increase in the amount adsorbed. This

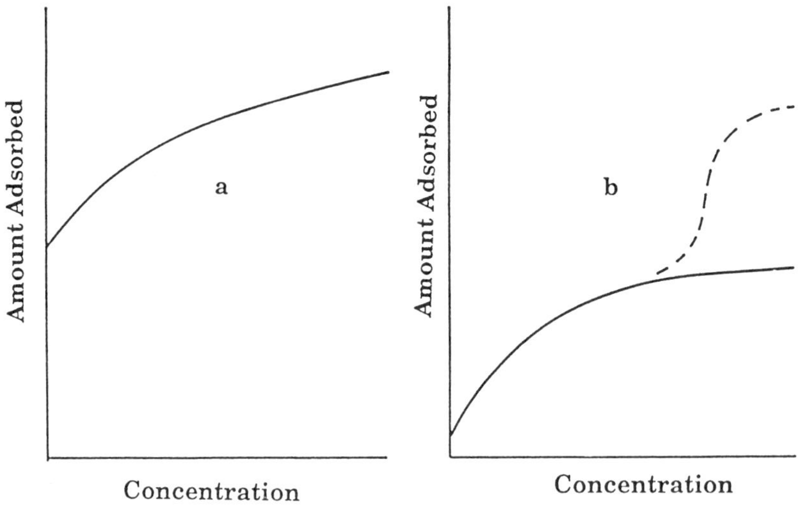

FIGURE 10.1. Representative protein adsorption isotherms for (a) high-affinity adsorption of conformationally flexible protein and (b) adsorption of a conformationally rigid protein.

process is explained by conformational changes in the protein on adsorption, and these changes stabilize the protein in the adsorbed state. Smaller, more conformationally rigid proteins such as lysozyme tend to adsorb in a manner described by Figure 10.1b, where the initial slope of the isotherm is lower and a more well-defined plateau region is observed. The plateau often corresponds to a close-packed monolayer of adsorbed protein. The dashed line in Figure 10.1b illustrates a third type of adsorption behavior, where a "step" in the adsorption isotherm is present at higher protein concentrations. This result indicates different adsorption mechanisms at low and high protein concentration and could be caused by adsorption of a second monolayer, a structural change in the adsorbed layer, or nucleation of a two-dimensional protein "crystal" at higher concentration, leading to decreasing desorbability at higher concentration (*27*). The plateau value for monolayer adsorption usually falls within the range of 0.1 to 0.5 $\mu g/cm^2$, depending on the size of the protein as well as orientation of the protein at the surface.

From the standpoint of formulation development, adsorption of proteins to surfaces can usually be considered irreversible; rinsing the surface with a protein-free solution will not significantly decrease the quantity adsorbed unless a surfactant is present. Sometimes addition of a second, more surface active protein such as human serum albumin to the formulation will cause displacement of the previously adsorbed protein. Hysteresis in the adsorption–desorption isotherm is indicative of a conformational change on adsorption; these conformational changes may result in multipoint contact with the surface, which stabilizes the protein in its adsorbed state and makes it difficult to remove. Adsorption of proteins from solution at a liquid–solid interface was reviewed extensively by Norde (*27*).

Interfacial adsorption, or adsorption at the air–water interface, is frequently a bigger problem in formulation and process development than surface adsorption for at least two reasons. First, the air–liquid interface is continually renewed, either by agitation or by natural convection, so that the likelihood of a large fraction of the protein being exposed to the denaturing environment at the interface is much greater than at a liquid–solid interface. Second, Langmuir film balance experiments have shown much higher average areas per molecule for monolayer adsorption at the air–liquid interface than for adsorption on a solid surface from solution—about 1 m^2/mg versus 0.1 m^2/mg of protein (*27*). This result suggests that unfolding is more complete at the air–liquid interface, increasing the exposure of hydrophobic groups and the probability of further manifestations of physical instability.

Cold denaturation of proteins, which is usually reversible, refers to unfolding of proteins at low temperature. It is caused by the strongly temperature-dependent interaction between water and nonpolar groups of the protein. Hydration of these groups is often favored at low temperature. Consequently, the polypeptide chain, which is tightly packed in its native structure, unfolds at temperatures low enough to favor hydration of nonpolar groups, thus exposing these groups to the aqueous environment. The reversible nature of this phenomenon would be expected to minimize its consequences in pharmaceutical formulation. However, cold denaturation can result in enhanced rates of reactions that could be responsible for protein instability. An accelerated rate of cross-linking of lysozyme was reported at low temperature due to cold denaturation (*28*). Cold denaturation, which is not to be confused with denaturation associated with freezing, was reviewed extensively by Privalov (*29*).

Aggregation of proteins to form ensembles that may be either soluble or insoluble, depending on the extent of aggregation, is one of the most frequent problems encountered in development of protein formulations. Aggregation has been reported for many protein drugs, including fibroblast growth factors (*30–32*), human growth hormone (*33*), interleukin-2 (*34*), β-galactosidase (*35*), immunoglobulins (*36*), and insulin (*37*). Protein aggregates are of particular concern in pharmaceutical formulations for several reasons (*11*):

1. Extraneous particulate matter is a potential hazard in any injectable product, and the U.S. Pharmacopeia places limits on such particulate matter.
2. Precipitated protein can cause clogging of administration devices such as iv catheters.
3. Protein aggregates have been shown to be responsible for an increased incidence of immunological adverse reactions.

Preformulation studies should examine the tendency of a protein to aggregate as well as to characterize the aggregates formed. Also, these studies should elucidate, as much as possible, the aggregation mechanism.

Aggregation is usually, for practical purposes, irreversible (dissolution of the aggregates requires a drastic change in the solvent environment, such as addition of a denaturant such as guanidinium hydrochloride) on a practical time scale. This characteristic distinguishes aggregation from protein association, a process that is generally reversible. Initial characterization should include distinguishing between covalent and noncovalent aggregates. Covalent aggregation is usually the result of intermolecular disulfide bridges formed by reaction between thiol groups that have been exposed to the aqueous environment by unfolding. Sodium dodecyl sulfate- polyacrylamide gel electrophoresis is a common analytical tool for characterizing aggregates with regard to covalent versus noncovalent nature as well as aggregate size of those ensembles small enough to enter the gel. If the sample containing aggregated protein that has been formed by intermolecular disulfide bridges is treated with a reducing agent such as 2-mercaptoethanol, the aggregates will dissociate, and the band arising from aggregates will be reduced in intensity relative to a sample containing no reducing agent. Treatment of a sample with strong denaturants such as 6 M urea or 8 M guanidinium hydrochloride will generally dissolve noncovalent aggregates, but not covalent aggregates.

Characterization of aggregated species and understanding of the mechanism of aggregation may offer useful clues to formulation and processing approaches to minimizing aggregation. The importance of characterizing all forms of aggregated protein, both soluble and insoluble, and establishing a mass balance for all species has been stressed by Shahrokh and co-workers (*32*). Relevant mechanisms include formation of noncovalent aggregates from either partially or fully unfolded proteins in solution via Smoluchowski-type kinetics (*38*), aggregation secondary to adsorption and denaturation at a surface, which is facilitated by a high local concentration of denatured protein, or aggregation secondary to interfacial denaturation, which is facilitated by excessive stirring that results in formation of foam and a high air–water interfacial area.

When aggregation proceeds to the extent that macroscopic assemblies are formed, the process is called precipitation. The nature of this precipitate is usually a floccule about 10–20 μm in diameter that is formed from primary particles of aggregated protein in the 0.1–0.2-μm size range. The size of the floccules is usually limited by shear forces (*39*).

A readily accessible method for quantitation of insoluble aggregates as a method for screening the effects of excipients on aggregate formation is to measure the absorbance of the sample by using a UV-vis spectrophotometer in the wavelength range of about 300 nm to about 400 nm (*40, 41*). Some formulation scientists use an aggregation index, for example $A_{340}/(A_{280} - A_{340}) \times 100$, where A is absorbance, to compensate for the effect of protein concentration. When the most promising excipients are identified, more careful examination of stability is generally done by using a wide variety of analytical techniques such as light scattering and light blockage instrumentation.

Formulation Development

A decision that must be made early in the formulation development process is whether to develop a solution or a freeze-dried formulation. Even though an aqueous solution is preferred from the standpoints of ease of use by the customer and cost of manufacture, a solution formulation is often not feasible because of unacceptable chemical stability, physical stability, or both. Prediction of a shelf life at the anticipated storage temperature, usually either room temperature or 4 °C, based on stability testing at elevated temperatures is much more uncertain for protein drugs than for low-molecular-weight compounds, because the temperature dependence of the relevant routes of degradation must be determined on a case-by-case basis (*8*).

Effect of Excipients on Stability–Solution Formulations

A critical aspect of formulation development is identifying the need for excipients to maintain the desired critical quality attributes of the formulation over the shelf life of the product. The number of excipients that are generally recognized as acceptable for administration by injection is limited, as illustrated by the excipients listed for representative protein formulations in Table 10.1.

Buffers

Buffers are used to maintain the desired pH range for optimum formulation stability, and they should be chosen so that the buffer system has an acceptable buffer capacity in the pH range of the formulation. This range is generally not more than one pH unit from the p*K* of the buffer. The most commonly used buffers are sodium or potassium phosphate and citric acid/sodium citrate. Sodium acetate, histidine, and tris(hydroxyethyl)aminomethane (TRIS, or THAM) have also been used.

Stabilizing Solutes

Solutes such as sugars and some amino acids can significantly enhance the solution stability of proteins by preferential exclusion, where the solute concentration is lower at the protein surface than in the bulk aqueous phase. The mechanisms of stabilization have been studied thoroughly by

TABLE 10.1. Representative Protein Formulations

Generic Name	*Trade Name*	*Physical Form*	*Administration*	*Excipients*
Interferon γ-1b	Actimmune	Aqueous solution	Subcutaneous	Mannitol Sodium succinate Polysorbate 20
Alteplase	Activase	Freeze-dried powder	Subcutaneous	L-Arginine Polysorbate 80 Phosphoric acid
Human growth hormone	Nutropin	Freeze-dried powder	Subcutaneous	Mannitol Glycine Sodium phosphates (monobasic and dibasic)
Human growth hormone	Protropin	Freeze-dried powder	Intramuscular or subcutaneous	Mannitol Sodium phosphates
Dornase α	Pulmozyme	Liquid aerosol	Inhalation	Calcium chloride dihydrate Sodium chloride
Algluconase	Ceredase	Aqueous solution	Intravenous infusion	Human serum albumin Sodium citrate
Interferon β-1b	Betaseron	Freeze-dried powder	Subcutaneous	HSA Dextrose
Imiglucerase	Cerezyme	Freeze-dried powder	Intravenous infusion	Mannitol Sodium citrate Polysorbate 80
Interferon α-2b	Intron A	Freeze-dried powder	Intramuscular Subcutaneous Intralesional	HSA Glycine Methyl- and propylparabens Phosphate
Sargramostim	Leukine	Freeze-dried powder	Intravenous infusion	Mannitol Sucrose TRIS
Filgrastim	Neupogen	Aqueous solution	Intravenous or subcutaneous	Sodium acetate Mannitol Polysorbate 80
Erythropoeitin	Epogen	Aqueous solution	Intramuscular or subcutaneous	Sodium chloride Sodium citrate HSA
Urokinase	Abbokinase	Freeze-dried powder	Intravenous and intracoronary infusion	Mannitol HSA Sodium chloride
Human chorionic gonadotropin	Pregnyl	Freeze-dried powder	Intramuscular injection	Sodium phosphates
Interferon α-2a	Roferon	Aqueous solution	Intramuscular or subcutaneous	Sodium chloride HSA Phenol

NOTES: TRIS is tris(hydroxyethyl)aminomethane. HSA is human serum albumin.

Timasheff and co-workers (*42*) and can be conceptually explained as follows. Even though preferential exclusion of the solute is thermodynamically unfavorable because it increases the entropy of the system, unfolding of the protein is even less favored thermodynamically, because unfolding of the protein would cause further exclusion of solute from the surface of the protein and increase the entropy of the system still further. Thus, the net effect of preferential exclusion is stabilization of protein structure.

Surfactants

Many ionic surfactants, such as sodium dodecyl sulfate (SDS), cause denaturation of proteins even at low concentrations. Nonionic surfactants, commonly polysorbates or polyethers, generally do not cause denaturation and are useful in helping to prevent protein adsorption to surfaces, interfacial denaturation, and subsequent aggregation by preferentially adsorbing to hydrophobic surfaces and to hydrophobic regions of proteins. The polysorbate compounds (Tweens) consist of a sorbitol–polyethylene oxide head group and a hydrocarbon tail. Polyethers typically are either PEG or polyethylene oxide–polypropylene oxide block copolymers (Pluronics). Selection of the best surfactant and the best concentration for a given protein formulation is largely empirical. Table 10.1 lists the types of surfactants and concentrations for representative protein formulations.

Human Serum Albumin

Human serum albumin (HSA) is one of the most common excipients in protein formulations, both solutions and freeze-dried formulations. Whereas not a surfactant in the traditional sense, HSA is surface active, which may partially explain its stabilizing effect. HSA has also been postulated to stabilize proteins by direct binding and by a preferential exclusion mechanism (*11*). Typical concentrations of HSA are shown in Table 10.1. The patent literature shows a wide range of concentrations, from as little as 0.003% to as high as 15% (*9*).

Preservatives

The use of antimicrobial preservatives in injectable formulations is a controversial issue in parenteral science for several reasons. Globalization within the pharmaceutical industry has resulted in emphasis on developing formulations for registration worldwide, yet international regulatory bodies have different views regarding acceptable preservatives and the conditions under which they are used. Even though multiple-dose parenteral products must contain a preservative, the real controversy arises with the use of preservatives in single-dose products. Arguments against the use of preservatives in single-dose products are as follows (*43*):

- well-documented cases of adverse reactions to preservatives,
- their contraindication for certain critical routes of administration, such as intraocular or intrathecal, and
- concern by regulatory agencies over the use of preservatives to cover up poor sterility assurance in processing.

Aqueous protein formulations present a particular dilemma with regard to the use of preservatives, because many protein formulations will promote the growth of accidentally introduced microorganisms. Adequate sterility assurance (a probability of not more than one in 10^6 that an individual vial contains a viable microorganism) dictates that the final product be manufactured by using advanced processing methodology that excludes interaction of people with the product, or a preservative be included in the formulation to assure that any accidentally introduced microorganism cannot grow.

Another source of confusion in the international registration of injectable products is that different compendia have different requirements for preservative efficacy. For example, a three-log reduction in the bacterial challenge level must occur in 1–7 days to meet the British/European pharmacopeial requirement, whereas 1 day will satisfy the U.S. Pharmacopeia preservative challenge test (*43*).

As with most excipients in injectable formulations, the list of acceptable preservatives is small and includes phenolic compounds (phenol, meta-cresol, and benzyl alcohol), chlorobutanol, *p*-aminobenzoic acids (methyl- and propylparabens), and mercurials (thimerosal). Representative protein formulations incorporating preservatives and the levels used are shown in Table 10.1.

Interaction between the preservative and the protein and its possible effect on structure and stability of the protein should not be overlooked. For example, the presence of phenol has been shown to stabilize the B chain of insulin (*44*).

Amino Acids

Amino acids probably have multiple mechanisms of stabilization of proteins. Stabilization by preferential exclusion was discussed, and this mechanism is not restricted to any class of compound. Amino acids also may stabilize by more specific mechanisms. For example, glutamic and aspartic acids have been shown to inhibit the aggregation of insulin, and chelation of zinc by the dicarboxylic amino acids was postulated as the mechanism of stabilization (*11*). Acidic amino acids also were demonstrated to inhibit surface adsorption of proteins, but the mechanism for this process is not clear (*11*). Glycine is the most commonly used amino acid in protein formulations and has concentrations ranging from about 0.01% to about 3%.

EDTA. Because oxidation of some proteins is catalyzed by trace metals, EDTA may be included as a stabilizer to chelate any trace metals introduced as a contaminant. For example, contamination of albumin by trace metals has been shown to correlate with instability with respect to oxidation.

Metals. Metals are essential to stabilizing the native conformation of some proteins. Calcium is a common example and has been shown to have a protective effect on proteases, amylases, fibrinogen, and Factor VIII.

Freeze-Dried Protein Formulations

The instability of a protein drug in aqueous solution often precludes a solution formulation. In this case, the next best option is a freeze-dried, or lyophilized, product. Advantages of freeze-drying over other drying methods are

1. Water is removed at low temperature, thereby avoiding damage to heat-sensitive materials.
2. The dried material generally has a high specific surface area, which promotes rapid reconstitution.
3. When properly done, freeze-dried products are pharmaceutically elegant.

Freeze-drying offers an advantage in manufacturing by allowing filling of vials with a liquid. This characteristic makes fill control more precise and avoids the cross-contamination concerns inherent in powder-filling operations.

The approach to developing a freeze-dried protein formulation requires separately determining the sensitivity of the protein to both freeze-thaw damage and freeze-drying damage. As pointed out by Crowe, Carpenter, and co-workers (*45–47*), the stresses involved in freezing and dehydration are different, and solutes that stabilize against damage by freezing (cryoprotectants) will not necessarily stabilize against damage by drying (lyoprotectant). Protein formulations can also be sensitive to the rate of freezing and thawing (*48, 49*).

The freeze-drying process takes place in three stages. During freezing, the product is frozen to a temperature where it is completely solidified (typically –40 °C to –50 °C). A vacuum is then applied to a level of about 0.1 mm Hg, and heat is applied to the shelves of the freeze-dryer to provide the heat of sublimation of ice (about 670 cal/g). Frozen water is removed by direct sublimation ice crystals by bulk flow of water vapor through the porous partially dried matrix (*primary* drying). For formulations that remain amorphous during the freezing process a significant amount of water remains unfrozen—up to about 20% of the weight

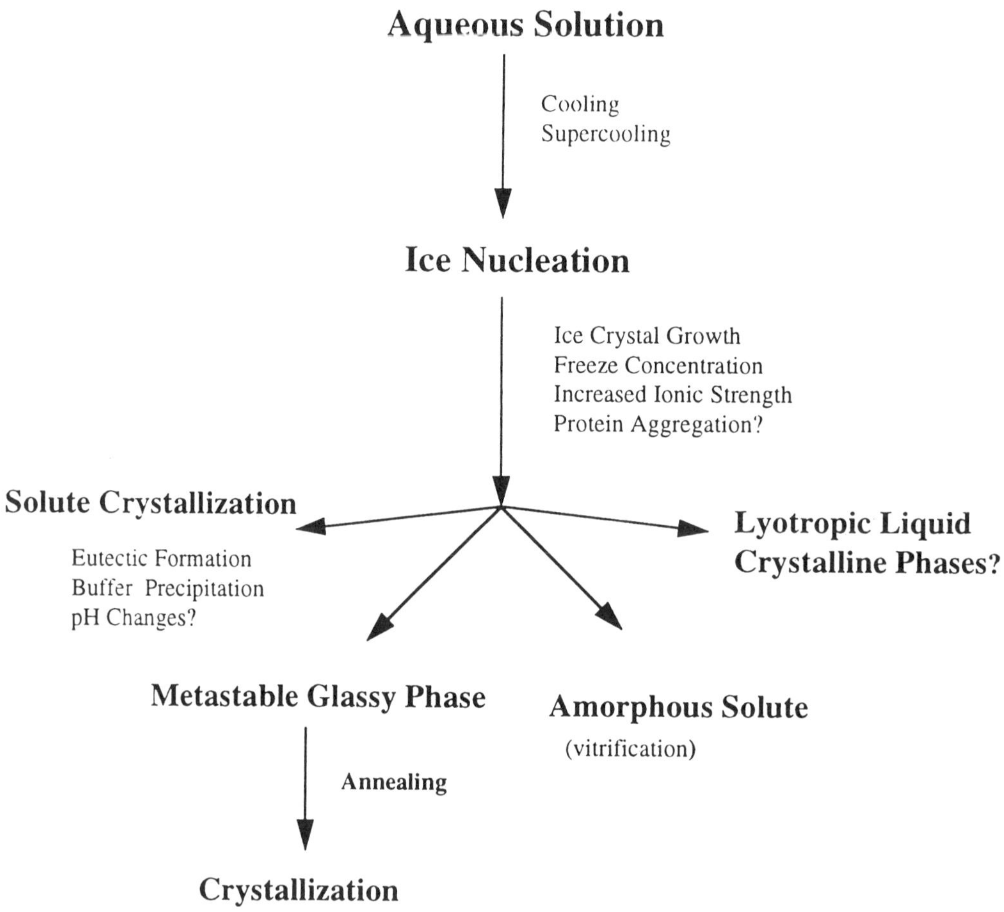

FIGURE 10.2. Schematic diagram of events occurring during freezing.

of the solids. This water must be removed during *secondary* drying.

Freezing of Protein Formulations

The freezing process is illustrated schematically by Figure 10.2. The product is usually frozen by placing filled vials on the shelf of the freeze-dryer, although in special situations where a very rapid freezing rate is required, a liquid nitrogen tunnel may be required. Freezing on the shelf is a relatively slow process. This process, along with the relatively particle-free nature of the solution, causes a considerable amount of supercooling—perhaps as much as 10–12 °C—before ice crystals nucleate. As the temperature is lowered and ice crystals grow, the solution becomes more concentrated. Freeze concentration, especially when combined with cold denaturation, can increase both the rate and extent of protein aggregation (*50*), as well as the rates of some chemical reactions (*51, 52*).

Crystallization of solute from the freeze-concentrated phase causes formation of a *eutectic* phase; that is, an intimate physical mixture of two (or more) crystalline solids that melts as though it were a pure compound. Inorganic salts such as sodium and potassium chloride, sodium and potassium phosphates, and organic compounds such as neutral glycine crystallize readily to form eutectic mixtures. Preferential crystallization of one buffer salt during the freezing process can cause significant pH changes in the freeze-concentrated solution, most notably for the phosphate buffer system, which has been investigated by van den Berg and co-workers (*53, 54*). Crystallization of the dibasic salt causes a decrease in pH, whereas crystallization of the monobasic salt causes an increase in pH. Crystallization, and the accompanying pH change, is also affected by the presence of other salts such as sodium and potassium chloride. Even though the effects of other solutes have not been investigated in detail, the formulation scientist should be aware that the presence of other solutes as well as supercooling effects (and subsequent rapid freezing) tend to inhibit crystallization. Nevertheless, the potential for changes in pH with freezing, particularly for proteins for which physical or chemical stabilities are highly pH dependent, should not be overlooked. Experimentally, changes in pH during freezing are generally measured directly by using a pH electrode that is designed to withstand the mechanical stresses associated with freezing.

The eutectic melting temperature of a frozen system is important with respect to its freeze-drying properties, because exceeding the eutectic melting temperature during freeze-drying would result in foaming, collapse of the partially dried matrix, and loss of a pharmaceutically acceptable dried product. Thus, for a formulation in which the major component crystallizes from the freeze concentrate, the eutectic temperature represents the maximum allowable product temperature during the primary drying process.

Solutes that do not crystallize during freezing form a glassy, or vitreous, phase at sufficiently low temperature. As the "frozen" system is cooled, the freeze-concentrated phase will undergo a glass transition, where the viscosity of the material increases by several orders of magnitude over a temperature range of a few degrees. Below the glass transition temperature (T_g), the amorphous freeze concentrate is mechanically a solid, and freeze-drying results in retention of the microstructure established by freezing. Exceeding T_g of the maximally freeze-concentrated solute (called T_g') can result in viscous flow of solute after the supporting ice matrix sublimes. This viscous flow results in collapse of the solute and loss of the microstructure established by freezing. T_g' is therefore a critical attribute of a formulation in which the major component does not crystallize during freezing, because it represents the maximum allowable product temperature during primary drying. Glass transition temperatures of freeze-concentrated systems tend to be lower than eutectic temperatures of organic compounds commonly used in freeze-drying. In the event of partial crystallization of a major component of a freeze-dried formulation, the freeze-drying characteristics are usually determined by the amorphous phase.

Compounds such as mannitol represent a special type of freezing behavior, because they tend to first form a glassy phase that is metastable; that is, the glass phase crystallizes when heated subsequent to the initial freezing process. This process is accelerated by heating above the T_g' of the freeze-concentrated solute, where increased molecular mobility allows the solute to attain a more stable state. This process is the basis for *annealing* in freeze-drying, where the product is first frozen, then heated to some temperature below its melting point to promote crystallization of one or more solutes. The product is then cooled again, and the drying process is initiated.

Freezing rate may be a critical processing variable with respect to protein aggregation. Fast freezing is generally thought to minimize aggregate formation, because the time during which the protein is exposed to the freeze-concentrated environment is minimized. However, this result must be determined on a case-by-case basis. Eckhardt, Hsu, and co-workers (*55, 56*) reported an inverse relationship between freezing rate and insoluble aggregate formation in human growth hormone. A

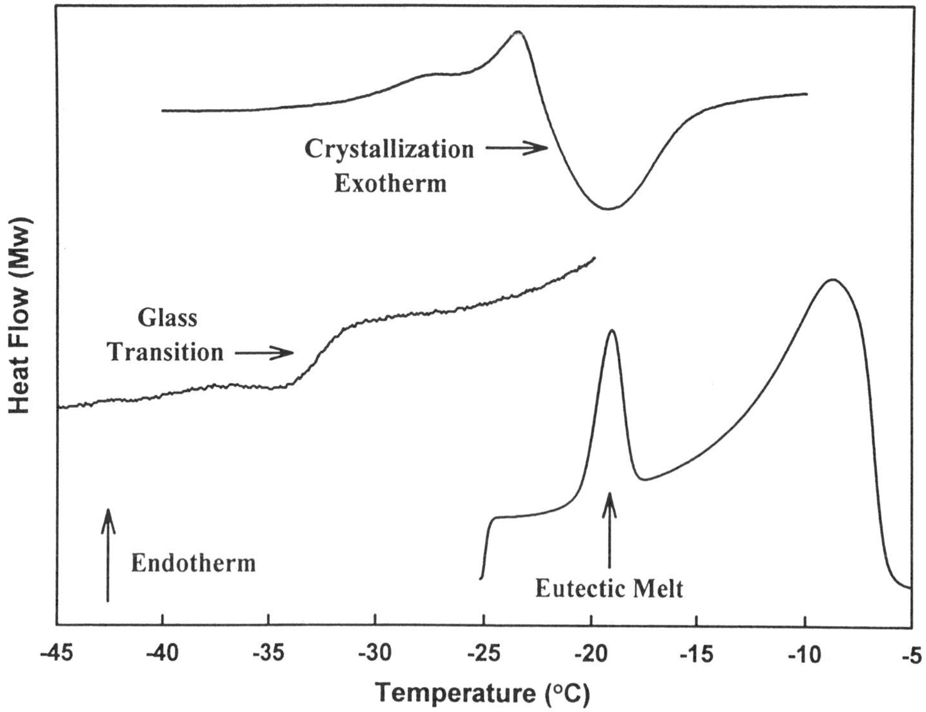

FIGURE 10.3. Representative events in thermal analysis of frozen solutions.

mechanism involving interfacial denaturation at the solid–vapor interface has been proposed, where the higher specific surface area created by rapid freezing promotes aggregation.

Characterization of Formulations

The driving force for freeze-drying is the vapor pressure of ice, which increases by about 75% for every 5 °C increase in temperature. This result makes the rate of freeze-drying very temperature dependent. The purpose of characterizing the formulation is to determine the maximum allowable product temperature during primary drying, so that the process can be carried out accordingly. This temperature is either a eutectic melting temperature (for a crystalline solute) or a glass transition temperature (for an amorphous solute). A second and equally important reason for characterizing the formulation is to determine the physical state of the solutes after freeze-drying. For traditional drug compounds, crystallization of the drug is desirable, because rates of degradation in the solid state, even if the solid is very dry, may be considerably faster for amorphous compounds (*57*).

The most common method for characterizing a formulation intended to be freeze-dried is thermal analysis, usually differential scanning calorimetry (DSC). Typical thermal transitions of interest in characterization of formulations are shown in Figure 10.3, which illustrates eutectic melting (neutral glycine), a glass transition (lactose), and a metastable glassy phase that crystallizes with heating after the initial freeze (mannitol). Interpretation of the thermograms is often not as easy as illustrated in Figure 10.3. Eutectic temperatures, when they occur, are often very close to the melting temperature of ice and may not be resolved from the ice melting endotherm. Glass transitions often involve very small heat capacity changes and may not be detected by the instrumentation. This characteristic is often a problem in formulations containing a relatively high concentration of protein. Experimentally, it is important to record the thermogram of the formulation at multiple conditions to maximize the information that can be obtained. These conditions include freezing rate, heating rate, solute concentration, and the effect of annealing (*58*).

Another method of characterizing freezing and freeze-drying behavior of a formulation is freeze-dry microscopy (*59*), where a freeze-drying stage allows direct observation of the sample during both freezing and freeze-drying. Freeze-dry microscopy is particularly useful for observation of collapse phenomena, as well as for characterization of the physical state of the solute after freeze-drying by observation of birefringence under polarized light.

Excipients in Freeze-Dried Formulations

All of the categories of excipients listed previously, with the usual exception of preservatives, are also used in freeze-dried formulations. Other excipients are used for these formulations as well. If the dose of protein to be administered is very small, as with a protein such as Factor VIII or interferon, the dried product must be dispersed in an inert, nontoxic matrix called a *bulking agent* to provide a pharmaceutically elegant dosage form and to avoid ejection of the drug from the vial during the freeze-dry process. Common bulking agents for protein formulations are glycine, lactose, mannitol, HSA, and combinations of these. A bulking agent may have multiple functions in that it may also be a cryo- or lyoprotectant. Even though many compounds have cryoprotective activity (largely explained by the preferential exclusion mechanism), disaccharides such as sucrose, lactose, maltose, and trehalose are the class of compounds that are most consistently effective in providing both cryoprotection and lyoprotection. A different mechanism than preferential exclusion must be operative to explain the protective effect of excipients during freeze-drying. Two theories have been proposed to explain the lyoprotective effect. One is the vitrification hypothesis (*60*), which states that solutes that form a stable glassy phase are the best protective solutes (i.e., solutes that form glasses with a high T_g) because formation of a glassy phase (vitrification) is necessary and sufficient to render the protein stable. The vitrification hypothesis assumes that all molecular motion that is relevant to stability ceases below T_g. Even though vitrification is considered necessary for stability, several investigations support the idea that vitrification alone is not sufficient. Pikal and co-workers (*21*) observed that dextran, which remains amorphous during freeze-drying and has a relatively high T_g, is an ineffective lyoprotectant for human growth hormone. Tanaka and co-workers (*61*), studying catalase, observed that the effectiveness of saccharides is inversely related to their molecular weight. This result is not what would be expected based on the vitrification hypothesis, because increasing the molecular weight of the solute should result in a higher T_g (a more stable glass). In addition, Tanaka and co-workers noted that the weight ratio of saccharide to protein, not the concentration of saccharide per se, correlates with protective efficacy. This result supports the idea that some type of direct interaction between excipient and protein is involved. The vitrification hypothesis supposes no such direct interaction, because glass formation is a bulk property of the solute.

The theory of lyoprotection that appears to be best supported by the available experimental data is the "water replacement" hypothesis (reviewed in reference 10), which states that lyoprotective solutes work by acting as a replacement for water in the hydration shell of the protein as water is removed. Carpenter, Crowe, and co-workers have used Fourier transform IR (FTIR) spectroscopy to study interactions between proteins and stabilizing solutes, and they have demonstrated hydrogen bonding between protein and stabilizing solutes (*62*). Such bonding appears to be necessary for lyoprotection. This result has been demonstrated by both protein-induced spectral changes in disaccharides and by the influence of disaccharides on the spectrum of proteins such as lysozyme. In the case of the influence of disaccharides, the solute preserves the spectrum of the fully hydrated protein in the dried state, whereas pronounced changes in the amide I and amide II regions are observed when the protein is dried with no protective solute. The use of FTIR for measurement of secondary structure of proteins was reviewed critically by Surewicz and co-workers (*63*).

The most common bulking agents for freeze-dried protein formulations are mannitol, glycine, and mixtures of the two. These compounds may have a multiple purpose in protein formulations as a bulking agent and as a stabilizing solute. Both compounds tend to crystallize during freezing and freeze-drying when they are the only solute. However, their crystallization behavior is more complex in multicomponent systems. The degree of crystallization is important not only from the standpoint of freeze-drying behavior, but also with respect to lyoprotective effect. Izutsu and co-workers (*64*) reported that three model proteins were protected by mannitol in a manner that was dependent on the concentration of amorphous mannitol, but that the stabilizing effect was decreased by crystallization of mannitol. Pikal and co-workers (*21*) reported that a combination of mannitol and glycine, where the glycine remains amorphous, provides the best protection for human growth hormone against decomposition and aggregation.

Stability of Freeze-Dried Formulations

In addition to the normal assays for potency loss and physical stability of freeze-dried protein formulations, it is important to characterize the physical state of the formulation both initially and periodically during the course of a stability study. Excipient crystallization during storage can lead not only to loss of activity resulting from a decreased effective concentration of lyoprotectant resulting from the crystallization process, but also from a redistribution of residual water within the dried matrix. The increase in water content of the amorphous phase, which constitutes the microenvironment of the active compound, can cause increased rates of reactions either from increased water activity or from the plasticizing effect of water, which increases molecular mobility (*65, 66*).

Stability studies of freeze-dried protein formulations should also include periodic assay of the total amount of residual water in the freeze-dried cake, because small but significant increases in residual moisture due to transfer of water vapor from the stopper to the dried solid are also common (*67*).

Processing Considerations

Processing of protein formulations can require particular care to avoid adverse effects on the final product. Preparation of the bulk solution may require that overly vigorous agitation be avoided, because this process can cause foaming and a subsequent drastic increase in air–liquid interface that can promote interfacial denaturation. For oxygen sensitive proteins, special precautions may be needed to carry out processing in an oxygen-free environment.

Sterile filtration generally requires special attention when dealing with proteins, primarily because of the tendency of filters to adsorb protein. Pitt (*68*) demonstrated significant differences between different membrane filters with regard to protein adsorption by using bovine serum albumin and immunoglobulin G (IgG) as model proteins (*68*). Process development should include examination of several different types of filters to determine the degree of protein adsorption. Filtration can also cause changes in protein conformation, perhaps after initial adsorption of protein to the filter surface. Truskey and co-workers (*69*) demonstrated that changes in the circular dichroism spectra of bovine alkaline phosphatase, bovine insulin, and human IgG were independent of shear rate during processing, but the changes correlated with the degree of adsorption to the filter.

Another aspect of pharmaceutical processing that is not a concern in the processing of low-molecular-weight drugs, but is often mentioned in connection with protein formulations, is shear denaturation. Charm and Wong (*70*) reported that loss of protein activity is related to the product of the average shear rate and the time of exposure to this shear. Measurable loss of protein activity was found only at very high shear rate-time products ($>10^5$). This result is perhaps two orders of magnitude higher than the highest shear rate–time product found in processing of protein formulations, because flow through a typical sterilizing membrane filter would have a shear rate–time product on the order of 10^3. Whereas more studies like the one reported by Charm and Wong are needed using different proteins, the available data suggest that shear denaturation is not likely to be significant in typical unit operations involved in manufacture of injectable protein formulations. It is likely that interfacial denaturation caused by, for example, vigorous stirring in a mixing tank, is often mistaken for shear denaturation.

Conclusion

The development of pharmaceutically acceptable formulations of therapeutic proteins is not trivial but is worthy of the best efforts of pharmaceutical scientists. A detailed understanding of routes and rates of chemical degradation as well as how these degradation pathways are affected by formulation variables is essential. Assessment of the physical stability of a protein drug is equally important and requires careful examination of the adsorption and aggregation characteristics of proteins. The presence of other compounds in the formulation is often essential to maintaining the integrity of the protein over a suitable shelf life. Understanding concepts such as preferential exclusion provides a useful framework for systematic evaluation of potential stabilizing solutes. Because many proteins must be supplied as a dried solid to be reconstituted before use, an understanding of the physical chemistry of the freeze-drying process is essential to minimizing the amount of empiricism associated with formulation and process development.

References

1. Center for Biologics Evaluation and Research, Information System, U.S. Food and Drug Administration, Bethesda, MD, 1995.
2. Kelley, W. S. *Biotechnology* **1996**, *14*, 28–31.
3. Jaenicke, R. *Biochemistry* **1991**, *30*, 3147–3161.
4. Dill, K. A. *Biochemistry* **1990**, *29*, 211–233.
5. Pusey, M.; Gernert, K. *J. Cryst. Growth* **1988**, *88*, 419–424.
6. Middaugh, C. R.; Volkin, D. B. In *Stability of Protein Pharmaceuticals. Part A, Chemical and Physical Pathways of Protein Degradation;* Ahern, T. J.; Manning, M. C., Eds.; Pharmaceutical Biotechnology Series; Plenum: New York, 1992; Vol. 2, p 112.
7. Trissel, L. A. In *Handbook on Injectable Drugs*, 8th ed.; American Society of Hospital Pharmacists: Bethesda, MD, 1994.
8. Manning, M. C.; Patel, K.; Borchardt, R. T. *Pharm. Res.* **1989**, *11*, 903–918.
9. Cleland, J. L.; Powell, M. F.; Shire, S. J. *Crit. Rev. Ther. Drug Carrier Syst.* **1993**, *10*, 307–377.
10. Arakawa, A.; Prestrelski, S. J.; Kenney, W. C.; Carpenter, J. F. *Adv. Drug Delivery Rev.* **1993**, *10*, 1–28.
11. Wang, J.; Hanson, M. A. *J. Parenter. Sci. Technol.* **1988**, *42*, 2S.
12. *Stability of Protein Pharmaceuticals. Part A, Chemical and Physical Pathways of Protein Degradation;* Ahern, T. J.; Manning, M. C., Eds.; Pharmaceutical Biotechnology Series; Plenum: New York, 1992; Vol. 2.
13. Brange, J.; Langkjaer, L.; Havelund, S.; Pand Volund, A. *Pharm. Res.* **1992**, *9*, 715–726.
14. Brange, J.; Langkjaer, L. *Acta Pharm. Nord.* **1992**, *4*, 149–158.
15. Gu, L. C.; Erdos, E. A.; Chiang, H. S.; Calderwood, T.; Faster, L. C. *Pharm. Res.* **1991**, *8*, 485–490.
16. Kroon, D. J.; Baldwin-Ferro, A.; Lulan, D. *Pharm. Res.* **1992**, *9*, 1386–1393.
17. Bhatt, N. P.; Patel, K.; Borchardt, R. T. *Pharm. Res.* **1990**, *7*, 593–599.

18. Strickley, R. G.; Brandl, M.; Chan, K. W.; Straub, K.; Gu, L. *Pharm. Res.* **1990,** *7,* 530–536.
19. Moss, J.; Bundgaard, H. *Pharm. Res.* **1990,** 7, 751–755.
20. Patel, K.; Borchardt, R. T. *J. Parenter. Sci. Technol.* **1990,** *44,* 300–301.
21. Pikal, M. J.; Dellerman, K. M.; Roy, M. L.; Riggin, R. M. *Pharm. Res.* **1991,** *8,* 427–436.
22. Kosen, P. A. In *Stability of Protein Pharmaceuticals. Part A, Chemical and Physical Pathways of Protein Degradation;* Ahern, T. J.; Manning, M. C., Eds.; Pharmaceutical Biotechnology Series; Plenum: New York, 1992; Vol. 2, p 40.
23. Quinlan, G. J.; Coudray, C.; Hubbard, A.; Gutteridge, J. M. *J. Pharm. Sci.* **1992,** *81,* 611–614.
24. Motto, M. G.; Hamburg, P. F.; Graden, D. A.; Shaw, C. J.; Cotter, M. L. *J. Pharm. Sci.* **1991,** *80,* 419–423.
25. Schrier, J. A.; Kenley, R. A.; Williams, R.; Corcoran, R. J.; Kim, Y.; Northey, R. P.; D'Augusta, D.; Huberty, M. *Pharm. Res.* **1993,** *10,* 933–944.
26. Chen, T. *Drug Dev.* **1992,** *18,* 1311–1354.
27. Norde, W. *Adv. Colloid Interface Sci.* **1986,** *25,* 267–340.
28. Sotelo, C. G.; Kurosky, A. *J. Agric. Food Chem.* **1994,** *42,* 1845–1849.
29. Privalov, P. L. *Biochem. Mol. Biol.* **1990,** *25,* 281–305.
30. Tsai, P. K.; Volkin, D. B.; Dabora, J. M.; Thompson, K. C.; Bruner, M. W.; Gress, J. O.; Matuszewska, B.; Keogan, M.; Bondi, J. V.; Middaugh, C. R. *Pharm. Res.* **1993,** *10,* 649–659.
31. Chen, B.; Arakawa, T.; Hsu, E.; Narhi, L.; Tressel, T. J.; Chien, S. L. *J. Pharm. Sci.* **1994,** *83,* 1657–1661.
32. Shahrokh, Z.; Stratton, P. R.; Eberlein, G. A.; Wang, Y. J. *J. Pharm. Sci.* **1994,** *83,* 1645–1650.
33. Brems, D. N.; Brown, P. L.; Becker, G. W. *J. Biol. Chem.* **1990,** *265,* 5504–5511.
34. Wang, P. L.; Johnston, T. P. *Int. J. Pharm.* **1993,** *96,* 41–49.
35. Yoshioka, S.; Aso, Y.; Izutsu, K.; Terao, T. *Pharm. Res.* **1993,** *10,* 687–691.
36. Levine, H. L.; Ransohoff, T. C.; Kawahata, R. T.; McGregor, W. C. *J. Parenter. Sci. Technol.* **1991,** *45,* 160–165.
37. Costantino, H. R.; Langer, R.; Kilbanov, A. M. *Pharm. Res.* **1994,** *11,* 21–29.
38. DeYoung, L. R.; Fink, A. L.; Dill, K. A. *Acc. Chem. Res.* **1993,** *26,* 614–620.
39. Glatz, C. E. In *Stability of Protein Pharmaceuticals. Part A, Chemical and Physical Pathways of Protein Degradation;* Ahern, T. J.; Manning, M. C., Eds.; Pharmaceutical Biotechnology Series; Plenum: New York, 1992; Vol. 2, p 135.
40. Hsu, C. C.; Ward, C. A.; Pearlman, R.; Nguyen, H. M.; Yeung, D. A.; Curley, J. G. *Dev. Biol. Stand.* **1992,** *74,* 255–271.
41. Eberlein, G. A.; Stratton, P. R.; Wang, Y. J. *PDA J. Pharm. Sci. Technol.* **1994,** *48,* 224–230.
42. Arakawa, T.; Timasheff, S. N. *Biophys. J.* **1985,** *47,* 411–415.
43. Akers, M. J.; Nail, S. L. *Pharm. Technol.* **1994,** 26–36.
44. Hardaway, L. A.; Brems, D. N.; Beals, J.; MacKenzie, N. E. *Biochim. Biophys. Acta* **1994,** *1208,* 101–103.
45. Crowe, J. H.; Carpenter, J. F.; Crowe, L. M.; Anchordoguy, T. J. *Cryobiology* **1990,** *27,* 219–231.
46. Carpenter, J. F.; Crowe, J. H. *Cryobiology* **1988,** *25,* 244–255.
47. Carpenter, J. F.; Crowe, J. H. *Cryobiology* **1988,** *25,* 459–470.
48. van den Berg, L. *Cryobiology* **1966,** *3,* 236–242.
49. Soliman, F. S.; van den Berg, L. *Cryobiology* **1971,** *8,* 73–78.
50. Hansson, U. B. *Acta Chem. Scand.* **1968,** *22,* 490–494.
51. Hatley, R.; Franks, F. *Biophys. Chem.* **1986,** 24, 41–46.
52. Hatley, R.; Franks, F. *Biophys. Chem.* **1986,** *24,* 187–192.
53. van den Berg, L.; Rose, D. *Arch. Biochem. Biophys.* **1959,** *84,* 305–315.
54. van den Berg, *Arch. Biochem. Biophys.* **1959,** *84,* 305–15.
55. Eckhardt, B. M.; Oeswein, T. W.; Bewley, T. A. *Pharm. Res.* **1991,** *81,* 1360–1364.
56. Hsu, C. C.; Nguyen, H. M.; Yeung, D. A.; Brooks, D. A.; Koe, G. S.; Bewley, T. A.; Pearlman, R. *Pharm. Res.* **1995,** *12,* 69–78.
57. Pikal, M. J.; Lukes, A. L.; Lang, J. E. *J. Pharm. Sci.* **1977,** *66,* 1312–1316.
58. Her, L. M.; Nail, S. L. *Pharm. Res.* **1994,** *11,* 54–59.
59. Nail, S. L.; Her, L. M.; Proffitt, C.; Nail, L. L. *Pharm. Res.* **1994,** *11,* 1098–1100.
60. Franks, F. *Cryo-Lett.* **1990,** *11,* 93–110.
61. Tanaka, R.; Tanaka, T.; Miyajima, R. *Chem. Pharm. Bull.* **1991,** *39,* 1091–1094.

62. Carpenter, J. F.; Crowe, J. H. *Biochemistry,* **1989,** *28,* 3916–3922.
63. Surewicz, W. K.; Mantsch, H. H.; Chapman, D. *Biochemistry* **1993,** *32,* 389–394.
64. Izutsu, K.; Yoshioka, S.; Terao, T. *Chem. Pharm. Bull.* **1994,** *42,* 5–8.
65. Ahlneck, C.; Zografi, G. *Int. J. Pharm.* **1990,** *62,* 87–95.
66. Herman, B. D.; Sinclair, B. D.; Milton, N.; Nail, S. L. *Pharm. Res.* **1994,** *11,* 1467–1473.
67. Pikal, M. J.; Shah, S. *Dev. Biol. Stand.* **1992,** *74,* 167–179.
68. Pitt, A. *J. Parenter. Sci. Technol.* **1987,** *41,* 110.
69. Truskey, G. A.; Gabler, R.; DiLeo, A.; Manton, T. *J. Parenter. Sci. Technol.* **1987,** *41,* 180–193.
70. Charm, S.; Wong, B. *Biotechnol. Bioeng.* *12,* 1103–1109.
71. *Physician's Desk Reference;* Arky, R., Medical Consultant; Medical Economics: Montvale, NJ, 1996.

11

Stability of Peptides and Proteins

Christian Schöneich, Michael J. Hageman, and Ronald T. Borchardt

Proteins and peptides exhibit both physical and chemical instability. This chapter focuses on the hydrolytic and oxidative pathways of chemical degradation and the aggregation and precipitation pathways of physical degradation. With respect to hydrolytic pathways, the endogenous factors (e.g., primary sequence and secondary and tertiary structures) and exogenous factors (e.g., pH, temperature, moisture, and excipients) that influence the rates and pathways of degradation of asparagine and aspartic acid residues in proteins are reviewed. For oxidation reactions, this chapter reviews the amino acid residues labile to oxidation, the influence of protein sequence and structure, the nature of the reactive oxygen species, autoxidations, and mechanisms of metal-catalyzed oxidations. With respect to physical instability, the role of protein unfolding and "molten-globule" intermediates on irreversible aggregation, gelation, and precipitation are discussed in light of the stresses most commonly encountered during storage, processing, and administration or delivery.

Although the pharmacological properties of proteins and peptides have long been recognized, the realization of their use as therapeutic agents came only recently with the advent of biotechnology and modern peptide synthetic chemistry. As a result, the number of recombinant proteins and synthetic peptides with potential clinical applications is ever increasing. However, these molecules represent a significant challenge to the pharmaceutical scientists who are responsible for developing stable and efficacious formulations or delivery systems of these molecules (*1–5*).

Unlike small organic molecules, peptides and proteins possess not only a primary structure (sequence of amino acids) but also, in the case of proteins, higher order structures (e.g., secondary, tertiary, and quaternary), which are required for biological activity. Instability of small synthetic peptides involves primarily chemical pathways of degradation. With proteins, the degradation pathways include chemical transformations, which are similar to those observed in peptides, as well as physical instability (*6–8*). Chemical instability can be defined as any process involving modification of the protein or peptide by bond formation or cleavage, yielding a new chemical entity. In contrast, physical instability does not necessarily involve covalent modifications of the protein.

Copyright © 1997 American Chemical Society

Instead, it generally refers to changes in the higher order structure (secondary and above). This physical instability results in denaturation, adsorption to surfaces, aggregation, and precipitation.

This chapter is not intended to be a comprehensive review of the physical and chemical instability of proteins and peptides. For a comprehensive treatment of these subjects, the reader is directed to several recent reviews (*5, 6*) and books (*7, 8*). Instead, this chapter will focus on the most prevalent types of chemical instability (hydrolysis and oxidation) and physical instability (irreversible aggregation, gelation, and precipitation) that occur during processing and storage of the bulk drug and during preparation and storage of the delivery system.

Chemical Instability

Hydrolytic Reactions

The most commonly observed hydrolytic reactions (spontaneous, nonenzymatic) that occur in proteins and peptides involve asparagine residues (e.g., –XAsnY– and –XAsn–OH) and aspartic acid residues (e.g., –XAspY–) (*9*). Less commonly observed reactions in proteins involve glutamine residues (e.g., H_2N–GlnY–, –XGlnY–) (*8*) and N-terminal sequences that have proline (Pro) in the penultimate position (H_2N–XProY–). Each of these reactions is discussed in detail.

Deamidation of –XAsnH– Sequences

Deamidation of Asn residues is perhaps the most commonly reported pathway of chemical degradation of proteins and peptides (*5, 6, 9*). Under neutral or basic conditions in solution, deamidation of Asn residues (–XAsnY–) proceeds through a five-membered cyclic imide intermediate formed by intramolecular attack of the following peptide bond nitrogen on the Asn side-chain carbonyl-carbon atom in a reaction that releases ammonia (Scheme 11.1). The resulting cyclic imide intermediate is prone to racemization at the α-carbon. The cyclic imide then hydrolyzes rapidly by attack of water to yield a mixture of peptides in which the polypeptide backbone is attached via an α-carboxyl linkage (D,L-Asp) or is attached via a β-carboxyl linkage (D,L-isoAsp) (Scheme 11.1) (*10, 11*). Gln residues (–XGlnY–) can undergo a similar deamidation reaction; however, the rate of the reaction is substantially slower than that observed for Asn residues (*9, 12*). Under acidic conditions, Asn residues (–XAsnY–) can also undergo deamidation, but this process involves a direct attack of water on the Asn side-chain carbonyl-carbon atom, releasing ammonia and generating Asp. Alternatively, the Asn residue can degrade (minor reaction) under acidic conditions via the same mechanism described previously to form the cyclic imide, which is stable under acidic conditions (Scheme 11.1) (*11*). Under acidic conditions, degradation of Asn residues does not lead to the formation of isoAsp.

The pathways of degradation of Asn residues shown in Scheme 11.1 obviously can generate multiple degradation products that are structurally similar (e.g., Asp vs. isoAsp or L-Asp vs. D-Asp)

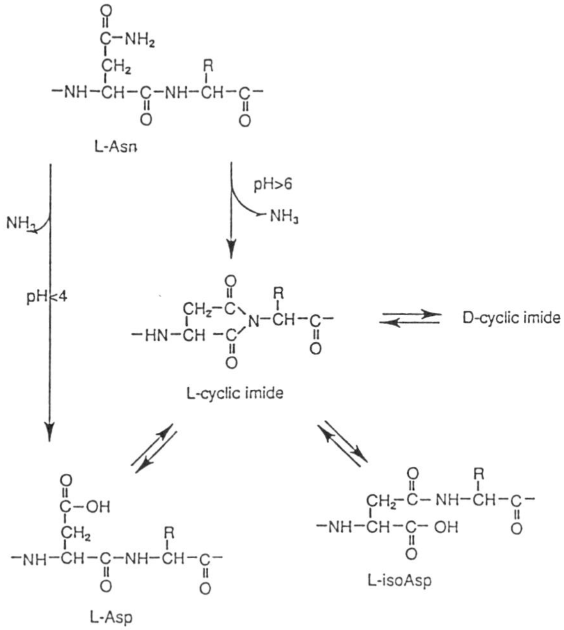

SCHEME 11.1. Pathways for Asn degradation.

and thus difficult to separate and detect (*13–18*). In addition, deamidation of Asn residues can lead to changes in protein conformation [e.g., ribonuclease (*19, 20*) and human growth hormone releasing factor (*21*)], ultimately resulting in self-association and aggregation [triosephosphate isomerase (*22*), adrenocorticotropin (ACTH) (*23*), and interleukin 1-β (*24*)]. However, deamidation of a protein does not always result in loss of biological activity [e.g., hirudin (*25*)].

Many factors, both endogenous (e.g., primary sequence and secondary, tertiary, and quaternary structures) and exogenous (e.g., pH, buffer concentration, buffer species, and excipient) can influence both the rate of deamidation and the distribution of the products. One of the major factors that influences the rate of deamidation is the nature of the residue (Y) on the C-terminal side of the Asn residue (e.g., –XAsnY–) (*10, 12, 26, 27*). For example, with the series of hexapeptides Val–Tyr–X–Asn–Y–Ala, the half-lives of deamidation in 0.1 M phosphate buffer (pH 7.4) at 37 °C were reported to be as follows: X = Pro, Y = Gly, 1.1–1.9 days; X = Pro, Y = Ser, 5.6–8.0 days; X = Pro, Y = Ala, 20.2 days; X = Pro, Y = Leu, 70 days; and X = Pro, Y = Val, 106 days (*10, 26, 27*). In contrast, changes in the amino acid residue (X) on the N-terminal side of the Asn residue (e.g., –XAsnY–) did not have a significant effect on the rate of deamidation (*27*). Another endogenous factor that affects the rate of deamidation of Asn residues (–XAsnY–) is the higher order structure of the protein. Clarke (*28*) has shown that Asn residues in native proteins generally exist in conformations where the peptide bond nitrogen atom cannot approach the side-chain carbonyl carbon without large-scale conformational changes. Therefore, certain proteins will not undergo deamidation unless they have been denatured and sufficient conformational flexibility exists to allow for the formation of the cyclic imide. A similar effect was recently observed by Stevenson et al. (*29*) using a series of synthetic growth-hormone-releasing factors.

Exogenous factors can also influence the rate of deamidation of Asn residues as well as the distribution of products. Perhaps the major exogenous factor that influences the deamidation of Asn residues is pH. Using a model hexapeptide whose primary sequence coincides with residues 22–27 of ACTH (Val–Tyr–Pro–Asn–Gly–Ala), Patel and Borchardt (*11*) have shown that Asn deamidation is pH dependent and follows pseudo-first-order kinetics. Under acidic conditions (pH 1–2) at 37 °C, this hexapeptide degrades to produce the corresponding Asp–hexapeptide via direct hydrolyses (Scheme 11.1). This Asp–hexapeptide, in turn, undergoes Asp–Gly amide bond hydrolysis (Scheme 11.2). The formation of the cyclic imide (Asu–hexapeptide) was also detected at acidic pH, although its appearance was much slower than the direct hydrolysis reaction, constituting only 10% of the total product. The Asu–hexapeptide was stable under acidic conditions and thus no isoAsp–hexapeptide was observed. Under neutral and basic conditions (pH 5–12), the Asn–hexapeptide degraded via the cyclic imide to form the Asp– and isoAsp–hexapeptides in a ratio of 4 to 1 (Scheme 11.1). The rate constants of deamidation in neutral and alkaline buffers were shown to be much faster than the rates observed in acidic buffers, and maximum stability was observed in the range of pH 3.0–5.0. The pH rate profile showed a unit negative slope in the acidic region (pH 1–2), indicating specific acid catalysis, whereas in the neutral and alkaline regions (pH 5–12), the rate was catalyzed by hydroxide ion.

Other exogenous factors reported to influence the rate of deamidation include temperature, buffer species, buffer concentration, and ionic strength. For example, Patel and Borchardt (*11*) showed that the deamidation of the hexapeptide Val–Tyr–Pro–Asn–Gly–Ala in solution obeyed the Arrhenius relationship within the temperature ranges studied (e.g., pH 7.5, 25 °C to 70 °C). Patel and Borchardt (*11*) also reported buffer catalysis for deamidation of this hexapeptide in the pH range 7–11. Recently, the deamidation of the tripeptide Boc–Asn–Gly–Gly was studied in the presence of different salts, buffer ions, and organic solvents (*30*). The addition of aprotic NaCl, $MgSO_4$, and Na_2SO_4 had little effect on the rate of

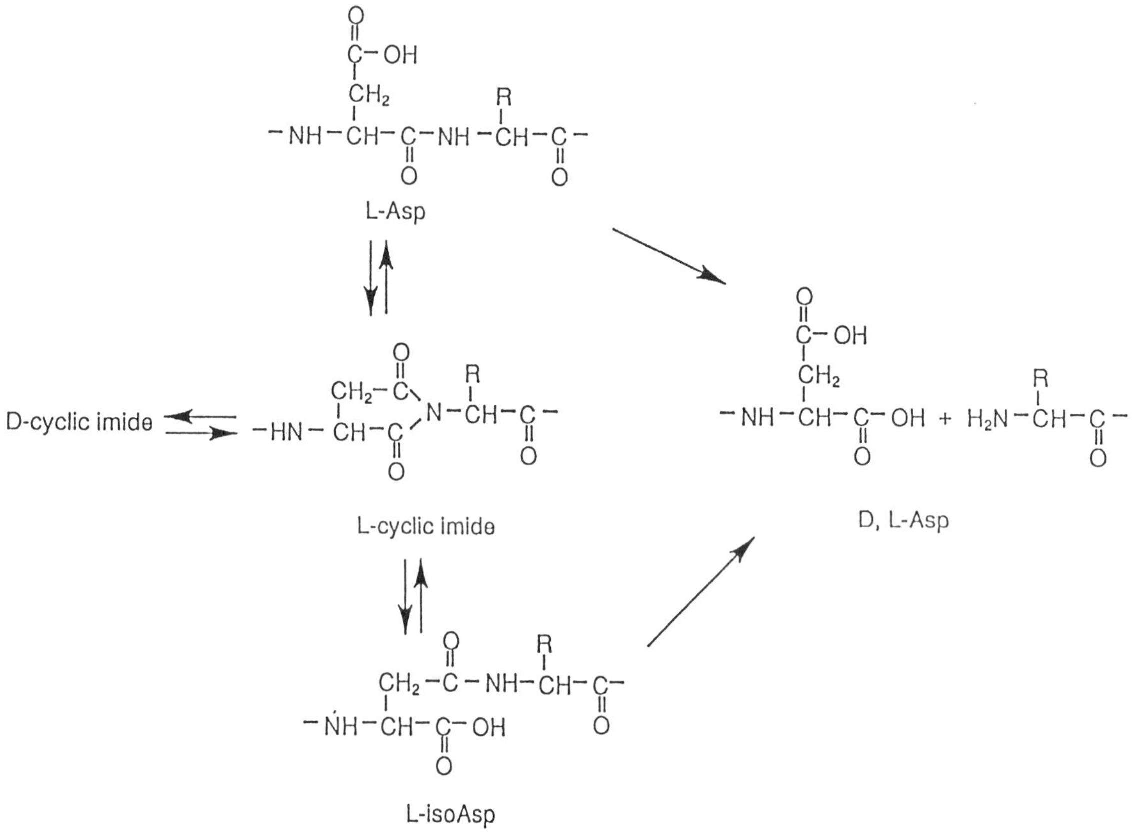

SCHEME 11.2. Pathways for Asp degradation.

deamidation, whereas $(NH_4)_2SO_4$ and K_2HPO_4/KH_2PO_4 greatly enhanced the reaction rate. The rank ordering of enhancement of the deamidation rate was $H_2PO_4^- >> CO_2^= =$ imidazole $> NH_3 >$ tris(hydroxymethyl)aminomethane. Capasso et al. (*30*) also reported that the deamidation rate of Boc–Asn–Gly–Gly decreased in the presence of solvents (e.g., ethanol) typically used in protein isolation and crystallization. Recently, Brennan and Clarke (*31*) have shown that the main effect on the deamidation rate of Val–Thr–Pro–Asn–Gly–Ala by addition of cosolvents occurs because of a decrease in the dielectric constant of the solution. Another factor that could influence the deamidation of Asn and Gln residues is ionic strength. For example, using the pentapeptides Gly–Leu–Gln–Ala–Gly and Gly–Arg–Gln–Ala–Gly, Scotchler and Robinson (*32*) showed that the rate of deamidation of the Gln residues increased with increasing ionic strength. Similar results were observed by McKerrow and Robinson (*33*) for the Asn-containing peptide Gly–Arg–Asn–Arg–Gly.

Whereas extensive research efforts have been devoted to understanding deamidation in aqueous medium, little is known about the instability of –XAsnY– sequences in the solid state. Deamidations in lyophilized formulations were reported for a model tripeptide (Gly–Asn–Gly) (*34*), a model hexapeptide (Val–Tyr–Pro–Asn–Gly–Ala) (*35–37*), glucagon (*38*), and human growth hormone (*39*). Factors important in solid-state stability of these Asn-containing peptides and proteins included the pH of the lyophilization solution, the moisture level of the lyophilized products, and the storage temperature. For example, Oliyai and co-workers (*36, 37*) using the hexapeptide Val–Tyr–Pro–Asn–Gly–Ala have shown that if the pre-lyophilization solution was pH 3.5, the lyophilized peptide deamidated to form the Asp–hexapeptide. This Asp–hexapeptide then underwent hydrolysis at the Asp–Gly amide bond to generate small quantities of the tetrapeptide Val–Tyr–Pro–Asp. In addition, significant quantities of the cyclic imide (Asu–hexapeptide) were detected in these formulations. The amounts of this cyclic imide increased significantly at pH 3.5 with increases in temperature or moisture level. While the propensity of the Asn–hexapeptide to cyclize to Asu–hexapeptide increased slightly when going from pH 3.5 to 5.0, the rate of hydrolysis of the Asn side chain decreased. The Asn–hexapeptide was most unstable when lyophilized from a pH 8.0 solution. If the pH of the prelyophilization solution was 8.0, the isoAsp–hexapeptide and Asp–hexapeptide were generated as the major and minor products, respectively.

In conclusion, deamidation of Asn residues is perhaps the most significant pathway of chemical degradation of peptides and proteins. The rates of degradation of Asn residues are very dependent on endogenous and exogenous factors, many of which have been characterized, particularly those in solution. However, our understanding of the factors that influence deamidation in lyophilized formulations or delivery systems are less well developed.

Hydrolytic Reactions of –XAspY– Sequences

In recent years, it has become increasingly apparent that –XAspY– sequences in peptides and proteins can be chemically unstable (*9, 10, 26*). Depending primarily on the pH of the solution, Asp residues can degrade via two different pathways (Scheme 11.2). One pathway involves the Asp residue degrading to form a cyclic imide by a mechanism similar to that observed in the deamidation of Asn residues. This cyclic imide may or may not (depending on pH) degrade further to form isoAsp-containing products (Scheme 11.2). The cyclic imide is prone to racemization at the α-carbon, resulting in the potential formation of D,L-cyclic imide-, D,L-Asp-, or D,L-isoAsp-containing degradation products. The second pathway for Asp degradation involves Asp–Y or X–Asp amide bond hydrolysis (Scheme 11.2). In dilute acid, this type of peptide bond hydrolysis reaction adjacent to an Asp residue is approximately 100 times faster than the hydrolysis of peptide bonds adjacent to other amino acid residues (*40*). The enhanced hydrolysis occurring at Asp–Y or X–Asp is attributed to intramolecular catalysis by the carboxyl group of the Asp side chain.

One of the major factors that prevented scientists from recognizing the importance of the cyclic imide pathway for Asp degradation in proteins was the lack of good analytical methodology to detect the formation of D-Asp and D,L-isoAsp. However, in recent years investigators have effectively used modern high-performance liquid chromatographic methodologies to study these reactions in small peptides (*10, 26, 36, 37, 41–43*) and the enzyme D-aspartyl/L-isoaspartyl methyltransferase and mass spectrometry to study these reactions in proteins (*15–18, 44–46*). Using this type of analytical methodology, investigators have shown that Asp residues in glucagon (*47*), calmodulin (*17, 48*), human growth hormone (*15*), tissue plasminogen activator (*18*), epidermal growth factor (*49*), daptomycin (*50*), and small model peptides (*10, 26, 36, 37, 41–43*) undergo this pathway of degradation. Obviously, as in the deamidation of Asn residues, the degradation of Asp residues can lead to changes in protein conformation and loss of biological activity (*15–18, 44–50*).

Many endogenous and exogenous factors influence the rate of Asp degradation and the distribution of the products. With respect to endogenous factors, the effect of primary structure on Asp degradation is important, but clearly not as obvious as that observed for Asn degradation. For example, Clarke and co-workers observed L-isoAsp formation from an Asp–Tyr sequence in glucagon (*47*), from Asp–Gln and Asp–Thr sequences in calmodulin (*48*), and from Asp–Gly sequences in human growth hormone (*15*) and human epidermal growth factor (*49*). Stephenson and Clarke (*26*) also studied a series of hexapeptides (Val–Tyr–Pro–Asp–Y–Ala) and observed that the half-lives of degradation to isoAsp were as follows: Y = Gly, 40.8 days; Y = Ala, 266 days; and Y = Ser, 168 days. Recently, Oliyai and Borchardt (*42*) studied the degradation of this same series of hexapeptides (Val–Tyr–X–Asp–Y–Ala; X = Pro, Gly; Y = Gly, Ser, Val) under both acidic (pH 1.1) and alkaline (pH 10.0) conditions. Under acidic conditions (pH 1.1), the Asp–Gly–hexapeptides and the C-terminally modified analogs (Asp–Ser– and Asp–Val–hexapeptides) underwent hydrolysis of the Asp–X amide bond, which constituted the major degradation route. With the Asp–Gly– and Asp–Ser–hexapeptide, the cyclic imides (Asu–Gly and Asu–Ser–hexapeptides) were also formed, but in lower amounts compared with the Asp–X amide bond hydrolysis products. Insignificant amounts of the Asu–Val–hexapeptide were formed from the Asp–Val–hexapeptide. With respect to the rate of hydrolysis of the Asp–Y amide bond, substituting larger, more bulky compounds (Y = Ser, Val) caused a diminutive, albeit still notable effect. The Asp–Gly–hexapeptide hydrolyzed only 1.6 times faster than the Asp–Ser peptide and 2.3 times faster than the Asp–Val peptide. In contrast, the formation of the cyclic imide was more sensitive to modifications made in the amino acid on the C-terminal side of Asp. For example, when Gly was replaced with Ser, the rate constant for the formation of cyclic imide decreased by nearly fourfold. Placing Val at the same position diminished the rate of cyclic imide formation even further, such that no cyclic imide was detected. Interestingly, replacement of the Pro residue with Gly on the N-terminal side of the Asp residue results in an eightfold increase in the rate of hydrolysis of the Y–Asp bond. However, this replacement of Pro with Gly did not affect the rates of Asp–X hydrolysis or cyclic imide formation.

When Oliyai and Borchardt (*42*) studied the stability of this same series of hexapeptides under alkaline conditions (pH 10.0), they observed that the Asp–Gly, Asp–Ser and Asp–Val–hexapeptides degraded to form isoAsp. As expected, the rates of isoAsp formation decreased when Gly was substituted with Ser and Val. With the Asp–Gly– and Asp–Ser–hexapeptides, degradation resulted in isoAsp/Asp ratios of approximately 4–5. In contrast, the ratio of the products arising from the Asp–Val–hexapeptide was less than 1. Also, the Asp–Val–hexapeptide underwent more rapid racemization. As expected, replacement of the Pro with Gly on the N-terminal side of the Asp residue had no significant impact on the rate of isoAsp formation or the distribution of the products. Clearly, from these studies by Stephenson and Clarke (*26*) and Oliyai and Borchardt (*42*), the effects of primary sequence on Asp degradation are not as obvious as those observed for Asn degradation. Clarke et al. (*9*) hypothesized that the reactivity of an Asp residue may be very dependent on the local chemical environment provided by the neighboring residues and its effect on the pK_a of the Asp residue. If this hypothesis is correct, then predictions of reactivity based on primary sequence alone will be difficult.

Exogenous factors can also influence the rate of Asp degradation as well as the distribution of products. The exogenous factor that most influences Asp degradation is pH. Using a model hexapeptide (Val–Tyr–Pro–Asp–Gly–Ala), Oliyai and Borchardt (*41*) have shown that the rate of Asp degradation and the distribution of the products are highly dependent on pH. Asp–hexapeptide degraded predominantly via intramolecular hydrolysis of the Asp–Gly amide bond. However, the peptide also degraded to a minor extent to the cyclic imide (Asu–hexapeptide), which was stable under strongly acidic conditions. As the pH of the solution was increased (e.g., pH 4.0–5.0), the Asp–hexapeptide favored cyclization over peptide bond hydrolysis. Under these conditions, the Asu–hexapeptide further degraded to isoAsp. For example, at pH 4.0, the majority of the Asp–hexapeptide (approx. 60%) rearranged to produce the Asu–hexapeptide, which then hydrolyzed to form the isoAsp–hexapeptide, whereas approximately 10% underwent Asp–X hydrolysis. In the pH range 6.0–10.0, the Asp–hexapeptide degraded completely to the isoAsp peptide. Above pH 8.0, the degradation of the Asp–hexapeptide was pH independent. The peptide was most unstable at approximately pH 4.0. Little or no buffer catalysis was observed for the degradation of the Asp–hexapeptides in the pH range 3.0–10.0.

Other exogenous factors reported to influence the rate of degradation of this Asp residue in peptides include temperature (*41*) and dielectric constant of the solvent (*31*). Brennan and Clarke (*31*) reported that the rates of isomerization of the Asp residue in the hexapeptide Val–Thr–Pro–Asp–Gly actually increased slightly with decreases in dielectric constant.

Even though progress has been made in recent years on understanding Asp degradation pathways in solution, our knowledge of these processes in the solid state is less well developed. Recently, Oliyai, Borchardt, and co-workers (*36, 43*) determined the individual and interactive effects of formulation variables such as the pH of prelyophilized solution, moisture level, temperature, and type of bulking agent on the stability of the hexapeptide Val–Tyr–Pro–Asp–Gly–Ala in the lyophilized state. The degradation pathways of the Asp–hexapeptide in the lyophilized state were dependent on the pH of the prelyophilization solutions and the moisture content of the freeze-dried formulations. Under acidic conditions (pH 3.5 and 5.0), the lyophilized Asp–hexapeptide predominantly decomposed to generate the Asu–hexapeptide, irrespective of the type of excipient (lactose or mannitol) present in the formulation. The hydrolysis of the Asp–Gly amide bond constituted a much less significant pathway under these conditions. At pH 6.5 and 8.0, the parent hexapeptide exclusively isomerized via formation of the Asu–hexapeptide to produce the isoAsp–hexapeptide. The extent of hydrolysis of the Asu–hexapeptide intermediate at pH 8.0 exceeded that at pH 6.5 and rendered the isoAsp–hexapeptide the major product of degradation in the basic environment. Although the type of excipient (lactose or mannitol) did not influence the degradation routes, the choice of excipient substantially affected the mean rate constant of peptide decomposition. Consistently, the amorphous lactose/peptide formulations were significantly more chemically stable at all temperatures, pH values, and moisture levels than formulations containing crystalline mannitol.

On examination, the product distribution in the solid state was significantly different from that in solution. Evidently the hydrolysis of the Asu–hexapeptide intermediate and the Asp–Gly peptide bond (formation of the tetrapeptide) was suppressed in a water-deficient environment. Thus, at lower pH values (3.5, 5.0, and 6.5), the major product observed was the Asu–hexapeptide, and only trace amounts of the tetrapeptide and isoAsp–hexapeptide were detected at pH 3.5 and 6.5, respectively. At pH 8.0, the base catalysis component of the hydrolysis of the Asu–hexapeptide intermediate compensated for the nearly inoperative water catalysis term. Consequently, the rate of decomposition of the Asu–hexapeptide was sufficient to afford the isoAsp–hexapeptide as a major degradation product at pH 8.0.

In conclusion, it is becoming increasingly evident that –XAsp–Y– sequences in proteins are a potentially important site for chemical degradation. In recent years, we have gained a greater understanding of the pathways by which –XAspY– sequences can degrade in proteins. However, our ability to predict chemically labile Asp residues in proteins is inadequate. In addition, we do not have a good understanding of how exogenous factors, particularly in lyophilized formulations or delivery systems (e.g., polymers), will influence the degradation of –XAspY– sequences in proteins.

Hydrolytic Reactions of Gln, Pro, and Asn Sequences

Several other hydrolytic reactions have been shown to be important degradation pathways in proteins and peptides. As discussed previously, a Gln residue in a primary sequence –XGlnY– deamidates roughly 10-fold slower than an Asn residue in a similar sequence (*12, 51*). However, Gln residues at the N-terminal end of a peptide or protein (H_2N–GlnX–) react more rapidly than

Asn residues to generate a pyroglutamate-containing product (*52*). This type of reaction was recently reported to occur in human relaxin A chain (*53*). Instability at the N-terminus of a peptide or protein can also occur when a H_2N–XProY– sequence exists. This sequence is labile to hydrolysis of the Pro–Y peptide bond through an intramolecular reaction generating X–Pro diketopiperazine (*54–57*).

Finally, Asn residues at the C-terminal end of a peptide or protein (–XAsn–OH) may deamidate to generate a cyclic anhydride through nucleophilic attack by the C-terminal carboxylate on the carbonyl carbon of the Asn side chain (*58*). This reaction has been extensively studied in insulin, where hydrolysis of the A-21 Asn results in formation of desamido A-21 insulin (*59–60*).

Oxidative Protein Degradation

Oxidation constitutes an additional major degradation pathway of protein pharmaceuticals. However, in contrast to the extensive information available about the hydrolytic degradation pathways (*see* the section in this chapter called *Hydrolytic Reactions*), there is, at present, considerably less mechanistic understanding of the underlying oxidation mechanisms. This difference is caused, in part, by the lack of information about the involved oxidizing species that might be present or formed in pharmaceutical formulations. Much of our knowledge of protein oxidation originates from the biochemical literature, because protein oxidation in vivo has been identified as a major covalent modification accompanying conditions of oxidative stress or aging (*61–65*).

Some recent reviews (*5, 66*) have identified a number of protein pharmaceuticals that are susceptible to oxidative degradation. However, we need to understand the mechanisms to rationally develop stabilization strategies of protein formulations (instead of trial-and-error approaches). Therefore, this chapter will focus on the mechanistic concepts governing the oxidation of amino acid residues in proteins and the potential products and their detection rather than providing a comprehensive summary of oxidation labile proteins.

Oxidation Labile Amino Acid Residues

The lability of an amino acid residue toward oxidation depends on both the nature of its side chain and the nature of the oxidizing species. For example, the hydroxyl radical ($HO^{\bullet}$) reacts rather nonselectively with all amino acids because of its high reactivity (*67*). On the other hand, less reactive species such as singlet oxygen (O_2) (*68*), ozone (O_3) (*69*), or hydrogen peroxide (H_2O_2) (*70, 71*) would preferentially oxidize methionine (Met), cysteine (Cys), histidine (His), tryptophan (Trp), and tyrosine (Tyr). An additional determining factor is introduced when reactive oxygen species are formed through transition metal catalysis. In that case, amino acids directly involved in metal chelating or located near a metal-binding site through the distinct structure of a protein would oxidize predominantly via so-called "site-specific" mechanisms (*61*). Metal contamination constitutes a problem for many pharmaceutical formulations, and transition metal-catalyzed oxidation may be assumed to play a major role during the manufacturing and storage of such formulations.

Influence of Protein Sequence and Structure

The protein structure influences the sensitivity of particular amino acid residues toward oxidation by the following:

- accessibility of a protein domain by exogenous oxidants,
- presence of metal-binding sites, and
- potential neighboring group effects.

Shechter et al. (*72*) defined three distinct classes of *exposed*, *partially exposed*, and *buried* methionine residues in proteins, based on their reactivity with added *N*-chlorosuccinimide (NCS). In general, buried methionines are least reactive toward exogenous oxidants. In accord with this proposal, hydrogen peroxide was able to oxidize the two exposed methionine residues Met_{14} and Met_{125} in native human growth hormone (hGH), but not the buried Met_{170} (*73*). In human relaxin, the more exposed Met-B_{25} reacted more efficiently with hydrogen peroxide than did the sterically less accessible Met-B_4 (*74*). On the other hand, small flexible model peptides composing the sequences around Met-B_{25} and Met-B_4 reacted equally well with exogenous hydrogen peroxide (*74*). These results were reversed on oxidation of human relaxin via a transition metal-catalyzed pathway (Cu(II)/ascorbate/O_2) (*75*). Here, the less exposed Met-B_4 reacted considerably better than Met-B_{25}, possibly because of its location in the vicinity of metal-binding sites such as Lys-A_{17} and several Glu residues. This selectivity was maintained even for small model peptides composing the sequences around Met-B_{25} and Met-B_4, respectively. In addition to the two Met residues, there was significant oxidation of the only His residue of the protein, His-A_{12} (*75*). This finding is in accord with the general sensitivity of His toward the oxidation by transition-metal/pro-oxidant systems as demonstrated for various proteins such as glutamine synthetase (*76, 77*), superoxide dismutase (*78*), and glycated insulin (*79*). However, although the site-specific mechanisms clearly exist, there is little mechanistic information available with regard to the reactive species involved and the physicochemical and structural parameters that influence their respective reactions. We are currently elucidating some of these factors employing small model peptides (*see* subsequent discussion).

At present, only a few specific examples exist for neighboring group effects other than metal-binding influencing protein oxidation. One illustrative example may emerge from the recent investigation of calmodulin oxidation (*80*). The exposure of calmodulin to H_2O_2 or peroxynitrite ($ONOO^-$) resulted in the predominant oxidation of vicinal Met–Met subdomains over the oxidation of other Met residues. A potential rationale for this observation may be provided by the closeness of two thioether functionalities in these subdomains, together with the general high flexibility of Met side chains (*81*). The one-electron (*82*) and two-electron oxidation (*83*) of a thioether can be facilitated by the proximity of a second thioether, capable of stabilizing cationic intermediates by the donation of lone-pair electron density. Such neighboring group effects may lower the corresponding oxidation potentials for the involved thioether groups by as much as 1.0 V, as determined electrochemically for a series of organic model compounds (*84*). Such effects are not only restricted to neighboring thioether groups, however. Similarly, the oxidation of a thioether can be facilitated by the proximity of other functional groups providing electron density, such as carboxylate, hydroxyl, and amino groups (*82, 85*).

Both parameters, metal-binding and neighboring group effects, are closely related to the secondary and tertiary structure of a respective protein, orienting functional groups into the conformation required for action. Neighboring group effects will probably be somehow less sensitive to protein structure as long as an interaction between the involved functional groups can occur for a finite time necessary for catalysis. On the other hand, metal binding may require the protein to arrange in a defined conformation even before metal binding can take place. In addition, this particular conformation will eventually change with the redox state of the metal.

Reactive Species Involved in Protein Oxidation

In pharmaceutical formulations as well as under biological conditions of oxidative stress, proteins may encounter a large variety of oxidizing species. Each of these species will react with the individual amino acid residues via distinct mechanisms. These mechanisms will experience additional influences from neighboring groups, pH, and the nature of the buffer or other components (e.g., excipients) present in solution. In the following section some characteristic features of these mechanisms will be outlined, preceded by some general considerations of the potential occurrence of true autoxidation processes in pharmaceutical protein formulations.

Autoxidation

By definition, the term autoxidation describes the direct oxidation of a substrate by molecular oxygen in the absence of any catalytic process. The direct oxidation of an organic molecule in its singlet ground state (i.e., presenting paired electron spins) with molecular oxygen in its naturally abundant triplet state ($^3O_2$) represents a spin-forbidden process. Thus, during the lifetime of an encounter complex between $^3O_2$ and an organic molecule, there will be no significant bond formation. Any potential oxidation reaction, therefore, is expected to proceed initially via an outer-sphere electron transfer (forward eq 1).

$$S + O_2 \underset{k_{-1}}{\overset{k_1}{\rightleftharpoons}} S^{\bullet +} + O_2^{\bullet -} \quad (1)$$

This electron transfer was demonstrated experimentally by Merényi et al. (*86*) for a number of organic mono- and dianions for which a plot of k_1 vs. K_1 could be fitted to the Marcus equation, where K_1 represents the equilibrium constant for equation 1, that is $K_1 = k_1/k_{-1}$. An interesting question is now whether autoxidation processes such as eq 1 may account for protein degradation in pharmaceutical formulations. Let us consider a hypothetical formulation of 1 mg/mL of a standard protein of molecular weight 22,000 (i.e., a protein of the size of monomeric hGH) in air-saturated, metal- and peroxide-free aqueous buffer. Let us further assume that the forward eq 1 represents the rate-determining step of an overall oxidation process of the standard protein in this system. The potential oxidation-sensitive amino acids of this protein will be Cys ($E^\circ_{CysS^-/CysS^\bullet}$ = 0.75 V (*87*)), Trp ($E^\circ_{Trp/Trp^\bullet,H^+} \approx$ 1.1 V, pH 7.0 (*88*)), Tyr ($E^\circ_{TyrOH/TyrO^\bullet,H^+} \approx$ 0.93 V, pH 7.0 (*88*)), Met ($E^\circ_{MetS/MetS^\bullet,H^+} \approx$ 1.3 V (*89*)), and His ($E^\circ_{His/His^\bullet,H^+}$ = 0.93 V (*90*)), where E° is the one-electron redox potential. In this series the amino acid most labile toward autoxidation via the outer-sphere mechanism 1 should be the deprotonated Cys residue (denoted as $CysS^-$). By comparison with electron donors of similar redox potentials and their measured or predicted forward rate constants k_1 (*86*), we obtained an estimate for k_1 for S = $CysS^-$ on the order of 10^{-6} M^{-1} s^{-1}. Under hypothetical conditions of a pseudo-first-order process, that is taking 1 mg/mL protein in an air-saturated aqueous solution ([protein] $\approx 4.5 \times 10^{-5}$ M, excess $[O_2] = 2.5 \times 10^{-4}$ M), we can then approximate a half-life for the autoxidation of one $CysS^-$ residue within our standard protein according to eq 1 of $t_{1/2} \approx$ 32,090 days. Such a hypothetical value for Cys oxidation is unreasonably high and would still appear too high even if it were overestimated by a factor of 100: that is, its half-life were 320 days. Moreover, these calculations would indicate that Met residues would by no means autoxidize according to eq 1 on an experimentally accessible time scale ($t_{1/2} > 8.8 \times 10^7$ years). In reality, however, the Cys and the Met residues have been characterized as the most important oxidation sites in pharmaceutical protein formulations (*5*, *66*). From these considerations we may exclude a true autoxidation of a protein as an important process of protein degradation.

Does this picture change when proteins are formulated in the presence of excipients and antioxidants? From the published pseudo-first-order rate constant of ascorbate monoanion (AH^-) oxidation in chelex-treated buffer under conditions of excess oxygen and pH 7.0 (*91*), we extrapolate an upper limit for the overall rate constant for the autoxidation process of $k_2 \approx 2.4 \times 10^{-3}$ M^{-1} s^{-1}.

$$AH^- + O_2 \rightarrow \text{products} + O_2^{\bullet -} + H_2O_2 \quad (2)$$

This value of k_2 represents an overall rate constant and cannot directly be related to k_1. However, in a hypothetical formulation containing 5×10^{-5} M ascorbate at pH 7.0 in metal- and peroxide-free water ($[O_2] = 2.5 \times 10^{-4}$ M), the ascorbic acid monoanion will autoxidize with a half-life of $t_{1/2} \approx$ 13.4 days. In addition to dehydroascorbic acid and other products, this process will yield superoxide radical anion and hydrogen peroxide. Superoxide oxidizes thiol groups (*92*), whereas hydrogen peroxide efficiently oxidizes thiols (*71*) and converts Met into Met sulfoxide (eq 3) (*70*).

$$H_2O_2 + (Met)>S \rightarrow H_2O + (Met)>S{=}O \quad (3)$$

Thus, if ascorbate autoxidation were to take place in yet another hypothetical liquid formulation containing 5×10^{-5} M ascorbate, pH 7.0, and 5×10^{-5} M protein, the generated superoxide and hydrogen peroxide would subsequently oxidize the protein. Taking $k_3 \approx 1 \times 10^{-2}$ M^{-1} s^{-1} (*70*), we derive that significant protein oxidation could take place during the half-life of ascorbate autoxidation.

In conclusion, these model calculations show the feasibility of antioxidant autoxidation to initiate protein oxidation via the formation of reactive oxygen species. On the other hand, protein autoxidation alone appears to be of little importance as a potential mechanism for protein degradation. This picture may change dramatically with the additional presence of redox-active transition metals in a pharmaceutical formulation, as described subsequently.

Mechanisms of Metal-Catalyzed Oxidation

The spin restrictions applying to the reaction of triplet oxygen with organic molecules in their singlet states can be minimized by the complexation of O_2 to transition metals (*93*). The mode of interaction of molecular oxygen with a transition metal will depend largely on its ligand sphere (*93*). The first elementary step of the reaction of ferrous iron with $^3O_2$ will be the formation of an adduct **1** (eq 4). This adduct can be regarded as a peroxyl-type radical, which itself should show considerable reactivity toward many amino acids by analogy to organic peroxyl radicals such as $CCl_3OO^{\bullet}$ (*90, 94, 95*). However, adduct **1** will easily oxidize a second ferrous iron to yield hydrogen peroxide and two equivalents of ferric iron (eq 5). Hydrogen peroxide can then either oxidize amino acids (such as Cys and Met) directly or enter additional reaction pathways. The complexation of ferrous iron by oxygen acids may lower its redox potential sufficiently to permit the direct reduction of molecular oxygen to superoxide anion (eq 6).

$$Fe(II) + O_2 \rightarrow Fe(II){-}O{-}O^{\bullet}\ (\mathbf{1}) \quad (4)$$

$$Fe(II){-}O{-}O^{\bullet}\ (\mathbf{1}) + Fe(II) + 2H^+ \rightarrow 2Fe(III) + H_2O_2 \quad (5)$$

$$Fe(II){-}O{-}O^{\bullet}\ (\mathbf{1}) \rightarrow Fe(III) + O_2^{\bullet -} \quad (6)$$

$$Fe(II) + H_2O_2 \rightarrow Fe(II)(H_2O_2)\ (\mathbf{2}) \quad (7)$$

$$Fe(III) + H_2O_2 \rightarrow Fe(II){-}O{-}O{-}H\ (\mathbf{3}) + H^+ \quad (8)$$

$$Fe(II)(H_2O_2)\ (\mathbf{2}) \rightarrow Fe(IV){=}O + H_2O \quad (9)$$

$$Fe(IV){=}O + S \rightarrow Fe(II) + SO \quad (10)$$

$$Fe(IV){=}O + H_2O_2 \rightarrow Fe(II) + O_2 + H_2O \quad (11)$$

$$Fe(II)(H_2O_2)\ (\mathbf{2}) \rightarrow Fe(III) + HO^{\bullet} + HO^- \quad (12)$$

$$Fe(II){-}O{-}O{-}H\ (\mathbf{3}) \rightarrow Fe(V){=}O + HO^- \quad (13)$$

$$Fe(II){-}O{-}O{-}H\ (\mathbf{3}) \rightarrow Fe(II) + H^+ + O_2^{\bullet -} \quad (14)$$

Much emphasis has been placed on the characterization of the reaction of hydrogen peroxide and organic peroxides with complexes of ferrous (*96–98*) and ferric iron (*99*), primarily chelates of ethylenediaminetetraacetic acid, ethylenediaminediacetate, nitrilotriacetic acid, and diethylenetriaminepentaacetic acid. The initial step appears to be the addition of hydrogen peroxide to the complex (eqs 7 and 8), followed by a series of reactions yielding various specific reactive intermediates, all of which may oxidize amino acid targets in proteins.

Are these mechanisms relevant to pharmaceutical formulations? Pharmaceutical formulations often contain contaminations of hydrogen peroxide originating from sterilization procedures, or organic hydroperoxides introduced through excipients such as polyols. On the other hand, frequently buffers and proteins are contaminated by transition metals. Even though transition metals cannot be expected to be present initially in their reduced forms [i.e., Fe(II) or Cu(I)], the reduction of Fe(III) and Cu(II) can occur through peroxides as delineated in eqs 8 and 14 (representatively for Fe(III)). In addition, proteins contain amino acids that themselves can function as electron donors, such as Trp, Cys, and Tyr.

Depending on the specific target amino acids affected during protein oxidation, there is the additional possibility of further oxidation either through metal-catalysis or chain reactions of protein-bound peroxyl radicals. For example, the hydroxylation of protein tyrosyl residues yields protein-bound 3,4-dihydroxyphenylalanine (DOPA) (*100*). DOPA is a potent reductant and chelator of Fe(III) and Cu(II) and was found to initiate further protein damage. Hydrogen abstraction from any amino acid residue of a protein in the presence of molecular oxygen yields protein-bound peroxyl radicals (*101*). These radicals abstract hydrogen atoms from neighboring amino acid residues to initiate protein-chain oxidations (*102*) and generate protein-bound hydroperoxides, which represent additional substrates for metal-catalyzed processes.

Oxidation by Reactive Oxygen Species

Apart from reacting in a metal-bound form, the activation of molecular oxygen can result in the formation of various reactive oxygen species, which subsequently oxidize protein amino acids through non-metal-catalyzed processes. *N*-Formylkynurenine, a common oxidation product of Trp, can be detected by means of its characteristic fluorescence spectrum (λ_{em} = 435 nm) (*103*). In the first elementary step of its formation by means of the oxidation by different reactive oxygen species, the addition of singlet oxygen to the indole ring yields an indole dioxetane (*104*), whereas ozone produces a Criegee intermediate (*69*). Peroxyl radicals form a radical adduct that preferentially decomposes into epoxide and alkoxyl radical by means of breakage of the peroxide bond, and only a minor fraction yields an indole radical cation via electron transfer (*105*).

The oxidation of Tyr may yield DOPA (*100*) as well as various bityrosine species (*106*). These bityrosine species exhibit a characteristic fluorescence band with $\lambda_{em} \approx$ 420 nm. However, lack of bityrosine fluorescence in an oxidized protein does not necessarily preclude Tyr oxidation because bityrosine formation requires the spatial neighborhood of two Tyr residues (*103*).

The most prominent oxidation products of His have been shown to be 2-oxo-His (*78*), Asn (*76*), and Asp (*107*). However, the associated reaction mechanisms have not been studied in detail. A recently developed analytical method for 2-oxo-His, employing reversed-phase chromatography with electrochemical detection (*78*), should be beneficial for further investigations targeted at the mechanisms of 2-oxo-His formation. In addition, this product may well be characterized by enzymatic protein digestion followed by sensitive mass spectrometric analysis of the peptide fragments.

The oxidation of Met to Met sulfoxide requires a formal two-electron oxidation of the thioether side chain. This oxidation is easily achieved by peroxides and singlet oxygen. Recent findings also have indicated the possibility of sulfoxide formation from thioethers via initial one-electron oxidation, followed by reaction of an initially formed sulfur-centered radical cation with hydroxide and molecular oxygen (*108*). Evidence for a potential intermediary sulfur-based peroxyl radical $R_2S(OH)OO^{\bullet}$ was presented (*109*).

Studies Involving Model Peptides

To gain insight into the potential reaction mechanisms by which Met residues in proteins are oxidized by reactive oxygen species, we initiated a series of studies with Met-containing model peptides for the elucidation of the following contributing factors:

- peptide sequence,
- presence of metal-binding sites,
- nature of transition metals,
- nature of pro-oxidants,
- buffer, and
- pH.

The presence of metal-binding sites such as His (*110–112*) or deprotonated N-terminal amino groups (Zhao, F.; Schöneich, C. H., unpublished results) within the peptide sequences promoted the oxidation of Met to Met sulfoxide. Further studies with specifically generated reactive oxygen species (e.g., $HO^•$, $SO_4^{•-}$, and triplet benzophenone derivatives) revealed that complex neighboring group effects may significantly influence such oxidation yields and material balances. For example, the initial addition of the hydroxyl radical ($HO^•$) to Met in Thr–Met resulted in only a small fraction of Met oxidation, and the major products originated from the cleavage of the Thr side chain (*113*). The characterization of short-lived intermediates permitted the proposal of a potential mechanism involving intramolecular proton and electron transfer reactions. Similar experimental observations were made when the oxidation of Thr–Met was initiated by Fe(II)EDTA/H_2O_2 (Schöneich, C. H.; Yang, J., unpublished results), an oxidation system of potential pharmaceutical relevance. In a separate study, sulfur-centered radical cations at C-terminal Met residues were investigated for their propensity to undergo C-terminal decarboxylation (*114*). This decarboxylation mechanism was influenced by the pK_a of the N-terminal amino group in the peptide radical cations, a result revealing another interesting example of neighboring group interactions controlling peptide oxidation.

We are still far from predicting the stability of protein pharmaceuticals toward oxidation. However, further mechanistic information, generated by various physicochemical and biochemical approaches (e.g., site-directed mutagenesis at proteins), will potentially lead us to conclusive information, enabling formulation scientists to search more effectively for oxidation-resistant protein formulations.

Physical Instability

The physical instability of peptides and proteins generally is manifested as a loss of active species from solution via processes such as adsorption to surfaces, formation of soluble or insoluble aggregates, and sometimes, formation of highly structured gels (*3–8, 115*). Even though the deleterious effects of physical instability are well recognized, the understanding of the dynamics and kinetics of such physical changes is only beginning to approach the quantitative descriptions currently used for chemical modifications. Unlike covalent modifications, physical instability is generally much more sensitive to stresses encountered during processing, such as freezing, lyophilization, and exposure to interfaces during agitation, filtration, and release from a polymeric delivery system. Similar to covalent modifications, but for somewhat different reasons, factors such as thermal fluctuations, ionic strength, pH, or excipient composition may all influence the sensitivity of the protein to physical instability. To appreciate the role that such a myriad of seemingly unconnected environmental stresses has on the physical stability of proteins and peptides, a general model of physical instability will be discussed.

Model for Physical Instability

Historically, the stability of proteins has been described in terms of thermodynamic equilibria between a native state and a denatured state (*115–117*). Lumbry and Eyring's (*118*) inclusion of irreversible modifications of the denatured protein, at the expense of a loss in native protein (eq 15), has found widespread application in describing the overall loss of protein (*5–6, 19–20, 118–123*).

$$\text{native} \leftrightarrow \text{denatured} \rightarrow \text{irreversibly modified} \quad (15)$$

Reversible Components

Although equilibrium unfolding from the native to the denatured state has generally been regarded as a single-step process, it has become increasingly clear through studies of folding kinetics and thermodynamic equilibria of perturbed systems that partially folded intermediates can exist and the two-state equilibrium is an over-simplification for many proteins (*5, 116, 123–132*). The observed equilibrium intermediates have been termed "molten globules" and are characterized as having a nativelike secondary structure held together in a rather compact, nonspecific form through hydrophobic interactions but lacking the tertiary structure typically attributed to native proteins (*124, 133, 134*). Kinetic intermediates observed during refolding appear to have structural properties very similar to those observed in equilibrium studies (*130, 133, 134*). Mildly denaturing conditions such as acidic pH and low-level chaotropes can be used to generate conditions where intermediates are more highly populated and thus more easily characterized than the low-level steady-state kinetic intermediates. The importance of these molten-globule intermediates to physical stability of proteins lies in their propensity to associate or aggregate, leading to off-pathway products of folding and formation of insoluble aggregates (eq 16) (*5, 123, 125–129, 131*).

$$\begin{array}{ccccc} \text{native} & \leftrightarrow & \text{intermediate(s)} & \rightarrow & \text{denatured} \\ & & \updownarrow & & \downarrow \\ & & \text{associated} & \rightarrow & \text{irreversibly} \\ & & \text{intermediate(s)} & & \text{modified} \end{array} \qquad (16)$$

An important implication of the presence of these molten-globule intermediates lies in the potential for decoupling of thermodynamic stability and overall physical stability of the protein toward irreversible modification. This decoupling argues against the generally accepted adage that the more conformationally stable a protein is, the greater will be its stability toward irreversible hydrophobic aggregation. This decoupling occurs with α-interferon, where a change in pH from 6 to 5 decreases conformational stability but decreases irreversible aggregation as well (*116, 123*). Also, disulfide cross-linked multiple mutants of T4 lysozyme, which have global stabilities less than the wild type, are actually stabilized toward irreversible aggregation (*116*).

These native molten-globule and denatured forms are actually an ensemble of rapidly fluctuating species of similar free energy for which the energetic distribution of species would be expected to increase in going from the native to molten-globule to denatured state (*115, 135*). Unlike native proteins, smaller polypeptides (i.e., less than 40 amino acids (AA)) generally exist in an ensemble of structures with similar free energies. The existence of a more subtle structural or energetic hierarchy in proteins also raises questions regarding the relative importance of global (i.e., along the pathway from native to molten globule to denatured) versus localized conformational stability in a protein (i.e., manifested in flexibility or surface chain mobility) (*136, 137*). Even though the global and local stabilities of basic pancreatic trypsin inhibitor appear to be correlated (*138*), other cases exist where local and global stability are not necessarily coupled (*136, 137*). The high degree of surface flexibility on a globular protein may be an important factor in its interfacial properties, irrespective of the effect that flexibility has on structural global stability (*139*).

This dynamic picture of the native protein or peptide is crucial to understanding the availability of "activated" protein/peptide susceptible to irreversible interactions through exposure of either fleeting hydrophobic surfaces in the native subpopulation, much longer-lived hydrophobic surfaces in molten-globule intermediates, hydrophobic residues in the random coil structure of the unfolded protein, or previously buried or unreactive hot spots now vulnerable to covalent modification.

Irreversible Components

Although potential reactions contributing to the irreversible portion of this model include chemical modification of the primary structure (*6, 19–20, 115, 119, 121, 140*), the emphasis here will be placed on physical interactions generally attributed to hydrophobic aggregation (*5, 6, 115, 116, 119–121, 140–143*). Even though these hydrophobic aggregates are considered to be irreversible, they usually can be reversed by chaotropic solvents and then reactivated by subsequent refolding under appropriate conditions (*142*). However, it is not uncommon for covalent modifications to occur following aggregation: for example, insulin undergoes intermolecular disulfide exchange following B-elimination of the intact disulfide in the unfolded species (*140*), disulfide cross-linking occurs in soluble aggregates of keratinocyte growth factor (*143*), and unfolding of interleukin-1β leads to the autoxidation of free cysteines followed by hydrophobic aggregation and precipitation (*24*).

The morphological characteristics of aggregated or precipitated protein are largely controlled by kinetics of formation and specific properties of the protein or peptide involved. In general, the lower the steady-state concentration of the aggregating species, the more time will exist for greater orientation to occur during aggregate formation. Aqueous solutions of small peptides such as luteinizing-hormone-releasing hormone analogs become more viscous with time and yield associated species and birefringent gels or liquid crystals, which can be solubilized by increasing the temperature (*144*). Larger peptides such as glucagon (29 AA) (*145*), calcitonin (32 AA) (*146*), insulinotropin (31 AA) (*147*), and amyloid peptides (20–40 AA) (*148, 149*) undergo conversion to an intermolecular β-sheet structure of limited solubility. Glucagon and calcitonin can undergo further supramolecular structuring, resulting in formation of fibrils or highly viscous gels (*145, 146*). Calcitonin fibrils are composed of both α-helical and β-sheet components (*146, 150*). In aqueous solutions, insulinotropin can undergo self-association into an α-helical rich tetramer, which forms easily solubilized precipitates. However, on exposure to interfacial stresses of agitation, irreversible precipitates rich in β-sheet character are formed (*147*).

The morphological and structural aspects of insulin and sickle-cell hemoglobin aggregates were studied extensively (*151, 152*). The type of stress, concentration of hemoglobin, and ionic strength dictate whether liquid crystals, spherulites, fibrils, or gels are formed (*151*). In the case of insulin, it appears that the major morphological form is that of fibril nuclei, which can be manifested either as increases in viscosity, amorphous flocculates, or a "frost" on the sides of glass vials (*151*). Aggregates of erythropoietin can vary from a limited size of 20 molecules at neutral pH to aggregates of unlimited size at acidic pH (*153*). In general, both the kinetics and the morphological characteristics of the aggregates are highly dependent on the balance between attractive and repulsive forces of the particles. Hence, manipulations of protein concentration, pH, ionic strength, type of salt, or presence of surfactant can result in the formation of opaque gels, translucent gels, gellike precipitates, or amorphous white precipitates (*154, 155*).

Induction of Physical Instability

Similar to chemical instability, physical instability can be induced by changes in temperature, pH, ionic strength, and levels of various excipients that alter solvent properties. Unlike chemical instability, less commonly considered stresses such as interfacial tension and dehydration by freezing or lyophilization can initiate physical instability.

Thermal Fluctuations

Typical protein molecules have conformational stabilities of 5–15 kcal/mol near room temperature, which is roughly equivalent to the presence of one-billionth to one-hundredth of 1% of the protein molecules existing in the denatured form (i.e., equilibrium constant of denatured/native $K_{D/N} = 10^{-11}$ to 10^{-4}) (*117*). However, with every 5 °C increase in temperature the amount of denatured protein can easily increase 10-fold or more. At 55 °C, anywhere from 0.01% to 90% of the protein can exist in the unfolded form (i.e., $K_{D/N} = 10^{-4}$ to 10^{0}). Therefore, the impact of temperature on the amount of unfolded protein susceptible to subsequent irreversible aggregation can be quite extensive. The drastic temperature effect on these unfolding equilibria is probably responsible for the apparent energies of activation exceeding 100 kcal/mol that are observed for aggregation/ precipitation of bovine somatotropin (*156*) and antithrombin III (*157*). An unexpected decrease in off-path aggregation of lysozyme during refolding was observed with increasing temperature (*125*). This decrease in aggregation was explained by the greater thermal destabilization of the folding intermediates (from which aggregation occurs) relative to the native structure. Temperature effects on all equilibria and irreversible reactions must be considered when describing the impact of thermal stress on physical instability.

The potential for cold denaturation of proteins, as recently reviewed by Privalov (*158*), should not be overlooked, especially if storage or processing of systems near 0 °C is being contemplated. *Streptomyces subtilisin* inhibitor can exist in three distinct forms at pH 2.5—cold-denatured (0 °C), native (20 °C), and heat-denatured (50 °C) (*159*). Similarly, cytochrome C is present in at least four different states at low pH, depending on ionic strength and temperature (*132*). The clear structural differences between cold- and heat-denatured subtilisin inhibitor and cytochrome C also point out the different states of proteins that can be encountered using different denaturants. The practical significance of cold denaturation to irreversible aggregation is not yet clear.

The irreversible process could be extremely fast and consume the partially or completely unfolded material as fast as it is formed. In such cases, the rate-limiting process would then become the unfolding step, which is typically a first-order reaction. However, given milliseconds to days as the typical rates of unfolding (*117*), there are not too many practical situations wherein this process would be rate-limiting. For example, half-lives for bovine somatotropin unfolding (induced by guanidine HCl) at 3 °C are still less than about a minute (*127*).

Solvent Conditions

The effect of solvent conditions and excipients on the thermodynamic equilibrium of proteins has been studied extensively, and recent reviews in the area provide an excellent overview (*4–8, 115, 160*). Additives such as sucrose or other polyols, which are known to stabilize the conformational structure of proteins through preferential hydration (*160*), decrease the rate of thermally induced aggregation of bovine somatotropin (*156*) and basic fibroblast growth factor (*161*).

Unlike its effect on chemical stability, the effect of pH on physical stability may be quite unpredictable and highly protein dependent. A small increase in pH from 5 to 6 results in a rapid increase in the irreversible aggregation of α-interferon, despite an increase in conformational stability (i.e., decoupling of conformational and overall physical stability) (*123*). Effects of pH on an amyloid peptide (*116*) and on α-interferon (*116, 123*) implicate histidine and its ionization as playing a key role in the aggregation step. A decrease in pH from 9 to 7 led to a change from chemical to physical instability as the predominant decomposition pathway for bovine somatotropin (*162, 163*), whereas pH impacted the type of erythropoietin (*153*) and human growth hormone aggregate formed (*164*).

Ionic strength can drastically affect the impact of pH (or vice versa) on aggregation of ovalbumin (*154*). Increases in ionic strength from 0.01 to 0.15 at pH 2 greatly increase the formation of interferon-β1 oligomers (*142*). Ionic strength can actually be important because of differential effects on folding intermediates and native structures, wherein molten-globule states can be stabilized by binding of ions (*132, 133*).

Specific ligands can have a significant impact on protein stability. The physical stability of the metalloprotein fibrolase was increased by the addition of zinc and destabilized by the presence of complexing agents or decreases in pH, which lead to dissociation of zinc from fibrolase (*165*). Similarly, the stability of insulin is enhanced through the presence of zinc, which induces formation of the hexamer, thus minimizing the concentration of the monomeric insulin that is responsible for aggregation (*141, 152, 166*). The physical stability of various growth factors can be greatly improved by the presence of anionic ligands (*129, 143, 167, 168*).

Even though nonionic surfactants can minimize interfacial adsorption and induce denaturation of insulin (*169*), interleukin-2 (*170*), human growth hormone (*171*), and urease (*170*), they have been shown to enhance thermally induced aggregation of bovine somatotropin (*156*), basic fibroblast growth factor (*161*), and urease (*172*). The use of polyethylene glycol (PEG) during refolding of bovine carbonic anhydrase minimizes off-path dimerization and subsequent aggregate formation by binding to hydrophobic folding intermediates prone to dimerization (*173*).

Interfacial Tension

Physical instability of proteins at interfaces arises from the inherent surface activity of proteins and their tendency to concentrate at interfaces (*174, 175*). The complexity of factors influencing protein adsorption at interfaces is exemplified by the use of 12 different properties of a protein to describe its tendency to interact at an interface (*174*). Obviously, the properties of the surface (i.e., hydrophilicity, hydrophobicity, surface charge density, potential for acid/base character, surface dynamics, heterogeneity, and topography) are also important (*174*). One of the simplest and most commonly encountered interfaces is the air–water interface (*175, 176*). More complex interfaces are often encountered during manufacturing, packaging, and product administration (including controlled delivery systems) (*115, 169, 177*). The interface interaction of greatest complexity is encountered between implantable biomaterials and endogenous proteins, actually providing the impetus for most of the fundamental work in this area (*174*).

Even though the collisional frequency at the interface is dictated by concentration and diffusion coefficient, the ability of the interaction to provide sufficient residence time at the interface for subsequent interfacial processes to occur becomes dependent on the external surface properties of the protein and the dynamics of change in the protein structure at the interface (*175, 178*). Efficiencies of binding on initial collision were as high as 60–80% for the adsorption of lysozyme and α-lactoglobulin to hydrophobic surfaces, whereas very low efficiencies (<5%) were noted for hydrophilic surfaces (*179*). The specific orientation of the molecules appears to be much less restrictive for hydrophobic than for hydrophilic interfaces, whereas electrostatic and desolvation effects tend to play a more critical role in adsorption to hydrophilic surfaces (*179*). Because of the very rapid diffusion-dominated nature of the initial interaction step, adsorption from concentrated protein solutions can actually result in weaker protein binding (*180*). In such situations, binding occurs more rapidly than spreading and conformational alteration, a characteristic thus enhancing the probability of desorption with minimal conformational change.

The diffusion-controlled nature of the interaction manifests itself in a relatively temperature-independent process wherein the rates of adsorption are relatively fast on a practical timescale, even when efficiencies of collision are exceedingly low (*175*). Steady-state levels of adsorption are generally reached in less than 1–2 h, and in many cases, within minutes, depending on the degree of surface structuring that occurs and the initial concentration of protein (*175, 179*). Typical saturation levels of binding are generally less than 1 μg/cm^2, and they reach steady-state level with protein concentrations of less than 100 μg/mL. Consequently, significant losses due to irreversible adsorption are encountered only in very low concentration solutions of highly potent proteins such as tissue necrosis factor (0.25 μg/mL), interleukin-2 (1.7 μg/mL), and erythropoietin (0.5 μg/mL) (*115, 175*).

The more important role of the interface lies in its ability to generate or kinetically trap partially unfolded proteins. These proteins are then accessible to act as nucleation sites for further aggregation (eq 17). The air–water interface is rapidly renewed on agitation, permitting new molecules to be adsorbed to the interface and partially unfolded molecules to be desorbed into the media, where they are then subject to the competitive nature of refolding versus intermolecular nucleation and aggregation. Irreversibly adsorbed insulin appears to form a nucleation site on which aggregates are formed until a critical size of ~100–150 nm is reached and the aggregate breaks away from the surface (*141, 166*).

$$\begin{array}{ccccc} \text{native} & \leftrightarrow & \text{desorbed altered} & \rightarrow & \text{irreversibly aggregated particulates} \\ \updownarrow & & \updownarrow & & \uparrow \\ \text{adsorbed to interface} & \leftrightarrow & \text{conformationally altered} & \rightarrow & \text{irreversibly adsorbed} \end{array} \qquad (17)$$

The difficulties in obtaining reproducible kinetics from the air–water interface recently were addressed by using hydrophobic beads to mimic the air–water interface when studying the physical instability of insulin solutions (*166*). In spite of the tremendous amount of work that has gone into attempts to understand and modify the interfacially induced denaturation of insulin, it still remains a significant problem, especially when in extended contact with various drug delivery systems (*141, 152, 166, 175*).

The nebulization of recombinant human granulocyte colony stimulating factor, lactate dehydrogenase, and human growth hormone resulted in aggregate formation (*176*). In many cases, the formation of aggregates during nebulization can be alleviated by the addition of nonionic surfactants or PEG 1000, presumably through the preferential interaction of the surface active stabilizers at the air–water interface (*171, 176*). In many cases, shear forces have been implicated in the instability of proteins during high speed agitation of solutions, sterile filtration, and tangential flow filtration. For human somatotropin and other proteins, the shear-induced denaturation in solution is due to denaturation at rapidly renewed membrane–water or air–water interfaces rather than to actual shear forces encountered (*39, 171*).

Dehydration

The dehydration of proteins by freezing or lyophilization is generally undertaken to provide enhanced storage stability. However, these processes can often result in physical instability if appropriate precautions are not taken (*181–185*). Rates of freezing affect formation of soluble versus insoluble aggregates of human growth hormone (*164*). Freezing of human growth hormone in the presence of preservatives such as benzyl alcohol, *m*-cresol, or phenol results in formation of a significant amount of insoluble aggregate (*171*). The physical instabilities induced by freezing are not necessarily the same as those induced by lyophilization (*181*). Through the use of solid-state Fourier transform IR (FTIR) spectroscopy it is possible to probe the effects of freezing, primary and secondary drying stages of lyophilization, and in some cases, the rehydration or reconstitution steps on the conformational structure of the protein (*181–182, 186*). The stabilization of proteins through these steps has been approached from both a kinetic and a thermodynamic perspective (*183*).

The kinetic approach assumes that some degree of mobility is required for unfolding to occur. Consequently, by vitrification of a protein in a glassy amorphous matrix below its glass transition temperature (T_g), the kinetic barrier to unfolding will be limiting (*183–185*). The effects of excipients and moisture on the denaturation temperatures of bovine somatotropin and lysozyme are indicative of the importance of the proposed kinetic mechanism (*121, 187*). The physical stability of lyophilized tissue-type plasminogen activator (t-PA) during long-term storage at 50 °C correlated with the degree of porosity of the lyophilized cake (*188*). It was proposed that those t-PA molecules at the solid–gas interface (originally at the ice crystal–amorphous concentrate interface) were less well protected by the amorphous stabilizer, lysine, and therefore more susceptible to denaturation than those imbedded in the amorphous glassy matrix.

The thermodynamic approach assumes that an excipient provides conformational stabilization via "solute exclusion" during freezing and acts as a water substitute during dehydration (*160, 181–183*). The evidence for the water substitute mechanism is based largely on FTIR measurements (*181–182, 186*) and to some extent on recent calorimetric evaluations of lysozyme/sucrose mixtures (*187*). Regardless of the specific mechanism of stabilization by these glass-forming and hydrogen-bond-donating excipients, some proteins clearly show significant changes in conformational structure during lyophilization, which may or may not be reversible on reconstitution, whereas some proteins are seemingly unaffected by the lyophilization process. The irreversibility of the conformationally altered proteins in the dehydrated solid may be largely due to the competitive process of refolding and aggregation, which occurs during the reconstitution step (*182*). Moisture levels are critical for stability of proteins toward unfolding and aggregation in the solid state (*121, 140, 185, 187*).

Principles of Colloid Science and Kinetics of Aggregation

The aggregating form of the protein, irrespective of how it is formed (whether by temperature fluctuations, solvent changes, or interfacial interactions), reaches a point of supersaturation, initiating the formation of a nucleus for aggregate growth. In some cases, this nucleation core may be very specific in nature and may actually exist as dimeric or stoichiometrically dictated complexes as seen with bovine growth hormone (*128*), insulin aggregation (*152*), α-chymotrypsin (*189*), γ-interferons (*123*), or amyloid fibrils (*149*). Despite their diverse nature and potential to undergo very specific interactions, the general solution behavior of polyelectrolytes such as proteins is still primarily governed by their colloidal nature. Colloidal science principles such as the Derjaguin–Landau–Verwey–Overbeck theory have been used to better understand the intermolecular interactions that can lead to irreversible states of interaction (*120*). In most cases, rapid aggregation occurs only after a lag time during which a sufficient number of these nucleation cores accumulate and reach a critical level to support rapid aggregate growth (*123, 141, 148–150, 161, 166, 189, 190*).

Nucleation rate-controlled reactions and their lag times can be very sensitive to the concentration of the aggregating species, depending on the size of the nucleating core. Assuming an octameric nucleus for the amyloid peptide, it might be expected that a change in lag time from 100 years to 3 h can occur with only a sixfold increase in concentration (*149*). Furthermore, the kinetics of nucleation rate-limited reactions can be greatly accelerated by the addition of seeding aggregates (*148–151*). Covalently linked aggregates of bovine somatotropin, either as initial impurities or generated with time in situ, greatly enhance the rate of aggregation and precipitation of monomeric bovine somatotropin (*156*). Double nucleation models have been invoked to describe the aggregation kinetics of sickle-cell hemoglobin and calcitonin fibril growth (*150, 151*). In such models, not only does homogeneous nucleation occur with the formation of nodules or spherulite domains leading to fiber growth, but heterogeneous nucleation also occurs on the surface of growing fibrils, resulting in a branching of aggregates and ultimately gel formation (*150, 151*). Even though the lag times for calcitonin fibrillation are temperature dependent (apparent activation energies of ~20 kcal/mol), the rate of rapid growth following the lag period is relatively independent of temperature (*150*).

In cases where aggregate growth is rate-limiting, that is, beyond the lag time or under conditions where the concentration of aggregating species is very high (lag phase undetectable), kinetic descriptions of aggregate growth also employ colloidal science models. For example, the Smoluchowski aggregation theory of colloids successfully described the rates of precipitation of denatured proteins (*189–191*) and of isoelectrically precipitated proteins (*190*). From such diffusion rate-limited models, based on collisional dynamics, the often-observed second-order kinetics can be rationalized during periods of rate-limited aggregate growth. Factors that influence diffusional rates or efficiency of collision (i.e., viscosity, pH, surface charge, and ionic strength) would be expected to have a large impact on aggregate growth rates and morphology, whereas direct temperature effects would be minimal, as seen with calcitonin fibril growth (*150*).

In spite of the acknowledged importance of physical instability of proteins and peptides, which leads to deleterious effects of aggregation, precipitation, or gelling, the ability to quantitatively describe or predict kinetics of such processes is emerging very slowly. This evolution, of necessity, will be required to entwine the effects of numerous stress variables on the dynamics of

- the availability of "activated" protein/peptide through exposure of fleeting hydrophobic surfaces in the native subpopulation, exposure of much longer-lived hydrophobic surfaces in molten-globule intermediates, or exposure of hydrophobic residues in the random coil structure of the unfolded protein; and
- the subsequent irreversible processes governed largely by lag phases of nucleation and rapid phases of aggregate growth as delineated by colloidal science principles.

Even though many trends may be consistent among proteins, the quantitative effects of specific variables on such irreversible processes most likely will be highly peptide- and protein-dependent.

References

1. *Peptide and Protein Delivery;* Lee, V. H., Ed.; Marcel Dekker: New York, 1991.
2. *Biological Barriers to Protein Delivery;* Audus, K. L.; Raub, T. J., Eds.; Plenum: New York, 1993.
3. *Stability and Characterization of Protein and Peptide Drugs;* Wang, Y. J.; Pearlman, R., Eds.; Plenum: New York, 1993.
4. *Formulation and Delivery of Proteins and Peptides;* Cleland, J. L.; Langer, R., Eds.; ACS Symposium Series 567; American Chemical Society: Washington, DC, 1994.
5. Cleland, J. L.; Powell, M. F.; Shire, S. J. *Crit. Rev. Ther. Drug Carrier Syst.* **1993,** *10,* 307–377.
6. Manning, M. C.; Patel, K.; Borchardt, R. T. *Pharm. Res.* **1989,** *6,* 903–917.
7. *Stability of Protein Pharmaceuticals, Chemical and Physical Pathways of Protein Degradation;* Ahern, T. J.; Manning, M. C., Eds.; Plenum: New York, 1992.
8. *Stability of Protein Pharmaceuticals, In Vivo Pathways of Degradation and Strategies for Protein Stabilization;* Ahern, T. J.; Manning, M. C., Eds.; Plenum: New York, 1992.
9. Clarke, S.; Stephenson, R. C.; Lowenson, J. D. In *Stability of Protein Pharmaceuticals, Chemical and Physical Pathways of Protein Degradation;* Ahern, T. J.; Manning, M. C., Eds.; Plenum: New York, 1992; pp 1–29.
10. Geiger, T.; Clarke, S. *J. Biol. Chem.* **1987,** *262,* 785–794.
11. Patel, K.; Borchardt, R. T. *Pharm. Res.* **1990,** *7,* 703–711.
12. Robinson, A. B.; Rudd, C. J. *Curr. Top. Cell. Regul.* **1974,** *8,* 247–295.
13. Stevenson, C. L.; Williams, T. D.; Anderegg, R. J.; Borchardt, R. T. *J. Pharm. Biomed. Anal.* **1992,** *10,* 567–575.
14. Stevenson, C. L.; Anderegg, R. J.; Borchardt, R. T. *J. Pharm. Biomed. Anal.* **1993,** *11,* 367–373.
15. Johnson, B. A.; Shirokawa, J. M.; Hancock, W. S.; Spellman, M. W.; Basa, L.; Aswad, D. W. *J. Biol. Chem.* **1989,** *264,* 14262–14271.
16. Johnson, B. A.; Aswad, D. W. *Anal. Biochem.* **1991,** *192,* 384–391.
17. Potter, S. M.; Henzel, W. J.; Aswad, D. W. *Protein Sci.* **1993,** *2,* 1648–1663.
18. Paranandi, M.; Guzzetta, A. W.; Hancock, W. S.; Aswad, D. W. *J. Biol. Chem.* **1994,** *269,* 243–253.
19. Ahern, T. J.; Klibanov, A. M. *Science (Washington, D.C.)* **1985,** *228,* 1281–1284.
20. Zale, S. E.; Klibanov, A. M. *Biochemistry* **1986,** *26,* 5432–5444.
21. Stevenson, C. L.; Donlan, M. E.; Friedman, A. R.; Borchardt, R. T. *Int. J. Pept. Protein Res.* **1993,** *42,* 24–32.
22. Yuan, P. M.; Talent, J. M.; Gracy, R. W. *Mech. Ageing Dev.* **1981,** *17,* 151–162.
23. Bhatt, N. P.; Patel, K.; Borchardt, R. T. *Pharm. Res.* **1990,** *7,* 593–599.
24. Gu, L. C.; Erdos, E. A.; Chiang, H.-S.; Calderwood, T.; Tsai, K.; Visor, G. C.; Duffy, T.; Hsu, W.-C.; Foster, L. C. *Pharm. Res.* **1991,** *8,* 485–490.
25. Bischoff, R.; Lepage, P.; Jaquinod, M.; Cauet, G.; Acker-Klein, M.; Clesse, D.; Laporte, M.; Bayol, A.; Van Dorsselaer, A.; Roitsch, C. *Biochemistry* **1993,** *32,* 725–734.
26. Stephenson, R. C.; Clarke, S. *J. Biol. Chem.* **1989,** *264,* 6164–6170.
27. Patel, K.; Borchardt, R. T. *Pharm. Res.* **1990,** *7,* 787–793.
28. Clarke, S. *Int. J. Pept. Protein Res.* **1987,** *30,* 808–821.
29. Stevenson, C. L.; Friedman, A. R.; Kubiak, T. M.; Donlan, M. E.; Borchardt, R. T. *Int. J. Pept. Protein Res.* **1993,** *42,* 497–503.

30. Capasso, S.; Mazzarella, L.; Zagari, A. *Pept. Res.* **1991**, *4,* 234–241.
31. Brennan, T. V.; Clarke, S. *Protein Sci.* **1993**, *2,* 331–338.
32. Scotchler, J. W.; Robinson, A. B. *Anal. Biochem.* **1974**, *59,* 319–322.
33. McKerrow, J. H.; Robinson, A. B. *Anal. Biochem.* **1971**, *42,* 565–568.
34. Lou, S.; Lroa, C.; McClelland, J. F.; Graves, D. J. *Int. J. Pept. Protein Res.* **1987**, *29,* 728–753.
35. Patel, K.; Borchardt, R. T. *J. Parenter. Sci. Technol.* **1990**, *44,* 300–301.
36. Oliyai, C.; Borchardt, R. T. In *Formulation and Delivery of Proteins and Peptides;* Cleland, J. L.; Langer, R., Eds.; ACS Symposium Series 567; American Chemical Society: Washington, DC, 1994; pp 46–58.
37. Oliyai, C.; Patel, J.; Carr, L.; Borchardt, R. T. *J. Parenter. Sci. Technol.* **1994**, *48,* 167–173.
38. Gearhart, D. A.; Kirsch, L. E. *Pharm. Res.* **1988**, *5,* S244.
39. Pikal, N. J.; Dellerman, K. M.; Roy, M. L.; Riggin, R. M. *Pharm. Res.* **1991**, *8,* 427–436.
40. Perseo, G.; Forino, R.; Galantino, M.; Gioia, B.; Malatesta, V.; DeCastiglione, R. *Int. J. Pept. Protein Res.* **1986**, *27,* 51–60.
41. Oliyai, C.; Borchardt, R. T. *Pharm. Res.* **1993**, *10,* 95–102.
42. Oliyai, C.; Borchardt, R. T. *Pharm. Res.* **1994**, *11,* 751–758.
43. Oliyai, C.; Patel, J. P.; Carr, L.; Borchardt, R. T. *Pharm. Res.* **1994**, *11,* 901–908.
44. Clarke, S. *Annu. Rev. Biochem.* **1985**, *54,* 479–506.
45. Lowenson, J. D.; Clarke, S. *J. Biol. Chem.* **1991**, *266,* 19396–19406.
46. Lowenson, J. D.; Clarke, S. *J. Biol. Chem.* **1992**, *267,* 5985–5995.
47. Ota, I. M.; Ding, L.; Clarke, S. *J. Biol. Chem.* **1987**, *262,* 8522–8531.
48. Ota, I. M.; Clarke, S. *J. Biol. Chem.* **1989**, *264,* 54–60.
49. George-Nascimento, C.; Lowenson, J.; Borissenko, M.; Calderon, M.; Medina-Selby, A.; Kuo, J.; Clarke, S.; Randolph, A. *Biochemistry* **1990**, *29,* 9584–9591.
50. Kirsch, L. E.; Molloy, R. M.; Debonon, M.; Baker, P.; Farid, K. Z. *Pharm. Res.* **1989**, *6,* 387–393.
51. Terwilliger, T. C.; Clarke, S. *J. Biol. Chem.* **1981**, *256,* 3067–3076.
52. Melville, J. *Biochem. J.* **1935**, *29,* 179–186.
53. Stults, J. T.; Bourell, J. H.; Canova-Davis, E.; Ling, U. T.; Laramee, G. R.; Winslow, J. W.; Griffin, P. R.; Rinderknecht, E.; Vandlen, R. L. *Biomed. Environ. Mass Spectrom.* **1990**, *19,* 655–658.
54. Marsden, B. J.; Nguyen, T. M. D.; Schiller, P. W. *Int. J. Pept. Protein Res.* **1993**, *41,* 313–316.
55. Strickley, R. G.; Visor, G. C.; Lin, L. H.; Gu, L. *Pharm. Res.* **1989**, *6,* 971–975.
56. Mazurov, A. A.; Andronati, S. A.; Korotenko, T. I.; Gorbatylk, V. Y.; Shapiro, Y. E. *Int. J. Pept. Protein Res.* **1993**, *42,* 14–19.
57. Battersby, J. E.; Hancock, W. S.; Conova-Davis, E.; Oeswein, J.; O'Connor, B. *Int. J. Pept. Protein Res.* **1994**, *44,* 215–222.
58. Leach, S. J.; Lindley, H. *Trans. Faraday Soc.* **1953**, *49,* 921–928.
59. Slobin, L. I.; Carpenter, F. H. *Biochemistry* **1963**, *2,* 22–28.
60. Darrington, R. T.; Anderson, B. D. *Pharm. Res.* **1994**, *11,* 784–793.
61. Stadtman, E. R. *Free Radical Biol. Med.* **1990**, *9,* 315–325.
62. Stadtman, E. R. *Biochemistry* **1990**, *29,* 6323–6331.
63. Stadtman, E. R. *Science (Washington, D.C.)* **1992**, *257,* 1220–1224.
64. Stadtman, E. R. *Annu. Rev. Biochem.* **1993**, *62,* 797–821.
65. Dean, R. T.; Gieseg, S.; Davies, M. J. *Trends Biochem. Sci.* **1993**, *18,* 437–441.
66. Nguyen, T. H. In *Formulation and Delivery of Proteins and Peptides;* Cleland, J. L.; Langer, R., Eds.; ACS Symposium Series 567; American Chemical Society: Washington, DC, 1994; pp 59–71.
67. Buxton, G. V.; Greenstock, C. L.; Helman, W. P.; Ross, A. B. *J. Phys. Chem. Ref. Data* **1988**, *17,* 513–886.
68. Foote, C. S. *Science (Washington, D.C.)* **1968**, *162,* 963–969.
69. Pryor, W. A.; Uppu, R. M. *J. Biol. Chem.* **1993**, *268,* 3120–3126.
70. Sysak, P. K.; Foote, C. S.; Ching, T.-Y. *Photochem. Photobiol.* **1977**, *26,* 19–27.
71. Barton, J. P.; Packer, J. E.; Sims, R. J. *J. Chem. Soc. Perkin Trans. 2* **1973**, 1547–1549.
72. Shechter, Y.; Burstein, Y.; Patchornik, A. *Biochemistry* **1975**, *14,* 4497–4503.
73. Pearlman, R.; Nguyen, T. H. *J. Pharm. Pharmacol.* **1992**, *44,* 178–185.

74. Nguyen, T. H.; Burnier, J.; Meng, W. *Pharm. Res.* **1993,** *10,* 1563–1571.
75. Li, S.; Nguyen, T. H.; Schöneich, C. H.; Borchardt, R. T. *Biochemistry* **1995,** *34,* 5762–5772.
76. Levine, R. L. *J. Biol. Chem.* **1983,** *258,* 11823–11827.
77. Farber, J. M.; Levine, R. L. *J. Biol. Chem.* **1986,** *261,* 4574–4578.
78. Uchida, K.; Kawakishi, S. *J. Biol. Chem.* **1994,** *269,* 2405–2410.
79. Cheng, R.-Z.; Kawakishi, S. *Eur. J. Biochem.* **1994,** *223,* 759–764.
80. Yao, Y.; Yin, D.; Squier, T. C.; Schöneich, C. H. *Biophys. J.* **1994,** *66,* A73.
81. Gellman, S. H. *Biochemistry* **1991,** *30,* 6633–6636.
82. Asmus, K.-D. In *Sulfur-Centered Reactive Intermediates in Chemistry and Biology;* Chatgilialoglu, C.; Asmus, K.-D., Eds; NATO ASI Series A; Plenum: New York, 1990; Vol. 197, pp 155–172.
83. Doi, J. T.; Musker, W. K. *J. Am. Chem. Soc.* **1981,** *103,* 1159–1163.
84. Wilson, G. S.; Swanson, D. D.; Klug, J. T.; Glass, R. S.; Ryan, M. D.; Musker, W. K. *J. Am. Chem. Soc.* **1979,** *101,* 1040–1042.
85. Glass, R. S. In *Sulfur-Centered Reactive Intermediates in Chemistry and Biology;* Chatgilialoglu, C.; Asmus, K.-D., Eds; NATO ASI Series A; Plenum: New York 1990; Vol. 197, pp 213–226.
86. Merényi, G.; Lind, J.; Jonsson, M. *J. Am. Chem. Soc.* **1993,** *115,* 4945–4946.
87. Surdhar, P. S.; Armstrong, D. A. *J. Phys. Chem.* **1987,** *91,* 6532–6537.
88. Wardman, P. *J. Phys. Chem. Ref. Data* **1989,** *18,* 1637–1756.
89. Sanaullah; Wilson, G. S.; Glass, R. S. *J. Inorg. Biochem.* **1994,** *55,* 87–99.
90. Jovanovic, S. V.; Simic, M. G. In *Oxygen Radicals in Biology and Medicine;* Simic, M. G.; Taylor, K. A.; Ward, J. F.; von Sonntag, C., Eds.; Plenum: New York, 1988; pp 115–122.
91. Buettner, G. R. *J. Biochem. Biophys. Methods* **1988,** *16,* 27–40.
92. Zhang, N.; Schuchmann, H.-P.; von Sonntag, C. *J. Phys. Chem.* **1991,** *95,* 4718–4722.
93. Feig, A. L.; Lippard, S. J. *Chem. Rev.* **1994,** *94,* 759–805.
94. Mönig, J.; Göbl, M.; Asmus, K.-D. *J. Chem. Soc. Perkin Trans. 2* **1985,** 647–651.
95. Packer, J. E.; Mahood, J. S.; Willson, R. L.; Wolfenden, B. S. *Int. J. Radiat. Biol.* **1981,** *39,* 135–141.
96. Rush, J. D.; Koppenol, W. H. *J. Am. Chem. Soc.* **1988,** *110,* 4957–4963.
97. Rahal, S.; Richter, H. W. *J. Am. Chem. Soc.* **1988,** *110,* 3126–3133.
98. Yamazaki, I.; Piette, L. H. *J. Am. Chem. Soc.* **1991,** *113,* 7588–7593.
99. Walling, C. *Acc. Chem. Res.* **1975,** *8,* 125–131.
100. Gieseg, S. P.; Simpson, J. A.; Charlton, T. S.; Duncan, M. W.; Dean, R. T. *Biochemistry* **1993,** *32,* 4780–4786.
101. Gebicki, S.; Gebicki, J. M. *Biochem. J.* **1993,** *289,* 743–749.
102. Neuzil, J.; Gebicki, J. M.; Stocker, R. *Biochem. J.* **1993,** *293,* 601–606.
103. Guptasarma, P.; Balasubramanian, D.; Matsugo, S.; Saito, I. *Biochemistry* **1992,** *31,* 4296–4303.
104. Adam, W.; Ahrweiler, M.; Peters, K.; Schmiedeskamp, B. *J. Org. Chem.* **1994,** *59,* 2733–2739.
105. Shen, X.; Lind, J.; Eriksen, T. E.; Merenyi, G. *J. Chem. Soc. Perkin Trans. 2* **1989,** 555–562.
106. Giulivi, C.; Davies, K. J. A. *Methods Enzymol.* **1994,** *233,* 363–371.
107. Dean, R. T.; Wolff, S. P.; McElligott, M. A. *Free Radical Res. Commun.* **1989,** *7,* 97–103.
108. Schöneich, C. H.; Aced, A.; Asmus, K.-D. *J. Am. Chem. Soc.* **1993,** *115,* 11376–11383.
109. Schöneich, C. H.; Bobrowski, K. *J. Phys. Chem.* **1994,** *98,* 12613–12620.
110. Schöneich, C. H.; Zhao, F.; Wilson, G. S.; Borchardt, R. T. *Biochim. Biophys. Acta* **1993,** *1158,* 307–322.
111. Li, S.; Schöneich, C. H.; Wilson, G. S.; Borchardt, R. T. *Pharm. Res.* **1993,** *10,* 1572–1579.
112. Zhao, F.; Yang, J.; Schöneich, C. H. *Pharm. Res.* **1996,** *13,* 931–938.
113. Schöneich, C. H.; Zhao, F.; Madden, K. P.; Bobrowski, K. *J. Am. Chem. Soc.* **1994,** *116,* 4641–4652.
114. Bobrowski, K.; Schöneich, C. H.; Holcman, J.; Asmus, K.-D. *J. Chem. Soc. Perkin Trans 2* **1991,** 353–362.
115. Wang, Y.-C. J. In *Pharmaceutical Dosage Forms: Parenteral Medications;* Avis, K. E.; Lieberman, H. A.; Lachman, L., Eds.; Marcel Dekker: New York, 1992; Vol. 1, pp 283–319.

116. Wetzel, R. *Trends Biotechnol.* **1994,** *12,* 193–198.
117. Shirley, B. A. In *Stability of Protein Pharmaceuticals. Part A, Chemical and Physical Pathways of Protein Degradation;* Ahern, T. J.; Manning, M. C., Eds.; Plenum: New York, 1992; pp 167–194.
118. Lumbry, R.; Eyring, H. *J. Phys. Chem.* **1954,** *58,* 110–120.
119. Zale, S. E.; Klibanov, A. M. *Biotechnol. Bioeng.* **1983,** *25,* 2221–2230.
120. DeYoung, L. R.; Fink, A. L.; Dill, K. A. *Acc. Chem. Res.* **1993,** *26,* 614–620.
121. Bell, L. N.; Hageman, M. J.; Bauer, J. M. *Biopolymers* **1995,** *35,* 201–209.
122. Owusu Apenten, R. K.; Berthalon, N. *Food Chem.* **1994,** *51,* 15–20.
123. Mulkerrin, M. G.; Wetzel, R. *Biochemistry* **1989,** *28,* 6556–6561.
124. Kuwajima, K. *Proteins Struct. Funct. Genet.* **1989,** *6,* 87–103.
125. Fischer, B.; Sumner, I.; Goodenough, P. *Arch. Biochem. Biophys.* **1993,** *306,* 183–187.
126. Brems, D. N.; Plaisted, S. M.; Kauffman, E. W.; Havel, H. A. *Biochemistry* **1986,** *25,* 6539–6543.
127. Brems, D. N.; Plaisted, S. M.; Dougherty, J. J., Jr.; Holzman, T. F. *J. Biol. Chem.* **1987,** *262,* 2590–2596.
128. Brems, D. N. *Biochemistry* **1988,** *27,* 4541–4546.
129. Mach, H.; Ryan, J. A.; Burke, C. J.; Volkin, D. B.; Middaugh, C. R. *Biochemistry* **1993,** *32,* 7703–7711.
130. Hughson, F. M.; Wright, P. E.; Baldwin, R. L. *Science (Washington, D.C.)* **1990,** *249,* 1544–1548.
131. Cleland, J. L.; Wang, D. I. C. *Biochemistry* **1990,** *29,* 11072–11078.
132. Kuroda, Y.; Kidokoro, S.; Wada, A. *J. Mol. Biol.* **1992,** *223,* 1139–1153.
133. Xie, D.; Fox, R.; Freire, E. *Protein Sci.* **1994,** *3,* 2175–2184.
134. Xie, D.; Bhakuni, V.; Freire, E. *Biochemistry* **1991,** *30,* 10673–10678.
135. Ansari, A.; Berendzen, J.; Bowne, S. F.; Frauenfelder, H.; Iben, I. E. T.; Sauke, T. B.; Shyamsunder, E.; Young, R. D. *Proc. Natl. Acad. Sci. U.S.A.* **1985,** *82,* 5000–5004.
136. Clarke, J.; Hounslow, A. M.; Bycroft, M.; Fersht, A. R. *Proc. Natl. Acad. Sci. U.S.A.* **1993,** *90,* 9837–9841.
137. Schejter, A.; Luntz, T. L.; Koshy, T. I.; Margoliash, E. *Biochemistry* **1992,** *31,* 8336–8343.
138. Wagner, B.; Wuthrich, K. *J. Mol. Biol.* **1979,** *130,* 31–37.
139. Kato, A.; Tanimoto, S.; Muraki, Y.; Kobayashi, K.; Kumagai, I. *Biosci. Biotechnol. Biochem.* **1992,** *56,* 1424–1428.
140. Costantino, H. R.; Langer, R.; Klibanov, A. M. *Pharm. Res.* **1994,** *11,* 21–29.
141. Dathe, M.; Gast, K.; Zirwer, D.; Welfle, H.; Mehlis, B. *Int. J. Pept. Protein Res.* **1990,** *36,* 344–349.
142. Utsumi, J.; Yamazaki, S.; Kawaguchi, K.; Kimura, S.; Shimizu, H. *Biochim. Biophys. Acta* **1989,** *998,* 167–172.
143. Chen, B.; Arakawa, T.; Morris, C. F.; Kenney, W. C.; Wells, C. M.; Pitt, C. G. *Pharm. Res.* **1994,** *11,* 1581–1587.
144. Rogerson, A.; Sanders, L. *Proc. Int. Symp. Controlled Release Bioact. Mater.* **1987,** *14,* 97–98.
145. Moran, E. C.; Chou, P. Y.; Fasman, G. D. *Biochem. Biophys. Res. Commun.* **1977,** *77,* 1300–1306.
146. Bauer, H. H.; Muller, M.; Goette, J.; Merkle, H. P.; Fringeli, U. P. *Biochemistry* **1994,** *33,* 12276–12282.
147. Yesook, K.; Rose, C. A.; Yongliang, L.; Ozaki, Y.; Datta, G.; Tu, A. T. *J. Pharm. Sci.* **1994,** *83,* 1175–1180.
148. Come, J. H.; Fraser, P. E.; Lansbury, P. T., Jr. *Proc. Natl. Acad. Sci. U.S.A.* **1993,** *90,* 5959–5963.
149. Jarrett, J. T.; Berger, E. P.; Lansbury, P. T., Jr. *Biochemistry* **1993,** *32,* 4693–4697.
150. Arvinte, T.; Cudd, A.; Drake, A. F. *J. Biol. Chem.* **1993,** *268,* 6415–6422.
151. Samuel, R. E.; Salmon, E. D.; Briehl, R. W. *Nature (London)* **1990,** *345,* 833–835.
152. Brange, J.; Langkjoer, L. *Pharm. Biotechnol.* **1993,** *5,* 315–350.
153. Endo, Y.; Nagai, H.; Watanabe, Y.; Ochi, K.; Takagi, T. *J. Biochem. (Tokyo)* **1992,** *112,* 700–706.
154. Hegg, P.-O.; Martens, H.; Lofqvist, B. *J. Sci. Food Agric.* **1979,** *30,* 981–993.
155. Gimel, J.-C.; Durand, D.; Nicolai, T. *Macromolecules* **1994,** *27,* 583–589.
156. Hageman, M. J.; Admiraal, S. J.; Bauer, J. M. *Pharm. Res.* **1994,** *11,* S-71.
157. Mitra, G.; Schneider, P. M.; Lundblad, J. L. *Biotechnol. Bioeng.* **1982,** *24,* 97–107.
158. Privalov, P. L. *Crit. Rev. Biochem. Mol. Biol.* **1990,** *25,* 281–305.

159. Tamura, A.; Kimura, K.; Akasaka, K. *Biochemistry* **1991**, *30*, 11313–11320.
160. Timasheff, S. N. *Annu. Rev. Biophys. Biomol. Struct.* **1993**, *22*, 67–97.
161. Eberlein, G. A.; Stratton, P. R.; Wang, Y. J. *PDA J. Pharm. Sci. Technol.* **1994**, *48*, 224–230.
162. Bauer, J. M.; Hageman, M. J.; Vidmar, T. J. *Pharm. Res.* **1993**, *10*, S-86.
163. Davio, S. R.; Hageman, M. J. In *Stability and Characterization of Protein and Peptide Drugs;* Wang, Y. J.; Pearlman, R., Eds.; Plenum: New York, 1992; pp 59–89.
164. Eckhardt, B. M.; Oeswein, J. Q.; Bewley, T. A. *Pharm. Res.* **1991**, *8*, 1360–1364.
165. Pretzer, D.; Schulteis, B. S.; Vander Velde, D. G.; Smith, C. D.; Mitchell, J. W.; Manning, M. C. *Pharm. Res.* **1992**, *9*, 870–877.
166. Sluzky, V.; Tamada, J. A.; Klibanov, A. M.; Langer, R. *Proc. Natl. Acad. Sci. U.S.A.* **1991**, *88*, 9377–9381.
167. Volkin, D. B.; Tsai, P. K.; Dabora, J. M.; Gress, J. O.; Burke, C. J.; Linhardt, R. J.; Middaugh, C. R. *Arch. Biochem. Biophys.* **1993**, *300*, 30–41.
168. Vemuri, S.; Beylin, I.; Sluzky, V.; Stratton, P.; Eberlein, G.; Wang, Y. J. *J. Pharm. Pharmacol.* **1994**, *46*, 481–486.
169. Thurow, H.; Geisen, K. *Diabetologia* **1984**, *27*, 212–218.
170. Wang, P.; Johnston, T. P. *J. Parenter. Sci. Technol.* **1993**, *47*, 183–189.
171. Pearlman, R.; Bewley, T. A. In *Stability and Characterization of Protein and Peptide Drugs;* Wang, Y. J.; Pearlman, R., Eds.; Plenum: New York, 1992; pp 1–58.
172. Wang, P.; Johnston, T. P. *Int. J. Pharm.* **1993**, *96*, 41–49.
173. Cleland, J. L.; Randolph, T. W. *J. Biol. Chem.* **1992**, *267*, 33147–33153.
174. Andrade, J. D.; Hlady, V.; Wei, A.-P.; Ho, C.-H.; Lea, A. S.; Jeon, S. I.; Lin, Y. S.; Stroup, E. *Clin. Mater.* **1992**, *11*, 67–84.
175. Horbett, T. A. In *Stability of Protein Pharmaceuticals. Part A, Chemical and Physical Pathways of Protein Degradation;* Ahern, T. J.; Manning, M. C., Eds.; Plenum: New York, 1992; pp 195–214.
176. Niven, R. W.; Ip, A. Y.; Mittelman, S. D.; Farrar, C.; Arakawa, T.; Prestrelski, S. J. *Int. J. Pharm.* **1994**, *109*, 17–26.
177. Burke, C. J.; Steadman, B. L.; Volkin, D. B.; Tsai, P. K.; Bruner, M. W.; Middaugh, C. R. *Int. J. Pharm.* **1992**, *86*, 89–93.
178. Kato, A.; Yutani, K. *Protein Eng.* **1988**, *2*, 153–156.
179. Norde, W. In *Stability and Stailization of Enzymes;* van den Tweel, W. J. J.; Harder, A.; Buitelaar, R. M., Eds.; Elsevier Science: Amsterdam, Netherlands, 1993; pp 3–11.
180. Nygren, H.; Stenberg, M. *Biophys. Chem.* **1990**, *38*, 77–85.
181. Carpenter, J. F.; Prestrelski, S. J.; Anchordoguy, T. J.; Arakawa, T. In *Formulation and Delivery of Proteins and Peptides;* Cleland, J. L.; Langer, R., Eds.; ACS Symposium Series 567; American Chemical Society: Washington, DC, 1994; pp 134–147.
182. Prestrelski, S. J.; Arakawa, T.; Carpenter, J. F. In *Formulation and Delivery of Proteins and Peptides;* Cleland, J. L.; Langer, R., Eds.; ACS Symposium Series 567; American Chemical Society: Washington, DC, 1994; pp 148–169.
183. Pikal, M. J. In *Formulation and Delivery of Proteins and Peptides;* Cleland, J. L.; Langer, R., Eds.; ACS Symposium Series 567; American Chemical Society: Washington, DC, 1994; pp 120–133.
184. Franks, F.; Hatley, R. H.; Mathias, S. F. *BioPharm (Eugene, Oreg.)* **1991**, *4*, 38–42, 55.
185. Hageman, M. J. In *Stability of Protein Pharmaceuticals. Part A, Chemical and Physical Pathways of Protein Degradation;* Ahern, T. J.; Manning, M. C., Eds.; Plenum: New York, 1992; pp 273–309.
186. Remmele, R. L.; Stushnoff, C.; Carpenter, J. F. In *Formulation and Delivery of Proteins and Peptides;* Cleland, J. L.; Langer, R., Eds.; ACS Symposium Series 567; American Chemical Society: Washington, DC, 1994; pp 170–191.
187. Bell, L. N.; Hageman, M. J.; Muraoka, L. M. *J. Pharm. Sci.* **1995**, *84*, 707–712.
188. Hsu, C. C.; Nguyen, H. M.; Yeung, D. A.; Brooks, D. A.; Koe, G. S.; Bewley, T. A.; Pearlman, R. *Pharm. Res.* **1995**, *12*, 69–77.
189. Przybycien, T. M.; Bailey, J. E. *AIChE J.* **1989**, *35*, 1779–1790.
190. Glatz, C. E. In *Stability of Protein Pharmaceuticals. Part A, Chemical and Physical Pathways of Protein Degradation;* Ahern, T. J.; Manning, M. C., Eds.; Plenum: New York, 1992; pp 135–166.
191. Patro, S. Y.; Przybycien, T. M. *Biophys. J.* **1994**, *66*, 1274–1289.

12

Peptide, Protein, and Vaccine Delivery from Implantable Polymeric Systems

Progress and Challenges

Steven P. Schwendeman, Henry R. Costantino, Rajesh K. Gupta, and Robert Langer

*Controlled release is an important alternative to conventional methods of delivery of therapeutic peptides and proteins, in particular those used as vaccine antigens. In this mini-review, we describe the fundamentals of this mode of delivery, and we emphasize biodegradable injectables, such as microspheres from the poly(lactide-*co*-glycolide)s, and their role as vaccine adjuvants. Progress toward stabilizing peptides and proteins within polymeric carriers is a major factor limiting the development of these dosage forms. Several examples from the literature are used to examine the status of this field and to define the challenges associated with this form of drug delivery.*

The controlled release of drugs from polymers has developed into an extremely important area of drug delivery. The concept of delivering small molecules (molecular weight, MW < 600 Da) slowly and continuously from polymer systems has been applied to a long list of diseases such as cancer (*1*), heart disease (*2*), and parkinsonism (*3, 4*). Likewise, an increasing number of peptide drugs has been formulated into controlled-release devices for use in prostate cancer (*5, 6*), diabetes (*7*), and cell transplantation (*8*). Even though the physical description of the mechanisms of protein release from polymers has been highly developed (*9–14*), the stability of the encapsulated protein, in terms of the native three-dimensional structure required for pharmacological activity (protein drugs) or conformational epitopes (vaccine antigens) required to elicit a protective immune response has been relatively unexplored. This focus is particularly important for injectables based on biodegradables such as poly(lactide-*co*-glycolide)s (PLGA), which are among the most promising new controlled-release systems (*15*). On the basis of our most recent

Copyright © 1997 American Chemical Society

research efforts (*15–20*) we believe that understanding protein stability within the biodegradable polymer systems is one of the most critical factors in launching these dosage forms into the next generation of controlled drug delivery devices.

The great flexibility and general low cost of polymer materials have led to their wide-spread use in controlled-release devices. For the purpose of controlled-release systems, polymers can be classified into three groups: nondegradable synthetic (e.g., poly(ethylene-*co*-vinyl acetate)(EVA)), degradable synthetic (e.g., PLGA), and degradable naturally derived (e.g., starch). Most notable examples of the delivery of peptides and proteins over extended time periods have been achieved with synthetic polymers (*1, 6, 21*). Thus, the focus of this chapter will be on this polymer class (particularly synthetic biodegradable polymers).

Several vital characteristics of polymers may be relevant to the stability of peptides and proteins at the time of release. These characteristics were reviewed recently (*15*) and consist of the following:

- polymer water content,
- factors relating to protein adsorption (e.g., polymer hydrophobicity),
- (if degradable) chemical nature of degradation products (e.g., if acidic pH may drop), and
- the ability of the polymer to retain low-molecular-weight stabilizing excipients.

When evaluating the stability of the peptide or protein of interest, it is essential to identify which of the above conditions (or other destabilizing factor) is primarily responsible for irreversible inactivation, so that logical experimental methodologies can be employed to determine the inactivating mechanism. This rational approach can then provide clues to how one may alter the formulation to inhibit or bypass the mechanism and thus stabilize the protein (*22*).

Categorizing peptides and proteins into at least three types is useful:

- type I, small peptides (including peptide isosteres) with little or no higher order structure that exhibit good chemical stability under relevant conditions for their release (e.g., pH, physiological temperature, and ionic strength);
- type II, small peptides like type I except those that are chemically labile under relevant conditions; and
- type III, larger peptides or proteins that have substantial higher order structure and have both physical and chemical stability limitations.

For type I, the peptide can be formulated with minimal stability considerations. Type II typically does not exhibit aggregation, whereas chemical inactivation is the primary obstacle (*23*). Type III would be a protein (roughly the size of insulin or larger) that in addition to having chemical lability has an exquisite three-dimensional structure, giving rise to several physical stability limitations (*22*). Furthermore, it is important to distinguish between a typical therapeutic peptide or protein and a vaccine antigen, because in the vaccine antigen case chemical characterization may be tedious (due to heterogeneity) and biological activity difficult to determine and interpret.

Unlike low-molecular-weight drugs, polypeptides and proteins cannot permeate films of hydrophobic polymers (i.e., which generally take up little water) (*13*). For example, a protein cannot diffuse through a nonporous film of silicone rubber. Channels or pores of sufficient size must be present at some stage for drug release. These channels may be created during manufacture (air pockets) or during release (dissolution and release of next-neighbor protein molecules, polymer rearrangement and microcrack formation under osmotic pressure, or erosion (for biodegradable polymers)). The percolation of channels formed due to release of protein powder from solvent-cast EVA films has been described (*9, 10, 24*). In addition to these percolation effects, the geometry of the pore has been predicted to be important (*25*). For biodegradable polymers, the transport problem can become simple to interpret, as in the case of "erosion-controlled" systems whose drug release kinetics closely correlate with the mass released from the polymer (*26*), or much more difficult, as in systems where multiple release mechanisms are involved (e.g., PLGA) (*5, 11*).

One growing application of biodegradable polymers is for delivery of peptide and protein vaccine antigens. These new "polymer adjuvants" may soon offer an alternative to the aluminum-based adjuvants and reduce the number of injections required for complete immunization (*27*). The elimination of the need for booster injections has immediate importance because diseases that have been eradicated in most developed countries are now present in countries where the logistical difficulty of delivering 2–3 doses of vaccine required for protection still remains (*28*). PLGA, which is known to be quite safe (*29, 30*), is the most commonly studied polymer adjuvant for protein vaccines. Biocompatibility is an extremely important characteristic for adjuvants, because side effects in even the smallest group of immunized population are largely unacceptable (*31*). Research in this area has focused largely on the release kinetics of the vaccine antigen and eliciting a good immune response relative to a positive control (i.e., two doses of antigen made into an adjuvant with alum). More definitive research remains ahead to provide new insights on how to release antigen in a single dose that has intact conformational antigenic determinants and can elicit a persistent and high-affinity immune response.

The objectives of this chapter are to examine the fundamentals of controlling the release of stable therapeutic peptides and proteins and vaccine antigens from polymer implants and to discuss a few of the important examples of recent approaches from our group and elsewhere in the development of this area of research.

Background of Polymer Systems for Delivery of Peptides and Proteins

Implantable controlled-release systems for peptides and proteins have emerged largely because of the tremendous difficulty in delivering these drugs into the body intact by noninvasive routes (e.g., oral and transdermal) and the poor patient comfort and compliance that accompany delivery of immediate-release dosage forms (e.g., daily injections). For protein and peptides, the environment in the polymer may improve drug stability relative to an ordinary aqueous environment (*26, 32, 33*). The potential disadvantages of implantable polymer delivery systems include their invasiveness, possible immunogenic and inflammatory responses, difficulty in removing the implants should drug delivery need to be ceased, and possible device failure (e.g., drug dumping).

Design of Implantable Controlled-Release Systems

Typical specific design parameters for implantable drug delivery devices include

- systemic, local, or targeted drug therapy;
- time of release and release profile (release rate as a function of time); and
- degradability or nondegradability.

For situations where release from the device is the rate-limiting step in drug delivery, a rough estimate of the desired delivery rate (dm/dt) can be obtained by its equivalence with the rate of drug elimination at steady state:

$$\frac{dm}{dt} = C_{ther} \cdot CL \qquad (1)$$

where C_{ther} and CL are the therapeutic drug concentration in, and drug clearance from, the blood circulation (i.e., systemic therapy) or local tissue (for local therapy), respectively. For local therapy, the decline in drug concentration as a function of position away from the implant site may also need to be considered (*34*).

The total drug mass required in the device may then be estimated by the product of the release time and dm/dt or by the integral of the release profile (in the case where the release rate is not nearly constant over time). These two design criteria, release rate as a function of time and mass of drug to be released, provide a good starting point for designing implantable drug delivery systems.

Synthetic Polymers

Nondegradable Polymers

Chemically inert, biocompatible polymers with high mechanical strength such as EVA and silicone rubber were the most widely used materials for long-term delivery of therapeutic agents before the use of biodegradables. These polymers were effective in illustrating the flexibility and potential of controlled drug delivery devices (*21, 35–37*). Currently, they are used more commonly for delivery of smaller molecules in implantables where a nondegradable polymer is required for a specific property that is unavailable in a biodegradable polymer (i.e., to house incompatible items) (*38, 39*) or in cases where long-term delivery is required (i.e., 5 years for the Norplant System, Wyeth-Ayerst (*40*)). The two likely sources for protein inactivation inside the nondegradables (besides those that are ubiquitous or irrespective to the presence of the polymer, e.g., oxygen) are the presence of moisture in the polymer (*19*) and the polymer surface (i.e., inactivation on adsorption) (*41*).

Synthetic Biodegradables

In the event that the body is able to eliminate the polymer (excluding water-soluble polymers) or its erosion products via metabolism or excretion, the polymer is said to be degradable. Long lists of biodegradable polymers under development are given elsewhere (*42–44*). Some of the more promising polymeric carriers are listed in Table 12.1 (*45–51*). The most common is the polyester based on lactic and glycolic acids (or more precisely, poly(lactide-*co*-glycolide) when prepared by ring-opening polymerization). Moreover, these degradables, like most, erode following chemical hydrolysis of their backbone to form acids and alcohols. This mode of degradation is highly desirable, because it requires only water for degradation (as opposed to enzyme-catalyzed hydrolysis). However, during degradation of the backbone, water is sorbed by the polymer and new chemical species are formed that dramatically change the polymer environment. For example, as PLGA degrades, the environment can become more hydrophilic (acids and alcohols are more hydrophilic than esters) and more acidic (due to the formation of new carboxylic acids) (*52*). Increases in water content of PLGA specimens will accompany the increased hydrophilicity during release: originally dry samples may reach 60 wt% water by later stages of erosion (*45, 46*). These characteristics of polymer erosion provide added challenges to the delivery of polypeptides and proteins, particularly because these events remain, to some extent, ambiguous. In addition to the effects of moisture and surface adsorption (hypothesized as the key protein destabilizing source for the nondegradables), pH, physical–chemical character of the soluble products formed, and reactivity of the polymer with the protein are now plausible sources of irreversible physical and chemical inactivation of peptides and proteins (*15*).

Preparing Implantable Devices

Device Configuration

With multiple types of polymers (Table 12.1), copolymer ratios, biocompatible excipients, geometries, and possible sizes available to incorporate into the design of a device, an enormous number of possible configurations can provide the desired release profile. The most basic configurations are those that provide a predetermined rate of drug release and typically are of the reservoir or monolithic (or matrix) type. Furthermore, more than one type of polymer may be used in a mixture or solution, or as a coating to reduce the release rate at initial times. More advanced configurations are those implantable self-regulated systems such as the insulin delivery/glucose detection systems (*53*).

Encapsulation

After a configuration has been selected, methods of microencapsulation and device fabrication must be examined to meet the design criteria. Methods

TABLE 12.1. Some Synthetic Biodegradable Polymers

Polymer Class	*Example*
Polyesters	poly(lactide-*co*-glycolide) 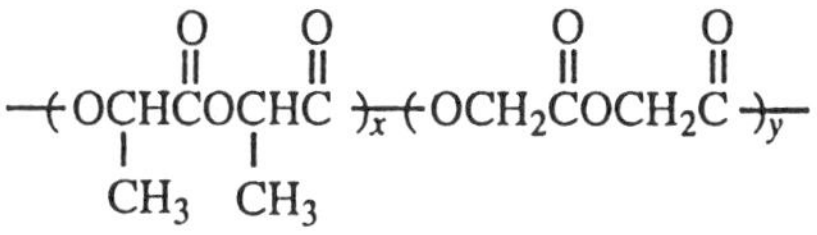
Polyanhydrides	poly[1,3-bis(*p*-carboxyphenoxy)propane-*co*-sebacic acid] 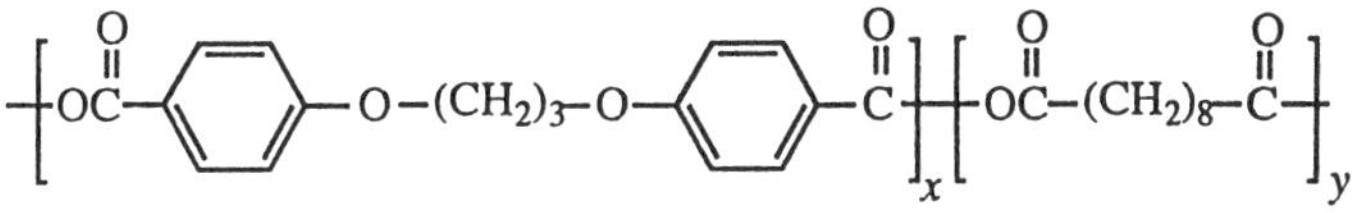
Poly(*ortho*-esters)	poly(*ortho*-esters) copolymerized with *trans*-cyclohexanedimethanol and 1,6-hexanediol monomers 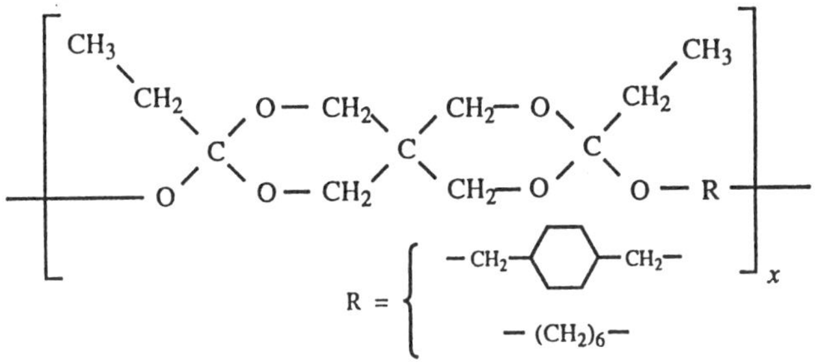
Polyiminocarbonates	poly(DAT-Tyr-Hex iminocarbonate) 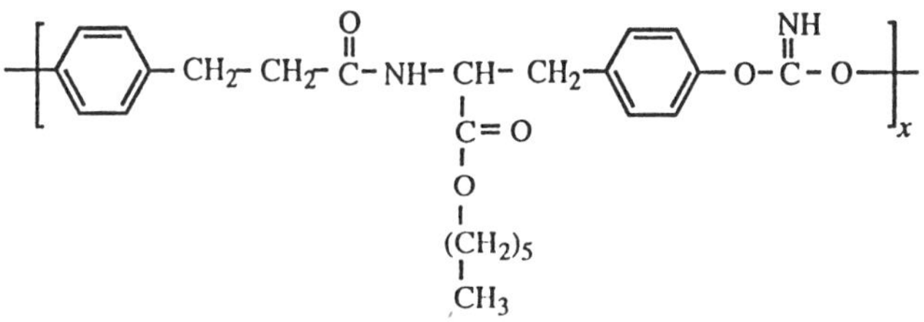
Polyphosphazenes	poly[bis(carboxylatophenoxy)phosphazene] 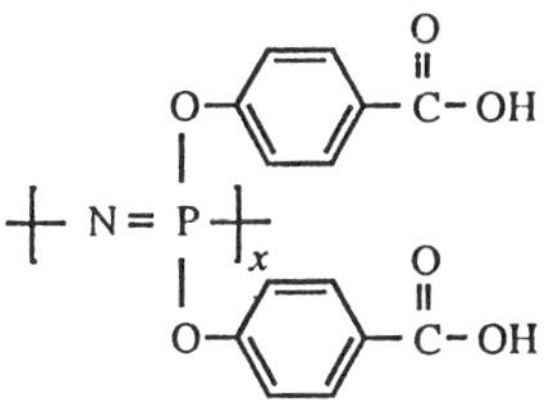

of preparing microspheres or microcapsules include emulsion and in-liquid drying methods (*54*), spray drying (*55*), coacervation and other phase-separation methods (*56*), and hot-melt encapsulation (*57*). For preparing larger geometries such as cylinders or discs, commonly used polymer processing techniques are employed such as those based on extrusion (*11*), molding (*58*), and solvent-casting (*59*). To process the polymer, it must be brought through its glass transition by some means such as by increasing temperature or pressure, or by plasticization. We have brought forward the concept that future encapsulation methods should process the polymer while maintaining the protein/excipient phase as a glass, because proteins are known to be most stable when maintained "rigidified" (*15*). An example is processing PLGA with organic solvents such as acetonitrile or ethyl acetate. If water has been removed from the system, solid proteins are known to be extremely stable in such organic solvents that do not dissolve them (*60*).

Such preparation techniques are relatively harsh with respect to peptides and proteins, potentially inducing unfolding and further deleterious covalent and noncovalent events. This result may be ameliorated by the presence of sugars, which increase protein stability during exposure to organic solvents (*61*). This stabilization is probably due to the preferential hydration effect; sugars increase the activity of water on the protein surface, favoring the folded structure thermodynamically (*62*).

Future alternatives to the simultaneous exposure of the protein to aqueous and organic solvents are methods that expose the solid protein to the organics and harden in the absence of water (e.g., o/o emulsion methods). Hot-melt encapsulation is a possibility because organic solvents are not used, although typically high temperatures are required to soften the polymer, which can also increase the flexibility or "unrigidify" the solid protein to promote unfolding (e.g., as observed during differential scanning microcalorimetry (*63*)).

Microspheres and Microcapsules. Microparticles of monolithic (microspheres) or capsular (microcapsules) configuration, with their smooth surfaces, can be dispersed readily in vehicles for injection. Despite their small size, these particles can release their contents and remain in the body for more than a month when prepared with slow degrading PLGA (*6*). Most methods of preparation of microspheres expose the protein or peptide drug to water and the organic solvent simultaneously, which is not a good combination (e.g., small amounts of organics dissolve in the water phase and become an unwanted denaturant to the protein), as described previously. For example, in the double-emulsion method, the aqueous drug-containing phase is dispersed in an organic solution containing the polymer to form a single water-in-oil (w/o) emulsion before formation of the particles. In coacervation, an aqueous protein solution is often emulsified in a dilute polymer solution before forming the coacervate on addition of a nonsolvent. In spray drying, a w/o emulsion is also commonly used.

Larger Geometries. Even though commonly used methods of preparing larger cylindrical or slab geometries are not typically studied with the rigor of standard polymer processing techniques (*64*), some very useful, simple, and reproducible methods have been developed. One of the first was a solvent casting method of preparing protein-loaded EVA films that homogeneously distributed the particles by freezing the polymer solution/protein suspension in liquid nitrogen before the particles could sediment (*59*). In another method, the polymer and dry protein powder were compressed in the absence of an organic solvent (*58*). A third method involving extrusion of the polymer solution/protein powder dispersion yielded impressive release profiles (*11*). In this method, the PLGA (100% D,L-lactide, MW 46,000–512,000 Da) was dissolved in acetone (25 to 55% w/w) and mixed with albumin particles (~20–150 μm, 10–30% albumin loading). The suspension was extruded at room temperature with a syringe pump, dried, and die-coated with pure polymer leaving the ends of the cylinders open. Most preparations released albumin continuously for more than a month (*11*) without the multiphasic behavior typical of PLGA mixed diffusion and erosion-controlled release mechanisms (*see* subsequent discussion) (*65*).

Physical Factors That Govern Release and Release Mechanisms

A central question in evaluating the release mechanism of peptides and proteins is that if these molecules cannot diffuse through a continuous and rigid (or glassy) polymer film itself, how are the channels created for the peptide and protein to be transported out? For example, in the classical case of EVA, the channels are believed to form by the next-neighbor solid drug particles themselves, which have since dissolved and diffused out of the polymer. Thus, a "percolation cluster" (or network) of drug particles throughout a polymer matrix are formed during preparation of the polymer–drug film (*10*). A second critical question is why does it take so long for a protein to diffuse out of a polymer matrix such as EVA, when the time of release should be orders of magnitude less (based on estimates of protein diffusion coefficients in, and tortuosities of, aqueous polymer pores)? The geometry of the channels in the percolating cluster and their spatial relationship to one another have provided a clue to the answer to this question (*25*). During the erosion of biodegradable polymers (which we define as the loss of polymer mass), new channels can form (and potentially close) as oligomers and monomers diffuse out of erosion zones. In addition, the polymer may exist in a rubbery state allowing changes in its morphological pore structure (as opposed to the rigid case, which does not). Swelling pressure can also create channels in polymers by local crack formation or by causing the polymer to approach or move through its glass transition, and thus deform in favor of a reduction in overall osmotic pressure. A physical description of these four important aspects of release of peptides and proteins from synthetic polymers is given in Figure 12.1. Correct interpretations of these physical phenomena are essential to enable mathematical models of the release mechanism to be formulated, which in turn will provide a mathematical estimate of the critical parameter, the release rate.

Percolating Drug Clusters

When EVA matrices were loaded with bovine serum albumin (BSA) below a critical drug loading (w/w), some fraction of the protein was not releasable (e.g., entrapped by the polymer) and this critical loading was dependent on protein particle size (*59*). The morphology of matrices before and after release was characteristic of a connected pore network of the remnants of drug particles (*59*) (Figure 12.1A). A stochastic description of the percolation of drug particles through a cubic lattice geometry (system size = matrix thickness/particle size) could predict with good agreement the experimentally observed entrapment of BSA in EVA (*10*). Namely, as the drug loading was reduced, the probability of the drug particle being in contact with a cluster of particles having access to the surface of the matrix also decreased.

Pore Geometry

Even though this percolation phenomena offered a reasonable explanation as to why the drug release was incomplete, it did not explain why drug release was so slow. Scanning electron micrographs of EVA matrices revealed large constrictions between pores in the percolating drug clusters, which were present before water penetration (eliminating the possibility that these pores were formed by osmotic forces). Monte Carlo simulations of diffusion through constricted geometries like those observed in the polymer (Figure 12.1B) helped explain the additional "retardation" of diffusion (*25*). The retardation factor was proportional to the empirical expression, $[(w/W)^{-\beta}]/n_T$, where w/W is the ratio of the width of the constricted throat exit or entry to the width of the central pore body, n_T is the number of outlet throats, and β is an empirical exponent dependent on the ratio of throat length to the width of the pore body. The constricted pore model as described by Siegel and Langer (*25*) is like a "trap" for the blind molecule, which must spend extra time reflecting off the walls of the pore

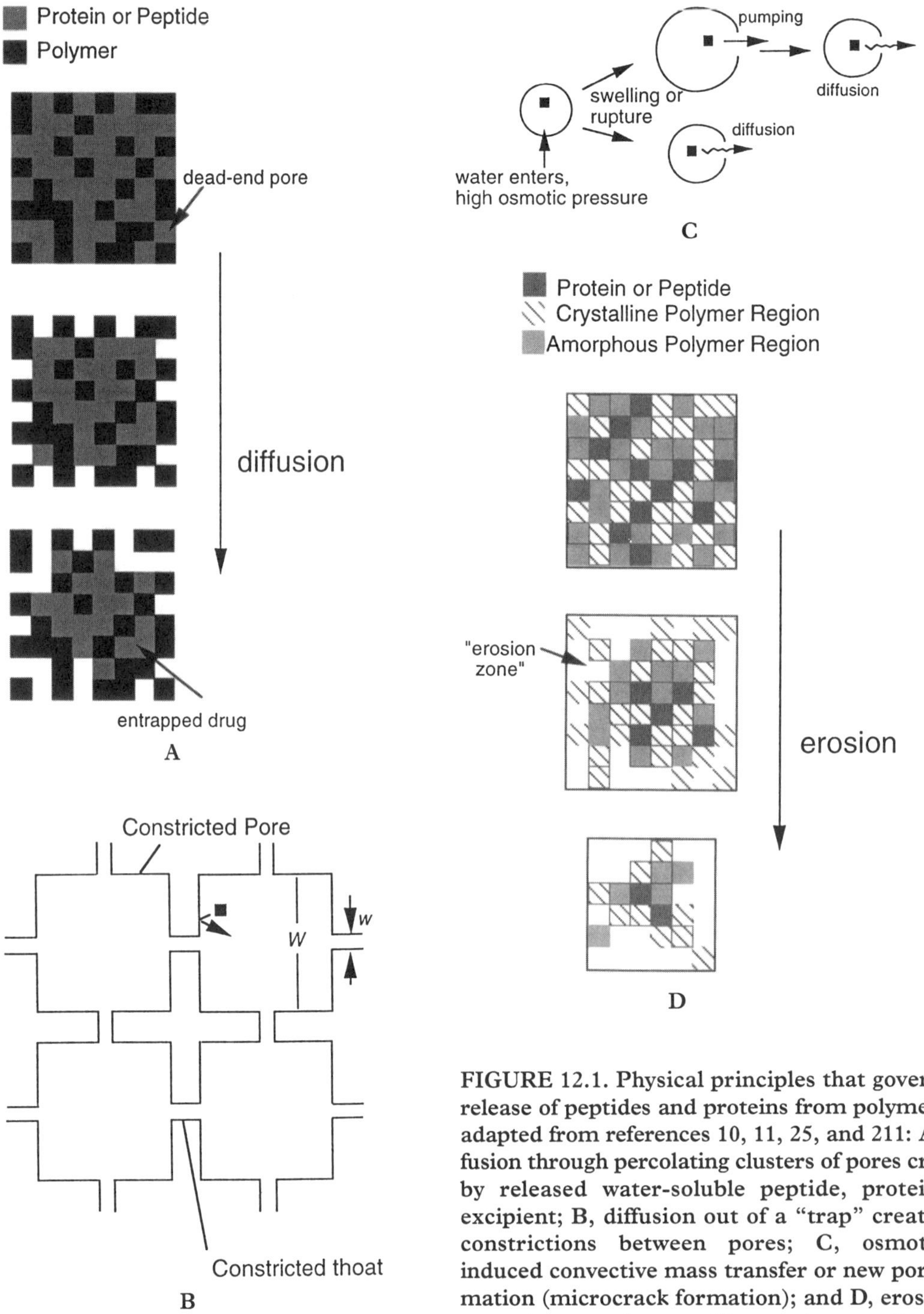

FIGURE 12.1. Physical principles that govern the release of peptides and proteins from polymers, as adapted from references 10, 11, 25, and 211: A, diffusion through percolating clusters of pores created by released water-soluble peptide, protein, or excipient; B, diffusion out of a "trap" created by constrictions between pores; C, osmotically induced convective mass transfer or new pore formation (microcrack formation); and D, erosion of amorphous and crystalline regions to form an "erosion zone" (a new pore for release).

until it finds the exit throat (these events occurring as the particle exercises its random walk during Brownian motion within the polymer matrix).

Osmotically Induced Structural Changes and Osmotic Forces

Water has the ability to permeate most polymers used in controlled release, and therefore, it can have an effect on each of the release mechanisms. Thus, encapsulated peptides and proteins will sorb water until they eventually dissolve (and typically refold or aggregate). Proteins, because of their large molecular weight and relatively low solubility, do not introduce very large osmotic pressures (12 atm has been determined for saturated albumin (*66*)) into the polymer matrix. However, low-molecular-weight salts, which induce enormous osmotic pressures, can deform otherwise glassy polymers (e.g., EVA) to the point of microscopic crack formation (*67*). In addition to the formation of new channels, the osmotic pressure can provide an additional driving force (beside the drug's own chemical potential gradient) for the release of peptide and protein drugs from polymers (*11*) (Figure 12.1C).

Polymer Erosion

In the new era of synthetic biodegradable polymers, the physical chemistry of erosion (i.e., the interrelationship between the various physical properties of the polymer as it chemically degrades) becomes central not only to the release kinetics but also to the environment in which the protein or peptide must retain its stability over the duration of release incubation. Additional details of erosion processes can be found in recent reviews (*42, 68*). For the typical case of biodegradables with hydrolyzable polymer backbones such as PLGA, the first stage of erosion occurs on sorption of water. The polymer backbone begins to degrade (if hydrolyzable), and changes in the microstructure may be observed immediately by microscopy (*69*). The glass transition temperature (T_g) of PLGA may decrease to below physiological temperature (*70*). During hydrolysis, the polymer may become more hydrophilic as new chemical species are formed until it becomes soluble in extracellular fluid, typically as an oligomer or monomer, which can be eliminated by the body. These events can lead to steady or dramatic increases in polymer permeability (Figure 12.1D), resulting in release that is either continuous or abrupt, often in one or two pulses (*65*).

Because water content, pH, and hydrophobicity appear to be very important factors in the stability of peptides and proteins (*22*), the key question is how to accurately measure and manipulate these variables within polymeric systems to promote the most stable environment. For example, the pH inside PLGA cylinders has been reported to reach less than 2 during erosion in aqueous media (*52*). In that study, the soft internal phase was measured by removing it by compression and directly inserting a pH electrode. Is such a result relevant for the typically smaller and more porous PLGA microspheres? How can one determine the pH inside microspheres? In addition, many factors may affect erosion such as device size, porosity, polymer characteristics, and method of device fabrication. The "reversed" heterogeneous erosion observed in PLGA systems as described by Li and co-workers (*45, 46*) (i.e., increased loss of molecular weight from the center of the matrix relative to the surfaces) is not necessarily observable in micrographs of matrices prepared by other investigators (e.g., *see* Hutchinson and Furr (*5*)). Hence, there is much work to be done to better define polymer erosion if the controlled release of stable proteins from these polymers is to become common practice.

Case Studies on Delivery of Therapeutic Peptides and Proteins

Release from Biodegradable Microspheres (PLGA)

The most common of the new biodegradable injectables is based on monolithic microparticles, known as microspheres. Probably the most notable example of the development of these controlled-release devices is the 1-month delivery of luteinizing-hormone-releasing hormone analog, leuprolide acetate (a synthetic nonapeptide) (*6, 71–74*). The Lupron Depot (TAP Pharmaceuticals) is currently on the market for treatment of prostate cancer. The PLGA (75:25 D,L-lactide:glycolide, weight-average MW 14,000 Da) microspheres are prepared by a double emulsion/solvent evaporation method using methylene chloride as a carrier solvent (*71*). Several principles were established by a series of rational formulation studies (*71, 72*). The erosion-diffusion mechanism from these particles could be adjusted to provide near zero-order drug release by meticulously adjusting process parameters (particularly glycolide content and polymer molecular weight). A reasonable explanation of the zero-order effect is that as the polymer erodes with time the increasing porosity within the microspheres (which increases release rate) compensates for the increasing diffusion path and decreasing area normal to transport (both of which decrease release rate) as the source of peptide "shrinks" within the degrading matrix.

During microencapsulation of leuprolide, losses of water-soluble peptide from the inner-water phase (i.e., reducing encapsulation efficiency) could be minimized by increasing the viscosity of the inner-water phase with gelatin, which was cooled down before formation of the second emulsion in the second aqueous phase (*72*). Sterilization was performed by filtration of starting materials, because gamma radiation degraded the PLGA and the peptide (*75*).

The release of the peptide from the microspheres was determined indirectly by the drug remaining in the polymer (by extraction), because the peptide was unstable in the release medium (*72*). Analysis of the release profile can be particularly difficult for proteins, which can adsorb strongly onto release vessel surfaces and to PLGA itself, thus yielding lower concentrations in the release medium. Therefore, a suitable control (e.g., peptide concentration the equivalent of 20% drug release) to verify the physical and chemical stability in the release medium should be established before extensive release studies are performed. In addition, nonionic surfactants often used in release vehicles can interfere in many protein assays (*76*), so investigators should exercise caution when developing release protocols using proteins to avoid experimental artifacts. The drug release in vitro (in 33 mM phosphate buffer pH 7.0/0.05% Tween 80) was very similar to that observed in vivo, a result illustrating the utility of in vitro analysis (*6, 71*).

Another example of delivery of a larger peptide from PLGA microspheres, epidermal growth factor (EGF, 54 amino acids), was developed to enhance the engraftment of transplanted liver cells (*8*). In this study, microspheres of roughly the same size as liver cells (~20 μm) were prepared and co-incorporated with the cells before the mixture was seeded onto porous poly-L-lactide. Following an initial burst, EGF was released from microspheres continuously over a month (Figure 12.2).

Release from Biodegradable Microcylinders

The less frequently studied type of biodegradable injectable is that based on cylindrical geometry.

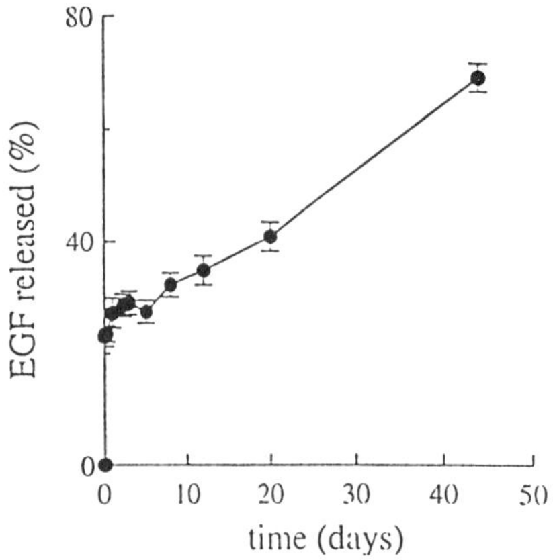

FIGURE 12.2. Release of mouse ^{125}I-labeled epidermal growth factor from PLGA microspheres in phosphate-buffered saline, pH 7.4, 0.1% Tween 20 at 37 °C (data from reference 8). Microspheres were 10–30 μm and were prepared by a double-emulsion/solvent evaporation method using ethyl acetate as the solvent carrier. The encapsulation efficiency was 53%, and the polymer was poly(D,L-lactide-*co*-glycolide) 75:25, 0.2 dL/g inherent viscosity.

One example is a competitor of the Lupron Depot, the controlled-release preparation of Zoladex (goserelin acetate, Zeneca Pharmaceuticals). This cylindrical formulation from PLGA is injectable through an applicator provided with the product (trocar). The depot achieves systemic delivery for one month at a total dose of 3.6 mg (130 μg/day) (*5*). The polymer selected was PLGA (50:50 D,L-lactide:glycolide), and drug release was diffusion and erosion controlled (*65*). The erosion of model implants was characterized by significant induction periods before device mass loss when higher molecular weight polymers were used (inherent viscosity > 0.33 dL/g), whereas no lag time was observed for low-molecular-weight PLGA (inherent viscosity = 0.1 dL/g) (*65*). For low-molecular-weight PLGA, degradation began immediately in a first-order manner until dramatic structural changes took place.

Several potential advantages of the larger cylindrical geometry exist relative to microparticles discussed previously, including: higher attainable peptide loading, simpler encapsulation methodology, and the ease of implant removal (e.g., the Zoladex implant can be localized by ultrasound (*40*)). However, significant discomfort may accompany administration of such large implants. For example, the Zoladex implant must be applied through a large 16-gauge needle, often requiring an anesthetic (*40*).

Release from Multipolymeric Systems

Even though the design criteria in the previous two important examples can be met with a single polymer, in some cases the addition of a second polymer to form a solution, mixture, or coating can provide an effect that could not otherwise be attained. For example, Zhang and co-workers (*11*) prepared microcylinders of PLGA (100% D,L-lactide) at very high protein loading, that is, up to 30% w/w. They were able to control the release slowly and continuously over 40 days via diffusion and osmotic mechanisms by coating the cylinders with a thin layer of a second PLGA (also 100% D,L-lactide, MW 410,000 Da) while leaving the ends open. Thus, transport occurred through the ends, largely independent of the erosion of the polymer (which often causes bi- and triphasic release kinetics). Another example of a coating was the encapsulation of microspheres with a second polymer to reduce the burst of drug release. In this case the microspheres were placed in the polymer solution as if they were drug powder, and the emulsion in the aqueous phase was prepared with subsequent hardening of the coating (*77*).

Recently, spherical particles of submicrometer size (i.e., nanospheres) were prepared from diblock polyethylene glycol (PEG)–PLGA copolymers (10% by wt PEG). PLGA nanospheres, which can be injected directly into the blood stream, are rapidly taken up by the reticuloendothelial system (RES) resulting in very short circulation times (e.g., minutes). Gref and co-workers (*78*) demonstrated that when nanospheres are prepared from a PEG–PLGA diblock copolymer, long circulating times in the blood can be obtained. The protection from the RES is due to the PEG, which was shown to occupy the surface of the copolymer. Indeed, no surfactant in the continuous phase was necessary to prepare nanospheres by standard oil-in-water solvent evaporation methods, indicating that the copolymer has a built-in surfactant (*78*). Because peptides and proteins typically have very short half-lives in vivo, this technology may offer a substantial advantage in their short-term delivery, and it has potential for drug targeting.

The encapsulation of peptides and proteins in nanoparticles is not as straight-forward as with microspheres, because multiple emulsions are too large to form such small particles. One approach is to find mixed solvent systems (e.g., methylene chloride and acetone) that dissolve both the peptide and the polymer and form nanoparticles by typical single-emulsion protocols (*79*). However, new approaches will be required in the future to maximize encapsulation efficiency and minimize the burst (due to the tremendous surface area/volume ratio of nanospheres).

Another example of a novel copolymer is formed with lactide and lysine monomers (*80*). The ε-amino group is a very useful moiety for conjugation of pharmacological or immunologically active peptides, immunoglobulins, and polysaccharides (for drug targeting). Thus, numerous possibilities exist to modify the properties of PLGA while retaining, to a large extent, its biodegradable controlled-release function.

Protein Stability Related to Polymer Systems

The structural complexity of peptides and proteins offers formidable challenges to their development as controlled-release pharmaceutical agents from polymeric systems. Pharmaceutical proteins, unlike most other low-molecular-weight drugs, possess exquisite three-dimensional structures that are critical to their bioactivity. Furthermore, the multiple functional groups that proteins possess are prone to a slew of deleterious processes under conditions relevant to their formulation, storage, and delivery from polymeric systems.

To address these issues, the possible sources of protein inactivation during the preparation, storage, and administration of protein from polymeric systems must be considered. Additionally, understanding and identifying the mechanisms of protein inactivation are critical. Once the instability pathways (and their sources) are elucidated, one can develop a rational approach to stabilizing proteins within polymeric systems. Although general guidelines are presented, the uniqueness of each protein/polymer system demands individual study to attain maximum stability. At the end of this section, several case histories are presented illustrating how protein stability is compromised under conditions relevant to their formulation within polymer systems, and how this situation can be judiciously ameliorated by a mechanistic approach.

Identifying Sources of Protein Inactivation

Three basic stages exist at which a protein can be inactivated pertaining to its delivery from a polymeric system:

- during incorporation of the protein in the polymeric system,
- during its storage, and
- during its release from the polymeric system in vivo.

Preparation of Polymeric Systems

Various approaches have been developed for the controlled release of polypeptides and proteins from polymer matrices, as discussed previously in this chapter and reviewed elsewhere (*81, 82*). For any polymeric protein delivery system, one must consider the conditions that the protein experiences during its preparation, and how these conditions may influence protein stability. Depending on the system in question, in most cases the therapeutic protein experiences relatively severe conditions during preparation. The potential problems that may lead to protein inactivation include the presence of organic phases (which are necessary to dissolve hydrophobic monomers and polymers) and mixing (typically via sonication) resulting in an increase in the air–liquid interface and local increases in temperature. For example, all factors occur during the preparation of peptides and proteins within biodegradable polymer microspheres (*15*). For a more detailed description of potential protein stability issues related to preparation of polymeric systems, *see* the recent review by Schwendeman et al. (*15*).

Storage and Shelf Life of Polymer Systems

Like other pharmaceutical protein formulations, polymeric-based ones must be designed to exhibit superior stability during storage. The U.S. Food and Drug Administration (FDA) requires less than 10% deterioration over a two-year storage period (*83*). To increase the stability of polymeric protein formulations, they are processed as solids: that is, excess solvent is removed, typically by lyophilization.

Stability of lyophilized (or freeze-dried) proteins is critically dependent on several factors including T_g and the level of moisture in the powder. The reactivity of proteins in the glassy state (below T_g) is typically lower than in the rubbery state (above T_g) (*84, 85*). As a rule of thumb, it is advantageous to store solid protein formulations at a temperature of at least 20 °C below their T_g (*85*).

Instead of controlling the storage temperature to ensure it is below T_g, one may also alter the protein formulation to raise its T_g. To this end, it has been proposed that the storage stability of protein formulations may be enhanced by incorporation of excipients with a higher T_g than the protein (*84, 86–88*). One relationship that can be used to describe the T_g of protein:excipient systems is the Gordon–Taylor (G-T) equation (*89*):

$$T_{g,mixture} = \frac{w_1 T_{g,1} + k w_2 T_{g,2}}{w_1 + k w_2} \quad (2)$$

where $T_{g,mixture}$ is the observed glass transition for the mixture consisting of components 1 and 2, their mass fractions are denoted by w_1 ands w_2, their pure glass transition temperatures are denoted by $T_{g,1}$ and $T_{g,2}$, and k is a constant.

Another critical factor in shelf stability of solid therapeutic proteins is the level of moisture. The role of water in the stability of solid proteins has been previously established as discussed in several reviews (*19, 88, 90, 91*). The presence of water in solid protein formulations may be harmful for several reasons:

- water increases protein molecular motions (the "molecular lubricant" effect), which increases protein reactivity;
- water is a reactant in many deleterious reactions (*see* subsequent discussion); and
- water is a mobile phase for reactants.

This first (and perhaps most important) point, that water acts as a plasticizer, may be understood in terms of a significantly lowered T_g of the hydrated versus dry formulation. (*84, 85, 91, 92*). The T_g of hydrated protein formulations has been successfully described by the G-T equation (*93*); note that $T_{g,water}$ = −139 °C (*94*), which is far lower than that of proteins.

Thus, another general rule for proper storage conditions, in addition to keeping solid protein formulations below the T_g, is to maintain low levels of moisture. However, it is not necessarily the case that the lower the moisture content the greater the stability. Hsu and co-workers (*95*) have shown that for tissue-type plasminogen activator stored at 50 °C, aggregation occurred more rapidly for the protein below monolayer water coverage compared with that at or above the monolayer coverage.

In Vivo Delivery from Polymeric Systems

Ultimately, protein or peptide loaded within a polymer depot must exhibit bioactivity on its delivery in vivo. However, the conditions are such that protein stability may be compromised. For example, the water content within the device will increase as water diffuses into the matrix. Thus, the protein will become increasingly hydrated, which decreases its stability (as discussed previously). The kinetics and ultimate level of hydration will depend on the system, but may be quite dramatic; for example, the water content within PLGA can reach high levels (*70*).

If the water content within the polymer is high enough, T_g will fall below the physiological temperature. For most proteins, this result occurs at a hydration of approximately 20 g water/100 g dry protein (*88, 93, 96*). The consequence is that deleterious reactions are accelerated in the rubbery state above the T_g. Furthermore, it has been hypothesized (*84*) that the kinetics of deleterious reactions (specifically those processes dependent on diffusion) for proteins in the rubbery state near the T_g are governed not by Arrhenius behavior but instead by the Williams–Landel–Ferry (WLF) (*97*) relationship:

$$\ln k = \frac{-c_1(T - T_g)}{c_2 + (T - T_g)} \quad (3)$$

where k is the rate constant, and c_1 and c_2 are constants. Because rate constants increase with the term $(T - T_g)$, because T_g is lowered by increasing water content, kinetics increase at a given temperature (i.e., 37 °C). The WLF theory has been successful in describing the kinetics of deleterious processes occurring to solid proteins (*98, 99*). For example, WLF behavior has been confirmed for the Maillard reaction occurring in hydrated food proteins at temperatures up to 10 °C above their T_g (*99*).

Mechanisms of Protein Inactivation Relevant to Polymer Systems

Recent reviews of the various modes of protein deterioration can be found elsewhere (*22, 23, 100*). Briefly, the deleterious processes occurring to proteins can be divided into chemical (covalent) instability pathways:

- deamination,
- hydrolysis,
- oxidation,
- β-elimination,
- incorrect disulfide formation (thiol–disulfide interchange and thiol-catalyzed disulfide interchange),
- Maillard reaction, and
- transamidation;

and physical (conformational or noncovalent) instability pathways:

- reversible unfolding and
- irreversible aggregation via noncovalent interactions.

Some of these processes are intermolecular, and some are intramolecular; some can occur in both fashions. Many of these processes have been observed occurring to therapeutic proteins either within polymeric matrices or under conditions directly relevant to their formulation in such a system.

Protein Conformation

The structural sequence of amino acids of a polypeptide chain is referred to as its primary sequence. According to the permitted backbone bond torsion angles, φ and ϕ, the amino acid chain can form various secondary structures such as α-helices and β-sheets (*101*). Typically, more than half of the native (folded) protein structure is composed of these elements; the remainder may be in other, less ordered secondary structures such as loop, turn, and random coil. The tertiary structure, or conformation, of a protein refers to its overall three-dimensional structure. Finally, associations of peptide chains into oligomeric forms is referred to as quaternary structure.

Physical Instability Pathways

Protein structure is altered by its unfolding, or denaturation, as described by the following (reversible) equation:

$$N \leftrightarrow U \qquad (4)$$

where N and U represent native (folded) and unfolded (denatured) conformations. In general, the protein surface has a higher density of charged and polar residues, which can interact with water and other polar solvents, whereas hydrophobic residues are tightly packed within the protein core, and therefore excluded from the polar solvent. Usually, the unfolded species has little or no bioactivity relative to the native one. Unfolding can also accelerate other, covalent pathways of deterioration (*22, 23, 100*). At least some proteins also exhibit an intermediate species on unfolding:

$$N \leftrightarrow I \leftrightarrow U \qquad (5)$$

where the intermediate species, I, contains both folded and unfolded domains and may be involved in other destabilization pathways. However, such intermediate species are difficult to isolate and identify.

Various forces have been shown to be involved in stabilizing the native state. Briefly, these forces include electrostatic interactions, hydrogen bonding, van der Waals interactions, and hydrophobic interactions (*102*). The native structure is stabilized slightly over the denatured one; typically, the Gibbs free energy for the folding equation is only about 5–50 kcal/mol (*103, 104*). Thus, only a moderate change in the protein environment can lead to the disruption of the delicate balance of forces involved in stabilizing protein conformation.

Protein unfolding can occur because of an environmental change such as an extreme of pH, temperature, pressure, ionic strength, and the solvent medium. All of these factors may occur during the processing or release of protein in polymeric matrices. Additionally, the removal of water (usually a required step in the production of controlled-release protein–polymer formulations) should lead to considerable changes in protein conformation, as hypothesized by Kuntz and Kauzman (*105*). Indeed, several recent investigations have confirmed the view that removal of water from proteins leads to significant (reversible) structural rearrangement (*106–109*). For example, Griebenow and Klibanov (*108*) uncovered, by Fourier transform IR (FTIR) spectroscopy, that for some dozen proteins, the removal of water results in a significant decrease in α-helical structure with a rise in β-sheet structure.

Protein unfolding usually leads to the exposure of hydrophobic moieties previously buried in the protein interior. This unfavorable situation (of hydrophobic groups in contact with aqueous solvent) can be ameliorated by the association of unfolded molecules via noncovalent interactions to form aggregates. Ultimately, these aggregates can reach such a large molecular weight that they fall out of solution because of their low solubility (*23*). Even though unfolding is a reversible process, aggregation is practically irreversible and usually leads to a loss of soluble, folded monomeric protein. In addition to binding to each other, unfolded proteins may bind to hydrophobic surfaces such as that present by a hydrophobic polymer, and this event may also promote noncovalent aggregation (*110, 111*).

Chemical Instability Pathways

In addition to the unfolding event and the ensuing related events of adsorption and aggregation, proteins also may exhibit a variety of chemical (covalent) processes leading to their loss of activity. Several of these processes may be relevant to proteins formulated within polymeric systems, including deamidation, hydrolysis, oxidation, deleterious reactions involving cysteine and cystine residues, the Maillard reaction, and transamidation.

Deamidation. Among the various deleterious intramolecular chemical processes occurring in proteins, deamidation is perhaps the most common. Deamidation of asparagine (Asn) and glutamine (Gln) residues has been observed for numerous therapeutic proteins under a variety of conditions (*112*). Asn residues, particularly those adjacent to glycine (Gly) residues, are most susceptible. This process occurs most often at neutral to alkaline conditions. Under these conditions, the proposed mechanism of deamidation is via an intramolecular rearrangement to form a five-membered succinimide ring with the liberation of ammonia (Scheme 12.1). This succinimide ring is rather unstable and undergoes hydrolysis and racemization. Thus, the possible products are L- and D-aspartyl and L- and D-isoaspartyl (β-aspartyl) peptides (Scheme 12.1). In addition to deamidation occurring via the succinimide intermediate, deamidation may occur at acidic conditions (pH 1–2) by direct hydrolysis of the side-chain amide; however, this process occurs at a much slower rate than the succinimide-intermediate pathway at neutral to alkaline conditions (Scheme 12.1).

Regardless of which product (either aspartyl or isoaspartyl) is generated, deamidation has the consequence of introducing a new negative charge (e.g., a free carboxylic acid) on the protein. This result may or may not affect the bioactivity, depending on the location of the event on the protein. For example, for the peptide growth hormone releasing factor, deamidation leading to aspartyl and isoaspartyl forms reduces the bioactivity by 25- and 500-fold, respectively, compared with native (asparaginyl-containing) peptide (*113*). In contrast, deamidation occurring at residue Asn-149 in another therapeutic protein, recombinant human growth hormone, did not significantly affect the biological activity (*114*).

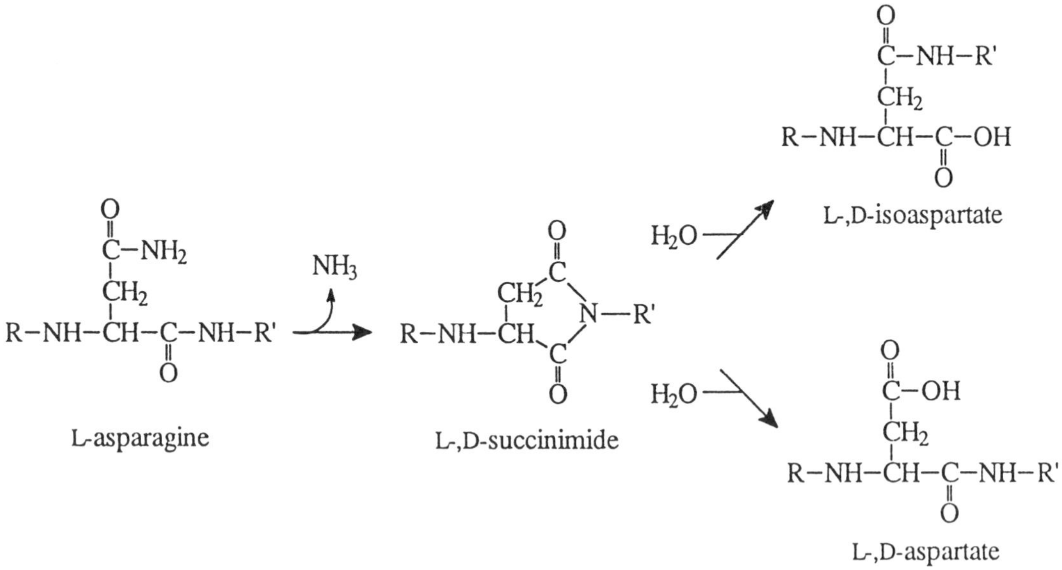

SCHEME 12.1. Pathway for deamidation of asparaginyl residues. A similar pathway may be envisioned for glutaminyl residues, except that the reaction proceeds via a six-membered succinimide ring intermediate, rather than the five-membered one shown.

Deamidation has been observed occurring to solid protein and peptides on exposure to moisture and elevated temperatures, conditions that are relevant to their delivery from polymeric systems (*15, 19*). For example, Hageman and co-workers (*92*) found some deamidation occurring to lyophilized recombinant growth hormone incubated at 30 °C and 96% relative humidity. In another study, for a lyophilized model peptide, deamidation occurred under various solid-state conditions (*115*). These studies indicate that deamidation may be a serious problem for solid proteins or peptides suspended within polymeric matrices intended for controlled release in vivo.

Hydrolysis. The peptide backbone can undergo hydrolysis under acidic conditions. This event is particularly prevalent occurring adjacent to Asp residues, which hydrolyze about 100 times faster than other peptide bonds (*116*). When Asp is adjacent to Pro, the hydrolysis is particularly prevalent (*117*). Cleavage can occur at either the C-terminal (via a five-membered ring intermediate) or the N-terminal (via a six-membered ring intermediate) peptide bonds of Asp. The mechanism for the C-terminal case is shown in Scheme 12.2. Note that the carboxyl group of Asp acts as an intramolecular catalyst. A similar mechanism is proposed for N-terminal peptide bond hydrolysis.

Hydrolysis may be a concern for peptides and proteins loaded within polymeric systems, especially if such systems bioerode to produce acidic monomeric units (e.g., PLGA). Work done with enzymes such as ribonuclease A and lysozyme has shown that the significant loss of protein structure accompanied by hydrolysis will result in significant loss of therapeutic protein biological activity (*118, 119*). An additional concern is that the lower molecular weight hydrolysis products will probably release more readily from polymeric systems because their smaller size relative to native protein. Also, therapeutic proteins, such as recombinant bovine growth hormone (*88*), as well as model peptides (*115*) undergo hydrolysis in the solid state.

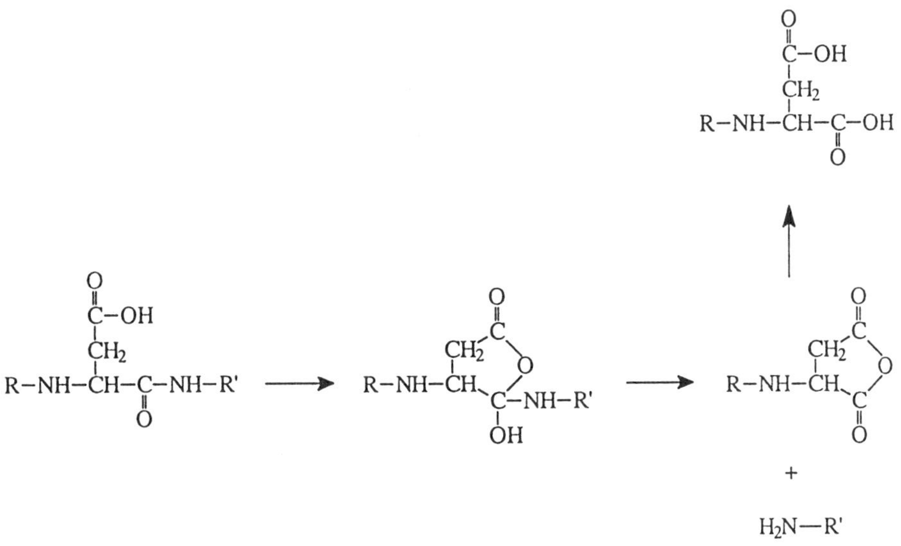

SCHEME 12.2. Hydrolysis of the peptide backbone. Shown is hydrolysis of the C-terminal peptide bond adjacent to an aspartyl residue. A similar scenario may be envisioned for peptide hydrolysis of the N-terminal peptide bond adjacent to an aspartyl residue via a six-membered ring intermediate.

Oxidation. Various residues such as methionine (Met), cysteine (Cys), histidine (His), tryptophan (Trp), and tyrosine (Tyr) are all conceivably susceptible to oxidation, particularly in the presence of oxidizing agents such as hydrogen peroxide. Of these, serine (Ser) and Cys are probably the most likely to experience oxidation under conditions relevant to protein loaded within polymeric systems.

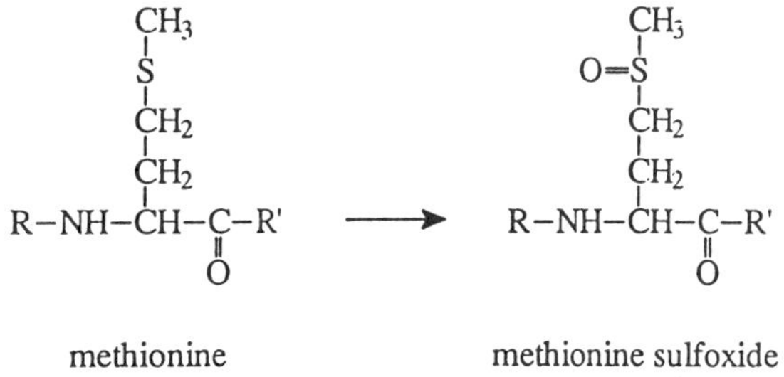

SCHEME 12.3. Oxidation of methionyl residues to methionine sulfoxide.

At mildly acidic conditions in the presence of hydrogen peroxide, Met is the primary residue experiencing oxidation. The product of the reaction is Met sulfoxide (Scheme 12.3). Like deamidation of Asn or Gln, the oxidation of Met may or may not affect the biological activity of a therapeutic protein. Presumably, this result depends on whether the Met oxidation disrupts the protein structure near the active site. For example, oxidation of Met^{14} and Met^{125} in human growth hormone does not significantly alter its biological potency (*114, 120*). The oxidation of Met^{125} to the sulfoxide in human growth hormone has been observed for the lyophilized powder exposed to light (*121*). In contrast to the effect on human growth hormone, for another related protein, chorionic somatomammotropin, oxidation occurring at Met^{64} and Met^{179} resulted in drastically decreased biological potency (*120*).

Residues of Cys are also prone to oxidation, yielding various forms such as disulfide, sulfenic, sulfinic, and sulfonic acids, depending on the conditions (*23*). Because oxidation rates are greater for the thiolate ion (deprotonated) compared with the undissociated thiol, oxidation of Cys is more prevalent at alkaline conditions. Oxidation of Cys, particularly when it leads to inter- or intramolecular disulfide formation, can be deleterious with respect to protein activity (*see* subsequent discussion).

β-Elimination of Disulfide (Cystine). Disulfide bonds, or cystines, are susceptible to β-elimination. This process occurs via a nucleophilic attack of a hydroxide ion on a carbon atom in a carbon–sulfur bond (Scheme 12.4). The two new resulting residues, dehydroalanine and thiocysteine, are both relatively unstable (*22*): dehydroalanine may react with Lys, forming a lysinoalanine cross-link, and thiocysteine decomposes to yield various products such as cysteine and hydrosulfide ion (Scheme 12.4).

Because β-elimination is catalyzed by the hydroxide ion, it is greatly accelerated under alkaline conditions. Because disulfides play a role in maintaining the three-dimensional structure of a protein (*122*), their scission via β-elimination may have dire consequences for the bioactivity of therapeutic proteins. A further deleterious consequence of β-elimination is that the thiols that are generated may subsequently catalyze disulfide exchange. Indeed, this process was shown to be responsible for the solubility loss of lyophilized insulin exposed to elevated temperature and moisture (*18*).

Incorrect Disulfide Formation (Disulfide Scrambling). Disulfide bonds constitute an important structural motif in proteins. The formation of non-native, or incorrect intra- and intermolecular disulfides, may lead to drastic changes in protein structure and function (*122*). In addition, the dramatic changes in protein structure concomitant with incorrect disulfide formation may also contribute to other destabilization pathways such as adsorption and precipitation (noncovalent aggregation).

Intra- or intermolecular disulfide scrambling can occur by various pathways, depending on the conditions. For example, at acidic conditions disulfides can be cleaved by attack of a proton yielding a sulfenium ion that can subsequently react with a sulfur atom in another disulfide (*123*) (Scheme 12.5). The formation of disulfides at neutral to alkaline conditions requires the presence of sulfide ion (*124*). This ion can be present on the protein itself, in the form of Cys residues; in this case the pathway of disulfide formation is thiol–disulfide interchange (Scheme 12.6).

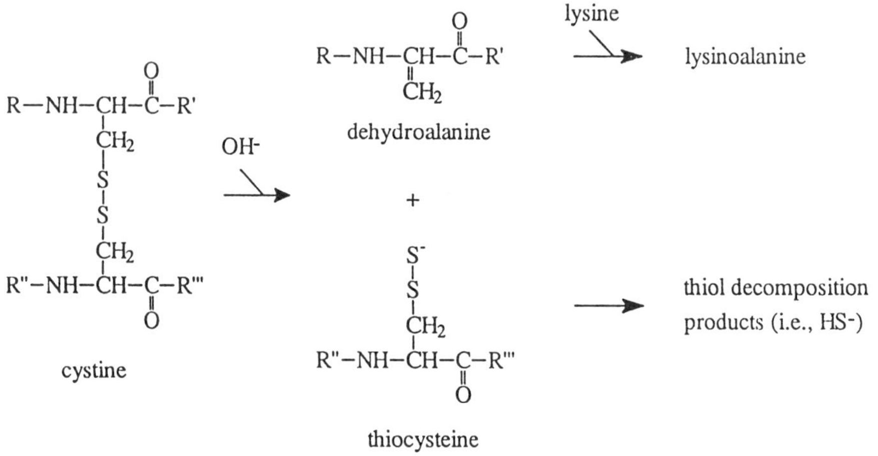

SCHEME 12.4. Mechanism of β-elimination of disulfide bonds.

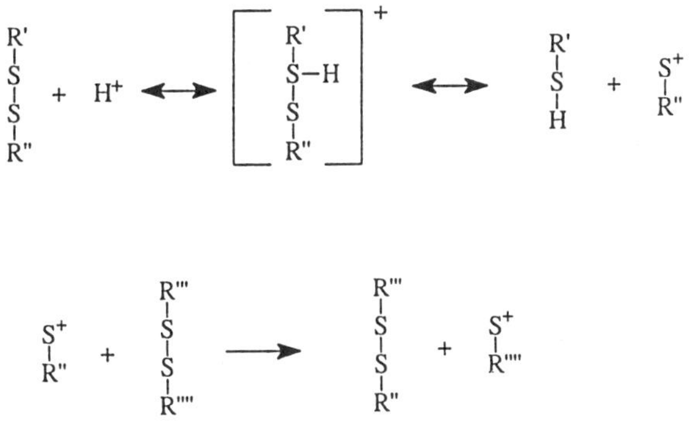

SCHEME 12.5. Disulfide interchange in acidic media.

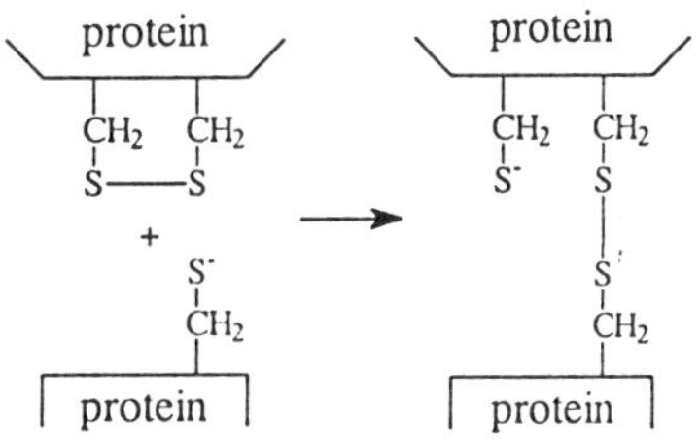

SCHEME 12.6. The mechanism of thiol–disulfide interchange.

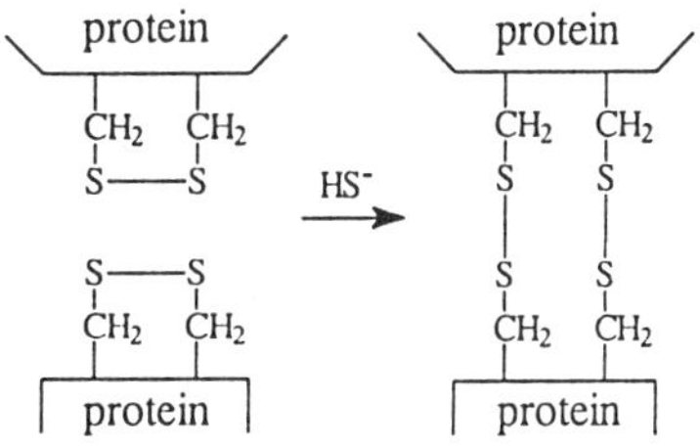

SCHEME 12.7. Thiol-catalyzed disulfide exchange.

The reactive thiolate ion is conserved in this process. Alternatively, the catalytic ion can be of small molecular weight, for instance, hydrosulfide ion, which was generated by a β-elimination of an intact disulfide. This scenario can be denoted as thiol-catalyzed disulfide exchange (Scheme 12.7).

Intermolecular disulfide bonding is responsible for moisture-induced aggregation of a number of lyophilized therapeutic proteins such as insulin (*18*), albumin (*20, 125*), and tetanus toxoid (*16, 126*), a characteristic suggesting it may be an important consideration in formulation of polymer protein delivery depots. The low solubility, low bioactivity, and potential immunogenicity of such aggregates make their controlled delivery problematic (*18*).

Maillard Reaction. Another important and deleterious reaction involving proteins is the Maillard reaction (Scheme 12.8). In this process a Schiff base is formed as a result of the condensation of a Lys ε-amino group and the carbonyl group of a reducing sugar or aldehyde. This Schiff base is thought to undergo Amadori rearrangement to a more stable product (*127*). Maillard-type reactions occur most readily in neutral to weakly alkaline conditions (*128*).

The Maillard reaction may be relevant to proteins and peptides loaded within polymer systems. Potential reactants include the reducing ends of various oligo- or polysaccharides (which may be candidates as matrices for controlled-release formulations) and the reducing ends of monosaccharides, which are commonly used as excipients in therapeutic protein formulations (*23, 100, 129–131*). A related reaction may occur with the carbonyl of formaldehyde, which is used in the preparation of vaccines (*16*).

The Maillard reaction has been observed occurring in the solid state in the presence of moisture to therapeutic proteins, such as insulin (*132*). However, because most investigations of solid-state Maillard-type reactions have focused on hydrated food proteins (*133*), it is not certain whether, in general, the bioactivity of pharmaceuticals will be adversely affected. A further consequence of this reaction is that the Schiff base may also be involved in cross-linking (*16*), as in the proposed aggregation pathway for the moisture-triggered solubility loss of tetanus toxoid.

Transamidation. Yet another deleterious covalent process involving Lys is transamidation. In this reaction, the ε-amino group of Lys forms an

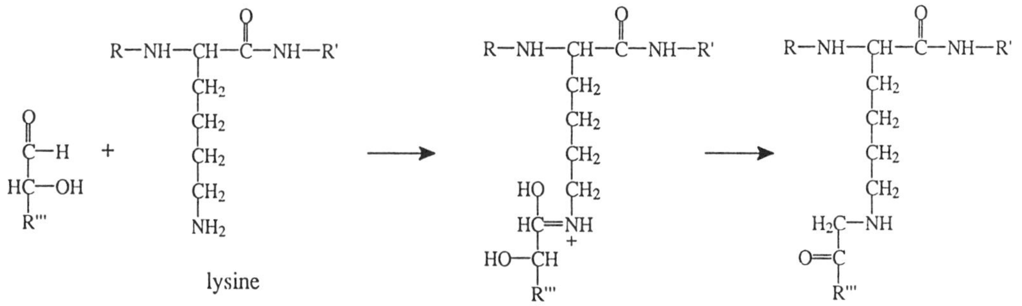

Schiff base

SCHEME 12.8. The Maillard reaction.

isopeptide bond with the carbonyl group of either Asn or Gln (Scheme 12.9), or dehydroalanine following β-elimination (Scheme 12.4). The cross-links may be either intra- or intermolecular.

Intermolecular transamidation is proposed to be responsible for the moisture-induced aggregation of the lyophilized therapeutic protein somatotropin (*90, 92*) as well as for the aggregation of lyophilized ribonuclease A (*134*). Transamidation also was exhibited by solid enzymes suspended in organic solvents at elevated temperatures (*60*). Given this susceptibility of solid proteins, aggregation via transamidation may be a problem for solid pharmaceutical proteins loaded within polymeric matrices intended for controlled release.

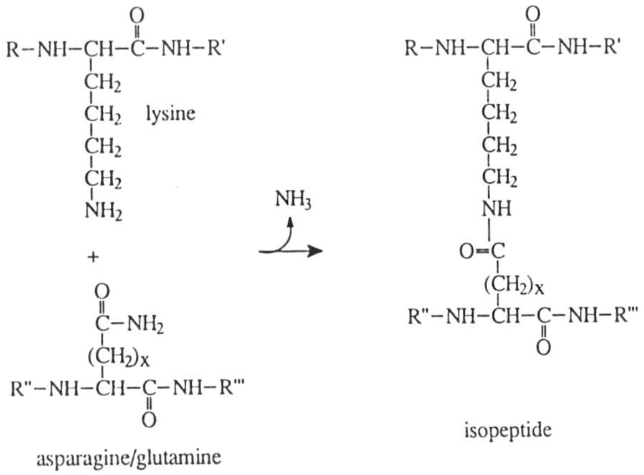

SCHEME 12.9. Isopeptide bond formation between lysine and asparagine (x = 1) or glutamine (x = 2) by means of transamidation.

Rational Stabilization Approaches for Proteins in Polymer Systems

Given the myriad of deleterious processes relevant to therapeutic proteins loaded within polymer delivery system, how can one develop a rational approach to stabilize them? Initially, one must choose pharmaceutically relevant conditions under which to study protein stability. Because the stability of therapeutic proteins varies widely, it is important to elucidate the mechanisms of inactivation that occur under such conditions. Using guidelines established for these pathways, one can then improve protein stability in polymeric systems.

Mechanisms of Protein Inactivation Under Pharmaceutically Relevant Conditions

There are various methodologies for investigating the stability of therapeutic proteins within polymeric systems. The initial consideration is whether to study the stability of the protein alone or incorporated within the polymeric system intended for use. The protein-alone method is more common because it is generally quicker, less costly, and more convenient in terms of the biochemical tools used to study protein stability. Because protein loaded within a polymeric device is usually in the solid form (i.e., suspended within the polymer matrix), a common preparation method to obtain therapeutic proteins for study is lyophilization. Discussions of this technique and the implications for protein stability can be found elsewhere (*84–87, 135*).

The next consideration is the conditions one chooses to study. For example, which is more relevant, the solution stability or the solid-state stability? This decision depends on the system (e.g., the kinetics of water uptake in and release from the polymer in vivo and the solubility of the protein). In either case, one must also choose additional variables such as pH and ionic strength. Although physiological conditions are often chosen, it is by no means definite that these conditions exist in the microenvironment around the protein within a polymer system. For example, proteins loaded within bioerodible polymers composed of acidic units, such as PLGA, may exhibit a local pH far more acidic than the physiological case during release in vivo. In this case, it is appropriate to study stability under acidic conditions.

Even though the protein is ultimately fully hydrated at release, at some earlier point it exists in a partially hydrated state. Thus, it is appropriate to study the stability of proteins under solid-state conditions. This study is done using lyophilized, or freeze-dried proteins. (One still needs to choose the conditions of the aqueous solution before lyophilization with the considerations discussed previously.) Next, one needs to choose the relative humidity at which to study stability. Again, this choice will depend on the polymer nature. The most thorough approach would be to investigate a number of relative humidities to mimic stability at various hydrations as the protein evolves from dry to fully hydrated.

Finally, one must choose the temperature of study. Of course, 37 °C is directly relevant, but reactivities at this temperature may be too low to make measurements practical. In the pharmaceutical industry, it is common to choose an elevated temperature and extrapolate data back to physiological temperature (*83*). The advantages and disadvantages of such accelerated storage stability have been discussed elsewhere (*84, 100*). One must exercise caution in extrapolating data from elevated to lower temperatures because such systems may experience WLF, and not Arrhenius behavior (*84*). Ultimately, for a pharmaceutical protein formulation the FDA does require real-time stability data (*83*).

Once the conditions have been chosen, the next step is to incubate the protein and uncover possible deleterious mechanisms. Herein, we will focus on study of the solid-state stability. In this case, following incubation, one must reconstitute the protein in aqueous media and physiological conditions that are most appropriate.

On the basis of the amount of protein dissolved against an unincubated standard, one can determine how much aggregation (i.e., solubility loss) had occurred. In addition, one can investigate the nature of soluble oligomeric species by analytical techniques such as size-exclusion chromatography and light scattering.

To investigate the nature of the cross-links responsible for the loss of solubility, one can attempt to dissolve the aggregates in various aqueous solutions. If the aggregates are held together by noncovalent forces (hydrophobic interactions), the aggregates should dissolve in high concentrations of denaturing agents such as urea and guanidine hydrochloride. If the aggregates are held together by disulfide bonds, the addition of a disulfide reducing agent, such as dithiothreitol, should be sufficient to dissolve the aggregates. If neither a denaturing nor a thiol-reducing agent dissolves the aggregates, then another covalent mechanism is responsible for the aggregation. In this case, the elucidation of the aggregation is more difficult. One possibility is the lysinoalanine cross-link, which can be identified by acid hydrolysis and amino acid analysis of the aggregates. Others have implicated Lys as a member of a cross-link in transamidation or other pathway; loss of Lys residues can also be followed via partial acid hydrolysis and amino acid analysis or by the liberation of NH_3. The aggregation picture becomes more complex when more than one mechanism of cross-linking is occurring, which may be a common case for solid pharmaceutical proteins (*18, 19, 125, 136*).

As for the intramolecular processes, they can be elucidated by various analytical techniques (*see* the book edited by Creighton (*137*), for example). Briefly, we shall discuss a few of these techniques. For example, the change in charge (as the result of a deamidation) can be revealed by a technique such as isoelectric focusing and hydrolysis can be uncovered by size-exclusion chromatography. Intramolecular cross-linking may be more difficult to detect and identify, but if the protein three-dimensional structure is perturbed, which is likely, potential tools of investigation include circular dichroism and FTIR spectroscopy. This FTIR technique has great promise because of its ability to quantify secondary structural elements in both solution and solid states (*108*).

Some Guidelines for Rational Stabilization Approaches

Once elucidated, the deleterious mechanisms found to occur to proteins loaded within polymeric systems may be targeted specifically. This targeting can be done by altering the formulation or altering the protein itself either chemically, or via genetic engineering. Herein, we will concentrate on the approaches of altering the formulation or the protein. Eventually, genetic engineering of therapeutic proteins may prove the most beneficial strategy. However, because of the current state of the art, this procedure usually requires screening of many mutants, and thus it can be costly and time-consuming. Additionally, the prospect of creating a new pharmaceutical agent has increased regulatory considerations.

A recent review outlines strategies for improving solid protein stability against various aggregation mechanisms (*19*). If aggregation occurs via thiol–disulfide interchange, stability may be improved by reducing the concentration of reactive thiolate ion species either by lowering the pH or by S-alkylation. (*125*). If aggregation occurs via β-elimination followed by thiol–disulfide interchange, the aggregation may be avoided either by lowering pH to avoid the β-elimination (Scheme 12.4) or by oxidation of low-molecular-weight thiols by means of Cu^{2+} (*18*). If Lys is a participant in the cross-linking, aggregation may be decreased by blocking its amino groups via succinylation or by including excipients that can compete for the reaction (*16, 17*). More detailed descriptions of modification of potentially reactive amino acids can be found elsewhere (i.e., *see* Lundblad and Noyes (*138, 139*)).

Similarly, one can improve stability against intramolecular processes as well. For intramolecular cross-linking involving disulfides or Lys residues, one can invoke strategies discussed previously for the intermolecular case. If hydrolysis occurs because of a low pH, one can attempt to control the pH at a higher level by either lyophilizing from a more basic pH, or adding buffering agents to the formulation. Because deamidation is more prevalent at high pH, in this case a similar strategy can be invoked to lower the pH level. Because of the potential reaction with reducing ends of sugars, their use as excipients should be avoided; nonreducing sugars, such as sorbitol or trehalose, are preferred.

A general approach to improve the stability of pharmaceutical proteins may be to control the physical or conformational stability. For solid proteins (the most relevant formulation for biopharmaceuticals within polymer systems) there is a significant structural rearrangement on dehydration (*106–108*). Such a loss in native structure may result in aggregation via hydrophobic interactions or increased rates of other deleterious mechanisms if there is an increased exposure of reactive groups (*22, 23, 100*).

Carpenter and co-workers (*106, 140–142*) as well as others (*107, 108*) demonstrated that addition of sugars and other polyols can stabilize protein structure on lyophilization. Such an agent may act as a "water substitute" in the solid-state hydrogen bonding with the protein (*129*). In addition to improving the physical stability of protein, the presence of such an excipient may also provide protection (against intermolecular processes) by decreasing protein intermolecular contacts (the "dilution" effect (*125*)). Successful examples of sugars as stabilizers for solid proteins include the use of mannitol to stabilize growth hormone against aggregation in the dry state at elevated temperature (*143*) and the use of sucrose and a sucrose polymer, Ficoll 70, to protect ribonuclease against aggregation under various solid-state conditions (*144*).

In addition to these aforementioned mechanisms, excipients may also increase the stability of solid proteins by raising the T_g of the formulation (*84, 86–88*). The magnitude of this effect depends on the T_g values of the protein and excipient, as illustrated by the G-T equation.

Yet another general approach to stabilize proteins against various deleterious mechanisms within polymeric systems is to control the amount of water associated with the protein at optimal levels. As a general rule, the lower the water content on the protein, the higher the stability; this rule has been elucidated for a number of pharmaceutical proteins, as reviewed by Costantino et al. (*19*). For example, the moisture-induced aggregation of insulin, bovine serum albumin, recombinant human albumin, and tetanus toxoid were all critically dependent on the level of moisture on the protein (Figure 12.3). In general, the dependency of aggregation with moisture was bell-shaped. Plasticization is thought to be the major factor in the ascending portion of the curve, whereas dilution and hydration-induced conformational changes are involved in the descending portion. It may not be true *in all cases* that lower water ensures an improved stability, as discussed above for the case of tissue-type plasminogen activator (*95*). In fact, "overdrying" may decrease stability, perhaps via denaturation at the solid–void interface (*145*).

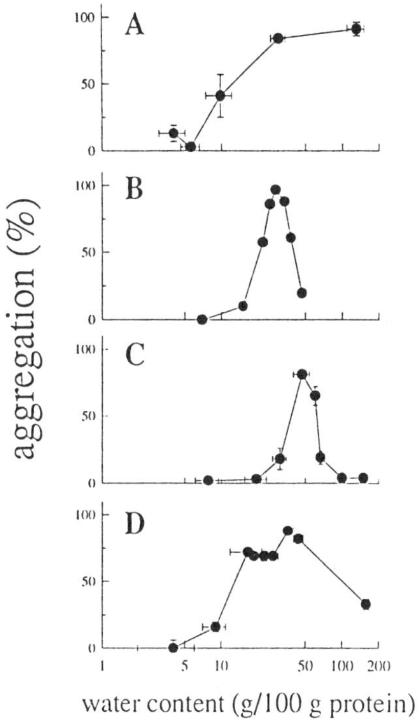

FIGURE 12.3. Effect of water content on the solid-phase aggregation of various proteins: A, aggregation of lyophilized insulin at 50 °C (data from reference 18); B, aggregation of lyophilized bovine serum albumin at 37 °C (data from reference 125); C, aggregation of lyophilized recombinant human albumin at 37 °C (data from reference 20); D, aggregation of tetanus toxoid at 37 °C (data from reference 16).

Case Studies on Inactivation of Therapeutic Proteins

Several case studies will illustrate the utility of our approach to rational stabilization of therapeutic proteins under conditions relevant to their delivery from polymeric systems. The model therapeutic proteins we will consider herein are all type III and consist of albumin, insulin, tetanus toxoid, and somatotropin.

Albumin

Albumin is a therapeutically useful protein and is a potential vehicle for sustained delivery of low-molecular-weight drugs that bind to it (*146–148*). In addition, albumin is commonly used as a model therapeutic protein to evaluate controlled-release formulations (*149*). However, this protein exhibits moisture-induced aggregation following its lyophilization and incubation at 37 °C and elevated moisture levels (*125*).

By dissolution in various aqueous solutions, the mechanism responsible for the aggregation was determined to be thiol–disulfide interchange (Scheme 12.6) among the lone free thiol, Cys^{34}, and the 17 native intramolecular disulfide bonds. On the basis of this mechanism, several approaches were devised and tested for stabilization. These approaches included lyophilization from acidic pH (to protonate reactive thiols), S-alkylation of the reactive Cys^{34} with either iodoacetamide or iodoacetic acid, and dilution in the solid phase with the co-lyophilized water-soluble polymeric excipients dextran, diethylaminoethyldextran, carboxymethylcellulose, and PEG.

Aggregation of albumin was revealed to be strongly dependent on the moisture content (*20, 125*). The bell-shaped dependency is shown in Figures 12.3B and 12.3C. Thus, another stabilization approach for albumin is incubation at lower water activities, resulting in lower amounts of water bound to the protein. In practice, this process could be accomplished by storing the solid protein–polymer formulation at optimal humidity levels or choosing a low water-sorbing polymer for delivery in vivo.

Another successful approach in stabilizing lyophilized albumin against aggregation was the addition of strongly water-sorbing co-lyophilized excipients (*20*). A plot of the water uptake for various water-sorbing excipients against their ability to stabilize recombinant human albumin against moisture-induced aggregation at 37 °C and 96% relative humidity (Figure 12.4) illustrates the correlation. This extra water, when brought in proximity to the protein, induced stabilization as expected from the behavior at various moisture levels (Figure 12.3C).

Insulin

Another protein that is both very useful pharmaceutically and is a common model for study is insulin (*150*). Insulin exhibits a number of deleterious pathways under conditions relevant to its release from polymer systems. For a recent review of the stability of insulin, *see* Brange (*151*).

In aqueous solution, insulin is known to exhibit noncovalent aggregation as a direct result of unfolding at a hydrophobic interfaces, which is a serious problem for its delivery from not only polymeric devices but also implantable pumps (*150–154*). This effect may be important because insulin may aggregate within certain polymeric matrices, resulting in its incomplete release (*7*). On the basis of this elucidated mechanism, Sluzky and co-workers (*41, 111*) devised rational approaches for its stabilization in aqueous solution. One highly successful strategy was to add nonionic detergents to the insulin solution; these detergents have high affinities for the hydrophobic interface. Compared with insulin in aqueous solution, which aggregated completely in less than 24

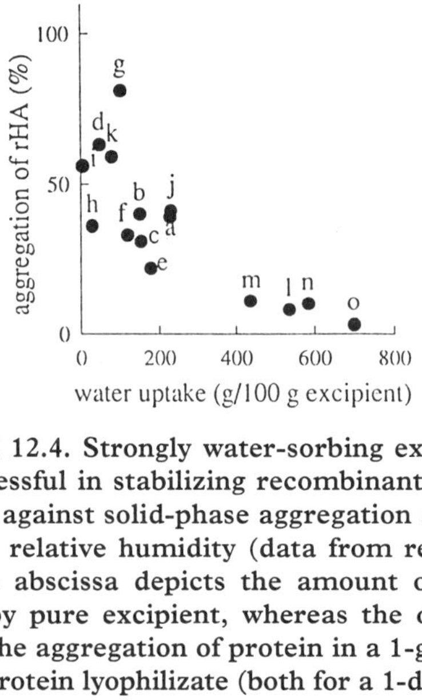

FIGURE 12.4. Strongly water-sorbing excipients are successful in stabilizing recombinant human albumin against solid-phase aggregation at 37 °C and 96% relative humidity (data from reference 20). The abscissa depicts the amount of water sorbed by pure excipient, whereas the ordinate depicts the aggregation of protein in a 1-g excipient:6-g protein lyophilizate (both for a 1-day incubation). Excipients were as follows: a, sorbitol; b, sucrose; c, trehalose; d, dextran; e, carboxymethyldextran; f, diethylaminoethyldextran; g, poly(ethylene glycol); h, D-glucosaminic acid; i, D-glucose diethyl mercaptal; j, L-serine; k, L-alanine; l, D-lactic acid; m, glycolic acid; n, propionic acid; o, sodium chloride.

h, insulin in the presence of *n*-dodecyl-β-D-maltoside or *n*-octyl-β-D-glucoside was remarkably stable and exhibited no aggregate formation even after 6 weeks. Such an approach may be useful in stabilizing insulin in polymeric systems.

Insulin also undergoes deleterious processes in the solid form. For example, Fisher and Porter (*155*) observed deamidation and cross-linking by an undetermined mechanism occurring in insulin crystals. In addition, Costantino et al. (*18*) have shown that lyophilized insulin undergoes moisture-induced aggregation via noncovalent and intermolecular disulfide bonding. However, in the case of insulin there are no free thiols present in the native protein (*150*). During the preparation and high-temperature, high-humidity incubation of lyophilized insulin, native disulfides were cleaved via β-elimination (Scheme 12.4), yielding thiols that subsequently catalyzed disulfide scrambling (Scheme 12.7).

Several rational strategies were devised to inhibit moisture-induced insulin aggregation. These strategies included controlling the pH and moisture at optimal levels and addition of trace amounts of Cu^{2+} to the lyophilized insulin to catalyze the oxidation of low-molecular-weight thiols to unreactive species (*18*). This catalysis was particularly notable; only a 1:10 cation-to-insulin molar ratio resulted in a complete stabilization under the conditions studied.

Tetanus Toxoid

One highly studied controlled-release vaccine is that for tetanus (*16, 17, 126, 156–161*). This disease remains a killer in developing countries even though it can be prevented by immunization. The major difficulty has been the failure to administer the 2–3 doses required for protection (*28*). Thus, tetanus toxoid (TT) and other protein vaccines are good candidates for controlled release from polymer systems (i.e., injectable PLGA microspheres).

However, incomplete release of total (*159*) and antigenically active (*158*) TT from one such system, PLGA microspheres, indicated that deterioration occurred within the polymer. Thus, several investigations were initiated by our group to understand these phenomena and to enhance the solubility and antigenicity of the protein released from PLGA microspheres.

During exposure to moisture, aggregation of lyophilized TT was consistent with three mechanisms depending on the water content within the lyophilized protein powder: noncovalent interactions, thiol–disulfide interchange, and the formaldehyde-mediated conversion of labile intramolecular bonds to stable and labile intermolecular cross-links (*16*). In the third pathway, which is the principal mechanism of aggregation, formaldehyde is stored in labile linkages in TT (i.e., hydroxymethylene- and aminal-type) as described in Scheme 12.10 following the formaldehyde exposure during the vaccine preparation (i.e., detoxification by formalinization). Both the alteration in the composition of the strongly formaldehyde-interacting amino acid residues (i.e., Lys, His, and Tyr (*162*)) in the moisture-induced aggregates relative to the unincubated TT control, and the insolubility of aggregates in combined denaturing (6 M urea) and reducing solvents (10 mM dithiothreitol) were used to formulate this hypothesis (*16*).

The formaldehyde-mediated aggregation pathway was tested by devising rational means of stabilization under conditions where this mechanism dominates (~86% relative humidity and 37 °C). As depicted in Figure 12.5, two methods were successful in disrupting the aggregation pathway and in retaining the solubility of TT (*16*). The first method was by reacting TT extensively with succinic anhydride, which is known to irreversibly react with Lys residues and reversibly react with Tyr residues (*162*), which participate in the mechanism in Scheme 12.10. The second method of inhibiting aggregation was by cyanoborohydride reduction of TT before its lyophilization. Borohydrides are known to reduce reactive imines and stabilize the bound formaldehyde (*163*). Further evidence for the formaldehyde-mediated mechanism was provided by the following:

- suggestion of its generality with another, very different formalinized protein vaccine, diph-

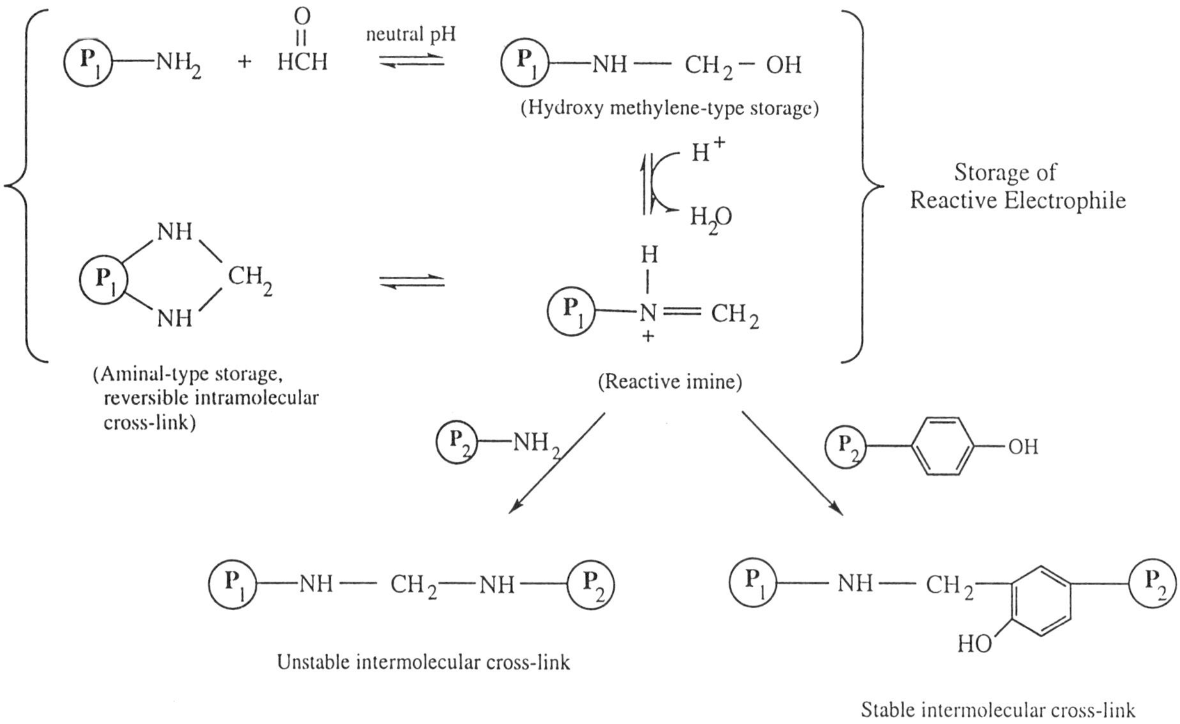

SCHEME 12.10. Proposed formaldehyde-mediated aggregation pathway occurring to formalinized vaccine antigens such as tetanus toxoid.

theria toxoid (DT), which had solid-state aggregation kinetics superimposable with those of TT (*17*);

- stabilization of DT by succinylation (*16*); and
- stabilization by co-lyophilization with free Gly, whose α-amino groups can compete with those on TT (*17*).

In addition to aggregation triggered by moisture, TT also forms soluble and insoluble aggregates as a result of organic solvents and other microencapsulation conditions (i.e., sonication and homogenization) (*17, 159*). Also, the acidic pH within PLGA may cause irreversible inactivation of TT during erosion of the polymer matrix (*164*). We have determined that broad isoelectric precipitation of TT occurs rapidly between pH 4 and 5.5 when solutions of the protein (1 mg/mL) are lightly agitated at 45 °C (*161*). Furthermore, a dramatic loss in antigenicity of TT is rapidly observed between pH 3 and 4 at the same temperature and protein concentration (*161*).

Somatotropin

Growth hormone, or somatotropin, is yet another therapeutic protein (*165*) that exhibits various destabilization pathways under conditions relevant to its release from polymeric systems. The storage stability of this protein was studied by Pikal and co-workers (*143*). In that investigation, it was revealed that human growth hormone is susceptible to Met oxidation (Scheme 12.3), deamidation (Scheme 12.1), and irreversible aggregation by an undetermined mechanism when stored at elevated temperatures in the dry state. In addition, Pearlman and Nguyen (*121*) have shown that when exposed to light at 30 °C, human growth hormone

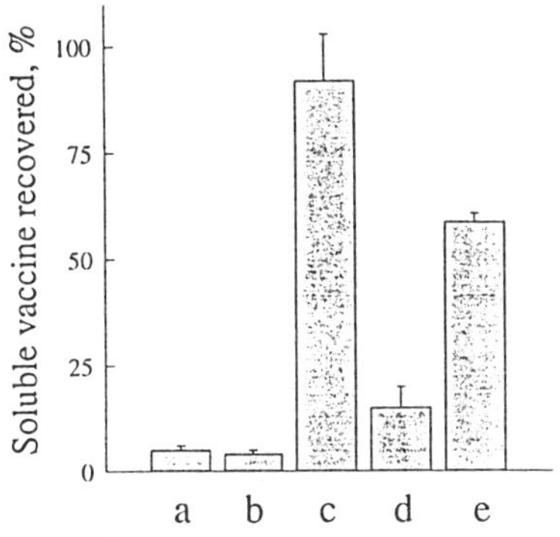

FIGURE 12.5. Stabilization of tetanus (TT) and diphtheria (DT) toxoids against moisture-induced aggregation (data from reference 16). Following a 9-day exposure of the lyophilized proteins to 86% relative humidity at 37 °C, the soluble protein recovered in phosphate buffer was determined for a, TT; b, dithiothreitol-reduced TT; c, reduced and succinylated TT; d, DT; and e, succinylated DT.

exhibits oxidation at Met^{125} in addition to covalent and noncovalent dimerization.

Hageman and co-workers (*90, 92*) studied the stability of various forms of growth hormone (bovine, porcine, and human) when the lyophilized powder was exposed to elevated temperature and moisture. Various deleterious mechanisms were observed including covalent, non-disulfide-bonded dimer- and oligomerization, deamidation, and hydrolysis. The authors observed that intermolecular processes appeared to predominate over intramolecular ones in the solid state. A transamidation pathway (Scheme 12.9) was hypothesized as a possible mechanism for the cross-linking.

Aggregation of growth hormone also was reported for protein loaded within various polymeric systems in vitro including porcine growth hormone in PLGA microspheres (*166*) and bovine growth hormone in ethyl cellulose pellets (*167*), EVA slabs (*167*), and polyanhydride matrices (*33*). In the polyanhydride matrices (*33*), unreleased growth hormone could be recovered following addition of a denaturing agent, indicating that aggregates had formed by noncovalent interactions and not by transamidation.

Because the aggregation of growth hormone is triggered by moisture (*90, 92*), a potential approach for improving the stability of this protein within polymeric systems is to employ more hydrophobic polymers. This process would presumably limit the water content within the system during its use. Indeed, Ron et al. (*33*) found an increased cumulative release of growth hormone from poly[1,3-bis(*p*-carboxyphenoxy)hexane] (90%) versus a less hydrophobic polymer, poly[1,3-bis(*p*-carboxyphenoxy)propane-*co*-sebacic anhydride] (<50%).

Another approach that was tested for stabilizing growth hormone in polymer systems is the use of sugars. For example, the addition of sucrose to growth hormone loaded within EVA (*167*) and polyanhydride matrices (*33*) increased the cumulative protein release in vitro.

Controlled-Release Polymers as Vaccine Adjuvants

Vaccination has been the most cost-effective way of controlling infectious diseases. Vaccines comprised of non-living antigens, particularly from purified inactivated toxins, subunits, and synthetic components (all mostly peptides and proteins), are poor immunogens and require adjuvants to achieve early, high, and long-lasting immunity (*168, 169*). The only adjuvants that have been widely used with routine human vaccines are aluminum adjuvants, which have several limitations. These limitations consist of local reactions, production of immunoglobulin (Ig) E antibodies, ineffectiveness for some antigens, and inability to elicit cell-mediated immune responses (especially cytotoxic T-cell responses) (*168–170*).

Even with the use of aluminum adjuvants, the logistics of delivering at least 2–3 doses for primary immunization to achieve protection are difficult and compliance is frequently inadequate, particularly in developing countries (*28*). In recent years, adjuvants and delivery systems for vaccines have received much attention because of their promise in

- modulating the immune response selectively to the type required for prevention against a particular disease,
- reducing the number of doses required for primary immunization with the ultimate goal of developing single-dose vaccines, and
- developing combination vaccines by reducing the amounts of vaccine required to achieve protection, thus avoiding antigenic competition, antigenic suppression, and side effects due to large amounts of antigens (*169*).

Objectives for Polymers as Vaccine Adjuvants

The potential of controlled-release polymers for the delivery of vaccine antigens was shown by the late 1970s by demonstrating sustained antibody responses to a single dose of antigen encapsulated in EVA (*35, 171, 172*). Since that time, there have been numerous studies evaluating the immune response to antigen-loaded polymer formulations. Some of these investigations are listed in Table 12.2. The two major areas of investigations are the targeting of antigens to microfold (M) cells on mucosal surfaces after oral administration (or to antigen presenting cells after parenteral inoculations) and the controlled release of vaccines with the aim of reducing the number of doses required for primary immunization to even a single dose (*27–30, 51, 156–159, 173–198*).

The use of polymers for the controlled release of vaccines is now focused on biodegradable polymers and those that can be prepared as microspheres and injected through a syringe needle. Microspheres can be made from a variety of polymers such as PLGA, polyphosphazenes, and polyanhydrides (Table 12.1). PLGA has the advantage of its proven safety as surgical sutures in humans (*27*).

Evaluation of Immune Response

For preclinical studies, the immune response to vaccine antigens encapsulated in controlled-release polymer microspheres is evaluated in animal models with the antigens either adsorbed onto aluminum adjuvants or given with Freund's adjuvants. It is important to use optimal formulations of aluminum adsorbed antigens because their adjuvanticity depends on a number of factors including the degree of antigen adsorption and adjuvant dose (*168, 170*). Animal species and strain as well as antigen type and dose all affect the results of immunogenicity studies (*168*). In several studies, model proteins such as ovalbumin or bovine serum albumin were evaluated as vaccine antigens in the controlled-release polymer systems (Table 12.2) (*35, 171, 172, 174, 179, 188, 195*). These model antigens may be useful for preliminary studies to optimize the formulation conditions. However, there are several limitations of using such antigens:

- These proteins have no clinical significance.
- Very high doses of these proteins are usually employed in animals (*188*) that would be clinically unacceptable and may be so high that it is not possible to discriminate differences between various formulations (it is possible to reach an upper limit of the immune response; doses administered to animals should be scaled relative to clinical doses in humans accordingly, e.g., dose in mice ≤ 1/10 dose in humans).
- Formulation conditions worked out with model antigens usually do not work with the actual human vaccine antigens because of their differences in physicochemical properties and purity.
- No functional antibody assays can be performed because these model proteins do not have distinct biological activities.

Antibody responses in animals to antigens encapsulated within polymer microspheres are usually measured by immunoassays such as enzyme-linked immunosorbent assay (ELISA). A major limitation in the use of ELISA is that different investigators assign antibody concentrations by

TABLE 12.2. Model Proteins and Vaccine Antigens Encapsulated in Controlled-Release Polymers and Evaluated in Animals for Immune Response

Antigen	*Ref.*	*Polymer*	*Use*	*Immunogenicity*
Bovine serum albumin	35	EVA	Subcutaneous implant, mice	Antibody response similar to FCA
	172	EVA	Subcutaneous implant, rabbits	Antibody response similar to FCA
	51	Tyrosine-based poly(iminocarbonate)	Subcutaneous implant, mice	Antibody response similar to FCA Adjuvant effect of tyrosine
Ovalbumin	174	Polyacrylamide	Oral boost, rats	Strong secretory IgA memory
	179	PLGA	Subcutaneous, microspheres, mice	Antibody response similar to FCA
	188	PLGA	Oral, microspheres, mice	Higher IgA than soluble antigen
	188	PLGA	Subcutaneous, microspheres, mice	High IgG antibody levels for 1 year
	195	PLGA	Subcutaneous, intraperitoneal microspheres, oral, mice	High IgG; DTH similar to ISCOMs; intestinal IgA; CTL
Staphylococcal enterotoxin B	175	PLGA	Oral, microspheres, mice	Circulating and local antibodies
	30	PLGA	Subcutaneous, microspheres, mice	<10 μm More immunogenic
	30	PLGA	Intraperitoneal, oral boost, mice	More immunogenic than soluble antigen
Human chorionic gonadotrophin	178	PLGA	Intramuscular, microsphere, rabbits	Sustained antibody response
E. coli CFA/I	190	PLGA	Oral, microsphere, rabbits	High serum IgG antibodies
E. coli CFA/II	191	PLGA	Oral, microsphere, rabbits	Highly immunogenic and safe
E. coli adhesin	184	PLGA	Intradermal, microspheres, rabbits	Protection against oral challenge
Tetanus toxoid	156	PLGA–glucose	Subcutaneous, microspheres, mice	Immunogenicity preserved
	194	PLGA	Intranasal, adsorbed, guinea pigs	Greater response than free antigen
	186	PLGA	Microspheres, rats[a]	Antibodies similar to 2 doses of alum
	158	PLGA	Subcutaneous, microspheres, mice	Antitoxin levels higher than soluble antigen
	157, 159	PLGA	Subcutaneous, microspheres, mice	Antitoxin levels similar to $AlPO_4$
Diphtheria toxoid	180, 181	PLGA	Subcutaneous, microspheres, mice	IgG similar to 3 doses of $CaPO_4$
Influenza virus	176, 185	PLGA	Intraperitoneal, oral, microspheres, mice	Good priming and boosting
	193	Poly(amino acid)	Oral, microspheres, rats	High IgG similar to 3 doses of $CaPO_4$
Human parainfluenza virus	189	PLGA	Intraperitoneal, microspheres, mice	High serum antibodies
			Intraperitoneal, oral, intranasal, hamsters	Reduction in virus after challenge
Rota virus	197	Anionic polymers and amines	Oral, microspheres, mice	Enhanced humoral immune responses
Mycoplasma hypneumoniae	182	Cellulose acetate phthalate	Oral boost, pigs	Protection against vaginal challenge
SIV	192	PLGA	Oral boost, monkeys	Protection against vaginal challenge
Hepatitis B surface antigen	183	PLGA	Intraperitoneal, microspheres, guinea pigs	Antibody levels higher than alum
Bordetella pertussis antigens	198	PLGA	Intranasal, microspheres, mice	More immunogenic and better protection

NOTE: FCA is Freund's complete adjuvant, alum is aluminum-based adjuvant, $AlPO_4$ is aluminum phosphate adjuvant, $CaPO_4$ is calcium phosphate adjuvant, CFA is colonization factor antigen, SIV is Simian immunodeficiency virus, ISCOMs are immunostimulating complexes, DTH is delayed-type hypersensitivity, CTLs are cytotoxic T-lymphocytes, Ig is immunoglobulin, and challenge is exposure of animal to pathogen.

[a]Injection route not specified.

different methods, including titers (reciprocal of the highest serum dilution giving a specified absorbance), absorbance (optical density) units (usually multiplication of absorbance and reciprocal of the dilution showing that absorbance), arbitrary units, international units, or weight-based units (µg/mL) (*169*). In most reports, ELISA results are not compared with results from a functional assay of antibodies such as neutralization or bactericidal and opsonophagocytic assays. This fact makes comparison of ELISA results from various research laboratories extremely difficult. Results in titers and absorbance units are not usually determined against an ELISA reference serum, and these results are affected by environmental factors such as temperature, time, and humidity, particularly at the last step involving enzyme and substrate interaction. Results obtained on different days on the same samples in the same laboratory are sometimes difficult to reproduce. However, results in arbitrary units determined against an ELISA reference serum can be reliably compared in a laboratory obtained on different days, but still are difficult to compare among different laboratories.

In certain cases, the ELISA reference serum is calibrated in international units by a functional antibody assay against an international or national standard. In this case, results can be compared among different laboratories, although the differences in avidities (i.e., strength of antibody–antigen binding) of the ELISA reference serum and unknown sera cause discrepancies between ELISA titers and titers obtained by functional antibody assays (*199*). Therefore, the use of an ELISA reference serum with similar avidity or from animals with similar immunization status (i.e., reference animals bled at similar time as test animals) as the test sera was proposed (*199–201*). This proposal might result in the use of several ELISA reference sera, one at every time point at least during early course of immunization (*201*).

In addition to measuring antibodies by ELISA or any other immunoassay, measurement of functional antibodies by neutralization or bactericidal assays, for example, is desirable because encapsulation methods may modify epitopes on the antigens, which may elicit antibodies that bind in the immunoassays, yet have no functional activity.

Mucosal Immunization

Because most human pathogens enter the body through mucosal surfaces (e.g., the intestinal mucosa) (*202, 203*), development of vaccines, adjuvants, or delivery systems for this route is highly desirable. However, the majority of the vaccines available today are given parenterally because of the problems in achieving effective mucosal immunization (*202*). The major problems in development of effective mucosal vaccines include

- instability of antigen delivered orally due to unfavorable conditions of pH, presence of bile salts, and proteolytic enzymes;
- induction of tolerance to soluble antigens (*204*);
- inefficiency of antigen uptake; and
- (in infants) interference by maternal antibodies (*203*) obtained through breast-feeding.

In recent years, biodegradable polymer microspheres have been used for effective mucosal immunization (Table 12.2) (*29, 174, 175, 182, 185, 190–195*).

Many advantages of oral delivery of antigens from microspheres exist, as reviewed by Walker (*202*). For example, the antigen is protected from the severe conditions in the gut and maternal antibodies. In addition, encapsulated antigens are delivered efficiently to M cells without inducing tolerance. Other advantages of oral immunization over immunization by the parenteral route include the following:

- easier production and quality control (*205*),
- increased safety of toxic components, such as lipopolysaccharides,
- easier delivery of vaccines without the need for personnel and equipment required for injections,

- increased vaccine effectiveness in the elderly,
- control of infection possibly leading to eradication of certain diseases, and
- larger number of antigens delivered due to the vast surface area of the mucosal immune system (*205*).

For certain pathogens entering through mucosal surfaces, effective induction of mucosal immunity has led to prevention of the disease. Marx and co-workers (*192*) showed protection in monkeys against simian immunodeficiency virus (SIV) when they elicited effective mucosal immune response by intratracheal booster with formalinized SIV encapsulated inside PLGA microspheres. In contrast, monkeys injected parenterally with microspheres containing SIV showed high levels of circulating antibodies but were not protected against vaginal challenge. PLGA microspheres containing staphylococcal enterotoxin B antigen elicited both systemic and mucosal immunity when delivered orally (*29, 30, 175, 203*). Microspheres less than 10 µm are taken up by the M cells of the Peyer's patches and transported to the T- and B-cell zones. Microspheres less than 5 µm are ingested by the cells and are taken to systemic lymphoid tissues such as spleen where the released antigen elicits a serum antibody response. Microspheres greater than 5 µm remain in Peyer's patches and provide a sustained release of antigen eliciting IgA response. Therefore, mucosal delivery of antigens from biodegradable polymer microspheres may be very useful for booster doses of routine vaccines such as TT and DT, which will increase immunization coverage. Additionally, mucosal immunization with microspheres will be beneficial for new vaccines targeting the pathogens at entry into the host, which may be the only protective method for certain diseases.

Controlled Release of Vaccine Antigens

Biodegradable microspheres also were evaluated for controlled release of antigens and vaccines by parenteral immunizations (Table 12.2) (*29, 171, 173*). Two approaches have been followed to develop controlled release of vaccines based on the release kinetics from the polymer adjuvant. The first "pulsatile-release" approach involves selecting polymers and encapsulation conditions to achieve little or no antigen release for some period (e.g., a month) followed by a large booster dose of antigen. The microspheres made up from two or more different formulations with varying pulsing times are mixed. The multiple pulses mimic the conventional multiple injections used for vaccination.

The second "continuous-release" approach results in high levels of antibodies, normally only observed after multiple injections. Also in this approach, the characteristics of polymer formulation affect the release profile of antigen. Depending on the molecular weight, ratio of lactide to glycolide, and device geometry and dimensions, PLGA polymers can degrade over a few days or weeks to several months or a year. Several human vaccine antigens encapsulated in microspheres elicited antibody levels in animals similar to those obtained with potent adjuvants such aluminum compounds and complete Freund's adjuvant (Table 12.2). The antibodies elicited in mice and guinea pigs by TT encapsulated in the PLGA microspheres were of high affinity as shown by neutralizing activity (*157–159*). This approach has the potential advantage of reducing the number of injections required for primary immunization, thus increasing vaccine coverage.

A number of issues such as antigen stability (*15*) need to be resolved before the controlled release of vaccines becomes competitive with more conventional aluminum adjuvants. The importance of preserving conformational epitopes to elicit a protective immune response is well documented. For example, the S1 subunit of pertussis toxin contains a protective B-cell epitope that is noncontiguous in the primary sequence, but contiguous in the assembled molecule (*206*). In addition, acellular vaccines developed against pertussis require the native assembly of the five individual subunits that make up the native pertussis toxin. Administration of vaccines whose quaternary structure was not assembled correctly following expression or was disrupted with chemicals such as formaldehyde resulted in less than protective immune responses (*206*). Other potential limitations with the controlled-release vaccines may be immunologic hypersensitivity to vaccine antigen. Once the vaccine antigen is delivered via the microsphere vehicle, it will be released slowly over time and will be difficult to remove. This characteristic may result in perpetual hypersensitivity reactions. However, the possibility of IgE-mediated hypersensitivity appears low because slow release of small amounts of antigens would elicit high affinity IgG antibodies replacing IgE.

Mechanisms of Adjuvant Activity of Controlled-Release Systems

The mechanisms of enhanced immunogenicity of proteins delivered through controlled-release systems may be grouped into two major categories based on the size of microspheres. These categories consist of depot formation at the site of injection and targeting antigens to antigen-presenting cells (Figure 12.6). Depot formation at the site of injection with continuous release of antigen has been the most important mechanism for several adjuvants. Availability of low levels of antigen in its native conformation due to its resistance to extracellular proteases in the depot leads to development of high affinity antibodies (*207*). Because the antigen is continuously available, it will bind with the antibodies. The antigen–antibody complex then persists on the surface of follicular dendritic cells in lymphoid tissues leading to generation of a pool of memory cells (*208*). With the further continuous supply of antigen, the memory B cells are recruited to become antibody-secreting cells leading to generation of high levels of high-affinity antibodies. Aluminum adjuvants act by depot formation, even though the adjuvant effect of aluminum adjuvants continued even after surgical removal of depot from the site of injection (*170*). Microspheres greater than 10 μm can act as depots at the site of injection, and if the protein is stabilized by rational approaches as discussed previously, it may lead to continuous supply of antigen in its native conformation.

Microspheres less than 10 μm can be efficiently taken up by antigen-presenting cells, thus leading to efficient presentation of proteins to immune cells as compared with soluble proteins. This process leads to the appearance of a higher level of antibodies compared with those elicited by large microspheres (>10 μm) (*30*).

Furthermore, the adjuvant activity of continuous-release microspheres is expected to be mechanistically distinct from that of pulsatile-release preparations. The controlled release of antigen over prolonged periods would give rise to immune response of a moderate magnitude with high affinity that would persist or even increase slowly for long periods. Pulsatile release of antigen would be expected to elicit a response similar to the conventional anamnestic response exhibiting high levels of antibodies after the second pulse. The antibody levels would then begin to decline after remaining above protective levels for long periods. It remains an open question as to which of these release kinetics will provide the optimal response for a specific antigen.

Another possible method of enhancing the immunogenicity of proteins delivered via polymeric systems is to activate antigen-presenting cells by coincorporating an adjuvant or immunomodulator in the microspheres with the protein antigen. Because most immunomodulators are toxic when administered at soluble doses required for adjuvanticity, delivery through microspheres at continuous low levels may stimulate the immune cells without toxicity. This concept has already been shown to be effective with liposomes, the administration of which enhanced adjuvanticity without adverse side effects (*209, 210*).

Thorough examination of these mechanisms by which the immune response can be enhanced by polymers will require the use of formulations that release antigens possessing appreciable levels of their native antigenic determinants. Hence, it has become clear how the challenges of preparing

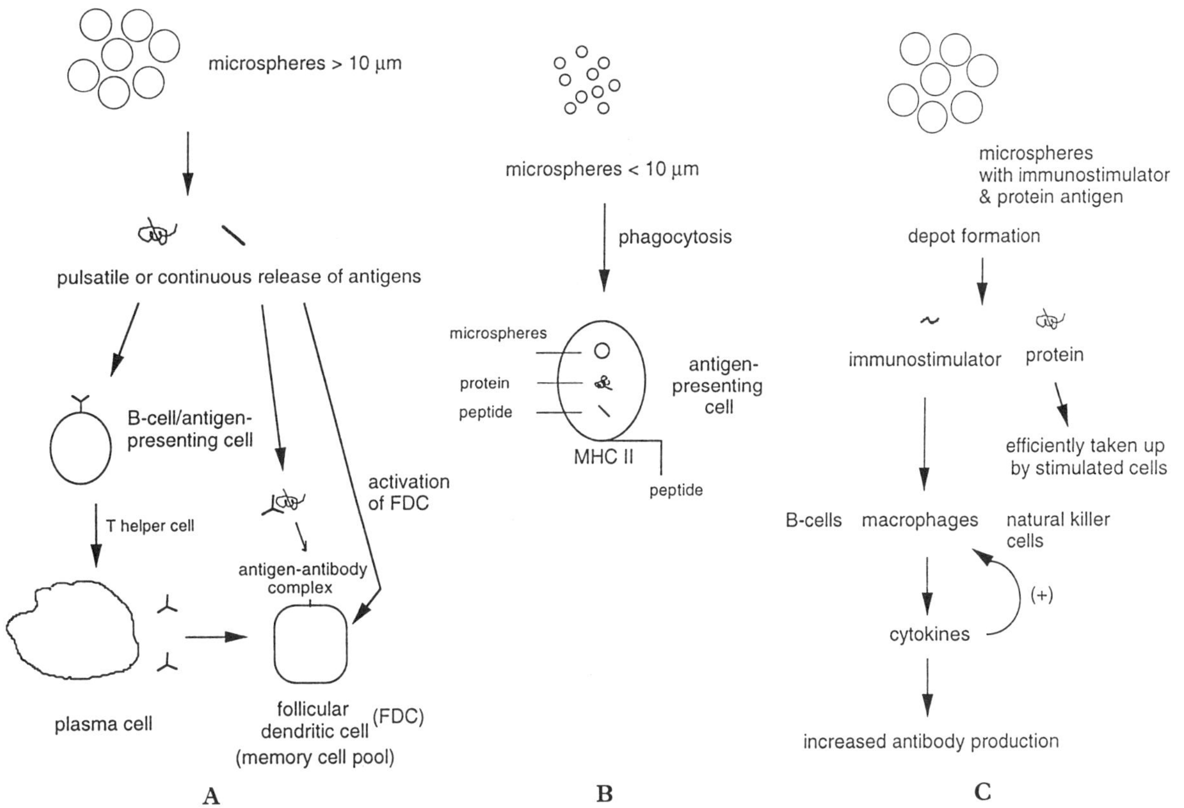

FIGURE 12.6. Potential mechanisms of adjuvant activity of antigen-loaded controlled-release biodegradable polymer microspheres administered parenterally: A, repeated antigen exposure and control over release profile; B, targeting to antigen-presenting cells via efficient uptake of microspheres by antigen-presenting cells; C, activation of antigen-presenting cells by immunostimulator, which leads to recruitment of cells of the immune system (B-cells, macrophages, and natural killer cells) and stimulation of natural killer cells to secrete cytokines, thus recruiting T helper cells.

stable formulations, which we have discussed in detail in earlier sections, impact directly the evaluation of these important biological questions.

Concluding Remarks

The enormous potential of polymers for the delivery of peptides and proteins is yet to be fully realized. Combination of several disciplines will be necessary for

- accurate assessment of the stability of the peptide or protein drug during conditions that mimic its release environment,
- formulation studies that take into account this stability, and
- definitive in vitro and in vivo characterization of the formulation, particularly when the encapsulated protein is a vaccine antigen.

Developing methods of characterizing both encapsulated proteins and the environment within degradable polymer devices are among the important experimental obstacles to overcome.

Acknowledgments

This work was supported by grants from the National Institutes of Health (GM 26698 and AI 33575), the Biotechnology Process Engineering Center at MIT, and the World Health Organization. Individual support was provided to Steven P. Schwendeman by a National Institutes of Health postdoctoral fellowship (AI 08965).

References

1. Brem, H.; Walter, K. A.; Langer, R. *Eur. J. Pharmacol.* **1993,** *39,* 2–7.
2. Levy, R. J.; Wolfrum, J.; Schoen, F.; Hawley, M.; Lund, S. A.; Langer, R. *Science (Washington, D.C.)* **1985,** *228,* 190–192.
3. Becker, J. B.; Robinson, T. W.; Barton, P.; Sintov, A.; Siden, R.; Levy, R. J. *Brain Res.* **1990,** *508,* 60–64.
4. Freese, A.; Sabel, B. A.; Saltzman, W. M.; During, M. J.; Langer, R. *Exp. Neurol.* **1989,** *103,* 234–238.
5. Hutchinson, F. G.; Furr, B. J. A. *J. Controlled Release* **1990,** *13,* 279–294.
6. Ogawa, Y.; Okada, H.; Yamamoto, M.; Shimamoto, T. *Chem. Pharm. Bull.* **1988,** *36,* 2576–2581.
7. Brown, L.; Munoz, C.; Siemer, L.; Edelman, E.; Langer, R. *Diabetes* **1986,** *35,* 692–697.
8. Mooney, D.; Kaufmann, P. M.; Sano, K.; Schwendeman, S. P.; McNamara, K.; Schloo, B.; Vacanti, J. P.; Langer, R. *Biotechnol. Bioeng.* **1996,** *50,* 422–429.
9. Siegel, R. A.; Langer, R. *J. Controlled Release* **1990,** *14,* 153–167.
10. Siegel, R. A.; Kost, J.; Langer, R. *J. Controlled Release* **1989,** *8,* 223–236.
11. Zhang, X.; Wyss, U. P.; Pichora, D.; Amsden, B.; Goosen, M. F. A. *J. Controlled Release* **1993,** *25,* 61–69.
12. Zhang, X.; Wyss, U. P.; Pichora, D.; Goosen, M. F. A. *J. Controlled Release* **1994,** *31,* 129–144.
13. Langer, R. S.; Peppas, N. A. *Biomaterials* **1981,** *2,* 201–214.
14. Göpferich, A.; Langer, R. *J. Controlled Release* **1995,** *33,* 55–69.
15. Schwendeman, S. P.; Cardamone, M.; Brandon, M. R.; Klibanov, A.; Langer, R. In *Microparticulate Systems for the Delivery of Proteins and Vaccines;* Cohen, S.; Bernstein, H., Eds.; Marcel Dekker: New York, 1996; pp 1–49.
16. Schwendeman, S. P.; Costantino, H. R.; Gupta, R. K.; Siber, G. R.; Klibanov, A. M.; Langer, R. *Proc. Natl. Acad. Sci. U.S.A.* **1995,** *92,* 11234–11238.
17. Schwendeman, S. P.; Costantino, H. R.; Gupta, R. K.; Tobio, M.; Chang, A.-C.; Alonso, M. J.; Siber, G. R.; Langer, R. *Dev. Biol. Stand.* **1996,** *87,* 293–306.
18. Costantino, H. R.; Langer, R.; Klibanov, A. M. *Pharm. Res.* **1994,** *11,* 21–29.
19. Costantino, H. R.; Langer, R.; Klibanov, A. M. *J. Pharm. Sci.* **1994,** *83,* 1662–1669.
20. Costantino, H. R.; Langer, R.; Klibanov, A. *Bio/Technology* **1995,** *13,* 493–496.
21. Langer, R.; Folkman, J. *Nature (London)* **1976,** *263,* 797–800.
22. Volkin, D. B.; Klibanov, A. M. In *Protein Function: A Practical Approach;* Creighton, T. E., Ed.; Oxford University: Oxford, England, 1989; pp 1–24.
23. Manning, M. C.; Patella, L.; Borchardt, R. T. *Pharm. Res.* **1989,** *6,* 903–918.
24. Bawa, R.; Siegel, R.; Marasca, B.; Karel, M.; Langer, R. *J. Controlled Release* **1985,** *1,* 259–268.
25. Siegel, R. A.; Langer, R. *J. Colloid Interface Sci.* **1986,** *109,* 426–440.
26. Tabata, Y.; Gutta, S.; Langer, R. *Pharm. Res.* **1993,** *10,* 487–496.
27. Morris, W.; Steinhoff, M. C.; Russell, P. K. *Vaccine* **1994,** *12,* 5–11.
28. Aguado, M. T.; Lambert, P. H. *Immunobiology (New York)* **1992,** *184,* 113–125.
29. Eldridge, J. H.; Staas, J. K.; Meulbroek, J. A.; McGhee, J. R.; Tice, T. R.; Gilley, R. M. *Mol. Immunol.* **1991,** *28,* 287–294.
30. Eldridge, J. H.; Staas, J. K.; Meulbroek, J. A.; Tice, T. R.; Gilley, R. M. *Infect. Immun.* **1991,** *59,* 2978–2986.
31. Hilleman, M. R. *Prog. Med. Virol.* **1966,** *8,* 131–182.
32. Cohen, S.; Yoshioka, T.; Lucarelli, M.; Hwang, L. H.; Langer, R. *Pharm. Res.* **1991,** *8,* 713–720.
33. Ron, E.; Turek, T.; Mathiowitz, E.; Chasin, M.; Hageman, M.; Langer, R. *Proc. Natl. Acad. Sci. U.S.A.* **1993,** *90,* 4176–4180.
34. Krewson, C. E.; Klarman, M. L.; Saltzman, W. M. *Brain Res.* **1995,** *680,* 196–206.

35. Preis, I.; Langer, R. S. *J. Immunol. Methods* **1979,** *28,* 193–197.
36. Hsieh, S. T.; Rhine, W. D.; Langer, R. *J. Pharm. Sci.* **1981,** *72,* 17–22.
37. Edelman, E. R.; Kost, J.; Bobeck, H.; Langer, R. *J. Biomed. Mater. Res.* **1985,** *19,* 67–83.
38. Schwendeman, S. P.; Labhasetwar, V.; Levy, R. J. *Pharm. Res.* **1995,** *12,* 790–795.
39. Labhasetwar, V.; Underwood, T.; Schwendeman, S. P.; Levy, R. J. *Proc. Natl. Acad. Sci. U.S.A.* **1995,** *92,* 2612–2616.
40. *Drug Facts and Comparisons;* Olin, B. R., Ed.; Facts and Comparisons: St. Louis, MO, 1994.
41. Sluzky, V.; Tamada, J. A.; Klibanov, A. M.; Langer, R. *Proc. Natl. Acad. Sci. U.S.A.* **1991,** *88,* 9377–9381.
42. Göpferich, A. In *Handbook of Biodegradable Polymers;* Domb, A.; Kost, J.; Wiseman, D., Eds.; Haarwood Academic: New York, in press.
43. Kalpana, R. K.; Park, K. *Adv. Drug Delivery Rev.* **1993,** *11,* 59–84.
44. Ron, E.; Langer, R. In *Treatise on Controlled Drug Delivery;* Kydonieus, A., Ed.; Marcel Dekker: New York, 1992; pp 199–224.
45. Li, S.; Garreau, H.; Vert, M. *J. Mater. Sci. Mater. Med.* **1990,** *1,* 123–130.
46. Li, S. M.; Garreau, H.; Vert, M. *J. Mater. Sci. Mater. Med.* **1990,** *1,* 131–139.
47. Leong, K. W.; Kost, J.; Mathiowitz, E.; Langer, R. *Biomaterials* **1986,** *7,* 364–370.
48. Heller, J. *Biomaterials* **1990,** *11,* 659–665.
49. Lin, W.-J.; Flanagan, D. R.; Linhardt, R. J. *Pharm. Res.* **1994,** *11,* 1030–1034.
50. Cohen, S.; Bano, M. C.; Visscher, K. B.; Chow, M.; Allcock, H. R.; Langer, R. *J. Am. Chem. Soc.* **1990,** *112,* 7832–7833.
51. Kohn, J.; Niemi, S. M.; Albert, E. C.; Murphy, J. C.; Langer, R.; Fox, J. G. *J. Immunol. Methods* **1986,** *35,* 609–612.
52. Herrlinger, M. Ph.D. Thesis, University of Heidelberg, Germany, 1994.
53. Fischel-Ghodsian, F.; Brown, L.; Mathiowitz, E.; Brandenburg, D.; Langer, R. *Proc. Natl. Acad. Sci. U.S.A.* **1988,** *85,* 2403–2406.
54. Thies, C. In *Microcapsules and Nanoparticles in Medicine and Pharmacy;* Donbrow, M., Ed.; CRC Press: London, 1992; pp 47–71.
55. Gander, B.; Wehrli, E.; Alder, R.; Merkle, H. P. *J. Microencapsulation* **1995,** *12,* 83–97.
56. Donbrow, M. In *Microcapsules and Nanoparticles in Medicine and Pharmacy;* Donbrow, M., Ed.; CRC Press: London, 1992; pp 17–45.
57. Mathiowitz, E.; Langer, R. *J. Controlled Release* **1987,** *5,* 13–22.
58. Cohen, J.; Siegel, R.; Langer, R. *J. Pharm. Sci.* **1984,** *73,* 1034–1037.
59. Rhine, W. D.; Hsieh, D. S. T.; Langer, R. *J. Pharm. Sci.* **1980,** *69,* 265–270.
60. Volkin, D. B.; Staubli, A.; Langer, R.; Klibanov, A. M. *Biotechnol. Bioeng.* **1991,** *37,* 843–853.
61. Cleland, J. L.; Jones, A. J. S. *Proc. Int. Symp. Controlled Release Bioact. Mater.* **1995,** *22,* 514–515.
62. Timasheff, S. N.; Arakawa, T. In *Protein Structure: A Practical Approach;* Creighton, T. E., Ed.; IRL Press : New York, 1990; pp 331–345.
63. Privalov, P. L.; Potekhin, S. A. *Methods Enzymol.* **1986,** *131,* 4–51.
64. Fenner, T. *Principles of Polymer Processing;* Chemical Publishing: New York, 1979.
65. Hutchinson, F. G.; Furr, B. J. A. *Biochem. Soc. Trans.* **1985,** *13,* 520–523.
66. Vilker, V. L.; Colton, C. K.; Smith, K. A. *J. Colloid Interface Sci.* **1981,** *79,* 549–565.
67. Amsden, B. G.; Cheng, Y.-L. *J. Controlled Release* **1994,** *31,* 21–32.
68. Göpferich, A. *Biomaterials* **1996,** *17,* 103–114.
69. Göpferich, A.; Langer, A. *J. Polym. Sci. Part A* **1993,** *31,* 2445–2458.
70. Shah, S. S.; Cha, Y.; Pitt, C. G. *J. Controlled Release* **1992,** *18,* 261–270.
71. Ogawa, Y.; Yamamoto, M.; Takada, S.; Okada, H.; Shimamoto, T. *Chem. Pharm. Bull.* **1988,** *36,* 1502–1507.
72. Ogawa, Y.; Yamamoto, M.; Okada, H.; Yashiki, T.; Shimamoto, T. *Chem. Pharm. Bull.* **1988,** *36,* 1095–1103.
73. Ogawa, Y.; Okada, H.; Heya, T.; Shimamoto, T. *J. Pharm. Pharmacol.* **1989,** *41,* 439–444.
74. Okada, H.; Inoue, Y.; Heya, T.; Ueno, H.; Ogawa, Y.; Toguchi, H. *Pharm. Res.* **1991,** *8,* 787–791.
75. Toguchi, H. *J. Int. Med. Res.* **1990,** *18,* 35–41.
76. Boolag, D. M.; Edelstein, S. J. *Protein Methods;* John Wiley & Sons: New York, 1991.
77. Göpferich, A.; Alonso, M. J.; Langer, R. *Pharm. Res.* **1994,** *11,* 1568–1574.
78. Gref, R.; Minamitake, Y.; Peracchia, M. T.; Trubetskoy, V.; Torchilin, V.; Langer, R. *Science (Washington, D.C.)* **1994,** *263,* 1600–1603.

79. Niwa, T.; Takeuchi, H.; Hino, T.; Kunou, N.; Kawashima, Y. *J. Pharm. Sci.* **1994,** *83,* 727–732.
80. Barrera, D. A.; Zylstra, E.; Lansbury, P. T., Jr.; Langer, R. *J. Am. Chem. Soc.* **1993,** *115,* 11010–11011.
81. Pitt, C. G. *Int. J. Pharm.* **1990,** *59,* 173–196.
82. Langer, R. *Acc. Chem. Res.* **1993,** *26,* 537–542.
83. Cleland, J. L.; Langer, R. In *Formulation and Delivery of Proteins and Peptides;* Cleland, J. L.; Langer, R., Eds.; ACS Symposium Series 567; American Chemical Society: Washington, DC, 1994; pp 1–19.
84. Franks, F.; Hatley, R. H. M.; Mathias, S. F. *BioPharm (Eugene, Oreg.)* **1991,** *4*(9), 38–55.
85. Franks, F. *Bio/Technology* **1994,** *12,* 253–256.
86. Pikal, M. J. *BioPharm (Eugene, Oreg.)* **1990,** *3*(8), 18–27.
87. Pikal, M. J. *BioPharm (Eugene, Oreg.)* **1990,** *3*(9), 26–30.
88. Hageman, M. J. In *Stability of Protein Pharmaceuticals. Part A, Chemical and Physical Pathways of Protein Degradation;* Ahern, T. J.; Manning, M. C., Eds.; Plenum: New York, 1992; pp 273–309.
89. Gordon, M.; Taylor, J. S. *J. Appl. Chem.* **1952,** *2,* 493–500.
90. Hageman, M. J. *Drug Dev. Ind. Pharm.* **1988,** *14,* 2047–2070.
91. Ahlneck, C.; Zografi, G. *Int. J. Pharm.* **1990,** *62,* 87–95.
92. Hageman, M. J.; Bauer, J. M.; Possert, P. L.; Darrington, R. T. *J. Agric. Food Chem.* **1992,** *40,* 348–355.
93. Kalichevsky, M. T.; Blanshard, J. M. V.; Tomarczuk, P. F. *Int. J. Food Sci. Technol.* **1993,** *28,* 139–151.
94. Sugisaki, M.; Suga, H.; Seki, S. *Bull. Chem. Soc. Jpn.* **1968,** *41,* 2591–2599.
95. Hsu, C. C.; Ward, C. A.; Pearlman, R.; Nguyen, H. M.; Yeung, D. A.; Curley, J. G. *Dev. Biol. Stand.* **1991,** *74,* 255–271.
96. Kakivaya, S. R.; Hoeve, C. A. J. *Proc. Natl. Acad. Sci. U.S.A.* **1975,** *72,* 3505–3507.
97. Williams, M. L.; Landel, R. F.; Ferry, J. D. *J. Am. Chem. Soc.* **1955,** *77,* 3701–3707.
98. Roy, M. L.; Pikal, M. J.; Rickard, E. C.; Maloney, A. M. *Dev. Biol. Stand.* **1991,** *74,* 323–340.
99. Karmas, R.; Buera, M. P.; Karel, M. *J. Agric. Food Chem.* **1992,** *40,* 873–879.
100. Cleland, J. L.; Powell, M. F.; Shire, S. J. *Crit. Rev. Ther. Drug Carrier Syst.* **1993,** *10,* 307–377.
101. Creighton, T. E. *Proteins: Structures and Molecular Properties;* W. H. Freeman: New York, 1993.
102. Dill, K. A. *Biochemistry* **1990,** *29,* 7133–7155.
103. Privalov, P. L. *Adv. Protein Chem.* **1979,** *33,* 167–241.
104. Pace, C. N. *Trends Biochem. Sci.* **1990,** *10,* 14–17.
105. Kuntz, I. D.; Kauzman, W. *Adv. Protein Chem.* **1974,** *28,* 239–345.
106. Prestrelski, S. J.; Tedeschi, N.; Arakawa, T.; Carpenter, J. F. *Biophys. J.* **1993,** *65,* 661–671.
107. Desai, U. R.; Osterhout, J. J.; Klibanov, A. M. *J. Am. Chem. Soc.* **1994,** *116,* 9420–9422.
108. Griebenow, K.; Klibanov, A. M. *Proc. Natl. Acad. Sci. U.S.A.* **1995,** *92,* 10969–10976.
109. Costantino, H. R.; Griebenow, K.; Mishra, P.; Langer, R.; Klibanov, A. M. *Biochim. Biophys. Acta* **1995,** *1253,* 69–74.
110. Thurow, H.; Geisen, K. *Diabetologia* **1984,** *27,* 212–218.
111. Sluzky, V.; Klibanov, A. M.; Langer, R. *Biotechnol. Bioeng.* **1992,** *40,* 895–903.
112. Clark, S.; Stephenson, R. C.; Lowenson, J. D. In *Stability of Protein Pharmaceuticals. Part A, Chemical and Physical Pathways of Protein Degradation;* Ahern, T. J.; Manning, M. C., Eds.; Plenum: New York, 1992; pp 1–29.
113. Friedman, A. R.; Ichpurani, A. K.; Brown, D. M.; Hillman, R. M.; Krabill, L. F.; Marin, R. A.; Zurcher-Neely, H. A.; Guido, D. M. *Int. J. Pept. Protein Res.* **1991,** *37,* 14–20.
114. Becker, G. W.; Tackitt, P. M.; Bromer, W. W.; Lefeber, D. S.; Riggin, R. M. *Biotechnol. Appl. Biochem.* **1988,** *10,* 326–337.
115. Oliyai, C.; Patel, J. P.; Carr, L.; Borchardt, R. T. *J. Pharm. Sci. Technol.* **1994,** *48,* 167–173.
116. Schulz, J. *Methods Enzymol.* **1967,** *11,* 255–263.
117. Piszkiewicz, D.; Landon, M.; Smit, E. L. *Biochem. Biophys. Res. Commun.* **1970,** *40,* 1173–1178.
118. Ahern, T. J.; Klibanov, A. M. *Science (Washington, D.C.)* **1985,** *228,* 1280–1284.
119. Zale, S. E.; Klibanov, A. M. *Biochemistry* **1986,** *25,* 5432–5444.
120. Teh, L.-C.; Murphy, L. J.; Huq, N. L.; Surus, A. S.; Friesen, H. G.; Lazarus, L.; Chapman, G. E. *J. Biol. Chem.* **1987,** *262,* 6472–6477.
121. Pearlman, R.; Nguyen, T. *J. Pharm. Pharmacol.* **1992,** *44,* 178–185.

122. Torchinski, Y. M. *Sulhydryl and Disulfide Groups of Proteins;* Consultants Bureau: New York, 1974.
123. Benesch, R. E.; Benesch, R. *J. Am. Chem. Soc.* **1958,** *80,* 1666–1669.
124. Cecel, R.; McPhee, J. R. *Adv. Protein Chem.* **1959,** *14,* 255–389.
125. Liu, W. R.; Langer, R.; Klibanov, A. M. *Biotechnol. Bioeng.* **1991,** *37,* 177–184.
126. Schwendeman, S. P.; Lee, J. H.; Gupta, R. K.; Costantino, H. R.; Siber, G. R.; Langer, R. *Proc. Int. Symp. Controlled Release Bioact. Mater.* **1994,** *21,* 54–55.
127. Hunt, S. In *Chemistry and Biochemistry of the Amino Acids;* Barrett, G. C., Ed.; Chapman and Hall: New York, 1985; pp 376–398.
128. Takahashi, K. *J. Biochem.* **1977,** *81,* 395–402.
129. Crowe, J. H.; Crowe, L. M.; Carpenter, J. F.; Aurell Wistrom, C. *Biochem. J.* **1987,** *242,* 1–10.
130. Hanson, M. A.; Rouan, S. K. E. In *Stability of Protein Pharmaceuticals. Part B, In Vivo Pathways of Degradation and Strategies for Protein Stabilization;* Ahern, T. J.; Manning, M. C., Eds.; Plenum: New York, 1992; pp 209–233.
131. Timasheff, S. N. In *Stability of Protein Pharmaceuticals. Part B, In Vivo Pathways of Degradation and Strategies for Protein Stabilization;* Ahern, T. J.; Manning, M. C., Eds.; Plenum: New York, 1992; pp 265–286.
132. Schwartz, H. M.; Lea, C. H. *Biochem. J.* **1952,** *50,* 713–716.
133. Duckworth, R. B. In *Water Activity. Influences of Food Quality;* Rockland, L. B.; Steward, G. F., Eds.; Academic: New York, 1981; pp 295–317.
134. Townsend, M. W.; DeLuca, P. P. *J. Pharm. Sci.* **1991,** *80,* 63–66.
135. Franks, F. *Cryo-Lett.* **1990,** *11,* 93–110.
136. Schwendeman, S. P.; Gupta, R. K.; Costantino, H. R.; Siber, G. R.; Langer, R. *Pharm. Res.* **1993,** *10,* S-200.
137. Creighton, T. E. *Protein Structure: A Practical Approach;* Oxford University: Oxford, England, 1990.
138. Lundblad, R. L.; Noyes, C. M. *Chemical Reagents for Protein Modification. Volume I;* CRC Press: Boca Raton, FL, 1984.
139. Lundblad, R. L.; Noyes, C. M. *Chemical Reagents for Protein Modification. Volume II;* CRC Press: Boca Raton, FL, 1984.
140. Carpenter, J. F.; Crowe, J. H. *Biochemistry* **1989,** *28,* 3916–3922.
141. Arakawa, T.; Kita, Y.; Carpenter, J. F. *Pharm. Res.* **1991,** *8,* 285–291.
142. Prestrelski, S. J.; Arakawa, T.; Carpenter, J. F. *Arch. Biochem. Biophys.* **1993,** *303,* 465–473.
143. Pikal, M. J.; Dellerman, K. M.; Roy, M. L.; Riggin, R. M. *Pharm. Res.* **1991,** *8,* 427–436.
144. Townsend, M. W.; DeLuca, P. P. *J. Parenter. Sci. Technol.* **1988,** *42,* 190–199.
145. Hsu, C. C.; Nguyen, H. M.; Yeung, D. A.; Brooks, D. A.; Koe, G. S.; Bewley, T. A.; Pearlman, R. *Pharm. Res.* **1995,** *12,* 69–77.
146. Peters, T. *Adv. Protein Chem.* **1985,** *37,* 161–245.
147. He, X. M.; Carter, D. C. *Nature (London)* **1992,** *358,* 209–215.
148. Geisow, M. J. *Bio/Technology* **1991,** *9,* 921–924.
149. Langer, R.; Rhine, W. D.; Hsieh, D. S. T.; Bawa, R. S. In *Controlled Release of Bioactive Materials;* Baker, R., Ed.; Academic: New York, 1980; pp 83–98.
150. Brange, J. *Galenics of Insulin: The Physico-Chemical and Pharmaceutical Aspects of Insulin and Insulin Preparations;* Springer-Verlag: Berlin, Germany, 1987.
151. Brange, J. *Stability of Insulin. Studies on the Physical and Chemical Stability of Insulin in Pharmaceutical Formulation;* Kluwer Academic: Boston, MA, 1994.
152. Feingold, V.; Jenkins, A. B.; Kraegen, E. W. *Diabetologia* **1984,** *27,* 373–378.
153. Grau, U. *Diabetologia* **1985,** *28,* 458–463.
154. Melberg, S. G.; Havelund, S.; Villumsen, J.; Brange, J. *Diabetic Med. (Chichester, Engl.)* **1988,** *5,* 243–247.
155. Fisher, B. V.; Porter, P. B. *Pharm. Pharmacol.* **1981,** *33,* 203–206.
156. Esparza, I.; Kissel, T. *Vaccine* **1992,** *10,* 714–720.
157. Alonso, M. J.; Cohen, S.; Park, T. G.; Gupta, R. K.; Siber, G. R.; Langer, R. *Pharm. Res.* **1993,** *51,* 945–953.
158. Gupta, R. K.; Siber, G. R. In *Vaccines 93;* Brown, F.; Chanock, R.; Ginsberg, H.; Lerner, R., Eds.; Cold Spring Harbor Laboratory: Cold Spring Harbor, NY, 1993; pp 391–396.
159. Alonso, M. J.; Gupta, R. K.; Min, C.; Siber, G. R.; Langer, R. *Vaccine* **1994,** *12,* 299–306.
160. Men, Y.; Thomasin, T.; Merkle, H. P.; Gander, B.; Corradin, G. *Vaccine* **1995,** *13,* 683–689.

161. Schwendeman, S. P.; Gupta, R. K.; Siber, G. R.; Langer, R. *Pharm. Res.* **1995**, *12,* S-80.
162. Means, G. E.; Feeney, R. E. *Chemical Modification of Proteins;* Holden-Day: San Francisco, CA, 1971.
163. Rice, R. H.; Means, G. E. *J. Biol. Chem.* **1971**, *246,* 831–832.
164. Gander, B., Eidgenössische Technische Hochschule, Zürich, Switzerland, personal communication, 1993.
165. Baumann, G.; Silverman, B. L. *Growth Regul.* **1991**, *1,* 43–50.
166. Wyse, J. W.; Takahashi, Y.; DeLuca, P. P. *Proc. Int. Symp. Controlled Release Bioact. Mater.* **1989**, *16,* 334–335.
167. Sivaramakrishnan, K. N.; Rahn, S. L.; Moore, B. M.; O'Neil, J. *Proc. Int. Symp. Controlled Release Bioact. Mater.* **1989**, *16,* 14–15.
168. Gupta, R. K.; Relyveld, E. H.; Lindblad, E. B.; Bizzini, B.; Ben-Efraim, S.; Gupta, C. K. *Vaccine* **1993**, *11,* 293–306.
169. Gupta, R. K.; Siber, G. R. *Vaccine* **1995**, *13,* 1263–1276.
170. Gupta, R. K.; Rost, B. E.; Relyveld, E.; Siber, G. R. In *Vaccine Design: The Subunit Approach;* Powell, M. F.; Newman, M. J., Eds.; Plenum: New York, 1995; pp 229–248.
171. Langer, R. S. *Methods Enzymol.* **1981**, *73,* 57–74.
172. Niemi, S. M.; Fox, J. G.; Brown, L. R.; Langer, R. *Lab. Anim. Sci.* **1985**, *35,* 609–612.
173. Wise, D. L.; Trantolo, D. J.; Marino, R. T.; Kitchell, J. P. *Adv. Drug Delivery Rev.* **1987**, *1,* 19–39.
174. O'Hagan, D. T.; Palin, K.; Davis, S. S.; Artursson, P.; Sjoholm, I. *Vaccine* **1989**, *7,* 421–424.
175. Eldridge, J. H.; Gilley, R. M.; Staas, J. K.; Moldoveanu, Z.; Meulbroek, J. A.; Tice, T. R. *Curr. Top. Microbiol. Immunol.* **1989**, *146,* 59–66.
176. Moldoveanu, Z.; Staas, J. K.; Gilley, R. M.; Ray, R.; Compans, R. W.; Eldridge, J. H.; Tice, T. R.; Mestecky, J. *Curr. Top. Microbiol. Immunol.* **1989**, *146,* 91–99.
177. Langer, R. *Science (Washington, D.C.)* **1990**, *249,* 1527–1533.
178. Stevens, V. C.; Powell, J. E.; Ricky, M.; Lee, A. C.; Lewis, D. H. In *Gamete Interaction: Prospects for Immunocontraception;* Alexander, N. J.; Griffin, D.; Spieler, M.; Waites, G., Eds.; Wiley-Liss: New York, 1990; pp 549–563.
179. O'Hagan, D. T.; Rahman, D.; McGee, J. P.; Jeffery, H.; Davies, M. C.; Williams, P.; Davis, S. S.; Challacombe, S. J. *Immunology* **1991**, *73,* 239–242.
180. Singh, M.; Singh, A.; Talwar, G. P. *Pharm. Res.* **1991**, *8,* 958–961.
181. Singh, M.; Singh, O.; Singh, A.; Talwar, G. P. *Int. J. Pharm.* **1992**, *85,* R5–R8.
182. Weng, C. N.; Tzan, W. L.; Lin, S. D.; Lin, S. Y.; Lee, C. J. *Res. Vet. Sci.* **1992**, *53,* 42–46.
183. Nellore, R. V.; Pande, P. G.; Young, D.; Bhagat, H. R. *J. Parenter. Sci. Technol.* **1992**, *46,* 176–180.
184. McQueen, C. E.; Boedeker, E. C.; Reid, R.; Jorbre, D.; Wolf, M.; Le, M.; Brown, W. R. *Vaccine* **1993**, *11,* 201–206.
185. Moldoveanu, Z.; Novak, M.; Huang, W. Q.; Gilley, R. M.; Staas, J. K.; Compans, R. W.; Mestecky, J. *J. Infect. Dis.* **1993**, *167,* 84–90.
186. Raghuvanshi, R. S.; Singh, M.; Talwar, G. P. *Int. J. Pharm.* **1993**, *93,* R1–R5.
187. Eldridge, J. H.; Staas, J. K.; Chen, D.; Marx, P. A.; Tice, T. R.; Gilley, R. M. *Semin. Hematol.* **1993**, *30,* 16–25.
188. O'Hagan, D. T.; Jeffery, H.; Davis, S. S. *Vaccine* **1993**, *11,* 965–969.
189. Ray, R.; Novak, M.; Duncan, J. D.; Matsuoka, Y.; Compans, R. W. *J. Infect. Dis.* **1993**, *167,* 752–755.
190. Edelman, R.; Russell, R. G.; Losonsky, G.; Tall, B. D.; Tacket, C. O.; Levine, M. M.; Lewis, D. H. *Vaccine* **1993**, *11,* 155–158.
191. Reid, R. H.; Boedeker, E. C.; McQueen, C. E.; Davis, D.; Tseng, L.-Y.; Kodak, J.; Sau, K.; Wilhelmsen, C. L.; Nellore, R.; Dalal, P.; Bhagat, H. R. *Vaccine* **1993**, *11,* 159–167.
192. Marx, P. A.; Compans, R. W.; Gettie, A.; Staas, J. K.; Gilley, R. M.; Mulligan, M. J.; Yamshchikov, G. V.; Chen, D.; Eldridge, J. H. *Science (Washington, D.C.)* **1993**, *260,* 1323–1327.
193. Santiago, N.; Milstein, S.; Rivera, T.; Garcia, E.; Zaidi, T.; Hong, H.; Bucher, D. *Pharm. Res.* **1993**, *10,* 1243–1247.
194. Almeida, A. J.; Alpar, H. O.; Brown, M. R. W. *J. Pharm. Pharmacol.* **1993**, *45,* 198–203.
195. Maloy, K. J.; Donachie, A. M.; O'Hagan, D. T.; Mowat, A. M. *Immunology* **1994**, *81,* 661–667.
196. Khan, M. Z. I.; Opdebeeck, J. P.; Tucker, I. G. *Pharm. Res.* **1994**, *11,* 2–11.
197. Offit, P. A.; Khoury, C. A.; Moser, C. A.; Clark, H. F.; Kim, J. E.; Speaker, T. J. *Virology* **1994**, *203,* 134–143.

198. Shahin, R.; Leef, M.; Eldridge, J.; Hudson, M.; Gilley, R. *Infect. Immun.* **1995,** *63,* 1195–1200.
199. Gupta, R. K.; Siber, G. R. *Biologicals* **1994,** *22,* 215–219.
200. Gupta, R. K.; Siber, G. R. *Dev. Biol. Stand.* **1996,** *86,* 207–215.
201. Gupta, R. K. *J. Immunol. Methods* **1995,** *179,* 277–279.
202. Walker, R. I. *Vaccine* **1994,** *12,* 387–400.
203. McGhee, J. R.; Mestecky, J.; Dertzbaugh, M. T.; Eldridge, J. H.; Hirasawa, M.; Kiyono, H. *Vaccine* **1992,** *10,* 75–88.
204. Tomasi, T. B., Jr. *Transplantation* **1980,** *29,* 353–361.
205. Holmgren, J.; Czerkinsky, C.; Lycke, N.; Svennerholm, A.-M. *Immunobiology* **1992,** *184,* 157–179.
206. Rappuoli, R.; Podda, A.; Pizza, M.; Covacci, A.; Bartoloni, A.; de Magistris, M. T.; Nencioni, L. *Vaccine* **1992,** *10,* 1027–1032.
207. Ada, G. L. In *Vaccination and World Health;* Cutts, F. T.; Smith, P. G., Eds.; John Wiley & Sons: Chichester, England, 1994; pp 67–80.
208. Burton, G. F.; Kapasi, Z. F.; Szakal, A. K.; Tew, J. G. In *Strategies in Vaccine Design;* Ada, G. L., Ed.; R. G. Landes Company: Austin, TX, 1994; pp 35–49.
209. Alving, C. R.; Verma, J. N.; Rao, U.; Krzych, U.; Amselem, S.; Green, S. M.; Wassef, N. M. *Res. Immunol.* **1992,** *143,* 197–198.
210. Alving, C. R. *Immunobiology* **1993,** *187,* 430–446.
211. Göpferich, A.; Langer, R. *Macromolecules* **1993,** *26,* 4105–4112.

13
Oral Immunization Using Microparticles

Terry L. Bowersock and Harm HogenEsch

Immunity at mucosal sites is important in preventing the attachment of pathogenic microorganisms and neutralizing toxins that initiate infectious disease. Oral vaccination is an effective means of stimulating an immune response at any mucosal site in the body. The dose of orally administered vaccines necessary to induce an immune response is much greater than parenterally administered vaccines because of dilution and inactivation by the low pH and enzymes in the gastrointestinal tract. Encapsulation within microparticles protects vaccines as they traverse the gastrointestinal tract. They also release antigen slowly over time resulting in a sustained mucosal immune response. In this review we discuss the use of polymeric microparticle delivery systems for the oral delivery of vaccines including ways in which they are being altered to enhance site-specific delivery to the gut-associated lymphoid tissue and to enhance or modulate mucosal immunity.

Immunization is an efficient and cost-effective technology for the prevention of various infectious diseases. In fact, immunization can be considered the best medicine for many diseases (*1*). For many years immunization has relied on the induction of immunity by parenteral administration of vaccines. Antibodies induced in this manner, however, do not necessarily reach mucosal surfaces where infectious agents enter the host and most infections start. At mucosal sites, secretory immunoglobulin A (sIgA) is the predominant antibody isotype present. Secretory IgA prevents the attachment of bacteria and viruses to mucosa and neutralizes viruses and toxins. Because mucosal immunity provides the first line of immunological defense, protective immunity should ideally consist of antibodies or cells active at mucosal sites. Induction of immunity at mucosal surfaces generally requires administration of antigen directly to the mucosal site. Alternatively, stimulating immune responses at multiple sites is possible by inoculation at one mucosal site using the common mucosal immune system (CMIS). The greatest source of lymphoid tissue

Copyright © 1997 American Chemical Society

in the CMIS is the gut-associated lymphoid tissue (GALT) in the intestinal tract. The presence of GALT makes oral vaccination possible for induction of immunity at many other mucosal sites in the body. Additional advantages of oral vaccination include decreased risk of painful adverse reactions associated with injected vaccines and less chance of immediate hypersensitivity reactions. These benefits would increase patient compliance in receiving vaccines. In the food-animal industry, oral vaccinations result in reduced injection-site abscesses that reduce carcass and hide quality, reduced labor needed to vaccinate large numbers of animals, and reduced stress on animals restrained individually for inoculation. Oral vaccines can be produced with less purification and with less expense. A greater number of different vaccines can be included in a dose because larger doses could be more easily ingested than injected.

Despite the advantages of the oral vaccination, the development of oral vaccines has been slow. Oral administration of antigens must overcome several challenges. Most oral vaccines are degraded by bacteria and enzymes present in the gastrointestinal tract (GIT). These factors limit absorption of antigens, resulting in insufficient immune stimulation (*2*). Some animal species have anatomically complex GITs, further complicating oral delivery. Ruminants have four stomachs; retention of antigens in the rumen (first stomach) and exposure to the resident microflora could degrade any antigen before it has a chance to reach the GALT. In birds, the crop and ventriculus (gizzard) can retain vaccines and break them down mechanically or enzymatically before they reach intestinal lymphoid tissue. To overcome these problems, delivery systems have been developed to deliver vaccine-relevant antigens to GALT.

Mechanism of Oral Vaccination

Mucosal immunity is stimulated by administering antigen directly to the mucosal site where an infection begins. Antigen is processed by the mucosa-associated lymphoid tissue (MALT). Major concentrations of MALT are found in the upper respiratory (nasal-associated lymphoid tissue), the lower respiratory (bronchus-associated lymphoid tissue), as well as in the GIT (the GALT). Administration of vaccines to MALT by either intranasal, intrabronchial, or oral administration results in protective immunity at other mucosal sites (*2, 3*). For this discussion, we will focus on uptake of antigen in MALT by the Peyer's patches, circumscribed areas of GALT in the small intestine. Peyer's patches are collections of lymphoid nodules containing B and T lymphocytes, dendritic cells, and macrophages. Each nodule consists of a dome region and a follicle, and the nodules are separated by interfollicular areas. Special cells in the epithelium overlying the dome, the microfold (M) cells, pick up antigen and transport it intact to underlying antigen-presenting cells. Antigen is processed and presented to T lymphocytes. The antigen stimulates antigen-specific plasma (B lymphocyte) precursors and memory cells in the germinal centers of follicles located beneath the Peyer's patches. These plasma cell precursors are influenced by $CD3^+$, $CD4^+$, and $CD8^-$ T helper cells located between follicles to preferentially produce IgA (*4–6*).

The IgA plasma cell precursors leave the site of antigen uptake and processing and traffic through lymph nodes, such as the mesenteric lymph nodes that drain the GIT. Further differentiation and maturation occur in the lymph nodes before the lymphocytes enter the efferent lymph. Plasma cell precursors then enter the blood and migrate to mucosal sites. Most plasma cells return to the organ of origin of the immune response. However, a significant number also end up at other mucosal sites. The homing to mucosal sites is not fully understood. However, special receptors on high endothelial venule cells and lymphocytes are thought to interact to allow extravasation of lymphocytes selectively into the lamina propria of mucosal sites (*7*). The lamina propria is the effector site of the mucosal immune system where IgA is produced and T lymphocyte mediated responses occur.

Uptake of Microparticles by Intestinal M Cells

Oral administration of vaccines generally requires much larger doses of antigen because of alteration or loss of antigens by the pH, enzymes, mucus lining, and motility of the GIT that prevent uptake by the Peyer's patches. To overcome these problems, various methods have been developed to deliver vaccine-relevant antigens to the GALT.

Antigen delivery systems can be divided into two classes: live vectors and nonreplicating antigen carriers. Live vectors have proven promising but face difficulties in acceptance by the public, and they may induce an immune response in the host that eliminates the delivery system resulting in immune stimulation of an uncertain length of time. Nonreplicating delivery systems offer a more ready acceptance by the public, long-term controlled release and immunostimulation, proven safety, and less chance of an immune response eliminating the delivery of antigens. Delivery systems have 2 primary purposes: protection of antigen through the pH and enzymes present in the GIT and delivery of intact antigen to the GALT at the M cells. Any sized device can protect an antigen, but stimulation of GALT depends on either release of the antigen in the vicinity of the M cells, or on the device (particle) being taken up. Uptake of particles ensures optimal protection (release is in the lymphoid tissue) from degradation in the GIT. To be taken up by cells, the delivery device (microparticles) must be small, 10 µm or less in diameter (*8*). Microparticles can then offer controlled release over time stimulating a longer lasting mucosal immune response. More information on microparticles in vaccines can be obtained in the recently published book on the subject (*9*).

The efficiency with which orally administered microparticles are taken up by M cells depends on several factors. The more hydrophobic a particle is, the better it is taken up by M cells (*8*). Size of particles is another important consideration. In general the smaller the particle, the better it is taken up (*8, 10–12*). Particles greater than 10 µm are usually considered to be the ones not taken up by M cells, the primary cells responsible for uptake of antigen from the intestinal lumen. Particles 5–10 µm in diameter are taken up but remain in the Peyer's patch, stimulating a local or mucosal immune response. Uptake of microparticle depends to some extent on the fed state and integrity of the intestinal epithelium (*13–15*), age (*16*), and morphology of the Peyer's patch of the host (*17, 18*). These factors must be considered in the design of orally administered microparticle vaccines.

The uptake of orally administered particulates is extremely low in most animal species; less than 0.01% of particles are taken up in mice (*13, 19*) and 1–5% in rabbits (*20, 21*). Therefore, one way to increase uptake of antigen is to increase the number of microparticles per dose. Differences exist in GI morphology between hosts that also affect the efficiency of uptake of particles. Because uptake is an inefficient process, there is now great emphasis on enhancing the uptake of microparticles. This increase can be done by conjugating lectins of plants or adhesins of bacteria like the pili of *Escherichia coli,* or cholera or heat-labile *E. coli* toxins that bind to specific receptors in the intestinal epithelium to the surface of microparticles (*22–25*). Not only does this process enhance uptake of the antigen, in the case of the cholera toxin and *E. coli* toxins, it appears to increase the mucosal immune response to antigen as well (*2, 24*).

Poly(lactide-*co*-glycolide) for Oral Delivery

The most studied polymer for oral administration, as well as by other routes, using primarily microparticles is poly(lactide-*co*-glycolide) (PLG). This polymer has received so much attention because it has been used for many years in resorbable sutures, bone plates, and for the controlled release of drugs (*26–28*). Safety of PLG has also been proven. PLG breaks down in vivo by hydrolysis of ester linkages to lactic and glycolic acid metabolites, constituents proven to be biocompatible (*29*). PLG does not stimulate damaging tissue reaction such as fibrous capsule formation. Any inflammatory response was completely resolved within a short period of time. Also, many companies make PLG microparticles, so availability and scale up of vaccines should not be difficult.

PLG microparticles are produced by one of several techniques. The most commonly used method is solvent evaporation (or emulsion polymerization), whereby an oil-in-water emulsion of antigen and copolymer is produced with an organic solvent. The solvent is removed by evaporation or extraction leaving microparticles. This technique produces very small particles but incorporates antigen at an efficiency of around 30% (*30, 31*). A water-in-oil-in-water emulsion technique has been developed with a much higher encapsulation efficiency (60–94%) (*32*). The second most common method is the phase-separation technique in which polyesters (dissolved in an organic solvent) and antigen are mixed with salts or a second polymer. This method causes a separation of the polyester phase and encapsulation of antigen. The polyester–antigen can then be solidified by adding a nonsolvent (*33*). This technique is limited to use with antigens that are insoluble in the organic solvent used to dissolve the polyester. A higher efficiency of antigen loading is possible with this technique. A less used technique is spray-drying of an aerosolized suspension of polyester and antigen (*34*). The small droplets solidify when exposed briefly to hot air. Other techniques are used but will not be discussed here.

The method of preparation of the PLG microparticles also affects the release profile of antigens. Production using an oil-in-water emulsion solvent-evaporation technique resulted in a triple phase release consisting of an immediate burst, a lag phase of minimal release, followed by a period of sustained release. When a phase-separation technique was used, antigen was released at a slow and steady rate without a burst or lag phase (*35*). Although it is not clear what effect this more predictable release rate will have on immune responses, it provides a more defined pattern of release for vaccines.

The effect of the production methods on the antigenicity of vaccines is a concern because many proteins may be denatured under the conditions of heat and solvents used in creating PLG microspheres. In this regard, the antigenicity of several encapsulated antigens has been studied. Immunogenicity of influenza vaccines was determined to be as good with encapsulation as without, with similar titers induced in mice (*36*). Sodium dodecyl sulfate–polyacrylamide gel electrophoresis of encapsulated and nonencapsulated bovine serum albumin showed similar bands indicating that degradation or aggregation did not occur (*32*). Analysis of ovalbumin (OVA) using Western blots, gel electrophoresis, and isoelectric focusing also demonstrated that this protein was unaffected by encapsulation in PLG (*37*).

Using OVA as a model antigen, release of antigen over time has been determined for PLG microparticles (*38*). These studies have demonstrated that polymer content—the proportion of lactide to glycolide—dramatically affects the rate of release of antigen in vitro. The greater the proportion of lactide, the slower the deterioration of the particle and subsequent release of antigen. The molecular weight of the copolymers also affects the degradation of the microparticles and subsequent rate of release of antigen. The higher the molecular mass, the slower the degradation and subsequent release of antigen. The concentration of antigen also affects the rate of release with a greater concentration causing a faster rate of release. Assuming the rate of degradation and release of antigen in vivo correlates with those in vitro, these factors can be used to design vaccines containing multiple preparations of microparticles that release antigen at different rates to stimulate an immediate as well as a sustained immune response.

Animal studies have been performed to examine these effects on the immune system. Oral administration of OVA in different copolymers resulted in different immune responses. Particles that released antigen quickly (50:50 lactide:glycolide) stimulated a greater secretory IgA response in the salivary secretions of mice, whereas particles that release antigen more slowly (75:25 lactide:glycolide) stimulated a greater serum immune response (*39*). This result demonstrates how the rate of release of antigen from microparticles in vitro correlates with the immune response in animals. Oral vaccines composed of multiple formulations of PLG copolymers have been tested in rodents using human immunodeficiency virus (HIV) antigens. Immune responses were demonstrated for up to a year following oral immunization using a mixture of 100% lactide, 50:50 lactide:glycolide, and 75:25 lactide:glycolide polymer microparticles (*40*). This vaccine shows promise for further development based on the good titers seen long term following oral administration.

PLG microparticles have an adjuvant effect because a greater immune response to antigen is seen than when antigen alone is administered (*41*). PLG microparticles have also been shown to induce cytotoxic T cell to OVA when administered either parenterally or orally to mice (*42, 43*). This process is mediated through the class I major histocompatibility complex (MHC) receptor because the target EG7.OVA cells express only MHC I and not MHC II. However, it is not clear how PLG microparticles target the endogenous antigen processing pathway. More frequent and larger doses of antigen in PLG microparticles are required to stimulate cytotoxic T-lymphocyte (CTL) response compared with liposomes.

PLG microparticles have been used to successfully immunize animals by oral administration using many different antigens from infectious agents. The first antigen used for oral administration with PLG was the enterotoxin toxoid of *Staphylococcus aureus* (SEB). Oral administration of SEB in PLG microspheres stimulated antigen-specific IgM, IgG, and IgA in serum, gut washings, saliva, and lung washings in mice (*41*).

For influenza, formalin-inactivated virus incorporated in PLG microspheres and administered orally to mice stimulated high immunoglobulin titers in serum, IgA in saliva, and reduced virus in nose and trachea of challenged mice (*36*). A similar vaccine was tested for use in parainfluenza. Mice vaccinated with PLG microspheres containing whole virus had nearly complete reduction in virus recovery postchallenge if vaccinated by intraperitoneal administration. However, oral or intranasal administration resulted in only a 10–15-fold reduction in virus (*44*). Protection against simian immunodeficiency virus (SIV) was induced in female macaques by systemic priming followed by either oral or intratracheal inoculation with formalin-killed SIV in PLG microspheres (*45*). The response to this virus is especially exciting because SIV is a model for HIV—a disease for which a vaccine is desperately needed.

Good oral vaccines for enteropathogenic bacteria are needed for use in infants and travelers. Pili from a diarrhea-causing strain of *E. coli* incorporated in PLG microspheres induced anti-pili biliary IgA, reduced bacterial adherence to intestinal epithelium, and increased anti-pili lymphocyte proliferation in spleen cells in orally vaccinated rabbits (*46*). The fimbrial adhesin, colonization factor I (CFA), of *E. coli* stimulates immunity to prevent the colonization of the gut by this organism. It has been incorporated in PLG microspheres and induced serum IgG and, less reliably, intestinal IgA when administered by intragastric tube to rabbits (*47*). When CFA II, another colonization factor, was incorporated in PLG, intraduodenal administration resulted in enhanced lymphocyte proliferation by lymphocytes from both the spleen and Peyer's patches and detection of antibody secreting cells from spleen (*48*). This use was the first reported of good manufacturing practices (GMP) to produce a PLG microsphere vaccine that met World Health Organization guidelines for oral vaccines. This fact is important because microparticles will be taken up into the lymphoid tissue of the intestine and will not merely release antigen into the intestinal lumen. A phase I clinical trial in human volunteers with CFA II in PLG microspheres administered orally was subsequently undertaken. Five of 10 volunteers developed antibody secreting cells to proteins contained in CFA II in peripheral blood mononuclear cells. Three of these vaccinates did not develop diarrhea when challenged with live enterotoxigenic *E. coli*, a result indicating a 30% protection for this oral vaccine (*49*). This procedure was the first reported human trial using PLG-based vaccines.

Not only have antigens from infectious agents been used for oral administration of antigens, but fertility vaccines also have been investigated. A recombinant protein antigen of fox sperm was incorporated within PLG microspheres and administered orally to rats. The number of antibody-secreting cells detected in the intestine of these rats was similar to that seen in rats in which the same antigen was injected into Peyer's patches and then injected intraduodenally (*50*). Whether antibodies will also be seen in the reproductive tract from such an immunization regimen remains to be seen.

Use of microspheres allows incorporation of other adjuvants or immunomodulators along with antigens. It is necessary to first determine whether such components can retain their biological activity in PLG or other polymers. In an early study, only 2–3% of interleukin-2 was released for up to 30 days (*51*). There is also evidence that the cytokine interferon can retain biological activity when encapsulated in PLG microparticles (*52*). More recent studies support earlier work on the effect of cytokines incorporated with an antigen on enhancing an immune response. Bovine serum albumin incorporated in PLG microspheres with IL-5 and IL-6 administered intraperitoneally or topically in the eye increased both the titer and duration of the IgA response in rats (*53*). These studies are important in that incorporation of cytokines or polypeptides in microparticles alone or with vaccines can enhance an immune response or direct a response toward a type I (cell-mediated) or type II (humoral) response as desired for immunity to specific pathogens. However, their efficacy when administered orally has not been determined as yet.

Even though PLG microparticle vaccines have many advantages, several disadvantages to the use of PLG exist. The most pressing issue is the use of organic solvents to prepare the microparticles. The solvents can adversely affect the antigenicity of the vaccine. Solvents will have to be carefully evaluated for each antigen prepared. A recent study showed that most solvents commonly used decrease the immunogenicity of antigens used in a parenteral vaccine (*54*). The solvents can also be retained within microparticles and pose a danger to the host when microparticles are taken up and solvent is released in vivo. In many of the preparations microparticles are subjected to heat and mechanical agitation that could also damage antigens. Polymers can also react with peptides of some antigens altering epitopes that could dramatically affect the immunogenicity of the vaccine. Stability of PLG microparticles, as for most other polymeric systems described here, on the shelf over long periods of time has not yet been determined. More work needs to be done in this regard. Postproduction sterilization must also be addressed. Gamma irradiation of PLG polymers has increased the degradation rate of the polymers. This result could have serious effects on the long-term stability of vaccines. In one report noted previously, microparticles made under GMP had few pathogenic bacteria as well as acceptable levels of solvent present (*48*). This study is encouraging but further studies are needed to support these findings.

Other Polymeric Oral Delivery Systems

Other polymers have been used for systemic immunization, but limited trials have been performed for oral administration of these polymers. Adsorption of antigen to the surface of poly(butylcyanoacrylate) particles induced secretory immunity to an antigen greater than when the antigen was administered orally alone (*55, 56*). One of the more hydrophobic polymers well taken up by Peyer's patches was poly(methyl methacrylic) acid (PMMA). PMMA has minimal tissue reactivity. Nanoparticles made of PMMA have adjuvant activity increasing the immune response to antigens administered parenterally better than aluminum hydroxide (*57*). The adjuvanticity of PMMA seems to be effective for orally administered vaccines. Discs of 3 × 5 mm of PMMA hydrogels were sorbed with culture supernatants of *Pasteurella haemolytica*. These larger hydrogels were produced to ensure that the vaccine would bypass the rumen (*58*). Cattle inoculated orally with these hydrogels lived longer and had less pneumonia than nonvaccinated calves when challenged intrabronchially with viable bacteria (*59*). Production of such vaccines with more efficient loading of antigen may make this delivery system more acceptable for use in animals.

Respiratory diseases are a major cause of concern to humans and of economic losses in swine as well. One organism for which a cellulose acetate phthalate (CAP)-based delivery of a vaccine has been developed is *Mycoplasma hyopneumoniae*. Pigs administered a whole organism–CAP vaccine orally after parenteral priming had reduced pneumonic lesions and a higher level of serum anti-*Mycoplasma* antibodies than nonvaccinated pigs (*60*). In this study the particles containing vaccine were 350–1400 μm in diameter. It will be interesting to see how long the protection from a vaccine with such large microparticles will last and whether continual or pulsed feeding of the vaccine will be needed to maintain protection, because it was unlikely that particles were taken up to release antigen over time.

Liposomes

Liposomes are particles composed of a bilayered phospholipid membrane or lamella. Liposomes vary in size from 0.01 to 150 μm and can be produced with one layer (unilamellar) or many, usually greater than five, layers (multilamellar). Multilamellar vesicles have a thin layer of water between layers. The single layered liposomes have an aqueous core surrounded by a single lipid bilayer and are described as small or large unilamellar vesicles. The amphipathic nature of the phospholipid molecules causes vesicle formation to occur spontaneously when produced. The hydrophobic part of the phospholipids is contained within the bilayer, whereas the hydrophilic part is directed into the water-soluble (hydrophilic) part of the vesicle. This unique arrangement allows either hydrophilic or hydrophobic antigens to be incorporated within liposomes (*61*). However, although most any type of antigen can be incorporated within liposomes, some materials lose their biological activity when incorporated within them. This characteristic can be attributed in part to the use of organic solvents in the preparation of liposomes.

Two primary methods of preparing liposomes exist: sonication followed by microemulsification of liposomes suspensions and dehydration–rehydration. In the sonication–microemulsion method, both multilamellar and unilamellar vesicles of varying size are produced, whereas the dehydration–rehydration method yields a more homogeneous, uniformly small, unilamellar product (*62*). Both methods are equally efficient and incorporate 75–80% of antigen. Leakage of material over time can be a problem with liposomes. This problem can be overcome by lyophilization of liposomes using trehalose as a cryoprotectant (*63*). The primary limiting factor is size. Liposomes 70–100 nm in diameter can be lyophilized and stabilized the best. Lyophilization serves another function: it doubles the antigen concentration within the liposomes, presumably by removing much of the moisture (*62*). Alternative methods of production of liposomes have shown promise for encapsulation of more labile antigens including whole live bacteria (*64*). Giant (5.5 μm in diameter) liposomes are made using a double emulsion technique. Such liposomes retain bacteria or toxoid antigens even in the presence of serum. Antigens in liposomes prepared in this manner are retained in tissue much longer than if given alone. This method has the potential for providing a means of delivering vaccines in a way that protects them from maternal antibodies, as well as from other degrading substances in the body.

The mechanism of delivery of antigen in liposomes relies on the fact that these vesicles readily adsorb to most mammalian cells. Liposomes may then release the antigen or be phagocytosed by macrophages or antigen-presenting cells. This result may explain how liposomes act as (immuno)-adjuvants, because they increase the uptake and processing of antigens (*65*). Part of the adjuvant effect can be explained by liposomes preventing elimination from the body as well as increasing the concentration of antigen in macrophages where antigen processing and presentation occur (*66*). Liposomes enhance cell-mediated responses, a role vital in protection from viral infections (*67*). More recently, liposomes have been shown to be effective in enhancing the immune response to B-cell dependent epitopes by providing co-immunization with T helper lymphocyte dependent epitopes. This result ensures an immune response, thereby avoiding the need for more toxic adjuvants (*68*).

The rapid uptake of liposomes makes them an ideal delivery system for vaccines. For oral administration, the stability of liposomes in bile salts and intestinal contents is important. Although this area remains controversial, phospholipids such as dipalmitoylphosphatidylcholine and distearoylphosphatidylcholine combined with cholesterol are resistant to bile salts and enzymes that have been used for oral vaccines. Determining whether liposomes can be taken up by the M cells of Peyer's patches is important. Transmission electron micrographs proved that intraintestinally administered liposomes were indeed taken up by M cells and transported to the lymphoid follicles underlying Peyer's patches (*69*). Subsequent studies by others have confirmed these results and shown that orally administered liposomes were selectively taken up by Peyer's patches in the lower ileum (*70, 71*). These studies support earlier reports where antigens administered orally in liposomes were detected in Peyer's patches up to 17 days after administration (*72*). Because liposomes are not as stable as other microparticles in intestinal contents, it is even more important to enhance uptake of liposomes into Peyer's patches. Attachment protein of reovirus, a compound shown to be specific for binding to M cells, was incorporated into liposomes and enhanced the uptake of liposomes in vitro (*73*). The relevance of this study is questionable because it used L929 cells, a fibroblast cell line that is considerably different from M cells. The effect the use of this attachment protein will have in vivo remains to be seen, not only concerning uptake, but whether it will also increase a mucosal immune response to antigens encapsulated within the liposomes.

The first use of liposomes for oral vaccines was in induction of immunity to dental caries caused by *Streptococcus mutans* (*74*) by using purified cell-wall antigens. This study and others (*61*) demonstrated that antigens from cariogenic bacteria incorporated in liposomes could induce as much protection as antigens prepared with other adjuvants. These studies showed that antigens administered orally in liposomes increased intestinal, serum, and salivary IgA to subunit antigens as well as reduced caries and bacterial colonization. A variety of antigens have been used for oral administration using liposomes. These antigens have been reviewed in other places (*62*). A brief summary of these studies will be included here with emphasis on more recent studies. Cell-wall or enzyme antigens of *Ascaris suum* in liposomes administered intraintestinally or intragastrically stimulated protection from migrating ascarid larvae in pigs (*75*). Surface and metabolic antigens of the gastrointestinal helminthic parasite *Nippostrongylus brasiliensis* incorporated within liposomes conferred greater and more reliable protection than when antigen alone was administered orally to mice (*76*). Both systemic and mucosal immunity were demonstrated for HIV (*77*) and *Bordetella pertussis* (*78*). Incorporation of tetanus toxoid in liposomes required 10 times more toxoid than needed for parenteral inoculation but resulted in similar serum IgG titers. Titers also lasted as long as those induced by parenteral inoculation in this study. Unencapsulated toxoid did not induce an immune response when administered orally to other rats. Electron microscopy confirmed the presence of the liposomes trapped in intestinal tissue (ileum) of rats used in this study although the exact location was not identified (*79*). Culture supernatants of *Vibrio cholerae* incorporated within liposomes by dehydration–rehydration method induced strong immunity in the gut of rabbits.

The lowest fluid volume occurred in challenged gut loops, the highest protection compared to nonvaccinated controls, and the greatest IgA titer occurred in the intestine (*80*). Overall protection was similar to that seen with antigens administered by parenteral routes in liposomes or antigen administered intraintestinally although intestinal IgA titers were much lower in these rabbits. Although this study relied on intraintestinal administration of liposomes, the results are encouraging in that liposomes could be useful for stimulation of protective intestinal immunity.

Encouraging results for orally administered antigens in liposomes were recently reported in human studies. Glucosyltransferase of *Streptococcus mutans* incorporated in liposomes was administered orally in enteric coated capsules to human volunteers (*81*). Salivary IgA was increased in 5 of 7 individuals for 56 days after administration. In another study 4 volunteers ingested carbohydrate cell-wall antigens of *S. mutans*. Three of four individuals had increased anticarbohydrate antibodies in salivary secretions by 5 weeks after ingestion as well as following booster inoculation over 6 months later (*82*). Further studies are needed to determine whether a more consistent immune response is possible, and whether the responses are related to protection from caries.

Stability and efficacy of liposome vaccines following oral administration are still questionable. Not all studies using orally administered antigens incorporated in liposomes have proven successful. OVA, keyhole limpet hemocyanin, cell-wall antigens of *E. coli*, and cholera toxin did not induce significant mucosal immunity following oral administration in mice although serum IgG responses were induced to bacterial antigens (*83, 84*). Immune responses in animals and humans immunized orally using liposomes have been variable and of short duration. To be useful, longer lasting, and more reliable induction of immunity is needed. Liposomes can be loaded with multiple antigens and immunomodulators or adjuvants to enhance the mucosal immune response. Muramyl dipeptide derivatives from mycobacterium enhance the antibody response to *S. mutans* (*85*). Cytokines could also be useful. Interleukin-2 incorporated in liposomes administered intranasally enhanced the local immune response to bacterial antigens (*86*). Further studies could be done with other mucosally active adjuvants or cytokines to evaluate their efficacy when administered orally with antigens in liposomes.

Cochleates (Proteoliposomes)

Liposomes have demonstrated mixed success in stimulating mucosal immunity when administered orally. Part of this inconsistency may relate to stability in the intestinal tract. Technological improvements have resulted in the creation of protein cochleates, a stable, protein–phospholipid–calcium precipitate (*87*). Cochleates consist of a lipid bilayer, a unilamellar liposome enclosing a solid center. When calcium is added, these cochleates roll up on themselves to form cochleate structures with calcium bridges between layers. Glycoproteins can be incorporated as integral components into the lipid bilayers in the final configuration. When antigens are included in the preparation, cochleates can be used as vehicles for vaccine delivery. The mechanism of interaction with immune cells is not fully known, but it is hypothesized that as calcium is removed, the cochleate unrolls forming a proteoliposome. The proteoliposome then attaches to and fuses with cell membranes. This process not only results in strong mucosal and long-lasting humoral antibody formation, but cell-mediated (cytotoxic and proliferative) immunity in Peyer's patches and spleen due to processing of antigen in the MHC class I pathway as well. The inclusion of viral glycoproteins as integral components of the cochleates mediates the fusion with cell membranes needed for induction of a cytotoxic T-cell response to encapsulated subunit viral proteins that are not able to induce such a response by themselves.

A single oral inoculation can result in slowly increasing mucosal IgA and serum IgG and IgA antibody titers over several months. Two or three oral doses stimulated even greater titers when the glycoproteins of either influenza or Sendai virus were used as antigens. Oral administration of an influenza subunit cochleate vaccine protected mice from challenge. Such mice had 70–100% reduction in virus in trachea and lungs compared with non-vaccinated mice. Oral administration of the principal neutralizing domain of HIV-1 was strongly immunogenic and induced antigen-specific cytotoxic intraepithelial T-lymphocytes in the intestine, and secretory and circulating IgA in mice (*88*).

Cochleates are stable at room temperature, can be lyophilized or kept in calcium containing buffer for long-term storage, and appear to be as safe (noninfectious and noninflammatory) as liposomes. Cochleates induce both circulating and mucosal antibodies, cytotoxic and proliferating T cells, and long-term memory immune responses. They represent a readily producible, effective alternative delivery system for oral administration of vaccines.

Immune-Stimulating Complexes

Immune-stimulating complexes (ISCOMs) are negatively charged cagelike five-sided dodecahedral particles 30–70 nm in size composed of 12-nm subunits. Their size and shape depend on the type and ratio of lipids, saponin and cholesterol, as well as the nature of the antigen incorporated. They form spontaneously when cholesterol and Quil A (a saponin derived from the bark of the tree *Quillaja saponaria* Molina) are mixed with amphipathic proteins and other lipids such as phosphatidylcholine. Quil A is a potent adjuvant. Its use in ISCOMs is also associated with its unique capacity to penetrate into and complex with lipids, especially cholesterol (*89*). Highly amphipathic proteins are incorporated best, but techniques have recently been developed to allow encapsulation of a broader range of recombinant or purified proteins. These include palmitification or acidification of proteins before mixing, and the addition of phosphatidylcholine or ethanolamine to the reaction mixture when making micelles. These techniques initially resulted in relatively inefficient loading of antigen: 1–5% (*90*) to 10–15% (*91*). Unfortunately, this methodology could damage immunogenic epitopes. Recent studies indicate that the loading efficiency can be further improved to 33–64% for a protozoan outer membrane protein (*92*) and to greater than 95% (*93*) for OVA by attaching palmitic acid residues to proteins followed by addition of phosphatidyl groups.

Very low amounts of antigen (1 μg of protein) are required in ISCOMs to stimulate an immune response when administered parenterally. Quil A is toxic when administered in doses of 15–20 μg. However, only 0.5 μg is required in ISCOMs to stimulate an optimal immune response. The purpose of developing ISCOMs was to bind multiple antigens to one adjuvant. This process is necessary for stimulation of an immune response to highly purified proteins. Antigens presented in ISCOMs stimulate both a humoral and cell-mediated immune response to subunit antigens. This characteristic was demonstrated for several types of viruses including influenza, measles, and rabies (*94–96*). ISCOMs have proven efficacious in enhancing the serum IgG and IgM responses 10-fold compared with soluble antigen alone, equal to or greater than responses seen with classical adjuvants. ISCOMs stimulate a low IgE response in vivo (*97*). This fact is important because IgE may increase the pathology caused by infectious agents: for example, in pneumonia caused by respiratory syncytial virus.

ISCOMs are especially effective in stimulating cell-mediated immunity. They induce greater delayed type hypersensitivity, proliferative responses, cytotoxic T cell response, as well as cytokine production by T lymphocytes than vaccines based on complete Freund's adjuvant (*93, 97–99*). The cell-mediated responses require small doses of antigen and the presence of $CD4^+$ T cells. ISCOMs appear to stimulate primarily a TH_1 type response because large amounts of interferon gamma were produced, yet low levels of IL-4 were detected. The CTL response is unique because it is difficult for purified antigens to initiate a CTL response. Live influenza virus stimulates a strong

CTL response. However, killed-virus vaccines and subunit antigens do not. Incorporation of subunit hemagglutinin and neuraminidase antigens in ISCOMs stimulated a CTL and cytotoxic memory-cell response following nasal administration (*100*). These results offer hope for more effective and safe virus vaccines. The CTL response is mediated by $CD8^+$ T cells. This response is MHC class I restricted, a result indicating that ISCOMs cause protein antigens to enter the endogenous pathway of antigen processing. How ISCOMs stimulate the CTL response is not fully understood. ISCOMs may present purified antigens (of viruses) in a way that mimics that of living virus. They are processed by antigen-presenting cells in an unusual chloroquine-insensitive manner by the MHC class I (endogenous) and MHC II (exogenous) pathways. ISCOMs may also stimulate an MHC class II response.

When ISCOMs are administered at a mucosal site, both IgA and IgG responses are initiated. IgG_{2a} and IgA are needed for protective immunity to influenza. Inoculation with antigens not incorporated within ISCOMs did not stimulate IgG_{2a}, and mice were not protected as well as mice vaccinated with ISCOMs. Even more impressive is the possibility that immunity can be induced to internal viral proteins not usually recognized by the host inoculated with whole virus particles (*101*).

Many studies have investigated parenterally administered ISCOMs. However, advantages to the oral administration of ISCOMs also exist. ISCOMs are more stable than liposomes in acid pH and higher temperatures. They are stable in bile salts and stimulate an immune response when injected directly into the intestine (*102*). The size of the particles is also critical. The small size of the ISCOMs makes uptake by intestinal lymphoid cells possible. ISCOMs were suggested to be taken up by Peyer's patches (*103*). The small size allows uptake by other sites in the intestine as well including epithelial cells (*104*). The hydrophobicity of these particles also would favor interaction with intestinal cells necessary for uptake.

Another important consideration is safety. ISCOMs seem to increase lymphocyte proliferation. Spleen cells from inoculated mice incubated with ISCOMs in vitro had increased H^3-thymidine uptake and increased IL-2 production. Parenterally administered Quil A is toxic to mice. Therefore, care must be taken in removing excess Quil A from ISCOMs and carefully adjusting the amount in each dose. Also, saponins released in the lumen of the bowel may induce a bystander immune response to other (food) antigens, an undesirable effect. Control of release of saponins only in association with the vaccine antigen will be necessary to avoid undesired immune responses. To date, no adverse side effects have been seen in mucosally administered ISCOMs. Also, no immunopathology exists with orally administered ISCOMs even with high local priming (*93*). Orally administered Quil A greatly increases the serum response to an antigen and subsequent protection from challenge with an infectious agent (*105*). For future applications, it would be desirable to develop ISCOMs without any toxicity. Recent work has shown that saponins in Quil A can be separated into fractions that retain adjuvanticity without toxicity (*106*).

ISCOMs have been used to enhance immune responses to many viruses including rabies, pseudorabies, influenza, bovine viral diarrhea, feline leukemia virus, bacteria (*Brucella abortus*), and protozoa. Host species inoculated have included mice, monkeys, dogs, sheep, cattle, cats, and horses (reviewed in refs. 107 and 108). The efficacy of orally administered ISCOMs was investigated more recently. Oral administration of ISCOMs stimulated a strong mucosal and serum antibody response. The mucosal immune response depended on the presence of saponin and could not be reproduced by palmitified antigen alone (*107*). Oral administration of OVA in ISCOMs did not induce tolerance as shown by a positive delayed-type hypersensitivity response in mice. An OVA-specific CTL response was induced in these mice. The response was detectable both in splenic and mesenteric lymph node cell populations, demonstrating that a systemic and a local cell-mediated immune response were induced. The CTL response was dependent both on the amount of and number of doses given. Six administrations of 100 μg of OVA were superior to one day or three administrations of 50- or 10-μg doses. The total amount of intestinal IgA as well as the antigen-specific antibody also were increased, results suggesting that ISCOMs have some nonspecific adjuvant effect (*89*).

Fewer feeds (3 vs. 6) and lower doses (50 vs. 100 μg) of OVA administered orally in ISCOMs to mice stimulated a stronger OVA-specific CTL response than did OVA administered orally in PLG (*109*). Lymphocytes harvested from mice orally vaccinated with ISCOMs could be activated in vitro by IL-2 alone, whereas PLG-vaccinated mice required in vitro antigen stimulation before CTL could be detected. This difference may be due to ISCOMs priming a higher percentage of CTL cells than other delivery systems. This process will need to be studied further. This study supported the results seen by Mowat and further supported the idea that the added lipids and saponin present in ISCOMs stimulate MHC class I endogenous antigen processing.

More recently, ISCOMs administered orally were shown to stimulate a mucosal antibody response to an intestinal protozoan parasite. Mice inoculated orally with ISCOMs containing the surface protein (p27) of the sporozoite *Eimeria falciformis* had increased systemic IgG, intestinal IgA, and enhanced proliferation by mesenteric lymph node cells to p27 in vitro. Vaccinated mice had reduced oocyst shedding in feces and greater body weight following challenge than did nonvaccinated challenged mice or mice vaccinated orally but without ISCOMs (*92*). Not only is this study a demonstration of the usefulness of orally administered ISCOMs, it is the first use for an intestinal parasitic disease, and the first study whereby orally administered ISCOMs were shown to induce protective immunity to an infectious agent.

For all their advantages, ISCOMs have a few shortcomings. Optimal dosing schedules are yet to be determined. A single dose of ISCOMs stimulates poor intestinal and serum antibody responses. Cytotoxic T-cell reactivity is also low. Multiple doses are needed to stimulate optimal immune responses, much like other orally administered antigens. The immune response that is initiated is short in duration lasting only 3–4 weeks. Improved formulations also may be developed to slow the release over time, and other mucosally active immunomodulators such as heat-labile toxin of *E. coli* or IL-6 may be included to enhance the immune response in intensity and duration. Lectins or monoclonal antibodies may be incorporated in the surface to increase site-specific delivery to Peyer's patches.

ISCOMs provide a powerful way to enhance immune responses. Some people are poor responders to certain antigens such as the hepatitis surface antigen. When such patients are exposed to a parenteral inoculation of this antigen, a subsequent inoculation with the antigen in ISCOMs results in a strong secondary immune response. Patients inoculated with the antigen in a conventional adjuvant did not respond (*110*). Part of the adjuvant effect of orally administered ISCOMs may be due to the effect on intestinal absorption of proteins. The prior administration of ISCOMs increases the uptake of soluble antigen into the general circulation. The increased immune response to antigen may be related to this increased absorption of protein. This result also points to a need to better understand the mechanism of immune stimulation of orally administered ISCOMs. ISCOMs have many novel properties that make them good candidates for oral delivery of antigens. The binding of low weight proteins to the surface of preformed ISCOMs provides the means to stimulate immunity to haptens and the possibility of adding other immunomodulators or molecules to the surface to further enhance uptake by Peyer's patches. This process could result in a more efficient stimulation of mucosal immune response with lower doses of antigen and could also better direct the immune response toward a type 1 or type 2 T helper response depending on the need for stronger humoral or cell-mediated immunity. ISCOMs may be good candidates for delivery of multiple antigens within multiple preparations because the dose required to stimulate an immune response appears to be fairly low. Although orally administered

ISCOMs induce protective immunity against some intestinal parasites, it still remains to be proven that they can induce protective immunity at other mucosal sites. Earlier studies using influenza virus ISCOMs administered orally were disappointing in stimulating a respiratory antibody response (*111*). The need for a way to stimulate a longer lasting immunity, or memory response, with fewer doses is desirable for more practical applications.

Polysaccharide Microparticles of Sodium Alginate

Microparticle mucosal vaccines must be biocompatible, not only with the host, but with the antigen they contain for delivery. If production requires strong solvents or heat that alters the epitopes of the antigen, no matter how well the microparticles are taken up by the Peyer's patches, an effective immune response cannot be stimulated. Production of ISCOMs, liposomes, or PLG microparticles requires conditions that could damage fragile antigens. Delivery systems are needed that use milder conditions for encapsulation of antigens. We have investigated the use of another polymer for oral delivery, sodium alginate. Algin is a naturally occurring polysaccharide produced by many species of brown seaweed such as *Macrocystis pyrifera*. Alginic acid is an unbranched glycuronan composed of two monomeric units, mannuronic acid and guluronic acid. Sodium alginate is a viscous liquid that polymerized into a solid matrix when mixed with a divalent cation like calcium (*112*). The rigidity of the resultant gels depends on the ratio of guluronic and mannuronic acids. The greater the proportion of guluronic acid, the stronger yet more brittle the gel formed. The greater the amount of mannuronic acid, the more flexible the gel. Conditions for gel formation are very mild—there is no need for high temperatures or solvents for gelation. Therefore, alginate gels can be used to encapsulate any type of antigen without damage to the epitopes. Even live cells can be encapsulated. Alginate has been used in the wine industry to encapsulate yeast used in fermentation (*113*). It has also been used to encapsulate beta islet cells, and it has been injected intraperitoneally for insulin therapy in mice (*114*). Now, hybridoma cells are encapsulated within alginate gels for antibody formation (*115*).

Microparticles of alginate were formed for parenteral administration of immunomodulators to the lung by using an aerosol technique (*116*). We have used this technique to produce microparticles less than 30 μm in diameter (70% less than 10 μm) containing antigen for oral delivery of vaccines. Our goal is to develop a system to incorporate any antigen for oral administration. Uptake of microparticles by Peyer's patches depends to some degree on the hydrophobicity of the microparticles. To enhance the hydrophobicity and to stabilize the alginate microspheres to prevent breakdown in the acid environment of the stomach, microparticles are coated with poly(L-lysine). To test whether this treatment would permit adherence to Peyer's patches, fluorescent-labeled alginate microspheres were prepared and injected into the lumen of tied off sections of intestine containing a single Peyer's patch in a rabbit. After 20 min, the Peyer's patches were removed, snap-frozen, cut into thin sections, and observed under a fluorescent microscope. Microparticles were observed in close approximation with the surface of Peyer's patches (*117*). These results suggested that it was possible for alginate microparticles to adhere to and presumably be taken up into Peyer's patches to stimulate an immune response.

Once alginate microspheres were known to bind to Peyer's patches, studies were undertaken to determine whether antigens incorporated within them could induce an immune response in animals. OVA has been encapsulated as a test antigen and administered to mice (*118*). Mice inoculated with OVA in alginate microspheres by either oral or parenteral administration had increased humoral IgG and IgM as well as increased antibody secreting cells in spleen comparable with that seen in mice inoculated parenterally with OVA in complete Freund's adjuvant. These data not only suggest that alginate microparticles are useful for antigen delivery, but that they act as an adjuvant as well.

Cattle suffer many respiratory infections that cause an estimated $500 million per year to North American producers alone. Better vaccines are needed to control respiratory diseases in cattle. Administration of a vaccine to large numbers of animals is time consuming and laborious. An easy way to vaccinate animals is through the feed or water. To test the possible use of alginate for oral delivery of vaccines to cattle, alginate microspheres containing OVA were prepared. Microspheres were encapsulated in larger alginate gels for oral administration to calves to bypass the rumen, the first stomach of ruminants where feed is degraded by bacteria and protozoa. Calves inoculated orally with OVA microspheres were then challenged by an intrabronchial administration of OVA. A bronchoalveolar lavage was performed. Lavage fluids and serum were analyzed for antibodies to OVA. Lymphocytes were isolated from the lavage fluids and used in an enzyme-linked assay to detect antigen-specific antibody secreting cells in the lung. Calves inoculated with two regimens of five daily doses of 5 mg of OVA, or one regimen of three daily doses of 5 mg of OVA in microspheres had significantly greater OVA-specific IgA in the lung as well as IgA antibody secreting cells compared with either nonvaccinated or subcutaneously vaccinated calves. The response in orally vaccinated calves was similar to that seen in calves primed with a subcutaneous inoculation of OVA in incomplete Freund's adjuvant followed by three intrabronchial inoculations of OVA in saline (*119*). These results offer hope for an easy way to administer oral vaccine to prevent diseases such as pneumonia caused by *Pasteurella haemolytica*, the most pathogenic respiratory bacterium in cattle.

Bacterial antigens have also been incorporated within alginate. The oral administration of antigens in water would be an effective way to vaccinate rabbits for a bacterium that causes multiple diseases (pneumonia, sinusitis, middle ear infections, septicemia, and orchitis). Cell-wall proteins in a crude potassium thiocyanate (KSCN) extract of *Pasteurella multocida* were incorporated within alginate microspheres and administered to rabbits in their drinking water three times at weekly intervals. Rabbits were then challenged by an intranasal dose of virulent *P. multocida*. Rabbits vaccinated with bacterial extracts in microspheres had significantly greater serum and nasal antibodies and reduced numbers of bacteria in nasal, pulmonary, and tympanic bullae compared with rabbits vaccinated orally with unencapsulated antigens or nonvaccinated rabbits (*120*). Protection was similar to that of rabbits inoculated repeatedly with intranasal doses of KSCN extracts. Incorporation of cholera toxin within the microspheres with antigen did not offer any greater protection in rabbits vaccinated orally with *P. multocida* KSCN extracts.

Salmonella enteritidis causes bacteremia in chickens and is of interest because it can be passed vertically through the egg to infect future generations of birds and can enter the food chain causing food-borne disease in humans. The flagellar antigens of *S. enteritidis* have been incorporated into alginate microspheres and administered to chicken poults. Flagella incorporated within alginate microspheres retained antigenicity as shown by sodium dodecyl sulfate–polyacrylamide gel electrophoresis of antigens recovered from alginate microspheres (*119*). Studies are in progress to determine the efficacy of vaccines in preventing colonization of the intestinal tract of chicken poults.

Because of the mild conditions needed for encapsulation, alginate can be used for encapsulation of nucleic acids. Studies have shown that DNA can be encapsulated in alginate and released over time with no adverse effect on the structure of the DNA. The proposed use for DNA encapsulation would target nucleic acids or oligonucleotides for transfection of intestinal epithelial cells for synthetic peptide or pharmaceutical therapy (*121*). A similar approach for mucosal vaccination by targeting nucleic acids to Peyer's patches would be interesting.

Alginate has been used to encapsulate microparticles for oral delivery of drugs (*122*). Another novel use of alginate is the incorporation of liposomes or ISCOMs for oral delivery. Alginate has been used to encapsulate liposomes for parenteral inoculation of a vaccine. Such a vaccine resulted in immune responses 2–3-fold greater than the same antigen in complete Freund's adjuvant (*123*). Because one of the drawbacks of liposomal vaccines is the short-term immunity they induce, this incorporation process would be a way to prolong the immune response for orally administered liposomes. A similar approach may be useful for oral ISCOM vaccines by providing stability in the intestinal tract until the particles reach the Peyer's patches.

Other polysaccharide polymers could be used for oral administration. Hyaluronic acid would be biologically compatible because it is present in the body, but conditions for polymerization would not be as mild as for alginate. Cost would be much greater as well compared with alginate. Synthetic polymers such as polyphosphazene offer mild conditions similar to alginate for polymerization. This group of polymers gels when placed in bivalent cations similar to alginate (*124*). No oral vaccine studies have been reported to date using polyphosphazene microparticles.

Although the use of alginates for oral vaccination offers much promise, there is more to be learned. The duration of immune responses to antigens administered orally in alginate microspheres has not been determined fully as yet. Studies are in progress to determine how best to increase uptake by Peyer's patches. Certain lectins or bacterial adhesins bind to intestinal epithelium or M cells of Peyer's patches (*22, 125*). Alginate particles will be coated with lectins such as *Lycopersicon esulentum*, *Ulex europeus*, or cholera toxin, as well as poly(L-lysine) to determine whether adherence to and uptake by Peyer's patches can be enhanced. Controlled release to stimulate both an immediate and long-term immune response will also be attempted by altering the type of alginate used (altering the ratio of mannuronic acid:guluronic acid), altering the concentration of antigen incorporated within the gels, and by coating the gels with poly(L-lysine) of different molecular weights because this process affects pore size and subsequent release of antigen over time (*124*). Enhancement of the mucosal immune response will also be attempted by incorporating cytokines (such as IL-6) within the microspheres in association with antigens. Cell-mediated immune responses in animals inoculated with alginate microspheres still need to be evaluated.

Virus Particles

One of the advantages of polymeric microparticles is avoiding the use of recombinant live vectors. Live vectors may stimulate an immune response limiting their usefulness, and there are concerns about reversion to virulence. This concern reduces acceptance by the public, and practical application in the immediate future is unlikely. However, novel approaches have produced virus-like particle pseudotypes that combine recombinant technology to produce particles for oral delivery. Co-infection of a mammalian cell with a recombinant enveloped vaccinia virus expressing a glycoprotein with a viruslike particle (formed by co-infection of a cell with 2 different viruses resulting in a hybrid replicating virus) resulted in a noninfectious viruslike particle pseudotype expressing viral glycoprotein. These particle pseudotypes provide a means of producing soluble subunit proteins in particulate manner much as they would be expressed on a virus during infection, but without replication (*126*). These particle pseudotypes are relatively new and have not been studied in an oral presentation, but the right viruses that could replicate in the GIT and are therefore stable in these conditions would be primary candidates.

Another novel oral delivery system that has been investigated is using rabies ribonucleocapsid protein (R-RNP) as an immunoenhancer by coupling it to other antigens. When a synthetic peptide of HIV was coupled to R-RNP, oral administration resulted in enhanced serum antibody and splenic T-cell proliferation to HIV–peptide compared with nonvaccinated controls (*127*). R-RNP offers a potent immunoenhancer for use in mucosal immunostimulation.

Conclusions

The use of polymeric microparticles for delivery of oral vaccines is an area of intense interest. Demand is growing for long-acting vaccines that stimulate mucosal immunity, including cell-mediated activity. Ideally, such vaccines would require one dose, include multiple antigens for many infectious diseases, be stable at room temperature for reasonably long periods of time, and be given to infants as well as older children and adults. The polymers used must be nontoxic, both to the host and to the antigens encapsulated within them. Microparticles for oral vaccines may need to be produced under good manufacturing procedures to ensure sterility without adversely affecting antigen content or release kinetics of the product. Because microparticles are taken up into the body to stimulate an immune response, sterility will be more of an issue than for products that rely on delivery to intestinal surface with uptake of final product and not the delivery device across the intestinal epithelium. The microparticles should release antigen in a controlled manner to induce long-term as well as short-term immunity. The design of antigen release should be similar to the way natural infections expose the body to antigens, a relatively large immediate antigen release, with lower levels over time, to stimulate effective immunological memory for long-term protection. The size, hydrophobicity, and surface characteristics of the microparticles should be controlled for optimal uptake by M cells in Peyer's patches. Incorporation of lectins (*22*), or modification of surface with antibodies to M cells (*128*), may be helpful to enhance uptake and efficiency of stimulating an immune response. Co-administration of adjuvants or immunomodulators may be advantageous to enhance the desired immune response. This characteristic may be especially important for poorly immunogenic antigens. Interest is growing in using oral administration of antigens to control autoimmune diseases, or possibly to prevent rejection of tissue transplants by controlled stimulation of tolerance. This area is exciting and could make a major impact on diseases difficult to control. It also suggests an interesting way to further control allergies by controlled down regulation of the immune response. As the number of people with allergies increases, this area may gain importance. As understanding of mucosal immunity increases, microparticles will be tested for application to other mucosal sites such as intravaginally, rectally, and intranasally. These routes may change design requiring smaller volumes of dosage but with greater antigen load, increased contact time with mucosa for uptake, along with special means of administration to ensure that microparticles reach the site of uptake. How these microparticle vaccines will be packaged for storage, shipment, and usage is of interest. Most studies suggest lyophilization of particles is possible. How long will such vaccines be stable for administration following hydration? Will these microspheres be encapsulated in a gelatin capsule or hydrated in a suspension for oral administration? Will refrigeration be required? These considerations could affect how well they are accepted for use in children, in developing nations, and in farm animals.

The protection of antigens from harsh environments, controlled release of antigen over time, and incorporation of additional mucosal immunomodulators make the use of microparticles for oral vaccination an exciting prospect now and in the immediate future.

References

1. Beardsley, T. *Sci. Am.* **1995**, *272*, 88–95.
2. Holmgren, J.; Czerkinsky, C.; Lycke, N.; Svennerholm, A. M. *Am. J. Trop. Med. Hyg.* **1994**, *50*, 42–54.
3. Rudzik, R.; Perey, D.; Bienenstock, J. *J. Immunol.* **1975**, *114*, 1599–1604.

4. Mosmann, T. R.; Coffman, R. L. *Annu. Rev. Immunol.* **1989,** *7,* 145–173.
5. Mosmann, T. R.; Moore, K. W. *Immunol. Today* **1991,** *12,* A43–A53.
6. Street, N. E.; Mosmann, T. R. *FASEB J.* **1991,** *5,* 171–177.
7. Butcher, E. *Curr. Top. Microbiol. Immunol.* **1986,** *128,* 85–122.
8. Eldridge, J. H.; Gilley, R. M.; Staas, J. K.; Moldoveanu, A.; Meulbroek, J. A.; Bite, T. R. *Curr. Top. Microbiol. Immunol.* **1989,** *146,* 59–65.
9. *Novel Delivery Systems for Oral Vaccines;* O'Hagan, D. T., Ed.; CRC Press: Boca Raton, FL, 1994.
10. Jani, P. U.; Halbert, G. W.; Langridge, J.; Florence, A. T. *J. Pharm. Pharmacol.* **1990,** *42,* 821–826.
11. Jenkins, P. G.; Howard, K. A.; Blackhall, N. W.; Thomas, N. W.; Davis, S. S.; O'Hagan, D. T. *Int. J. Pharm.* **1994,** *102,* 261–266.
12. Uchida, T.; Goto, S. *Biol. Pharm. Bull.* **1994,** *17,* 1272–1276.
13. Ebel, J. P. *Pharm. Res.* **1990,** *7,* 848–851.
14. Worthington, B. S.; Boatman, E. S.; Kenny, G. E. *Am. J. Clin. Nutr.* **1974,** *27,* 276–283.
15. Isolauri, E.; Gotteland, M.; Heyman, M.; Pochart, P.; Desjeux, J. F. *Dig. Dis. Sci.* **1990,** *35,* 360–365.
16. LeFevre, M. E.; Boccio, A. M.; Joel, D. D. *Proc. Soc. Exp. Biol. Med.* **1989,** *190,* 23–27.
17. Pappo, J.; Steger, H. J.; Owen, R. L. *Lab. Invest.* **1988,** *58,* 692–695.
18. Smith, M. W.; Peacock, M. A. *Am. J. Anat.* **1980,** *159,* 167–172.
19. LeFevre, M. E.; Joel, D. D. *Proc. Soc. Exp. Biol. Med.* **1986,** *182,* 112–116.
20. Pappo, J.; Ermak, T. H. *Clin. Exp. Immunol.* **1989,** *76,* 144–150.
21. Jepson, M. A.; Simmons, J. L.; Savidge, T. C.; James, P. S.; Hirst, B. H. *Cell Tissue Res.* **1993,** *271,* 399–403.
22. de Aizpurua, H. J.; Russell-Jones, G. J. *J. Exp. Med.* **1988,** *167,* 440–451.
23. Roy, M. J. *Cell Tissue Res.* **1987,** *248,* 483–489.
24. Walker, R.; Clements, J. D. *Vaccine Res.* **1993,** *2,* 1–10.
25 Elson, C. O. *Immunol. Today* **1989,** *146,* 29–33.
26. Reul, G. J. *Am. J. Surg.* **1977,** *134,* 297–302.
27. Christel, P.; Chabot, F.; Leray, L. F.; Morin, C.; Vert, M. In *Biomaterials;* Winter, G. D.; Biobonnes, D. F.; Plenk, H., Eds.; John Wiley: New York, 1982; pp 271–294.
28. Langer, R. *Science (Washington, D.C.)* **1990,** *249,* 1527–1533.
29. Visscher, G. E.; Robison, R. L.; Argentieri, G. I. *J. Biomater. Appl.* **1987,** *2,* 118–122.
30. Watts, P. J.; Davies, M. C.; Melia, C. D. *Crit. Rev. Ther. Drug Carrier Syst.* **1990,** *7,* 235–259.
31. Thies, C.; Bissery, M. C. In *Biomedical Applications of Microencapsulation;* Lim, F., Ed.; CRC Press: Boca Raton, FL, 1984; pp 53–74.
32. Yan, C.; Resau, J. H.; Hewetson, J.; West, M.; Rill, W. R.; Kende, M. *J. Controlled Release* **1994,** *32,* 231–241.
33. Deasy, P. B. *Crit. Rev. Ther. Drug Carrier Syst.* **1988,** *2,* 99–139.
34. Morris, W.; Steinhoff, M. C.; Russell, P. K. *Vaccine* **1994,** *12,* 5–11.
35. McGee, J. P.; Davis, S. S.; O'Hagan, D. T. *J. Controlled Release* **1994,** *31,* 55–60.
36. Moldoveanu, Z.; Novak, M.; Huang, W. Q.; Gilley, R. M.; Staas, J. K.; Schafer, D.; Compans, R. W.; Mestecky, J. *J. Infect. Dis.* **1993,** *167,* 84–90.
37. Jeffery, H.; Davis, S. S.; O'Hagan, D. *Pharm. Res.* **1993,** *10,* 362–368.
38. O'Hagan, D. T.; Jeffery, H.; Davis, S. S. *Int. J. Pharm.* **1994,** *103,* 37–45.
39. O'Hagan, D. T.; Rahman, D.; Jeffery, H.; Sharif, S.; Challacombe, S. J. *Int. J. Pharm.* **1994,** *108,* 133–139.
40. Richardson, J. L.; McGee, J. P.; Gumaer, D.; Potts, B.; Wang, C. Y.; Koff, W.; O'Hagan, D. T. *Proc. Int. Symp. Controlled Release Bioact. Mater.* **1994,** *21,* 869–870.
41. Eldridge, J. H.; Staas, J. K.; Meulbroek, J. A.; McGhee, J. R.; Tice, T. R.; Gilley, R. M. *Mol. Immunol.* **1991,** *28,* 287–294.
42. Maloy, K. J.; Donachie, A. M.; O'Hagan, D. T.; Mowat, A. M. *Immunology* **1994,** *81,* 661–667.
43. O'Hagan, D. T.; Jeffery, H.; Davis, S. S. *Vaccine* **1993,** *11,* 965–969.
44. Ray, R.; Novak, M.; Duncan, J. D.; Matsuoka, Y.; Compans, R. W. *J. Infect. Dis.* **1993,** *167,* 752–755.
45. Marx, P. A.; Compans, R. W.; Gettie, A.; Staas, J. K.; Gilley, R. M.; Mulligan, M. J.; Yamshchikov, G. V.; Chen, D.; Eldridge, J. H. *Science (Washington, D.C.)* **1993,** *260,* 1323–1327.
46. McQueen, C. E.; Boedeker, E. C.; Reid, R.; Jarboe, D.; Wolf, M.; Le, M.; Brown, W. R. *Vaccine* **1993,** *11,* 201–206.

47. Edelman, R.; Russell, R. G.; Losonsky, G.; Tall, B. D.; Tacket, C. O.; Levine, M. M.; Lewis, D. H. *Vaccine* **1993,** *11,* 155–158.
48. Reid, R. H.; Boedeker, E. C.; McQueen, C. E.; Davis, D.; Tseng, L. Y.; Kodak, J.; Sau, K.; Wilhelmsen, C. L.; Nellore, R.; Dalal, P.; et al. *Vaccine* **1993,** *11,* 159–167.
49. Tacket, C. O.; Reid, R. H.; Boedeker, E. C.; Losonsky, G.; Naataro, J. P.; Bhagat, H.; Edelman, R. *Vaccine* **1994,** *12,* 1270–1275.
50. Muir, W.; Husband, A. J.; Gipps, E. M.; Bradley, M. P. *Immunol. Lett.* **1994,** *42,* 203–207.
51. Hora, M. S.; Rana, R. K.; Numberg, J. H.; Tice, T. R.; Gilley, R. M.; Hudson, M. E. *Biotechnology* **1990,** *8,* 755–758.
52. Eppstein, D. A.; van der Pas, M. A.; Schryver, B. B.; Felgner, P. L.; Gloff, C. A.; Soike, K. F. In *Delivery Systems for Peptide Drugs;* Davis, S. S.; Illum, L.; Tomlinson, E., Eds.; Plenum: New York, 1986; pp 277–295.
53. Rafferty, D. E.; Montgomery, P. C. *FASEB J.* **1995,** *9,* A214.
54. Martin, S.; Grimm, S.; Davidson, P.; Guimond, P.; Krinick, N.; Frank, K.; Dunn, R. *Proceedings of the 4th International Veterinary Immunology Symposium;* Davis, CA, July 16–21, 1995; p 123.
55. O'Hagan, D. T.; Palin, K.; Davis, S. S.; Artursson, P.; Sjoholm, I. *Vaccine* **1989,** *7,* 421–424.
56. O'Hagan, D. T.; Palin, K.; Davis, S. S. *Vaccine* **1989,** *7,* 213–216.
57. Steineker, F.; Kreuter, J.; Lower, J. *AIDS* **1991,** *5,* 431–435.
58. Bowersock, T. L.; Shalaby, W. S. W.; Levy, M.; Blevins, W. E.; White, M. R.; Borie, D. L.; Park, K. *J. Controlled Release* **1994,** *31,* 245–254.
59. Bowersock, T. L.; Shalaby, W. S. W.; Levy, M.; Samuels, M. L.; Lallone, R.; White, M. R.; Borie, D. L.; Lehmeyer, J.; Park, K. *Am. J. Vet. Res.* **1994,** *55,* 502–509.
60. Weng, C. N.; Tzan, Y. L.; Liu, S. D.; Lin, S. Y.; Lee, C. J. *Res. Vet. Sci.* **1992,** *53,* 42–46.
61. Childers, N. K.; Michalek, S. M. In *Novel Delivery Systems for Oral Vaccines;* O'Hagan, D. T., Ed.; CRC Press: Boca Raton, FL, 1994; pp 241–254.
62. Childers, N. K.; Michalek, S. M.; Denys, F.; McGhee, J. R. In *Mucosal Immunology;* Mestecky, J.; McGhee, J. R.; Bienenstock, J.; Ogra, P. L., Eds.; Plenum: New York, 1987; pp 1771–1780.
63. Harrigan, P. R.; Madden, T. D.; Cullins, P. R. *Chem. Phys. Lipids* **1990,** *52,* 139–149.
64. Animisiaris, S. G.; Jayasekera, P.; Gregoriadis, G. *J. Immunol. Methods* **1993,** *166,* 271–280.
65. Alving, C. R. *J. Immunol. Methods* **1991,** *140,* 1–13.
66. Therien, H.-M.; Shahum, E. *Immunol. Lett.* **1989,** *22,* 253–258.
67. Manesis, E. K.; Cameron, C.; Gregoriadis, G. *FEBS Lett.* **1979,** *102,* 107–111.
68. Gregoriadis, G.; Wang, Z.; Barenholz, Y.; Francis, M. J. *Immunology* **1993,** *80,* 535–540.
69. Childers, N. K.; Denys, F. R.; McGhee, J. R.; Michalek, S. M. *Reg. Immunol. (New York)* **1990,** *3,* 8–13.
70. Aramaki, Y.; Tomizawa, H.; Hara, T.; Yachi, K.; Kikuchi, H.; Tsuchiya, S. *Pharm. Res.* **1993,** *10,* 1228–1231.
71. Tomizawa, H.; Aramaki, Y.; Fujii, Y.; Hara, T.; Suzuki, N.; Yachi, K.; Kikuci, H.; Tsuchiya, S. *Pharm. Res.* **1993,** *10,* 549–552.
72. Wachmann, D.; Klein, J. P.; Scholler, M.; Ogier, J.; Ackermans, F.; Frank, R. M. *Infect. Immun.* **1986,** *52,* 408–413.
73. Rubas, W.; Banerjea, A. C.; Gallati, H.; Speiser, P. P.; Joklik, W. K. *J. Microencapsulation* **1990,** *7,* 385–392.
74. Michalek, S. M.; Morisaki, I.; Gregory, R. L.; Kiyono, H.; Hamada, S.; McGhee, J. R. *Mol. Immunol.* **1983,** *20,* 1009–1013.
75. Rhodes, M. B.; Baker, P. K.; Christensen, D. L.; Anderson, G. A. *Vet. Parasitol.* **1988,** *26,* 343–349.
76. Rhalem, A.; Bourdieu, C.; Luffau, G.; Pery, P. *Ann. Inst. Pasteur/Immunol.* **1988,** *139,* 157–161.
77. Thibodeau, L.; Constantineau, L.; Tremblay, C. *Vaccine Res.* **1992,** *1,* 233–239.
78. Guzman, D. A.; Molinari, F.; Fountain, M. W.; Rohde, M.; Timmis, K. N.; Walker, M. J. *Infect. Immun.* **1993,** *61,* 573–578.
79. Alpar, H. O.; Bowen, J. C.; Brown, M. R. W. *Int. J. Pharm.* **1992,** *88,* 335–344.
80. Chandrasekhar, U.; Sinha, S.; Bhagat, H. R.; Sinha, B. B.; Srivastava, B. S. *Vaccine* **1994,** *12,* 1384–1388.
81. Childers, N. K.; Zhang, S. S.; Michalek, S. M. *Oral Microbiol. Immunol.* **1994,** *9,* 146–153.
82. Childers, N. K.; Michalek, S. M.; Pritchard, D. G.; McGhee, J. R. *Reg. Immunol. (New York)* **1991,** *3,* 289–294.

83. Clark, C. J.; Stokes, C. R. *Vet. Immunol. Immunopathol.* **1992,** *32,* 125–138.
84. Clark, C. J.; Stokes, C. R. *Vet. Immunol. Immunopathol.* **1992,** *32,* 139–148.
85. Michalek, S. M.; McGhee, J. R.; Mestecky, J.; Arnold, R. R.; Bozzo, L. *Science (Washington, D.C.)* **1976,** *192,* 1238–1240.
86. Abraham, E.; Shah, S. *J. Immunol.* **1992,** *149,* 3719–3724.
87. Gould-Fogerite, S.; Edghill-Smith, Y.; Kheiri, M.; Wang, Z.; Das, K.; Feketeova, E.; Canki, M.; Mannino, R. J. *AIDS Res. Hum. Retroviruses* **1994,** *10,* S99–S103.
88. Edghill-Smith, Y.; Gould-Fogerite, S.; Mannino, R. J. Presented at Novel Vaccine Strategies for Mucosal Immunization, Genetic Approaches and Adjuvants, October 24–26, 1994; Rockville, MD.
89. Mowat, A. M.; Maloy, K. J. In *Novel Delivery Systems for Oral Vaccines;* O'Hagan, D. T., Ed.; CRC Press: Boca Raton, FL, 1994; pp 207–223.
90. Sundquist, B.; Lovgren, K.; Hoglund, S.; Morein, B. *Vaccine* **1988,** *6,* 44–48.
91. Thapar, M. A.; Parr, E. L.; Bozzola, J. J.; Parr, M. B. *Vaccine* **1991,** *9,* 129–133.
92. Kazanji, M.; Laurent, F.; Pery, P. *Vaccine* **1994,** *12,* 798–804.
93. Mowat, A. M.; Maloy, K. J.; Donachie, A. M. *Immunology* **1993,** *80,* 527–534.
94. Morein, B.; Sundquist, B.; Hoglund, S.; Dalsgaard, K.; Osterhaus, A. *Nature (London)* **1984,** *308,* 457–460.
95. Chavali, S. R.; Barton, L. D.; Campbell, J. B. *Immunology* **1988,** *74,* 339–343.
96. Lovgren, K.; Kaberg, H.; Morein, B. *Clin. Exp. Immunol.* **1990,** *82,* 435–439.
97. Villacres-Eriksson, M.; Bergstrom-Mollaoglu, M.; Kaberg, H.; Morein, B. *Scand. J. Immunol.* **1992,** *36, 421–426.*
98. Mowat, A. M.; Donachie, A. M.; Reid, G.; Jarrett, O. *Immunology* **1991,** *72,* 317–322.
99. Fossum, C.; Bergstrom, M.; Lovgren, K.; Watson, D. L.; Morein, B. *Cell. Immunol.* **1990,** *129,* 414–419.
100. Jones, P. D.; Tha-Hla, R.; Morein, B.; Lovgren, K.; Ada, G. L. *Scand. J. Immunol.* **1988,** *27,* 645–652.
101. Morein, B.; Fossum, C.; Lovgren, K.; Hoglund, S. *Semin. Virol.* **1990,** *1,* 49–55.
102. Kersten, G. F. A. Ph.D. Thesis, University of Utrecht, Netherlands, 1990.
103. Claasen, I.; Osterhaus, A.; Boersma, W.; Schellekens, M.; Claassen, E. *Adv. Exp. Med. Biol.* **1995,** *371,* 1485–1489.
104. O'Hagan, D. T. *Clin. Pharmacokinet.* **1992,** *22,* 1–10.
105. Chavali, S. R.; Campbell, J. B. *Immunobiology* **1987,** *174,* 347–351.
106. Kensil, D. R.; Patel, U.; Lennick, M.; Marciani, D. *J. Immunol.* **1991,** *146,* 4431–4435.
107. Mowat, A. M.; Donachie, A. M. *Immunol. Today* **1991,** *12,* 383–385.
108. Morein, B.; Lovgren, K.; Hoglund, S.; Sundquist, B. *Immunol. Today* **1987,** *11,* 333–338.
109. Maloy, K. J.; Donachie, A. M.; O'Hagan, D. T.; Mowat, A. M. *Immunology* **1994,** *81,* 661–667.
110. Howard, C. R.; Sundquist, B.; Allan, J.; Brown, S. E.; Chen, S.-H.; Morein, B. *J. Gen. Virol.* **1987,** *68,* 2281–2289.
111. Lovgren, K. *Scand. J. Immunol.* **1988,** *2,* 241–247.
112. Gacesa, P. *Carbohydr. Polym.* **1988,** *8,* 161–182.
113. Fumi, M. D.; Trioli, G.; Colagrande, O. *Biotechnol. Lett.* **1987,** *9,* 339–342.
114. Lim, F.; Sun, A. M. *Science (Washington, D.C.)* **1980,** *210,* 908–910.
115. Lee, G. M.; Chuck, A. S.; Palsson, B. O. *Biotechnol. Bioeng.* **1993,** *41,* 330–340.
116. Kwok, K. K.; Graves, M. J.; Burgess, D. J. *Pharm. Res.* **1991,** *8,* 341–344.
117. Bowersock, T. L.; HogenEsch, H.; Park, H.; Park, K. *Proceedings of the 21st International Symposium on Controlled Release of Bioactive Materials;* Controlled Release Society: Deerfield, IL, 1994; Vol. 21, pp 839–840.
118. Bowersock, T. L.; HogenEsch, H.; Suckow, M.; Davis-Snyder, E.; Borie, D.; Park, H.; Park, K. *Polym. Preprints (Am. Chem. Soc. Div. Polym. Chem.)* **1994,** *35,* 405–406.
119. Bowersock, T. L.; HogenEsch, H.; Suckow, M.; Porter, R. E.; Jackson, R.; Park, H.; Park, K. *J. Controlled Release,* in press.
120. Suckow, M. A.; Bowersock, T. L.; Park, K. *Proceedings of the 21st International Symposium on Controlled Release of Bioactive Materials;* Controlled Release Society: Deerfield, IL, 1994; Vol. 21, pp 843–844.

121. Smith, T. J. *Pharm. Technol.* **1994,** *18,* 26–30.
122. Bodmeier, R.; Chen, H.; Paeratakul, O. *Pharm Res.* **1989,** *6,* 413–417.
123. Cohen, S.; Bernstein, H.; Hewes, C.; Chow, M.; Langer, R. *Proc. Natl. Acad. Sci. U.S.A.* **1991,** *88,* 10440–10444.
124. Andrianov, A. K.; Cohen, S.; Visscher, K. B.; Payne, L. G.; Allcock, H. R.; Langer, R. *J. Controlled Release* **1993,** *27,* 69–77.
125. Gebert, A.; Hach, G. *Gastroenterology* **1993,** *105,* 1350–1361.
126 McGuigan, L. C.; Stallare, V.; Roos, J. M.; Payne, L. G. *Vaccine* **1993,** *6,* 675–678.
127. Hooper, D. C.; Pierard, I.; Modelska, A.; Otvos, L., Jr.; Fu, Z. F.; Koprowski, H.; Dietzschold, B. *Proc. Natl. Acad. Sci. U.S.A.* **1994,** *91,* 10908–10912.
128. Pappo, J.; Ermak, T. H.; Steger, H. J. *Immunology* **1991,** *73,* 277–280.

14

Peptide and Protein Delivery for Animal Health Applications

Thomas H. Ferguson

Advances in genetic engineering and biotechnology have led to the commercial production of a variety of recombinant polypeptides and proteins for use in the animal health and veterinary markets. Important applications for improved lactation response in dairy cows and enhanced growth in beef cattle, swine, and poultry have emerged with the improved supply of economical recombinant proteins. However, proteins are notoriously difficult to formulate into sustained- or controlled-release delivery systems. The complexities of developing a sustained- or controlled-release delivery system for proteins in the animal health industry are reviewed. Attempts to formulate somatotropins and other important peptides and proteins in sustained- or controlled-release delivery systems are reviewed. Protein formulation and manufacturing issues are discussed. Several areas for future research in protein delivery systems for animal health applications potentially leading to innovative products are outlined.

Agriculture will be the first industry to be revolutionized by biotechnology. Stronger and more productive herds of livestock, custom-designed crops better to grow in harsh environments, plants and animals better able to resist insects and disease, and food offering more nutritious protein are all on our immediate horizon.

—Senator Patrick Leahy, 1988 (*1*)

Has the revolution begun yet? Certainly! Important advances have been made in genetic engineering and biotechnology that have led to the practical production of a variety of polypeptides for use as human pharmaceuticals and for use in animal health or veterinary applications. Molecules such as insulin, insulin analogs, human growth hormone, tissue plasminogen activator, bovine growth hormone, porcine growth hormone, growth-hormone-releasing factors, and cytokines can now be economically manufactured in quantities only dreamed of even 20 years ago. Certainly, Asimov and Krouze in 1937 (*2*), using their "lactogenic preparation" (anterior pituitary extract) to improve milk production in dairy cows, could not have envisioned the possibility of treating 42 million dairy cows (estimated number of

Copyright © 1997 American Chemical Society

dairy cows in the old Soviet Union), at approximately 6 g activity each per year, and still have enough of the bovine species survive the pituitary extraction and purification process.

These relatively large and complex molecules represent extremely difficult challenges from the administration or delivery point of view because of their inherently difficult chemistry, the complex biology into which they integrate, and inherent susceptibility to chemical and physical degradation. These challenges are more than adequately addressed in other chapters in this book. These challenges are further complicated when molecules such as bovine growth hormone are desired to perform over extended periods of time (e.g., weeks from a single administration), necessitating sustained- or controlled-release delivery technology.

Table 14.1 shows the potential desired duration of release of a sustained- or controlled-release delivery system developed to enhance production efficiency for each major food producing species. The life span of a broiler in today's highly concentrated poultry industry is typically 6–7 weeks, and in some cases for heavy broilers, 9–10 weeks. The birds are generally handled twice, once after hatch to vaccinate and then again for slaughter. An injectable delivery system, therefore, needs to be deliverable and of appropriate size to be administered to the chick, yet be capable of delivering the protein drug for the period of time required, potentially as long as 6–7 weeks. Swine are typically finished for 90–120 days. The producers generally will not handle the hogs to administer an injection during this period of time for fear of reduced average daily gain and reduced feed efficiency caused by stress and because of increased labor costs unless a clear economic benefit can be shown. Cattle stay in feedlots for approximately 120–140 days, dependent on the cost of feed and the breed of cattle. Some of the larger frame breeds (e.g., Holstein) may stay in the feedlot for longer periods of time: for example, up to a year. Again, handling of the animals is generally kept to a minimum. In some cases, because of either management practices or biology, a delivery system that exhibits a latent period after administration and before initiation of release (e.g., broilers or swine) may be desirable, so that the protein drug is delivered to the animal for a period of time late in the finishing phase. Because dairy cows are handled either two or three times a day, the criteria for extended duration of release from a controlled-release delivery system in this application is not as severe. Dairy cows increase milk production very nicely from daily injections of bovine somatotropin. Sustained-release injections of bovine somatotropin reduce the labor associated with giving the injections and reduce stress to the animal.

TABLE 14.1. Desired Duration of Release for Controlled-Release Delivery Systems To Enhance Production Efficiency in Food-Producing Animals

Species	*Duration of Release (days)*
Broilers	35–43 (may need only the last 14 days)
Swine	90–120 (finishing phase only)
Feedlot cattle	120–140 (60 days may be acceptable)
Dairy cows	14–30

Needless to say, the great expectations of the late 1980s for biotechnology in animal health have yielded to date, at best, moderate results. Lack of effort is not to blame. Many creative, innovative, and yes, esoteric ideas have surfaced, destined to deliver proteins from controlled-release devices or by unique delivery routes, only to expire when evaluated because they were too impractical, too complicated to produce or administer, or too expensive. In almost all cases, proteins are not economically deliverable or bioactive when administered through oral or dermal routes, two very common drug administration routes in the animal health industry. Given that the parenteral route of administration is possible, the mere size of the administered drug delivery system necessary to achieve the required dosages over extended periods of time may then become the limiting factor for use.

The extrapolation of protein drug delivery technology from human pharmaceuticals to the animal health industry is not always direct. First, tissue residues of the protein drug or from the delivery system itself need to be addressed, because these animals enter the human food chain. Human food safety is a growing public and regulatory concern as the food-producing industry strives to produce a high quality product while maintaining the low cost to which the consumer is accustomed. The fate of the delivered protein drug and the formulation at the injection site and within the edible tissues of the carcass, as well as the oral activity of the protein drug, has to be defined. In addition, the human food safety of elevated levels of secondary endogenous proteins (e.g., insulin-like growth factor-1) stimulated by product administration (e.g., porcine somatotropin) are required to be addressed. Second, because the excreta of all meat-producing animals ultimately finds its way into the environment, the impact of or the presence of environmental residues needs to be addressed. Water contamination is a growing public concern and issue. Third, food-producing animals are intensively (e.g., 10,000–30,000 broilers per broiler house) and extensively (hands-off!) managed. Just by sheer numbers alone, a protein drug delivery system suitable for administration in a physician's office or in the comfort of one's home may be a nightmare to administer to 1000 feedlot cattle on the eastern plains of Colorado at –10 °F.

To be an effective and marketable protein delivery system in the animal health industry, the product must encompass at least some of the following attributes:

- ease of administration,
- improved control of therapy,
- improved pharmacokinetics of the drug,
- adaptability of the delivery system to different management practices,
- reduced discomfort or stress on the animal,
- safety to the handler and the environment,
- cost effectiveness, and
- patentability.

Many of the product attributes are synonymous with the development of pharmaceutical drug delivery systems. But, because the animal health and veterinary markets are value-driven, the most desired product attributes may not be scientific and technical uniqueness, but may focus on cost, ease of use, safety, and integration into current management practices. For example, Figure 14.1 shows the average weekly price for finished hogs from 1990 through the end of 1994 (*3*).

The average price for hogs from January 1992 through the end of 1994 was approximately \$42.59 per 100 lb of live weight. The cost to produce hogs over the same period of time ranged from \$34.54 to \$46.25 per 100 lb, averaging \$40–\$41 per 100 lb (*3*). Thus, for a 240-lb market hog, a producer in recent years may only realize on average a profit of approximately \$6.20 per hog. It is apparent that if a producer wants to maintain his business, the producer either needs to increase his volume (number of hogs sold) or reduce his production costs, or both. The products for the hog industry need to be, of course, safe and efficacious, but also cost effective enough and easy to integrate into the

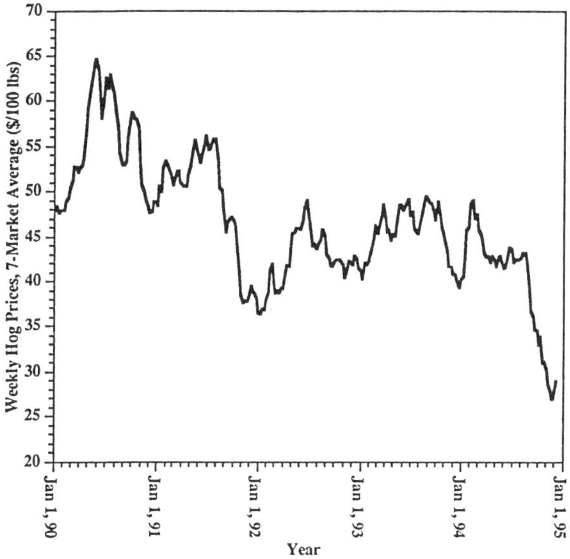

FIGURE 14.1. Weekly hog prices per 100 lb live weight, 1990–1994.

high-volume management system so that the hog producer still realizes a profit. The retail cost of a controlled-release protein product then becomes limited by the economics of the industry and ultimately defines the acceptable limits for the cost of producing the product.

The importance of cost of a controlled-release protein product is further strained by the fact that with many of the peptides and proteins of interest the cost of the protein drug is a very large component of the cost of manufacturing the product. With recombinant somatotropins, recent improvements in expression, isolation, and purification processes and economies of scale have driven the cost of the recombinant protein to approximately 50–60% of the manufacturing cost of the final product. Five years ago the cost of the recombinant protein may have been as much as 75–85% of the manufacturing cost of the final product. Even with these cost improvements for the protein drug, this change does not allow much room for the added cost of the formulation, manufacturing, and packaging, much of which are fixed costs (e.g., wages, benefits, and overhead). As additional improvements in recombinant protein manufacturing processes are implemented, more cost-competitive products for the animal health industry should become available.

Yet, with all the technical and socioeconomic issues surrounding the use of biotechnology, and in particular, recombinant proteins in animal health, tremendous potential exists for improving the quality and the cost of production of protein for human consumption. Table 14.2 lists some potential uses of peptides and proteins in the animal health industry. The remainder of this chapter will address the progress made toward achieving suitable controlled- or sustained-release delivery systems for potentially useful proteins in the animal health and veterinary industries.

Somatotropin Delivery Systems

Much of the protein drug delivery effort in animal health in the past 10 years has focused on the delivery of somatotropins (e.g., bovine growth hormone or porcine growth hormone). For good reason: economically viable markets have been identified, efficacy has been demonstrated, and through the use of recombinant biotechnology, large quantities of bulk protein are possible. One prediction is that by the year 2000, 45% of the agricultural biotechnology market share will be due to growth hormones (*4*). A variety of delivery systems have been used to achieve sustained release of somatotropins.

By far, diffusion-controlled systems are the most widely used controlled-release systems. They can be either reservoir systems (membrane systems, microcapsules, liposomes, or hollow fibers) or matrix systems, where the drug is distributed throughout the delivery system. Usually, the matrix systems are easier to manufacture compared with reservoir systems, but the drug release pattern is not generally zero-order unless geometries are controlled (*5–7*).

Hydrogels

Hydrogels from hydroxyethyl methacrylate–methyl methacrylate copolymer and hydroxyethyl methacrylate homopolymer cross-linked with ethylene glycol dimethacrylate have been used as rate-limiting membranes for the release of polypeptides and somatotropins (*8*). In this reference, ovine, equine, and bovine somatotropins were suspended in silicone oil and placed into cross-linked hydroxyethyl methacrylate cylindrical reservoirs. Sustained release was implied but no data were reported. In another reservoir system, small spheres of porcine somatotropin–chitosan (60:40 by weight) were coated with a hydrogel coating of polyvinyl alcohol (*9*). Significant growth of coho salmon over a period of 20 weeks was observed when these spheres were implanted in the peritoneal cavity. In a matrix system, biodegradable thermoplastic hydrogels consisting of ABA or AB block polymers where the B block is polyethylene oxide and the A block is a glycolide, or in the case of an ABA block polymer, a glycolide and trimethylene carbonate, have released bovine somato-

TABLE 14.2. Peptides and Proteins with Potential Uses in the Animal Health Industry

Peptide or Protein	*Use*
Somatotropins	Enhanced milk production Increased growth Reduced fat deposition Increased ovulation
Somatotropin releasing hormone	Increased growth Reduced fat deposition Increased milk production Increased ovulation
Somatostatin	Production enhancer by using finely ground feeds without gastric ulcer formation in swine
β-Lipotropin	Lipolytic agent Production enhancer
Gonadotropin releasing hormone (GnRH)	Induced molting in laying hens produces increased egg production
Luteinizing hormone releasing hormone (LHRH) and agonists	Agonists: promotion of weight gain in feedlot cattle prevention of estrus suppression of spermatogenesis LHRH: stimulation of ovulation stimulation of spermatogenesis stimulation of androgen production
LHRH antagonists	Opposite effects to LHRH agonists
Thymosins	Enhanced immune system
Interferons and cytokines	Shipping fever prevention in cattle Pseudorabies prevention in swine Prevention of equine infectious anemia Mastitis prevention Decreased bacterial infections in poultry
Prolactin antagonist	Increased egg production
Immunization against vasoactive intestinal peptide	Increased egg production
Glucagon	Ketosis management in dairy cows Lipolytic agent in poultry

tropin for periods up to 10 days in hypophysectomized rats (*10*). Copolymers of D,L-lactide–glycolide containing polyethylene glycol, which on absorbing water form a hydrogel, have been used to sustain the in vitro release of bovine somatotropin for periods of at least 12 days (*11*). Other matrix hydrogel systems, such as Pluronics (*12*), cross-linked poly(ethylene oxide) (*12*), phospholipid conjugated poly(*N*-iso-propylacrylamide (*13*), starch hydrogels (*14*), tetrahydrofurfuryl methacrylate–ethyl methacrylate gelled with hydroxyethyl methacrylate (*15*), gels based on polyethylene glycol (*16–19*), dextran gels (*20, 21*), hydroxyethyl methacrylate copolymer systems (*22, 23*), and hydrogel–liposome systems (*24*) have been evaluated for the delivery of polypeptides and proteins. With the inherent instability and aggregation problems with somatotropins in the presence of water, and the high water content of hydrogels after administration, the commercial use of hydrogel delivery systems for the delivery of somatotropins will be limited. Only by modifying the proteins themselves, so that they are stable to aggregation, will hydrogels find application.

Liposomes

Sustained release of bovine somatotropin has been demonstrated from several types of liposomes. Body weight gain in hypophysectomized rats lasting from 8 to 21 days has been observed when bovine somatotropin was formulated into egg phosphatidylcholine, ethanolamine, and α-tocopherol hemisuccinate (*25*) or phosphatidylcholine and cholesterol hemisuccinate (*26*). From similar preparations, elevated plasma levels of bovine somatotropin in dairy cows for approximately 21 days (*26*) and increased milk production (*27*) have been observed. These results are intriguing. However, liposomal delivery of somatotropins in animal health has languished largely due to low product loadings, difficulty and cost of manufacture, and stability issues. Recent efforts have focused on the use of liposomes in combination with other existing delivery technologies perhaps amenable to the delivery of somatotropins or other proteins. In one research group, liposomes containing bovine serum albumin (BSA) have been microencapsulated by using sodium alginate and coating with poly-L-ornithine (*28, 29*). In vitro release of BSA was observed for over 50 days. In yet another research group, cyclodextrin inclusion complexes of drugs were entrapped into liposomes (*30*). This concept offers some intriguing possibilities for the sustained delivery of somatotropins (and other proteins) for it is well documented that cyclodextrins improve the aqueous solubility and stability of bovine somatotropin (*26*) and ovine somatotropin (*31*) and decrease the thermal and interfacial degradation of porcine somatotropin (*32*), presumably by inhibiting aggregation. The expense of cyclodextrins and toxicology concerns have limited their use in controlled-release formulations to date.

Microparticles

Microparticulate formulations have been used to sustain the delivery of somatotropins. In my laboratory (*26*), bovine-somatotropin-containing microspheres composed of ethylene-vinyl acetate, cellulose acetate propionate, polyvinyl alcohol, polylactic-*co*-polyglycolic acid, D,L-polylactic acid, sodium alginate, polycaprolactone, waxes, hydrogenated oils, and 100% bovine somatotropin itself, each have been evaluated as sustained-release delivery systems. Various wax and triglyceride microspheres have been prepared with bovine somatotropin by other researchers and evaluated in dairy cows (*33–37*). These microspheres, when injected into dairy cows at 14-day intervals, yielded elevated plasma levels of bovine somatotropin and significant increases in milk production (*33–36*). In vitro release of porcine somatotropin was studied from microspheres of polyglycolic acid (*38*) where aggregation of the protein within the polymer was noted. Microcapsules of cellulose acetate butyrate containing bovine somatotropin (*26*) when injected into dairy cows surprisingly elevated plasma bovine somatotropin levels for approximately 28 days. However, no significant increase in milk production was noted from this formulation. This phenomenon of measurable quantities of elevated plasma bovine somatotropin with no increase in milk production points one to the issue of formulation-induced changes to protein molecules leading to immunoassayable proteins in the plasma, yet loss of protein biological activity.

Oleaginous Vehicles

Somatotropin release from oil suspensions has been evaluated extensively. The only three marketed somatotropin products to date use this technology (Posilac, bovine somatotropin injection, Monsanto Company; Optiflex, bovine somatotropin injection, Elanco; and Boostin, LG Chemicals, Ltd.) These types of delivery systems, in contrast to the aforementioned delivery systems, yield sustained release of bioactive somatotropin for periods of 7 to 28 days and are relatively inexpensive and easy to manufacture as sterile preparations, stable, and easy to administer. Thus, in light of the product requirements for animal health products, oil suspensions of somatotropins have made good business sense to pursue. Thickened biocompatible oils have been prepared by using glyceride derivatives, polyglycerol esters, waxes, absorption regulating agents, or the antihydration agent aluminum monostearate, and by using a suspension of either the somatotropin or a transition-metal salt of the somatotropin (*39–46*). The efficacy (*47–49*) and the safety (*50*) of bovine somatotropin when delivered from oleaginous formulations is well documented. Figure 14.2 shows typical effects of bovine somatotropin on milk production. The shaded area in Figure 14.2 represents the increase in milk production due to increasing doses of bovine somatotropin from a summary of 21 experiments, involving approximately 1,000 dairy cows, many of them injected without sustained-release preparations but with daily injections (*49*). Milk production increased ranging from 10 to 20%, dependent on the bovine somatotropin dose. The milk production responses from different oleaginous formulations of bovine somatotropin injected at intervals of either 14 or 28 days are also shown in Figure 14.2, expressed as a function of the average daily dose (injected dose divided by the injection time interval). Milk response profiles versus average daily dose of bovine somatotropin from the oleaginous formulations are similar to the target response curve (shaded area), albeit of somewhat lower magnitude. This result can be due to numerous biological and animal management variables, but also to aggregation, sequestering, or proteolysis of the bovine somatotropin, and thus loss of bioactivity, within the injection depot as well as the less than ideal release profiles of the protein from these types of formulations (Figure 14.3).

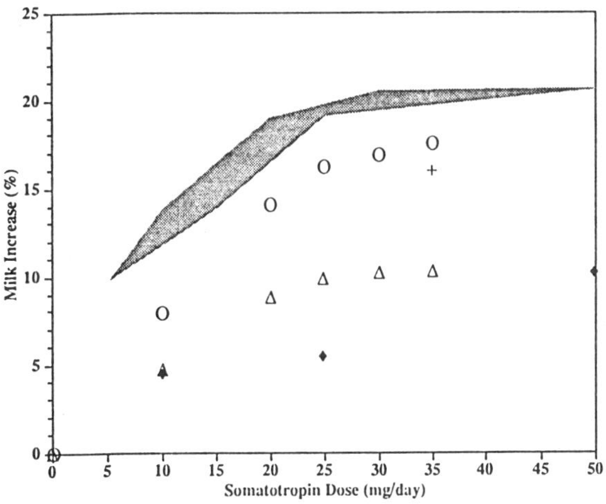

FIGURE 14.2. Effect of bovine somatotropin dose on milk production. Shaded area represents summary of 21 trials, many conducted with daily injections of bovine somatotropin: (O) summary lactation response from five U.S. trials at 320, 640, and 960 mg of bovine somatotropin (somidobove) every 28 days in oleaginous sustained-release injection; (△) summary lactation response from 21 European trials at 320, 640, and 960 mg of somidobove every 28 days in oleaginous sustained-release injection (adapted from reference 49); (♦) lactation response at 140, 350, and 700 mg of bovine somatotropin every 14 days in waxy microsphere suspended in oil sustained-release injection (adapted from reference 36); (+) lactation response at 500 mg of bovine somatotropin (sometribove) every 14 days in oleaginous sustained-release injection (adapted from reference 47).

Other Injectable Systems

Other injectable nonoleaginous delivery systems for somatotropins have been used. Multiple water-in-oil-in-water (WOW) emulsions have been shown to sustain the release of bovine somatotropin (*51, 52*). Multiple WOW emulsions using bovine somatotropin in the W_1 phase, sorbitan surfactants in light mineral oil (O phase), and polyoxyethylene sorbitans in the W_2 phase showed sustained growth in hypophysectomized rats through 10 days. Similar compositions showed elevated plasma bovine somatotropin levels in sheep through 17 days. In other research, adducts of activated polysaccharides with bovine somatotropin have been used to prolong the biological

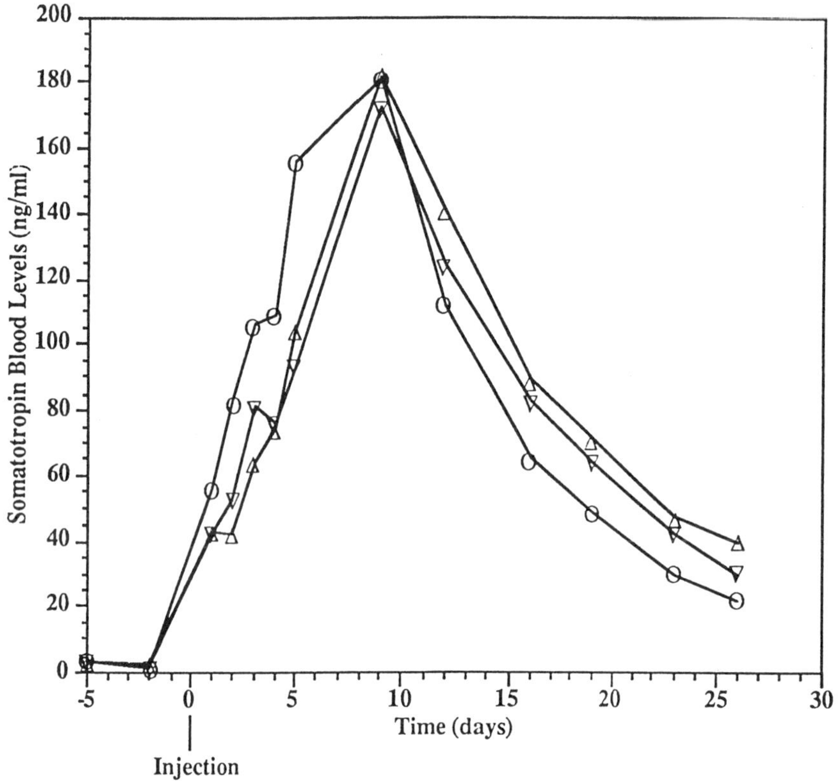

FIGURE 14.3. Somatotropin blood levels in dairy cows after injection of a suspension of recombinant bovine somatotropin in peanut oil thickened with beeswax. Key: △, 1.77 g of bovine somatotropin, 7.40 g of sesame oil, and 0.83 g of white beeswax; ○, 1.24 g of bovine somatotropin, 8.32 g of sesame oil, and 0.44 g of white beeswax; and ▽, 1.26 g of bovine somatotropin, 7.87 g of peanut oil, and 0.87 g of yellow beeswax. (Adapted from reference 44.)

effect of bovine somatotropin in vivo. Sustained growth of hypophysectomized rats over a period of 10 days was demonstrated from a single dose of bovine somatotropin–agarose adduct (*53*). I evaluated (*26*) other complexes of bovine somatotropin with Sephadex, oxycellulose, carboxymethylcellulose, polyvinylpyrrolidone, and polyvinylpyrrolidone–polyacrylic acid. Sustained release of bovine somatotropin up to 7 days in hypophysectomized rats was observed. Carbohydrate polymers have been used to elevate plasma bovine somatotropin levels and increase milk production in dairy cows (*54*). A dextrin solution containing bovine somatotropin maintained elevated plasma bovine somatotropin levels through 7 days from a single injection in dairy cows. Milk production was also increased by 10–20% by using the same injection in a 3-week study. By using a xanthan and bean locust gum injection, milk production in dairy cows was increased through 2 weeks from a single injection of 350 mg bovine somatotropin.

Implants

Various implant delivery systems, with and without polymers, have been used to deliver somatotropins. Elevated porcine somatotropin plasma levels for 8–10 days were observed from compressed implants of a leupeptin complex of porcine somatotropin (*55*). This process resulted in improved feed-to-gain ratios and average daily gain over that of the untreated pigs. Porcine somatotropin complexed with an aromatic aldehyde and compressed into implants showed an ~12% increase over control in average daily gain and a 21% increase over control in feed–gain ratio over a 21-day treatment period (*56*). Increased milk production and serum bovine somatotropin levels over a period of 56 days were reported in dairy cows from compressed tablets of bovine somatotropin implanted every 14 days (*57*). Similarly, compressed tablets of porcine somatotropin or copper-associated porcine somatotropin gave improved feed efficiency in hogs over a period of 4 weeks when implanted every 14 days. Also, these compressed porcine somatotropin tablets were placed into silicone rubber tubing so that release only occurred at the ends. Implanted at 4 weeks intervals, hogs demonstrated improved feed efficiency of 6–10% through 5 weeks (*57*). Compressed implants of zinc-associated bovine somatotropin or copper-associated porcine somatotropin coated with polyvinyl alcohol have been used for the parenteral administration of somatotropins (*58*).

Somatotropin implants surrounded by porous polymeric membranes have been evaluated as delivery systems (*59, 60*). In one case, cylindrical pellets containing 30% bovine or porcine somatotropin, 40% sucrose, and 30% ethylcellulose were encased within microporous polyethylene tubes (*59*). In other efforts to sustain the release of bovine somatotropin, bovine somatotropin implants with either polylactic acid or ethylcellulose were encased within microporous sleeves (*60*). In vitro release of bovine somatotropin was demonstrated over a period of 14 days. Stabilized porcine somatotropin implants, either stabilized with porcine serum albumin (*61*) or with sugars, amino acids, or choline derivatives, have been evaluated for their growth-promoting effects in pigs (*62*). Stabilized porcine somatotropin implants with sucrose and capable of release through microporous discs elevated serum porcine somatotropin levels in pigs for approximately 10 days (*63*). Yet, other implants have been made containing bovine somatotropin and either polylactic acid or polycaprolactone and then coated with either polylactic acid or polycaprolactone (*64*). Wax-coated implants of zinc complex of porcine somatotropin, when implanted weekly in pigs, depressed blood urea nitrogen and enhanced feed efficiency during a 21-day treatment period (*65*). Compressed implants of lactose, calcium phosphate, and ^{125}I-human somatotropin coated with Eudragit NE30 D released ^{125}I-human somatotropin in vitro over a period of 29 days (*66*). Implants of silicone rubber (*26*), ethylene vinyl acetate (*26*), polyanhydride (*26, 67*), and cholesterol (*68*) containing bovine somatotropin have been evaluated as sustained-release devices. Limited sustained release, for approximately 1 week in vitro, of bovine somatotropin from the polyanhydride devices in vitro was observed (*67*). Similar to the work with porcine somatotropin in polyglycolic acid microspheres (*38*) and in carbohydrate–porcine somatotropin pellets coated with polyvinyl alcohol (*9*), aggregation of the unreleased bovine somatotropin within the polyanhydride devices was suggested (*67*).

Osmotic Devices

Osmotic pumps have been studied extensively for the delivery of somatotropins (*69, 70*). In vitro release of porcine somatotropin over a 2-month period was demonstrated (*70*). In this case, the porcine somatotropin was stabilized in a glycerol, gelatin, L-histidine gel. Other stabilizing compositions for somatotropins suitable for osmotic devices have been suggested (*71*). Although not necessary to achieve efficacy, patterned or episodic release of porcine somatotropin by separating porcine somatotropin tablets with osmotic tablets has been done (*72*). Improved growth rate and feed efficiency in poultry roasters has been documented by using osmotic pumps to deliver recombinant chicken somatotropin (*73*). Although osmotic pump boluses are being used in the rumen of cattle, the commercial use of osmotic pumps for the parenteral delivery of proteins has been limited by the cost of the final product, final product size, cumbersome administration to the animal, and delivery system recovery issues after slaughter of the animal.

Other Important Peptides and Proteins

Not surprisingly, many of the same types of sustained- or controlled-release delivery systems that have been evaluated with the somatotropins have been evaluated for the delivery of other important peptides and proteins for the animal health industry. A common (but dwindling) concept is that if a delivery system works for somatotropins then it will perform the same for other proteins: for example, somatotropin-releasing hormone. However, because many delivery systems do not behave similarly for bovine somatotropin and porcine somatotropin, why would one think a priori that a delivery system for bovine somatotropin would be appropriate for somatotropin-releasing hormone? Researchers working in the area soon acknowledged that the equivalency concept breaks down, that all proteins are "not created equal", and that a firm appreciation of the physicochemical behavior of each specific protein is a necessary, albeit tedious and less glamorous, prerequisite to formulations evaluation.

Luteinizing-Hormone-Releasing Hormone

Luteinizing-hormone-releasing hormone (LHRH) is a neurohormone, a decapeptide, of hypothalamic origin that controls synthesis and release of gonadotropins from the pituitary. As such, it is of interest for controlling fertility and estrus in livestock and companion pet animals. A series of compressed implants containing lactose and calcium phosphate and coated with Eudragit NE30 D have been evaluated as controlled-release implants for LHRH, epidermal growth factor, gonadotropin-releasing hormone (GnRH), and luteinizing hormone (*66, 74, 75*). In vitro release of epidermal growth factor over 28 days, luteinizing hormone over 63 days, GnRH over 298 days, and LHRH over 168 days was measured. When evaluated in vivo, implants of LHRH induced ovulation and mating in anestrous ewes. Higher payload implants of LHRH released 10–17 mg of LHRH over a period of 25–30 days (*75*). Peptide release from this implant system was controlled by the compressed core excipient ratios, percent active present, and the coating thickness of Eudragit NE30 D. In another application, manipulation of ovulation and spawning in catfish and sea bream using an LHRH implant was evaluated (*76*). Ethylene-vinyl acetate coated hemispheres containing a single aperture showed constant in vitro release of LHRH of about 10 μg/day over a period of 20 days. When implanted in sea bream, spawning was induced within several days.

Luteinizing-Hormone-Releasing Hormone Agonists

Because of the low potency and short circulatory half-life of LHRH, only transitory and inefficient stimulation of the pituitary and gonads was achieved in early studies. Synthesis of substituted analogs, such as incorporation of D-amino acids in the sixth position and alkyl-amides in the tenth position (*77*), achieved increased potency and greater suppression of the hypophysial–gonadal axis. Even though LHRH usually acts most efficiently as short-lived, episodic pulses, stimulating ovulation and spermatogenesis, the high potency of the analogs due to increased receptor binding affinity and resistance to metabolic degradation and continuous administration made the analogs of interest for inhibition of fertility. Nafarelin acetate, an analog of LHRH, has been studied extensively (*77–83*). Estrus suppression in female rats was observed for 24 days and for 40 days with nafarelin acetate containing microspheres of 50–50 poly(D,L-lactide-*co*-glycolide) (PLGA) with intrinsic viscosities of 0.38 and 1.52 dL/g, respectively (*78, 79*). A triphasic estrus suppression was observed with nafarelin acetate containing microspheres of 69–31 PLGA, intrinsic viscosity of 0.97

dL/g, over a period of 90 days (*78, 79*). The effect of gamma irradiation on the triphasic estrus response was noted and attributed to the decrease in molecular weight of the polymer with increasing irradiation dose (*78, 79*). As the dose of nafarelin acetate containing microspheres increased, so to did the length of estrus suppression (*77*). Other studies have evaluated the effect of nafarelin acetate in PLGA microspheres on the clinical performance of the formulation (*80, 81*). Implants of PLGA and the effect of lactide monomer ratio and molecular weight of the polymer on the in vitro release of nafarelin acetate were evaluated (*82, 83*). Another analog, leuprolide acetate, has been effective in suppressing luteinizing hormone (LH), follicle-stimulating hormone (FSH), and testosterone secretion in rats and dogs over a period of 1 month from a single administration in PLGA microspheres (*84*). In other work, a silicone rubber implant containing an LHRH agonist (RS-49947) suppressed reproductive function in both female and male dogs for as long as 1 year (*85*). In vitro release of an LHRH analog from D,L-lactide–glycolide containing either vinylpyrrolidone or polyethylene glycol methyl ether has been observed over a period of several days (*11*). In vitro release of an LHRH analog, D-Trp^6-LHRH, from PLGA microspheres was observed over a period of 56 days (*86*). NMR analysis of the peptide–polymer system during in vitro release suggested that the release of the peptide involved not only diffusion of the peptide through a porous structure created by degradation and dissolution of the polymer matrix, but also binding of the basic peptide to terminal polymer carboxylate groups. This "binding" phenomena stopped in vitro release of the peptide after about 15 days and only resumed once the molecular weight of the polymer reached some critical small molecular weight. By deliberately conjugating free COOH groups of biodegradable polyester polymers to ionogenic amines of polypeptides, researchers have demonstrated 28 days of in vitro release of D-Trp^6-LHRH (*87*). Moreover, this technology is claimed to be effective for sustained release of other acid-stable peptides, such as glucagon or growth-hormone-releasing hormone. By using another biodegradable polymer, the cross-linked poly(ortho ester)s, estrus suppression in rats has been documented for up to 140 days by using the LHRH analog, nafarelin (*88, 89*). A more rapidly hydrolyzing polymer was prepared by varying the polymer hydrophilicity by the addition of the acidic comonomer, 9,10-dihydroxystearic acid. With this modification, in vitro release of nafarelin of 1 month was achieved (*89*).

Somatotropin-Releasing Hormone

Somatotropin-releasing hormone (more commonly called growth-hormone-releasing hormone, GHRH or GRF) has been evaluated extensively as a next generation to the somatotropins presumably because the same biological effects of increased growth or milk production can be achieved with a substantially lower dose of GHRH. Japanese researchers have evaluated implants of GHRH in silicone rubber (*90*) and GHRH in collagen (*91, 92*). GRF(1-29)NH_2 was mixed with silicone elastomer, sodium chloride, or sodium chloride and glycerol and cured to form matrix-type implants. In vitro release was monitored over 6 days. In another formulation, hGRF(1-44)NH_2 was lyophilized from a solution of gelatin and suspended in sesame oil to give a sustained-release oleaginous injection (*91*). Pellets of hGRF(1-29)NH_2 and collagen were fabricated and gave 3 days of sustained release of hGRF(1-29)NH_2 (*91*). Atelocollagen implants of GRF(1-29)NH_2 containing various amounts of acidic compounds, such as citric acid, glycine, aspartic acid, or glutamic acid, were evaluated for their sustained-release properties (*92*). Sustained in vitro release of the GRF(1-29)NH_2 was observed for 3 days by using citric acid in the implant and for 14 days by using glycine in the atelocollagen implants.

Similar to the cholesterol implants for somatotropins (*68*), cholesterol acetate implants of GHRH have been made demonstrating in vitro release of the GHRH over a period of 20 days (*93*). Elevated growth hormone levels in sheep for approximately 14 days from a single subcutaneous injection of a GHRH analog, *p*-methyl hippuroyl pGRF (2-76)OH, suspended in an oleaginous vehicle containing polyglycerol esters to prolong the release, was observed (*46*). The in vitro sustained release of GRF(1-29)NH_2, the biologically active fragment of the 44 amino acid GRF, has been evaluated in PLGA pellets (*94*). In vitro release was influenced by pH and ionic strength of the release media. Release of the peptide was dependent more on the solubility characteristics of the peptide than on the diffusion through channels in the polymer matrix or polymer morphology. Significant gelling of the peptide was noted in salt solutions and in the presence of plasma. Similar solubility and gelling phenomena with the GHRH analog, *p*-methyl Hippuroyl pGRF(2-76)OH, affected and inhibited the in vitro and in vivo sustained release of the peptide from various formulations (*26, 95*).

Other Peptides

Controlled-release implants of a growth-hormone-releasing hexapeptide for ruminants were reported (*96*). Fatty acid salts of the hexapeptide were prepared and compressed into implants and coated with a variety of polymers, leaving the cylindrical implant ends coating-free for drug release. Release of the hexapeptide was controlled by the fatty acid salt chain length ($C_{18} > C_{14} > C_{13} > C_{12} > C_{10}$) and the type of implant coating. Approximately 42 days of hexapeptide release in steers was measured. Polylactic acid implants of porcine pituitary extracts rich in FSH and LH have been used to induce superovulation in cows (*97*) to overcome the arduous and stressful task of twice daily injections for 4 days. Adequate plasma FSH concentrations to induce superovulation were obtained from this formulation. Other researchers have evaluated the release of a superactive agonist of gonadotropin-releasing hormone, D-Phe^6-GnRH, in rats, rabbits, and guinea pigs from PLGA implants (*98*). Release of the peptide was monitored by the presence of ^{125}I -labeled peptide at the subcutaneous injection site and was complete within 4–7 weeks.

Formulation and Manufacturing Issues

One would think, what with all the efforts and investigations during the past 10–15 years into the sustained delivery of somatotropins (and other important peptides and proteins for the animal health industry), that there would be several manageable, cost-effective delivery technologies ready to propel these important peptides and proteins into the animal health marketplace. Of course, the regulatory approval process contributes heavily to the timing of market entry, but one very important reason these products have not reached their full potential for use in the animal health industry is the inherent difficulties in maintaining either the chemical stability or bioactivity of the peptides and proteins either during manufacturing, or within the formulation itself before and after administration to the animal. The literature (*9, 38, 67, 94, 95, 99*) points to the problems associated with protein aggregation within various sustained-release formulations and within the biological milieu, resulting in decreased product performance. Design questions that govern the acceptance of a formulation's behavior in vivo also are common to the development and design of the manufacturing process for the formulation. Thus, it is imperative that the individual characteristics of the formulation and the protein drug and the interactions of the formulation and protein drug during manufacturing and in the in vivo environment be documented, well understood, and controlled in the manufacturing process.

Several recent reviews on formulation and manufacturing concerns for protein drugs (*99–102*), and more specifically on the somatotropins (*103, 104*), have been published. These reviews and the chapters within this book should be consulted for additional information.

As shown in Figure 14.4, several factors influence the design of a protein drug formulation and manufacturing process. In general, the physicochemical properties of the candidate protein are considered first: isoelectric point, molecular weight, amino acid composition, aqueous solubili-

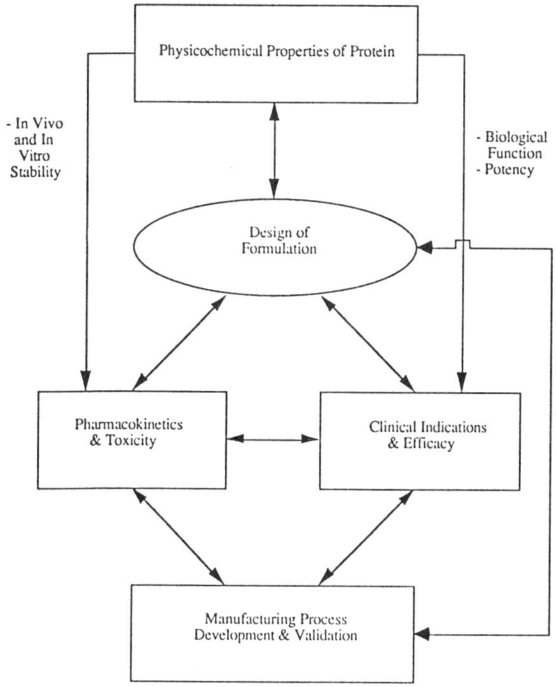

FIGURE 14.4. Key factors affecting the development of protein drug formulations. Note that the manufacturing process development effort can have effects on the performance of the protein delivery system and thus ultimately may affect the final design of the protein delivery system. (Adapted from reference 102.)

ties at different pH levels and salt concentrations, and organic solubilities.

Second, definition of chemical degradation routes and aggregation phenomena are critical to the design of the formulation, lyophilization of the protein, and the manufacturing process. Three of the most common routes of protein degradation are aggregation, deamidation, and oxidation; and protein aggregation is the most critical to the success of sustained- or controlled-release systems. Correlation of the chemical degradation products (e.g., deamidation or oxidation) to biological activity is desirable. Also, it must be demonstrated that these degradation products (both in vitro as well as in vivo) do not have any adverse effect on safety and efficacy of the product. Many proteins can degrade extensively without affecting their safety or efficacy, such as recombinant human growth hormone (*102*). However, the U.S. Food and Drug Administration usually requires that a commercial product of this type not degrade more than 10% during manufacturing and maintain at least 90% of the original drug composition after 2 years of storage. One often overlooked factor that pervades the development process is the generation of appropriate analytical procedures and techniques to assess protein stability and degradation.

Third, after potential sustained-release formulations have been evaluated against proposed product specifications, desired pharmacokinetics, and toxicity, the formulation is chosen and the development of the manufacturing process begins. It is desirable that the pivotal field trials be done with a product that is made by a manufacturing process identical to, in the same type of equipment as, and at least 1/10 the scale of the proposed final manufacturing process. These trials are when one really determines how well the physicochemical properties, the degradation routes of the candidate protein, and the behavior of the sustained-release formulation are understood. Manufacturing issues arise such as

- method of drying the protein,
- storage before processing,
- formulation processing conditions (e.g., time, heat, shear, light, and oxidation),
- intermediate storage during processing steps,
- sterilization, and
- compatibility in contact with equipment surfaces and final packaging materials.

All of these manufacturing variables have the capability to affect and to be deleterious to the product. And of course, the cost of manufacturing the product needs to be kept at a minimum. As commonly happens, the formulation changes because of the development of the manufacturing process for one reason or another, and additional studies are required. To provide continuity throughout the product development process, the formulation scientist needs to be an integral part of not only the discovery research efforts, but also the development and scale-up of the manufacturing process.

Freeze-drying, also termed lyophilization, has been the drying method of choice for pharmaceutical therapeutic proteins, and for proteins for animal health use as well. Another method of drying proteins, spray-drying, is of interest largely because it is more economical, especially for the large quantities of therapeutic proteins needed in the animal health industry. Because lyophilization is a low temperature process, it is normally considered less destructive to proteins, and thus it currently is the drying method most used. However, it is not without its disadvantages. For instance, pH and ionic strength around the protein can increase during the freezing step, leading to protein denaturation, aggregation, and precipitation, even before formulation into a sustained-release system. Yet, recent studies of lyophilizing proteins in the presence of sugars such as trehalose have shown that the stability of lyophilized proteins can be dramatically improved (*105*). Another recent study demonstrated that spray-drying of therapeutic proteins may be a potential alternative to lyophilization (*106*). Spray-drying of mannitol-formulated human growth hormone resulted in extensive aggregation, either from surface-induced denaturation at the air–liquid interface during atomization or from thermal degradation. Tissue-type plasminogen activator in the presence of a surfactant was not degraded by atomization. Although not optimized, we have seen only minimal aggregation and maintenance of a lactation response in dairy cows with recombinant bovine somatotropin spray-dried in a small production-size dryer (*26*). Clearly, additional work in this area is needed.

Sustained-release delivery systems for proteins offer special problems from a sterilization perspective, because protein formulations generally are perceived to promote growth of microorganisms. Sterility of the final product is a manufacturing and regulatory issue. Both Posilac and Optiflex, bovine somatotropin sustained-release injections, are manufactured aseptically. However, not all delivery systems for proteins are amenable to aseptic processing either because of a multitude of complex processing steps or loss of protein potency. Even though there has been extensive approved use of nonsterile cattle ear implants in the United States with few adverse effects to the treated animals and with no human food safety implications, the Center for Veterinary Medicine has proposed that veterinary injectable products, including protein-containing implants, be sterile. However, if on a case-by-case basis there is adequate data and information to support that sustained-release delivery systems for proteins cannot be sterilized and that there are no safety or efficacy issues and the bioburden in the product can be maintained at appropriately low levels, waiver from the sterility requirement should be possible.

In general, therapeutic proteins are considered unstable to terminal sterilization techniques, such as heat or irradiation. One example where this is not true is with gamma irradiation of LHRH analogs (*78, 79*). Substantial degradation of recombinant bovine somatotropin was noted with dry heat sterilization (*104*). Approximately a 30% loss of lyophilized recombinant bovine somatotropin measured by reversed-phase high-performance liquid chromatography was noted at an irradiation dose of 2.5 Mrad (*104*). I noted (*26*) similar responses of recombinant bovine somatotropin to heat, gamma irradiation, and E-beam irradiation. These data suggest that traditional terminal sterilization techniques of delivery systems containing proteins that are the size of somatotropins are probably not feasible, unless the protein can be stabilized by some means, or if bioburdens are maintained at very low levels so that dramatically reduced irradiation doses are required for sterility.

Future of Peptide and Protein Delivery for Animal Health Applications

The future of peptide and protein delivery systems (sustained or controlled) in the animal health industry is bright provided the protein drugs and final products (e.g., delivery systems) can be made economically, the protein drug can be delivered efficiently, and the efficacy and safety of the products can be demonstrated. Several areas of additional research need to be addressed:

- improving the stability of the protein drugs during drying (e.g., lyophilization), formulation, manufacturing, and postadministration;
- exploring different routes of administration;
- designing improved polymeric materials, perhaps mimicking biomaterials more closely;
- addressing the efficiency of pulsatile release or self-regulated drug release versus continuous drug release;
- targeted delivery, perhaps with peptidomimics;
- gene therapy;
- improved analytical procedures; and
- improvements in manufacturing processes and plant design, scale-up, and sterilization.

Some of these areas should be consulted in other chapters of this book.

The one area that has plagued formulation scientists most persistently to date in the development of sustained- or controlled-release systems for peptides and proteins is the area of protein stabilization. Protein aggregation and its effects on controlled-release delivery system performance has been noted and discussed substantially with the somatotropins (*9, 38, 67, 99, 103, 104*) and more recently with GHRH (*94, 95*). The role of moisture in protein stability has been studied (*107*) and of course needs to be understood if protein delivery systems are to be used in vivo. For example, one salt of recombinant bovine somatotropin lost approximately 50% of bovine somatotropin monomer (presumably lost to aggregation) after 2 weeks of storage at 75% relative humidity (*107*). If this protein molecule were to be used in a controlled-release delivery system similar to those discussed, what would be the expected in vivo duration of product performance? Presumably, not very long unless the aggregation could be inhibited. Other researchers have pursued chemical modifications to somatotropins such as porcine somatotropin to improve solution solubility and stability or to provide sites to link to polyethylene glycol, with the goal of inhibiting aggregation and extending sustained-release product performance (*108, 109*).

Of potential interest to this area of protein stabilization is the recent work on thermodynamically stable intermediates (molten globules) of proteins (*110, 111*), particularly in light of the suggestion that the molten-globule state of proteins plays an important role in a number of physiological processes, such as the transfer of proteins across membranes and protection of certain signal peptides against proteolytic degradation (*111, 112*). Also of interest is the finding that bovine α-lactalbumin binds to phospholipid vesicles at low pH (e.g., in the denatured state) (*112*). The unfolding of bovine somatotropin by guanidine hydrochloride has been studied by several researchers and found to be a multistate process with at least one stable intermediate (*113–116*) (Figure 14.5). The intermediates were suggested to be the molten-globule state for bovine somatotropin (*117*). The molten-globule state of bovine somatotropin was characterized as follows (*110, 114*):

- largely α-helical,
- retaining a compact hydrodynamic radius,
- having the aromatic side chains randomly oriented, and
- possessing a solvent-exposed hydrophobic surface, leading to protein association.

Similar behavior for human somatotropin was observed at high protein concentrations (*110*). Because the molten globule is fully reversible to the native state (presumably once in contact with the in vivo milieu), this characteristic opens some intriguing possibilities for improved protein delivery systems. For somatotropins, the proteins may be able to be captured in the intermediate or molten-globule state potentially leading to improved resistance to degradation during lyophilization, spray-drying, formulation, manufacturing,

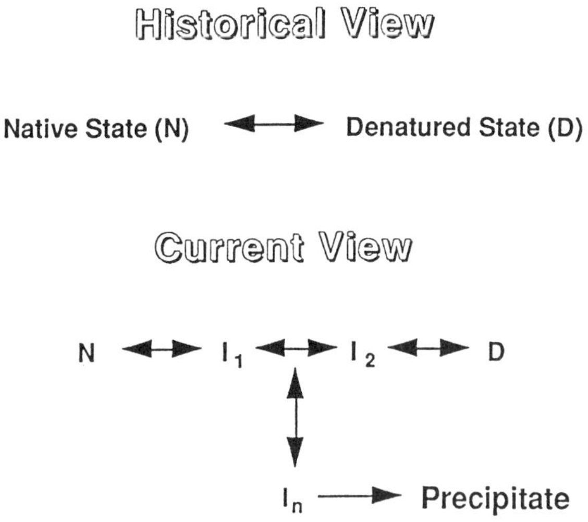

FIGURE 14.5. Historical and current bovine somatotropin unfolding schemes showing that the current scheme, in contrast to the historical scheme where the folding was thought to consist of two states, native (N) and denatured (D), is a multistate process with at least one stable intermediate (I). I_n is the self-associated intermediate species, where n is between 3 and 5. (Adapted from references 111, 113, and 115.)

or terminal sterilization (heat), or after administration to the animal. Also, alternative routes of administration, such as oral delivery, may be allowed.

Different routes of administration for peptide and protein drugs are continually being addressed in pharmaceutical applications. The more common delivery routes include nasal, pulmonary, buccal, rectal–vaginal, transdermal–iontophoretic–sonophoretic, oral, and parenteral (subcutaneous or intramuscular). Most of these administration routes are not common commercial routes for protein delivery for human health much less for the animal health industry. The efficiency of protein delivery via alternative routes (e.g., nasal, transdermal, buccal, or rectal) is generally low even with the use of permeation enhancers. The use of permeation enhancers and other formulation excipients (e.g., bioadhesives), more complex product manufacturing processes, and poor protein delivery efficiency add expense to the already cost-constrained product for the animal health industry.

In addition, the regulatory issues surrounding residues of the protein or the formulation or both either at the administration site or within the edible tissues of the animal result in a complicated scenario for many routes of administration. The administration site residue issue is further complicated across species. For example, several controlled-release implants for growth promotion are used in cattle. The implants are typically implanted subcutaneously in the ear, which is classified as inedible tissue. Thus, at slaughter, the ears are removed and disposed of with other inedible portions of the carcass. However, for swine, the ears are classified as edible portions of the carcass. There is a saying in the swine industry that "Everything in the swine carcass is used, except for the squeal!" This fact presents an interesting scientific and regulatory dilemma for those pursuing controlled-release porcine somatotropin implants, desired to be administered subcutaneously in the ear. Does one develop a completely biodegradable implant so no residues are left at the administration site after slaughter? What happens if the producer decides to market his hogs a week earlier than anticipated? Does one excise the implant site before slaughter? Does one reclassify swine ears as inedible tissue and remove the ears at slaughter? How do the inspection personnel oversee this process? Does the efficiency of the slaughter operation then decrease? Is there an economic loss, either due to loss in slaughter efficiency or due to the value of the excised ears, to the producer or packer? As biotechnological advances continue and novel anatomical delivery pathways, such as nasal and pulmonary, continue to be addressed, hopefully some of these concerns can be resolved.

Most proteins of interest in the animal health industry are secreted in a pulsatile or episodic fashion, as shown in Table 14.3 (*118*), usually under the control of other neuroendocrine proteins. The somatotropins, for example, have 9–29 episodic releases per day under the control of somatotropin-releasing hormone and the somatostatins. Researchers are continuing to evaluate pul-

TABLE 14.3. Examples of Pulsatile Secretion of Various Protein Hormones Useful in Animal Health Industry

Protein Hormone	*Pulses/Day*
Somatotropin	9–29
Prolactin	4–22
Thyroid stimulating hormone	6–13
Luteinizing hormone	7–121
Follicle stimulating hormone	4–19
Insulin	108–144
Somatostatin	72
Glucagon	103–144

SOURCE: Adapted from reference 18.

satile (or sometimes called feedback) delivery systems for insulin, with just cause because insulin is more effective when delivery is pulsatile. Nature's scheme for delivering neuroendocrine proteins and steroid hormones has biological purpose because the pulses reduce down-regulation of receptors, overwhelm proteolytic degradation processes, and are metabolically more efficient for the animal (*118*). Thus, from the point of view of efficacy and efficient protein use, feedback delivery systems are an intriguing goal for proteins in the animal health industry. To date, no pulsatile delivery systems for animal health proteins or even insulin are optimized and reproducible enough to be marketed. In addition, even if pulsatile delivery technology were available, delivery systems using this technology may be too expensive for use in the animal health industry. Fortunately, many of the useful proteins to date, such as the somatotropins and somatotropin-releasing hormones, are efficacious when delivered continuously, and with the issues of protein stabilization, it is just as well. Other chapters in this book discuss feedback delivery systems more fully.

One last intriguing direction toward which drug delivery technology in the animal health industry is progressing is the administration of active compounds to affect the growing animal via treatment of the embryo. This means of administration to affect growth of the animal has potential advantages of having to be treated only once and reducing drug residues in edible tissues. This process can readily be done in the poultry industry where the embryo is in a neat, defined package, the egg. The intensive management and large number of birds limit access to administer sustained-release delivery systems. Most medicaments today to poultry are administered via the feed or through the water. Vaccines (*119*), insulin-like growth factors or their active analogs (*120*), tissue-specific stem cells (*121, 122*), and thyroid-stimulating hormone (*123, 124*) have been injected into eggs while the embryo is developing to provide for disease control in the hatched bird (*119*) or improved growth rate of the bird (*120–124*). This intriguing administration route to treat growing birds offers the formulation scientist interested in protein controlled-release delivery systems opportunities to develop targetable delivery systems, systems with latent periods before drug release, and even stable delivery systems within one of the most rapidly changing environments known to humankind, the egg!

Acknowledgments

I wish to thank Greg Needham, Mark Heiman, Leo Richardson, and Snehlata Mascarenhas for their comments and suggestions to improve the manuscript.

References

1. Leahy, P. J. *Issues Sci. Technol.* **1988**, 26–29.
2. Asimov, G. J.; Krouze, N. K. *J. Dairy Sci.* **1937**, *20,* 289–306.
3. Elam, T., Elanco Animal Health, A Division of Eli Lilly and Company, personal communication, 1995.
4. *Animal Pharm. Rep. (Westbury, N.Y.)* **1994**, *309,* 10.
5. Brooke, D.; Washkuhn, R. J. *J. Pharm. Sci.* **1977**, *66,* 159–162.
6. Lipper, R. A.; Higuchi, W. I. *J. Pharm. Sci.* **1977**, *66,* 163–164.
7. Hsieh, D.; Rhine, W.; Langer, R. *J. Pharm. Sci.* **1983**, *72,* 17–22.

8. Sanders, L. M.; Domb, A. U.S. Patent 4,959,217, 1990.
9. Younsik, C.; Sittner, R. T.; Pitt, C. G.; Donaldson, E. M.; McLean, E. *Proc. Int. Symp. Controlled Release Bioact. Mater.* **1991,** *18,* 595–596.
10. Casey, D. J.; Rosati, L. U.S. Patent 4,882,168, 1989.
11. Churchill, J. R.; Hutchinson, F. G., inventors; Imperial Chemical Industries PLC, assignee; Eur. Patent EP809482, 1983.
12. Gombotz, W.; Pankey, S.; Braatz, J.; Ranchalis, J.; Puolakkainen, P. *Proc. Int. Symp. Controlled Release Bioact. Mater.* **1992,** *19,* 108–109.
13. Wu, X. S.; Hoffman, A. S.; Yager, P. *Proc. Int. Symp. Controlled Release Bioact. Mater.* **1992,** *19,* 192–193.
14. Tefft, J.; Roskos, K. V.; Heller, J. *Proc. Int. Symp. Controlled Release Bioact. Mater.* **1992,** *19,* 371–372.
15. Downes, S.; Clifford, C. J.; Davy, K.; Braden, M. *Proc. Int. Symp. Controlled Release Bioact. Mater.* **1992,** *19,* 204–205.
16. Hubbell, J. A.; Hill-West, J. L.; Pathak, C. P.; Sawhney, A. S. *Proc. Int. Symp. Controlled Release Bioact. Mater.* **1993,** *20,* 137–138.
17. Sawhney, A. S.; Stading, E.; Roth, L.; Ron, E. S. *Proc. Int. Symp. Controlled Release Bioact. Mater.* **1994,** *21,* 521–522.
18. Fortier, G.; Gayet, J.-C. H. *Proc. Int. Symp. Controlled Release Bioact. Mater.* **1994,** *21,* 640–641.
19. d'Urso, E. M.; Fortier, G. *J. Bioact. Compat. Polym.* **1995,** *9,* 367–387.
20. Kamath, K. R.; McPherson, T.; Park, K. *Proc. Int. Symp. Controlled Release Bioact. Mater.* **1993,** *20,* 111–112.
21. Borchert, J. C. H.; van Soest, M. J.; Hennink, W. E. *Proc. Int. Symp. Controlled Release Bioact. Mater.* **1994,** *21,* 306–307.
22. Nizuka, T.; Nakamae, K.; Hoffman, A. S. *Proc. Int. Symp. Controlled Release Bioact. Mater.* **1994,** *21,* 308–309.
23. Moo-Young, A. J.; Kuzma, P.; Bardin, C. W.; Sundaram, K.; Moro, D.; Quandt, H.; Nash, H. A. *Proc. Int. Symp. Controlled Release Bioact. Mater.* **1994,** *21,* 312–313.
24. Wu, X. S.; Hoffman, A. S.; Yager, P. *Proc. Int. Symp. Controlled Release Bioact. Mater.* **1992,** *19,* 192–193.
25. Janoff, A. S.; Bolcsak, L. E.; Weiner, A. L.; Tremblay, P. A.; Bergamini, M. V. W.; Suddith, R. L. U.S. Patent 4,861,580, 1989.
26. Ferguson, T. H., Animal Science Product Development, Elanco Animal Health, A Division of Eli Lilly and Company, unpublished data.
27. Weiner, A. L.; Estis, L. F.; Janoff, A. S., inventors; The Liposome Company, Inc., assignee; World Office Patent WO8905151, 1989.
28. Feeser, T. M.; Wheatley, M. A. *Proc. Int. Symp. Controlled Release Bioact. Mater.* **1993,** *20,* 32–33.
29. Feeser, T. M.; Wheatley, M. A. *Proc. Int. Symp. Controlled Release Bioact. Mater.* **1994,** *21,* 196–197.
30. Gregoriadis, G.; McCormack, B. *Proc. Int. Symp. Controlled Release Bioact. Mater.* **1994,** *21,* 89–90.
31. Simpkins, J. W. *J. Parenter. Sci. Technol.* **1991,** *45,* 266–269.
32. Charman, S. A.; Mason, K. L.; Charman, W. N. *Pharm. Res.* **1993,** *10,* 954–962.
33. Cady, S. M.; Steber, W. D.; Fishbein, R. *Proc. Int. Symp. Controlled Release Bioact. Mater.* **1989,** *16,* 22–23.
34. Steber, W. D.; Fishbein, R.; Cady, S. M. U.S. Patent 4,837,381, 1989.
35. Steber, W. U.S. Patent 5,213,810, 1993.
36. Downer, J. V.; Patterson, D. L.; Rock, W. W.; Chalupa, W. V.; Cleale, R. M.; Firkens, J. L.; Lynch, G. L.; Clark, J. H.; Brodie, B. O.; Jenny, B. F.; De Gregorio, R. *J. Dairy Sci.* **1993,** *76,* 1125–1136.
37. Domb, A. J.; Maniar, M., inventors; Nova Pharmaceutical Corporation, assignee; World Office Patent WO9214449, 1992.
38. Wyse, J. W.; Takahashi, Y.; DeLuca, P. P. *Proc. Int. Symp. Controlled Release Bioact. Mater.* **1989,** *16,* 334–335.
39. Mitchell, J. W. U.S. Patent 4,985,404, 1991.
40. Mitchell, J. W. U.S. Patent 5,013,713, 1991.
41. Martin, J. L.; Kraemer, J. F Australian Patent 764,400, 1987.
42. Thakkar, A. L.; Harrison, R. G. U.S. Patent 4,775,659, 1988.
43. Ferguson, T. H.; McGuffey, R. K.; Moore, D. L.; Paxton, R. E.; Thompson, W. W.; Wagner, J. F.; Dunwell, D. *Proc. Int. Symp. Controlled Release Bioact. Mater.* **1988,** *15,* 55.
44. Ferguson, T. H.; Harrison, R. G.; Moore, D. L. U.S. Patent 4,977,140, 1990.
45. Bramley, M. R.; Carter, A. B.; Dunwell, D. W., inventors; Lilly Industries Ltd., assignee; Eur. Patent EP314421, 1989.
46. Brooks, N. D.; Needham, G. F. U.S. Patent 5,352,662, 1994.
47. Bauman, D. E.; Hard, D. L.; Crooker, B. A.; Partridge, M. S.; Garrick, K.; Sanders, L. D.; Erb, H. N.; Franson, S. E.; Hartnell, G. F.; Hintz, R. L. *J. Dairy Sci.* **1989,** *72,* 642–651.

48. McGuffey, R. K.; Basson, R. P.; Snyder, D. L.; Block, E.; Harrison, J. H.; Rakes, A. H.; Emery, P. W.; Muller, L. D. *J. Dairy Sci.* **1991**, *74*, 1263–1276.
49. Cunningham, E. P. *Ir. Vet. J.* **1994**, *47*, 207–210.
50. Juskevich, J. C.; Guyer, C. G. *Science (Washington, D.C.)* **1990**, *249*, 875–884.
51. Tyle, P. U.S. Patent 4,857,506, 1989.
52. Tyle, P.; Cady, S. M. *Proc. Int. Symp. Controlled Release Bioact. Mater.* **1990**, *17*, 49–50.
53. Arendt, V. D., inventor; American Cyanamid Company, assignee; Eur. Patent EP281809, 1988.
54. Cady, S. M.; Fishbein, R.; Schroder, U.; Eriksson, H.; Probasco, B. L. U.S. Patent 5,266,333, 1993.
55. Lindsey, T. O.; Clark, M. T. U.S. Patent 5,015,627, 1991.
56. Clark, M. T.; Gyurik, R. J.; Lewis, S. K.; Murray, M. C.; Raymond, M. J. U.S. Patent 5,198,422, 1993.
57. Azain, M. J.; Eigenberg, K. E.; Kasser, T. R.; Sabacky, M. J., inventors; Monsanto Company, assignee; Eur. Patent EP403032, 1990.
58. Castillo, E. J.; Eigenberg, K. E.; Patel, K. R.; Sabacky, M. J., inventors; Monsanto Company, assignee; Eur. Patent EP462959, 1991.
59. Janski, A. M.; Yang, R. D. U.S. Patent 4,786,501, 1988.
60. Sivaramakrishnan, K. N.; Rahn, S. L.; Moore, B. M.; O'Neil, J. *Proc. Int. Symp. Controlled Release Bioact. Mater.* **1989**, *16*, 14–15.
61. DePrince, R. B.; Viswanatha, R. U.S. Patent 4,765,980, 1988.
62. Hamilton, E. J.; Burleigh, B. D., inventors; International Minerals & Chemical Corporation, assignee; Eur. Patent EP303746, 1989.
63. Viswanatha, R.; DePrince, R. B. U.S. Patent 4,917,685, 1990.
64. Shalati, M. D.; Viswanatha, R. U.S. Patent 4,761,289, 1988.
65. Sivaramakrishnan, K. N.; Miller, L. F., inventors; Pitman-Moore, Inc., assignee; World Office Patent WO9011070, 1990.
66. Williams, A. H.; Staples, L. D.; Thiel, W. J.; Oppenheim, R. C.; Clarke, I. J., inventors; The State of Victoria, Monash Medical Centre, and Victorian College of Pharmacy Limited, assignees; World Office Patent WO8706828, 1987.
67. Ron, E.; Turek, T.; Mathiowitz, E.; Chasin, M.; Langer, R. *Proc. Int. Symp. Controlled Release Bioact. Mater.* **1989**, *16*, 338–339.
68. Kent, J. S. U.S. Patent 4,452,775, 1984.
69. Magruder, J. A.; Peery, J. R.; Eckenhoff, J. B., inventors; Alza Corporation, assignee; World Office Patent WO9200728, 1992.
70. Eckenhoff, J. B.; Magruder, J. A.; Cortesse, R.; Peery, J. R.; Wright, J. C. U.S. Patent 4,959,218, 1990.
71. Azain, M. J.; Kasser, T. R.; Sabacky, M. J., inventors; Monsanto Company, assignee; Eur. Patent EP374120, 1990.
72. Wong, P. S. L.; Theeuwes, F.; Eckenhoff, J. B.; Larsen, S. D.; Huynh, H. T. U.S. Patent 5,023,088, 1991.
73. Scanes, C. G.; Ricks, C. A., inventors; Rutgers University and American Cyanamid Company, assignees; Eur. Patent EP353045, 1990.
74. Oppenheim, R. C.; Thiel, W. J.; Staples, L. D.; Williams, A. H.; Clarke, I. J. *Proc. Int. Symp. Controlled Release Bioact. Mater.* **1988**, *15*, 54–55.
75. Thiel, W. J.; Tsui, K. C. *Proc. Int. Symp. Controlled Release Bioact. Mater.* **1991**, *18*, 207–208.
76. Gombotz, W. R.; Grizzle, J. M.; Goodwin, A.; Brown, L. R.; Healy, M. S.; Shaver, K. M.; Zohar, J. *Proc. Int. Symp. Controlled Release Bioact. Mater.* **1989**, *16*, 253–254.
77. Vickery, B. H.; McRae, G. I.; Tallentire, D.; Foreman, J.; Nerenberg, C.; Kushinsky, S.; Sanders, L. M. *Proc. Int. Symp. Controlled Release Bioact. Mater.* **1983**, *10*, 97–101.
78. Sanders, L. M.; McRae, G. I.; Kent, J. S.; Vickery, B. H. *Proc. Int. Symp. Controlled Release Bioact. Mater.* **1983**, *10*, 91–96.
79. Sanders, L. M.; Kent, J. S.; McRae, G. I.; Vickery, B. H.; Tice, T. R.; Lewis, D. H. *J. Pharm. Sci.* **1984**, *73*, 1294–1297.
80. Burns, R.; Sanders, J. *Proc. Int. Symp. Controlled Release Bioact. Mater.* **1988**, *15*, 452–453.
81. Sanders, L.; Burns, R.; Vitale, K.; Hoffman, P. *Proc. Int. Symp. Controlled Release Bioact. Mater.* **1988**, *15*, 62–63.
82. Sanders, L. M.; Kell, B. A.; McRae, G. I.; Whitehead, G. W. *Proc. Int. Symp. Controlled Release Bioact. Mater.* **1985**, *12*, 177–178.
83. Sanders, L. M.; Kell, B. A.; McRae, G. I.; Whitehead, G. W. *J. Pharm. Sci.* **1986**, *75*, 356–360.
84. Okada, H. *Proc. Int. Symp. Controlled Release Bioact. Mater.* **1989**, *16*, 12–13.
85. Burns, R.; McRae, G.; Sanders, L. *Proc. Int. Symp. Controlled Release Bioact. Mater.* **1988**, *15*, 64–65.
86. Lawter, J. R.; Brizzolara, N. S.; Lanzilotti, M. G.; Morton, G. O. *Proc. Int. Symp. Controlled Release Bioact. Mater.* **1987**, *14*, 99–100.

87. Shalaby, S. W.; Jackson, S. A.; Moreau, J. P., inventors; Kinerton Limited, assignee; World Office Patent WO9415587, 1994.
88. Heller, J.; Sanders, L. M.; Mishky, P.; Ng, S. Y. *Proc. Int. Symp. Controlled Release Bioact. Mater.* **1986,** *13,* 69–70.
89. Heller, J.; Ng, S. Y.; Penhale, D. W. H.; Fritzinger, B. K.; Sanders, L. M.; Burns, R. M.; Bhosale, S. S. *J. Controlled Release* **1987,** *6,* 217–224.
90. Fujioka, K.; Sato, S.; Tamura, N.; Takada, Y., inventors; Sumitomo Pharmaceuticals Company, assignee; Eur. Patent EP219076, 1987.
91. Fujioka, K.; Sato, S.; Takada, Y. Australian Patent 8,655,983, 1986.
92. Fujioka, K.; Sato, S.; Tamura, N.; Takada, Y.; Sasaki, Y.; Maeda, M., inventors; Sumitomo Pharmaceuticals Company, assignee; Eur. Patent EP326151, 1989.
93. Leonard, R. J.; Harman, S. M. U.S. Patent 5,039,660, 1991.
94. Mariette, B.; Coudane, J.; Vert, M.; Gautier, J.-C.; Moneton, P. H. *J. Controlled Release* **1993,** *24,* 237–246.
95. Needham, G. F.; Pekar, A. H.; Havel, H. A. *J. Pharm. Sci.* **1995,** *84,* 437–442.
96. Cady, S. M.; Fishbein, R.; SanFilippo, M. *Proc. Int. Symp. Controlled Release Bioact. Mater.* **1988,** *15,* 56–57.
97. Demoustier, M. M.; Beckers, J.; Taper, H.; Vert, M.; Gillard, J. *Proc. Int. Symp. Controlled Release Bioact. Mater.* **1988,** *15,* 350–351.
98. Heinrich, N.; Fechner, K.; Berger, H.; Lorenz, D.; Albecht, E.; Rafler, G.; Schafer, H.; Mehlis, B. *J. Pharm. Pharmacol.* **1991,** *43,* 762–765.
99. Pitt, C. G. *Int. J. Pharm.* **1990,** *59,* 173–196.
100. Chen, T. *Drug Dev. Ind. Pharm.* **1992,** *18,* 1311–1354.
101. Cady, S. M.; Langer, R. *J. Agric. Food Chem.* **1992,** *40,* 332–336.
102. *Formulation and Delivery of Proteins and Peptides*; Cleland, J. L.; Langer, R., Eds.; ACS Symposium Series 567; American Chemical Society: Washington, DC, 1994.
103. Hageman, M. J. *Proc. Int. Symp. Controlled Release Bioact. Mater.* **1992,** *19,* 76–77.
104. Hageman, M. J.; Bauer, J. M.; Possert, P. L.; Darrington, R. T. *J. Agric. Food Chem.* **1992,** *40,* 348–355.
105. Fox, K. C. *Science (Washington, D.C.)* **1995,** *267,* 1922–1923.
106. Mumenthaler, M.; Hsu, C. C.; Pearlman, R. *Pharm. Res.* **1994,** *11,* 12–20.
107. Hageman, M. J. *Drug Dev. Ind. Pharm.* **1988,** *14,* 2047–2070.
108. Buckwalter, B. L.; Cady, S. M.; Shieh, H.-M.; Chaudhuri, A. K.; Johnson, D. F. *J. Agric. Food Chem.* **1992,** *40,* 356–362.
109. Chaleff, D. T., inventor; American Cyanamid Company, assignee; Eur. Patent EP488279, 1992.
110. Wicar, S.; Mulkerrin, M. G.; Bathory, G.; Khundkar, L. H.; Karger, B. L. *Anal. Chem.* **1994,** *66,* 3908–3915.
111. Ptitsyn, O. B.; Uversky, V. N. *FEBS Lett.* **1994,** *341,* 15–18.
112. Bychkova, V. E.; Pain, R. H.; Ptitsyn, O. B. *FEBS Lett.* **1988,** *238,* 231–234.
113. Holladay, L. A.; Hammonds, R. G.; Puett, D. *Biochemistry* **1974,** *13,* 1653–1661.
114. Brems, D. N.; Plaisted, S. M.; Havel, H. A.; Kauffman, E. W.; Stodola, J. D.; Eaton, L. C.; White, R. D. *Biochemistry* **1985,** *24,* 7662–7668.
115. Havel, H. A.; Kauffman, E. W.; Plaisted, S. M.; Brems, D. N. *Biochemistry* **1986,** *25,* 6533–6538.
116. Brems, D. N.; Plaisted, S. M.; Kauffman, E. W.; Havel, H. A. *Biochemistry* **1986,** *25,* 6539–6543.
117. Brems, D. N.; Havel, H. A. *Proteins Struct. Funct. Genet.* **1989,** *5,* 93–95.
118. Siegel, R. A.; Pitt, C. G. *J. Controlled Release* **1995,** *33,* 173–188.
119. Gore, A. B.; O'Connell, T. O.; Phelps, P. V.; Tyczkowski, J. K. U.S. Patent 5,339,766, 1994.
120. Ballard, F. J.; Francis, G. L.; McMurtry, J. P.; Phelps, P. V.; Walton, P. E., inventors; Embrex Inc., Gropep Proprietary Ltd., U.S. Department of Agriculture, assignees; World Office Patent WO9406445, 1994.
121. Petitte, J. M.; Ricks, C. A.; Spence, S. E., inventors; Embrex Inc., North Carolina State University, assignees; World Office Patent WO9315185, 1993.
122. Petitte, J. M.; Ricks, C. A., inventors; Embrex Inc., North Carolina State University, assignees; World Office Patent WO9314629, 1993.
123. Phelps, P. V.; Gilderslee, R. P., inventors; Embrex Inc., assignee; World Office Patent WO9112016, 1991.
124. Christense, V. L., inventor; North Carolina State University, Embrex Inc., assignees; Eur. Patent EP167381, 1986.

Tissue Engineering and Gene Therapy

15
Protein Delivery by Microencapsulated Cells

*Julia E. Babensee and Michael V. Sefton**

Microencapsulation is a means of isolating cells from the immune system, thereby enabling the transplantation of mammalian cells without immunosuppression and the use of xenogeneic or genetically engineered cells. The cells are transplanted to correct a disease state by the delivery of a cell product, typically a protein. Insulin from pancreatic islets or a neurotrophic factor from genetically engineered fibroblasts are examples of this mode of therapy. Unlike conventional drug delivery devices, the cells have an inexhaustible supply of the protein (pending cell viability) in an intrinsically stable form and without the problem and expense of protein purification. Furthermore, the protein is delivered at a rate determined by the normal physiology of the cells, which might involve regulation by glucose level (islets) or potassium concentration (dopamine-secreting cells) or cytokine levels (antitrypsin). Successful microencapsulation and cell transplantation requires high cell number and viability; control of cell function (through the extracellular matrix, for example); and maintenance of function for extended duration. These properties of encapsulated cells are illustrated with the use of hydroxyethyl methacrylate–methyl methacrylate copolymer.

Microencapsulation of Cells

Microencapsulation is the process of placing a protective polymer membrane around a cellular core. Microcapsules are prepared by different methods based on the underlying mechanism of membrane formation and can be 300–1500 μm in diameter. Membrane formation by interfacial adsorption, interfacial polymerization, polyelectrolyte complexation, and simple coacervation or precipitation has been used to prepare cell-containing capsules.

Microencapsulation of mammalian cells within a synthetic polymer membrane is proposed as a means of using these cells as a source of therapeutic biomolecules, the absence or abnormality of which cause a disease state. Alternatively, they may replace diseased organ function. The polymer wall prevents contact between the encapsulated cells and the host immune system (Figure

*Corresponding author.

Copyright © 1997 American Chemical Society

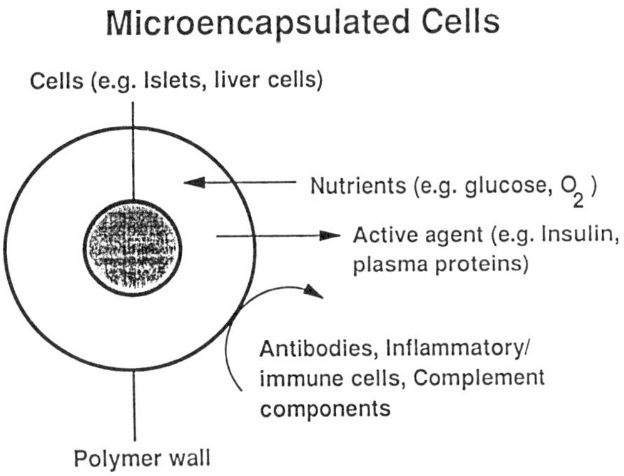

FIGURE 15.1. The microcapsule and immunoisolation concept.

15.1) allowing for the transplantation of allogeneic and even xenogeneic cells or cell lines without immunosuppressive therapy. The permeability of the polymer membrane allows the diffusion of nutrients into the capsule to sustain cells and the diffusion of the secreted therapeutic product out of the capsule. The need for immunosuppression therapy and the limited availability of suitable donor tissue are the primary limitations of cell transplantation. Immunoisolation is an alternative to immune suppression (*1–4*) or induction of tolerance (*5–7*) of the host or immunomodulation of the graft (*8, 9*).

Microcapsules are one of a number of cell-based delivery systems that may be used for protein delivery but are usually used for cell transplantation. A variety of cell types have been encapsulated to demonstrate the success of the various techniques in maintaining encapsulated cell viability, product secretion, and applicability to a variety of diseases. Alginate-poly-L-lysine microcapsules have been prepared containing the following:

- islets for diabetes treatment (reviewed in refs. 10 and 11),
- PC12 cells (*12*) and bovine chromaffin cells (*13*) for Parkinson's disease,
- hepatocytes for liver disease (*14–17*), and
- parathyroid tissue for hypoparathyroidism (*18, 19*).

Microencapsulation of genetically modified cells has added other potential applications including the following:

- Factor IX secretion for the treatment of hemophilia B (*20*),
- growth hormone secretion for the treatment of pituitary dwarfism (*21–23*),
- replacement of adenosine deaminase activity (*24*), and
- cytokine delivery for modulation of specific immune responses (*25*).

For a review of the cellular encapsulation in alginate poly-L-lysine microcapsules and other materials such as agarose, *see* reference 26.

Macrocapsules prepared from poly(acrylonitrile-*co*-vinyl chloride) (PAN-*co*-VC) copolymer have been used for the transplantation of islets (*27–29*), PC12 cells (*30–32*), rat fibroblasts genetically modified to secrete nerve growth factor (NGF) (*33*), baby hamster kidney (BHK) cells genetically modified to secrete human NGF for the treatment of Alzheimer's disease (*34–36*) and Huntington's disease (*37*), BHK cells genetically modified to secrete glial cell line-derived neurotrophic factor (*38, 39*), and bovine adrenal chromaffin cells for the treatment of chronic pain (*40, 41*). Planar chambers comprised of laminated membranes with an inner cell-impermeable poly(tetrafluoroethylene) (PTFE, Biopore) membrane (pore size, 0.45 μm) for immunoisolation and an outer cell-permeable expanded PTFE (Gore) membrane (pore size, typically 5 μm) of a microarchitecture that induces close vascular structures (*42*) have been used for the transplantation of various cells (*43, 44*) with a view toward gene therapy applications. Intravascular devices require anastomoses to the blood supply and have been used for the transplantation of islets with immunoisolation provided by the membrane (*45, 46*). However, device failure due to loss of islet viability and thrombosis frequently occurred.

Microencapsulated cells have also been used in vitro in biotechnology applications such as antibody production by encapsulated hybridomas (*47*). The polymer membrane protects the encapsulated cells from high shear stresses created as the bioreactor contents are agitated. The secreted proteins are easily recovered from the bioreactor medium and separated from the cells by removing the microcapsules. As another example, alginate poly-L-lysine microencapsulation of human bone marrow combined with rapid medium exchange has been used as a method for clonal expansion of single hematopoietic progenitor cells in vitro (*48*).

Microcapsules are comprised of the polymer membrane enclosing mammalian cells, which may be embedded in an immobilization matrix or a cell-attachment matrix. The function of each of these components and common examples of each microcapsule component are listed in Table 15.1.

Microencapsulated cells function as a renewable source of secreted therapeutic product in response to physiological stimuli as long as the cells remain viable and differentiated. By using cells in the protein delivery device, we are able to take advantage of the physiological control mechanism by which these cells function. Glucose stimulation of insulin secretion by islets of Langerhans is perhaps the best known mechanism (*49*). However, there are many other features of cells that at least in principle could be exploited in cell-based delivery systems. Cell-based biomolecule delivery can be modulated by factors that influence the phenotypical gene expression of the cells such as secreted paracrine (*50*) and autocrine growth factors or cytokines (*51*), positive/negative feedback loops of these factors, hormones, neurotransmit-

TABLE 15.1. Microcapsule Components

Microcapsule Component	*Function*	*Examples*
Polymer membrane	Enhance permeability Enhance permselectivity Enhance immunoisolation Sequester cells Prevent encapsulated cell contact with host cells Prevent tumor formation Enhance encapsulated cell compatibility Enhance biocompatibility Enhance mechanical structure	HEMA–MMA polyacrylate Alginate–poly-L-lysine Agarose
Living mammalian cells	Act as a renewable source of secreted therapeutic product in response to physiological stimuli as long as the cells remain viable and differentiated	Primary cells (e.g., islets and hepatocytes) Cell lines (e.g., PC12 and HepG2) Genetically engineered cells
Immobilization matrix	Distribute cells within capsule core Prevent clumping of cells and central necrosis due to diffusion limitations	Alginate Agarose Chitosan Synthetic polymers (e.g. thermoresponsive hydrogels)
Cell-attachment matrix	Provide sites for cell attachment (differentiation, physiological environment) for anchorage-dependent cells Function as an immobilization matrix	Collagen Matrigel Laminin

ters, extracellular matrix (ECM) (*52*), and associated cytokine effects (*53, 54*). One of the main difficulties to be overcome in using living cells in microcapsule applications is their characteristic sensitivity to their external environment, which restricts the insults that the cells will endure and still remain viable and functional. Therefore, the encapsulation process must be carefully designed because the cells cannot be exposed to pH, osmotic, thermal, shear, or solvent stresses.

Cell-based protein delivery systems differ from conventional drug delivery systems in that the amount of protein available is not restricted by the finite amount preloaded into the device or delivered by a pump. The cells provide a continuous source of protein. Conventional drug delivery systems, in general, are not responsive to physiological stimuli as are cells, although there are a few exceptions (*55*). Even though most conventional drug delivery systems secrete a single protein, cell-based protein delivery devices may supply more than one protein depending on the cell type. For example, microencapsulated hepatocytes could secrete a variety of plasma proteins that may be deficient due to liver disease. It has been suggested that Parkinson's disease could effectively be treated by microcapsules containing two cell types, one genetically modified to secrete a neurotrophic factor and one producing dopamine to treat the neurodegenerative component of the disease at the same time as supplying the depleted neurotransmitter (*39*). Fabrication of the protein delivery system in which cells are the source of the therapeutic protein, in some sense, is simplified over conventional protein delivery systems. Because the protein is produced by viable cells from within the capsules, protein degradation due to heat, enzymes, solvents, pH, and water during device fabrication is not likely an issue (*56*). Furthermore, the protein does not need to be prepared [e.g., by using recombinant technology (*57*)], and protein separation, purification, concentration, and sterilization steps are not required (*58, 59*).

Microencapsulated cells may be implanted into a location that is dependent on the protein and its route of delivery. Microencapsulated cells implanted into the orthotopic (physiological) location would deliver the biomolecule at the site of its action. Microencapsulated cells containing cells secreting neurotrophic factor or neurotransmitter implanted into the brain would circumvent biomolecule passage across the blood–brain barrier, and the secreted factor or transmitter activity would be regulated by other neurotrophic factors, cofactors, or enzymes at the site. Lack of systemic administration of such molecules obviates their deleterious action on peripheral receptors of the nervous system. In other cases, microcapsules need not be placed in the orthotopic site, but there may be an indication for protein delivery via the physiological route (*60*). Insulin delivered to the peritoneal cavity would be absorbed by the liver (*61*), and insulin absorbed by omental tissue vasculature would be transported to the liver (*62*), both of which deliver insulin to the systemic circulation via its physiological route. If the route of protein administration is not important, placement of capsules in the subcutaneous site may be preferred because of the minimal invasive survey (*63*).

Even though the typical site for microcapsule implantation is intraperitoneal, as discussed previously, other sites may better satisfy issues of physiological product delivery route, endogenous vascularizing capabilities [e.g., omentum (*64*)] or ease of implantation surgery (e.g., subcutaneous). Tissue implantation of capsules avoids chronic, direct contact with flowing blood and circumvents problems of coagulation and vascular anastomoses. Capsules of small enough size may be injected, thus eliminating the need for invasive surgery. The issues of retrievability, the number of cells required, and the capsule size will influence the choice of implant site [e.g., kidney capsule, liver (*65*), omental pouch (*66*), or epididymal fat pad (*67*)] and the mode of access. Microcapsules may also be seeded in a prevascularized implantation bed generated in response to a porous polymer scaffold (*68, 69*) or a vascularized planar device (*70, 71*). Problems in the retrievability of free floating capsules implanted intraperitoneally would suggest the need to localize the capsules in a receptacle such as a "tea bag" (*72*). The retrievability and localization of PAN-*co*-VC macrocapsules were enhanced by attaching a tether at the end of the device to remove the implanted macrocapsule after localization with the aid of a radio-opaque marker (*41*).

Microcapsules have a high surface-to-(total) volume ratio and an optimal geometry for diffusion and are thus expected to result in implants of lower total volume than is possible with hollow fibers with faster response times to physiological stimuli. The implant volume is strongly dependent on capsule size, on the number of capsules that need to be implanted, and on how many cells are contained in each capsule. This calculation is particularly important for pancreatic islets because more than 3×10^5 islets are thought to be necessary for human application (*73*). With one islet per capsule and a low 300,000 islets per transplant, the transplant volume would need to be an unmanageable 1150 mL for 900-μm capsules, a reasonable 30 mL with 400-μm capsule, and only 2.3 mL with 170-μm "capsules". Loss of islet efficacy or viability due to transport limitations or cell damage increases the implant volume needed. This effect may be compensated by packing more than 1 islet per capsule. The geometric limitations of hollow fiber or planar geometries are worse and result in even higher implant volumes. Recent modifications to microencapsulation processes have added greater flexibility in the size of microcapsules by making possible the preparation of small-diameter (*74, 75*) or conformally coated cellular aggregates (*76–78*).

An intact polymer membrane sequesters the transplanted cells to the capsule core, and in the case of potentially tumorigenic cell lines such as PC12 cells, it acts as a physical barrier limiting cell growth and preventing tumor formation. A mechanically strong polymer membrane is essential for the maintenance of capsule integrity, and the tough thermoplastic hydroxyethyl methacrylate–methyl methacrylate (HEMA–MMA) microcapsules may be preferred in this context to the mechanically weaker alginate–poly-L-lysine capsules. Microcapsules are tougher and mechanically more stable in vivo, without failures due to bending and breakage as observed with tubular structures (*29*). With certain cell lines and host combinations, tumor formation would not occur because the host's immune system would recognize the leaked transplanted cells as being foreign (e.g., because of the presence of tumor-specific antigens) and resorb them. Microencapsulation may protect the transplanted cells from the original disease pathology, which in the case of diabetes may be affected by autoantibodies against islet antigens (*79, 80*) or insulin (*81*).

The clinical applicability of microencapsulation of living cells for the treatment of disease or drug delivery brings up other issues in addition to those already discussed. Scale-up of the microencapsulation techniques is needed to produce the large number of capsules required of reproducible quality. The isolation of enough cells from a suitable source (e.g., cadaver islets or porcine islets) on a large scale with a high degree of purity and sterility remains an issue that is under development.

Intracapsule Cell Behavior

Encapsulated cells are in a nonphysiological environment in which the polymer wall acts as a diffusion barrier to nutrient transport into and secreted biomolecule and cellular waste transport out of the capsule. The intracapsular environment may not support the attachment of anchorage-dependent cells, limiting their proliferation. Furthermore, in the absence of a co-encapsulation immobilization matrix or attachment matrix, cells attach to each other or to cell-secreted ECM components such as laminin, fibronectin, and collagen (*82, 83*) to arrange into a spheroidal shape (*83*). Cellular arrangement into aggregates and changes in cell morphology are illustrated by anchorage-dependent cells such as Chinese hamster ovary cells (*84*) and HepG2 cells (*85*) and anchorage-independent PC12 (*86*) cells placed into the nonadherent environment of the HEMA–MMA capsule. This result is consistent with several reports on unencapsulated cell (e.g., hepatocyte) culture under nonadherent conditions (*87*).

Diffusion limitations associated with the capsule because of its wall and related to its size affect the transport of secreted biomolecules and metabolites (nutrients and waste products) with corresponding changes in the intracapsular environment (e.g., oxygen, nutrient, or metabolite concentrations and presence of products from dying or dead cells) that influence cell behavior, morphology, or three-dimensional arrangement and protein secretion. Minimization of the intracapsular and intercellular diffusion limitations is critical to the success of microencapsulation of living cells. For example, depletion of a particular nutrient (e.g., oxygen) may cause encapsulated cells to grow at a lower, diffusion-limited rate, or the center of a cell cluster may become necrotic (*85*), similar to what occurs with tumour spheroids (*88*). Because the cells in spheroids are at a distance from the surrounding medium, gradients of critical nutrients and growth factors and metabolites (and also pH) are set up (*89*). As a spheroid grows, the number of proliferating cells decreases, the proportion of quiescent cells increases, and eventually due to nutrient deprivation and waste-product accumulation, necrosis develops at the center of spheroids (*89*). The diffusion of larger molecular weight species (such as growth factors, hormones, and cytokines) into spheroid regions is affected by their receptor-mediated uptake by cells in the outer few layers of an aggregate such that inner spheroid regions receive less of these signals for proliferation and gene expression (*90*). Cellular association in a spheroidal arrangement increases cell–cell contact and contact with cell-derived ECM molecules, which maintain the aggregate arrangement and provide signals for control of phenotypical gene expression. Furthermore, protein secretion by encapsulated cells may be altered because of limitations in nutrient delivery to those cells located at a distance from the nutrient source [e.g., insulin secretion by islets is sensitive to oxygen concentration (*91*)]. Another consequence of the intracapsule microenvironment and cell morphology is the modulation of the sensitivity of encapsulated cells to exogenous influences such as cytokines (*92*) from that of the monolayer culture (*93–95*).

The microenvironment of spheroidal aggregates appears to maintain the in vivo differentiated cell characteristics for longer times than in monolayer culture (*83, 96*), presumably because of the three-dimensional arrangement, the corresponding diffusion gradients, and ECM–cell and high cell–cell contact. With the capsule wall controlling the aggregate size and strongly influencing the diffusion gradients, such differentiated cells with prolonged life spans would presumably be ideally suited for microencapsulated cell applications. Microcapsules are also characterized by a locally high cell density, and this too is expected to influence intracapsule cell behavior (*97, 98*).

The nonadherent intracapsular environment has been altered by the co-encapsulation of extracellular components such as a cell-attachment substrate, Matrigel (Collaborative Research, Bedford, MA) (*85, 99–102*), or the co-encapsulation of an immobilization matrix, such as agarose or chitosan (*100, 103–105*). Whereas Matrigel provides a substrate for cell attachment (which the capsule membrane does not), agarose or chitosan may simply improve the distribution of cells within the capsule and minimize intercellular diffusion limitations. The benefits of Matrigel also include the more uniform distribution of individual cells and aggregates in which the cells attached not only to each other but also to ECM components such that cell–cell association was limited. A uniform distribution of individual cells and aggregates affects the local cell density and hence consumption of nutrients, production of wastes, and cellular metabolic state, reducing cellular necrosis in central regions of aggregates.

The ECM, through its binding to cell-surface integrins, has a strong influence on cell behavior. Accordingly, adding ECM components to cells before encapsulation or otherwise altering the substrate on or within which the cells are grown is a means of controlling the phenotypical behavior of encapsulated cells to influence protein synthesis. For example, hepatocytes grown on Matrigel maintain protein mRNA synthesis, whereas cells grown on plastic or in collagen type I gel do not (*52*). Liver-specific cytochrome enzyme P-450 activities were also induced by culturing hepatocytes on Matrigel but not on collagen type I (*106*).

The ECM is a dynamic environment constantly undergoing remodeling (e.g., through the indirect effect of cytokines and growth factors) (*107*). The capsule wall will likely affect the extent of remodeling by impeding the inward transport of interstitial proteases that otherwise cause ECM turnover and consequently stabilizing the intracapsular ECM. On the other hand, necrotic cells may release significant amounts of lysosomal enzymes that can degrade biomolecules. The capsule wall will impair their outward diffusion and could cause a buildup of degrading enzymes or of the acidic lysosomal environment.

Hydroxyethyl Methacrylate–Methyl Methacrylate

Polyacrylates

The suitability of a polymer for microencapsulation of cells is determined by its processibility (e.g., its viscosity and solubility especially in solvents that are tolerated by the cells), its permselectivity in the form of the microcapsule wall, and its biocompatibility. The capsule wall must have a high permeability to nutrients and cell-derived biomolecules, yet it must exclude antibodies, complement components, and inflammatory and immune cells. Because the surface chemistry has a large role in defining the capsule biocompatibility, it too is significant.

The polyacrylate on which our current research is based is a thermoplastic HEMA–MMA (~75 mol% HEMA) copolymer prepared by solution polymerization after careful monomer purification to reduce the cross-linker content (*108*). This copolymer is hydrophilic and has a ~25–30% (w/w) water uptake (*109*) consistent with the poly(HEMA) content, but it has mechanical strength, toughness, and elasticity imparted by the poly(MMA) component. These properties lead to adequate permeability of the polymer capsules to aqueous solutes for cellular sustenance (*110*) and sufficient mechanical durability to tolerate normal handling and stresses in vivo. The critical requirement of microcapsule biocompatibility was considered likely due to the common use of the homopolymers poly(MMA) (*111*) and poly(HEMA) (*112*) in biomedical applications (e.g., bone cement and intraocular and contact lenses, respectively). The water insolubility of the HEMA–MMA polymer provides stability in the aqueous physiological environment but necessitates the use of an organic solvent to prepare the polymer solution. The ultimate success of this material depended on the ability to select a tolerable solvent and to design a gentle encapsulation process.

Microencapsulation Processes

We prepared microcapsules by using three different water-insoluble polyacrylates: commercially available polyacrylate (EUDRAGIT RL) (*113–115*), copolymers of dimethylaminoethyl methacrylate and methyl methacrylate (DMAEMA–MMA) (*116*), and most often, HEMA–MMA. Common to all these polymers was the use of coaxial extrusion of cell suspension and polymer solution, shearing of the capsule droplet, and polymer-wall formation by interfacial precipitation on nonsolvent contact. However, each polymer has different properties such as its solubility in solvents that would be suitable for microencapsulation with cells [(diethyl phthalate or polyethylene glycol (PEG)], the solution viscosity, and the precipitation characteristics in an appropriate nonsolvent. Therefore, their application to microencapsulation required appropriately selected polymer solvents and nonsolvents and droplet shearing conditions.

The early microencapsulation studies involving commercially available EUDRAGIT RL or DMAEMA–MMA (reviewed in ref. 117) indicated that encapsulation of living cells within water-insoluble polyacrylate polymers was feasible but that improvement of polymer properties was necessary in the areas of biocompatibility and permeability. The expected properties of HEMA–MMA, as supported by the use of the parent monomers (HEMA and MMA) in biomedical applications, suggested its suitability for microencapsulation purposes. Modifications to the coaxial extrusion, interfacial precipitation process used for EUDRAGIT RL were required to accommodate HEMA–MMA (*110, 118*). Microcapsule preparation using HEMA–MMA has been described in more detail elsewhere (*26*).

Large-diameter microcapsules were prepared by using a coaxial extrusion submerged jet, interfacial precipitation process (*110*). Microcapsules were produced as capsule droplets consisting of the cellular core surrounded by the polymer solution by pumping the HEMA–MMA/PEG-200 polymer solution (10%, w/v) and the mammalian cell suspension in their complete tissue culture medium to the tip of a coaxial needle assembly. The polymer solution flowed through the outer needle, and the cell suspension flowed through the inner needle. The cell suspension was augmented with the viscosity–density enhancer 20% (w/v) Ficoll-400 (*119*). Each capsule droplet was sheared from the needle assembly as its tip was withdrawn from the hexadecane overlayer. The capsule then passed through the hexadecane overlayer into the phosphate-buffered saline (PBS) precipitation bath, which contained 100 ppm of the Pluronic surfactant L101. Pluronic surfactant L101 was added to facilitate the passage of the droplet through the hexadecane–PBS interface. In this precipitation bath, the polymer solvent was extracted leaving behind a polymer wall surrounding the cellular core. Microcapsules produced by this process were spherical, opaque, and uniform in their diameters of 750–900 µm.

A more recent development has been the submerged nozzle–liquid jet extrusion process (Figure 15.2) to produce HEMA–MMA microcapsules of even smaller diameter (300–600 µm) (*120*). In this process, the polymer solution and cell suspension are delivered to the tip of a coaxial needle assembly as in the process described above. However, the needle assembly remains stationary while the hexadecane (or dodecane) is recirculated by a peristaltic pump and flows uniformly, coaxially to the needle assembly, to shear off each capsule

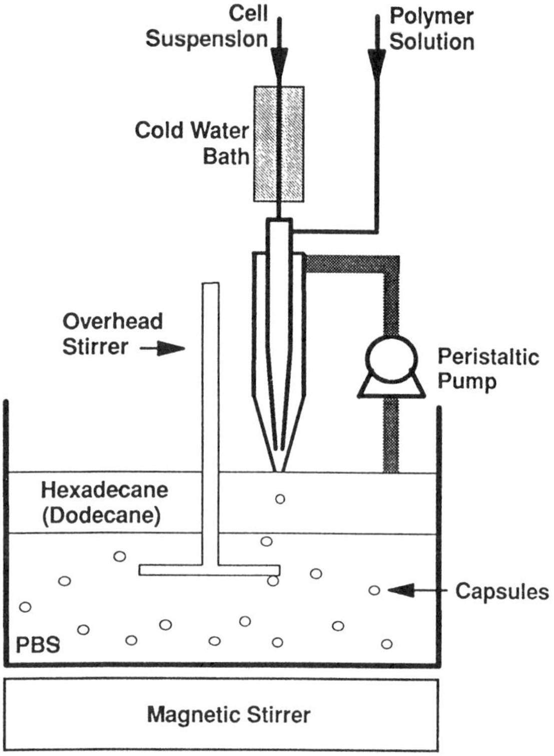

FIGURE 15.2. Schematic drawing of the encapsulation apparatus. (Reproduced with permission from reference 120. Copyright 1994 Wiley.)

droplet. The capsules are kept in suspension by a magnetic overhead stirrer. This inherently higher capsule shearing force produces the smaller capsules.

Following the trend toward diminishing even further the stagnant water layer by reducing the microcapsule diameter and at the same time, the membrane thickness, a conformal coating process that results in cellular aggregates (e.g., HepG2 cells of a diameter of <500 µm) or islets of Langerhans that are coated with a very thin layer of HEMA–MMA (<10 µm) is currently under development (*76*). These ultrasmall "capsules" are expected to have virtually no low-molecular-weight biomolecule diffusion limitations.

A consequence of the biocompatibility of HEMA–MMA is its failure to support the attachment, spreading, and growth of anchorage-dependent cells. The nonadherent HEMA–MMA intracapsule environment was modified by co-encapsulating cell attachment and growth substrates such as the commercially available Matrigel (*85, 99*) or hydrated cell-preloaded Cytodex beads (*121*) for anchorage-dependent cells or cell-immobilization matrices such as agarose or chitosan (*100*). Whereas agarose and chitosan would not present sites for cellular attachment, their presence would distribute the cells within the capsule core preventing their aggregation in this nonadherent environment and the consequent necrosis due to nutrient diffusion limitations. The cell attachment and growth substrates also provide these functions. Modifications to the microencapsulation process described previously were made to accommodate these additives in the capsule core solution as noted in corresponding references.

HEMA–MMA Microcapsules

Membrane Structure and Formation

HEMA–MMA polymer capsules prepared by either submerged jet process resulting in large- or small-diameter microcapsules were morphologically similar. The polymer walls were of an asymmetric morphology similar to ultrafiltration membranes (*122, 123*), but in a spherical geometry instead of the typical hollow fiber or flat-sheet geometry of ultrafiltration membranes. The capsule walls were ~150- and ~50-μm thick for large-diameter (Figure 15.3A) or small-diameter (Figure 15.3B) capsules, respectively. The wall consisted of a thin outer skin; a macroporous sublayer; a thick, seemingly dense layer; and an inner skin (Figure 15.3C). Capsules were frequently eccentric, although this characteristic was diminished by the addition of the viscosity and density enhancer, Ficoll-400, to the capsule core solution (*119*). The inclusion of this additive in the capsule core presumably resulted in a reduction in the mixing of the core and polymer solutions, and the denser capsule core was more completely surrounded by polymer.

Each region of the HEMA–MMA capsule wall has a particular function. The skin layers are expected to provide the selectivity to molecule permeation. The inner skin as well as an outer skin may provide a permselective barrier even if mechanical or chemical change occurs at the outer skin on implantation. The macroporous region is effective for the permeation of nutrients, cellular waste products, and cell-derived biomolecules through the capsule polymer wall, whereas this region and the "dense" layer provide mechanical support. Capsules with well-centered cores are expected to be mechanically more stable because weak spots can be avoided. These capsules would withstand any internal pressure exerted by proliferating cells or by a swollen immobilization matrix

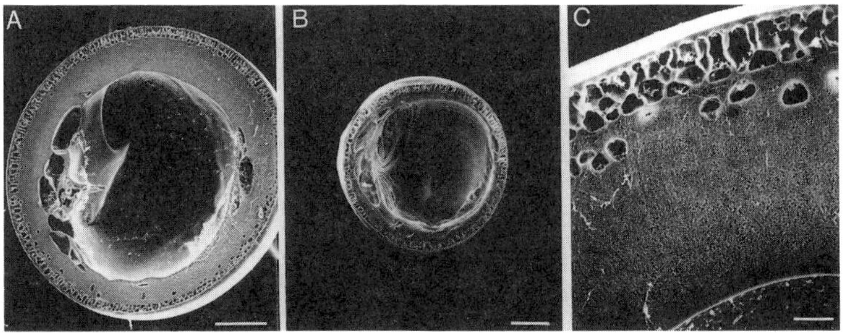

FIGURE 15.3. Scanning electron micrographs of HEMA–MMA microcapsules: A, large-diameter capsule (o.d., 660 μm); B, small-diameter capsule (o.d., 450 μm); C, wall structure of a large capsule. Magnification bars: A, 100 μm; B, 100 μm; C, 25 μm.

and would protect cells more completely, after their implantation, from cellular infiltration as well as normal handling.

Permeability

To provide an environment that is favorable for the maintenance of cellular viability and function, the capsules must be permeable to small molecules (nutrients and metabolites such as glucose, oxygen, and lactate) and intermediate or large molecules (growth factors, ~13–30 kDa; transferrin, 80 kDa). Yet following implantation, the capsules must be impermeable to components of the host immune system (IgG antibodies, 150 kDa; complement components such as C1q, 410 kDa) such that the encapsulated cells are not attacked by the host immune system on a molecular level. Cytokines such as IL-1β, released by inflammatory and immune cells in the tissue reaction to polymer microcapsules containing cells, may be present in sufficiently high concentrations intracapsularly to synergistically affect cell viability and differentiated function, possibly resulting in graft failure (*124*). For the microcapsule polymer membrane to lessen the concentration of these cytokines within the capsule, a molecular-weight cutoff of ~20 kDa would be necessary, but this cutoff would likely compromise nutrient delivery for cellular sustenance. Furthermore, for a functioning cell-based biomolecule delivery system, the polymer wall must be permeable to the therapeutic product such as dopamine (153 Da), insulin (6000 Da, if a monomer), or growth hormone (23 kDa). We developed techniques to determine the molecular-weight cutoff and the permeability of HEMA–MMA capsules and to modify these membrane characteristics.

The permeability of HEMA–MMA capsules was estimated by determining the time-dependent release into an extracapsular PBS sink of one of several molecules of different molecular weights [^{3}H-glucose (180 Da), ^{3}H-inulin (5.2 kDa), albumin (69 kDa), or alcohol dehydrogenase (150 kDa)] from an aliquot of 200–300 one-week-old capsules following their 3-day incubation in the respective PBS solution (*110*). A decrease occurred in the mass transfer coefficient with increasing molecular weight and a significant drop between 69 kDa and 150 kDa (log–log plot). This result suggested a molecular-weight cutoff for the membrane on the order of 100 kDa.

Capsule permeability can be altered by changing the normal polymer precipitation conditions to modify the degree of macroporosity of the polymer wall. For example, the replacement of the normal PBS/L101 precipitation bath with 0.3M aqueous glycerol resulted in membranes of unchanged wall thicknesses but that were more macroporous and had a corresponding doubling of the mass transfer coefficient (*117*). Capsule permeability increased with capsule diameter for small-diameter capsules prepared by using the submerged nozzle–liquid jet extrusion process, presumably reflecting a difference in rate of precipitation associated with the respective differences in diameter, wall thickness, and eccentricity (*125*).

More recent capsule permeability studies have been directed at examining the permeability of capsules at the single-capsule level because of the observation of significant fibrinogen release from a subset of presumably defective capsules containing HepG2 cells (*see* subsequent discussion and ref. 126). The broad distribution in fibrinogen release was consistent with the distribution in the permeability of similar capsules (without cells) to the model protein, horseradish peroxidase (HRP; M_w, 40 kDa). The enzyme was loaded by diffusion from an external incubation solution, and its release from a single capsule was detected by using a sensitive method based on 3,3′,5,5′-tetramethylbenzidine as a chromogenic substrate. Therefore, the heterogeneity of capsules in the delivery of cell-derived proteins is probably partly due to variability in the permeability of the capsules to proteins of similar molecular weight. The variation in HRP permeability was dependent on the care taken in setting up the encapsulation assembly. This variability could be reduced by careful processing. However, re-engineering of the co-extrusion nozzle has begun to make the process more reproducible on a regular basis.

Control over microcapsule permeability was extended by replacing the polymer solvent, PEG-

200, with triethylene glycol (TEG; M_w, 150 Da), which has a higher diffusivity to enhance the rate of solvent removal on precipitation. Microcapsules prepared from a 9% (w/v) HEMA–MMA solution in TEG had an increased permeability (*127*). An increase in capsule permeability was also noted on the addition of a pore-forming additive poly(vinyl pyrrolidone) (PVP; M_w, 10,000) to the 10% HEMA–MMA in TEG solution at a ratio of 0.3 parts PVP to 1 part HEMA–MMA on a weight basis. This improvement in permeability was attributed to a significant increase in the apparent surface pore density in the presence of PVP. It remains to be seen how these changes affect the molecular-weight cutoff.

Biological Properties: Specific Examples and Issues

Microencapsulation processes using HEMA–MMA have been developed solely for living cells. Initially, it was necessary to ensure that enough cells survived the encapsulation process, and as appropriate, that they proliferated within the capsule and remained viable long term. The maintenance of differentiated functions by a variety of encapsulated cells was also studied, for this maintenance was the means by which the potentially therapeutically active biomolecule would be produced. Also important has been the effect of the intracapsular environment, determined by the polymer wall on cell arrangement and consequent viability and function of the cells. Results to illustrate these key points as an indication of the success of HEMA–MMA microencapsulation of living cells and their performance in vitro as protein delivery devices will be outlined in the following sections.

Cell Survival and Proliferation

The success of encapsulation is initially dependent on the number of viable cells that are entrapped within the capsule on an absolute basis or compared with the theoretical number of cells fed to each capsule—the encapsulation efficiency. In certain situations, such as when few cells are encapsulated, cell proliferation occurring within the microcapsule may be desirable. Such proliferation also indicates a reasonable microenvironment within the capsule core for the cells. The rate of proliferation, however, may be lower than under normal tissue conditions, in part because the environments inside the capsule and the normal tissue culture dish are different. On the other hand, cells in an unproliferative state may function in a more differentiated manner with enhanced production of the desired product. For other cell types such as pancreatic islets, which do not proliferate in culture, the initial encapsulation efficiency must be high to provide adequate cell mass within the capsule.

A variety of cell types, including the Chinese hamster ovary fibroblast (*119*), rat pheochromocytoma, PC12 (*86*), human hepatoma, HepG2 (*85, 99*), murine fibroblast, L929 (*128*) cell lines, primary rat hepatocytes (*129*), and rat islet tissue (*130*) have been encapsulated within HEMA–MMA microcapsules. These studies have shown that the cells survive the encapsulation procedure, despite the exposure to shear forces and organic solvents and nonsolvents, and that the cells grow or function afterward in vitro for periods from 2 to at least 6 weeks. More recently we have microencapsulated transfected cell lines, specifically mouse fibroblasts (2A-50), which had been engineered to secrete human growth hormone and β-glucuronidase (*131*), as a potential application of gene therapy for the delivery of therapeutic proteins using genetically engineered cells.

Generally, only a fraction of the cells delivered to the needle assembly actually become enclosed by the polymer wall. For example, when HepG2 cells were encapsulated with medium augmented with 20% Ficoll-400 (regular capsules), the encapsulation efficiency was ~20–25% (*119*). However, inclusion of the cell immobilization matrix, Matrigel, in the capsule core increased the encapsulation efficiency for HepG2 cells to ~50%. The gelling properties of this cell attachment matrix may have protected the cells from the forces inherent in the encapsulation process (e.g., shear or

PEG200 contact), although fluid-mechanic differences associated with two-component droplet formation were likely more important. The encapsulation efficiency was also dependent on the sensitivity of the cells to the encapsulation process and the quality of the encapsulation run; better centered capsules had higher efficiencies because more cells were completely enclosed by the polymer wall (*84*).

Cellular proliferation within microcapsules has been examined directly by counting the number of cells from an aliquot of ~25 capsules and indirectly by examining the metabolic activity of cells within a single capsule by using the 3-(4,5-dimethylthiazol-2-yl)-2,5-diphenyltetrazolium bromide (MTT) assay (*132*). PC12 (*86*) and HepG2 (*99*) cells proliferated within HEMA–MMA microcapsules to an extent that was dependent on the initial encapsulation density and hence the number of cells per capsule. The results suggest that the intracapsular environment (without a matrix) could only maintain cellular proliferation and metabolic activity up to a certain, limited extent (approximately 500 cells/capsule or 17×10^6 cells/mL internal volume). Intracapsular space limitations may limit the number of cells within a capsule. Alternatively, the arrangement of cells within the capsule core (e.g., spheroids in the nonadherent capsule environment) may limit diffusion of nutrients and metabolites, such that more cells per capsule would have a lower metabolic activity per cell.

However, when the cells were encapsulated along with an attachment substrate (Matrigel), cells proliferated steadily over the 2-week period to reach higher cell number values per capsule than in the absence of the attachment matrix (Figure 15.4A) (*99*). Cell numbers in excess of 4000 per capsule were obtained. However, even though the number of cells per capsule increased, the cellular metabolic activity per capsule did not parallel this increase but rather decreased slightly after 2 weeks (Figure 15.4B). The relatively constant metabolic activity may indicate that the cells are correspondingly more differentiated while within Matrigel. HepG2 cells encapsulated at the high density within Matrigel in small-diameter capsules (~400 μm) maintained a constant level of metabolic activity over 3 weeks, similar to the large-diameter capsules but not decreasing after 2 weeks as for large-diameter capsules (*120*). Cell numbers per small-diameter capsule showed that there was a similar increase in cell number for small-diameter capsules as for large-diameter capsules (but fewer cells per capsule).

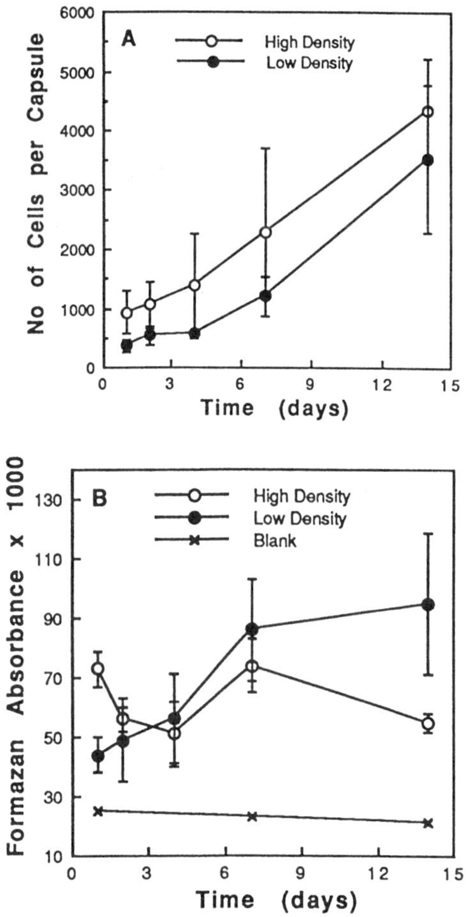

FIGURE 15.4. Time-dependent changes in the mean ± standard deviation cell number (A) and metabolic activity (B) of the HepG2 cells in Matrigel capsules when encapsulated at low (1×10^6 cells/mL, closed symbols) or high (5×10^6 cells/mL, open symbols) density. (Reproduced with permission from reference 99. Copyright 1993 Wiley.)

The picture that is emerging is that cells grow to fill the capsule space, but to the extent limited by the nutrient supply and intracapsule and intercellular diffusion gradients. They then may maintain a roughly constant cell number and cell activity level thereafter, provided the intracapsule environment is suitable: for example, anchorage-dependent cells have the necessary attachment substrate. The rate of proliferation is likely less than that in normal tissue culture, as is the rate of metabolic activity on a per cell basis (as exemplified by MTT conversion). How long this steady cell number and activity is maintained is not known, but MTT conversion has been observed for several months after encapsulation. Interestingly, cells are not found outside of the capsule, except on rare occasions, a result suggesting the capsule wall is strong enough to limit proliferation to the intracapsular space. Cells that do not receive adequate nutrients or that are apoptotic will die within the capsule. These products of necrosis remain in close proximity to viable cells and may affect their viability. Eventually, proteases released at cell death will have sufficiently degraded the cellular debris such that it could be removed from the capsule core by diffusive transport. However, the cell cycle and cell death due to an inadequate nutrient supply are dynamic processes, implying that the capsule core would presumably never be free of cellular debris. The implications on cell viability and function in vitro and as shed antigens eliciting an immune and inflammatory response in vivo are not yet understood.

Protein Release

HepG2 Cells and Islets of Langerhans

For the ultimate application of encapsulated cells as a bioartificial organ or physiologically controlled biomolecule-delivery system, the cells must not only survive the encapsulation process and remain viable, but they must also express their differentiated functions. The ability of the HEMA–MMA microcapsule to support the functional state of the cells has been assessed by the quantitation of encapsulated cell-derived biomolecule release into the extracapsular milieu.

As a protein-release example, human hepatoma HepG2 cells were used as a model for hepatocytes (*99*). Four plasma proteins that span the molecular-weight range of interest [α_1-acid glycoprotein (AG; M_w, 42 kDa), α_1-antitrypsin (AT; M_w, 52 kDa), haptoglobin (Hap; M_w, 98 kDa), and fibrinogen (Fbg; M_w, 340 kDa)] were released into the surrounding medium by HepG2 cells from an aliquot of ~100 Matrigel capsules. Protein release curves over a 2-week period (Figure 15.5) showed higher amounts of secreted protein paralleling the increase in the number of cells per capsule (*see* previous discussion); significantly more protein was secreted at day 7 as compared with day 3. No further increase occurred in the amount secreted at day 14, except for AT. The amount of Fbg released, relative to AG, was lower for encapsulated cells compared with control unencapsulated cells within Matrigel in tissue culture, results consistent with a sieving effect of the

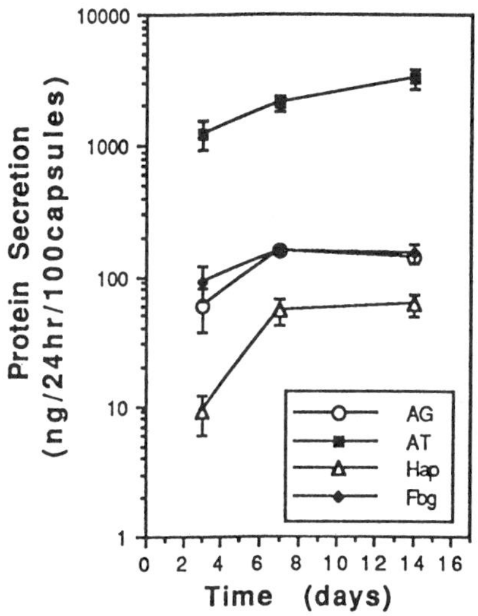

FIGURE 15.5. Mean ± standard deviation release rate of AG, AT, Hap, and Fbg by encapsulated HepG2 cells on days 3, 7, and 14 in ng per 24 h per 100 capsules (n = 3). (Reproduced with permission from reference 99. Copyright 1993 Wiley.)

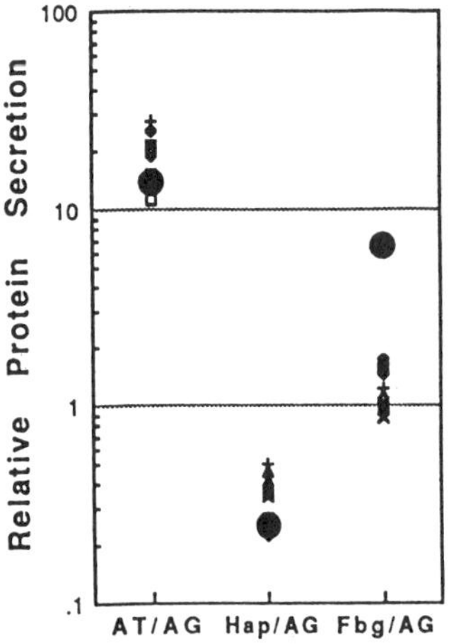

FIGURE 15.6. Release of AT, Hap, and Fbg with respect to AG (AT/AG, Hap/AG, and Fbg/AG, respectively) by the control (i.e., unencapsulated) and encapsulated HepG2 cells (large capsules). The relative release rates were obtained by dividing specific release rates by the AG release rate. Closed circles correspond to the average relative release rate of three independent control cultures (in duplicate), whereas the other symbols correspond to individual relative release rates from three batches of capsules on days 3, 7, and 14. Note that only the Fbg/AG was significantly lower for encapsulated as compared with control cells. (Reproduced with permission from reference 99. Copyright 1993 Wiley.)

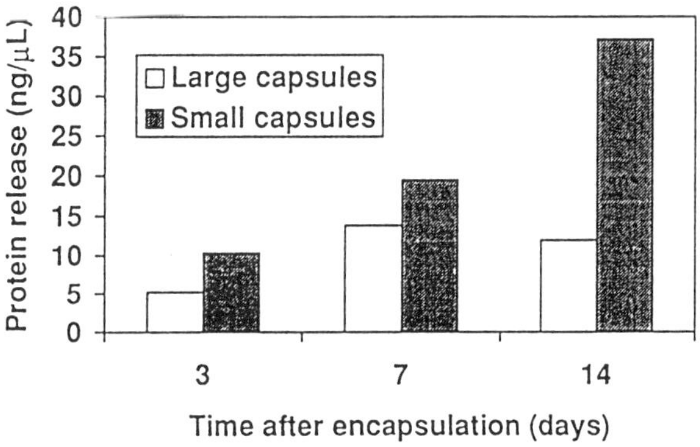

FIGURE 15.7. Comparison of AG release by small (~400 μm) and large (~900 μm) HEMA–MMA microcapsules containing HepG2 cells (initial encapsulation density, 5 × 10^6 cells/mL) in Matrigel. Large-diameter capsules were prepared by withdrawing nozzle assembly from stagnant hexadecane (*99*); small-diameter capsules were prepared as in Figure 15.2 by pumping dodecane past a stationary nozzle. Capsules were compared on the basis of equal intracapsule space: 11.3 μL/100 large capsules; 2.83 μL/200 small capsules.

polymer membrane on the higher molecular weight protein, Fbg, relative to the smaller AG for large-diameter (~800-μm) capsules (Figure 15.6) (*99*). The corresponding release rates for AT and Hap (relative to AG) from encapsulated and control cells were similar, a result suggesting that the cells from these two different environments may be compared in this manner and that the lower Fbg release from encapsulated cells was not likely because of an altered cell phenotype.

HepG2 cells within small-diameter capsules of ~400 μm also showed significantly lower Fbg:AG secretion rate ratio as compared to unencapsulated cells just as the aforementioned large-diameter capsules of ~800 μm had (*120*). A comparison of the amount of protein released from small- and large-diameter capsules containing HepG2 on the basis of the internal capsule volume is shown in Figure 15.7. At all time points after encapsulation, especially after 14 days, there was more AG secreted by small microcapsules than by large ones when compared on the basis of equivalent core volume. This result was attributed to a more efficient use of internal capsule volume for and a lower degree of cellular necrosis at the center of the core of the smaller capsules.

The dynamics of insulin release by microencapsulated pancreatic islets on glucose stimulation, although demonstrating the ability to encapsulate viable tissue fragments consisting of nonproliferating cells, also illustrated the effect of the polymer wall on stimulus-elicited protein release. The glucose enhancement of insulin secretion, expressed as a stimulation index (the ratio of insulin secretion for high to low glucose), was similar for both encapsulated and unencapsulated control islets.

However, the rates of insulin release were significantly lower (3.5–5 times) for encapsulated islets as compared with the control unencapsulated islets. This reduction was attributed to the polymer capsule wall acting as a barrier to insulin diffusion, because islets freed from within the capsule secreted insulin at rates similar to that of control islets. As a result of the diffusion resistance associated with the capsule wall, insulin was retained within the capsule and was released after breaking open the capsules. This high intracapsular insulin concentration was expected to downregulate insulin secretion by the encapsulated islets, exacerbating the diffusion barrier effect of the polymer wall. Because the stagnant water layer surrounding the islets in these 800-μm-diameter capsules was a significant part of the diffusion resistance, the smaller capsules with thinner polymer membranes prepared by using the submerged nozzle–liquid jet extrusion process (*120*) or the conformally coated islets (*76*) are expected to be preferred.

Genetically Engineered Cells

Mouse fibroblasts engineered to secrete human growth hormone (M_w, 48 kDa; cell line, LhGH-1) and β-glucuronidase (M_w, ~300 kDa; cell line, 2A-50) were encapsulated within HEMA–MMA capsules (*131*). Prior in vivo studies with these cells encapsulated within alginate poly-L-lysine capsules (*133*) demonstrated the delivery of human growth hormone with a circulating serum concentration of 0.1–1.5 ng/mL within the first two weeks and increasing antibody titer against human growth hormone for more than three months. The persistent expression of the transgene and survival of the transfected cells were verified when the microcapsules were retrieved periodically to show that the encapsulated cells remained viable, proliferative, and able to secrete human growth hormone even after 78–111 days (*21*). However, by 3 months many of the alginate–polylysine capsules had degenerated, a result precluding their long-term use. As an alternative, the cells were encapsulated in HEMA–MMA microcapsules as a potential means of extending their in vivo life span.

In vivo experiments have not yet been undertaken, but in vitro results have shown that the genetically modified mouse fibroblasts survived HEMA–MMA microencapsulation and proliferated within the capsules. Protein release studies showed that neither β-glucuronidase (M_w, ~300 kDa) nor hexosaminidase (a constitutive product molecular weight, M_w, 120 kDa) could diffuse out of the capsule, whereas human growth hormone (M_w, 48 kDa) was freely diffusible and recovered in the media at a rate similar to unencapsulated cells (Figure 15.8). Therefore, the capsules appeared to have a molecular-weight cutoff of <120 kDa. The limitation to β-glucuronidase diffusion from these capsules is in contrast to the release of Fbg (Figure 15.5) of a similar molecular weight from capsules containing HepG2 cells. All proteins were present in the intracapsule (extracellular) space at similar levels as determined by breaking open the capsules and assaying the extracellular supernatant. The intracellular protein levels (per 10^6 cells) were the same at all time points, with or without Matrigel. These results indicate that encapsulation did not affect cell behavior but that it did act as a diffusion barrier.

A significant amount of growth hormone was retained in the intracapsular but extracellular space when Matrigel was included in the encapsulation. This effect appeared to be saturable, because the amount retained did not increase with time in culture. The level of expression of the gene product accumulating in the intracellular fractions was not affected by the presence of Matrigel. The presence of this ECM material did not appear to pose a physical barrier to the exit of the recombinant gene product because the rates of human growth hormone secretion from the microcapsules were similar in the presence and absence of Matrigel. Presumably, there was an interaction between the nonglycosylated protein and the Matrigel, and once binding sites were saturated, no further binding could take place, allowing unrestricted efflux of the recombinant product.

Epidermal growth factor (EGF) was released by 19-3 cells (FR3T3 fibroblasts, transfected with the human EGF gene, obtained courtesy of R. A.

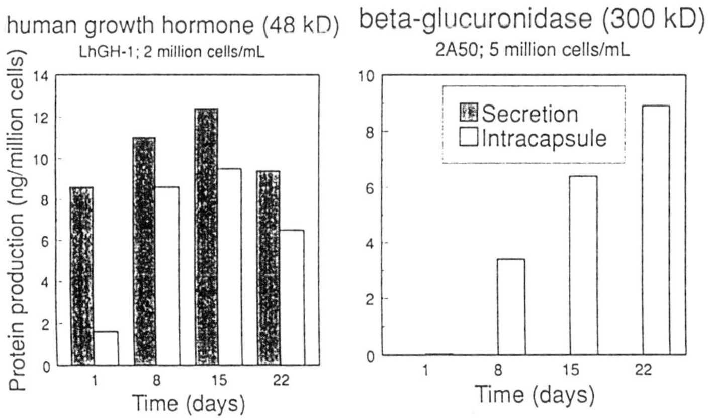

FIGURE 15.8. Protein production by genetically engineered cells encapsulated in HEMA–MMA microcapsules without an ECM. Cells were prepared and protein assays performed in the laboratory of P. L. Chang at McMaster University. Human growth hormone was secreted by transfected mouse Ltk$^-$ cells (LhGH initial encapsulation density, 5×10^6 cells/mL); β-glucuronidase was secreted by 2A-50 cells with an initial density of 2×10^6 cells/mL.

Weinberg, Massachusetts Institute of Technology), which were grown on Cytodex 2 microcarriers within HEMA–MMA microcapsules (Figure 15.9) (*121*). The microcarriers were used as a cell-attachment substrate in lieu of an ECM such as collagen or Matrigel. Capsules (large-diameter) were intended to contain either 20 beads/capsule and ~20 cells/bead (standard loading) or 60 beads/capsule and ~10 cells/bead (high loading). EGF activity was measured indirectly by using NRK-49F cells as a target. Encapsulated 19-3 cells were incubated in Dulbecco's minimum essential medium with 2% fetal calf serum for 10 h, and the effect of this conditioned medium (containing 19-3 secreted EGF) after various dilutions with fresh medium on enhancing the growth of NRK cells was measured. The dilution required to obtain no growth enhancement was compared with the concentration of pure human recombinant EGF required for the same "null" effect on growth in the same unconditioned medium; results were reported in terms of EGF equivalency. EGF secretion increased with cell loading, but the absence of an effect of time after encapsulation was attributed, in part, to the complexity of the mixture of growth factors, not only EGF, inhibitors that may be secreted by the 19-3 cells, and the crudeness of the assay procedure.

Capsule Variability: Fibrinogen Release and HRP Permeability

The release of Fbg, whose molecular weight is higher than that of molecules that would be expected to permeate the polymer membrane of capsules with a molecular-weight cutoff of ~100 kDa into the medium surrounding capsules containing HepG2 cells, was thought to have originated from a subset of these capsules that were defective. To test this hypothesis, protein secretion was quantified from individual capsules (*126*). Measurements of protein released from individual capsules were enabled by the sensitivity of the enzyme-linked immunosorbent assay techniques and the relatively high number of cells within the capsules that produced the protein. Both AT and Fbg were secreted in variable amounts from capsules with a significant proportion of capsules from which no protein release was detected. This proportion of nonprotein-releasing capsules decreased

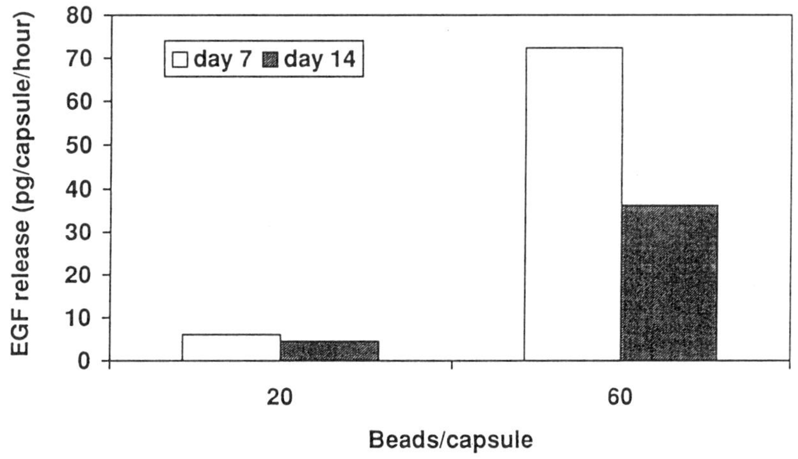

FIGURE 15.9. Epidermal growth factor (EGF) release by HEMA–MMA encapsulated 19-3 cells at two different bead loadings and 7 or 14 days after encapsulation. EGF concentration was inferred by bioassay of diluted, conditioned medium by using NRK cells as target.

with time, presumably as cellular proliferation occurred within the capsules and a detectable level of protein was secreted by these cells. Furthermore, the proportion of nonprotein-releasing capsules was higher for proteins of higher molecular weight. The distribution of the ratio of Fbg to AT secretion (Fbg/AT) (Figure 15.10) showed that 60% of the capsules secreted detectable amounts of AT but not Fbg. Only 10% of all capsules had a Fbg/AT release ratio greater than or equal to that of unencapsulated cells. This distribution of secretion rates was consistent with the distribution in the permeability of similar capsules without cells to the model protein HRP (M_w, 40 kDa) (*126*).

The subset of capsules that were detected as being more permeable to Fbg was considered to be defective. The cause of these defects has been blamed on the poor integrity of the polymer capsule wall due to thin areas, pin holes, or tears. However, the variability in the amount of protein released from capsules containing HepG2 cells may be only partly explained by the permeability to proteins of that molecular weight. There may also be contributions to this protein delivery variability due to a variability in the number of protein-secreting cells per capsule.

Future Directions

We have reached the stage where microparticles containing immobilized or encapsulated cells can be produced with natural (e.g., alginate) or synthetic polymers (e.g., HEMA–MMA) by using aqueous or organic solvents. Cell survival during the process and preserving cell viability or function afterward, at least in vitro, have been demonstrated. There is much to do to optimize the processes that are used, to improve their reproducibility, to be able to control permeability or molecular-weight cutoff, or simply to reduce the capsule diameter. These problems are primarily technical and will be resolved as the processes are scaled up and as further understanding of the processes is achieved.

There is much to do as well in understanding how the capsule wall affects the encapsulated cell behavior. The complexities of the interrelationships among wall permeability, ECM, cell density, etc., are beginning to be recognized, but we are still a long way from knowing how to design an encapsulated cell system with applications including cell transplantation and protein delivery. We have learned not to expect cell behavior that is

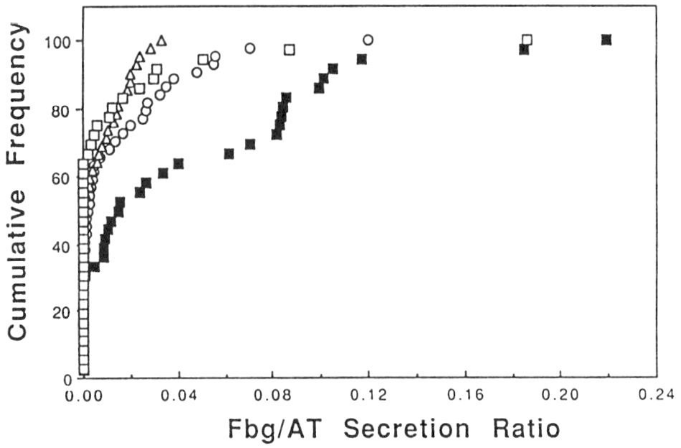

FIGURE 15.10. Cumulative distribution of Fbg to AT secretion ratio (Fbg/AT) among HepG2 containing microcapsules. Different symbols represent capsules from different batches. Mean ± standard deviation of Fbg/AT for unencapsulated HepG2 cells was 0.15 ± 0.07 (n = 3). The distributions for three batches were similar, whereas one batch was distinctly different. (Adapted from reference 126).

comparable to what is seen in conventional tissue culture: the three-dimensional arrangement of cells inside a capsule creates a unique microenvironment and may even be more physiological than that on a polystyrene dish. We are at the stage where any cell can be encapsulated, and provided the conditions are reasonable, its differentiated function can be retained in vitro.

On the other hand, we are only beginning to understand what happens in vivo. After implantation, the tissue reaction, the implant site, the presence of cytokines, the release of antigenic proteins or cell debris, the presence of adsorbed protein, the capsule chemistry, and durability all intervene to affect the encapsulated cell behavior, its ability to secrete the desired product, and the duration of viability. Even though considerable effort will be required to sort out these interrelationships, the therapeutic benefit of cell transplantation warrants the effort.

Acknowledgments

We acknowledge the financial support of the Medical Research Council of Canada, the Natural Sciences and Engineering Research Council, the Canadian Parkinson's Foundation, and the Canadian Liver Foundation, which provided a fellowship to Julia E. Babensee. We also acknowledge the assistance of Vlad Horvath and the advice of U. De Boni.

References

1. Kahan, B. D. *Curr. Opin. Immunol.* **1992**, *4,* 553–560.
2. Meiser, B. M.; Reichart, B. *Transplant. Proc.* **1994**, *26,* 3181–3183.
3. Thomson, A. W.; Forrester, J. V. *Clin. Exp. Immunol.* **1994**, *98,* 351–357.
4. Cosimi, A. B. *Dig. Dis. Sci.* **1995**, *40,* 65–72.
5. Oluwole, S. F.; Jin, M.-X.; Chowdhury, N. C.; Ohajekwe, O. A. *Transplantation* **1994**, *58,* 1077–1081.
6. Coulombe, M.; Gill, R. G. *Transplantation* **1994**, *57,* 1195–1200.
7. Coulombe, M.; Gill, R. G. *Transplantation* **1994**, *57,* 1201–1207.
8. Benhamou, P. Y.; Stein, E.; Hober, C.; Miyamoto, M.; Watanabe, Y.; Nomura, Y.; Watt, P. C.; Kenmochi, T.; Brunicardi, F. C.; Mullen, Y. *Horm. Metab. Res.* **1995**, *27,* 113–120.

9. Habibullah, C.M.; Ayesha, Q.; Khan, A. A.; Naithani, R.; Lahiri, S. *Transplantation* **1995**, *59,* 1495–1497.
10. Colton, C. K.; Avgoustiniatos, E. S. *Trans. ASME* **1991**, *113,* 152–170.
11. Clayton, H. A.; James, R. F. L.; London, N. J. M. *Acta Diabetol.* **1993**, *30,* 181–189.
12. Winn, S. R.; Tresco, P. A.; Zielinski, B.; Greene, L. A.; Jaeger, C. B.; Aebischer, P. *Exp. Neurol.* **1991**, *113,* 322–329.
13. Aebischer, P.; Tresco, P. A.; Sagen, J.; Winn, S. R. *Brain Res.* **1991**, *560,* 43–49.
14. Bruni, S.; Chang, T. M. S. *Biomater. Artif. Cells Artif. Organs* **1989**, *17*(4), 403–411.
15. Cai, Z.; Shi, Z.; Sherman, M.; Sun, A. M. *Hepatology* **1989**, *10*(5), 855–860.
16. Dixit, V.; Darvasi, R.; Arthur, M.; Brezina, M.; Lewin, K.; Gitnick, G. *Hepatology* **1990**, *12,* 1342–1349.
17. Dixit, V.; Gitnick, G. *J. Biomater. Sci. Polym. Ed.* **1995**, *7,* 343–357.
18. Fu, X. W.; Sun, A. M. *Transplantation* **1989**, *47,* 432–435.
19. Hasse, C.; Schrezenmeir, J.; Stinner, B.; Schark, C.; Wagner, P. K.; Neumann, K.; Rothmund, M. *World J. Surg.* **1994**, *18,* 630–634.
20. Liu. H.-W.; Ofosu, F. A.; Chang, P. L. *Hum. Gene Ther.* **1993**, *4,* 291–301.
21. Chang, P. L.; Shen, N.; Westcott, A. J. *Hum. Gene Ther.* **1993**, *4,* 433–440.
22. Chang, P. L.; Hortelano, G.; Tse, M.; Awrey, D. E. *Biotechnol. Bioeng.* **1994**, *43,* 925–933.
23. Al-Hendy, A.; Hortelano, G.; Tannenbaum, G. S.; Chang, P. L. *Hum. Gene Ther.* **1995**, *6,* 165–175.
24. Hughes, M.; Vassilakos, A.; Andrews, D. W.; Hortelano, G.; Belmont, J. W.; Chang, P. L. *Hum. Gene Ther.* **1994**, *5,* 1445–1455.
25. Savelkoul, H. F. J.; Van Ommen, R.; Vossen, A. C. T. M.; Breedland, E. G.; Coffman, R. L.; van Oudenaren, A. *J. Immunol. Methods* **1994**, *170,* 185–196.
26. Babensee, J. E.; Sefton, M. V. In *Microparticulates: Preparation, Characterization, and Application to Medicine;* Cohen, S.; Bernstein, H., Eds.; Drugs and the Pharmaceutical Sciences 77; Marcel Dekker: New York, 1996; pp 477–519.
27. Lacy, P. E.; Hegre, O. D.; Gerasimidi-Vazeou, A.; Gentile, F. T.; Dionne, K. E. *Science (Washington, D.C.)* **1991**, *254,* 1782–1784.
28. Lanza, R. P.; Borland, K. M.; Staruk, J. E.; Appel, M. C.; Solomon, B. A.; Chick, W. L. *Endocrinology* **1992**, *131,* 637–642.
29. Lanza, R. P.; Lodge, P.; Borland, K. M.; Carretta, M.; Sullivan, S. J.; Beyer, A. M.; Muller, T. E.; Solomon, B. A.; Maki, T.; Monaco, A. P.; Chick, W. L. *Transplant. Proc.* **1993**, *25*(1), 978–980.
30. Aebischer, P.; Goddard, M.; Signore, A. P.; Timpson, R. L. *Exp. Neurol.* **1994**, *126,* 151–158.
31. Kordower, J. H.; Liu, Y.-T.; Winn, S.; Emerich, D. F. *Cell Transplant. (Pittsburgh, Pa.)* **1995**, *4*(2), 155–171.
32. Emerich, D. F.; McDermott, P. E.; Krueger, P. M.; Winn, S. R. *Prog. Neuro-Psychopharmacol. Biol. Psychiatry* **1994**, *18,* 935–946.
33. Hoffman, D.; Breakefield, X. O.; Short, M. P.; Aebischer, P. *Exp. Neurol.* **1993**, *122,* 100–106.
34. Winn, S. R.; Hammang, J. P.; Emerich, D. F.; Lee, A.; Palmiter, R. D.; Baetge, E. E. *Proc. Natl. Acad. Sci. U.S.A.* **1994**, *91,* 2324–2328.
35. Emerich, D. F.; Winn, S. R.; Harper, J.; Hammang, J. P.; Baetge, E. E.; Kordower, J. H. *J. Comp. Neurol.* **1994**, *349,* 148–164.
36. Kordower, J. H.; Winn, S. R.; Liu, Y.-T.; Mufson, E. J.; Sladek, J. R.; Hammang, J. P.; Baetge, E. E.; Emerich, D. F. *Proc. Natl. Acad. Sci. U.S.A.* **1994**, *91,* 10898–10902.
37. Emerich, D. F.; Hammang, J. P.; Baetge, E. E.; Winn, S. R. *Exp. Neurol.* **1994**, *130,* 141–150.
38. Hammang, J. P.; Emerich, D. F.; Winn, S. R.; Lee, A.; Lindner, M. D.; Gentile, F. T.; Doherty, E. J.; Kordower, J. H.; Baetge, E. E. *Cell Transplant. (Pittsburgh, Pa.)* **1995**, *4,* S27–S28.
39. Lindner, M. D.; Winn, S. R.; Baetge, E. E.; Hammang, J. P.; Gentile, F. T.; Doherty, E.; McDermott, P. E.; Frydel, B.; Ullman, D.; Schallert, T.; Emerich, D. F. *Exp. Neurol.* **1995**, *132,* 62–76.
40. Sagan, J. *ASAIO J.* **1992**, *38,* 24–28.
41. Joseph, J. M.; Goddard, M. B.; Mills, J.; Padrun, V.; Zurn, A.; Zielinski, B.; Favre, J.; Gardaz, J. P.; Mosimann, F.; Sagen, J.; Christenson, L.; Aebischer, P. *Cell Transplant. (Pittsburgh, Pa.)* **1994**, *3,* 355–364.
42. Brauker, J. H.; Carr-Brendel, V. E.; Martinson, L. A.; Crudele, J.; Johnston, W. D.; Johnson, R. C. *J. Biomed. Mater. Res.* **1995**, *29,* 1517–1524.
43. Brauker, J.; Martinson, L. A.; Hill, R. S.; Young, S. K.; Carr-Brendel, V. E.; Johnson, R. C. *Cell Transplant. (Pittsburgh, Pa.)* **1992**, *1*(2/3), 163.

44. Hill, R. S.; Young, S. K.; Martinson, L. A.; Dudek, R. W. In *Pancreatic Islet Transplantation. Volume III: Immunoisolation of Pancreatic Islets;* Lanza, R. P.; Chick, W. L., Eds.; R. G. Landes Company: Austin, TX, 1994; pp 147–155.
45. Maki, T.; Ubhi, C. S.; Sanchez-Farpon, H.; Sullivan, S. J.; Borland, K.; Muller, T. E.; Solomon, B. A.; Chick, W. L.; Monaco, A. P. *Transplantation* **1991,** *51,* 43–51.
46. Maki, T.; Lodge, J. P. A.; Carretta, M.; Ohzato, H.; Borland, K. M.; Sullivan, S. J.; Staruk, J.; Muller, T. E.; Solomon, B. A.; Chick, W. L.; Monaco, A. P. *Transplantation* **1993,** *55,* 713–718.
47. Jarvis, A.; Grdina, T. *BioTechniques* **1983,** *1,* 22–27.
48. Levee, M. G.; Lee, G.-M.; Paek, S.-H.; Palsson, B. O. *Biotechnol. Bioeng.* **1994,** *43,* 734–739.
49. Malaisse, W. J. In *Nutrient Regulation of Insulin Secretion;* Flatt, P. R., Ed.; Portland Press: London, 1992; pp 83–100.
50. Otonkoski, T.; Beattie, G. M.; Rubin, J. S.; Lopez, A. D.; Baird, A.; Hayek, A. *Diabetes* **1994,** *43,* 947–953.
51. Kan, M.; Huang, J.; Mansson, P.-E.; Yasumito, H.; Carr, B.; Mckeehan, W. L. *Proc. Natl. Acad. Sci. U.S.A.* **1989,** *86,* 7432–7436.
52. Bissell, D. M.; Caron, J. M.; Babiss, L. E.; Friedman, J. M. *Mol. Biol. Med.* **1990,** *7,* 187.
53. Folkman, J.; Klagsbrun, M.; Sasse, J.; Wadzinski, M.; Ingber, D.; Vlodavsky, I. *Am. J. Pathol.* **1988,** *130,* 393–400.
54. Panayotou, G.; End, P.; Aumailley, M.; Timpl, R.; Engel, J. *Cell* **1989,** *56,* 93–101.
55. Kost, J.; Horbett, T. A.; Ratner, B. D.; Singh, M. *J. Biomed. Mater. Res.* **1984,** *19,* 1117–1133.
56. Davis, G. C. *Biologicals* **1993,** *21,* 105.
57. Evans, R. P.; Witcher, M. *Ther. Drug Monit.* **1993,** *15,* 514–520.
58. Andersson, L. *Cancer Invest.* **1992,** *10,* 71–84.
59. Leser, E. W.; Asenjo, J. A. *J. Chromatogr.* **1992,** *584,* 43–57.
60. Luck, R.; Klempnauer, J.; Ehlerding, G.; Kuhn, K. *Transplantation* **1990,** *50,* 394–398.
61. Flessner, M. F.; Dedrick, R. L.; Reynolds, J. C. *Am. J. Physiol.* **1992,** *262,* 275–287.
62. Koten, J. W.; Den Otter, W. *Lancet* **1991,** *338,* 1189–1190.
63. Johnson, P. C., University of Pittsburgh, personal communication, 1992.
64. Bikalvi, A.; Alterio, J.; Inyang, A. L.; Dupuy, E.; Laurent, M.; Hartmann, M. P.; Vigny, L.; Raulais, D.; Courtois, Y.; Tobelem, G. *J. Cell. Physiol.* **1990,** *144,* 151–158.
65. van Suylichem, P. T. R.; Strubbe, J. H.; Houwing, H.; Wolters, G. H. J.; van Schilfgaarde, R. *Diabetologia* **1992,** *35,* 917–923.
66. Ao, Z.; Matayoski, K.; Lakey, J. R. T.; Rajoote, R. V.; Warnockm, G. L. *Transplantation* **1993,** *56,* 524–529.
67. Pariseau, J.-P.; Leblond, F. A.; Harel, F.; Lepage, Y.; Halle, J.-P. *J. Biomed. Mater. Res.* **1995,** *29,* 1331–1335.
68. Mikos, A. G.; Sarakinos, G.; Lyman, M. D.; Ingber, D. E.; Vacanti, J. P. *Biotechnol. Bioeng.* **1993,** *42,* 716–723.
69. Yang, M. B.; Vacanti, J. P.; Ingber, D. E. *Cell Transplant. (Pittsburgh, Pa.)* **1994,** *3,* 373–385.
70. Johnson, R. C.; Carr-Brendel, V.; Martinson, L.; Neuenfeldt, S.; Young, S.; Vergoth, C.; Hodgett, D.; Maryanov, D.; Jacobs, S.; Hill, R.; Thomas, T. J.; Brauker, J. *Cell Transplant. (Pittsburgh, Pa.)* **1994,** *3,* 221.
71. Colton, C. K. *Cell Transplant. (Pittsburgh, Pa.)* **1995,** *4*(4), 415–436.
72. Aebischer, P.; Panol, G.; Galletti, P. M. *Trans. Am. Soc. Artif. Intern. Organs* **1986,** *32,* 130–133.
73. Warnock, G. L.; Rajotte, R. V. *Diabetes* **1988,** *37,* 467.
74. Halle, J.-P.; Leblond, F. A.; Pariseau, J.-F.; Jutras, P.; Brabant, M.-J.; Lepage, Y. *Cell Transplant. (Pittsburgh, Pa.)* **1994,** *3,* 365.
75. Uludag, H.; Horvath, V.; Black, J. P.; Sefton, M. V. *Biotechnol. Bioeng.* **1994,** *44,* 1199–1204.
76. May, M. H. Ph.D. Thesis, in preparation.
77. Cruise, G. M.; Sawhney, A. S.; Pahak, C. P.; Luther, K. M.; Hubbell, J. A. *Transactions 19th Annual Meeting Society for Biomaterials;* Society for Biomaterials: Birmingham, AL, 1993; p 205.
78. Zekron, T.; Siebers, U.; Horcher, A.; Schnettler, R.; Zimmermann, U.; Bretzel, R. G.; Federlin, K. *Acta Diabetol.* **1992,** *29,* 41–45.
79. Schaapherder, A. F. M.; Daha, M. R.; Van Der Woude, F. J.; Bruijn, J. A.; Gooszen, H. G. *Transplantation* **1993,** *56,* 739–741.
80. Vaughan, H. A.; McKenzie, I. F. C.; Sandrin, M. S. *Transplantation* **1995,** *59,* 102–109.
81. Petersen, K.-G.; Khalaf, A.-N.; Naithani, V.; Fabry, M.; Gattner, H. *Acta Diabetol.* **1994,** *31,* 66–72.
82. Varani, J.; Fligiel, S. E. G.; Inman, D. R.; Helmreich, D. L.; Bendelow, M. J.; Hillegas, W. *Biotechnol. Bioeng.* **1989,** *33,* 1235–1241.

83. Landry, J.; Bernier, D.; Ouellet, R.; Goyette, R.; Marceau, N. *J. Cell Biol.* **1985,** *101,* 914.
84. Uludag, H.; Sefton, M. V. *Biotechnol. Bioeng.* **1992,** *39,* 672–678.
85. Babensee, J. E.; De Boni, U.; Sefton, M. V. *J. Biomed. Mater. Res.* **1992,** *26,* 1401–1418.
86. Roberts, T.; De Boni, U.; Sefton, M. V. *Biomaterials* **1996,** *17,* 267–276.
87. Koide, N.; Sakaguchi, K.; Koide, Y.; Asano, K.; Kawagughi, M.; Matsushima, H.; Takenami, T.; Shinji, T.; Mori, M.; Tsuji, T. *Exp. Cell Res.* **1990,** *186,* 227.
88. Sutherland, R. M. *Science (Washington, D.C.)* **1988,** *240,* 177–184.
89. Sutherland, R. M. *Cancer* **1986,** *58,* 1668–1680.
90. Erlanson, M.; Daniel-Szolgay, E.; Carlsson, J. *Cancer Chemother. Pharmacol.* **1992,** *29,* 343.
91. Dionne, K. E.; Colton, C. K.; Yarmush, M. L. *Diabetes* **1993,** *42,* 12–21.
92. Ung, D. Y.-P. Bachelor of Applied Science Thesis, University of Toronto, 1993.
93. Matthews, N.; Neale, M. L. In *Lymphokines and Interferons: A Practical Approach;* Clemens, M. J.; Morris, A. G.; Gearing, A. J. H., Eds.; IRL Press: New York, 1987; p 221.
94. Ding, A. H.; Porteu, F.; Sanchez, E.; Nathan, C. F. *J. Exp. Med.* **1990,** *171,* 712.
95. Neale, M. L.; Matthews, N. *Br. J. Cancer* **1990,** *6,* 831.
96. Van Der Schueren, B.; Denef, C.; Cassiman, J.-J. *Endocrinology* **1982,** *110,* 513.
97. Nakumura, T.; Yoshimoto, K.; Nakayama, Y.; Tomita, Y.; Ichihara, A. *Proc. Natl. Acad. Sci. U.S.A.* **1983,** *80,* 7229–7233.
98. Yao, A.; Rubin, H. *Proc. Natl. Acad. Sci. U.S.A.* **1994,** *91,* 7712–7716.
99. Uludag, H.; Sefton, M. V. *J. Biomed. Mater. Res.* **1993,** *27,* 1213–1224.
100. De Castro, A. Bachelor of Applied Science Thesis, University of Toronto, 1994.
101. Dixit, V.; Arthur, M.; Reinhardt, R.; Gitnick, G. *Artif. Organs* **1991,** *15,* 272.
102. Hoffman, D.; Breakefield, X. O.; Short, M. P.; Aebischer, P. *Exp. Neurol.* **1993,** *122,* 100–106.
103. Aung, T.; Inoue, K.; Kogire, M.; Sumi, S.; Fujisato, T.; Gu, Y. J.; Shinohara, S.; Hayashi, H.; Doi, R.; Imamura, M.; Mitsuo, M.; Nakai, I.; Maetani, S.; Ikada, Y. *Transplant. Proc.* **1994,** *26,* 790–791.
104. Zielinski, B. A.; Aebischer, P. *Biomaterials* **1994,** *15,* 1049–1056.
105. Aung, T.; Inoue, K.; Kogire, M.; Doi, R.; Kaji, H.; Tun, T.; Hayashi, H.; Echigo, Y.; Wada, M.; Imamura, M.; Fujisato, T.; Maetani, S.; Iwata, H.; Ikada, Y. *Transplant. Proc.* **1995,** *27,* 619–621.
106. Schuetz, E. G.; Li, D.; Omecinski, C. J.; Muller-Eberhard, U.; Kleinman, H. K.; Elswick, B.; Guzelian, P. S. *J. Cell. Physiol.* **1988,** *134,* 309.
107. Alexander, C. M.; Werb, Z. *Curr. Opin. Cell Biol.* **1989,** *1,* 974.
108. Stevenson, W. T. K.; Evangelista, R. A.; Broughton, R. L.; Sefton, M. V. *J. Appl. Polym. Sci.* **1987,** *34,* 65–83.
109. Stevenson, W. T. K.; Sefton, M. V. *J. Appl. Polym. Sci.* **1988,** *36,* 1541–1553.
110. Crooks, C. A.; Douglas, J. A.; Broughton, R. L.; Sefton, M. V. *J. Biomed. Mater. Res.* **1990,** *24,* 1241–1262.
111. *Biocompatibility of Orthopedic Implants;* Williams, D. F., Ed.; CRC Press: Boca Raton, FL, 1982.
112. Wichterle, O.; Lim, D. *Nature (London)* **1960,** *185,* 117.
113. Boag, A. H.; Sefton, M. V. *Biotechnol. Bioeng.* **1987,** *30,* 954–962.
114. Sugamori, M. E.; Sefton, M. V. *ASAIO Trans.* **1989,** *35,* 791–799.
115. Broughton, R. L.; Sefton, M. V. *Biomaterials* **1989,** *10,* 462–465.
116. Mallabone, C. L.; Crooks, C. A.; Sefton, M. V. *Biomaterials* **1989,** *10,* 380–386.
117. Sefton, M. V.; Stevenson, W. T. K. *Adv. Polym. Sci.* **1993,** *107,* 145–197.
118. Dawson, R. M.; Broughton, R. L.; Stevenson, W. T. K.; Sefton, M. V. *Biomaterials* **1987,** *8,* 360–366.
119. Uludag, H.; Sefton, M. V. *Cell Transplant. (Pittsburgh, Pa.)* **1993,** *2,* 175–182.
120. Uludag, H.; Horvath, V.; Black, J. P.; Sefton, M. V. *Biotechnol. Bioeng.* **1994,** *44,* 1199.
121. Patterson, W. J. Masters of Applied Science Thesis, University of Toronto, 1992.
122. Koenhen, D. M.; Mulder, M. H. V.; Smolders, C. A. *J. Appl. Polym. Sci.* **1977,** *21,* 199–215.
123. Doi, S.; Hamanaka, K. *Desalination* **1991,** *80,* 167–180.
124. Cole, D. R.; Waterfall, M.; McIntyre, M.; Baird, J. D. *Diabetologia* **1992,** *35,* 231.
125. Hwang, J. R.; Sefton, M. V. *J. Controlled Release* **1995,** *33,* 273–283.
126. Uludag, H.; Hwang, J. R.; Sefton, M. V. *J. Controlled Release* **1995,** *33,* 273–283.

127. Hwang, J. R.; Sefton, M. V. *J. Membr. Sci.* **1995,** *108,* 257–268.
128. Ung, D. Y.-P. Bachelor of Applied Science Thesis, University of Toronto, 1993.
129. Wells, G. D. M.; Fisher, M. M.; Sefton, M. V. *Biomaterials* **1993,** *14,* 615–620.
130. Sefton, M. V.; Kharlip, L. In *Pancreatic Islet Transplantation. Volume III: Immunoisolation of Pancreatic Islets;* Lanza, R. P.; Chick, W. L., Eds.; R. G. Landes Company: Austin, TX, 1994; pp 107–117.
131. Tse, M.; Uludag, H.; Sefton, M. V.; Chang, P. L., submitted for publication in *Biotechnol. Bioeng.*
132. Uludag, H.; Sefton, M. V. *Biomaterials* **1990,** *11,* 708–712.
133. Chang, P. L.; Tse, M.; Hortelano, G.; Awery, D. E., submitted for publication in *Biomaterials.*

16

Tissue Engineering

Integrating Cells and Materials To Create Functional Tissue Replacements

D. J. Mooney and J. A. Rowley

Tissue engineering integrates cells, scaffolds, and specific signals (e.g., growth factors) to create new tissues. Matrices fabricated from natural or synthetic materials guide the development of a desired tissue structure from cells. These cells may be transplanted cells or cells that migrate to the matrix from the adjacent host tissue. Tissue engineering offers the possibility of replacing the complete metabolic and structural functions of virtually any lost or deficient tissues, and this field may revolutionize the practice of medicine.

The loss or malfunction of tissues or organs is one of the most significant classes of health problems in the United States. This problem affects a large number of people (Table 16.1), and also imposes a large monetary cost. Costs relating to these problems are estimated to be greater than $400 billion per year in the United States, approximately half of the total health care costs (*1*). A common approach to treat these types of problems (e.g., diabetes) is to replace the missing biochemical function of the tissue or organ by using a drug treatment. This approach has been enormously successful, as indicated by the worldwide pharmaceutical industry. However, it is difficult to replace the multitude of functions performed by certain organs (e.g., liver), and tissue replacement is the only cure in these situations. Although whole organ or tissue transplantation has become an established therapy for many diseases, these therapies are limited by the shortage of organs and tissues available for transplantation. For example, 3000 liver transplants are performed in the United States every year. However, over 30,000 people still die each year in the United States from liver failure (*2*).

Tissue engineering represents a strategy between tissue or organ transplantation and drug therapies. Tissue and organ transplantations replace the entire range of functions of a tissue, whereas conventional drug therapies treat diseases with single drugs or molecules. Tissue engineering creates functional new tissues with transplanted

Copyright © 1997 American Chemical Society

TABLE 16.1. Number of Patients or Procedures Relating to Specific Organ or Tissue Diseases or Injuries Each Year in the United States

Organ or Tissue Involved	*Approximate Procedure or Patients per Year (× 10^3)*
Skin	4,750
Neurologic or muscle	240
Bone or cartilage	2,475
Tendon or ligaments	123
Blood vessels	754
Gastrointestinal (liver, pancreas, intestine)	1,033
Urologic	739
Breast	261
Blood (transfusions)	18,000
Oral	10,000

cells, or induces new tissue formation from cells already existing in the patients' body (*1*). A variety of cell types can be isolated from tissues and greatly expanded in vitro, potentially providing an unlimited supply of therapeutic cells. These cells can provide specific missing factors (e.g., insulin) in a manner that maintains the normal tissue response to metabolic signals. Alternatively, the cells can be used to engineer tissues that provide a multitude of biochemical (e.g., liver) or structural (e.g., cartilage) functions. The first significant efforts in tissue engineering were directed toward creating skin tissue for burn victims (*3–9*) and engineering tissues containing β-islet cells to treat diabetics (*10–16*). These approaches have subsequently expanded to encompass virtually every tissue in the body.

Three major issues must be addressed in tissue engineering. First, the appropriate cell types must be accessed. Cells can be isolated from a variety of different tissue sources and expanded in culture. Second, a suitable matrix must be developed to transplant these cells or induce them to grow into the matrix from the surrounding tissue. The matrices are fabricated in a variety of configurations from naturally occurring extracellular matrix (ECM) molecules, ceramics, and synthetic polymers. Finally, the cells and materials must be integrated with a suitable strategy to create a tissue replacement with the appropriate structure and function. For example, cells can be encapsulated within a semipermeable membrane to protect them from the host immune system. Alternatively, macroporous matrices are used to engineer tissues structurally integrated with the host tissue. The current technology and major challenges in each of these three areas are discussed in detail in the next three sections of this chapter.

Cell Sourcing and Expansion

Induction–Conduction Versus Transplantation

A central question in tissue engineering is whether to transplant cells to a desired site or to provide an environment inductive or conductive to the formation of the desired tissue from cells already present in the host tissue. The growth of certain tissues (e.g., bone and blood vessels) can be induced by specific growth factors. Factors (e.g., bone morphogenetic proteins) have been isolated, expressed in recombinant form, and delivered by using conventional drug delivery technologies to induce the growth of desired tissues (*17, 18*). This approach has the advantage of simplicity, because no cell transplantation is involved, and one does not need to worry about tissue rejection. However, factors such as these have not been isolated, or may not even exist, for the majority of tissues in the body.

Two processes may occur following the injury of a tissue: repair or regeneration. Regeneration of the desired tissue structure is desirable, and the delivery of inducing factors may aid the regeneration process. The body also attempts to repair the injury. This attempt can lead to the formation of scar tissue and the loss of normal tissue structure and function. If regeneration occurs slowly, the body will repair the tissue. The desired tissue structure will then be lost (*19, 20*). Anything that speeds up the regeneration process (e.g., preseeding the site of injury with transplanted cells) may be beneficial. The race between regeneration and repair and the lack of inducing factors for the majority of tissue types make cell transplantation an attractive means to engineer most tissues.

Cell transplantation and induction of host cells may be combined in certain applications. It may be desirable for the mesenchymal element of an engineered tissue to be derived from the host tissue. For example, specific cell types (e.g., hepatocytes) are transplanted to the desired anatomic location, and blood vessels from the host tissue are induced to invade the forming tissue (*21–24*). Angiogenesis can be promoted by incorporating growth factors in the matrix (*25*) and by controlling the microstructure of the matrix (*22, 26*).

Cell Expansion and Maintenance of Tissue-Specific Function

The possibility that a great number of patients can be treated with a small tissue supply (Figure 16.1) has fueled interest in tissue engineering. The challenge in tissue engineering is to expand a small number of isolated cells to a clinically useful cell mass. A variety of bioreactors have been developed to expand and culture cells at high densities (*27, 28*). However, the specifics of each system must be altered for every new cell type, and strategies to optimize cell culture are lacking (*27*). A practical limitation to cell expansion is the finding that a number of cultured cell types will undergo a limited number of divisions in culture, and these cells may lose their ability to provide a desired tissue-specific function (*29*). These constraints have fueled interest in the use of stem cells.

Stem cells are largely responsible for replacing senescent cells within tissues, and they are potentially capable of multiplying unlimited times. Stem cells have been identified in a number of tissues, including bone marrow (*30*), intestine (*31*), liver (*32*), and mesenchymal tissues (*33*). Systems to reproducibly isolate and culture some of these stem cell populations, including bone marrow (*30*) and intestine (*31*), have been developed. Stem cell populations may provide a virtually limitless supply of cells to engineer tissues. A challenge in this approach is the identification of the stem cells in a given tissue, if they exist at all. Defining the conditions required to reproducibly differentiate the daughter cells will be another significant biological challenge (*34*).

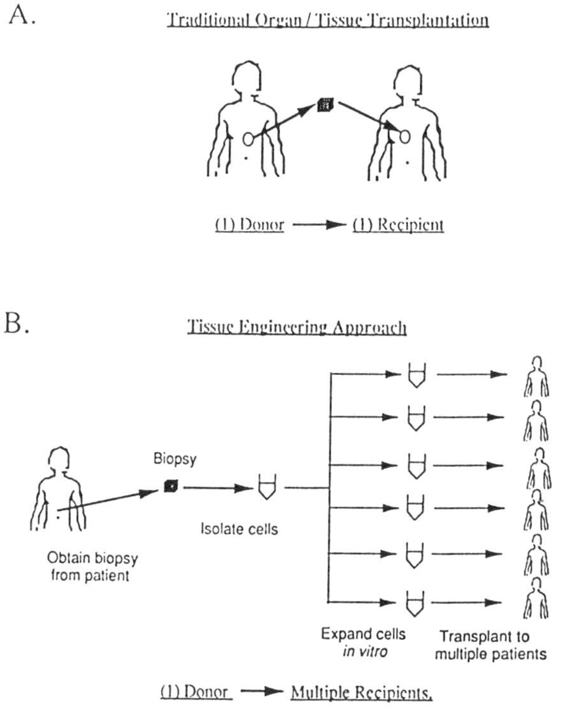

FIGURE 16.1. (A) One organ or tissue is transplanted from a single donor to a single recipient in traditional transplantation approaches. (B) In contrast, a tissue biopsy from a donor can potentially be used for multiple patients in a tissue engineering approach. This approach is possible because the cells isolated from the donor may be expanded in culture, greatly increasing the mass of cells available to treat patients.

Tissue Sources for Transplanted Cells

Cells for transplantation can be obtained from another site on the patient (autologous tissue), from another person (allogeneic tissue), or from animal tissue (xenogeneic tissue). Autologous cells are obtained from a biopsy, expanded or altered in culture, and transplanted to the desired site. This cell source eliminates concern over cell rejection and the need for immunosuppressants and is attractive in a variety of scenarios. However, this approach has a number of practical disadvantages. Two procedures on the patient would be required. The first procedure is performed to access the cells, and the second to transplant them. The time required to expand or alter the cells would create a time lapse before the patient could be treated. Autologous cells could not be used to treat acute diseases or injuries requiring immediate therapy. Also, it is not possible to expand certain cell types (e.g., hepatocytes) in culture at the present time.

Allogeneic tissue (unmatched same species) is used in whole organ transplantation and is an attractive source for cells. A variety of cell types can be multiplied in culture, and these cells are immediately available to a patient in large quantities (*8*). However, cell rejection is possible and immunosuppressive drugs may be required if this cell source is used. Also, multiplication of some cell types is not currently possible. The small supply of donated organs that can be used to access these cell types may again limit the number of patients who can be treated. To bypass this problem, one can use a transformed line of one of these cell types. These cell lines readily multiply in culture, even though the original cells did not. Several cell lines are being used in tissue engineering applications (*35, 36*). The uses for these cells are limited because they do not exhibit normal growth control and may be cancerous. Transformed cell lines are typically encapsulated within a semipermeable membrane to prevent them from directly interacting with the host tissue.

The use of xenogeneic (different species) cells could potentially alleviate the problem of a limited tissue supply for cell isolations. Animals could be bred specifically for this purpose, ensuring a constant and reproducible cell supply. However, these cells are rejected by the host immune system. They can only be used in situations where the host is heavily immunosuppressed or where the transplanted cells are protected from the host immune system (*11–16, 37–40*). Additionally, there may be differences between the metabolic regulation of gene expression in xenogeneic and human cells.

A variety of cell sources can be used in tissue engineering, and all have advantages and disadvantages. The requirements of a particular application guide the choice of cell source. This choice will then dictate the requirements of the matrix and the strategy used to form a functional tissue from the cells and the matrix.

Matrix Materials

Matrices are used in tissue engineering to first localize cells to a desired anatomic location, and to then serve as a scaffold for tissue development. The materials used to fabricate these matrices can be broken down into three types. Naturally derived materials, including ECM molecules and polysaccharides such as alginate, have been extensively used in tissue engineering. Matrices are also fabricated from a variety of synthetic materials. An exciting area of research is the synthesis of new materials that incorporate specific cell-recognition signals found in ECM molecules. These three types of matrices are discussed.

Naturally Derived Materials

Tissue engineering matrices are fabricated from a variety of materials that are isolated from human, animal, or plant tissue. Potential advantages of these types of materials are their biocompatibility and their biological activity. Because many of these molecules are found within tissues, they may not induce any foreign body reactions, and they are presumably receptive to the cell-mediated remodeling that occurs during tissue repair and regeneration (*19, 20*).

The primary, secondary, and tertiary structures of a variety of ECM molecules have been defined, as have specific cellular molecules involved in ECM recognition (Figure 16.2). Cells contain transmembrane receptors that specifically bind defined amino acid sequences present in ECM molecules (*41*). The binding of cellular

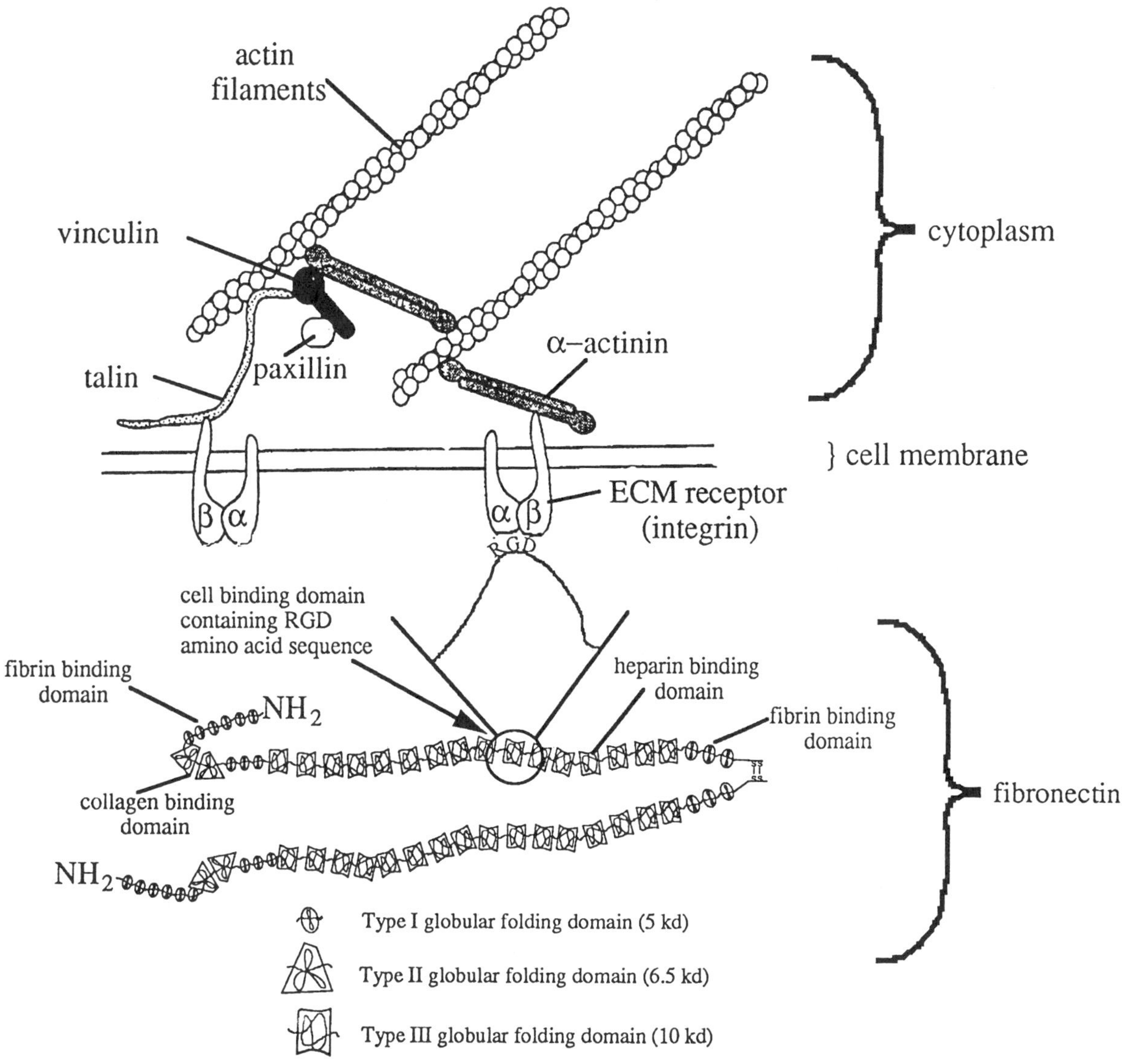

FIGURE 16.2. The structure of fibronectin and cell ECM receptors that specifically bind fibronectin. Fibronectin consists of two identical chains (220 kDa each) joined by two disulfide bonds. Each chain contains three types of globular folding domains (Type I, II, and III). Regions of the fibronectin with specific biological recognition properties have been identified (e.g., heparin binding). The cell binding ability of fibronectin has been localized to a three amino acid sequence (arginine–glycine–aspartate [RGD]), which interacts with a specific integrin receptor. These transmembrane receptors are composed of α and a β subunits and are physically continuous (via cytoplasmic proteins including talin, vinculin, α-actinin, and paxillin) with the actin cytoskeleton of the cell. (Adapted from references 88 and 89.)

receptors to ECM molecules starts a sequence of events that can alter both cell growth and tissue-specific gene expression (*42*). A variety of growth factors are also known to associate with the ECM in tissues. ECM molecules may serve as a repository of these factors for cells adherent to the ECM (*25, 32, 43*). One may precisely regulate where transplanted or induced cells adhere to the matrix and their exposure to specific growth factors by using specific ECM molecules to fabricate a matrix. This control may allow the gene expression of cells present in the matrix to be tightly regulated.

Type I collagen, the most prevalent ECM molecule in the body, is readily isolated from animal tissues and has been extensively used to fabricate cell delivery devices (*3–6, 44*). This material can be processed into a wide variety of structures (e.g., films, sponges, and fibers) (*3–6, 44*). The structure and resultant mechanical properties of collagen-based scaffolds can be regulated by the process used to extract the collagen from tissues (*44*) and by various cross-linking processes. Collagen molecules may be cross-linked physically by dehydrothermal (*45*) or UV radiation treatments, or chemically by using various chemical agents (*44–46*). However, the inflammatory response to these materials and their erosion rates are dependent on the specific cross-linking treatment that is used (*44–46*). Type I collagen may also be combined with glycosaminoglycans to form gels that mimic native dermal tissue (*4, 6, 7*). Xenogeneic collagen may induce an immune response in the host, but it is not clear if this response leads to an adverse clinical result (*46*). A variety of other ECM molecules, including laminin (*37, 47*), have also been used as cell-delivery matrices.

Alginate, a polysaccharide isolated from seaweed, is used as a cell-delivery vehicle. Water-soluble sodium alginate readily binds calcium, forming an insoluble calcium alginate hydrocolloid (*48*). These gentle gelling conditions have made alginate a popular material to encapsulate cells for transplantation (*10–16, 37, 38*) and as an injectable cell-delivery vehicle (*49*).

The potential advantages of these natural materials have made them popular for fabricating tissue engineering matrices. However, these materials also have a number of disadvantages. Many of these materials are isolated from human or animal tissue and are not available in large quantities. They suffer from large batch-to-batch variations and are typically expensive. Additionally, these materials exhibit a limited range of physical properties (e.g., mechanical strength and erosion times). These drawbacks have fueled interest in using synthetic materials to fabricate matrices.

Synthetic Matrices

A wide variety of polymers and calcium phosphate ceramics have been used to fabricate matrices. Both biodegradable and nondegradable materials have been used. Matrices fabricated from biodegradable materials theoretically will erode over time in the body to yield a completely natural tissue. These matrices will not induce any chronic inflammatory responses and cannot serve as a long-term site for infection. Nondegradable materials are clearly preferred when the matrix is intended to function as a barrier (e.g., protecting transplanted cells from the host immune system).

Biodegradable polymers have been extensively used to engineer tissues that will be structurally integrated with the host tissue (*1, 7, 8, 22–24, 37, 38, 50*). Even though a variety of biodegradable polymers exists (*51, 52*), polymers composed of monomers naturally present in the body (e.g., lactic acid and α-amino acids) are preferred. Polymers of lactic acid, glycolic acid, and copolymers of the two have been widely used to fabricate tissue engineering matrices (*7, 8, 22–24, 50*). These polymers are readily processed into a variety of configurations, including fibers (*53*), porous sponges (*22, 24, 50*), and tubular structures (*54*). The regular structure of homopolymers of lactic (**1**) and glycolic (**2**) acid results in a crystalline structure (*51*). The pendant methyl group on lactic acid introduces an asymmetry into the polymer structure. Polylactic acid can be synthesized from the L-isomer or a combination of the D- and L-isomers. Poly(L-lactic acid) and polyglycolic acid are crystalline, whereas poly(D,L-lactic acid) and

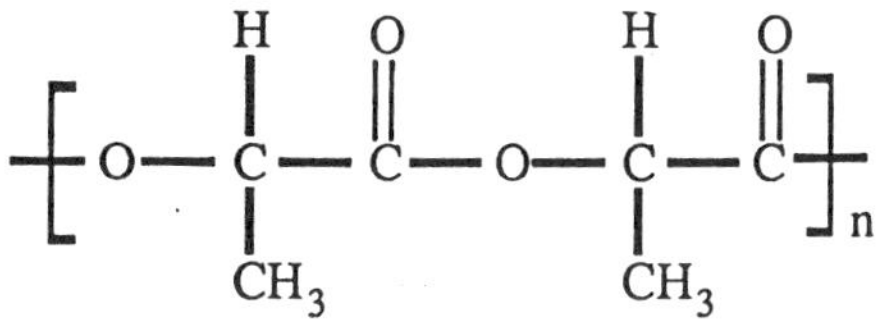

1, poly(lactic acid)

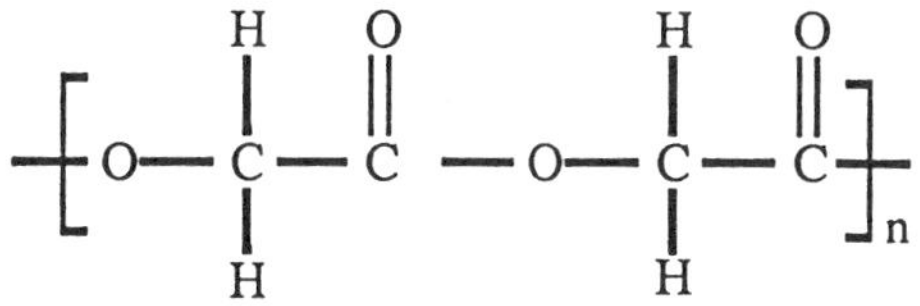

2, poly(glycolic acid)

TABLE 16.2. Typical Yield Stress Values and Erosion Times for Polymers of Lactic and Glycolic Acid

Polymer	*Yield Stress (psi × 10^3)*[a]	*Time for 50% Erosion*[b] *(weeks)*
Poly(glycolic acid)	11.2	4
50/50 Poly(D,L-lactic-*co*-glycolic acid)	7.7	6
85/15 Poly(D,L-lactic-*co*-glycolic acid)	6.3	20
Poly(D,L-lactic acid)	6.6	35
Poly(L-lactic acid)	8.5	>56

[a]Values represent the mean of five measurements obtained by using Instron testing with ASTM methods.

[b]Time at which half of the polymer has eroded (polymer mass = 1/2 initial mass) following immersion in a buffered saline solution maintained at 37 °C.

copolymers of lactic and glycolic acid containing significant amounts of both monomers (>10%) are typically amorphous (*51*). All polymers in this family degrade by hydrolysis of the ester bond. This polymer family's widely varying mechanical and erosion properties (Table 16.2) result both from the varying crystallinity and the differing hydrophobicity of lactic and glycolic acid (*51*). Cell-delivery matrices with a wide range of predefined degradation times and mechanical properties can be fabricated with this family of polymers (*55*). Other biodegradable polymers, including polyanhydrides, poly(ortho ester)s, and poly(amino acid)s (*52*) can also be used to fabricate biodegradable polymer matrices with controlled properties.

Nonbiodegradable polymers can provide a permanent barrier that prevents host cells or large-molecular-weight proteins (e.g., immunoglobulins) from directly contacting transplanted cells or cells in extracorporeal devices (*10–16, 37, 38, 40, 47, 56*). Semipermeable membranes are formed from a variety of polymers, including poly(acrylonitrile-*co*-vinyl chloride) (*39, 40, 47*), polylysine (*10–16*), cellulose acetate (*56*), and polysulfone (*56*). Cells can be immobilized within these semipermeable membranes by using a variety of gels, including alginate and polyphosphazenes. Polyphosphazenes are synthetic polymers, and aqueous solutions of polyphosphazenes will gel in the presence of specific ions (*57*). These polymers can be used in the same manner as alginate. The exceedingly stable backbone of these synthetic polymers allows significant alterations in side-group functionality without losing the gentle, physiologic gelling conditions (*58*).

The surface properties of synthetic materials can be easily and reproducibly altered to promote a superior biological response. Plasma modification and grafting of relatively inert substances such as polyethylene oxide or polyvinyl alcohol can mask the chemistry of the bulk matrix (*52*). The specific structure of adsorbed polymer coatings can be controlled by varying the chemical structure and molecular weight polydispersity of the coating polymer (*59*). Nanometer-scale molecular arrays also can be built at the material–cell interface by using molecular self-assembly strategies, defining the protein, and cellular interactions with material surfaces (*60*). For example, self-assembled monolayers of ω-functionalized alkane thiolates can be used as model systems for studying protein interactions with organic surfaces (*61*).

Calcium phosphate ceramics are used extensively in engineering bone tissue (*62*). Both hydroxyapatite and tricalcium phosphate, and mixtures of the two, are used. These materials can be coated over metal implants (*63*) or used alone as bone-conductive materials (*62, 64*) or as cell-delivery vehicles (*65*). These materials only release calcium and phosphate as breakdown products. They display no local or systemic toxicity, and they may become directly bonded to adjacent bone tissue with no intervening fibrous capsule (*62*). The erosion and mechanical properties of these materials are controlled by the specific chemical composition and processing conditions (*63*). However, applications of these materials are limited by their brittle nature and generally poor mechanical properties (*64*).

Synthetic Matrices That Mimic Natural Materials

Advantages and disadvantages exist for both natural materials (e.g., type I collagen) and synthetic materials (e.g., polyglycolic acid). Synthetic materials that incorporate design concepts or specific biological activities of natural biomaterials may combine the advantages of both types of materials. The reproducible, large-scale synthesis and flexible properties of synthetic polymers can be combined with the biocompatibility and biological activity of natural materials.

Amino acid sequences in ECM molecules that are responsible for specific biological activities (e.g., cell binding) have been identified in recent years (*41, 66*). This discovery has revolutionized the ability of researchers to design synthetic materials that are capable of precise cellular interactions. Genetic engineering approaches are being used to prepare artificial proteins with a desired backbone structure and amino acid side chains that promote cell adhesion (*67*). These artificial proteins can be expressed in bacterial cells, isolated and purified, and used to form matrices or coat other surfaces (*68*). This approach offers tremendous control over both the properties (bulk and surface) of the material and its ability to interact with cells. Issues of immunogenicity and purification from contaminants during large-scale production, however, must be addressed.

Traditional synthetic routes are also being used to develop biodegradable polymers that contain cell-recognition peptides as side chains (*69*). The advantages of synthetic polymers such as polylactide can be combined with the specific biological activity of ECM molecules with this approach (**3**). Amino acid side chains containing cell recognition peptides can be covalently bonded to the lysine residue (*69*). This polymer can be synthesized with varying ratios of lactide–lysine, or with glycolide in place of lactide. A family of polymers with a wide range of mechanical, degradative, and cell recognition properties can thus be synthesized. A similar approach is the synthesis of short amino acid chains containing a desired functional group that can be covalently bonded or adsorbed onto matrices fabricated from other synthetic materials (*70*).

Bone achieves its high mechanical moduli by combining organic and inorganic materials, and this principle is being used to synthesize new ceramic materials. Apatite crystals are being synthesized by nucleation and growth around poly(amino acid)s (*71*). This process results in an intimate dispersion of the organic molecules within the ceramic and improves the mechanical properties of the ceramic (*72*). This process may mimic the process of natural bone formation, and these materials show promise in engineering bone tissue (*73*).

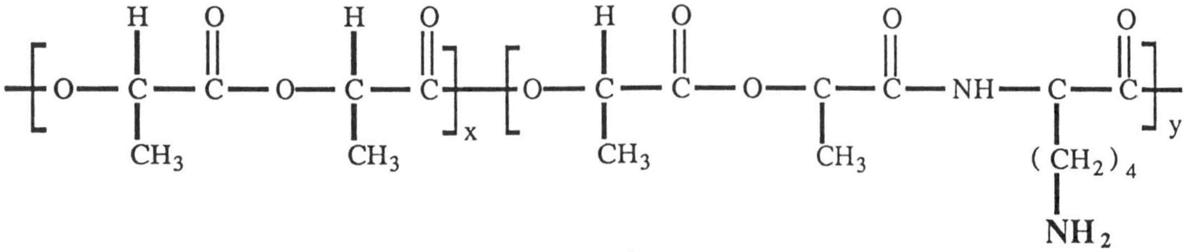

3, poly(lactic acid-*co*-lysine)

Integrating Cells and Matrices

Strategies

The goal of tissue engineering is to induce a group of cells to form a new tissue that replaces a lost or deficient tissue function. These cells could be a group of transplanted cells with no initial organization, or a group of cells already present in the host tissue at another site. The strategy to achieve this goal depends on the specific function one is attempting to replace and the available cell types. The first decision is whether an extracorporeal device will be used to replace lost function, or whether the new tissue will be engineered in vivo. A macroporous matrix is used to engineer a new tissue structurally integrated within the surrounding host tissue. Alternatively, the cells of interest may be encapsulated within a semipermeable membrane to protect them from the host immune system.

Extracorporeal support devices may be useful in treating acute diseases and diseases that only require intermittent treatment. Extracorporeal devices containing hepatocytes have been developed to treat acute liver failure and as a bridge to whole liver transplantation (*35, 37, 38*). This work has been motivated by the failure of hemodialysis and other non-cell-based extracorporeal devices to treat liver failure (*38*). These systems are unable to duplicate the numerous biochemical functions of the liver. Hepatocyte-based devices, however, show promise in limited clinical trials (*35*). Hepatocytes are typically localized to one compartment in these systems, whereas serum from the patient flows through a separate compartment (*35, 37, 38*). A semipermeable membrane separates the cells from the serum, allowing low-molecular-weight molecules to be exchanged but preventing the hepatocytes from directly contacting the patient's serum. This separation is necessary because several systems use a hepatoma-derived line (*35*) or porcine hepatocytes (*37*). The shortage of human livers and our inability to expand hepatocytes in culture limits the use of human hepatocytes.

Replacement tissues must be engineered in vivo if they are intended to replace a structural tissue (e.g., cartilage) or to replace a biochemical function on a continuous basis. Allogeneic or xenogeneic cells can be used without immunosuppression if limited biochemical functions will be performed by the transplanted cells. In this situation the cells are typically implanted within a semipermeable membrane to prevent their destruction by the host's immune system (*10–16, 35, 37–40, 74*). The semipermeable membrane allows the desired protein products (e.g., proteins or other low-molecular-weight species) to diffuse to the surrounding tissue while preventing cells from the host from directly contacting the implanted cells. The semipermeable membranes are preferably fabricated from nonbiodegradable materials.

Cells can be immunoprotected within semipermeable membranes using micro- or macroencapsulation techniques (*39*). Individual cells or small cell clusters are surrounded by a semipermeable membrane and delivered as a suspension in microencapsulation systems (*10–16, 37, 38*). Macroencapsulation systems typically use hollow fibers fabricated from semipermeable membranes to deliver multiple cells or cell clumps (*39, 40*). The small size, thin walls, and spherical shape of microcapsules optimize transport of molecules to and from the cells after implantation. Transport is not optimized in macroencapsulation devices, but these devices can exhibit high mechanical integrity. These devices can also be easily retrieved after implantation, in contrast to microcapsules. In both cases the cells can be immobilized within one matrix material (e.g., alginate), whereas the semipermeable membrane is formed of another material (e.g., poly[acrylonitrile-*co*-vinyl chloride]) (*39, 40*). The micro- and macroencapsulation materials and processes can be chosen to minimize cell death or damage during the process (*74*) while yielding a high density of encapsulated cells (Figure 16.3). Microencapsulated cells have cured animal models of human diseases, including single-gene liver defects (*37, 38*) and diabetes (*12–15*). Recently, microencapsulated islet cells have cured diabetics of insulin dependency in small-scale clinical trials (*16*). Macroencapsulated cells have been used to treat chronic pain and Parkinson's disease

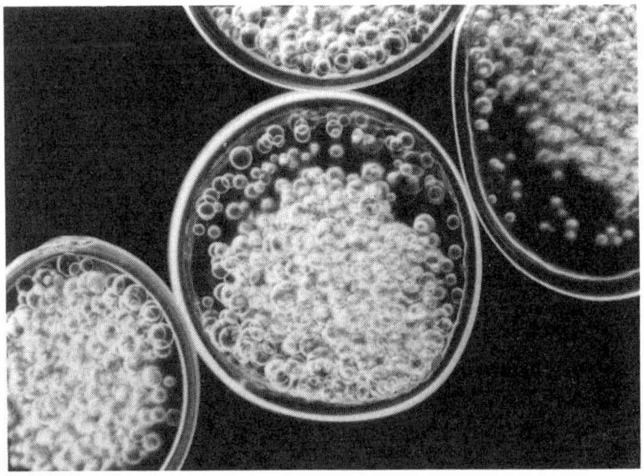

FIGURE 16.3. Calcium alginate microencapsulation matrices. The microencapsulation process can be used to encapsulate a variety of cell types at high densities, or agarose beads in this case, under physiologic conditions (H. Wang, unpublished data).

in animal models, and they may be effective in treating a variety of human diseases (*39, 40*).

The ultimate goal of tissue engineering is to replace a lost or deficient tissue with the same tissue. To achieve this goal the engineered tissue must be structurally continuous and integrated within the host. Matrices that allow free and unhindered communication and cellular transport between the new tissue and the surrounding tissue are used in these applications. Only autologous or allogeneic cells can be used with these matrices. Immunosuppression of the host may be required if allogeneic cells are used. The preferred matrix materials are biodegradable, because a completely natural tissue will result after matrix erosion.

A variety of structurally integrated tissues, including skin, bone, liver, ligament, intestine, cartilage, and tendon are currently being tested in animal or clinical studies (*1, 3–9, 21–24, 44, 50, 54, 75, 76*). Numerous attempts are being made to engineer skin tissue for burn victims (*3–9*). Skin tissue has been engineered using matrices fabricated from a variety of synthetic and naturally derived materials, including polyglycolic acid (*7, 8*) and type I collagen (*3–6, 9*). These matrices are being used to both transplant specific cell types (*3–5, 7–9*) and to serve as inductive matcrials for tissue regeneration by the host's own cells (*6*). Transplantation of chondrocytes on polyglycolic acid scaffolds leads to the formation of new cartilage tissue grossly and histologically resembling adult tissue (*75*). Chondrocytes proliferate on the polymer fibers in vitro and synthesize and deposit ECM molecules around themselves as they begin to form the cartilaginous matrix (*77*) (Figure 16.4). Liver tissue has also been engineered with cell-delivery matrices fabricated from synthetic polymers (*21–24*). The pore structure of these matrices has been controlled to promote rapid vascularization of the engineered liver tissue (*22, 24*).

Design Constraints

The microenvironment of an engineered tissue must be tightly regulated during the process of tissue development and perhaps well beyond this time. It will likely be critical that specific ECM molecules, growth factors, mechanical signals, and mass transport conditions be present to ensure the development of tissues with appropriate structure and function.

The tissue-specific function of cells in an engineered tissue must be maintained. The function of cultured cells is strongly dependent on the presence of specific growth factors and ECM molecules (*43*). For example, cells can be switched from a phase of tissue-specific gene expression to one of proliferation simply by altering the ECM presentation to the cell (*78*). Transplanted cells will probably require a defined microenvironment to maintain a desired program of gene expression. A variety of delivery matrices incorporate specific ECM molecules to provide the correct signaling to transplanted cells (*3–6, 9, 37, 38, 44–47*). Synthetic materials that incorporate specific peptides to enhance cell adhesion (*67, 69, 70*) likely will need to incorporate a variety of different peptides to mimic the multifunctional nature of ECM molecules (*66*). Growth factors critical to tissue development may not be present in the host tissue one is using to engineer a new tissue. To address this concern, traditional controlled drug delivery tech-

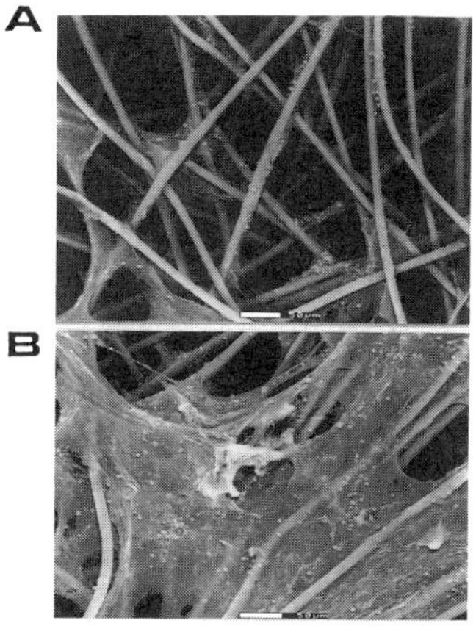

FIGURE 16.4. Scanning electron microscopy photomicrographs of bovine chondrocytes adherent to a macroporous matrix composed of polyglycolic acid fibers. The cells adhere to the fibers, multiply, and secrete and deposit ECM molecules within the matrix (*75, 77*). Numerous individual cells are adherent to the polymer fibers in photomicrograph (A), whereas the chondrocytes are largely embedded in the cell-generated cartilaginous matrix (B). Implantation of these chondrocyte–polyglycolic acid matrices leads to the formation of a new tissue that is histologically identical to adult cartilage following polymer erosion (*75*). These samples were glutaraldehyde-fixed and dehydrated and visualized by using an environmental scanning electron microscope. Size bars are shown on the photos (D. J. Mooney, unpublished data).

nology is being integrated with tissue engineering to provide transplanted cells with specific growth factors in their local environment (*79*).

Mechanical signals are known to regulate the development of a variety of tissues, including muscle (*80*) and bone (*81*). Mechanical stimuli (e.g., strain or shear) also clearly regulate the gene expression of cultured cells (*82*). To engineer a completely functional tissue the correct mechanical stimuli will probably have to be provided during the process of tissue development. For example, engineered tendons that are not subjected to mechanical loading do not develop mechanical moduli as high as normal tendons, even though they appear to be histologically identical (*76*). A thorough review recently was compiled on the regulation of mammalian cell phenotype by mechanical stimuli (*82*).

Mass transport between the engineered tissue and the host will be critical at several levels. First, the metabolic requirements (e.g., oxygen) of the developing tissue must be met, or the cells within the tissue will die. Micro- and macroencapsulated cells are dependent on diffusion of nutrients from the surrounding tissue to survive. Cells are typically encapsulated in small diameter devices (100–300 μm) to avoid mass-transport limitations on cell survival (*39*). However, uneven distribution of cells within matrices can lead to cell necrosis. Also, the development of mass-transport barriers (e.g., fibrous tissue encapsulation of matrices) decreases diffusive transport. Cells transplanted in macroporous matrices will also survive on diffusion until neovascularization of the new tissue occurs. Prevascularization of these devices before cell implantation improves cell engraftment (*21*). Transplanted cells also often perform some type of biochemical function that requires the transport of molecules to and from the vascular network (e.g., byproducts of metabolism that are present in blood must be transported to hepatocytes for modification or clearance). The transplanted cells may be unable to perform their desired function if the convective transport of blood to engineered tissues is low or significant diffusion barriers exist between capillaries and the transplanted cells.

Future Directions

A variety of approaches have been developed to engineer tissues by using cells and materials. The identification and isolation of stem cell populations may provide an inexhaustible source of cells for engineering tissues in the future. These cells may be capable of unlimited expansion in culture without losing their ability to differentiate. Using certain cell lines with any patient may also be possible by either genetically manipulating the cells to render them nonimmunogenic or by inducing host tolerance (*83, 84*). Techniques to precisely engineer materials and surfaces by using materials science, nanotechnology, and molecular biology approaches (*52, 60, 85*) are being developed. These techniques may allow the structure and function of engineered tissues, and individual cells within the tissues, to be precisely regulated. A significant challenge will involve engineering tissues (e.g., intestine and blood vessels) that contain multiple cell types organized in specific patterns (*86*). An important focus of this work will be the development of appropriate vascular and neural networks (*47*) in these complex tissues.

Tissue engineering represents the convergence of the pharmaceutical, biomaterials, and organ transplantation fields. Advances in cryopreservation and tissue banking may allow off-the-shelf replacement tissues to be routinely available to replace virtually any tissue that is damaged or that malfunctions. This ability would profoundly impact the manner in which medicine is practiced.

References

1. Langer, R.; Vacanti, J. P. *Science (Washington, D.C.)* **1993,** *260,* 920–926.
2. *Vital Statistics of the United States;* American Liver Foundation: Cedar Grove, NJ, 1988, Vol. 2.
3. Green, H.; Kehinde, O.; Thomas, J. *Proc. Natl. Acad. Sci. U.S.A.* **1979,** *76,* 5665–5668.
4. Yannas, I. V.; Burke, J. F.; Orgill, D. P.; Skrabut, E. M. *Science (Washington, D.C.)* **1981,** *215,* 174–176.
5. Bell, E.; Ehrlich, H. P.; Buttle, D. J.; Nakatsuji, T. *Science (Washington, D.C.)* **1981,** *211,* 1052–1054.
6. Stern, R.; McPherson, M.; Longaker, M. T. *J. Burn Care Rehabil.* **1990,** *11,* 7–13.
7. Heimbach, D.; Luterman, A.; Burke, J.; Cram, A.; Herndon, D.; Hunt, J.; Jordan, M.; McManus, W.; Solem, L.; Warden, G.; Zawacki, B. *Ann. Surg.* **1988,** *208,* 313–320.
8. Hansbrough, J. F.; Cooper, M. L.; Cohen, R.; Speilvogel, R.; Greenleaf, G.; Bartel, R. L.; Naughton, G. *Surgery (St. Louis)* **1992,** *4,* 438–446.
9. Compton, C.; Gill, J. M.; Bradford, D. A. *Lab. Invest.* **1989,** *60,* 600–612.
10. Lim, F.; Sun, A. M. *Science (Washington, D.C.)* **1980,** *210,* 908–910.
11. O'Shea, G. M.; Goosen, M. F. A.; Sun, A. M. *Biochim. Biophys. Acta* **1984,** *804,* 133–136.
12. Ricordi, C.; Scharp, D. W.; Lacy, P. E. *Transplantation* **1988,** *45,* 994–996.
13. Sullivan, S. J.; Maki, T.; Borland, K. M.; Mahoney, M. D.; Solomon, B. A.; Muller, T. E.; Monaco, A. P.; Chick, W. L. *Science (Washington, D.C.)* **1991,** *252,* 718–721.
14. Lacy, P. E.; Hegre, O. D.; Gerasimidi-Vazeou, A.; Gentile, F. T.; Dionne, K. E. *Science (Washington, D.C.)* **1991,** *253,* 1782–1784.
15. Levesque, L.; Brubaker, P. L.; Sun, A. M. *Endocrinology* **1992,** *130,* 644–650.
16. Soon-Shiong, P.; Sandford, P. A.; Heintz, R. *Abstracts of Papers,* Society for Biomaterials Annual Meeting, Boston, MA; Society for Biomaterials: Minneapolis, MN, 1994; Abstract 356.
17. Toriumi, D. M.; Kotler, H. S.; Luxenberg, D. P.; Holtrop, M. E.; Wang, E. A. *Arch. Otolaryngol. Head Neck Surg.* **1991,** *117,* 1101–1112.
18. Yasko, A. W.; Lane, J. M.; Fellinger, E. J.; Rosen, V.; Wozney, J. M.; Wang, E. A. *J. Bone Jt. Surg.* **1992,** *74,* 659–670.
19. Murphy, G. F.; Orgill, D. P.; Yannas, I. V. *Lab. Invest.* **1990,** *63,* 305–313.
20. Yannas, I. V.; Lee, E.; Orgill, D. P.; Skrabut, E. M.; Murphy, G. F. *Proc. Natl. Acad. Sci. U.S.A.* **1989,** *86,* 933–937.
21. Uyama, S.; Kaufmann, P. M.; Takeda, T.; Vacanti, J. P. *Transplantation* **1993,** *55,* 932–935.
22. Mooney, D. J.; Kaufmann, P. M.; Sano, K.; McNamara, K. M.; Vacanti, J. P.; Langer, R. *Transplant. Proc.* **1994,** *26,* 3425–3426.
23. Johnson, L. B.; Aiken, J.; Mooney, D.; Schloo, B. L.; Griffith-Cima, L.; Langer, R.; Vacanti, J. P. *Cell Transplant. (Elmsford, N.Y.)* **1994,** *4,* 273–281.
24. Mooney, D. J.; Park, S.; Kaufmann, P. M.; Sano, K.; McNamara, K.; Vacanti, J. P.; Langer, R. *J. Biomed. Mater. Res.* **1995,** *29,* 959–965.
25. Folkman, J.; Klagsbrun, M. *Science (Washington, D.C.)* **1987,** *235,* 442–447.
26. Mikos, A. G.; Sarakinos, G.; Lyman, M. D.; Ingber, D. E.; Vacanti, J. P.; Langer, R. *Biotechnol. Bioeng.* **1993,** *42,* 716–723.
27. Hu, W. S.; Peshwa, M. V. *Can. J. Chem. Eng.* **1991,** *69,* 409–420.
28. Nerem, R. M. *Ann. Biomed. Eng.* **1991,** *19,* 529–545.
29. Bruckner, P.; Horler, I.; Mendler, M. et al. *J. Cell Biol.* **1989,** *109,* 2537–2545.

30. Koller, M. R.; Palsson, B. O. *Biotechnol. Bioeng.* **1993,** *42,* 909–930.
31. Gordon, J. I.; Schmidt, G. H.; Roth, K. A. *FASEB J.* **1992,** *6,* 3039–3050.
32. Reid, L. M. *Curr. Opin. Cell Biol.* **1990,** *2,* 121–130.
33. Caplan, A. I. *J. Orthop. Res.* **1991,** *5,* 641–650.
34. Sigal, S. H.; Brill, S.; Fiorino, A. S.; Reid, L. M. *Am. J. Physiol.* **1992,** *263,* G139–G148.
35. Sussman, N. L.; Gislason, G. T.; Conlin, C. A.; Kelly, J. H. *Artif. Organs* **1994,** *18,* 390–396.
36. Grampp, G. E.; Sambanis, A.; Stephanopoulos, G. *Adv. Biochem. Eng./Biotechnol.* **1992,** *46,* 35–62.
37. Dixit, V. *Artif. Organs* **1994,** *18,* 371–384.
38. Kasai, S.; Sawa, M.; Mito, M. *Artif. Organs* **1994,** *18,* 348–354.
39. Emerich, D. F.; Winn, S. R.; Christenson, L.; Palmatier, M. A.; Gentile, F. T.; Sanberg, P. R. *Neurosci. Biobehav. Rev.* **1992,** *16,* 437–447.
40. Sagan, J.; Wang, H.; Tresco, P. A.; Aebischer, P. *J. Neurosci.* **1993,** *13,* 2415–2423.
41. Hynes, R. O. *Cell* **1987,** *48,* 549–554.
42. Schwartz, M. A.; Ingber, D. E. *Mol. Biol. Cell* **1994,** *5,* 389–393.
43. Stoker, A. W.; Streuli, C. H.; Martins-Green, M.; Bissell, M. J. *Curr. Opin. Cell Biol.* **1990,** *2,* 864–874.
44. Cavallaro, J. F.; Kemp, P. D.; Kraus, K. H. *Biotechnol. Bioeng.* **1994,** *43,* 781–791.
45. Koide, M.; Osaki, K.; Oyamada, K.; Konishi, J.; Katakura, T.; Takahashi, A.; Yoshizato, K. *J. Biomed. Mater. Res.* **1993,** *27,* 79–87.
46. DeLustro, F.; Dasch, J.; Keefe, J.; Ellingsworth, L. *Clin. Orthop. Relat. Res.* **1990,** *260,* 265–279.
47. Guenard, V.; Kleitman, N.; Morrissey, T. K.; Bunge, R. P.; Aebischer, P. *J. Neurol.* **1992,** *12,* 3310–3320.
48. Sutherland, I. W. In *Biomaterials: Novel Materials from Biological Sources;* Byron, D., Ed.; Stockton: New York, 1991; pp 308–331.
49. Atala, A.; Kim, W.; Paige, K. T.; Vacanti, C. A.; Retik, A. B. *J. Urol. (Baltimore)* **1994,** *152,* 641–643.
50. Mooney, D. J.; Vacanti, J. P. *Transplant. Proc.* **1993,** *7,* 153–162.
51. Gilding, D. K. In *Biocompatibility of Clinical Implant Materials;* Williams, D. F., Ed.; CRC Press: Boca Raton, FL, 1981; pp 209–232.
52. Peppas, N. A.; Langer, R. *Science (Washington, D.C.)* **1994,** *263,* 1715–1720.
53. Frazza, E. J.; Schmitt, E. E. *J. Biomed. Mater. Res.* **1971,** *1,* 43.
54. Mooney, D. J.; Organ, G.; Vacanti, J. P.; Langer, R. *Cell Transplant. (Elmsford, N.Y.)* **1994,** *3,* 203–210.
55. Mooney, D. J.; Mazzoni, C. L.; Breuer, C.; McNamara, K.; Hern, D.; Vacanti, J. P.; Langer, R. *Biomaterials* **1996,** *17,* 115–124.
56. Yang, M. B.; Vacanti, J. P.; Ingber, D. E. *Cell Transplant. (Elmsford, N.Y.)* **1994,** *3,* 373–385.
57. Cohen, S.; Bano, M. C.; Visscher, K. B.; Chow, M.; Allcock, H. R.; Langer, R. *J. Am. Chem. Soc.* **1990,** *112,* 7832–7833.
58. Allcock, H. R.; Kwon, S.; Riding, G. H.; Fitzpatrick, R. J.; Bennett, J. L. *Biomaterials* **1988,** *9,* 509–513.
59. Dan, N.; Tirrell, M. *Macromolecules* **1993,** *26,* 6467–6473.
60. Whitesides, G. M.; Mathias, J. P.; Seto, C. T. *Science (Washington, D.C.)* **1991,** *29,* 1312–1319.
61. Prime, K. L.; Whitesides, G. M. *Science (Washington, D.C.)* **1991,** *252,* 1164–1167.
62. *Bioceramics: Materials Characteristics and In Vivo Behavior;* Ducheyne, P.; Lemons, J., Eds.; New York Academy of Science: New York, 1988; Vol. 523.
63. Lemons, J. E. *Clin. Orthop. Relat. Res.* **1988,** *235,* 220–228.
64. Jarcho, M. *Clin. Orthop.* **1981,** *157,* 259–278.
65. Goshima, J.; Goldberg, V. M.; Caplan, A. I. *Clin. Orthop. Relat. Res.* **1991,** *269,* 274–283.
66. Hynes, R. O. *Fibronectins;* Springer-Verlag: New York, 1990.
67. McGrath, K. P.; Fournier, M. J.; Mason, T. L.; Tirrell, D. A. *J. Am. Chem. Soc.* **1992,** *114,* 727–733.
68. Anderson, J. P.; Cappello, J.; Martin, D. C. *Biopolymers* **1994,** *34,* 1049–1058.
69. Barrera, D. A.; Zylstra, E.; Lansbury, P. T.; Langer, R. *J. Am. Chem. Soc.* **1993,** *115,* 11010–11011.
70. Hubbell, J. A. *Trends Polym. Sci. (Cambridge, U.K.)* **1993,** *2,* 20–25.
71. Stupp, S. I.; Ciegler, G. W. *J. Biomed. Mater. Res.* **1992,** *26,* 169–183.
72. Stupp, S. I.; Mejicano, G. C.; Hanson, J. A. *J. Biomed. Mater. Res.* **1993,** *27,* 289–299.
73. Stupp, S. I.; Hanson, J. A.; Eurell, J. A.; Ciegler, G. W.; Johnson, A. *J. Biomed. Mater. Res.* **1993,** *27,* 301–311.
74. Wang, H. Y.; Eisfeld, T.; Sakoda, A. In *Biologics from Recombinant Microorganisms and Animal Cells;* White, M. D.; Reuveny, S.; Shafferman, K., Eds.; VCH Publishers: New York, 1994; pp 173–182.

75. Vacanti, C.; Langer, R.; Schloo, B.; Vacanti, J. P. *Plast. Reconstr. Surg.* **1991**, *88,* 753–759.
76. Cao, Y.; Vacanti, J. P.; Ma, X.; Paige, K. T.; Upton, J.; Chowanski, Z.; Schloo, B.; Langer, R.; Vacanti, C. A. *Transplant. Proc.* **1994**, *26,* 3390–3392.
77. Freed, L. E.; Vunjak-Novakovic, G.; Langer, R. *J. Cell. Biochem.* **1993**, *51,* 257–264.
78. Mooney, D.; Hansen, L.; Vacanti, J.; Langer, R.; Farmer, S.; Ingber, D. *J. Cell Physiol.* **1992**, *151,* 497–505.
79. Mooney, D. J.; Kaufmann, P. M.; Sano, K.; Schwendeman, S. P.; Majahod, K.; Schloo, B.; Vacanti, J. P.; Langer, R. *Biotechnol. Bioeng.* **1996**, *50,* 422–429.
80. Vandenburgh, H. H.; Swasdison, S.; Karlisch, P. *FASEB J.* **1991**, *5,* 2860–2867.
81. Carter, D. R.; Orr, T. E.; Fyhrie, D. P. *J. Biomechanics (Elmsford, N.Y.)* **1989**, *22,* 231–244.
82. *Physical Forces and the Mammalian Cell;* Frangos, J. A., Ed.; Academic: San Diego, CA, 1993.
83. Posselt, A. M.; Barker, C. F.; Tomaszewski, J. E.; Markmann, J. F.; Choti, M. A.; Naji, A. *Science (Washington, D.C.)* **1990**, *249,* 1293–1295.
84. Sykes, M.; Sachs, D. H.; Nienhuis, A. W.; Pearson, D. A.; Moulton, A. D.; Bodine, D. M. *Transplantation* **1993**, *55,* 197–202.
85. Ratner, B. D. *J. Biomed. Mater. Res.* **1993**, *27,* 837–850.
86. Atala, A. et al. *J. Urol. (Baltimore)* **1993**, *150,* 608–612.
87. Mooney, D. J.; Breuer, C.; McNamara, K.; Vacanti, J. P.; Langer, R. *Tissue Eng.* **1995**, *1,* 107–118.
88. Erickson, H. P. In *Plasma Fibronectin;* McDonagh, J., Ed.; Hematology Series; Marcel Dekker: New York, 1985; Vol. 5, p 38.
89. Simon, K. O.; Burridge, K. In *Integrins Molecular and Biological Responses to the Extracellular Matrix;* Cheresch, D. A.; Mecham, R. P., Eds.; Biology of Extracellular Matrix Series; Academic: San Diego, CA, 1994; p 59.

17

Drug Delivery Using Genetically Engineered Cell Implants

Richard L. Eckert, Daniel J. Smith, and Irwin A. Schafer

The therapeutic potential of somatic cell gene therapy has been investigated extensively in recent years; however, efforts to apply the techniques have been hampered by an inability to achieve sustained gene expression, difficulties in transferring the genes to the recipient cells, and problems with immune rejection of the cells. In this chapter we review plasmid-based and retrovirus-based methodologies for gene transfer. We describe efforts to deliver biomolecules to the central nervous system, epidermis, and pancreas by using cell implants. We also discuss an important new area that is evolving, the implantation of genetically engineered cells that are encapsulated within a semipermeable barrier. The encapsulation of genetically engineered cells is the focus of our research to develop a bandage to facilitate wound healing.

Genetic engineering is a process whereby the genetic composition of the cell is modified by addition of exogenous DNA. The process involves the physical transfer of DNA into a recipient cell and is generally performed for the purpose of programming the recipient cell to produce a new protein. The most efficient method of delivering DNA is by the use of transfer vectors, generally either plasmids or disabled viruses (*1, 2*).

Plasmids as Gene-Transfer Vectors

Plasmids are double-stranded, closed circular DNA molecules that are about 3–4 kb in length. Plasmids encode two features that are important for their propagation in bacteria, an origin of replication and a gene that confers antibiotic resistance (*1, 3*). These "prokaryotic" plasmid segments permit the production of large quantities of a given plasmid in bacteria (*3*). The expression plasmid shown in Figure 17.1 is designed to function in bacteria and in eukaryotic cells. It is grown in bacteria for purposes of amplification, purified, and then transferred into eukaryotic cells by transfection. In this example, the prokaryotic segment (upper half) contains the gene that encodes resistance to the antibiotic, ampicillin (AMP-R). The origin of replication is also indicated (ORI). The prokaryotic origin of replication is a specific DNA

Copyright © 1997 American Chemical Society

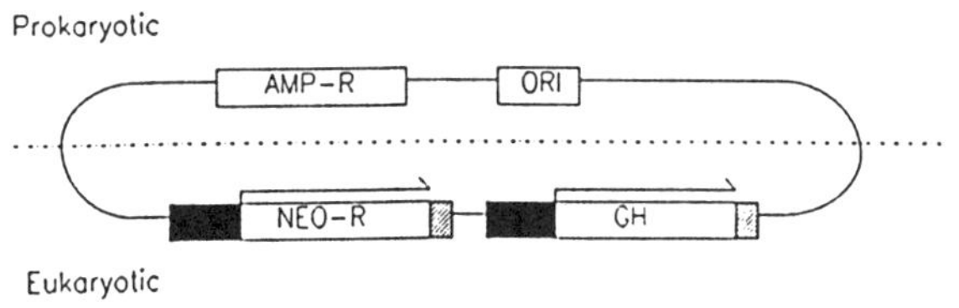

FIGURE 17.1. Plasmid expression vector. The narrow line represents the plasmid backbone. The solid boxes represent the promoter elements, and the slashed-line boxes represent the transcription terminator elements. The arrows indicate the direction of transcription: AMP-R, ampicillin resistance gene; NEO-R, gene that confers resistance to G418 in eukaryotic cells; GH, growth hormone; ORI, bacterial origin of replication. The dashed line divides the prokaryotic and eukaryotic plasmid segments.

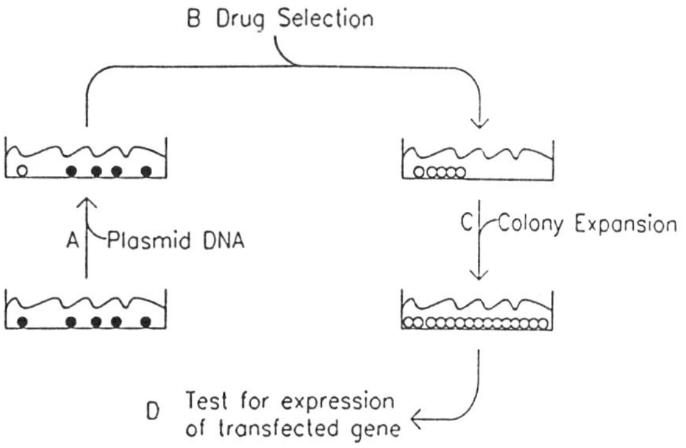

FIGURE 17.2. Transfer of DNA into cells by transfection. The target cells are indicated by the black circles. A cell that has taken up the plasmid is shown by the open circle. The transfection (A), drug selection (B), colony expansion (C), and testing for gene expression (D) steps are shown.

sequence that binds to factors that regulate replication of plasmid DNA and, in turn, control the number of copies of plasmid per bacterium. In general, the goal is to produce the largest possible number of copies of plasmid DNA per bacterium. In addition to the prokaryotic segments, expression plasmids contain two transcription units designed to function in eukaryotic cells (Figure 17.1). Each eukaryotic transcription unit includes a promoter and terminator. These elements flank the gene of interest. In Figure 17.1, the promoter of each transcription unit is indicated by a solid box, and the termination signal is indicated by the slashed-line boxes. The arrows indicate the direction of transcription. One transcription unit encodes the gene to be expressed (in this example growth hormone, GH), and the second encodes a marker that can be used to select those cells that have taken up the plasmid (i.e., the gene that encodes aminoglycoside phosphotransferase, NEO-R). The selection marker is necessary, because only a fraction of the cells (1–20%) in the population will take up the plasmid.

Transfer of Plasmids to Target Cells

Transfection is a process whereby naked DNA is incubated with agents that coat it and make it more amenable for uptake into cells. Figure 17.2 shows a typical experiment in which the eukaryotic recipient cells (growing in a culture dish) are transfected with plasmid DNA in the presence of lipids, calcium phosphate, or Polybrene (A) (*3*). After the transfection, the cells are allowed to recover and are then incubated for several days to several weeks in the presence of drug selection (B). Only those cells that have taken up the plasmid survive the drug selection, and eventually, clonal populations of cells appear (B). These colonies are then expanded (C) and tested for expression of the transferred gene (D).

Advantages and Disadvantages

The use of plasmids as vectors for gene transfer has several advantages and disadvantages. A major disadvantage is the low transfection efficiency (i.e., only 1–20% of the transfected cells take up and retain the plasmid). For this reason, transfection with plasmids works well with immortalized cell lines, where the cells can be passaged for an extended period of time in the selection medium. However, this process is more difficult with normal cells, because the lifetime in culture is limited. Major advantages include the fact that plasmid-based systems can be designed to produce large quantities of protein, and unlike viruses (*see* subsequent discussion), plasmids probably will not rearrange to form an infectious entity.

Retroviruses as Gene-Transfer Vectors

Retroviral Structure

The most commonly used type of virus in gene-transfer research is the retrovirus (Figure 17.3). The form of the virus used in gene therapy is a disabled retrovirus in which viral genes required for packaging or transcription are removed from the virus (*4*). In the recombinant virus, the product to be expressed replaces these segments of the viral genome. Figure 17.3 shows a map of the wild-type retrovirus showing the positions of the viral gag, pol, and env genes (A). A recombinant virus is shown in B. It is a growth hormone expression vector in which the three viral genes (gag, pol, and env) have been replaced by the genes encoding drug resistance (NEO-R) and the GH gene. The black boxes at each end are the retroviral long terminal repeats (LTR), which are responsible for mediating viral integration into the target-cell DNA and for starting and terminating transcription (i.e., the left LTR starts and the right LTR stops transcription). In this example, the left viral LTR is responsible for transcribing the NEO-R gene, and an additional promoter (central solid box) has been added to initiate transcription of the GH gene. Both transcription units terminate at the right LTR.

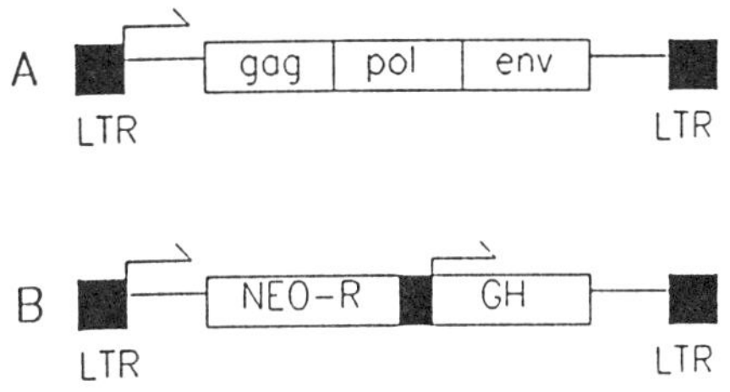

FIGURE 17.3. Retroviral genomes. A wild-type retrovirus is shown in A. The gap, pol, and env genes are proteins encoded by the virus that are required for packaging or synthesis of the viral genome. A recombinant retrovirus is shown in B. In this example, the virus has been engineered to express the drug resistance gene, NEO-R, and to produce growth hormone (GH). The solid box adjacent the GH gene is a promoter added to transcribe the GH protein. Arrowheads indicate the direction of transcription from the viral LTR and the internal promoter.

Generating Retroviral Infectious Particles

The gag, pol, and env genes encode viral proteins that are essential for synthesis and packaging of the viral genome. Because these genes are absent from the recombinant virus, recombinant viruses cannot be packaged (i.e., form infectious viral particles) without help. This assistance is provided by a packaging cell line (*4, 5*) (Figure 17.4). The recombinant virus is transfected (A) into the packaging line by using the scheme outlined in Figure 17.2. The packaging cells then begin to produce and package viable, infectious recombinant virus. These viral particles are released from the cells and collect in the culture medium. They are concentrated and are then used for infecting the target cells (B). The packaging cell can package the recombinant virus because it provides trans-functions (gap, pol, and env) via a helper virus that is integrated into its DNA. However, when the recombinant virus is transferred to its final destination (via infection), the target cell (i.e., liver, skin, spleen cell, etc.) does not possess helper virus functions. Therefore, the virus enters the target cell (Figure 17.4B) and becomes permanently integrated into the host cell DNA (Figure 17.4C). No new virus is produced; however, the viral transcriptional units continue to produce the NEO-R and GH gene products after integration. The cells are then assayed to ascertain the level of GH production (D). The advantage of this system over plasmids is that the transfer to the recipient cell takes place by infection, which is a highly effi-

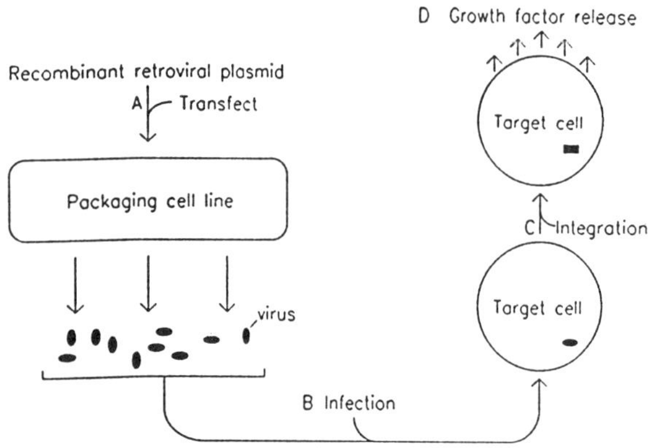

FIGURE 17.4. Packaging of the recombinant retrovirus. The retrovirus shown in Figure 17.3B exists as a plasmid and is grown in bacteria. The plasmid is purified from bacteria and then transferred to the packaging cell line by transfection (A). The packaging cell line converts the plasmid into a packaged virus, which is then released from the packaging line into the culture medium. The virus is collected from this medium, concentrated, and used to infect the target cell (B). The virus enters the cell and becomes permanently integrated into the target cell DNA (C). The cells are then assayed for growth factor production (D). The integrated virus transcribes the transferred gene, and the gene product is released. The virus is indicated by the small solid oval, and the integrated virus is indicated by the solid rectangle.

cient process. Thus, transfer to >90% of the targeted cells can sometimes be attained. However, two significant disadvantages of this system exist. First, the level of protein produced by the virus is generally relatively low, and second there is a small chance that the viral genome will be mobilized from its integration site in the host genome, which could lead to undesired side effects such as tumor formation.

Other Gene-Transfer Vectors

Two other types of vectors are worth mentioning. Adenovirus is a virus that exhibits tropism for respiratory and oral epithelial cells. It has been used to deliver the cystic fibrosis transmembrane conductance regulatory gene to airway epithelial cells (*6*).

Vectors derived from human papillomavirus (HPV), which target epithelial cell types, are also possible delivery vectors. HPV-based vectors remain as nuclear, extrachromosomal episomes following transfection (*7–9*). Because vector DNA can be altered (mutated) during the process of integration into the host cell genome, the lack of integration of the HPV vector can be an advantage. A disadvantage is that the episome may be lost from the cells during cell division.

Promoter Elements and Gene Delivery

Constitutive Promoters

A major factor when considering a gene-therapy application is the selection of the correct promoter to regulate expression of the gene. High-level expression can be achieved by using viral promoters (*1, 10, 11*) such as the SV40 promoter. These promoters yield constitutively high levels of expression in target cells in vitro. However, these promoters do not always maintain high-level in vivo expression over long periods of time. In addition, it is difficult to envision the use of these constitutive promoters for delivery of gene products that must be tightly regulated in response to physiological agents (e.g., glucose regulation of insulin release).

Tissue-Specific Promoters

An alternative is the use of tissue-specific promoters. These promoters are only expressed in particular tissues and only at specific stages during differentiation. For example, a gene encoding a clotting factor could be linked to a promoter that is only expressed in skin cells. Once skin cells had been engineered to produce the factor, they would be transplanted into the skin. The skin cells would then deliver the factor to the blood via the vessels that feed the skin and correct the blood deficiency. Thus, the delivery implant could be localized to a tissue different from that normally responsible for production.

Complete genes may also be used. This method is especially important when a complex pattern of regulation must be maintained. For a structural defect in a protein, the normal gene (complete with promoter) can be transferred into the cells containing the defective gene. In this case, the transferred gene contains the appropriate regulatory sequences and should be regulated appropriately by physiological agents. This process should correct the problem. However, in other situations, lack of expression may be due to a regulatory defect. The DNA sequence encoding the protein is correct, but the gene is not appropriately regulated. These problems can be more difficult to correct and may require the replacement or supplementation of the deficient cells with healthy cells.

Hybrid Promoters

Hybrid promoters are also a possibility. For example, in the case of skin cells that have been engineered to express a particular gene by using a tissue-specific promoter, expression could be controlled by adding to the promoter a regulatory element that responds to a topically applied agent. Topical application of the agent could then upregulate the expression of the target gene and increase delivery of the product to the bloodstream.

Diseases as Targets for Engineered Implant Therapy

Targeting Defective Cells

As previously mentioned, one form of therapy involves correcting a defect in a cell that normally provides the function. In these types of diseases, the endogenous gene is producing a defective product that is the cause of the disease or is producing too little of a product. Some examples of these types of diseases and how they can be corrected are listed below. Correction of defects in bone-marrow progenitor cells include replacing the adenosine deaminase gene in severe combined immune deficiency, the hypoxanthine phosphoribosyl transferase gene in Lesch–Nyhan disease, the β-globin gene in thalassemia, and the glucocerebrosidase in Gaucher's disease (*12–16*). Low-density lipoprotein receptor targeting in hepatocytes may be a possibility for treatment of hypercholesterolemia (*17*).

Targeting Heterologous Cell Types

A second form of gene therapy consists of targeting heterologous cell types. Various cell types can be used for this purpose, including fibroblasts, endothelial cells, lymphocytes, keratinocytes (epidermal cells), glial cells, and others. In this form of therapy, the gene is introduced so that a particular level of production is obtained. Because these cells do not ordinarily produce the protein that is expressed, regulated expression of the gene is not always possible. However, for genes that need to be expressed at a continuous basal level, this form of targeting may be useful.

Other disorders that are being studied as candidates for engineered implant therapy include blood vessel diseases. Engineering endothelial cells to produce tissue plasminogen activator or inhibitors of smooth muscle proliferation and placing them into vessels or prostheses may reduce blood vessel closure due to cell hyperproliferation (*18*). Clotting diseases may be treated by engineering cells to produce factors VIII and IX (*19, 20*). Pulmonary diseases that may be treated by gene therapy include cystic fibrosis (*6*). Endocrine or metabolism-related diseases include engineering cells to produce insulin to treat diabetes mellitus and glucocerebrosidase for Gaucher's disease (*21, 22*). Cancers can also be treated in this manner by engineering, for example, tumor-infiltrating lymphocytes to produce interferon or tumor necrosis factor (*23*). Finally, treatment of chronic wounds, as discussed subsequently, is a candidate application for the use of genetically engineered cells (*1*).

Gene Therapy Using Genetically Engineered Cells

Genetically engineered cells have been used in two distinct ways. The most common method of gene delivery has been the use of genetically engineered cells without a surrounding membrane. The second method is delivery of genetically engineered cells that have in some way been confined within a membrane or matrix. Although it is not a perfect description, for the sake of simplicity, we will refer to the process of enclosure of a cell inside of any semipermeable membrane or sphere as encapsulation.

Genetically Engineered Cell Implants

In some situations, the implanted cells should not be confined inside any type of membrane. For example, cells derived from a subject's bone marrow can be engineered in vitro to produce a particular factor by infection with a retrovirus and then replaced into the patient. These cells then circulate in the bloodstream and eventually take up residence in the bone marrow. In this case, encapsulation of the cells may serve no useful purpose and may hinder the cells from localizing in the marrow. A disadvantage of this type of approach is that the engineered cells come into direct contact with the patients tissue, and in general, this characteristic confines the technique to the use of normal cells.

Encapsulated Genetically Engineered Cell Implants

Efforts to replace pancreatic cell types (*24*) or to deliver proteins to the epidermal surface (*1*) have used cell encapsulation methods. These methods have a variety of advantages that will be discussed further, including maintaining the cells in a single location and protecting the cells from immune surveillance.

Encapsulation methods must meet several important criteria (*5*). Shape and size are of obvious importance, because the shape can influence where the device can be implanted. Size is important, because the absolute size of the device limits the number of cells that can be implanted and how efficiently the biomolecule is released. Porosity and macrostructure are important, because they affect the transport of nutrients to the implanted cells. Finally, surface microstructure and chemistry affect the ability of cells to attach, survive, and function normally in the implant.

Wound Healing and Delivery of Drugs to Skin

The epidermis has been proposed as a site for the delivery of drugs to the skin surface and for delivery to the bloodstream. The skin has several advantages as a site for a drug delivery system. The skin is readily accessible, and the total surface area of the skin is large and permits replacement of a substantial area with growth-factor-producing cells. Investigators have taken several approaches to exploiting this possibility, including repopulating the dermis with genetically engineered fibroblasts, or repopulating the epidermis with sheets of engineered keratinocytes (*10, 25*). This approach is feasible, for example, in the case of cultured keratinocytes, because sheets of epidermal keratinocytes can be cultured, engineered to express the protein of interest, and then transferred back onto the patient. A major problem with this approach is that if a cell type is engineered to overexpress a growth factor and these cells are placed on a patients skin, the continued overproduction may cause unwanted side effects. The possibility that this result may occur is a major consideration in all types of gene therapy.

These problems can be overcome by application of the engineered cells in a closed bandage system from which the cells cannot escape. The genetically engineered biological bandage (GEBB) is a concept in which genetically engineered cells are confined within a polymeric envelope to provide a versatile drug delivery system (*1*) (Figure 17.5). Such an encapsulated system can be used to deliver biological products to the bloodstream for purposes of gene therapy or to deliver products to the skin surface to improve wound healing.

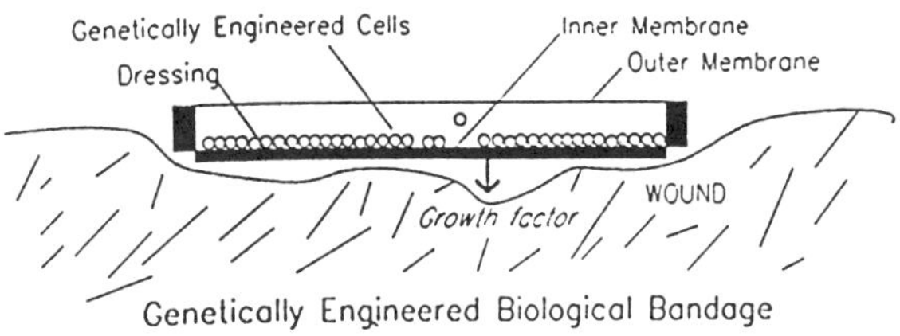

FIGURE 17.5. The genetically engineered biological bandage. The bandage is shown in cross-section. The solid rectangles indicate the circular silicon rubber gasket that provides structural support for the bandage. The outer membrane consists of a thin sheet of silicon rubber, and the inner membrane consists of a polymer that permits release of biomolecules. The dressing is a hydrogel material that provides an interface between the bandage and the wound. The genetically engineered cells are indicated by small circles.

We recently developed a prototype system designed to deliver biologicals to the skin surface for wound healing or gene therapy (*1*). In this system, a human epidermal keratinocyte cell line is engineered to produce a high level of a specific growth factor by using a plasmid-based expression vector. The engineered cells are then placed into a polymeric envelope consisting of a silicon rubber gasket bordered on one side by a thin film of silicon rubber and on the other by a membrane that is permeable to proteins of ≤40 kDa. The bandage is then placed onto the wound as shown in Figure 17.6. The interface between the wound and the bandage is a hydrogel dressing. Our initial results indicate that this system is able to deliver recombinant growth factor to the wound over a period of several days (Figure 17.6).

The GEBB concept has numerous advantages:

1. The bandage is completely reversible (i.e., the cells can be completely and quantitatively removed as desired).
2. The cells producing growth factor are isolated in an envelope and cannot escape into the wound.
3. The bandage can provide for uniform delivery of the biological over the entire wound surface.
4. The released growth factor will be biological active, because it is synthesized "on site" for immediate release.
5. The bandage can be changed as required to permit the delivery of different growth factors.
6. Bandages can be designed that produce multiple growth factors simultaneously.
7. The system is expected to be much less expensive than delivery of recombinant, purified growth factors.
8. The bandage can be used to deliver a wide variety of biological molecules that are important in tissue remodeling (i.e., proteases, protease inhibitors, growth factors, etc.).
9. The cells within the bandage are hidden from the patient's immune system.

Potential disadvantages with this system include problems with maintaining cell viability while the bandage is in place and problems associated with plugging of the pores in the bandage membrane (*1*).

Genetically Engineered Cells and the Nervous System

Encapsulated cell implants are ideally suited for use in situations where small changes in the local growth factor balance can produce a change in function. The central nervous system (CNS) is one such area. In this context, the cells can be confined so that release can be restricted to a small area. Frim et al. (*26*) used a fibroblast cell line that had been engineered to produce nerve growth factor (NGF) by using retroviral transfer technology. These cells, when implanted in the corpus callosum, reduced the maximal cross-sectional area of a excitotoxic lesion by 80% compared to a nonengineered cell control. Sagen et al. (*27*) have shown that xenogeneic chromaffin cells, isolated by semipermeable membranes, can survive and reduce pain when transplanted into the CNS. In this case a major concern of the study was to isolate the grafted cells from the immune system to prevent rejection. These cells, when implanted into the rat

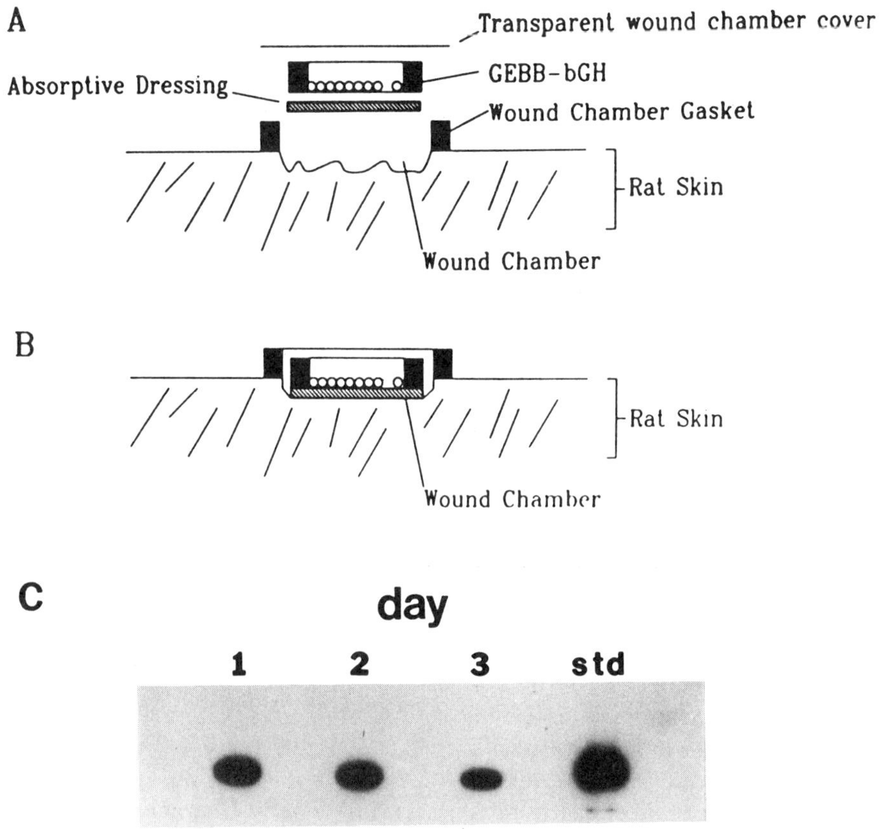

FIGURE 17.6. Genetically engineered bandage-mediated delivery of bovine growth factor to a rat wound. Panel A is a schematic of the bandage and wound model. The wound chamber is applied directly to the back of the rat, a full thickness wound is generated inside the wound chamber, and the bandage is placed atop the wound (B). After 1, 2, and 3 days, the bandage was removed and the hydrogel dressing was assayed for the presence of bGH by immunoblot (C). A known amount (1 μg) of authentic bovine growth hormone was electrophoresed in the std lane as a control. (Reproduced with permission from reference 1. Copyright 1993 John Wiley and Sons.)

spinal subarachnoid space, reduced pain sensitivity for intervals up to 3 months after implantation. These results suggest that cells can be successfully isolated from the immune system and can maintain functional integrity for prolonged periods of time. Alzheimer's disease is another target for genetic therapy. Recent studies show that primary fibroblasts can be engineered to express choline acetyltransferase. Implantation of these cells into the hippocampus of rats showed that the cells continued to produce acetylcholine after grafting, and interestingly, the level of acetylcholine produced by the cells could be regulated by local infusion of choline (*28*).

Other Delivery Systems

Several investigators have been working to develop a bioartificial pancreas by using microencapsulated islet cells (*24, 29, 30*). In most cases these studies used normal, nonengineered cells. An important finding with these studies is that the cells are protected from the immune system and continue to function for prolonged periods of time. Successful implantation of liver and cartilage cells also was reported (*31*). Some experiments used engineered cells. Tai and Sun (*32*) produced an encapsulated cell delivery system in which mouse fibroblasts, engineered to produce human growth hormone, were encased in an alginate–poly(L-lysine)–alginate (APA) sphere. When this system was placed in mice by intraperitoneal injection, growth hormone was continuously present in the blood over a period of several months (*32*). At the end of the treatment, the cells were recovered, explanted in culture, and still produced growth hormone.

Encapsulation of Permanent Immortalized Cell Lines

Although in some circumstances, it may be desirable to use normal primary cell cultures for use in gene-therapy applications, a lesson from these outlined studies is that immortal cell lines can be successfully used in therapy. For example, in the study described of mouse cells engineered to produce growth hormone, the cells were encapsulated at a low density in APA spheres and then implanted into mice (*32*). Although the cells grew to high densities in the spheres, the cells remained viable and there was no suggestion that the cells had escaped the encapsulation medium and populated other body areas. Andreatta-van Leyen and co-workers (*1*) showed that genetically engineered cells could be confined within a bandage, survive in vivo for several days in the hostile wound environment, and continue to produce the engineered factor. Additional studies (Andreatta-van Leyen, Smith, Schafer, and Eckert, unpublished) suggest that this length of time can be extended to several weeks.

Areas for Future Work

When designing a system for use in gene therapy, several factors must be considered (*33*). First, to achieve sustained delivery of the gene product, the promoter that is selected to drive transcription must maintain activity. Second, if the cells to be used are xenogeneic, they must be shielded from the immune system. Third, the cells must remain viable in vivo over extended time periods. Fourth, although constitutive gene expression can be readily achieved in vitro and sometimes observed in vivo, achieving correct *regulation* of gene expression of targeted genes (e.g., regulation of insulin gene expression by blood glucose) will require more extensive characterization of the gene regulatory elements of interest. For this reason, the diseases that are more amenable to correction by programming constitutive production of the engineered gene product are presently more amenable for gene therapy.

Progress has been made in these areas, especially in maintaining cell viability in vivo and in shielding the cells from the immune system by using semipermeable membranes. We can expect that the future will see increased use of xenogeneic cells and immortalized cell lines that are contained within polymeric structures. Care will need to be taken to assure that these systems guarantee 100% retention of the engineered cells. Although this result should not be hard to accomplish, some encapsulation mediums that are currently in use (e.g., APA spheres) may not survive this certification.

Acknowledgment

This work was supported by a grant from the Edison Biotechnology Center of Ohio.

References

1. Andreatta-van Leyen, S.; Smith, D. J.; Bulgrin, J. P.; Schafer, I. A.; Eckert, R. L. *J. Biomed. Mater. Res.* **1993**, *27*, 1201–1208.
2. Temin, H. M. In *Gene Transfer;* Kucherlapati, R., Ed.; Plenum: New York, 1986; pp 149–187.
3. *Current Protocols in Molecular Biology;* Ausubel, F. M.; Brent, R.; Kingston, R. E.; Moore, D. D.; Seidman, J. G.; Smith, J. A.; Struhl, K., Eds.; Current Protocols: New York, 1994; Vols. 1 and 2.

4. Miller, A. D.; Rosman, G. J. *BioTechniques* **1989,** *7,* 980–990.
5. Nerem, R. M. *Med. Biol. Eng. Comput.* **1992,** CE8–CE12.
6. Rosenfeld, M. A.; Yoshimura, K.; Trapnell, B. C.; Yoneyama, K.; Rosenthal, E. R.; Dalemans, W.; Fukayama, M.; Bargon, J.; Stier, L. E.; Stratford-Perricaudet, L.; Perricaudet, M.; Guggino, W. B.; Pavirani, A.; Lacocq, J. P.; Crystal, R. G. *Cell* **1992,** *68,* 143–155.
7. DiMaio, D.; Treisman, R.; Maniatis, T. *Proc. Natl. Acad. Sci. U.S.A.* **1982,** *79,* 4030–4034.
8. Broker, T. R.; Botchan, M. *Cancer Cells* **1986,** *4,* 17–36.
9. Mungal, S.; Steinberg, B. M.; Taichman, L. B. *J. Virol.* **1992,** *66,* 3220–3224.
10. Morgan, J. R.; Barrandon, Y.; Green, H.; Mulligan, R. *Science (Washington, D.C.)* **1987,** *237,* 1276–1480.
11. Miller, A. D.; Bender, M. A.; Harris, E. A. S.; Kaledo, M.; Gelinas, R. E. *J. Virol.* **1988,** *62,* 4337–4345.
12. Lim, B.; Williams, D. A.; Orkin, S. *Mol. Cell. Biol.* **1987,** *7,* 3459–3465.
13. Wilson, J. M.; Danos, O.; Grossman, M.; Raulet, D. H.; Mulligan, R. C. *Proc. Natl. Acad. Sci. U.S.A.* **1990,** *87,* 439–443.
14. Gruber, H. E.; Finley, K. D.; Luchtman, L. A.; Hershberg, R. M.; Katzman, S. S.; Laikind, P. K.; Meyers, E. N.; Seegmiller, J. E.; Friedman, T.; Yee, J. K. *Adv. Exp. Med. Biol.* **1986,** *195,* 171–175.
15. Bender, M. A.; Gelinas, R. E.; Miller, A. D. *Mol. Cell Biol.* **1989,** *9,* 1426–1434.
16. Nolta, J. A.; Sender, L. S.; Barranger, J. A.; Kohn, D. B. *Blood* **1990,** *75,* 787–797.
17. Wilson, J. M.; Johnston, D. E.; Jefferson, D. M.; Mulligan, R. C. *Proc. Natl. Acad. Sci. U.S.A.* **1988,** *85,* 4421–4425.
18. Dichek, D.; Neville, R.; Zwiebel, J. A.; Freeman, S. M.; Leon, M. B.; Anderson, W. F. *Circulation* **1989,** *80,* 1347–1353.
19. Palmer, T. D.; Thompson, A. R.; Miller, A. D. *Blood* **1989,** *73,* 438–445.
20. Gerrard, A. J.; Hudson, D. L.; Brownlee, G. G.; Watt, F. M. *Nat. Genet.* **1993,** *3,* 180–183.
21. Seldon, R. F.; Skoskiewicz, M. J.; Russell, P. S.; Goodman, H. M. *N. Engl. J. Med.* **1987,** *332,* 1067–1076.
22. Choudary, P. V.; Barranger, J. A.; Tsuji, S.; Mayor, J.; LaMarca, M. E.; Cepko, C. L.; Mulligan, R. C.; Ginns, E. I. *Mol. Biol. Med.* **1986,** *3,* 293–299.
23. Kasid, A.; Morecki, S.; Aebersold, P.; Cornetta, K.; Culver, K.; Freeman, S.; Director, E.; Lotze, M. T.; Blaese, R. M.; Anderson, W. J.; Rosenberg, S. A. *Proc. Natl. Acad. Sci. U.S.A.* **1990,** *87,* 473–477.
24. Lim, F.; Sum, A. M. *Science (Washington, D.C.)* **1980,** *210,* 908–910.
25. St. Louis, D.; Verma, I. M. *Proc. Natl. Acad. Sci. U.S.A.* **1988,** *85,* 3150–3154.
26. Frim, D. M.; Short, M. P.; Rosenberg, W. S.; Simpson, J.; Breadefield, X. O.; Isacson, O. *J. Neurosurg.* **1993,** *78,* 267–273.
27. Sagen, J.; Wang, H.; Tresco, P. A.; Aebischer, P. *J. Neurosci.* **1993,** *13,* 2415–2423.
28. Fisher, L. J.; Raymon, H. K.; Gage, F. H. *Ann. N. Y. Acad. Sci.* **1993,** *695,* 278–284.
29. Colton, C.; Augoustiniatos, E. S. *J. Biomech. Eng.* **1991,** *113,* 152–170.
30. Freidman, E. A. *Diabetes Care* **1989,** *12,* 415–420.
31. Cima, L. G.; Vacanti, J. P.; Vacanti, C.; Ingber, D.; Mooney, D.; Langer, R. *J. Biomech. Eng.* **1991,** *113,* 143–151.
32. Tai, I. T.; Sun, A. M. *FASEB J.* **1993,** *7,* 1061–1069.
33. Zwiebel, J. A.; Freeman, S. M.; Newman, K.; Dichek, D.; Ryan, U. S.; Anderson, W. F. *Ann. N. Y. Acad. Sci.* **1991,** *618,* 394–404.

18
Ribozymes as Antiviral Agents

Akira Wada, Takashi Shimayama, De-Min Zhou, Masaki Warashina, Masaya Orita, Tetsuhiko Koguma, Jun Ohkawa, and Kazunari Taira

The hammerhead ribozyme belongs to a class of antisense molecules. However, because of its pocket that captures magnesium ions (catalytic site) and because magnesium ions can cleave phosphodiester bonds, it can act as an enzyme (metalloenzyme). The hammerhead ribozyme is one of the smallest RNA enzymes. Because of its small size and potential as an antiviral agent, it has been extensively investigated in terms of its mechanism of action and its application in vivo. Although naked ribozymes are unstable in vivo, they can be completely protected from attacks by RNases in human serum, once they form complexes with poly-L-(lysine:serine) random copolymer. This kind of complex formation provides a potentially powerful methodology for delivering any nucleic acid drugs including antisense molecules into cells.

The term *ribozyme* is derived from the terms *ribo*nucleic acid (RNA) and en*zyme*, and it denotes a type of RNA molecule with catalytic properties. Until the last decade researchers believed that RNA merely acted as an intermediary in the process of genetic information transfer from DNA to protein molecules. It was not until the publication of work by Altman (*1*) and Cech (*2*) that RNA was shown to play a catalytic role in the cell. A number of other ribozymes (*3*) have been discovered since the original discovery of RNase P and the *Tetrahymena* ribozyme, but from certain standpoints the most important discovery has been that of the hammerhead-type ribozyme (*4–7*). This ribozyme can act within a single molecule (cis-acting) but has also been engineered in such a way that it acts against other molecules (trans-acting). The trans-acting hammerhead ribozyme developed by Haseloff and Gerlach (*6*) consists of an antisense section (stems I and III) and a catalytic domain with a flanking stem II–loop section (Figure 18.1).

Hammerhead Ribozymes

Cleavage of RNA Molecules

The general secondary structure of cis-acting hammerhead ribozyme is shown in Figure 18.1a.

Copyright © 1997 American Chemical Society

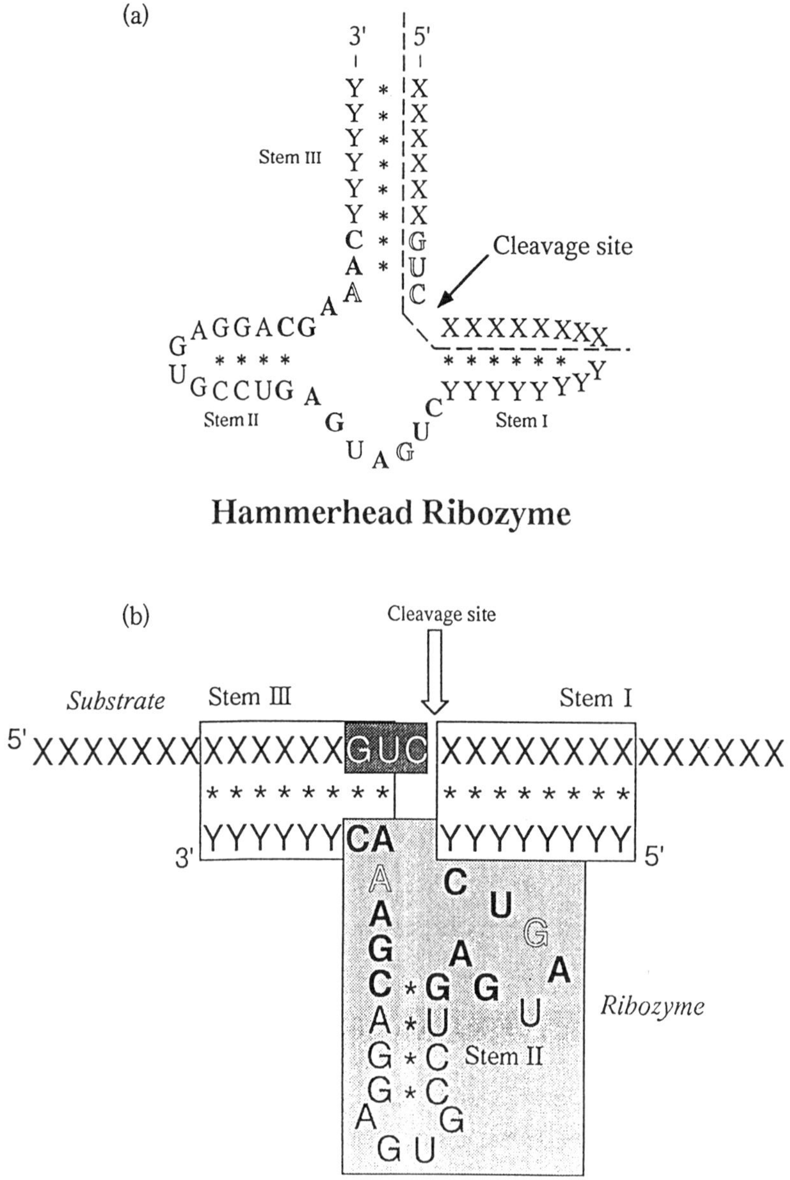

FIGURE 18.1. Naturally occurring cis-acting (a) and engineered trans-acting (b) hammerhead ribozymes.

This cis-acting molecule can be engineered in two ways such that it can act in trans (*5, 6*). In one case, the hammerhead domain was divided between stems I and II (*5*), and in another it was divided between stems I and III as shown by the dotted line in Figure 18.1a (*6*). These two trans-reaction systems are, hereafter, designated trans 1 and trans 2, respectively. To define the sequence requirements for the hammerhead structure, extensive mutagenesis studies of the conserved region have been performed in cis (*8*), trans 1 (*9, 10*), and trans 2 (*11–13*) reaction systems.

Shown in Figure 18.1b is the trans 2 system in that the binding sites indicated by Ys are complementary to the X regions of the substrate RNA (stem I and stem III). The arrow after GUC shows the site of cleavage. To examine the importance of the conserved trinucleotide GUC at the cleavage site, several mutagenesis studies were carried out (*8–13*). Earlier results revealed that G at the third base of the triplet, which might extend stem I by forming a G_{17}:C_3 pair [the numbering of the bases is in accord with the nomenclature of Hertel et al. (*14*); *see* Figure 18.2 of the structure used in our study (*12*)], inhibited the cleavage reaction (**b**, **c**, and **d** in Figure 18.3A) except in one case (**a** in Figure 18.3A). Also, a U residue in the central position was required for efficient cleavage (*9–11*). These observations led to the generally accepted NUX rule (N is A, U, G, or C; X is A, U, or C), which states that a substrate with a NUX triplet can be cleaved by a hammerhead ribozyme.

However, some inconsistencies existed among the results reported by different groups (Figure 18.3A). For example, a substrate that contained

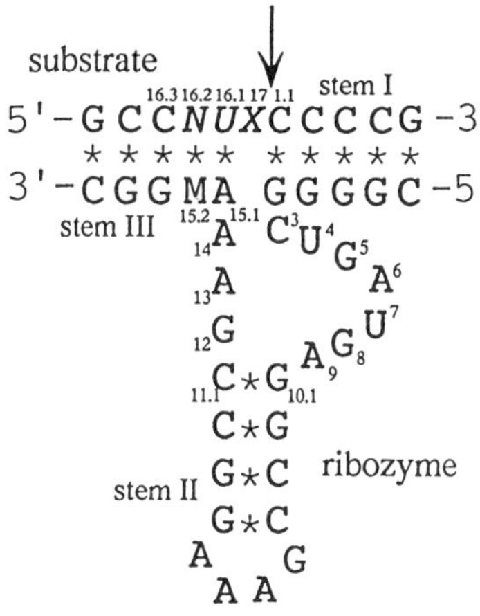

FIGURE 18.2. Secondary structure of hammerhead ribozymes used in our studies. Unless specified, N = G, X = C, and M = C. Wild-type ribozyme (M = C) is designated "R32". Mutagenesis results based on this sequence are summarized in Table 18.1 and Figure 18.3B.

the AUC triplet was cleaved in one study with efficiency comparable with cleavage of the wild type [**c** in Figure 18.3A (*10*)], whereas in other studies, that substrate was cleaved with much lower efficiency [**b** in Figure 18.3A (*9*)] or no cleavage was observed [**d** in Figure 18.3A (*11*)]. The differences could have been due not only to differences among reaction systems, which include the type of reaction (cis or trans and also trans 1 or trans 2), the sequence of the hammerhead complex, and the reaction conditions, but also to the experimental design. In every case only a simple comparison of cleavage activities of certain substrates at fixed concentrations was performed. Even though previous results were determined with ribozyme at a 1.3- to 1.5-fold molar excess compared with the substrate, the observed rate of cleavage would be lower than the maximum rate (k_{cat}) unless the total concentration of substrate and ribozyme were high enough with respect to the Michaelis constant (K_m). In other words, the determined rate constants could reflect either k_{cat} or k_{cat}/K_m values, depending on the concentrations of substrate and ribozyme used in the experiments. Therefore, determining whether a reduction in rate is caused by a reduced k_{cat} value or by an increase in K_m value was important. Thus, we felt that the determination of individual k_{cat} and K_m values was essential for an objective discussion of differences in activity between various mutant forms of the cleavage site (*12*).

From a practical point of view, because trans 2 reactions are suitable for targeting certain RNAs (*6, 7*), a detailed examination of the NUX rule in trans 2 systems should help us to choose target sequences. Therefore, to investigate the range of cleavage activity of the NUX triplets, we performed detailed kinetic analysis of all possible NUX mutants by measuring individual k_{cat} and K_m values. Results of such studies are shown in Figure 18.3B and actual kinetic parameters are listed in Table 18.1. These results indicate that GUC can be cleaved most efficiently and CUC is next. Therefore, in choosing a target site in trans-acting systems, GUC or CUC may be selected preferentially. However, in cis-acting systems in

A

a

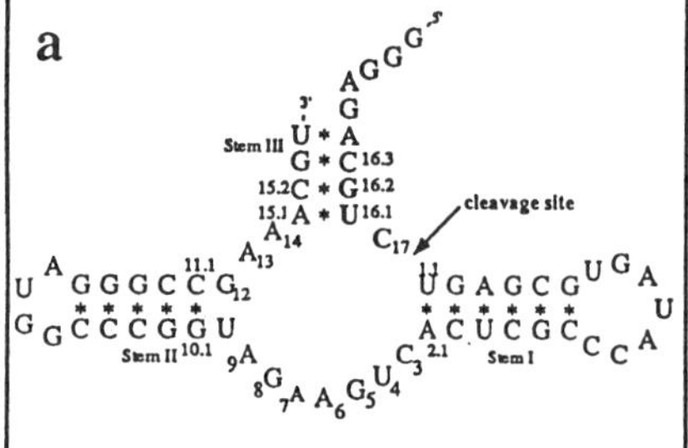

b

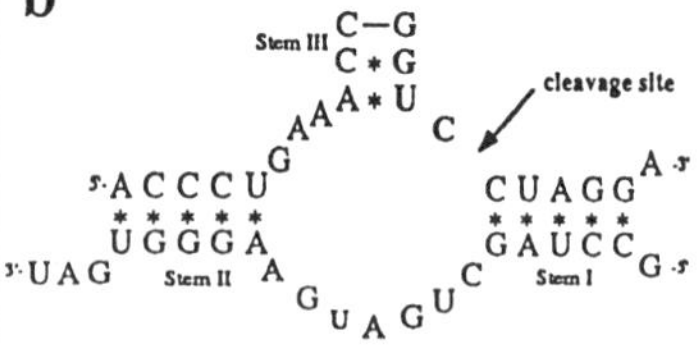

c

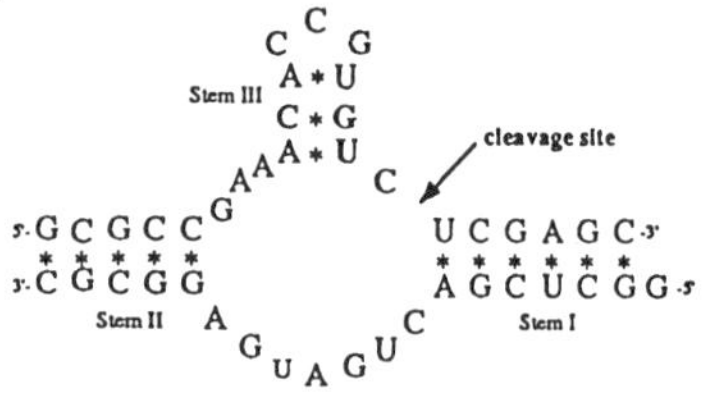

d

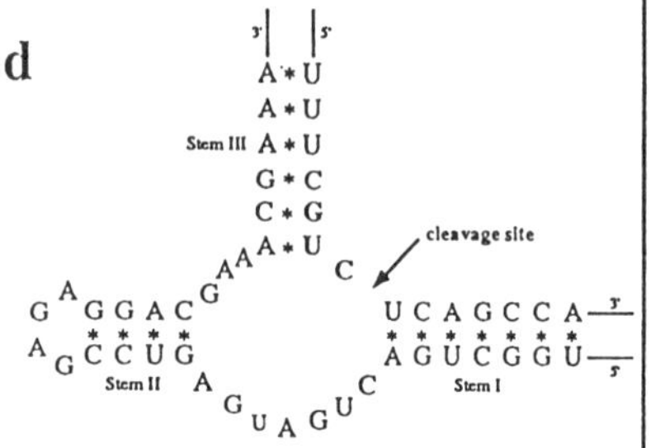

B

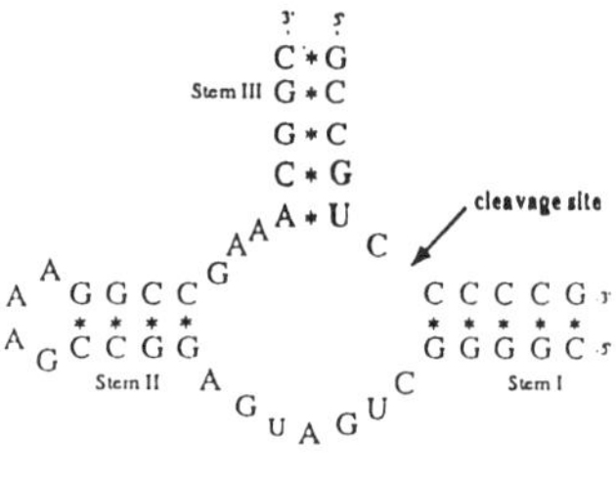

		AUC	UUC	CUC	GUA	GUU	GUG
A	a	NM	NM	NM	NM	1	0.3
A	b	0.1	0.4	1	1	1	0
A	c	0.57	0.25	0.083	0.7	0.05	<0.003
A	d	0	1.2	0.95	0.93	0.68	0
B	rel. k_{cat}	1.1	0.13	0.48	0.60	0.065	0
B	rel. K_m	35	1.7	2.5	30	1.7	

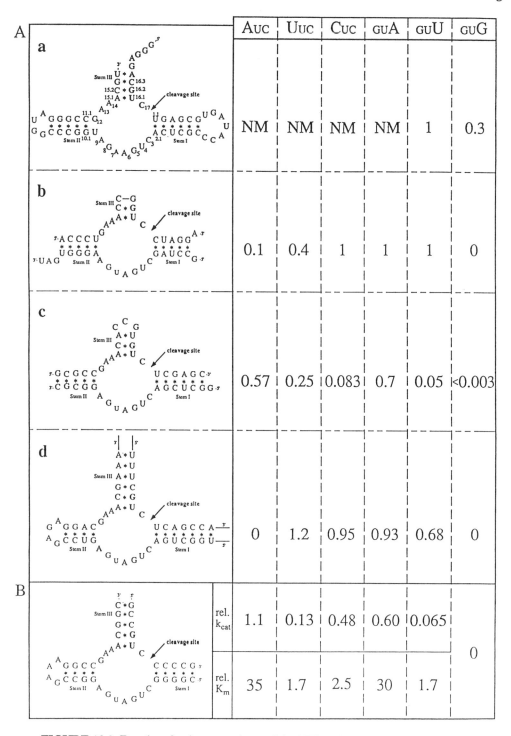

FIGURE 18.3. Results of point mutations of the NUX triplet at the cleavage site.

TABLE 18.1. Kinetic Parameters of Mutant Substrates with Changes in NUX Triplet

Triplet	k_{cat} *(min⁻¹)*	K_m *(nM)*	k_{cat}/K_m *(μM⁻¹·min⁻¹)*	*Relative* k_{cat}/K_m
GUC	4.0	20	200	1
AUC	4.4	700	6.3	0.032
UUC	0.52	33	16	0.080
CUC	1.9	50	38	0.19
GUA	2.4	600	4.0	0.020
GUU	0.26	34	7.6	0.038
AUA	2.1	320	6.6	0.033
AUU	0.53	320	1.7	0.0085
CUA	0.16	64	2.5	0.013
CUU	0.050	45	1.1	0.0055
UUA	0.39	140	2.8	0.014
UUU	0.10	33	3.3	0.015

which K_m values are irrelevant, other triplets such as AUC, GUA, and AUA may be chosen (Table 18.1). In fact, the minus strand of the virusoid of Lucerne transient streak virus, (-)vLTSV, and plus strand of satellite RNA of barley yellow dwarf virus, (+)sBYDV, use the GUA and the AUA triplets, respectively, for hammerhead cleavage during their replicating processes. The triplet (NUX)-dependent reactivity difference summarized in Table 18.1 appears general (*12*) because another well-controlled experiment that examined a completely different flanking sequence showed nearly the same result (*13*).

Minimum Reaction Scheme for trans-Acting Hammerhead Ribozymes

The trans-acting hammerhead ribozyme consists of an antisense section (stems I and III) and a catalytic domain with a flanking stem II–loop section (Figure 18.1b). The minimum reaction scheme can be described by Figure 18.4. First, the substrate (and Mg^{2+} ions) binds to the ribozyme to form a Michaelis–Menten complex via formation of base pairs with stems I and III (k_{assoc}). Then, a specific phosphodiester bond in the bound substrate is cleaved by the action of Mg^{2+} ions [k_{cleav}; the ribozyme is recognized to function as a metalloenzyme (*15–22*)]. This cleavage produces products with 2′,3′-cyclic phosphate and 5′-hydroxyl groups. Finally, the cleaved fragments dissociate from the ribozyme, and the liberated ribozyme is now available for a new series of catalytic events (k_{diss}). As can be seen in Figure 18.5, the nucleophile in the reaction catalyzed by a hammerhead ribozyme is 2′-alkoxide, similar to the cases catalyzed by RNases. Therefore, hammerhead ribozymes are incapable of cleaving DNA substrates.

The activation energy for a reaction can be determined by measuring the reaction rate constant (*k*) at different temperatures (*T*) and plotting log *k* versus 1/*T* to yield a so-called Arrhenius plot. The Arrhenius plot itself may be nonlinear if different steps become the rate-determining step at different temperatures. In some cases, the plot may show a sharp change in slope at some temperature (transition temperature) at which the rate-determining step changes from one to another. An Arrhenius plot was used to detect such changes in a ribozyme-catalyzed reaction (*23*). As shown in Figure 18.6, distinct changes in the slope of the plot were recognized. The plot provides evidence for three different rate-determining steps in the reaction. Arrhenius activation energies were calculated to be 16.0 kcal/mol at

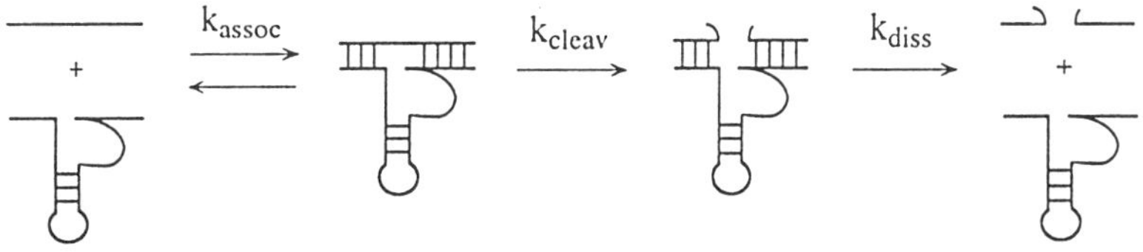

FIGURE 18.4. Minimal reaction scheme for hammerhead ribozymes.

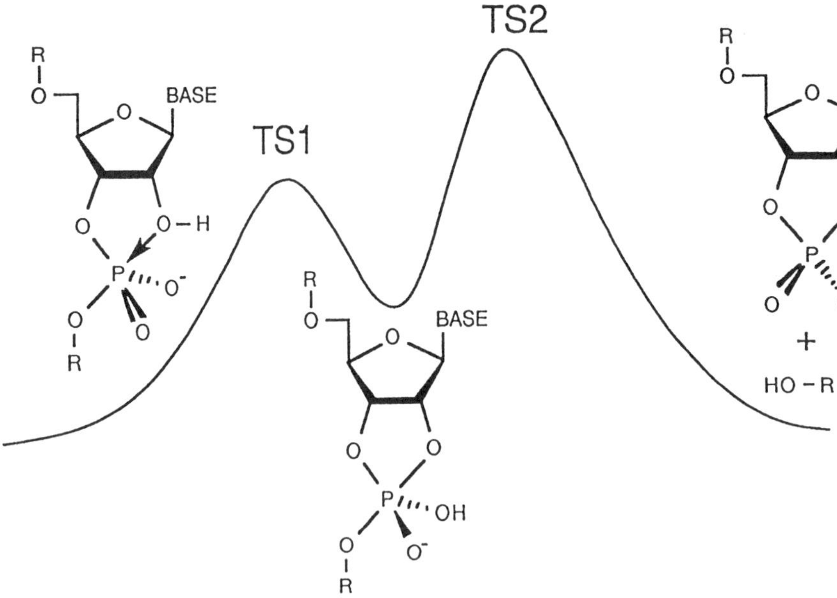

FIGURE 18.5. Schematic diagram for RNA-cleaving reaction.

midrange temperatures (25–50 °C) and 47.7 kcal/mol at lower temperatures (<25 °C). At 25–50 °C, the chemical cleavage step (k_{cleav}) was clearly the rate-determining step because no burst kinetics were detected at the measurement temperature of 37 °C (*24*). Therefore, at midrange temperatures, the cleaved fragments dissociated from the ribozyme at a higher rate than the rate of the chemical reaction ($k_{cleav} < k_{diss}$). When the temperature was below 25 °C, the cleaved fragments adhered to the ribozyme more tightly and the product-dissociation step became the rate-determining step. Above 50 °C, the rate of the reaction decreased, probably because at such high temperatures the formation of the Michaelis–Menten complex was hampered by thermal melting (therefore, the rate of the reaction above 50 °C does not reflect k_{cat}). The melting temperature (T_m) of stem II of this ribozyme was above 80 °C (*25*). This kind of analysis is useful in characterizing the reactions catalyzed by other types of ribozyme, including engineered ribozymes.

Transition temperatures heavily depend on the length of the binding arms (stems I and III). When longer binding arms are used, the product-dissociation step (k_{diss}) becomes the rate-determining step even at temperatures near 37 °C. Therefore, long binding arms are not always advantageous, especially in vitro. However, in vivo, proteins may exist that facilitate unwinding of the RNA duplex, and thus, longer binding arms might be used (*26, 27*).

Modified Ribozymes

As part of an effort to characterize structure–function relationships, several modified ribozymes including short ribozymes (minizymes) were chemically synthesized and their activities were examined.

Chimeric RNA–DNA Ribozymes

To elucidate structure–function relationships, several mixed RNA–DNA ribozymes and other modified ribozymes have been synthesized chemically. Examination of the mixed RNA–DNA ribozymes revealed that most deoxyribonucleotide substitutions can be tolerated. However, in the majority of

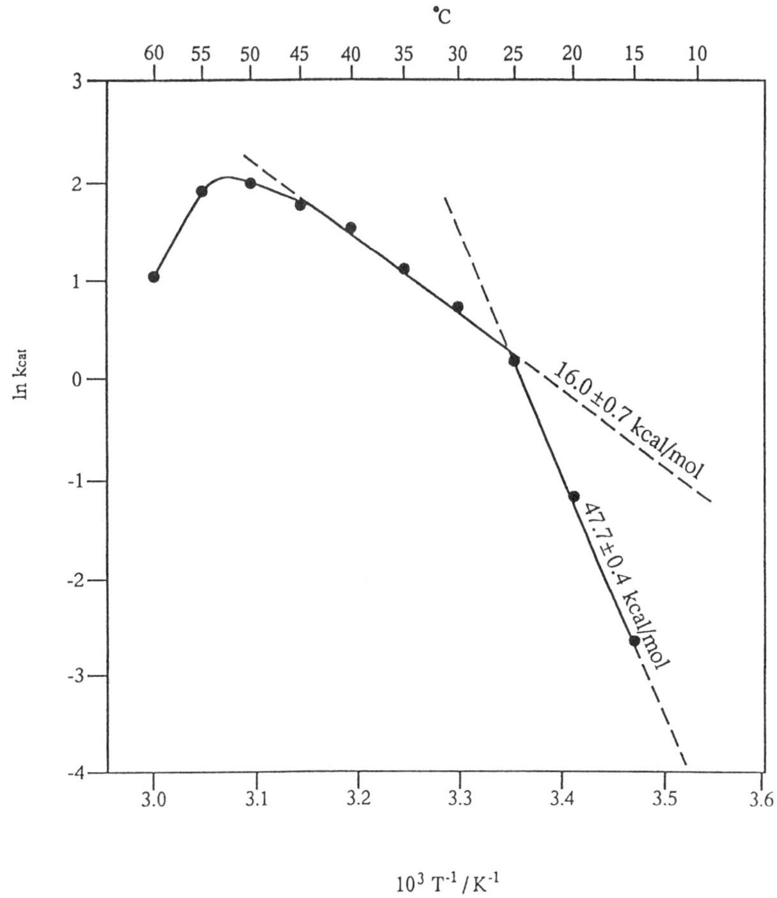

FIGURE 18.6. Arrhenius plot of (R32) ribozyme-catalyzed reaction.

cases, catalytic activities are somewhat reduced. When the rate-determining step is the product-dissociation step (k_{diss}, Figure 18.4), modifications that lower the affinity for substrate (and, thus, for products as well) enhance the catalytic activity. In fact, the first example of such an acceleration has been demonstrated for *Tetrahymena* ribozyme: The rate-limiting step in the reaction catalyzed by the *Tetrahymena* ribozyme is the product-dissociation step because replacement of one of the phosphoryl oxygens at the cleavage site by sulfur, which slows down the chemical cleavage reaction, has little effect on the overall rate of the reaction [the "no-thio effect" is apparent in this case (*28, 29*)]. Cech's group (*30*) successfully improved the activity of *Tetrahymena* ribozyme by mutating nonconserved sequences with resultant decreased affinity for RNA, and thus, higher turnover. However, mutations that increased the turnover number were associated with a slower chemical-cleavage step.

Enhancement in k_{cat} was also recognized for chimeric RNA–DNA hammerhead ribozymes in which binding arms (stem I and III) were made of DNA (*24, 31–34*). In all cases, the enhancement in k_{cat} was accomplished by the introduction of DNA arms and the origin of the acceleration was based on the two distinctly different phenomena.

In the case of the study by Rossi's group (*31*), the enhancement originated from the acceleration of dissociation of the products from the reaction complex. In general, DNA–RNA duplex is less stable than RNA–RNA duplex; thus, the DNA–RNA duplex enhances k_{diss} (Figure 18.4). In the case of the study by Hendry et al. (*32*), the rate-limiting step for the chimeric DNA–RNA ribozyme was the chemical cleavage step (k_{cleav}); although, for their all-RNA ribozyme, the rate-limiting step appeared still to be the product-dissociation step. In our case (*24, 33, 34*), the chemical cleavage step was clearly enhanced by the DNA arms.

As compared to that of the *Tetrahymena* ribozyme (*30*), the chemical step of the reaction catalyzed by hammerhead-type ribozymes is about two orders of magnitude slower, and thus, it can easily become the rate-determining step as shown in Figure 18.6 (at midrange temperatures). In fact, Uhlenbeck and others have demonstrated a thio effect for hammerhead ribozymes (*35–37*). Fedor and Uhlenbeck (*38*) also elegantly measured individual elementary rate constants for substrate-binding, cleavage, and product-dissociation steps, and they demonstrated a change of the rate-limiting step depending on the length of the substrate.

In designing the ribozymes in our case (Figure 18.2; N = G, X = C, M = C), we tried to minimize the substrate-binding sequence and thereby to increase the rate of the product-dissociation step (*12, 21–25, 33, 34*). Then, ambiguity regarding the rate-determining step should be minimized. Moreover, the binding sequence was designed so that self-complementation and inter-substrate complementarity could be avoided, resulting in complete cleavage of substrates without the formation of inactive complexes (*25*).

To compare the activities between all-RNA ribozyme (R32) and the corresponding DNA-armed ribozyme (DRD32), in which 5 nucleotides at the 5′-end and 3 nucleotides at the 3′-end were replaced by DNA [5′-dCdGdGdGdG-(sequence including the active site)-AC-dGdGdC-3′], cleavage reactions were carried out under substrate-saturating conditions (k_{cat} control) (*24, 33, 34*). Results are shown in Figure 18.7. In this figure, to demonstrate that the observation of a higher k_{cat} for the DNA-armed ribozyme (DRD32) is not an artifact caused by the degradation of the more ribonuclease-sensitive all-RNA enzyme (R32) during the preparation and reaction, both the ribozymes and the substrates were 5′-end-labeled with ^{32}P. Although the ratio of substrate to ribozyme used was roughly 300:1, for both to be visualized on the same gel (Figure 18.7), substrates were only partially end-labeled. In other

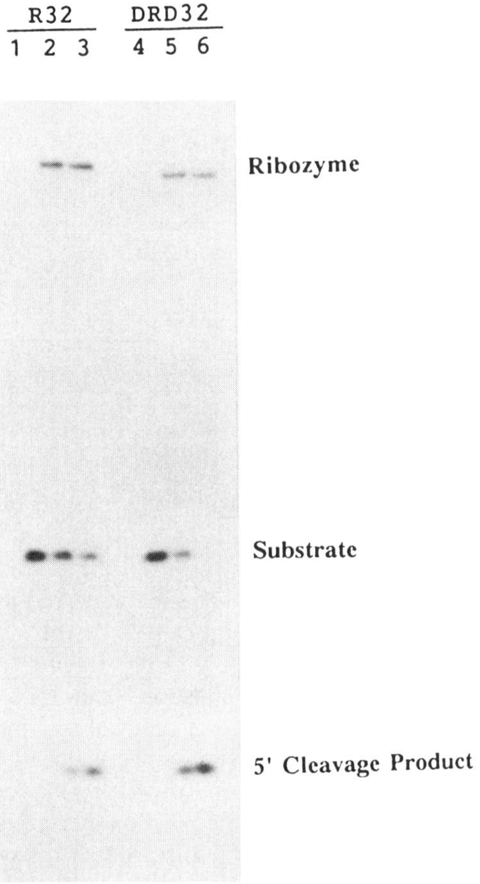

FIGURE 18.7. Reactions of fully ^{32}P-labeled all-RNA (R32) or DNA-armed (DRD32) ribozymes with partially ^{32}P-labeled substrates under k_{cat} conditions.

words, only a small percentage of substrate molecules were ^{32}P-labeled, and an excess of unlabeled cold substrate was mixed with the labeled substrate. Therefore, the intensities of the bands (ribozyme vs. substrate) in Figure 18.7 do not reflect the actual molar ratio of 1 to 300. In Figure 18.7, lanes 1 and 4 represent labeled substrate (R11; NUX = GUC in Figure 18.2) before the addition of ribozyme. To the substrate, either a ^{32}P-labeled all-RNA ribozyme (R32, lanes 2 and 3) or the DNA-armed DRD32 ribozyme (lanes 5 and 6) was added, and reactions were analyzed 10 min (lanes 2 and 5) and 20 min (lanes 3 and 6) after mixing at 37 °C. After 20 min of incubation, the substrate (87 μM) was cleaved almost completely by the chimeric DRD32 ribozyme (0.3 μM) (>95%, lane 6), whereas about 40% of the substrate remained intact when the substrate was treated with the all-RNA ribozyme (lane 3). No degradation of the ribozyme was observed. These findings clearly indicate that the DNA-armed ribozyme is indeed a better enzyme in terms of catalytic power under saturating conditions (k_{cat} conditions). Moreover, this kind of DNA-armed ribozyme is more stable in vivo than the corresponding all-RNA ribozyme against intracellular (ribo)nucleases (*31*).

Ribozymes Are Metalloenzymes

We examined thio effects for all-RNA (R32) and DNA-armed (DRD32) ribozymes (Figure 18.8). A single phosphorothioate linkage of defined Rp and Sp configuration (*35, 36*) was introduced at the cleavage site of the R11 substrate. In Figure 18.8, lanes 1 and 4 represent labeled substrates with the Rp and Sp configurations, respectively. These substrates were incubated either with the all-RNA R32 ribozyme (lanes 2 and 5) or with the chimeric DRD32 ribozyme (lanes 3 and 6), and the reactions were analyzed after 10 min of incubation at 37 °C, pH 8.0. The amount of 5′ cleavage product clearly indicates that, whereas the Sp isomer is nearly as reactive as the all-RNA substrate (lanes 5 and 6), the Rp isomer is cleaved only very slowly (lanes 2 and 3). The observations that when the pro-Rp oxygen was replaced with a sulfur, the hammerhead ribozymes lost their catalytic efficiencies (Figure 18.8), whereas they remain active when Mg^{2+} ions were replaced by Mn^{2+} ions (data not shown; Mn^{2+} is much more thiophilic than Mg^{2+}). These results appear to indicate that an Mg^{2+} ion is bound directly to the pro-R oxygen in both the all-RNA (*35*) and the DNA-armed ribozymes. In this arrangement, the bound metal ion can act as an electrophilic catalyst and thus, the proposed mechanism is very attractive as an explanation of the activities of metalloenzymes. However, more recently, a closer examination of the

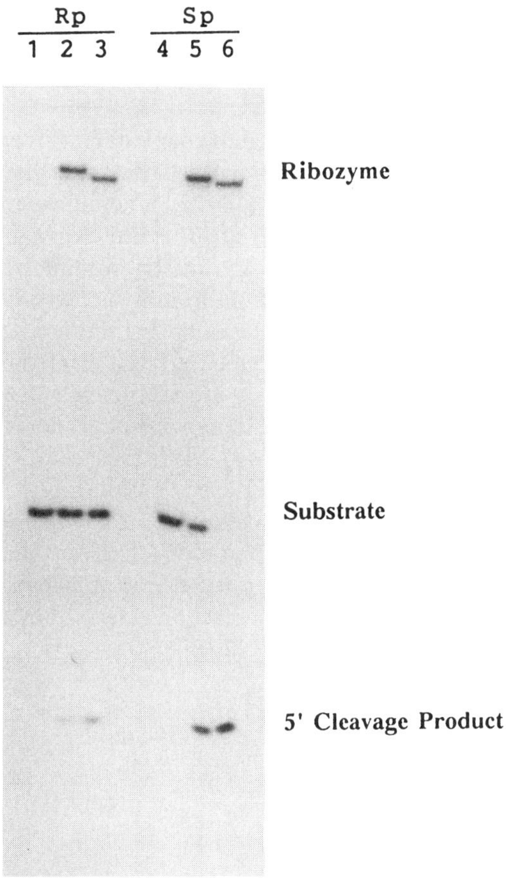

FIGURE 18.8. Cleavage of an RNA substrate with one phosphorothioate linkage of either Rp or Sp configuration by R32 (lanes 2 and 5) or DRD32 (lanes 3 and 6) ribozyme.

thio effects argues against the generally accepted mechanism of electrophilic catalysis, namely, the direct coordination of a metal ion with the pro-Rp oxygen (*36*).

The hammerhead ribozyme-catalyzed reaction can nevertheless be envisaged as a metalloenzyme-catalyzed reaction (*15–22*). Therefore, similar to the lead-catalyzed site-specific cleavage of tRNA (*3, 39–41*), the cleavage by hammerhead ribozymes may involve both intramolecular interactions among functional groups that reside within the central RNA loop region and true catalysis by Mg^{2+} ions (**1**). The possibility of true catalysis by Mg^{2+} ions is supported by our molecular orbital calculations (*18, 21, 42–49*). More importantly, in agreement with the proposed double-metal-ion mechanism shown in structure **1**, our kinetic measurements on solvent isotope effects clearly indicated that transfer of a proton does not take place in the transition state in a reaction catalyzed by the hammerhead ribozyme R32 (*22*).

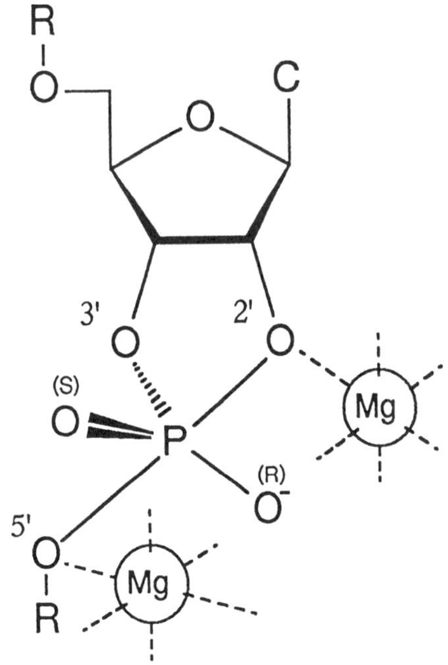

1

However, our NMR analysis indicated that Mg^{2+} ions are not only real catalysts but they also play a role in establishing an active form of a certain kind of ribozyme–substrate complex (Figure 18.9) (*25*). It remains to be answered whether the catalytically indispensable Mg^{2+} ions are the same as the structurally important Mg^{2+} ions. Recent X-ray data identified one Mg^{2+} ion that was bound to a site away from the cleavage site (*50–53*): an implication that a structurally important Mg^{2+} ion could be different from those used as catalysts.

Minizymes

Because conserved nucleotides are located within the catalytic loop and because we long have had a belief that ribozymes are metalloenzymes, we envisaged the function of the catalytic loop, at least in part, to be more like a crown ether that captures Mg^{2+} ions and brings them close to the cleavage site. To identify functional groups and to elucidate the role of the stem II region, several modifications and deletions have been made in the stem II region [*see* Figures 18.2 and 18.10a for the structure (*54–61*)]. Complete deletion of the stem-loop II abolished the catalytic activity (*55–59*). Although the conclusion was originally made that replacement of stem-loop II with a tetranucleotide (resulting in small ribozymes that were designated minizymes) affected hammerhead activity only moderately (*55*), more recent findings indicate that this type of alteration has significant consequences with respect to hammerhead activity (*56, 59*). In fact, activities of minizymes were reduced by 2–3 orders of magnitude, a result that led to the suggestion that minizymes might not be suitable as gene-inactivating reagents (*59*).

Tuschl and Eckstein (*56*) determined the minimal sequence requirements in the stem-loop II region for the full ribozyme activity. Stems with two base pairs (bp) had essentially unaltered catalytic activity, which was independent of the composition of the tetraloop, as long as the $G_{10.1}$–$C_{11.1}$ bp was maintained (*56*). Other results of mutagenesis tended to support this conclusion (*10, 33, 56–60*). NMR analysis and probing the hammer-

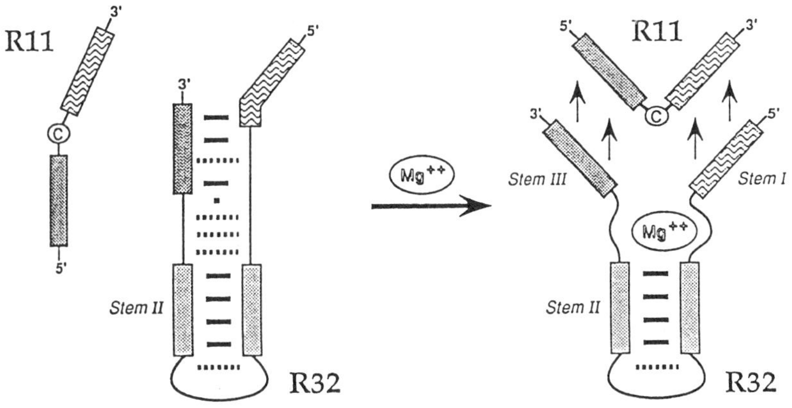

FIGURE 18.9. Magnesium-mediated formation of active ribozyme–substrate complex.

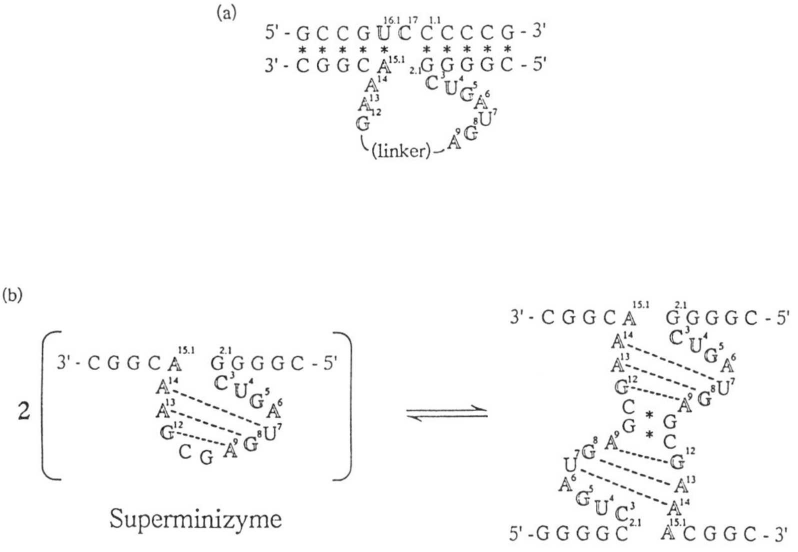

FIGURE 18.10. General structure of minizyme (a) and highly active superminizyme that performs catalysis in a form of dimer (b).

head ribozyme structure with ribonucleases also confirmed the $G_{10.1}$–$C_{11.1}$ base-pairing (*25, 62, 63*). Thus, a stem with 2 bp that includes the $G_{10.1}$–$C_{11.1}$ bp, in combination with an additional loop, is widely accepted as the minimal ribozyme structure to provide the necessary stability for high catalytic efficiency.

In a search for a small ribozyme that might be more amenable to structural and functional studies, as well as for a practical application as a gene-inactivating reagent, we reexamined the minimal sequence requirement for the formation of active minizymes because no evidence exists to indicate that the previously used tetra-loop is indeed the best sequence to join the 5′ and 3′ active sites. All previous investigations involved either complete deletion of the stem-loop II or maintenance of some base pairs, and no examination has been made of minizymes with mono-, di-, or trinucleotides that replace the stem-loop II. We found that some minizymes with appropriate linker sequences show extraordinary cleavage activities (*64*). Surprisingly, the cleavage activity of a good minizyme was 65% of that of the native ribozyme, R32, even though the molecular size of the good minizyme was only two-thirds of that of R32 (*64*). We designated the good minizyme a superminizyme (the exact structure is shown in Figure 18.10b) to distinguish this type of highly active minizyme from the conventional minizymes that have very low activity.

Further analysis indicated, however, that our superminizymes function in a form of a dimer (Figure 18.10b). In their dimer form, superminizymes can satisfy the minimal sequence requirement (the importance of bases $G_{10.1}$ and $C_{11.1}$) proposed by Tuschl and Eckstein, and they show extraordinary cleavage activities. If the dimer form could be stabilized by further modifications, superminizymes that have two active sites per dimer might be suitable as gene-inactivating reagents.

Not all minizymes form dimeric structures. In general, monomeric minizymes are significantly less active than the comparable full-sized ribozymes when cleaving short substrate. However, particular monomeric minizymes were shown to be more active than full-sized ribozymes when a long transcribed RNA was used as a substrate (*61*).

Delivery of Ribozymes into Cells

Specific association of the ribozyme with its target by means of base pairing and subsequent cleavage of the RNA substrate makes these catalytic RNA molecules attractive as antiviral agents. Basically, two ways introduce ribozymes into cells. One such technique is the drug delivery system in which the previously mentioned chemically synthesized ribozymes are encapsulated in liposomes or other related compounds, including cationic peptides as will be discussed later, and delivered to target cells. Another way to introduce ribozymes into cells is by transcription from the corresponding DNA template (gene therapy). Current gene therapy technology is primarily limited by the necessity for ex vivo manipulations of target tissues: namely, target cells must be removed from the body, engineered, and returned. Therefore, the limitations that determine which genetic diseases can potentially be treated are linked to the limits of current cell biology (*65*).

The ability of DNA and RNA tumor viruses to cause transformation by means of transfer of their genetic material led to their utility as gene-transfer vectors. Most commonly studied vectors include those of retroviruses, adenoviruses, and adeno-associated viruses (AAV). Each of these viruses has advantages and disadvantages as a gene-transfer vector (*65*). Retroviruses can integrate their genetic material directly into the chromosomes of the host cell. However, this system must be used in dividing cells. Adenoviruses are double-stranded DNA viruses that enter cells via receptor-mediated endocytosis, followed by disruption of the endosome and migration of the adenovirus genome to the nucleus where it remains extrachromosomal. Because adenoviral vector remains episomal and does not replicate, cell division leads to the eventual loss of the vector from the progenitor cells. AAV has the intermediary properties between the retrovirus and the adenovirus. AAV can only accept small inserts (< 5 kb), but it can be used in nondividing cells as well, as are adenoviruses, and it has the ability to undergo specific integration into a small region of human chromosome 19. However, not all recombinant AAVs exhibit this site specificity.

Retroviral Vector for Ribozyme–Gene Integration

An example of a ribozyme–gene transfer by means of the retroviral vector is illustrated in Figure 18.11. Retroviral-mediated gene transfer is dependent on a two-component system, the packaging cell and viral vector. To produce replication-incompetent retroviral particles, the packaging cell line is used, in which viral structural gene has

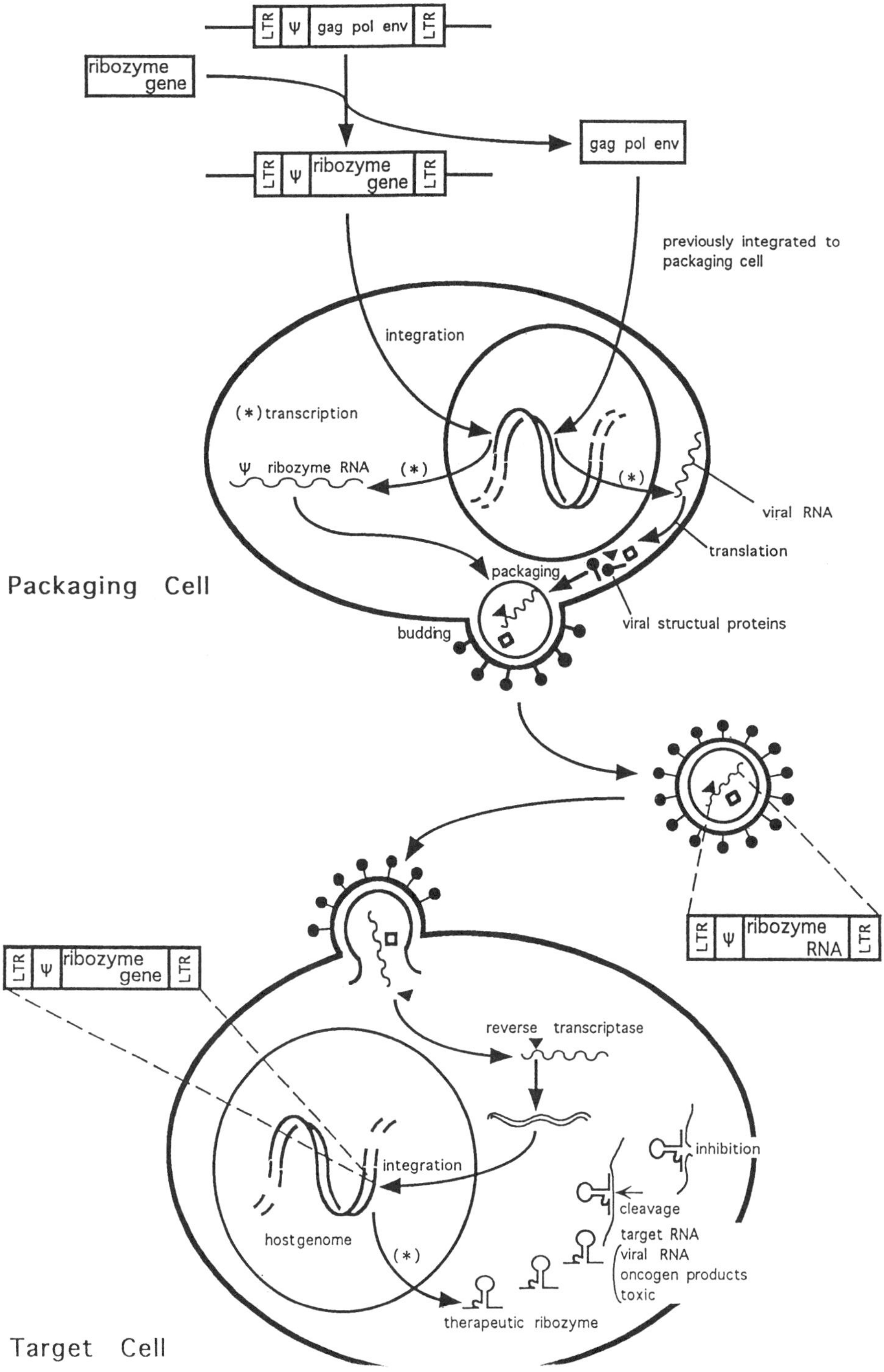

FIGURE 18.11. Retroviral vector and packaging cell.

already been integrated into the host genome. In such a cell line, the viral genome that encodes the viral proteins does not contain the packaging sequences (Ψ) that are responsible for the encapsidation of the viral genome into viral particles. As a result, the packaging lines do not efficiently package the viral genomic RNA into viral particles. Thus, the integrated viral genome in the chromosome of packaging cell line produces all the proteins necessary for virus replication and assembly, but it cannot create an infectious particle because of the lack of a transferable viral genome.

When one wants to produce viral particles that contain the ribozyme-expressing gene, this gene (ribozyme gene) is inserted into the region that had formally been occupied by the viral structural gene (gag, pol, env). When this retroviral vector is introduced into the packaging cell line, some of the ribozyme genes are integrated into the host genome. Then, the ribozyme RNA, expressed by host RNA polymerase, can be encapsidated to the viral particle because it contains the packaging sequence Ψ. This recombined viral particle has ribozyme gene in place of the natural viral genome (gag, pol, env), so it is replication incompetent but can infect to the specific target cells. Thus, the newly formed retrovirus is capable of an interaction between the viral envelope glycoprotein and the target-cell receptor. After infection, the ribozyme gene is reverse transcribed to double-stranded DNA, as are natural retroviruses, of which both 3′ and 5′ ends have long terminal repeat (LTR) sequences that are used to integrate the ribozyme gene to the host genome. The integrated ribozyme gene is transcribed by host RNA polymerases and thus therapeutic ribozyme RNAs are expressed in the target cell.

Ribozymes show their activities without any proteinaceous apparatus, and their specificity is very high, as is that of antisense molecules. So one can manipulate the target recognition sequence according to any target sequences (viral RNA, oncogene product RNA, toxic RNA, etc.). Ribozymes make a duplex with the target RNA and cleave at the predetermined site only, resulting in an inhibition of the expression of the target gene.

Ribozyme-Expression Vector

The finding that a hammerhead ribozyme can disarm human immunodeficiency virus-1 (HIV-1), at least in cells in culture, without any associated detrimental effects (*7*), has accelerated attempts at its application as an anti-HIV agent (*66*). HIV is infamous for its high mutation rate, which is caused by the low fidelity of its reverse transcriptase (*see* the life cycle of HIV-1 in Figure 18.12), an enzyme that lacks the proof-reading function (*67, 68*) and has a tendency to add an extra nucleotide when moving from one DNA template to another (*69*). This mutability of HIV not only makes it difficult to prepare vaccines against HIV but also hinders the application of ribozymes to cleavage of HIV RNA, because once the target site has undergone mutation, the ribozyme targeted to that specific site loses its effectiveness. One way to overcome this mutability of HIV would be to use ribozymes that target several conserved sites simultaneously. Then, even if one or more sites were to undergo mutation and avoid cleavage by the ribozyme, the other conserved sites could still potentially be cleaved by additional ribozymes targeted specifically to those sites. In fact, targeting several sites simultaneously by antisense DNAs prevented the development of escape mutants (*70*). We developed trimming vectors for liberation of multiple ribozymes, each with a different target site (Figure 18.13b) (*60, 71–73*). Other groups also constructed expression vectors for multitargeted ribozymes and demonstrated the importance of the multitargeting strategy (*74, 75*). At least two methods for expression of multitargeted ribozymes exist. The simpler way involves joining several sequences of ribozymes that are specific for different target sites in tandem, such that all the transcribed multitargeted ribozymes are connected in tandem as a single RNA (connected type; Figure 18.13a). The second strategy involves combining cis-acting ribozymes with

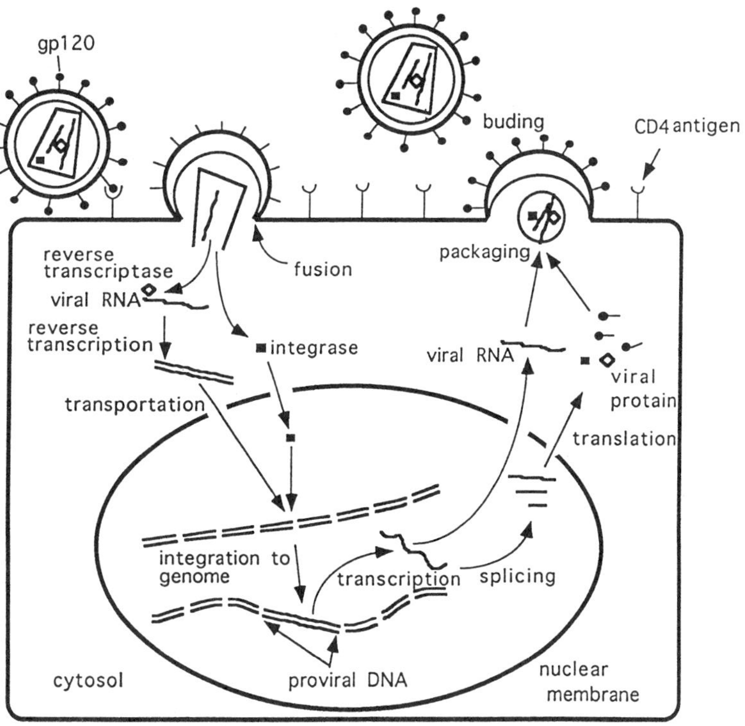

FIGURE 18.12. Life cycle of HIV-1.

trans-acting ribozymes so that several trans-acting ribozymes, targeted to HIV (or any other sequences), are trimmed at both their 5′ and 3′ ends by the actions of the cis-acting ribozymes, resulting in liberation of several trans-acting ribozymes that should function independently of one another (shotgun type; Figure 18.13b).

When levels of ribozyme expression were examined for the shotgun-type vector (*73*), the level of the ribozyme transcript was proportional to the number of units (n) connected in tandem. Accordingly, the activities of the shotgun-type ribozymes, in terms of the cleavage of HIV-1 RNA in vitro, also were proportional to the number of units connected in tandem (n). By contrast, the activities of the connected-type ribozymes reached plateau values at around $n = 3$. These results indicate that, when the shotgun-type expression system is used, generating various independent ribozymes is theoretically possible, each specific for a different target site, without sacrificing the activity of any individual ribozyme.

Activities of Ribozymes In Vivo

We investigated intensively, in vitro and in vivo, the ribozyme activity associated with the catalytic sequence against RNA coliphages (*76, 77*). RNA coliphages provided systems in which ribozyme activity could be rapidly evaluated in vivo. We found that a ribozyme designed to cleave the A2 gene (coding maturation enzyme) of RNA coliphage SP, when transcribed from a plasmid in *Escherichia coli,* caused failure of the proliferation of progeny phage. Also, inactive ribozymes with altered catalytic sequences, which might be expected to form a more stable RNA duplex than the active ribozyme, did not have significant inhibitory effects on phage growth. These results indicated that mainly the catalytic activity of the

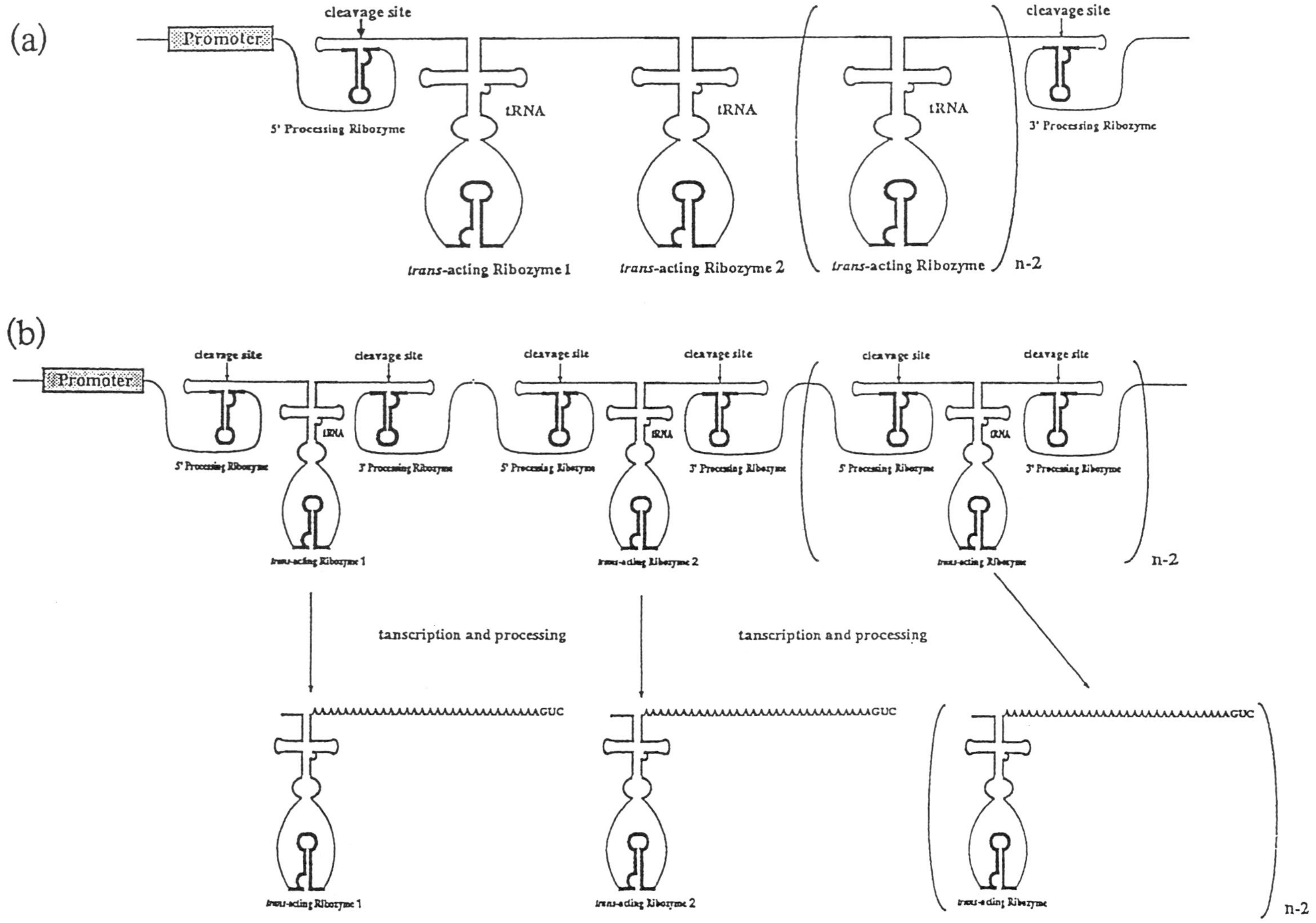

FIGURE 18.13. Multitargeting ribozyme-expression vector of either simply connected type (a) or shotgun type (b).

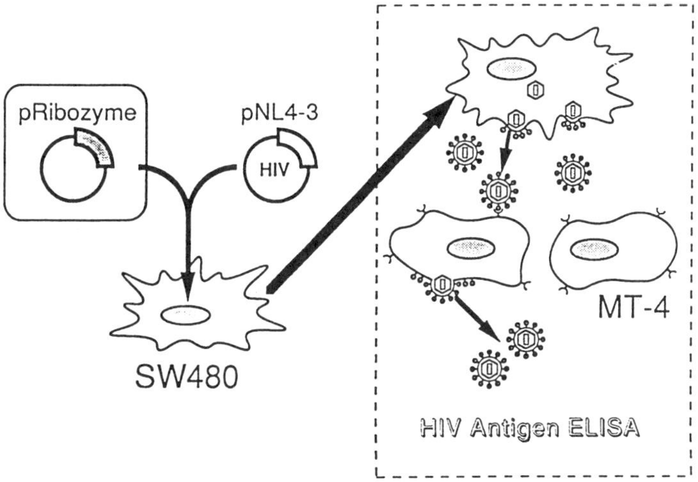

FIGURE 18.14. HIV-1 assay system used in our study.

ribozyme and not its function as an antisense molecule is responsible for suppressing the proliferation of the RNA phage (*77*).

Then, to examine the anti-HIV activity of tRNA-embedded 5′- and 3′-trimmed ribozymes in vivo, each ribozyme unit (Figure 18.13b) was connected to a mammalian expression vector under the control of SRα promoter (*78, 79*). Five relatively conserved regions on HIV-1 RNAs were chosen as the target sites of anti-HIV-1 ribozymes, and the activities of such ribozymes were examined by the procedure outlined in Figure 18.14 (*79, 80*). In short, human $CD4^-$ epithelioid colon carcinoma cells (SW480) were cotransfected with test plasmids and infectious proviral HIV-1-DNA pNL4-3 (*81*) by the Ca^{2+}-coprecipitation protocol. Subsequently, transfected SW480 cells were cocultivated with $CD4^+$ T-lymphoid MT-4 cells, which replicate HIV-1 efficiently, and then the p24 HIV-1 antigen was quantitated in supernatants from cocultures 4 days after transfection with a commercially available polyspecific enzyme-linked immunosorbent assay. As a control, a plasmid that encoded a ribozyme targeted to unrelated *lac Z* gene was used in place of the HIV-1 targeted ribozyme expression plasmid.

Although the extent of inhibition depended on the target site, all constructs caused significant inhibition of the HIV-1 p24 antigen production in culture supernatant (Figure 18.15). The ribozyme targeted to either 5′SS (5′ major splicing site) or *tat*1 site within *tat*-coding region had the highest inhibitory effect (>80%) when the molar ratio of template DNA for the target HIV-1 RNA to that

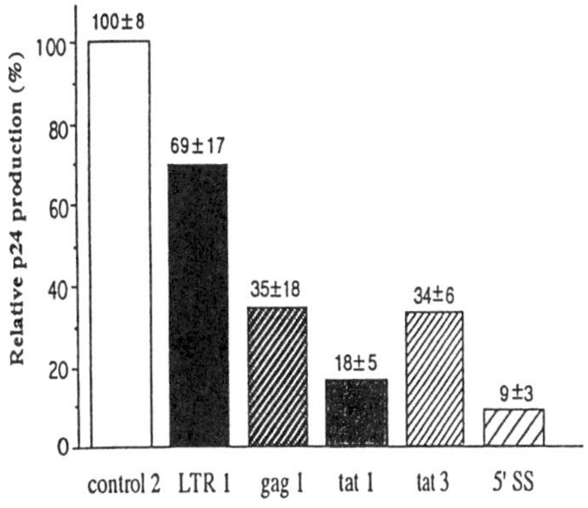

FIGURE 18.15. Inhibition of HIV-1 replication by various ribozymes targeted at a specific site; LTR is long terminal repeat.

for the ribozyme was only 1:8. Greater inhibition could be achieved by choosing a higher molar excess of ribozyme template. These results clearly demonstrate that our tRNA-embedded ribozyme-expression system under the control of the SRα promoter actively inhibited HIV-1 replication in vivo.

Multifunctionalization of Ribozyme-Expression Vector

As discussed previously (Figure 18.10), the stem II region of a hammerhead ribozyme (Figure 18.2) is not involved in catalysis. For synthetic ribozymes, it is advantageous to make ribozyme sequences as short as possible without sacrificing their enzymatic activities. However, for ribozymes transcribed from the corresponding DNA templates, the length of the ribozymes is less important. We wondered whether additional functions could be introduced into the stem II region without sacrificing the ribozyme activities (*60*).

In our original construct, the cis-acting ribozymes (Figure 18.13b) that had trimmed the 5′ and 3′ ends of each trans-acting ribozyme were designed merely to await degradation by RNases when they were used in vivo. Because several trans-activator proteins are essential for viral replication of HIV-1, we wondered whether a decoy function could be coupled with the cleavage activity of ribozymes. We therefore introduced the TAR and the RRE sequence, which are known to interact with a trans-activator protein such as Tat or Rev (Figure 18.16), respectively, into the stem II region of each cis-acting ribozyme. When the activity of each resulting cis-acting ribozyme that had been endowed with the decoy function was examined in vitro, it was found to retain almost full trimming activity (*60*). Moreover, cis-acting ribozymes with either the TAR or the RRE sequence were shown to be able to trap Tat or Rev protein successfully (*60*). Therefore, endowing the stem II region with a specific protein-binding function is possible without the loss of ribozyme function. Thus, cis-acting ribozymes, endowed with the decoy function, can first trim the 5′ and 3′ ends of each trans-acting ribozyme and possibly are still available for trapping trans-activator proteins before their degradation by RNases when they are to be used in vivo (Figure 18.16). Furthermore, the reduction in production of HIV RNA that is achieved by sequestering the trans-activator proteins might provide the trans-acting ribozymes, targeted to HIV RNA, with a better chance of eliminating the remaining HIV RNA (*60*).

Stabilization of Synthetic Ribozymes and Their Delivery into Cells

Among the nonviral gene-transfer techniques, cationic lipid- or cationic peptide-mediated delivery has received increasing attention in recent years (*82*). In the case of synthetic cationic lipids, hydrophobic effects provide a cationic glue that links the anionic nucleic acids and cell surfaces together. In the case of hemisynthetic polypeptides, nucleic acid binding polycationic peptide (protamine, polylysine) is chemically linked to another protein (e.g., asialo orosomucoid, ferritin, or insulin) whose recognition by a given cell surface receptor leads to active endocytosis (*82*). Synthetic ribozymes may be delivered into cells by these techniques if cationic carriers can sufficiently stabilize ribozymes in serum. Synthetic all-RNA ribozymes are unstable in serum. Although chemically modified chimeric ribozymes (*33*) gained some resistant against RNases in human serum, they were not stable enough for practical use. Therefore, we examined what kind of cationic carriers might stabilize ribozymes in serum (*83*).

*Delivery of Nucleic Acids Mediated by Poly(lysine-*co*-serine)*

Stability of ribozymes in human serum was examined after formation of complexes with various polycationic polypeptides by the method outlined in Figure 18.17. During the course of our investigation, we realized that although a ribozyme (R32, Figure 18.2) formed a complex with either poly-L-lysine (PLL) or poly-D-lysine (PDL), the effi-

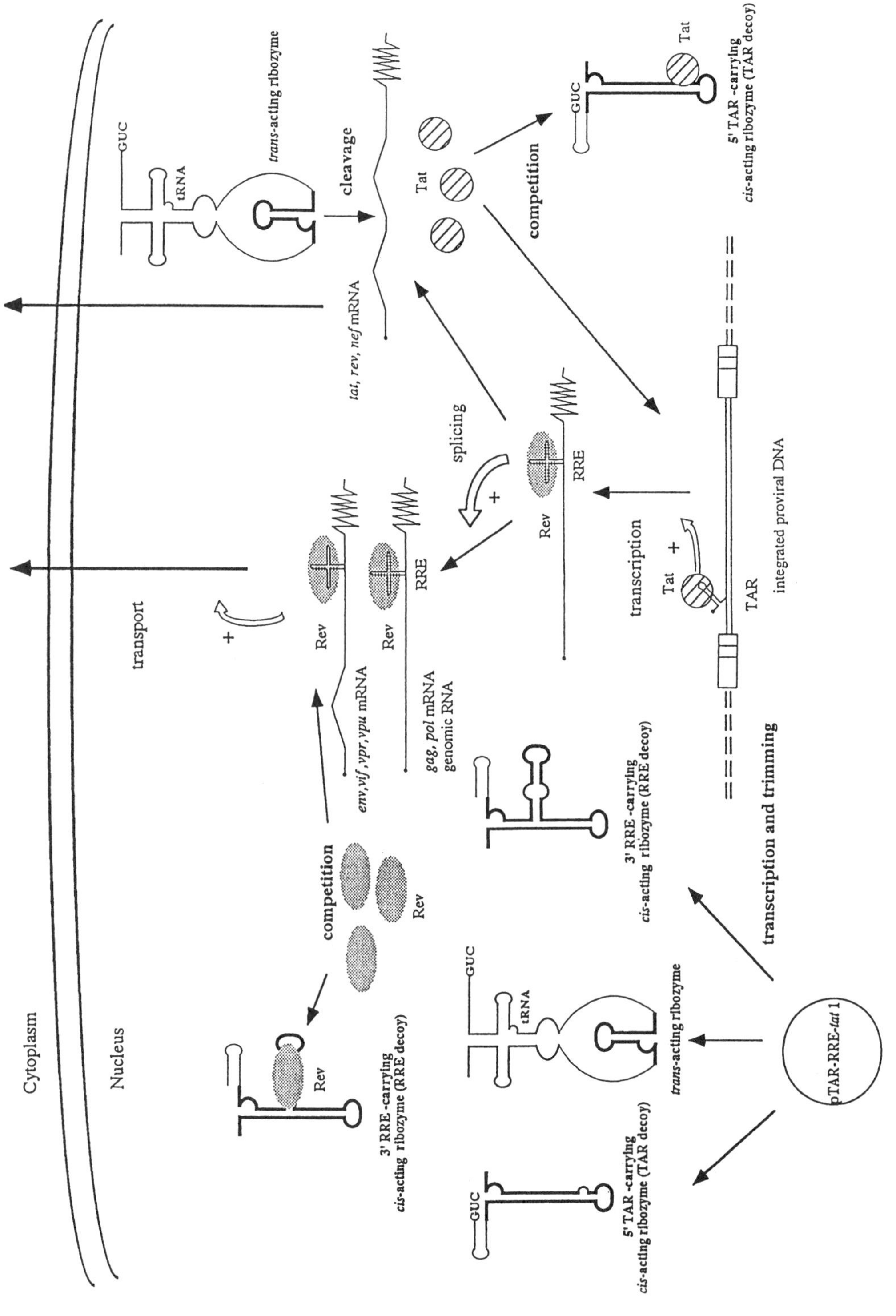

FIGURE 18.16. Multifunctional ribozyme-expression vector that produces a trans-acting-ribozyme and cis-acting ribozymes with decoy capability.

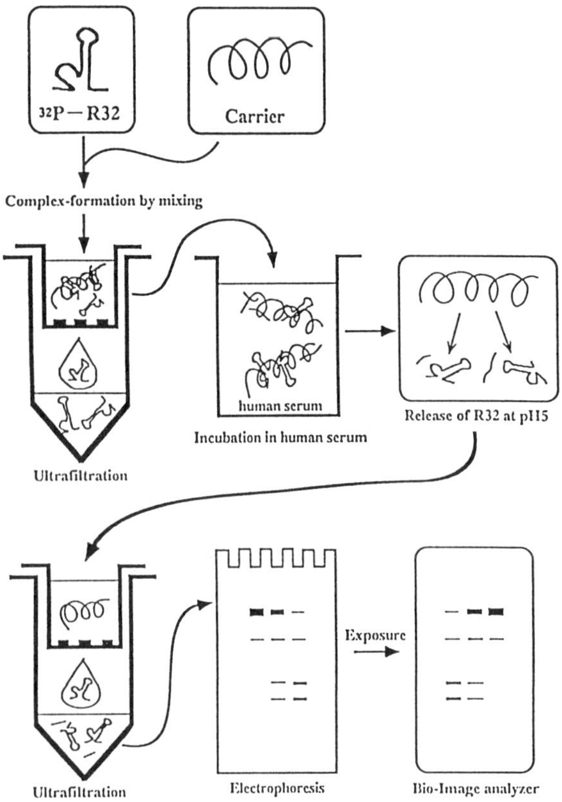

FIGURE 18.17. Assay system to detect binding of ribozymes (R32) to cationic peptides and to examine protection of R32 from attacks by RNases in serum.

ciency of the complex formation was lower with poly-(D,L)-lysine composed of racemic amino acids of D- and L-configurations. These results indicated that some structural elements are required for the favorable interaction between cationic peptides and nucleic acids (in addition to the cationic charges of the polypeptides). Following circular dichroism (CD) measurements hinted at the importance of some structural element such as α-helices in the formation of complexes between cationic peptides and ribozymes. We therefore replaced parts of the poly-L-lysine (PLL) chain (**2**) with L-serine residues (**3**). This random copolymer composed of L-lysine and L-serine (PLS) has less cationic charge but contains more α-helical character. When PLS was examined for its capacity to carry ribozymes, it indeed formed a tighter complex with R32 than did PLL (*83*).

Figure 18.18 shows the time courses of ribozyme degradation in undiluted human serum when complexes were formed with various cationic peptides. A hammerhead ribozyme (R32) complexed with poly(Lys:Ser = 1:1) survived in undiluted human serum more than 24 h, whereas naked ribozyme was degraded in seconds. Moreover, the higher the content of the serine residues, the longer the lifetime of R32 in the complex (Figures 18.18 and 18.19). Importantly, PLS also forms a complex with double-stranded plasmids and protects them from attack by nucleases in

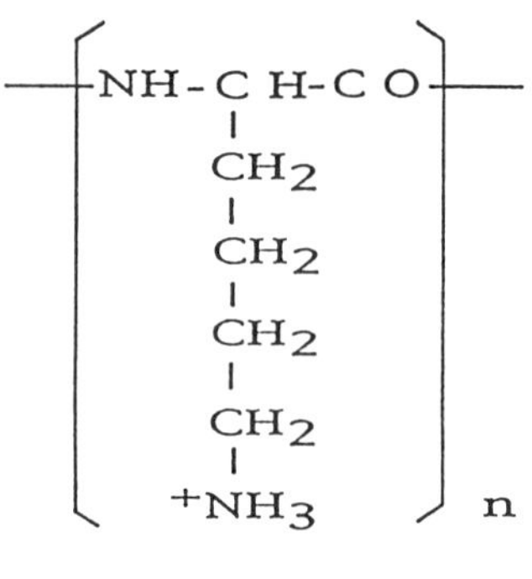

2, poly-L-lysine (PLL)

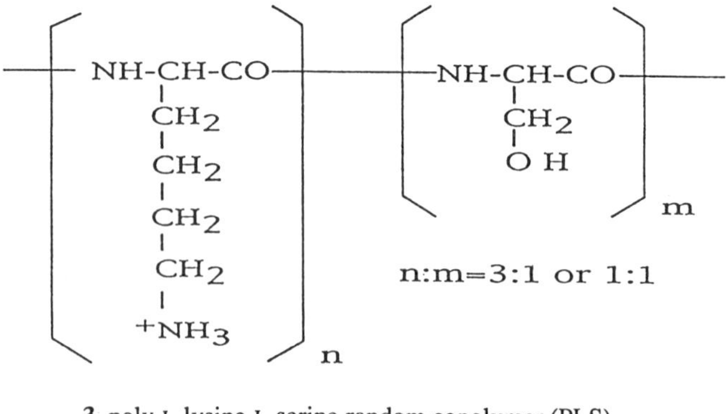

3, poly-L-lysine-L-serine random copolymer (PLS)

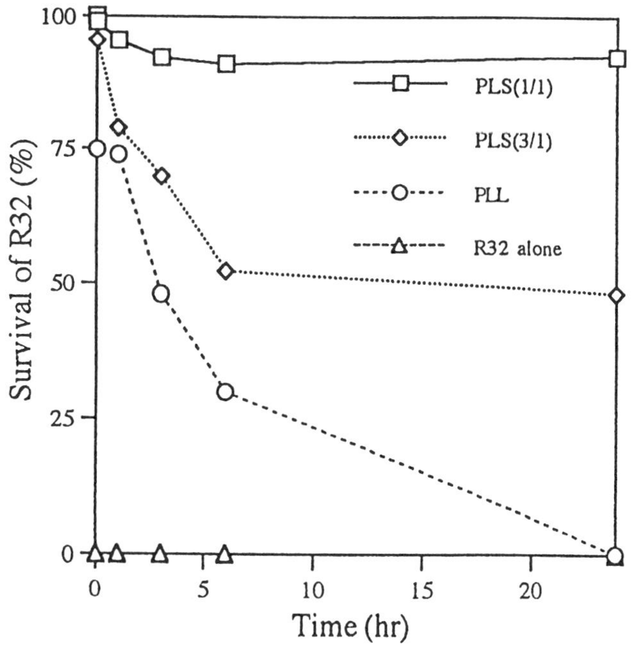

FIGURE 18.18. Survival of R32 in undiluted human serum when complexes were formed with various cationic peptides.

serum. The stability of nucleic acids varied with the purity of PLS: in general, commercially available PLS was less favorable than synthesized PLS in our laboratory. Examination of various types of peptides (Table 18.2) revealed that the higher the percentage of α-helical content of the complex (estimated from CD measurements after subtracting signals of R32), the better it is as a nucleic acid stabilizer (Figure 18.20).

Further analysis demonstrated that both ribozymes and plasmids can be released from the carrier (PLS) when pH becomes below 5.5. Most probably, this result is because α-helices will be destabilized at lower pH on protonation of more lysine residues. This character of PLS is advantageous as a carrier because the complex will be exposed to such a lower pH (about 5) on endocytosis (Figure 18.21) and the nucleic acids may escape from the endosome.

Our preliminary data (luciferase assay) indicate that complexes of nucleic acids and PLS by themselves cannot penetrate HeLa cells, requiring additional attachment of targeting molecules. However, an antisense DNA–PLS complex can exert an antisense effect in cells of macrophage origin (*84–86*). When DNAs with phosphorothioate linkages (S-oligo) were targeted at mRNA of proinflammatory cytokine, tumor necrosis factor (TNF)-α, approximately 40% inhibition of TNF-α secretion was observed. However, the inhibitory effect was not specific for the antisense S-oligo: sense S-oligo had similar inhibitory effect. On the other hand, natural antisense DNA for the same target site had higher than 90% inhibition of TNF-α secretion when it was complexed with PLS, and sense DNA–PLS complex or antisense DNA itself without formation of complex with PLS had no inhibitory effect. Moreover, PLS conjugated with polyethylene glycol

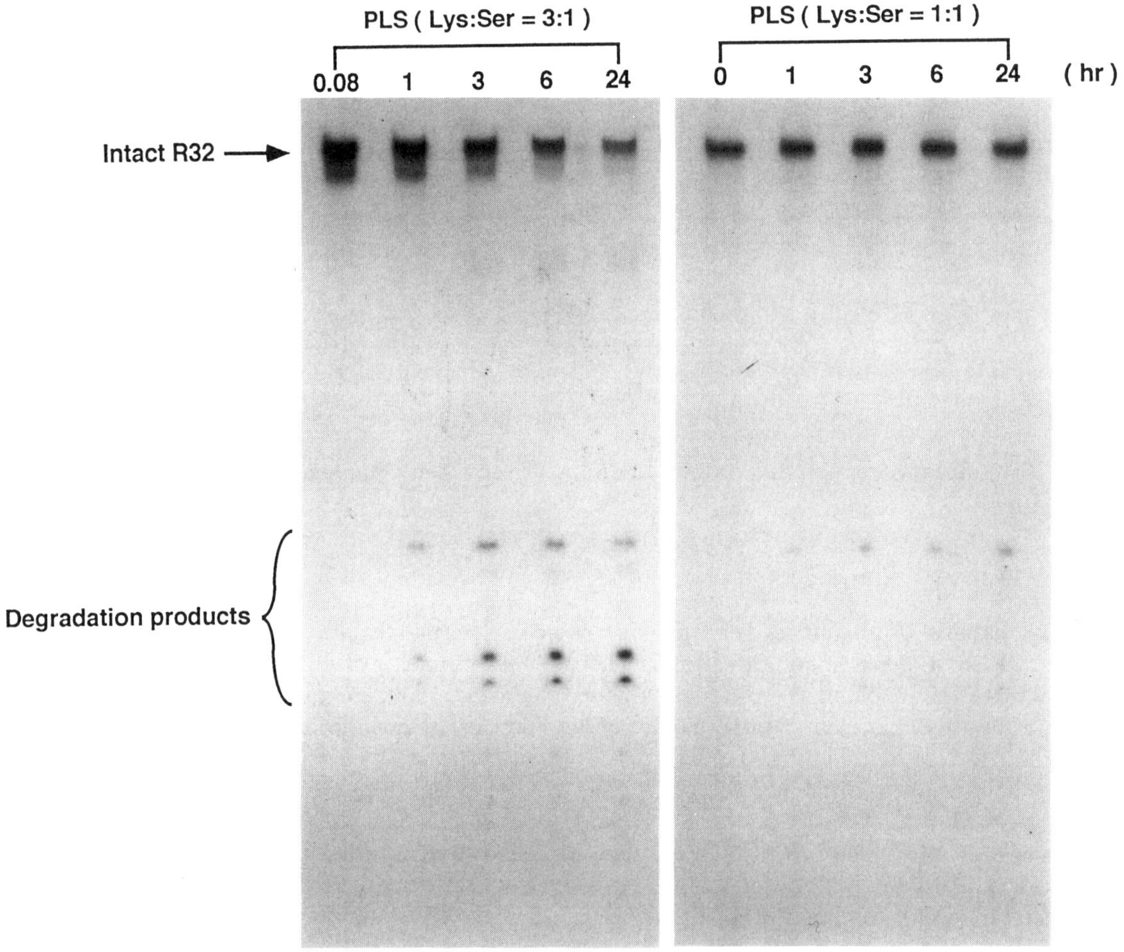

FIGURE 18.19. Results of gel electrophoresis to show the survival of R32 in undiluted human serum when complexes were formed with two kinds of PLS.

TABLE 18.2. Comparison of Polyamino Acids as Gene-Delivery Carriers

Polyamino Acids	*Solubility Against 0.1-M Tris Buffer*	*Complexity*		*Dissociation*		*Protect Effect (12 h)*		*Max. Helicity*	
		Rib.	*pUC*	*Rib.*	*pUC*	*Rib.*	*pUC*	*Rib.*	*pUC*
Homopolymers									
Poly(L-lysine) M_w = 1,000–4,000	Medium	85	55	75	80	73	NT	42	25
Poly(L-lysine) M_w = 4,000–15,000	Medium	80	42	70	75	68	NT	37	22
Poly(L-lysine) M_w = 15,000–30,000	Medium	85	35	75	60	62	NT	30	20
Poly(L-lysine) M_w = 30,000–70,000	Medium	45	20	54	25	44	NT	21	5
Poly(D-lysine) M_w = 15,000–30,000	Medium	80	47	72	68	65	NT	33	21
Poly(D,L-lysine) M_w = 15,000–30,000	Medium	0	0	ND	ND	NT	NT	ND	ND
Poly(L-arginine) M_w = 5,000–15,000	Low	25	NT	NT	NT	NT	NT	2	NT
Poly(L-serine) M_w = 5,000–10,000	Insoluble	ND	ND	ND	ND	ND	ND	ND	ND
Poly(L-ornithine) M_w = 15,000–30,000	Low	85	NT	25	NT	NT	NT	NT	NT
Poly(D,L-ornithine) M_w = 15,000–30,000	—	—	—	—	—	—	—	—	—
Random copolymers									
Poly(Arg,Trp) = (4/1) M_w = 20,000–50,000	Low	25	NT	5	NT	0	NT	0	NT
Poly(Arg,Tyr) = (4/1) M_w = 20,000–50,000	Low	12	NT	0	NT	0	NT	0	NT
Poly(Arg,Ser) = (3/1) M_w = 20,000–50,000	High	32	NT	10	NT	78	NT	45	NT
Poly(Arg,Pro,Thr) = (6/3/1) M_w = 10,000–30,000	Low	2	NT	NT	NT	0	NT	5	NT
Poly(Asp,Glu) = (1/1) M_w = 5,000–15,000	Low	0	NT	ND	NT	0	NT	0	NT
Poly(Glu,Lys) = 1/4 M_w = 150,000–300,000	Medium	42	NT	30	NT	15	NT	8	NT
Poly(Lys,Ala) = (1/1) M_w = 20,000–50,000	Low	51	NT	34	NT	28	NT	11	NT
Poly(Lys,Phe) = (1/1) M_w = 20,000–50,000	Medium	45	NT	34	NT	20	NT	8	NT
Poly(Lys,Ser) = (3/1) M_w = 20,000–50,000	High	100	NT	100	NT	78	65	39	NT
Poly(Lys,Tyr) = (1/1) M_w = 20,000–50,000	Low	43	NT	41	NT	33	NT	18	NT

Continued on next page

TABLE 18.2. Comparison of Polyamino Acids as Gene-Delivery Carriers—*Continued*

Polyamino Acids	*Solubility Against 0.1-M Tris Buffer*	*Complexity*		*Dissociation*		*Protect Effect (12 h)*		*Max. Helicity*	
		Rib.	*pUC*	*Rib.*	*pUC*	*Rib.*	*pUC*	*Rib.*	*pUC*
Poly(Lys,Ser)[a] = (3/1) M_w = 30,000	High	100	75	100	82	82	70	47	NT
Poly(Lys,Ser)[a] = (1/1) M_w = 30,000	High	100	85	100	90	100	91	51	NT
Conjugates									
PLL-PEG5000[a] M_w(Lys) = 5,000	High	90	75	92	80	100	91	50	NT
PLL-PEG5000-PLL[a] M_w(Lys) = 5,000	High	85	80	88	76	100	88	47	NT
PEG5000-PLS[a] M_w(PLS) = 5,000	High	100	NT	NT	NT	NT	NT	NT	NT

NOTES: All values are percents. Rib. is ribozymes. pUC is pUC series plasmids. NT indicates not tried. ND indicates not detectable.
[a]Samples were from Gunma University.

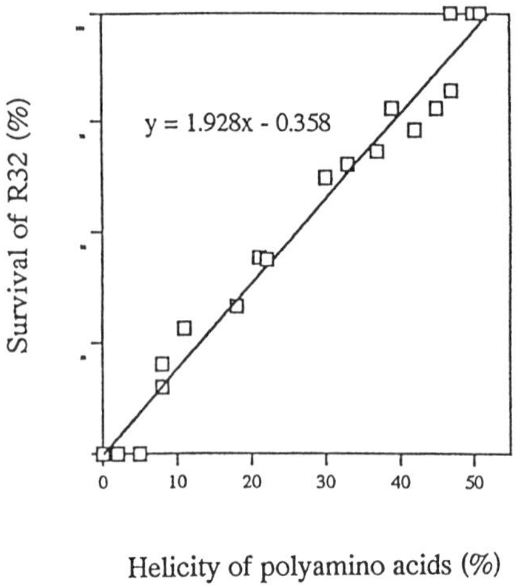

FIGURE 18.20. Relationship between the survival of R32 and the α-helical content of the complexes.

5000 (to avoid induction of antibody) worked well. These results suggest that PLS and its derivatives have great potential for protecting and delivering natural nucleic acids to target cells.

The PLL portion of cationic peptide carrier, which has been used favorably to date (*82*), can be replaced with PLS (Lys:Ser = 1:1). Another advantage of using PLS is that nucleic acids–PLS complex will not form precipitates even at high concentrations, and this result is in contrast to the case of nucleic acids–PLL complexes, which tend to precipitate out at high concentrations (*87*).

Reconstituted Histone-Mediated Delivery of Nucleic Acids

Nucleic acids can be stabilized similarly by being encapsulated into reconstituted histones (Figure 18.22). The order of addition of each subunit is very important in reconstituting soluble histone–R32 complex. Unless the exact order shown in Figure 18.22a is followed, the complex starts to precipitate out (Figure 18.23). The histone–R32 complex is also stable in undiluted human serum. However, the complex formed by the method described in Figure 18.22 was very stable, and R32 did not dissociate from the histone even at lower pH. Examination of various conditions revealed that the H1 subunit can be omitted to gain full protection of R32 in serum (Figure 18.24). Addition of R32 to original histones or reconstituted histones did not produce stable complexes, and R32 ribozymes on the surface of those histones were

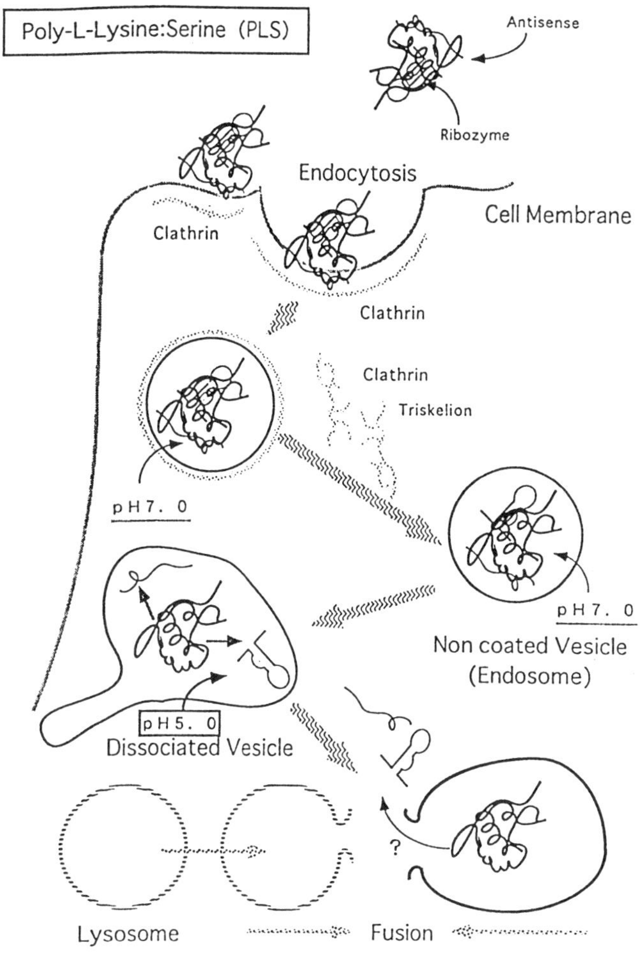

FIGURE 18.21. Hypothetical endocytosis for nucleic acids–PLS complex.

degraded in minutes in serum. Most importantly, R32 can be liberated from the $H1^{(-)}$-histone at pH below 5.5. Therefore, it is possible that $H1^{(-)}$-histone can function very similarly to PLS.

Detection of Undegraded Oligonucleotides In Vivo

In studies of possible therapeutic modalities, for example, after delivery of natural DNA–RNA into cells by means of PLS or $H1^{(-)}$-histone, the structural integrity of oligonucleotides merits careful consideration because of the dependence of specific binding activity on the length of the intact oligonucleotide sequence. The undesirable possibility always exists of depolymerization or cleavage by digestion by intracellular nucleases. Ribozymes are especially sensitive to intracellular ribonucleases because their active sites are composed of RNA (Figures 18.1 and 18.2). If we are to monitor the fate of ribozyme in vivo by use of conventional materials labeled with a fluorescent dye at one end, we are likely to detect degraded RNAs labeled with a fluorescent dye at their end. To

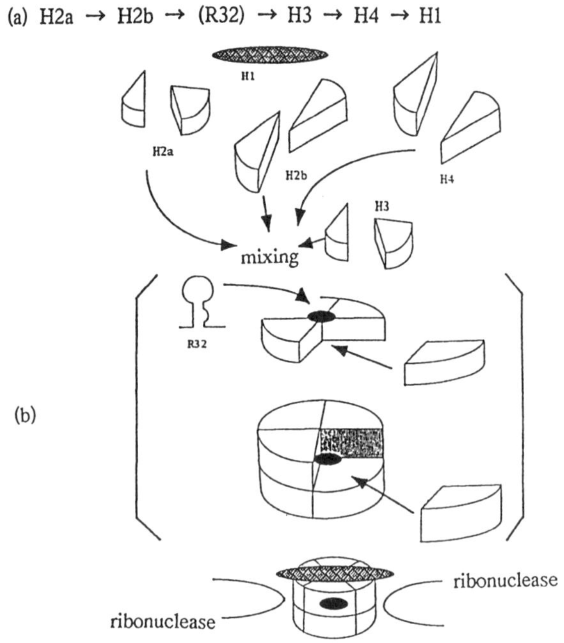

FIGURE 18.22. Reconstituting of histones. Exact order of addition of each subunit (a) and schematic representation (b).

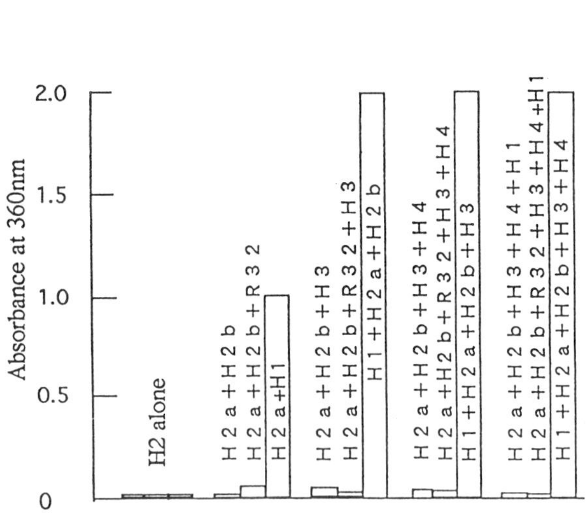

FIGURE 18.23. Examination of turbidity when the order of addition of each subunit was varied.

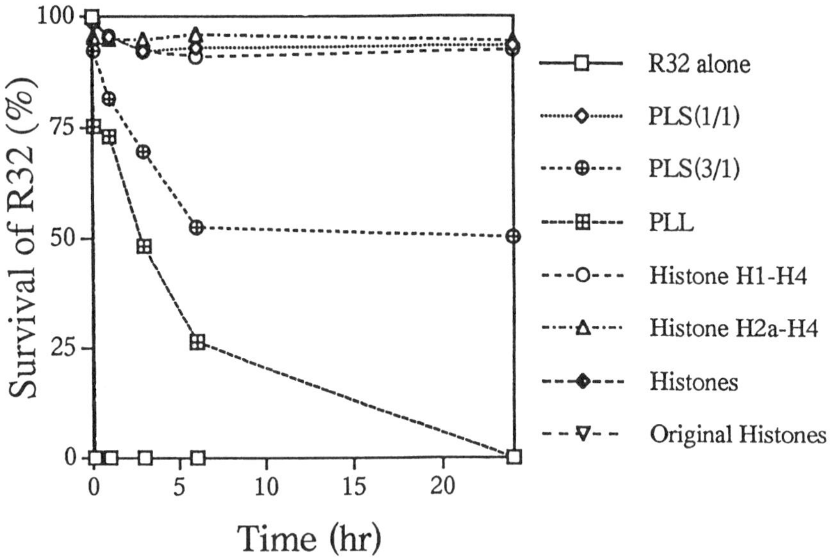

FIGURE 18.24. Survival of R32 in undiluted human serum when formed complexes with various carriers. Reconstituted histone (H1-H4) and $H1^{(-)}$-histone (H2a-H4) can protect R32 from attacks by RNases at the same level as PLS (Lys:Ser = 1:1).

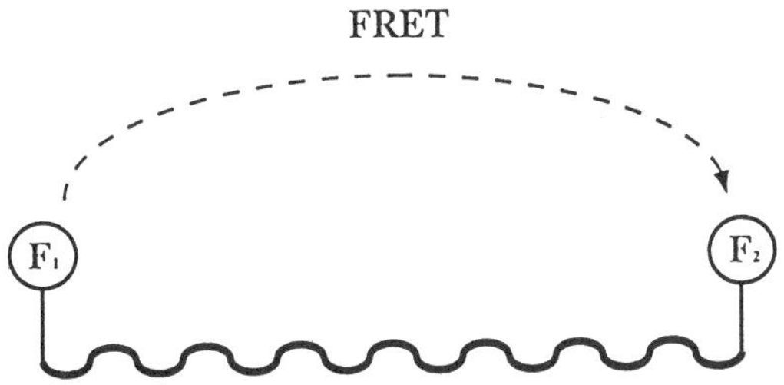

FIGURE 18.25. Detection of undegraded oligonucleotides by fluorescence resonance energy transfer (FRET). A single-stranded oligonucleotide was double labeled with fluorescein (F_1) at its 5′-end and rhodamine X (F_2) at its 3′-end to detect FRET within undegraded molecule.

address these issues, we investigated methods for estimating the integrity of injected oligonucleotides in living cells by using an oligonucleotide that had been double labeled with two fluorescent dyes (*88*). We examined the fluorescence characteristics of and the fluorescence resonance energy transfer (FRET) within this double-labeled oligonucleotide (Figure 18.25).

FRET is an interesting example of a fluorescence-related phenomenon. When the fluorescence spectrum of one fluorochrome, the donor, overlaps with the excitation spectrum of another fluorochrome, the acceptor, and when the donor and the acceptor are in close physical proximity, the excitation of the donor induces the emission of fluorescence from the acceptor as if the acceptor has been excited directly and the intensity of fluorescence from the donor decreases. The extent of FRET is extremely sensitive to the distance between the donor and the acceptor and is inversely proportional to the sixth power of the distance. This phenomenon can be explored for studies of intermolecular and intramolecular relationships in biophysical investigations and in cell biology.

To investigate the integrity of oligonucleotides in living cells, we synthesized a double-labeled fluorescent single-stranded oligodeoxyribonucleotide (Figure 18.25). FRET was demonstrated spectroscopically in a solution of this modified oligonucleotide and was visualized under the fluorescence microscope in sea urchin eggs after microinjection of this modified oligonucleotide. Enzymatic digestion eliminated FRET both in solution and in eggs. These results indicate that it is possible to detect only the intact oligonucleotide in living cells by monitoring FRET.

Conclusion

We have described in some detail how hammerhead ribozymes function as metalloenzymes. Because two ways exist to deliver ribozymes into cells, ribozyme-expression vectors and a method to stabilize ribozymes in serum that leads to a drug delivery system were discussed. Synthetic ribozymes can be completely stabilized against RNases in human serum (no degradation over 24 h), once ribozymes have formed a complex with cationic–hydrophilic peptides (PLS or $H1^{(-)}$-histone), and they can be released from the carrier at lower pH (in endosomes). This kind of complex formation is a potentially powerful methodology for delivering any nucleic acid drugs including antisense molecules into cells.

References

1. Altman, S. *Adv. Enzymol.* **1989,** *62,* 1–36.
2. Cech, T. R. *Angew. Chem. Int. Ed. Engl.* **1986,** *29,* 759–768.
3. *The RNA World;* Gasteland, R. F.; Atkins, J. F., Eds.; Monograph 24; Cold Spring Harbor Laboratory: Plainview, NY, 1993.
4. Symons, R. H. *Trends Biochem. Sci.* **1989,** *14,* 445–450.
5. Uhlenbeck, O. C. *Nature (London)* **1987,** *328,* 596–600.
6. Haseloff, J.; Gerlach, W. L. *Nature (London)* **1988,** *334,* 585–591.
7. Sarver, N.; Cantin, E. M.; Chang, P. S.; Zaia, J. A.; Ladne, P. A.; Stephens, D. A.; Rossi, J. J. *Science (Washington, D.C.)* **1990,** *247,* 1222–1225.
8. Sheldon, C. C.; Symons, R. H. *Nucleic Acids Res.* **1989,** *17,* 5679–5685.

9. Koizumi, M.; Iwai, S.; Ohtsuka, E. *FEBS Lett.* **1988,** *228,* 228–230.
10. Ruffner, D. E.; Stormo, G. D.; Uhlenbeck, O. C. *Biochemistry* **1990,** *29,* 10695–10702.
11. Perriman, R.; Delves, A.; Gerlach, W. L. *Gene* **1992,** *113,* 157–163.
12. Shimayama, T.; Nishikawa, S.; Taira, K. *Biochemistry* **1995,** *34,* 3649–3654.
13. Zoumadakis, M.; Tabler, M. *Nucleic Acids Res.* **1995,** *23,* 1192–1196.
14. Hertel, K. J.; Pardi, A.; Uhlenbeck, O. C.; Koizumi, M.; Ohtsuka, E.; Uesugi, S.; Cedergren, R.; Eckstein, F.; Gerlach, W. L.; Hodgson, R.; Symons, R. H. *Nucleic Acids Res.* **1992,** *20,* 3252.
15. Dahm, S. C.; Derrick, W. B.; Uhlenbeck, O. C. *Biochemistry* **1993,** *32,* 13040–13045.
16. Piccirilli, J. A.; Vyle, J. S.; Caruthers, M. H.; Cech, T. R. *Nature (London)* **1993,** *361,* 85–88.
17. Yarus, M. *FASEB J.* **1993,** *7,* 31–39.
18. Uchimaru, T.; Uebayasi, M.; Tanabe, K.; Taira, K. *FASEB J.* **1993,** *7,* 137–142.
19. Steitz, T. A.; Steitz, J. A. *Proc. Natl. Acad. Sci. U.S.A.* **1993,** *90,* 6498–6502.
20. Pyle, A. M. *Science (Washington, D.C.)* **1993,** *261,* 709–714.
21. Uebayasi, M.; Uchimaru, T.; Koguma, T.; Sawata, S.; Shimayama, T.; Taira, K. *J. Org. Chem.* **1994,** *59,* 7414–7420.
22. Sawata, S.; Komiyama, M.; Taira, K. *J. Am. Chem. Soc.* **1995,** *117,* 2357–2358.
23. Takagi, Y.; Taira, K. *FEBS Lett.* **1995,** *361,* 273–276.
24. Sawata, S.; Shimayama, T.; Komiyama, M.; Kumar, P. R.; Nishikawa, S.; Taira, K. *Nucleic Acids Res.* **1993,** *21,* 5656–5660.
25. Orita, M.; Vinayak, R.; Andrus, A.; Warashina, M.; Chiba, A.; Kaniwa, H.; Nishikawa, F.; Nishikawa, S.; Taira, K. *J. Biol. Chem.* **1996,** *271,* 9447–9454.
26. Tsuchihashi, Z.; Khosla, M.; Herschlag, D. *Science (Washington, D.C.)* **1993,** *262,* 99–102.
27. Homann, M.; Tzortzakaki, M.; Rittner, K.; Sczakiel, S.; Tabler, M. *Nucleic Acids Res.* **1993,** *21,* 2809–2814.
28. McSwiggen, J. A.; Cech, T. R. *Science (Washington, D.C.)* **1989,** *244,* 679–683.
29. Rajagopal, J.; Doudna, J. A.; Szostak, J. W. *Science (Washington, D.C.)* **1989,** *244,* 692–694.
30. Young, B.; Herschlag, D.; Cech, T. R. *Cell* **1991,** *67,* 1007–1019.
31. Taylor, N. R.; Kaplan, B. E.; Swiderski, P.; Li, H.; Rossi, J. J. *Nucleic Acids Res.* **1992,** *20,* 4559–4565.
32. Hendry, P.; McCall, M. J.; Santiago, F. S.; Jennings, P. A. *Nucleic Acids Res.* **1992,** *20,* 5737–5741.
33. Shimayama, T.; Nishikawa, F.; Nishikawa, S.; Taira, K. *Nucleic Acids Res.* **1993,** *21,* 2605–2611.
34. Shimayama, T.; Nishikawa, S.; Taira, K. *FEBS Lett.* **1995,** *368,* 304–306.
35. (a) Dahm, S. C.; Uhlenbeck, O. C. *Biochemistry* **1991,** *30,* 9464–9469. (b) Koizumi, M.; Ohtsuka, E. *Biochemistry* **1991,** *30,* 5145–5150. (c) Slim, G.; Gait, M. J. *Nucleic Acids Res.* **1991,** *19,* 1183–1188.
36. Zhou, D.-M.; Kumar, P. K. R.; Zhang, L.-H.; Taira, K. *J. Am. Chem. Soc.* **1996,** *118,* 8969–8970.
37. Zhou, D.-M.; Usman, N.; Wincott, F. E.; Matulic-Adamic, J.; Orita, M.; Zhang, L.-H.; Komiyama, M.; Kumar, P. K. R.; Taira, K. *J. Am. Chem. Soc.* **1996,** *118,* 5862–5866.
38. Fedor, M. J.; Uhlenbeck, O. C. *Biochemistry* **1992,** *31,* 12042–12054.
39. Brown, R. S.; Dewan, J. C.; Klug, A. *Biochemistry* **1985,** *24,* 4785–4801.
40. Pan, T.; Gutell, R. R.; Uhlenbeck, O. C. *Science (Washington, D.C.)* **1991,** *254,* 1361–1364.
41. Pan, T.; Uhlenbeck, O. C. *Biochemistry* **1992,** *31,* 3887–3895.
42. Taira, K.; Uebayasi, M.; Maeda, H.; Furukawa, K. *Protein Eng.* **1990,** *3,* 691–701.
43. Taira, K.; Uchimaru, T.; Tanabe, K.; Uebayasi, M.; Nishikawa, S. *Nucleic Acids Res.* **1991,** *19,* 2747–2753.
44. Uchimaru, T.; Tanabe, K.; Nishikawa, S.; Taira, K. *J. Am. Chem. Soc.* **1991,** *113,* 4351–4353.
45. Storer, J. W.; Uchimaru, T.; Tanabe, K.; Uebayasi, M.; Nishikawa, S.; Taira, K. *J. Am. Chem. Soc.* **1991,** *113,* 5216–5219.
46. Yliniemela, A.; Uchimaru, T.; Tanabe, K.; Taira, K. *J. Am. Chem. Soc.* **1993,** *115,* 3032–3033.
47. Taira, K.; Uchimaru, T.; Storer, J. W.; Yelimiena, A.; Uebayasi, M.; Tanabe, K. *J. Org. Chem.* **1993,** *58,* 3009–3017.
48. Uchimaru, T.; Tsuzuki, S.; Storer, J. W.; Tanabe, K.; Taira, K. *J. Org. Chem.* **1994,** *59,* 1835–1843.
49. Uchimaru, T.; Uebayasi, M.; Hirose, T.; Tsuzuki, S.; Yliniemela, A.; Tanabe, K.; Taira, K. *J. Org. Chem.* **1996,** *61,* 1599–1608.
50. Pley, H. W.; Flaherty, K. M.; McKay, D. B. *Nature (London)* **1994,** *372,* 68–74.
51. Pley, H. W.; Flaherty, K. M.; McKay, D. B. *Nature (London)* **1994,** *372,* 111–113.
52. Cech, T. R.; Uhlenbeck, O. C. *Nature (London)* **1994,** *372,* 39–40.

53. Scott, W. G.; Finch, J. T. *Cell* **1995,** *81,* 991–1002.
54. Goodchild, J.; Kohli, V. *Arch. Biochem. Biophys.* **1991,** *284,* 386–391.
55. McCall, M. J.; Hendry, P.; Jennings, P. A. *Proc. Natl. Acad. Sci. U.S.A.* **1992,** *89,* 5710–5714.
56. Tuschl, T.; Eckstein, F. *Proc. Natl. Acad. Sci. U.S.A.* **1993,** *90,* 6991–6994.
57. Thomson, J. B.; Tuschl, T.; Eckstein, F. *Nucleic Acids Res.* **1993,** *21,* 5600–5603.
58 Fu, D.; Benseler, F.; McLaughlin, L. W. *J. Am. Chem. Soc.* **1994,** *116,* 4591–4598.
59. Long, D. M.; Uhlenbeck, O. C. *Proc. Natl. Acad. Sci. U.S.A.* **1994,** *91,* 6977–6981.
60. Yuyama, N.; Ohkawa, J.; Koguma, T.; Shirai, M.; Taira, K. *Nucleic Acids Res.* **1994,** *22,* 5060–5067.
61. Hendry, P.; McCall, M. J.; Santiago, F. S.; Jennings, P. A. *Nucleic Acids Res.* **1995,** *23,* 3922–3927.
62. Heus, H. A.; Pardi, A. *J. Mol. Biol.* **1991,** *217,* 113–124.
63. Hodgson, R. J.; Shirley, N. J.; Simons, R. H. *Nucleic Acids Res.* **1994,** *22,* 1620–1625.
64. Amontov, S. V.; Taira, K. *J. Am. Chem. Soc.* **1996,** *118,* 1624–1628.
65. Morgan, R. A.; Anderson, W. F. *Annu. Rev. Biochem.* **1993,** *62,* 191–217.
66. Altman, S. *Proc. Natl. Acad. Sci. U.S.A.* **1993,** *90,* 10898–10900.
67. Preston, B. D.; Poiesz, B. J.; Loeb, L. A. *Science (Washington, D.C.)* **1988,** *242,* 1168–1171.
68. Roberts, J. D.; Bebenek, K.; Kunkel, T. A. *Science (Washington, D.C.)* **1988,** *242,* 1171–1173.
69. Peliska, J. A.; Benkovic, S. J. *Science (Washington, D.C.)* **1992,** *258,* 1112–1118.
70. Lisziewicz, J.; Sun, D.; Klotman, M.; Agrawal, S.; Zamecnik, P.; Gallo, R. *Proc. Natl. Acad. Sci. U.S.A.* **1992,** *89,* 11209–11213.
71. Taira, K.; Nakagawa, K.; Nishikawa, S.; Furukawa, K. *Nucleic Acids Res.* **1991,** *19,* 5125–5130.
72. Taira, K.; Nishikawa, S. In *Gene Regulation: Biology of Antisense RNA and DNA;* Erickson, R. P.; Izant, J. G., Eds.; Raven: New York, 1992; pp 35–54.
73. Ohkawa, J.; Yuyama, N.; Takebe, Y.; Nishikawa, S.; Taira, K. *Proc. Natl. Acad. Sci. U.S.A.* **1993,** *90,* 11302–11306.
74. Chen, C.-J.; Banerjea, A. C.; Harmison, G. G.; Haglund, K.; Schubert, M. *Nucleic Acids Res.* **1992,** *20,* 4581–4589.
75. Weizacker, F. V.; Blum, H. E.; Wands, J. R. *Biochem. Biophys. Res. Commun.* **1992,** *189,* 743–748.
76. Yuyama, N.; Ohkawa, J.; Inokuchi, Y.; Shirai, M.; Sato, A.; Nishikawa, S.; Taira, K. *Biochem. Biophys. Res. Commun.* **1992,** *186,* 1271–1279.
77. Inokuchi, Y.; Yuyama, N.; Hirashima, A.; Nishikawa, S.; Ohkawa, J.; Taira, K. *J. Biol. Chem.* **1994,** *269,* 11361–11366.
78. Takebe, Y.; Seiki, M.; Fujisawa, J.; Hoy, P.; Yokota, K.; Arai, K.; Yoshida, M.; Arai, N. *Mol. Cell. Biol.* **1988,** *8,* 466–472.
79. Ohkawa, J.; Koguma, T.; Kohda, T.; Taira, K. *J. Biochem. (Tokyo)* **1995,** *118,* 251–258.
80. Homann, M.; Tzortzakaki, M.; Rittner, K.; Sczakiel, G.; Tabler, M. *Nucleic Acids Res.* **1993,** *21,* 2809–2814.
81. Adachi, A.; Gendelman, H. E.; Koenig, S.; Folks, T.; Willey, R.; Rabson, A.; Martin, M. A. *J. Virol.* **1986,** *59,* 284–291.
82. Behr, J.-P. *Acc. Chem. Res.* **1993,** *26,* 274–278.
83. Wada, A.; Suzuki, Y.; Okayama, M.; Sato, S.; Shimayama, T.; Oya, M.; Uchida, C.; Koguma; T.; Nishikawa, S.; Kushitani, T.; Munekata, E.; Taira, K. *Nucleic Acids Res. Symp. Ser.* **1994,** *31,* 227–228.
84. Wada, A.; Suzuki, Y.; Kushitani, T.; Oya, M.; Munekata, E.; Koguma, T.; Nishikawa, S.; Taira, K. *Abstracts of Papers,* 5th Antisense Symposium, Tsukuba, Japan, 1995; No. 2–45.
85. Wada, A.; Kawai, S.; Goto, T.; Suzuki, Y.; Nishida, S.; Kato, M.; Oya, M.; Higaki, M.; Mizushima, Y. *Abstracts of Papers,* 5th Antisense Symposium, Tsukuba, Japan, 1995; No. 2–20.
86. Jie, S.; Kawai, S.; Wada, A.; Suzuki, Y.; Goto, T.; Oya, M.; Nishida, S.; Kato, M.; Narikawa, S.; Mizushima, Y. *Abstracts of Papers,* 5th Antisense Symposium, Tsukuba, Japan, 1995; No. 2–52.
87. Leonetti, J. P.; Rayner, B.; Lemaitre, M.; Gagnor, C.; Milhaud, P. G.; Imbach, J.- L.; Lebleu, B. *Gene* **1988,** *72,* 323–332.
88. Uchiyama, H.; Hirano, K.; Kashiwasake-Jibu, M.; Taira, K. *J. Biol. Chem.* **1996,** *271,* 380–384.

New Biomaterials for Drug Delivery

19

Pseudo-Poly(amino acid)s

Examples for Synthetic Materials Derived from Natural Metabolites

Kenneth James and Joachim Kohn

Pseudo-poly(amino acid)s consisting of α-L-amino acid building blocks linked by nonamide bonds have been used successfully as carriers in direct intracranial drug delivery, immunization systems, and the controlled release of anticoagulants. These natural metabolite-based polymers are biocompatible, degradable materials readily processed into microspheres, films, fibers, pins, and screws. In the last 5 years, most attention has been directed toward tyrosine-derived polycarbonates, polyiminocarbonates, and polyarylates. Pseudo-poly(amino acid) chemistry and synthesis, physicomechanical and degradation properties, biological response, immunological considerations, sterilization, and drug delivery applications thus far investigated are summarized.

Pseudo-poly(amino acid)s represent one of the newest classes of biomedical polymers. These materials are derived from naturally occurring amino acids. Unlike conventional poly(amino acid)s, where neighboring amino acids are linked by amide bonds to yield a peptide-like backbone, pseudo-poly(amino acid)s consist of α-L-amino acid building blocks linked by nonamide bonds such as urethane, ester, iminocarbonate, and carbonate bonds. Recently, polymers such as hydroxyproline-derived polyesters; serine-derived polyesters; and tyrosine-derived polyiminocarbonates, polycarbonates, and polyarylates have been synthesized. As will be described in detail, pseudo-poly(amino acid)s offer significant advantages over conventional poly(amino acid)s in terms of their inherent physicomechanical properties. For a comparison of the physicomechanical properties of commonly used biomedical polymers, the reader is referred to an earlier publication (*1*).

Studies of selected pseudo-poly(amino acid)s have shown them to be generally biocompatible, biodegradable materials that can be processed by a variety of means to yield microspheres, fibers, films, pins, and screws (*2*). These characteristics lend themselves to the formulation of a wide range of drug delivery systems. Because many of the pseudo-poly(amino acid)s under investigation

Copyright © 1997 American Chemical Society

have been synthesized only within the last 5 years, such research is still in its infancy.

Recently, a number of laboratories have prepared amino acid derived polymers whose backbones differ from conventional polypeptide backbone structures (*3–5*). Some confusion has entered the contemporary literature as investigators adopt existing terminology to new polymeric materials. Names such as "peptoids" (*3*), "isopeptide" (*4*), and "biopolymer" (*5*) are used without rigorous definition. In this chapter, the term "conventional poly(amino acid)" denotes synthetic polymers in which individual amino acids are linked by amide bonds derived from the N- and C-termini of neighboring amino acids. In contrast, "pseudo-poly(amino acid)s" are polymers where α-L-amino acids are linked by nonamide bonds. This convention is analogous to the use of the term "pseudo-peptide", which was introduced in 1983 (*6*) to designate a peptide containing nonamide backbone linkages.

A large number of poly(amino acid)s can be derived from different amino acid sequences. However, most conventional poly(amino acid)s cannot be used as industrial, engineering plastics because of their tendency to decompose in the molten state, to swell in moist environments, and to degrade, albeit slowly, by hydrolytic or enzymatic mechanisms. The high cost of conventional poly(amino acid)s is another obstacle for their commercial development.

During the 1970s, the exploration of conventional poly(amino acid)s shifted from an industrial focus to biomedical applications. Because poly(amino acid)s degrade in vivo to simple, naturally occurring amino acids, the application of these polymers in the manufacture of degradable, controlled-release drug systems was investigated. A small number of poly(γ-alkyl glutamate)s and other derivatives of poly(glutamic acid) were promising candidates for medical applications (*7*). For example, the solubilization of poly(γ-ethyl glutamate) on hydrolysis of the γ-ethyl pendant chain was used by Sidman et al. (*8, 9*) in the design of implantable controlled-release devices. Later, Bhaskar et al. (*10*) suggested the use of poly(γ-benzyl glutamate) as a matrix for controlled release. Exploiting the liquid-crystalline behavior of this polymer, the permeability of membranes could be controlled by the application of external magnetic fields. In spite of these early promising results, however, none of these approaches matured into clinically used systems.

The development of pseudo-poly(amino acid)s represents an attempt to circumvent some of the unfavorable material properties of conventional poly(amino acid)s and increase the range of amino acid derived polymers that can be considered for industrial or medical applications. Pseudo-poly(amino acid)s were first described in 1987 (*11*) and since have been evaluated for use in medical applications (*2*). This chapter reviews the synthesis, physical properties, biological response, and possible drug delivery applications of selected pseudo-poly(amino acid)s.

Chemistry and Polymer Synthesis

Since 1987, a wide range of new pseudo-poly(amino acid)s have been prepared (Table 19.1). Although several synthetic routes to pseudo-poly(amino acid)s exist, one of the most convenient ways is based on the polymerization of trifunctional amino acids or dipeptides (*2, 11*). This approach can be illustrated by the use of hydroxy amino acids such as serine, hydroxyproline, or tyrosine in the preparation of polyesters. For example, poly(*N*-acyl-L-serine ester) is a polymer

TABLE 19.1. Currently Available Pseudo-Poly(amino acid)s

Polymers	*Literature Citations*
Poly(*N*-acyl-*trans*-4-hydroxy-L-proline ester)	11, 13, 48
Poly(*N*-acyl-L-serine ester)	12, 49–51
Tyrosine-derived polyiminocarbonates	14, 23, 52, 53
Tyrosine-derived polycarbonates	15, 23, 54, 55
Tyrosine-derived polyarylates	17, 56, 57

obtained by the polymerization of N-protected α-L-serine by using the C-terminus and the hydroxyl group available at the side chain (Scheme 19.1) (*12*). First, the amino group of serine is protected. The protecting group (Z) can be a nontoxic acid (such as acetic acid or palmitic acid). Alternatively, Z can be one of the protecting groups used in peptide synthesis (such as benzyloxycarbonyl, *t*-butyloxycarbonyl, or triphenylmethyl). After formation of the lactone, poly(*N*-Z-L-serine ester) is obtained by a ring-opening polymerization.

A similar approach was used to obtain poly(*N*-acylhydroxy-L-proline ester), a polymer obtained when N-protected hydroxy-L-proline was used as the monomer in a polymerization reaction (Scheme 19.2) (*11, 13*). In this case, a simple melt transesterification yielded polymers of high molecular weight. The elimination of *N*-carboxyanhydrides as intermediates significantly reduced the cost of preparing poly(*N*-acylhydroxy-L-proline ester) as compared with the cost of preparing conventional poly(*O*-acylhydroxy-L-proline).

In some of these polymers, derivatives of dipeptides were used instead of individual amino acids as the monomeric repeat units. This approach can be illustrated by the preparation of tyrosine-based polyiminocarbonates (*14*). In these polymers, the basic, monomeric repeat unit is a tyrosyl–tyrosine dipeptide. First the N- and C-termini of the dipeptide were protected, followed by polymerization by means of the two phenolic hydroxyl groups located on the tyrosine side chains (Scheme 19.3). To optimize the polymer properties, the chemical structures of the N- and C-terminal protecting groups (R_1 and R_2) have to be designed carefully. One way to look at this reaction scheme is to regard tyrosine dipeptide as a replacement for industrially used diphenols such as bisphenol A.

A particularly promising series of tyrosine-based polycarbonates was obtained when the N-terminus of tyrosyl–tyrosine dipeptide was completely eliminated and ethyl, butyl, hexyl, or octyl esters were used as the C-terminus protecting groups (*15, 16*). The exact chemical structures of the polycarbonates obtained from desaminotyrosyl–tyrosine alkyl esters (designated as DTE, DTB, DTH, and DTO, respectively) are shown in structure **1.** These materials form a series of homologous polymers that differ only in the length

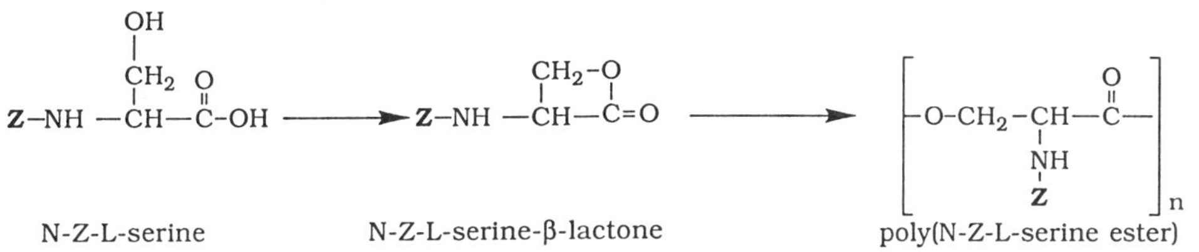

SCHEME 19.1. Reaction scheme for the preparation of poly(*N*-acyl-L-serine ester).

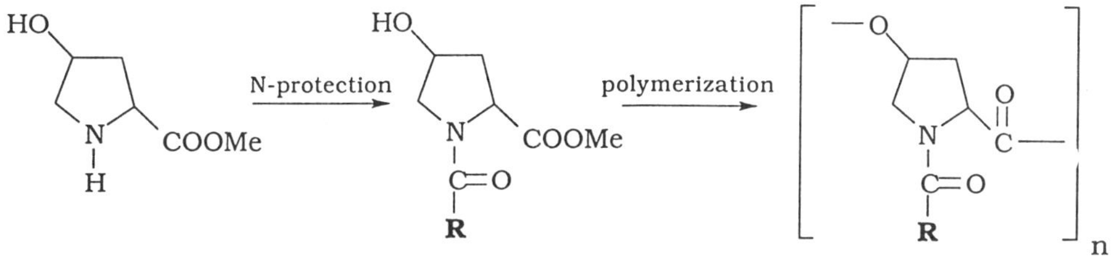

SCHEME 19.2. Reaction scheme for the preparation of poly(*N*-acylhydroxy-L-proline ester).

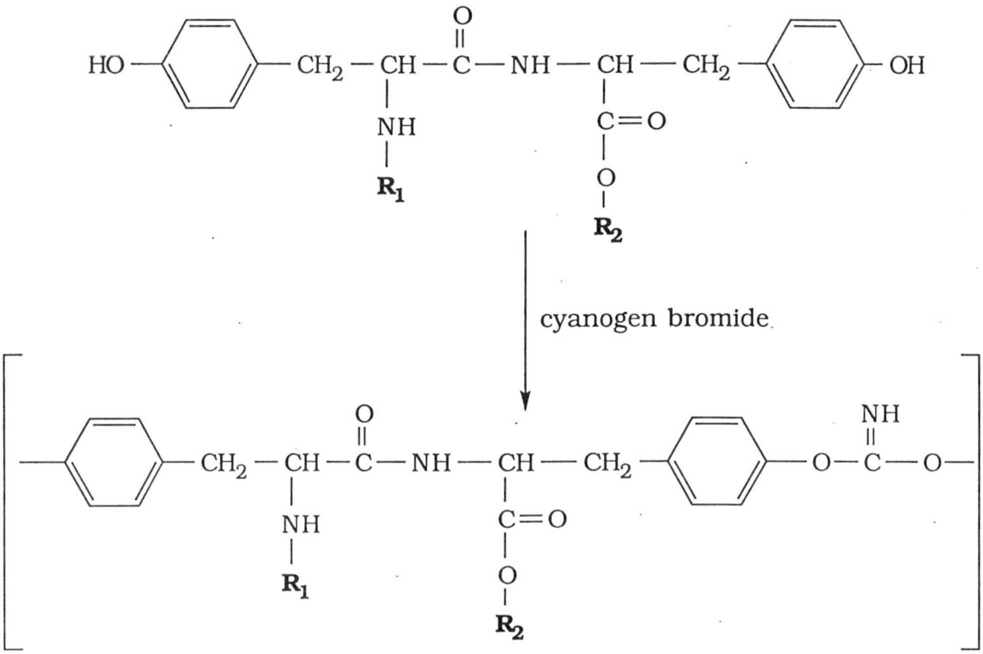

SCHEME 19.3. Preparation of tyrosine-derived polyiminocarbonates.

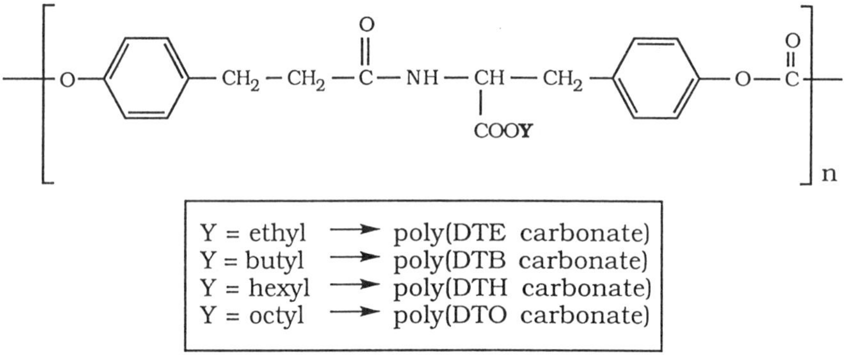

1

of their respective alkyl ester pendant chains. The ability to maintain the polymer backbone while changing the pendant chain structure is a powerful tool for the investigation of polymer structure–property relationships. For example, by varying the length of the alkyl ester pendant chain, properties such as the glass transition temperature (T_g), surface free energy, strength, stiffness, and the degradation rate can be controlled.

Among the pseudo-poly(amino acid)s listed in Table 19.1, the tyrosine-based polyarylates were synthesized most recently (*17*). Tyrosine-based polyarylates are made via the copolymerization of a diphenol and a diacid (**2**). Specifically, the

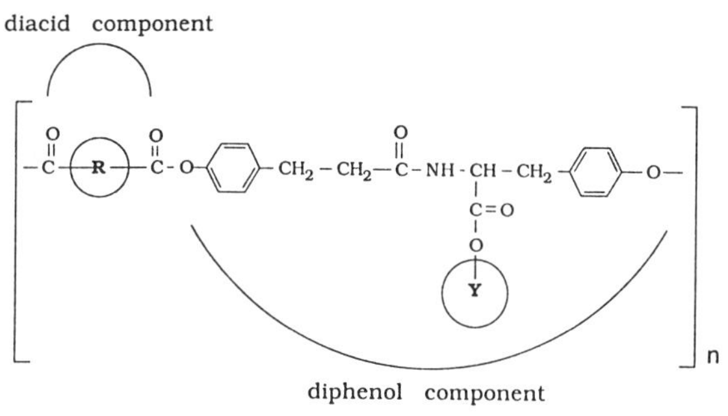

2

diphenol components selected were the desamino-tyrosyl–tyrosine alkyl esters previously described. The diacids include succinic, adipic, suberic, and sebacic acid, which have, respectively 2, 4, 6, or 8 methylene groups between two carboxylic acid groups. With this family of polymers it is possible to alter both the pendant chain length as well as the number of flexible methylene groups in the backbone. Hence, these materials are also a convenient model system for the investigation of structure–property relationships.

Tyrosine-Derived Pseudo-poly(amino acid)s as Drug Delivery Systems

Initial drug release research using pseudo-poly-(amino acid)s focused on hydroxyproline-derived polyesters (*13, 18*). By using poly(*N*-acyl-*trans*-4-hydroxy-L-proline ester)s (Scheme 19.2) as a model system, release times of commonly used dyes varied from several hours to many months, depending on the loading and the hydrophobicity of the dye. Poly(*N*-palmitoyl-*trans*-4-hydroxy-L-proline ester) in particular, which resembles candle wax and has long hydrophobic pendant chains, appeared promising as a long-lasting controlled-release device considering the favorable release profiles and initial biocompatibility evaluations. More recently, the physical properties, biodegradation, biocompatibility, and drug delivery applications of the tyrosine-derived polycarbonates, polyiminocarbonates, and polyarylates have been investigated. These studies will be the focus of this chapter.

Physicomechanical and Degradation Properties

One of the main differences between conventional poly(amino acid)s and pseudo-poly(amino acid)s is in their respective degree of order. Whereas conventional poly(amino acid)s are highly ordered, the disruption of the regular peptide backbone by nonamide backbone linkages reduces the inter- and intrachain forces in pseudo-poly(amino acid)s to the extent that all currently known pseudo-poly(amino acid)s are amorphous materials that are soluble in a variety of organic solvents. Consequently, these materials are processible using solvent-casting and evaporation techniques to yield polymer films, fibers, microspheres, and membranes (*19*). Furthermore, the lack of crystallinity reduces the processing temperatures of most pseudo-poly(amino acid)s to below 100 °C so that these materials can be extruded compression molded or fabricated into complex shapes by injection molding.

Tyrosine-Derived Polycarbonates

All members of this series of polymers are amorphous. Glass transition temperatures are a function of the pendant chain length (Table 19.2), and onset values range from 53 to 81 °C. These properties make it possible to formulate drug release systems by compression molding or extrusion at temperatures as low as 80 °C. The widely used poly(lactic acid), for example, usually requires processing temperatures upward of 170 °C, which often precludes thermal processing of drug–polymer mixtures.

Tyrosine-derived polycarbonates are characterized by their high mechanical strength and stiffness (Table 19.2). These polymers are stiffer than many other degradable polymers including polycaprolactone (0.5 GPa) and poly(ortho ester)s (<1 GPa) but are not as stiff as poly(lactic acid) (reported range of 2.4–10 GPa) or poly(glycolic acid) (6.5 GPa) (*20*). It is possible to obtain

TABLE 19.2. Physical Properties of Tyrosine-Derived Polycarbonates

Polymer	M_w ($\times 10^3$)	T_g[a] (°C)	T_d[b] (°C)	*Contact Angle* (°)	*Young's Modulus*[c] (GPa)	*Tensile Strength*[c] (MPa)	*Time Constant*[d] (weeks)
Poly(DTE carbonate)	176	81	290	73	1.5	67	12
Poly(DTB carbonate)	120	66	290	77	1.6	60	16
Poly(DTH carbonate)	350	58	320	86	1.4	62	21
Poly(DTO carbonate)	450	53	300	90	1.2	51	21

NOTE: Data are from Ertel and Kohn (*16*). *See* structure 1 for polymer structures. Weight-average molecular weights were determined by gel permeation chromatography.

[a]Glass-transition temperature determined by differential scanning calorimetry.

[b]Decomposition temperature determined by thermal gravimetric analysis. Measured at 2% weight loss.

[c]Unoriented samples. Properties measured at room temperature.

[d]Degradation time constant for thin, solvent-cast films under simulated physiological conditions (37 °C, pH 7.4 phosphate-buffered saline).

devices of sufficient mechanical strength at relatively low molecular weights (30–50 kDa). As with other polymer systems, the device degradation time can be controlled by changing the initial molecular weight. Considering the strength and stiffness of these materials, load-bearing drug delivery devices that may find application in orthopedics can conceivably be fabricated.

Another potential advantage of tyrosine-derived polycarbonates is their relatively high hydrophobicity, which can be controlled or modified by the pendant chain length (Table 19.2). At 37 °C, the most hydrophilic polymer in the series, poly(DTE carbonate) has an equilibrium water content of only 3–5%. For poly(DTO carbonate), the most hydrophobic member of the series, the equilibrium water content is well below 1%. Thus, even when hydrophilic drugs are incorporated into the polymer, the release profiles do not show strong burst effects. The length of the alkyl ester pendant chain has some effect on the intrinsic rate of polymer hydrolysis. For thin, solvent-cast films of high initial molecular weight, the degradation time constants are listed in Table 19.2. The time constants indicate that poly(DTE carbonate) degrades almost twice as fast as the more hydrophobic poly(DTO carbonate).

Under physiological conditions, tyrosine-derived polycarbonates degrade relatively slowly. As a first approximation, poly(DTE carbonate) degrades in vitro and in vivo about as fast as high-molecular-weight poly(L-lactic acid). On the basis of evidence obtained from electron spectroscopy for chemical analysis (ESCA), attenuated total reflection Fourier transform IR spectroscopy (ATR-FTIR), and gel permeation chromatography, an in vitro degradation mechanism was postulated (Scheme 19.4) (*16*). According to this mechanism, the ester bonds at the alkyl ester pendant chains are cleaved first, followed by the carbonate bonds in the polymer backbone. In vitro, the amide bonds are not hydrolyzed; thus, desaminotyrosyl–tyrosine is the final degradation product in vitro. In vivo, the cleavage of the amide bond by enzymatic or cellular mechanisms may lead to additional degradation products. No detailed mechanistic studies of the degradation products in vivo have so far been published.

Tyrosine-Derived Polyarylates

Polyarylates have relatively low T_g values and high decomposition temperatures. This characteristic allows processing of these materials via traditional means (compression molding, extrusion, and injection molding) at low molding temperatures without, in most cases, risking thermal degradation of the incorporated drug. The mechanical properties of the polyarylates differ substantially from those of the polycarbonates and can be controlled over a wide range by the selection of the

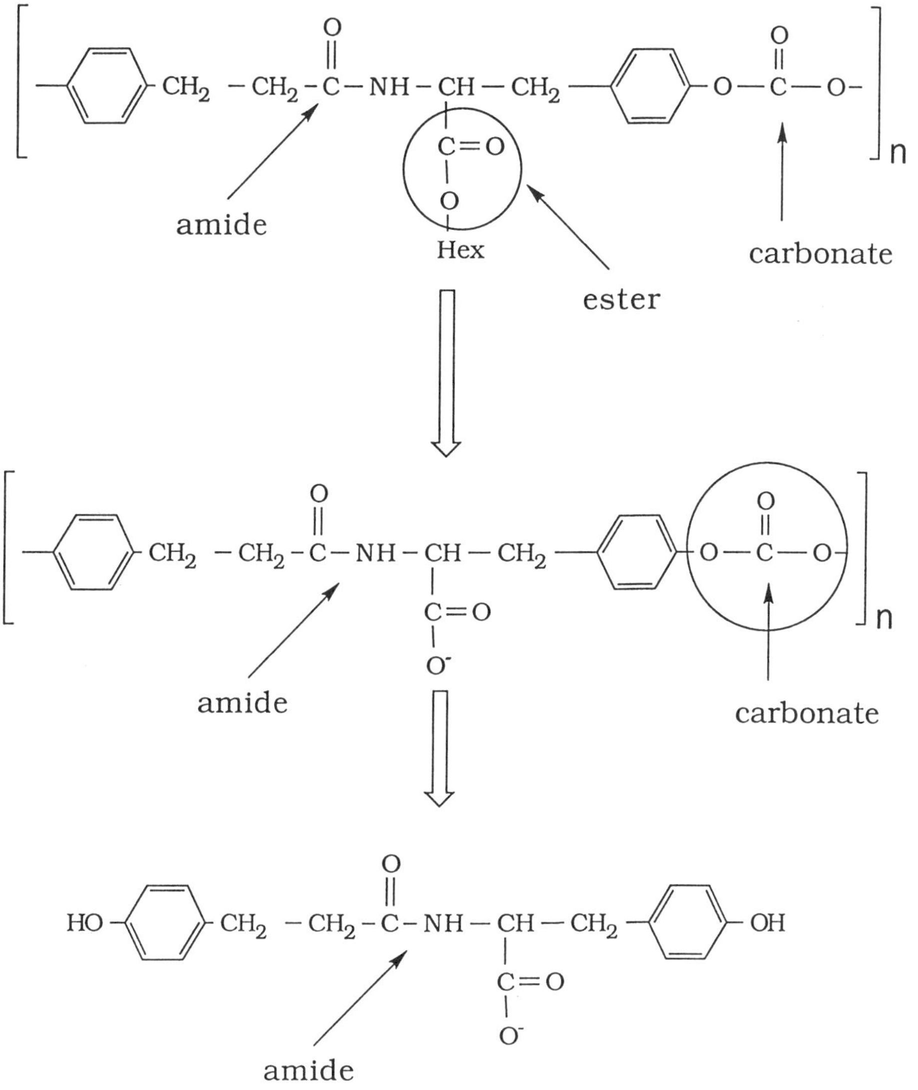

SCHEME 19.4. Suggested degradation mechanism of poly(DTH carbonate) based on surface analysis techniques. The degrading bond is circled.

alkyl pendant chain and diacid incorporated into the polymer structure.

Table 19.3 summarizes selected physical properties of poly(DTE adipate), poly(DTH adipate), and poly(DTO adipate). Increasing the length of the pendant chain decreases both T_g and polymer mechanical strength and stiffness. The stiffest and strongest of this series of polyarylates, poly(DTE adipate), approaches the mechanical strength of the tyrosine-derived polycarbonates. Increasing the pendant chain by 4 carbons decreases T_g to 34 °C and the associated modulus to below 0.5 GPa. If implanted into the body, both poly(DTH adipate) and poly(DTO adipate) will

Table 19.3. Physical Properties of Tyrosine-Derived Polyarylates

Polymer	M_w *(×10³)*	T_g[a] *(°C)*	T_d[b] *(°C)*	*Contact Angle (°)*	*Young's Modulus*[c] *(GPa)*	*Break Stress*[c] *(MPa)*	*Elongation at Break*[c] *(%)*
Poly(DTE adipate)	209	56	340	70	1.52	34	157
Poly(DTH adipate)	232	34	356	82	0.43	30	418
Poly(DTO adipate)	220	28	357	86	0.01	28	424

NOTE: Data are from Kohn et al. (*17, 57*). *See* structure **2** for polymer structures. Weight-average molecular weights determined by gel permeation chromatography.

[a]Glass-transition temperature determined by differential scanning calorimetry.

[b]Decomposition temperature as determined by thermal gravimetric analysis. Measured at 10% weight loss.

[c]Unoriented samples. Properties measured at room temperature.

be in a rubbery state. As indicated by the air–water contact angles, polyarylates become more hydrophobic as the length of the pendant chain length is increased.

In vitro degradation studies of thin films of poly(DTE adipate), poly(DTH adipate), and poly(DTO adipate) were performed (*17, 21*). Under simulated physiological conditions (37 °C in pH 7.4 phosphate-buffered saline (PBS)), changes in sample weight and molecular-weight loss over a 26-week period were assessed. The three tyrosine-derived polyarylates mentioned exhibited similar degradation profiles and retained about 30–40% of their initial molecular weight. No weight loss was observed during this test period. Although the time frame of degradation in vivo still needs to be determined, polyarylates probably will be comparable in their degradation rate to poly(lactic acid) and tyrosine-derived polycarbonates. Thus, these polymers appear most adept at addressing situations where a long-term degradation profile is required.

Tyrosine-Derived Polyiminocarbonates

In comparison to other pseudo-poly(amino acid)s, polyiminocarbonates have very low decomposition temperatures (approximately 150 °C) (*14, 22*). Consequently, during fabrication the temperature where thermal decomposition commences can be approached, and even exceeded. Because polyiminocarbonates are soluble in organic solvents, fabrication by standard solvent-casting methods is possible. Also, poly(DTH iminocarbonate) has a T_g of 55 °C; therefore, devices of this material can be compression molded at approximately 70 °C without significant decomposition.

The polyiminocarbonates are as stiff and as strong as the corresponding tyrosine-derived polycarbonates (*1, 14*). For example, poly(DTH iminocarbonate) (M_W = 101 kDa) has a tensile strength of 40 MPa and tensile stiffness of 1.6 GPa. A defining feature of tyrosine-derived polyiminocarbonates is the significant hydrolytic instability of the iminocarbonate bond, which is responsible for the rapid degradation of high-molecular-weight polymers (>100 kDa) to low-molecular-weight oligomers (1–6 kDa) within 1 week under simulated physiological conditions. For example, solvent-cast films exposed to an aqueous buffer solution swelled and became turbid due to the absorption of water, and these films crumbled within 5–7 days. The chemical mechanism of degradation leads to the formation of ammonia and carbon dioxide and the regeneration of the diphenolic monomer used in the synthesis of the polymer. The iminocarbonate bond ranks as one of the most rapidly degrading backbone structures, but because the low-molecular-weight oligomers formed are insoluble, the rapid hydrolysis of the backbone does not lead to equally rapid mass loss or resorption of the polymeric device.

Biological Response

Although studies concerning the biocompatibility of pseudo-poly(amino acid)s are still being actively pursued, several pertinent observations can be made from the results thus far. In vitro cytotoxicity assays and several in vivo tissue compatibility screening tests were conducted for poly(*N*-acylhydroxy-L-proline ester)s (*18*), tyrosine-derived polyiminocarbonates, and tyrosine-derived polycarbonates (*16, 23*). On the basis of in vitro assays, these materials were generally found to be noncytotoxic. The in vivo response was usually typical of a mild inflammatory response as seen for most medical polymers such as polyethylene or poly(lactic acid). Biocompatibility assays focusing on tyrosine-derived polyarylates have not yet appeared in the literature.

Tyrosine-Derived Polycarbonates

In a recent in vitro study, the number of fibroblasts attaching to polycarbonate surfaces and their subsequent proliferation over a 5-day period were measured (*16*). Surface hydrophobicity could be correlated with the cellular response. Consistent with the hypothesis that cells favor more hydrophilic and molecularly rigid polymer surfaces, the number of viable cells attached after 24 h decreased as the polymer pendant chain length increased. That is, cell attachment increased in the following order: poly(DTE carbonate) > poly(DTB carbonate) > poly(DTH carbonate) > poly(DTO carbonate). The same pattern was observed when the cells were allowed to proliferate for 5 days. By using cell proliferation on tissue culture polystyrene as a control (100% growth), the comparative cell growth values on poly(DTE carbonate), poly(DTB carbonate), poly(DTH carbonate), and poly(DTO carbonate) were 71, 59, 31, and 29% respectively (*16*).

In an in vivo pilot study (*24*), poly(DTH carbonate) pins (80 mg, 2 mm in diameter) were fabricated and compared to commercially available Orthosorb pins made of polydioxanone. The pins were implanted transcortically in the distal femur and proximal tibia of 10 New Zealand White rabbits for 1, 2, 4, and 26 weeks. In addition to routine histological evaluation of the implant sites, bone activity at the implant–tissue interface was visualized by UV illumination of sections labeled with fluorescent markers, and the degree of calcification around the implants was ascertained by backscattered electron microscopy.

The bone tissue response was characterized by a lack of fibrous capsule formation and an unusually low number of inflammatory cells at the bone–implant interface. Poly(DTH carbonate) exhibited close contact with bone tissue throughout the 6-month period of this initial study. A roughened interface was observed that was penetrated by new bone as early as 2 weeks postimplantation. Bone growth into the periphery of the implant material was clearly visible after 6 months (Figure 19.1) (*24*).

Tyrosine-Derived Polyiminocarbonates

In a comparative study (*23*), solvent-cast films of poly(DTH iminocarbonate) and poly(DTH carbonate) were evaluated in a subcutaneous rat model. In this study, high-density polyethylene (HDPE) and medical-grade poly(D,L-lactic acid) served as controls. Considering the significantly faster degradation rate of poly(DTH iminocarbonate), one would expect a different response from this material. Indeed, at 7 days postimplantation a greater cell density and inflammatory response was noted for this material. However, at time periods exceeding 1 month, the biological response to poly(DTH iminocarbonate) was not notably different from the response observed for poly(DTH carbonate), polyethylene, or poly(lactic acid). The tissue response was characterized by a thin tissue capsule, absence of giant cells, and a low inflammatory cell count (*23*).

Immunological Considerations

Although pseudo-poly(amino acid)s do not have the peptide backbone structure of conventional poly(amino acid)s, they are nevertheless polymers derived from amino acids. Thus, in the context of implantable materials, the immunological properties of pseudo-poly(amino acid)s become an important concern. Because of the current lack of detailed investigations of the immunological properties of pseudo-poly(amino acid)s, only prelimi-

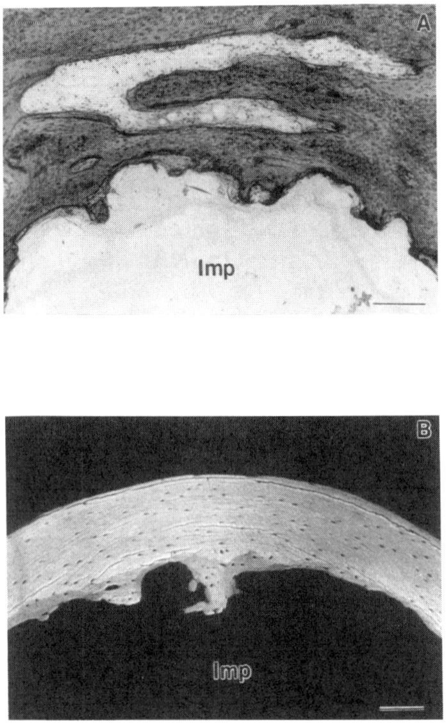

FIGURE 19.1. Bone tissue response at 26 weeks to poly(DTH carbonate) implant (Imp): A, histological section stained with Stevenel's Blue and Van Gieson picrofuchsin staining. (Note very thin fibrous capsule and bone penetrating into the outer surface of the implant.) B, backscattered electron image of the bone–implant interface. (Again note close juxtaposition of the bone and implant and evidence of calcified bone tissue penetrating into the implant.) Scale represents 100 μm.

nary conclusions can be drawn from the known immunological properties of conventional poly(amino acid)s and from some initial implantation studies in various animal models.

During the 1950s and 1960s, the exploration of poly(amino acid)s as synthetic antigens contributed to the elucidation of the molecular mechanisms of antigenicity. This work also provided insights into the possible use of poly(amino acid)s as biomaterials. In essence, homopolymers of amino acids were shown to be poor immunogens. For example, attempts to raise antibodies against poly-D,L-alanine, polysarcosine, poly-L-aspartic acid, poly-L-lysine, poly-L-hydroxyproline, poly-L-glutamic acid, poly-L-proline, and copolymers of L-tyrosine and L-aspartic acid in several animal species were consistently unsuccessful (*25, 26*).

However, the immunogenicity of poly(amino acid)s increases with increasing molecular complexity and with the number of different amino acid residues present in the copolymer (*27*). Immunogenicity is also strongly dependent on molecular structure and conformation (*28, 29*). For example, a multiresidue, branched copolymer of lysine, alanine, tyrosine, and aspartic acid is a powerful synthetic immunogen (*30*). A comparison of two such polymers with identical molecular weight, size, and composition, only differing in the order of the tyrosine and glutamine residues within the sequence of the tetrapeptide epitopes, revealed very different immunological properties (*29*). This result illustrates the possible pitfalls of trying to predict the immunogenicity of one material based on the known immunogenicity of a structurally related material.

The observations about the effect of structure, conformation, and molecular weight on the immunogenicity of amino acid derived polymers are in line with the experimental observation that some poly(amino acid)s are immunogenic and can serve as immunologically active carriers for haptens and other antigens, whereas other poly(amino acid)s can serve as immunologically inactive implant materials. The rule, historically, was that the search for such nonimmunogenic implant materials must be limited to unbranched homopolymers and copolymers containing no more than two different amino acid residues in random sequence (*7*). However, recent work by Urry et al. (*31*) indicated that polypeptides based on repeated sequences containing three different peptides did not induce systemic antigenicity nor dermal sensitization. Such observations enlarge the number of new polymeric materials derived from amino acids available for medical purposes.

Currently, no data specifically address the immunogenicity of pseudo-poly(amino acid)s.

However, as described, in all animal models tested (mouse, rat, rabbit, and dog) and all tissue sites evaluated (subcutaneous, intramuscular, and bone), the histological findings were consistent with a mild inflammatory response and not indicative of an immune response.

Sterilization

The effect of γ-irradiation on commonly used polymers such as poly(glycolic acid), polydioxanone, poly(methyl methacrylate), and poly(lactic acid) has been investigated by different groups (*32–37*). The most obvious effect of γ-radiation is a reduction in the average molecular weight caused by chain scission and sometimes accompanied by chain cross-linking. Copolymers of lactide–glycolide have exhibited a particularly high sensitivity to irradiation. To alleviate this problem, Dennis and Shalaby (*38*) explored the concept of incorporating aromatic components into the copolymer backbone to improve the resistance of the polymer to γ-irradiation. On the basis of their results and the known resistance of polycarbonates to γ-irradiation in general, tyrosine-derived polycarbonates were expected to be sterilizable by γ-irradiation.

In preliminary studies, somewhat conflicting observations were made. When γ-irradiation was performed in air, especially at higher doses, a considerable loss of molecular weight occurred. Exposure to 1, 2.5, and 4.0 Mrad reduced the respective weight-average molecular weight of tyrosine-derived polycarbonates by 25, 50, and 60% (*16*), irrespective of the structure of the pendant chain. However, when the same radiation doses were used to sterilize specimens immersed in aqueous buffer solutions, almost no reduction in molecular weight was observed and the radiation-induced damage seemed to increase with increasing length of the alkyl ester pendant chain (unpublished data, Kohn et al.). This result can in part be attributed to the shielding effect of water, and it may be possible to identify conditions that facilitate the effective sterilization of tyrosine-derived polycarbonates by γ-irradiation.

In several ongoing biocompatibility studies, treatment with ethylene oxide has been used successfully to sterilize films and rod specimens made of tyrosine-derived polycarbonates. Ethylene oxide treatment did not significantly alter the surface chemistry as documented by ESCA and ATR-FTIR, nor did it induce changes in molecular weight of more than 15% (*16*).

Controlled Drug Delivery: Example Applications

Tyrosine-Derived Polycarbonates

Poly(DTH carbonate) has been used in the design of an investigational long-term controlled-release device for the intracranial administration of dopamine (*39, 40*). For such applications, poly(DTH carbonate) has several potential advantages over other degradable polymers, which include the ease with which dopamine can be physically incorporated into the polymer (due to its relatively low processing temperature and the structural similarity between the drug and the polymer), the apparent protective action of the polymeric matrix on dopamine, the prolonged release of approximately 15% of the total load of dopamine over about 180 days (Figure 19.2), and the high degree of compatibility with brain tissue (unpublished results, Kohn et al.).

Preliminary results indicate an average dopamine release of about 1–2 μm/day over prolonged periods of time from poly(DTH carbonate) (Table 19.4). Although this release rate was within the therapeutically useful range, no in vivo release experiments have so far been reported in the literature. For more detailed information on other dopamine-releasing systems the reader is referred to several recent publications (*41–43*).

Because of its relatively high strength, poly(DTH carbonate) was also considered for low-weight-bearing orthopedic applications. Recent studies involving the transcortical implantation of compression molded poly(DTH carbonate) pins in the distal femur and proximal tibia of the rabbit indicated active bone remodeling at the surface of the degrading implant and excellent bone apposi-

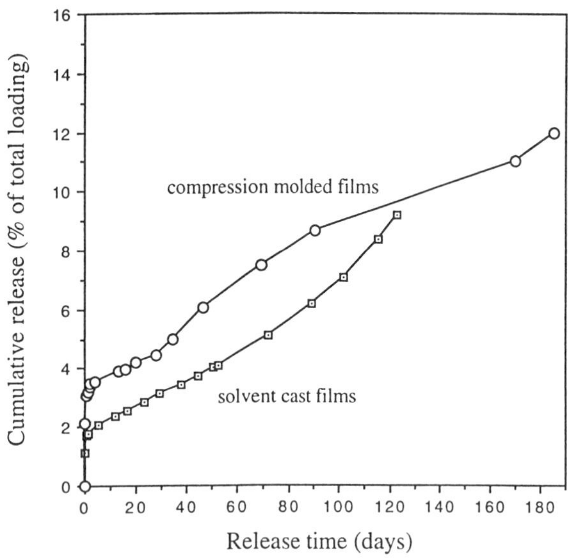

FIGURE 19.2. Release of dopamine from compression-molded disks and solvent-cast films containing a total loading of dopamine of 5 wt%. The devices were incubated in 0.9% saline solution containing 0.2% of ethylenediaminetetraacetic acid disodium salt as an antioxidant (pH 4.2) at room temperature.

tion without any intervening fibrous tissue layer throughout the 6-month test period (*24*). Although these observations indicate that poly(DTH carbonate) may find applications in the delivery of pharmaceutical agents to bone tissue, no publications describing this specific application of tyrosine-derived polycarbonates have so far been published.

Tyrosine-Derived Polyarylates

Model drug release studies have been performed by using the adipic acid series of polyarylates (*17, 21*). Solvent-cast devices incorporating 5% (w/w) *p*-nitroaniline dye were incubated in pH 7.4 PBS (37 °C) to simulate the release of low-molecular-weight drugs from these polymers. Over a 40-day period, all dye was released from devices made of poly(DTH adipate) and poly(DTO adipate), and over 70% of the drug was released from the polymers over the first week. These polymers were in

TABLE 19.4. Release of Dopamine from Devices Made of Poly(DTH carbonate)

Parameter	*Compression Molding*	*Solvent Casting*
Weight of device (mg)	55	22
Total dopamine loading (mg)	2.75	1.1
Observed release time (days)	180	120
Cumulative total release of dopamine (μg)	330	110
Average release of dopamine (μg/day)	1.8	0.92

NOTE: Data are from Dong (*40*). Although the release of dopamine continued, the studies were terminated at the indicated times.

the rubbery state at 37 °C. In contrast, poly(DTE adipate), which was in a glassy state at 37 °C, released only half of the incorporated dye during the 40-day incubation period. For each polymer, however, the dye-release profile was linear when plotted as a function of the square root of time (Figure 19.3). This result may indicate a diffusion-controlled-release profile.

In Europe, tyrosine-derived polyarylates (*17*) were tested as hemocompatible coatings for blood-contacting devices (*44*). Techniques were developed to incorporate anticoagulants into coatings made of tyrosine-derived polyarylates or lactide–glycolide copolymers. The ex vivo data obtained in human blood and the in vivo data obtained in pigs, supported by scanning electron microscopy, revealed that uncoated test materials (carbon fibers) were covered within minutes by a coagulation plug rich in fibrin and platelets. Degradable coatings without anticoagulants reduced the thrombogenicity of the test materials, but coatings releasing hirudin and prostacyclin inhibitors prevented the formation of thrombi at the coated surfaces.

Tyrosine-Derived Polyiminocarbonates

Delivery systems are also being considered for single-step immunization procedures. Such systems allow the sustained release of antigen to induce prolonged immunity without the need for repeated

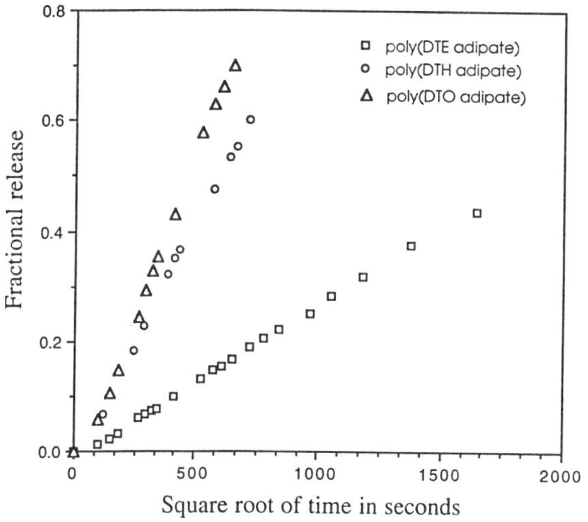

FIGURE 19.3. Release of *p*-nitroaniline dye from solvent-cast films of tyrosine-derived polyarylates under simulated physiologic conditions (pH 7.4 PBS at 37 °C). Note the curves are plotted as a function of the square root of time. The linear curves are indicative of a diffusion-controlled-release profile.

doses (*45*). Because many vaccines require the use of an adjuvant to reach an adequate level of antibodies, engineered carriers providing sustained adjuvancy at the same time as serving as an antigen repository have been considered. Because L-tyrosine is known for its adjuvant properties (*46*), the use of a degradable tyrosine-derived polyiminocarbonate as an antigen delivery device appeared promising (*47*). Such a device was tested by using a system that consisted of bovine serum albumin (BSA) released subcutaneously in mice from thin, solvent-cast films of a tyrosine-derived polyiminocarbonate. On degradation, the release of BSA and the concomitant presence of the tyrosine-derived polyiminocarbonate (or its degradation products) resulted in an anti-BSA antibody titer that was comparable to the titer observed when BSA was administered repeatedly in complete Freund's adjuvant. This system took advantage of the rapid degradation characteristics of tyrosine-derived polyiminocarbonates.

Summary

Pseudo-poly(amino acid)s use naturally occurring metabolites in the synthesis of new polymeric structures. By linking together amino acids with bonds other than amide linkages, the physicomechanical properties of such polymers were superior to conventional poly(amino acid)s. Tyrosine-derived polycarbonates, polyiminocarbonates, and polyarylates appear particularly promising as a new class of biodegradable, biocompatible materials. Offering a wide range of physical and degradation properties, pseudo-poly(amino acid)s allow for the selection and optimization of characteristics appropriate for specific applications. Controlled-delivery devices based on this new class of polymers have been implemented successfully in several animal models.

Acknowledgments

A portion of the scientific work reviewed here was supported by NIH grants GM39455 and GM49894. Joachim Kohn also acknowledges support from NIH Research Career Development Award GM00550.

References

1. Engelberg, I.; Kohn, J. *Biomaterials* **1991,** *12,* 292–304.
2. Kohn, J. *Trends Polym. Sci. (Cambridge, U.K.)* **1993,** *1,* 206–212.
3. Kessler, H. *Angew. Chem. Int. Ed. Engl.* **1993,** *32,* 543–544.
4. Sekiguchi, H.; Coutin, B.; Bechaouch, S.; Gachard I. *POLYMIX–93: International Symposium on Polymers;* Cancun, Mexico, 1993; pp 92–93.
5. Cho, C. Y.; Morgan, E. J.; Cherry, S. R.; Stephans, J. C.; Fodor, S. P. A.; Adams, C. L.; Sundaram, A.; Jacobs, J. W.; Schultz, P. G. *Science (Washington, D.C.)* **1993,** *261,* 1303–1305.
6. Spatola, A. F. In *Chemistry and Biochemistry of Amino Acids, Peptides, and Proteins;* Weinstein, B., Ed.; Marcel Dekker: New York, 1983; pp 267–357.

7. Anderson, J. M.; Spilizewski, K. L.; Hiltner, A. In *Biocompatibility of Tissue Analogs;* Williams, D. F., Ed.; CRC Press: Boca Raton, FL, 1985; Vol. 1, pp 67–88.
8. Sidman, K. R.; Schwope, A. D.; Steber, W. D.; Rudolph, S. E.; Poulin, S. B. *J. Membr. Sci.* **1980,** *7,* 277–291.
9. Sidman, K. R.; Steber, W. D.; Schwope, A. D.; Schnaper, G. R. *Biopolymers* **1983,** *22,* 547–556.
10. Bhaskar, R. K.; Sparer, R. V.; Himmelstein, K. J. *J. Membr. Sci.* **1985,** *24,* 83–96.
11. Kohn, J.; Langer, R. *J. Am. Chem. Soc.* **1987,** *109,* 817–820.
12. Zhou, Q. X.; Kohn, J. *Macromolecules* **1990,** *23,* 3399–3406.
13. Yu-Kwon, H.; Langer, R. *Macromolecules* **1989,** *22,* 3250–3255.
14. Pulapura, S.; Li, C.; Kohn, J. *Biomaterials* **1990,** *11,* 666–678.
15. Pulapura, S.; Kohn, J. *Biopolymers* **1992,** *32,* 411–417.
16. Ertel, S. I.; Kohn, J. *J. Biomed. Mater. Res.* **1994,** *28,* 919–930.
17. Fiordeliso, J.; Bron, S.; Kohn, J. *J. Biomater. Sci. Polym. Ed.* **1994,** *5,* 497–510.
18. Yu-Kwon, H. Ph.D. Thesis, Massachusetts Institute of Technology, 1988.
19. Kohn, J.; Yu, C.; Yeo, S. D.; Debenedetti, P. G. *Proceedings of the 21st International Symposium on Controlled Release of Bioactive Materials;* Controlled Release Society: Lincolnshire, IL, 1994; pp 132–133.
20. Daniels, A. U.; Chang, M. K. O.; Andriano, K. P.; Heller, J. *J. Appl. Biomater.* **1990,** *1,* 57–78.
21. Fiordeliso, J. M.S. Thesis, Rutgers University, 1993.
22. Li, C.; Kohn, J. *Macromolecules* **1989,** *22,* 2029–2036.
23. Silver, F. H.; Marks, M.; Kato, Y. P.; Li, C.; Pulapura, S.; Kohn, J. *J. Long-Term Eff. Med. Implants* **1992,** *1,* 329–346.
24. Ertel, S. I.; Kohn, J.; Zimmerman, M. C.; Parsons, J. R. *J. Biomed. Mater. Res.* **1995,** *29,* 1337–1348.
25. Maurer, P. H.; Subrahmanyam, D.; Katchalski, E.; Blout, E. R. *J. Immunol.* **1959,** *83,* 193–197.
26. Sela, M.; Katchalski, E. In *Advances in Protein Chemistry;* Anfinsen, C. B.; Anson, M. L.; Bailey, K.; Edsall, J. T., Eds.; Academic: New York, 1959; Vol. 14, pp 391–477.
27. Sela, M. *Proceedings of the Rehovot Symposium on Poly(Amino Acids), Polypeptides, and Proteins and Their Biological Implications;* Blout, E. R.; Bovey, F. A.; Goodman, M.; Lotan, N., Eds.; John Wiley: New York, 1974; pp 495–509.
28. Sela, M.; Katchalski, E.; Olitzki, A. L. *Science (Washington, D.C.)* **1956,** *123,* 1129.
29. Muszkat, K. A.; Schechter, B.; Sela, M. *Mol. Immunol.* **1992,** *29,* 1049–1054.
30. Sela, M.; Fuchs, S.; Arnon, R. *Biochem. J.* **1962,** *85,* 223–235.
31. Urry, D. W.; Parker, T. M.; Reid, M. C.; Gowda, D. C. *J. Bioact. Compat. Polym.* **1991,** *6,* 263–282.
32. Shintani, H.; Kikuchi, H.; Nakamura, A. *J. Appl. Polym. Sci.* **1990,** *41,* 661–675.
33. Martakis, N.; Niaounakis, M.; Pissimissis, D. *J. Appl. Polym. Sci.* **1994,** *51,* 313–328.
34. Lewis, D. A.; O'Donnell, J. H.; Hedrick, J. L.; Ward, T. C.; McGrath, J. E. In *The Effects of Radiation on High-Technology Polymers;* Reichmanis, E.; O'Donnell, J. H., Eds.; ACS Symposium Series 381; American Chemical Society: Washington, DC, 1989; pp 252–261.
35. Gupta, M. C.; Deshmukh, V. G. *Polymer* **1983,** *24,* 827–830.
36. Gorham, S. D.; Srivastava, S.; French, D. A.; Scott, R. *J. Mater. Sci. Mater. Med.* **1993,** *4,* 40–49.
37. Collett, J. H.; Lim, L. Y.; Gould, P. L. *Polym. Prepr. Am. Chem. Soc. Div. Polym. Chem.* **1989,** *30,* 468–469.
38. Jamiolkowski, D. D.; Shalaby, W. S. In *Radiation Effects on Polymers;* Clough, R. L.; Shalaby, S. W., Eds.; ACS Symposium Series 475; American Chemical Society: Washington, DC, 1991; pp 300–309.
39. Coffey, D.; Dong, Z.; Goodman, R.; Israni, A.; Kohn, J.; Schwarz, K. O. Presented at the 203rd National Meeting of the American Chemical Society, San Francisco, CA, April 1992; paper CELL 0058.
40. Dong, Z. M.S. Thesis, Rutgers University, 1993.
41. Sabel, B. A.; Freese, A.; During, M. J. *Adv. Neurol.* **1990,** *53,* 513–518.
42. Madrid, Y.; Langer, L. F.; Brem, H.; Langer, R. *Adv. Pharmacol. (San Diego)* **1991,** *22,* 299–324.
43. During, M. J.; Freese, A.; Deutch, A. Y.; Kibat, P. G.; Sabel, B. A.; Langer, R.; Roth, R. H. *Exp. Neurol.* **1992,** *115,* 193–199.
44. Stemberger, A.; Alt, E.; Schmidmaier, G.; Kohn, J.; Blümel, G. *Ann. Hematol.* **1994,** *68,* 48.
45. Preis, I.; Langer, R. S. *J. Immunol. Methods* **1979,** *28,* 193–197.

46. Wheeler, A. W.; Moran, D. M.; Robins, B. E.; Driscoll, A. *Int. Arch. Allergy Appl. Immunol.* **1982,** *69,* 113–119.
47. Kohn, J.; Niemi, S. M.; Albert, E. C.; Murphy, J. C.; Langer, R.; Fox, J. G. *J. Immunol. Methods* **1986,** *95,* 31–38.
48. Yu, H.; Lin, J.; Langer, R. *14th International Symposium on Controlled Release of Bioactive Materials;* Lee, P. I.; Leonhardt, B. A., Eds.; Controlled Release Society: Lincolnshire, IL, 1987; pp 109–110.
49. Gelbin, M. E.; Kohn, J. *Polym. Prepr. Am. Chem. Soc. Div. Polym. Chem.* **1991,** *32,* 241–242.
50. Gelbin, M. E.; Kohn, J. *J. Am. Chem. Soc.* **1992,** *114,* 3962–3965.
51. Fietier, I.; Le Borgne, A.; Spassky, N. *Polym. Bull.* **1990,** *24,* 349–353.
52. Haque, F.; Li, C.; Kohn, J. *16th International Symposium on Controlled Release of Bioactive Materials;* Perlman, R.; Miller, J. A., Eds.; Controlled Release Society: Lincolnshire, IL, 1989; pp 115–116.
53. Kohn J.; Li, C. U.S. Patent 5,140,094, 1992.
54. Ertel, S. I.; Parsons, R.; Kohn, J. *Abstracts of Papers,* 19th Annual Meeting of the Society of Biomaterials, Birmingham, AL; Society for Biomaterials; Minneapolis, MN, 1993; p 17.
55. Ertel, S. I.; Kohn, J. *Abstracts of Papers,* 19th Annual Meeting of the Society of Biomaterials, Birmingham, AL; Society for Biomaterials; Minneapolis, MN, 1993; p 133.
56. Bron, S.; Kohn, J. *Polym. Mater. Sci. Eng.* **1993,** *69,* 37–39.
57. Kohn, J. *Abstracts of Papers,* 20th Annual Meeting of the Society for Biomaterials, Boston, MA; Society for Biomaterials; Minneapolis, MN, 1994; paper number MN67.

20

Transductional Protein-Based Polymers as New Controlled-Release Vehicles

Dan W. Urry, Cynthia M. Harris, Chi Xiang Luan, Chi-Hao Luan, D. Channe Gowda, Timothy M. Parker, Shao Qing Peng, and Jie Xu

The transductional protein-based polymers of interest are composed of repeating peptide sequences that exhibit inverse temperature transitions of hydrophobic folding and assembly as the temperature is raised above a critical onset (transition) temperature, T_t. Importantly, innumerable means exist whereby T_t can be lowered from above to below physiological temperature to drive hydrophobic folding and assembly in the performance of various forms of mechanical and chemical work, such as lifting weights, uptake of protons, and construction of drug-laden controlled-release vehicles. This mechanism is called the ΔT_t-mechanism of free-energy transduction (energy conversion), and any of the inputs of mechanical, thermal, pressure, chemical, electrical, or electromagnetic energy with the proper polymer design can be employed to control drug loading and release.

One particularly useful approach to the construction of controlled-release vehicles is to design polymers with pK_a-shifted functional side chains such as the carboxyl moieties of glutamic acid and aspartic acid residues and the amino groups of lysine residues. The value for the onset of the transition temperature, T_t, under physiological conditions of pH and salt concentration for these multiply charged polymers can be above 100 °C, but on adding a drug with the opposite sign to the solution of the polymer, ion pairing dramatically lowers T_t to below 37 °C. A consequence is the formation of a drug-laden, more dense phase. The drug-laden phase acts like a precipitate with low solubility, such that, with regular renewal of the surrounding medium, there is a constant release of drug for a constant surface area with concomitant dispersal of the vehicle. The densities

Copyright © 1997 American Chemical Society

of charged sites on the polymer and in the resulting more dense phase determine the total drug to be released per unit volume of loaded vehicle, and the solubility product constant determines the level of release per day for a given surface area. Analogously, swollen cross-linked matrices composed of polymers with pK_a-shifted functional groups will contract on addition (i.e., loading) of the oppositely charged drug. Release of drug results in a return to the swollen state, which is generally perceived to be biodegradable.

The focus of this review is on transductional protein-based polymers: polymers composed of repeating peptide sequences that are capable of undergoing transitions between the gel and elastomer states or between the gel and plastic states. Transductional controlled release occurs as a result of the change of physical state of the polymer.

Transductional Protein-Based Polymers

Inverse Temperature Transitions and Energy Transduction

Protein-based polymers can be designed to perform free-energy transductions (energy conversions) involving each of the intensive variables of mechanical force (e.g., in dynes per square centimeter cross-sectional area), temperature, pressure, chemical potential, electrochemical potential, and electromagnetic radiation (*1*). Each of the free-energy transductions has the potential to be employed in the controlled release of therapeutic substances. The key property whereby protein-based polymers achieve transductionally controlled release is that of exhibiting a reversible inverse temperature transition (*1, 2*): that is, to hydrophobically fold and assemble as the temperature is raised through a critical temperature range, or conversely to hydrophobically unfold and disassemble as the temperature is lowered from above to below the transition range. Most importantly, however, there are the innumerable ways in which the transition range can be changed, from above to below physiological temperature to drive hydrophobic folding and assembly or from below to above physiological temperature to drive hydrophobic unfolding and disassembly (i.e., to drive dissolution or swelling for a cross-linked matrix).

The onset temperature for the inverse temperature transition on raising the temperature is designated as T_t, and the temperature interval over which the transition occurs varies from 5 to 20 °C depending on composition. Performing functions by controlling the value of T_t to change the folded and assembled state uses what is called the ΔT_t-mechanism.

Controlled-Release Modalities

The transductional protein-based polymers make possible diverse modalities for controlled release. The general categories can be listed as diffusional, degradational, and transductional. Variables for altering the rate of diffusional release include the relative hydrophobicity of polymer and drug, the water content of the polymer matrix (e.g., under physiological conditions), and the possibility of charges within hydrophobic matrices of varying degree of hydrophobicity. Also, moieties of opposite charge to that of the drug allow for ion pairing of designable affinity between drug and matrix. Degradational release from the matrix can be due to hydrolytic or enzymatic cleavage of backbone bonds under variable conditions of the state of matrix swelling, of the presence of proteolytic sites within the polymer sequence, and even of appropriate proteolytic enzymes within the matrix.

Transductional Release and Molecular Machines

The transductional release modalities themselves are many. One particularly attractive transductional approach uses polymers containing chemical clocks with programmable half-lives ranging from days to a decade (*3, 4*). In a related general statement of approach, polymers are designed as first-order molecular machines of the T_t-type in which stimuli or energy inputs of many different kinds can be used to effect swelling followed by diffusional release. In another approach, transductional polymers are designed as second-order molecular machines that can function to control the affinity of the drug for the matrix. A properly designed molecular machine has the potential to be a self-contained diagnostic–therapeutic pair, which would use the altered chemistry of the diseased state to signal release of the needed drug.

Strategies for Transductional Drug Loading and Modulation of Release

The state of the drug under physiological conditions of pH, temperature, and salt concentration becomes the key to matrix loading and to further modulation of release. More specifically, strategies for matrix design to modulate loading and release depend on whether the drug is neutral, positively charged, or negatively charged and how hydrophobic it is under physiological conditions.

In short, neutral drugs load into neutral matrices; they require swell-doping (*5, 6*) or organic solvents to load, and their rate of release depends on hydrophobicity, the hydrogel, elastic or plastic state, and chemical clocks within the matrix. Positively charged drugs load into negatively charged matrices; they use a dramatic decrease in T_t on their addition to the polymer solution or to the matrix to load, and their rate of release also depends on the hydrophobicity, the elastic or plastic state, polymer chain length (if in the viscoelastic, non-cross-linked state), and the chemical clocks of the matrix. Finally, negatively charged drugs load into positively charged matrices; they also can use a dramatic lowering of T_t on their addition to the polymer solution or matrix to load, and their release too depends on polymer or matrix hydrophobicity, the hydrogel, elastic or plastic state, and chemical clocks.

Chemical Clock as a Transductional Element

As used here, a chemical clock is a naturally occurring amino acid residue, namely an asparagine (Asn, N) or glutamine (Gln, Q). These residues contain a chemical moiety, in this case a carboxamide ($CONH_2$), which under physiological conditions spontaneously hydrolyzes to form a carboxylate (COO^-). The rate of hydrolysis depends on its nearest neighbor residues in the sequence (*3*), on the hydrophobicity of the protein-based polymer in which it resides, and on the physical state (e.g. hydrogel, elastic, or plastic) of the matrix in which it might occur. The formation of a COO^- moiety strikingly raises T_t, causing the polymer to go from a more dense viscoelastic phase to complete dissolution or, if in the form of a cross-linked matrix, to cause the matrix to swell dramatically and to degrade.

Next Generation of Controlled-Drug Delivery

The drug-delivery process to be discussed is very much under development; it is a case in which the potentialities are being demonstrated step by step as the principles of energy conversion become integral with the process of structure assembly and disassembly. In this perspective of drug delivery, loading of the monolith is a process of assembly, and controlled delivery becomes an integral part of the process of disassembly. Central to it all is the ΔT_t-mechanism of energy conversion. The matrix-loading and controlled-release approaches will be described in some detail after the biocompatibility and toxicity studies of basic representative, transductional protein-based polymers and their cross-linked matrices are reviewed, after the present status of protein-based polymer preparation and importance of sequence control are noted, and after the underlying physical processes are presented.

Biocompatibility of Representative Physical States

Foreign proteins are generally antigenic. Because of this characteristic, it is important at the outset to establish whether the classes of protein-based polymers of interest are biocompatible. They have as their origins repeating sequences in mammalian elastic fibers, and the elastic fiber and the core protein, elastin, are well-known for their low antigenicity. For the present, the two most relevant repeats of elastin are Gly–Val–Gly–Val–Pro (GVGVP) and Gly–Gly–Val–Pro (GGVP). Consistent with the low immunogenicity of elastin, even very aggressive efforts using multiple immunizations of highly sensitive and sensitized animals have not been successful in preparing monoclonal antibodies directly to poly(GVGVP).

Under physiological conditions, high-molecular-weight polymers of these repeats or of analogs of these repeats can exist in three different physical states: elastomeric, plastic, and hydrogel. Representative polymers for each of the states are poly(GVGVP), poly(AVGVP), and poly(GGAP), respectively. In terms of coarsely stated elastic moduli and water content, the cross-linked matrices obtained by using 20 Mrad of cobalt-60 γ-irradiation signified by the prefix X^{20}-, exhibit values of 10^4 to 10^5 dyne/cm^2 for the hydrogels with 90% water (*7*), 10^6 to 10^8 dyne/cm^2 for the elastomers with up to 60% water by weight (*8*), and $>10^8$ dyne/cm^2 for the plastics with little or no water (*7*, *8*). The entire range of values covering four orders of magnitude of accessible elastic moduli is possible by using mixed peptide compositions. Furthermore, the elastomers and the plastics can be reversibly converted to hydrogels depending on the interactions between polymer composition and the external perturbations.

Elastomers: Poly(GVGVP) and X^{20}-Poly(GVGVP)

The list of 11 biological tests employed for determining the biocompatibility of these elastic and plastic protein-based polymers is given in Table 20.1. This list includes the complete set of tests recommended in 1987 for nonmutagenic materials in contact with tissues, tissue fluids, and blood by the American Society for Testing and Materials (ASTM) Committee F-4 on Medical and Surgical Materials and Devices (ASTM Designation F748-87) in a publication entitled "Standard Practice for Selecting Generic Biological Test Methods for

TABLE 20.1. Biological Test Results for Poly(GVGVP) and X^{20}-Poly(GVGVP) and for Poly(AVGVP) and X^{20}-Poly(AVGVP)

Test	*Description*	*Test System*	*Results*
Ames mutagenicity	Determine reversion rate to wild type of histidine-dependent mutants	*Salmonella typhimurium*	Nonmutagenic
Cytotoxicity	Agarose overlay to determine cell death and zone of lysis	L-929 mouse fibroblast	Nontoxic
Systemic toxicity	Evaluate acute systemic toxicity from an intravenous or intraperitoneal injection	Mice	Nontoxic
Intracutaneous toxicity	Evaluate local dermal irritant or toxic effects by injection	Rabbit	Nontoxic
Muscle implantation	Evaluate effect on living muscle tissue	Rabbit	Favorable
Intraperitoneal implantation	Evaluate potential systemic toxicity	Rat	Favorable
Systemic antigenicity (British Pharmacopoeia Antigenicity Test)	Evaluate general toxicology	Guinea pigs	Nonantigenic
Sensitization (Kligman test)	Evaluate dermal sensitization potential	Guinea pigs	Nonsensitizing
Pyrogenicity	Determine febrile reaction	Rabbit	Nonpyrogenic
Clotting study	Evaluate whole blood clotting times	Dog	Normal clotting time
Hemolysis	Evaluate level of hemolysis in the blood	Rabbit blood	Nonhemolytic

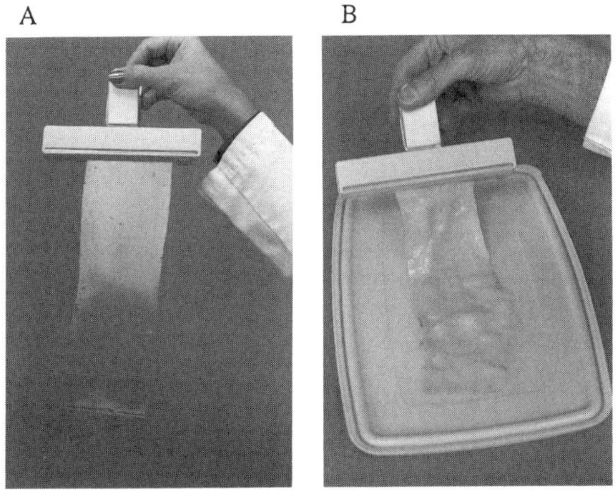

FIGURE 20.1. Cross-linked elastic matrices resulting from cobalt-60 γ-irradiation at 20 × 10⁶ radiation absorbed dose (20 Mrad) of the viscoelastic (coacervate) phase formed on raising the temperature above T_t,: A, X^{20}-poly(GVGVP); B, X^{20}-poly-(GVGIP).

Materials and Devices". More current guidelines have been proposed by the International Organization for Standardization (ISO Document No. 10993-1), and these guidelines were adopted by the U.S. Food and Drug Administration in July 1995.

A representative sheet of X^{20}-poly(GVGVP) is shown in Figure 20.1A, where it is seen as a transparent material in its hydrogel state containing some 90% water and has an elastic modulus of about 10^5 dyne/cm^2 at a temperature below its inverse temperature transition. On raising the temperature to physiological values, the matrix becomes opaque, expresses water, and contracts to about one-half the swollen dimensions of Figure 20.1A. On standing again it becomes transparent to form the elastomeric state containing about 50% water and has an elastic modulus of 10^6 dyne/cm^2. A sheet of X^{20}-poly(GVGIP) is shown in Figure 20.1B.

Table 20.1 contains the biocompatibility information for poly(GVGVP) and X^{20}-poly(GVGVP), the description of the test, the test animal or organism, and the results. In the publication reviewing the entire set of 11 tests (*9*), the abstract states, "Thus, this new elastomeric polypeptide biomaterial which is based on the most striking repeating sequence in the mammalian elastic fiber exhibits an extraordinary biocompatibility". Details for each of the test results should be sought in that publication or in the archival files of the testing laboratories of North American Science Associates, Inc. (NAmSA). In virtually every case, the test material was as innocuous as the negative control.

Plastics: Poly(AVGVP) and X^{20}-Poly(AVGVP)

Representative shapes of X^{20}-poly(AVGVP) are shown in Figures 20.2A and 20.2B on equilibration in water at 37 °C. This plastic state is a relatively hard, strong, yet deformable material that contains little or no water and exhibits an elastic modulus of 1.8×10^8 dyne/cm^2 as shown in Figure 20.2D at 37 °C (*7*). On lowering the temperature to 25 °C, the matrix swells to more than double the contracted dimensions to form the more delicate hydrogel state shown in Figure 20.2C, which has an elastic modulus of 1.5×10^5 dyne/cm^2 (*7*).

Table 20.1 summarizes the biocompatibility studies for poly(AVGVP) and X^{20}-poly(AVGVP) (*10*). Even though the short statements of results of Table 20.1 read identically for both the representative elastic and plastic compositions, a careful examination of each of the NAmSA reports shows the biocompatibility of poly(AVGVP) and X^{20}-poly(AVGVP) to be good but not as extraordinary as for poly(GVGVP) and X^{20}-poly(GVGVP). From these toxicity studies, the plastic protein-based polymers and their cross-linked matrices show substantial promise for medical applications in which the materials are in contact with tissues, tissue fluids, and blood.

Hydrogels: Poly(GGAP) and X^{20}-Poly(GGAP)

The hydrogel state, as represented by poly-(GGAP), presents the most extraordinary biocompatibility data (*6*). In the Ames mutagenicity test,

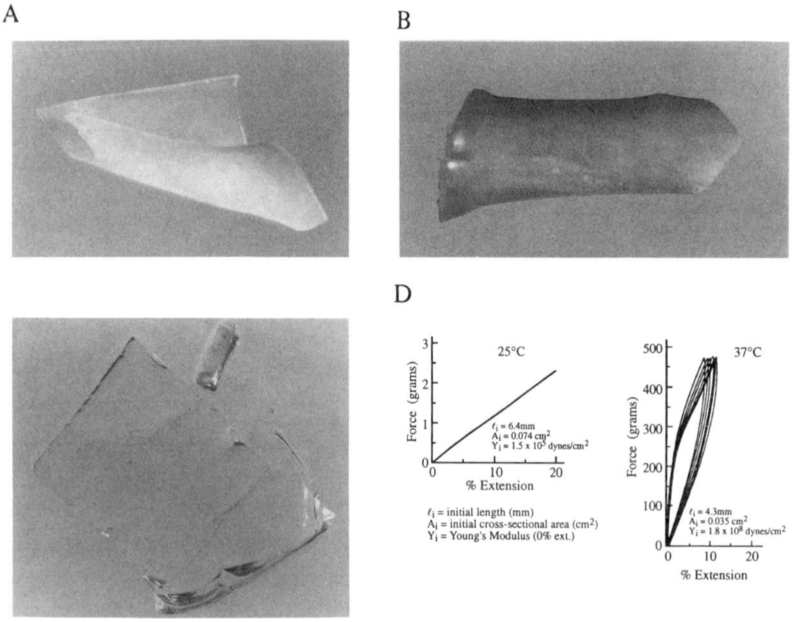

FIGURE 20.2. Matrices resulting from 20 Mrad γ-irradiation of phase-separated poly(AVGVP) formed on raising the temperature above T_t: A and B, plastic states of X^{20}-poly(AVGVP) above T_t; C, hydrogel states of X^{20}-poly(AVGVP) that on raising the temperature above T_t resulted in the plastic states of A and B; D, stress–strain curves of the hydrogel state at 25 °C and a repetitive series of stress–strain curves for the same matrix on converting to the plastic state by raising the temperature to 37 °C. (Part D is reproduced with permission from reference 7. Copyright 1991 Plenum.)

the test article, X^{20}-poly(GGAP), caused fewer reversions of the His-dependent mutant strains of *Salmonella typhimurium* than did the negative saline control, whereas a doubling of the control rate would be required before the test article could be considered mutagenic. Accordingly, poly(GGAP) is nonmutagenic and as such need not be tested for carcinogenicity. In the agarose overlay cytotoxicity test, confluent layers of L-929 mouse connective tissue cells are used. When the test article was placed in the overlay position, a 0-mm zone of cell lysis was observed, whereas the toxic positive control caused a 9-mm zone of cell lysis. In the acute systemic toxicity test, intravenous injections of 40-mg/mL solutions at a dose level of 50 mL/kg of mouse resulted in no signs of toxicity due to poly(GGAP). In the U.S. Pharmacopeia intracutaneous toxicity examination, injections resulted in the observations of no erythema and no edema. For the antigenicity studies, both the British Pharmacopoeia and the Kligman maximization method of dermal sensitization were used, and neither antigenicity nor sensitization was observed. In the Kligman test, the negative controls happened to exhibit more reaction than the test material. To evaluate thrombogenicity in terms of the effect on clotting time, the Lee–White clotting study was used with poly(GGAP) actually resulting in a small increase in clotting time. Clearly, poly(GGAP) is not thrombogenic. Finally, in the in vitro hemolysis test, poly(GGAP) caused 0% hemolysis. Thus, based on the set of tests carried out by the independent testing laboratory, poly(GGAP) exhibits extraordinary biocompatibility.

Another demonstration of the noninteractive nature of this polymer is seen in a series of cell attachment studies. In this case, X^{20}-poly(GGAP) is seen in Figure 20.3 to be nonadherent to bovine ligamentum nuchae fibroblasts and to human umbilical vein endothelial cells even in the presence of serum. Other more hydrophobic matrices appear to adsorb proteins, such as fibronectin, from the serum, which then result in suboptimal cell attachment (*11*). In the absence of serum, no significant attachment to any of the included matrices occurs, but in the presence of serum attachment, albeit poor, does occur to all but the X^{20}-poly(GGAP) matrix. Of course, the GRGDSP cell attachment sequence can be included within the sequence of the protein-based polymers, and the cross-linked matrices then support cell attachment, spreading, and growth to confluence (*11*).

Preparation and Importance of Sequence Control of Protein-Based Polymers

Because the chemical and microbial syntheses and the purifications of these elastic and plastic protein-based polymers have been described in detail elsewhere (*12–14*), they will only be briefly noted. Also, the importance of control of sequence has been emphasized previously (*14*), but because this aspect is central to achieving the desirable transductional properties reported in this account, more attention will be given to this issue.

Chemical Synthesis

The preparation of transductional elastic and plastic protein-based polymers by chemical synthesis has always been a challenging effort. Even achieving the fundamental property of the inverse temperature transition at the correct temperature (i.e., achieving reproducible T_t values) requires great care. Small amounts of racemization can cause T_t of poly(GVGVP) to vary from 25 °C to 40 °C (*14*). This issue is serious, of course, because controlling the temperature at which the inverse temperature transition occurs is central to the transductional properties of interest. The key to successful chemical syntheses is to stepwise build up the basic repeating sequence taking care to

A

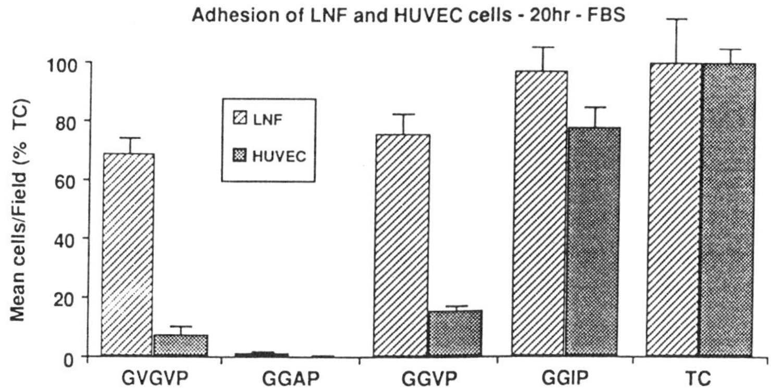

FIGURE 20.3. Cell-attachment studies of LNF (ligamentum nuchae fibroblasts) and HUVEC (human umbilical vein endothelial cells) to a series of 20 Mrad γ-irradiation cross-linked matrices formed from poly(GVGVP), poly(GGAP), poly(GGVP), and poly(GGIP) and to TC (tissue culture plastic). A, bar graph for both cell types showing no attachment to the matrix X^{20}-poly(GGAP) after 20 h in the presence of 20% FBS (fetal bovine serum) and even the counted cells attach only poorly to X^{20}-poly(GVGVP) and to X^{20}-poly(GGVP). *(Continued on next page)*

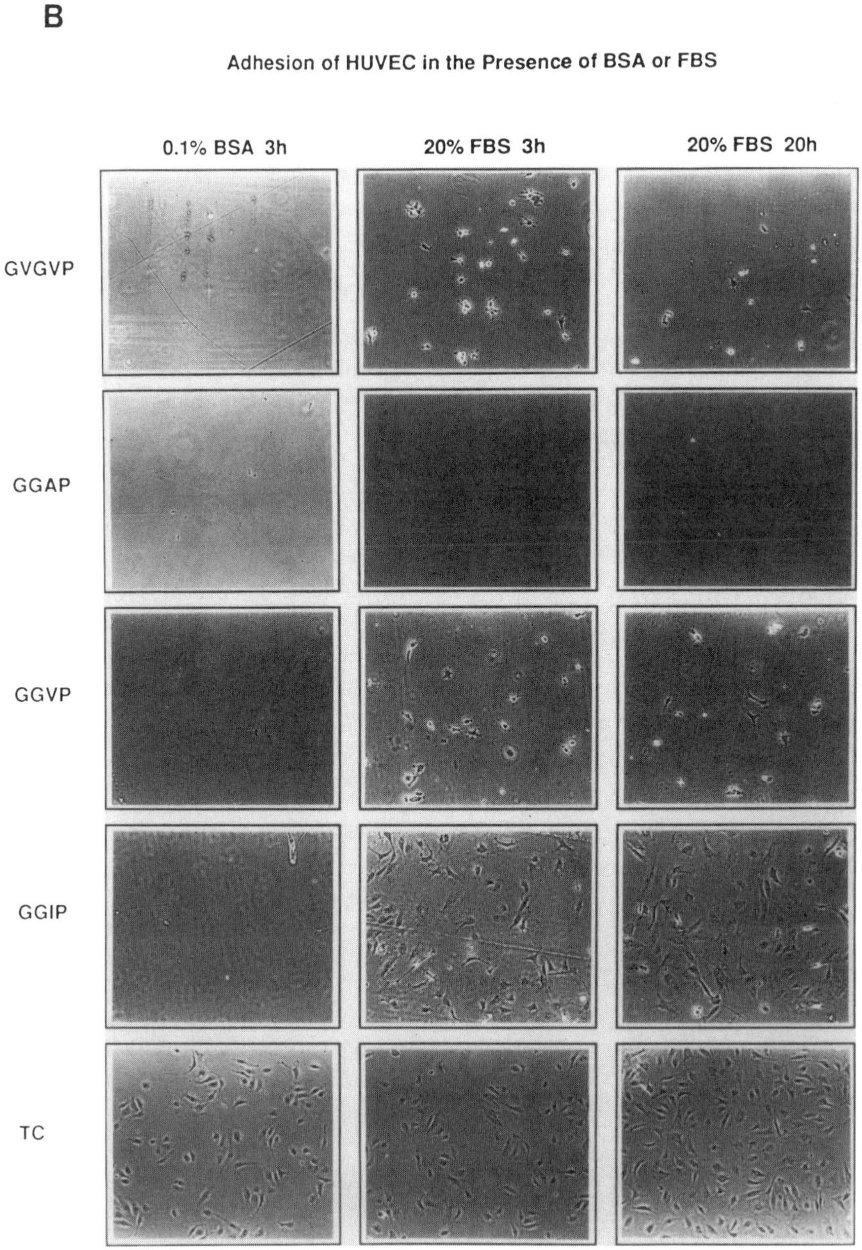

FIGURE 20.3—*Continued.* B, attachment to included matrices in the presence and absence of serum. (Reproduced with permission from reference 11. Copyright 1993 John Wiley.)

ensure that the composite peptides are pure and that they are coupled in a way that either allows for crystallization to remove impurities such as D-residues or would not give rise to racemization. The polymerization of the basic repeat itself, of course, must be done with either a Gly or Pro residue at the carboxyl terminus (*12, 13*). The definitive proof when such stringent purity is required actually comes from the microbial biosynthesis, as noted in the next section.

Microbial Biosynthesis, Gene Construction, and Expression

The protein-based polymers are characteristically prepared by the following sequential steps:

1. construction of a basic gene sequence (by using the polymerase chain reaction),
2. *E. coli* transformation using the pUC118 plasmid containing the basic gene,
3. verification of the basic gene sequence by sequence analysis on the pUC118 plasmid,
4. expression of an adequate concentration of the basic gene,
5. concatenation of the basic gene to obtain high-molecular-weight multimers of the basic gene fragment,
6. incorporation into the pET-11d expression vector, which uses the T_7 promoter, and
7. transformation of an appropriate *E. coli* strain followed by expression under favorable fermentation conditions.

For poly(GVGVP), the basic gene was $(GVGVP)_{10}$ in which there was no repeating nucleotide sequence due to the use of the available multiple codons for the repeating amino acid residues. The concatenation and transformation procedures resulted in the several genes encoding for $(GVGVP)_n$, where $n = 41$, 121, 141, and 251 (*6, 10, 14*). This result provided the opportunity to determine the effect of chain length on the value of T_t and to definitively determine the value of T_t for high-molecular-weight polymers. The values of T_t for $n = 41$, 141, and 251 were 34 °C, 28 °C, and 26 °C, respectively. The 26 °C value prompted the direct comparison of the chain lengths of the best chemically synthesized polymers by sodium dodecyl sulfate polyacrylamide gel electrophoresis. The best chemically prepared polymers were larger than the 102-kDa molecular weight of the 251-mer. In retrospect, this chemical synthesis of polypeptides of more than 1300 residues is a remarkable result.

Importance of Control and Diversity of Sequence

As will be discussed, the magnitude of the hydrophobic-induced pK_a shift is central to the loading into, and the sustained release of an ionizable drug from, a transductional coacervate or matrix. Accordingly, the capacity to shift the pK_a becomes the means whereby the successful use of these transductional polymers in controlled release is in large part realized.

Therefore, the synthesis of two families of polymers in which in each family there is a common set of pentamers was of interest. In one case six pentamers of varying composition were coupled in a specific order to form a 30-mer, and the 30-mer was polymerized. In the other case the same six pentamers were mixed at the same ratio as in the 30-mer and polymerized such that there was no control over the order of the pentamers in the resulting polymer. One family contained the Asp residue and the other the Glu residue. The polymers synthesized are listed in Table 20.2 along with their pK_a values and amino acid compositions (*14*). For the designed Asp-containing poly(30-mers), the largest pK_a value was 10.1, whereas the polymer that began with the same composite pentamers exhibited a pK_a of 4.6. For the designed Glu-containing poly(30-mers), the largest pK_a value was 8.1, whereas the polymer with essentially the same composition but with the pentamers not in the prescribed order exhibited a

TABLE 20.2. pK_a Values at 20 °C and Amino Acid Compositions

Polymer	pK_a	*Asp*	*Glu*	*Gly*	*Pro*	*Val*	*Phe*
Poly[(GDGFP),2(GVGVP),2(GVGFP),(GFGFP)] (random mix of pentamers)	4.6	0.32 ± 0.02	0	2.07 ± 0.10	1.0	0.90 ± 0.10	0.70 ± 0.08
Poly[GDGFPGVGVPGVGFPGFGFPGVGVPGVGFP]	10.1	0.17 ± 0.02	0	2.02 ± 0.10	1.0	0.92 ± 0.10	0.83 ± 0.08
Poly[(GDGFP)3(GVGVP)2(GFGFP)] (random mix of pentamers)	5.2	0.32 ± 0.02	0	2.05 ± 0.10	1.0	1.01 ± 0.10	0.76 ± 0.08
Poly[GDGFPGVGVPGVGVPGVGVPGFGFPGFGFP]	9.5	0.17 ± 0.02	0	2.00 ± 0.10	1.0	0.98 ± 0.10	0.82 ± 0.08
Poly[GDGVPGFGFPGFGVPGVGVPGFGFPGVGVP]	6.7	0.17 ± 0.02	0	2.00 ± 0.10	1.0	0.94 ± 0.10	0.85 ± 0.08
Normal pK_a for Asp(D)	3.8						
Poly[(GEGFP),2(GVGVP),2(GVGFP),(GFGFP)] (random mix of pentamers)	5.2	0	0.15 ± 0.02	1.88 ± 0.10	1.0	1.09 ± 0.10	0.90 ± 0.08
Poly[GEGFPGVGVPGVGFPGFGFPGVGVPGVGFP]	8.1	0	0.14 ± 0.02	1.98 ± 0.10	1.0	0.98 ± 0.10	0.81 ± 0.08
Poly[(GEGFP),3(GVGVP),2(GFGFP)] (random mix of pentamers)	4.7	0	0.15 ± 0.02	2.11 ± 0.10	1.0	1.01 ± 0.10	0.70 ± 0.08
Poly[GEGFPGVGVPGVGVPGVGVPGFGFPGFGFP]	7.7	0	0.16 ± 0.02	2.05 ± 0.10	1.0	0.90 ± 0.10	0.79 ± 0.08
Poly[GEGVPGFGFPGFGVPGVGVPGFGFPGVGVP]	7.8	0	0.18 ± 0.02	1.96 ± 0.10	1.0	0.93 ± 0.10	0.86 ± 0.08
Normal pK_a for Glu(E)	4.3						
Theoretical values	—	0.17	0.17	2.0	1.0	1.0	0.83

pK_a value of 5.2. Thus, control of sequence is essential for achieving large ΔpK_a values, and large ΔpK_a values are required to achieve sustained drug release levels.

Purification

The products are very effectively purified by using the inverse temperature transitional properties of the polymers. They are verified by the standard physical and chemical means (*15*).

Transductional Principles for Protein-Based Polymers

The transductional principles, as related to the use of protein-based polymers in controlled release, begin with a description of the inverse temperature transition of hydrophobic folding and assembly that results in a more ordered state as the temperature is raised. The T_t value is a measure of the functional hydrophobicity. More hydrophobic amino acid residues and other functional moieties lower T_t, whereas less hydrophobic, more polar residues raise T_t. The result is a T_t-based hydrophobicity scale for protein-based polymer engineering. Of particular interest is that energy inputs capable of changing T_t become the means with which to convert energy from one form to another. Finally in this section, the role of changing hydrophobicity in effecting large pK_a shifts is demonstrated and associated with the capacity to load with drug by ion pairing and to achieve a sustained release of drug.

Increase in Order with Increase in Temperature

When polymers are composed of the correct balance of hydrophobic and polar groups, they can be soluble in, or miscible in all proportions with, water at low temperatures. As the temperature is raised, they hydrophobically fold, self assemble, and settle out of solution to form a more dense phase called a coacervate. The coacervate can contain as much as 90% water or little or no water depending on the polymer structure.

A particularly striking example of a phase separation is given by the cyclic dodecapeptide, cyclo$(GVGVAP)_2$. This 12-residue cyclic peptide is soluble in water at a low temperature, but on raising the temperature, it reversibly crystallizes over a temperature range that depends on its concentration (*16*). Raising the concentration lowers the temperature range for crystallization. Thus, these polymeric molecules are randomly dispersed at low temperature, and they associate to form an ordered crystalline state on raising the temperature.

This increase in order with increase in temperature is referred to as *an inverse temperature transition* (*1, 2*), because the order of the polypeptide part of the water plus peptide system increases with increase in temperature. In the usual transition from one state to another, such as the melting of ice or the vaporization of water, the molecules become more disordered as the temperature is raised from below to above the transition. The reason for this inverse temperature behavior resides in the second component of the system, the water. The water, which surrounds the hydrophobic moieties (*17, 18*) of the dissolved polypeptide, is more ordered than bulk water. This more ordered water of hydrophobic hydration becomes less ordered bulk water as the polypeptides associate by hydrophobic assembly, in keeping with the second law of thermodynamics (*19–21*).

The linear high-molecular-weight polymers of this composition, poly(GVGVAP), are also soluble in water at low temperature, and similarly, depending on concentration, associate on raising the temperature to form an insoluble aggregate. Furthermore, high-molecular-weight polymers of the related composition, poly(GVGVP), are also soluble at low temperature and reversibly associate to form a viscoelastic coacervate phase. This poly(GVGVP) family of protein-based polymers provides significant advantages for controlled drug delivery.

Definition of T_t

A relatively quick means of determining the temperature at which the inverse temperature transition occurs is to follow spectrophotometrically the development of turbidity as the temperature is raised. This process is shown in Figure 20.4A for several compositions, where T_t is defined as the temperature at which 50% of maximal turbidity is reached.

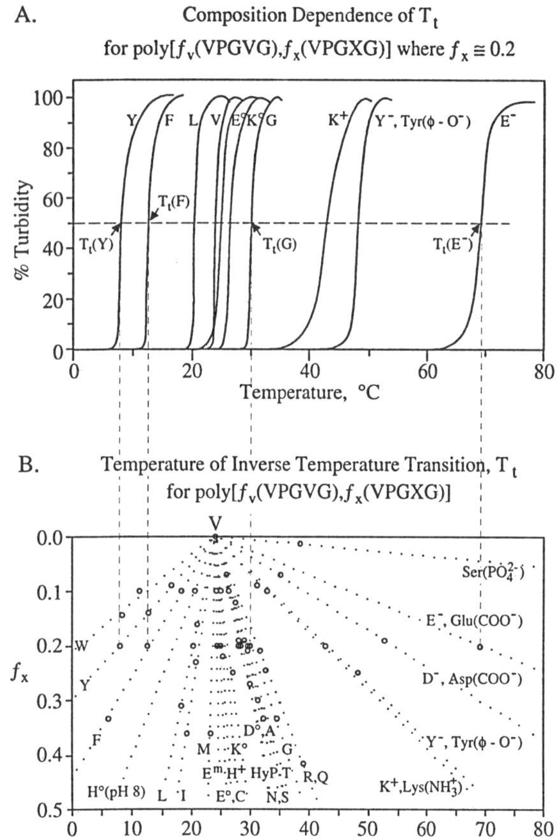

FIGURE 20.4. Dependence of T_t on the composition of the polymer, poly[f_V(GVGVP), f_X(GXGVP)], where f_V and f_X are mole fractions and $f_V + f_X = 1$. A, experimental determination of T_t for a series of protein-based polymers with $f_X = 0.2$; B, plots of f_X as a function of T_t that provide the experimental basis for a T_t-based hydrophobicity scale listed in Table 20.3. (Reproduced with permission from reference 1. Copyright 1993 VCH.)

Dependence of T_t on Chain Length and Polymer Concentration

The value of T_t decreases as the chain length increases (*14*), but the increment of decrease becomes smaller for each doubling of length until their is no longer a significant decrease. Above 100-kDa molecular weights, decreases in the value of T_t with increase in chain length become limiting.

For polymers at the high-molecular-weight limit, the value of T_t decreases as the concentration is raised. Again for molecular weights of the order of 100 kDa, this high concentration limit is reached by a concentration of 40 mg/mL or more (*1*). Accordingly, comparisons of how the value of T_t changes as the composition is changed are made with polymers that are of the order of 100 kDa or more and at concentrations of 40 mg/mL.

Effect of Hydrophobic Residues on T_t

By comparing T_t values of polymers of different compositions, more hydrophobic polymers like poly(GVGIP), which contains one more CH_2 moiety per pentamer, exhibit lower values of T_t by about 15 °C. Less hydrophobic polymers like poly(GAGVP), which contain two fewer CH_2 moieties per pentamer, exhibit higher values of T_t by about 30 °C. Clearly, increasing hydrophobicity lowers T_t, and decreasing hydrophobicity raises T_t. Accordingly, a hydrophobicity scale based on the value of T_t can be developed.

T_t-*Based Hydrophobicity Scale for Protein-Based Polymer Engineering*

For developing the T_t-based hydrophobicity scale, the model protein-based polymer, poly[f_V-(GVGVP), f_X(GXGVP)], is used where f_V and f_X are mole fractions, $f_V + f_X = 1$, and X is any naturally occurring amino acid residue or chemical modification thereof. The T_t values for several such polymers are demonstrated in Figure 20.4A for compositions where f_X is approximately 0.2, and their values are plotted in Figure 20.4B as f_X

versus T_t. Plots of f_X as a function of T_t form essentially straight lines. Therefore, relative hydrophobicities can be compared for any given value of f_X. A value of $f_X = 1$ is taken as the reference that would approximate the value of T_t for poly(GXGVP), if temperatures outside of the aqueous range could be reached. The more hydrophobic the residue the lower the value of T_t; the less hydrophobic the residue the higher the value of T_t. With this scale it becomes possible to engineer protein-based polymers to exhibit their inverse temperature transitions where desired after introducing the functionalities required for the intended function. The T_t-based hydrophobicity scale appears in Table 20.3 (*1, 22*).

Effect of Side-Chain Ionization

As seen in Table 20.3, the T_t for poly(GEGVP) is 30 °C when the Glu(E) residue is protonated (i.e., COOH), but it is 250 °C when the Glu residue is ionized (i.e., COO^-). The much more polar carboxylate dramatically increases the value of T_t; ΔT_t is 220 °C for deprotonation of the Glu side chain. This effect, the further capacity to shift hydrophobically the pK_a of the function, and the effect of ion pairing on lowering the value of T_t (*see* subsequent discussion) become important means of loading cationic drugs into the matrix or coacervate and of controlling the rate of release.

Effect of Change in State of Attached Prosthetic Group

When a prosthetic group, such as *N*-methyl nicotinate, is attached to a Lys(K) side chain by amide linkage, the oxidized nicotinamide has a T_t value of 120 °C, whereas on chemical or electrochemical reduction the T_t for this functional group becomes −130 °C (*see* Table 20.3). The ΔT_t for oxidation of this functional group is 250 °C. Thus, reduction can drive hydrophobic folding and assembly.

Effect of Side-Chain Phosphorylation

The most dramatic way to change the value of T_t is to phosphorylate a residue such as Ser(S). The T_t for poly(GSGVP) is 50 °C. On phosphorylation, however, the value of T_t, obtained by extrapolation to $f_X = 1$ becomes 1000 °C; the ΔT_t for phosphorylation is of the order of 1000 °C (*see* Figure 20.4A and Table 20.3). Thus, phosphorylation is the most effective way to drive hydrophobic unfolding and disassembly, or dephosphorylation becomes the most dramatic means of driving hydrophobic folding and assembly.

ΔT_t-*Mechanism for Free-Energy Transduction*

As has been appreciated for many years, an increase in temperature can cause hydrophobic folding; this phenomenon has long provided an explanation for the observation of "cold denaturation" of proteins. What has not been appreciated until more recently is that it is possible to cause hydrophobic folding and assembly by lowering the value of T_t from above to below physiological temperature or any available aqueous temperature. This process has been called the ΔT_t-mechanism of free-energy transduction, because the act of lowering T_t to effect hydrophobic folding can be used to perform such tasks as the mechanical work of lifting a weight.

Hexagon Representation of Energy Conversions

The introduction of the energies represented by the application of pressure, by chemical concentration changes, by electrochemical oxidation or reduction of an attached prosthetic group, or by the absorption of electromagnetic radiation to a suitably designed polymer can result in a change in the value of T_t and in the performance of mechanical work. This process is called the ΔT_t-mechanism of free-energy transduction or simply of energy conversion. The energy conversions possible by this mechanism are represented by using the hexagon shown in Figure 20.5 (*1*), and the six energies are at the apices. The energy conversions that have been achieved are given as the bold arrows.

Postulates for Energy Conversion

The energy conversions possible by the ΔT_t-mechanism can be stated in terms of three postulates (*23*).

TABLE 20.3. Hydrophobicity Scale Based on T_t Poly[f_V(VPGVG),f_X(VPGXG)] for Protein Engineering

Residue X	*Abbreviation*	T_t[a] *(°C)*	*Correlation Coefficient*
Lys(NMeN, reduced)[b]		–130	1.000
Trp	W	–90	0.993
Tyr	Y	–55	0.999
Phe	F	–30	0.999
His (pH 8)	H^0	–10	1.000
Pro	P[c]	(–8)	Calculated
Leu	L	5	0.999
Ile	I	10	0.999
Lys(HO–NMeN, reduced)[d]		15	1.000
Met	M	20	0.996
Val	V	24	Reference
Glu($COOCH_3$)	E^m	25	1.000
Glu(COOH)	E^0	30	1.000
Cys	C	30	1.000
His (pH 4)	H^+	30	1.000
Lys(NH_2)	K^0	35	0.936
Pro	P[e]	40	0.950
Asp(COOH)	D^0	45	0.994
Ala	A	45	0.997
HyP		50	0.998
Asn	N	50	0.997
Ser	S	50	0.997
Thr	T	50	0.999
Gly	G	55	0.999
Arg	R	60	1.000
Gln	Q	60	0.999
Lys(NH_3^+)	K^+	120	0.999
Tyr(ϕ-O^-)	Y^-	120	0.996
Lys(NMeN, oxidized)[b]		120	1.000
Asp(COO^-)	D^-	170	0.999
Glu(COO^-)	E^-	250	1.000
Ser(HPO_4^{2-})		1000	1.000

[a] T_t values are extrapolated linearly to $f_x = 1$.

[b] NMeN is for *N*-methyl nicotinamide pendant on a lysyl side chain (i.e., *N*-methyl nicotinate attached by amide linkage to the ε-NH_2 of Lys), and the reduced state is *N*-methyl-1,6-dihydro-nicotinamide.

[c] Calculated T_t value for Pro comes from poly(VPGVG) when the experimental values of Val and Gly are used. This hydrophobicity value of –8 °C is unique to the β-spiral structure, where there is hydrophobic contact between the $Val^1\gamma CH_3$ and $Pro^2\beta CH_2$ moieties.

[d] HO–NMeN, reduced, is for 6-hydroxy-*N*-methyl-1,4,5,6-tetrahydronicotinate attached by amide linkage to the ε-NH_2 of lysine; it is obtained on standing following reduction of *N*-methyl-nicotinamide.

[e] Value determined experimentally from poly[f_V(VPGVG), f_P(PPGVG)].

SOURCE: Adapted from reference 22.

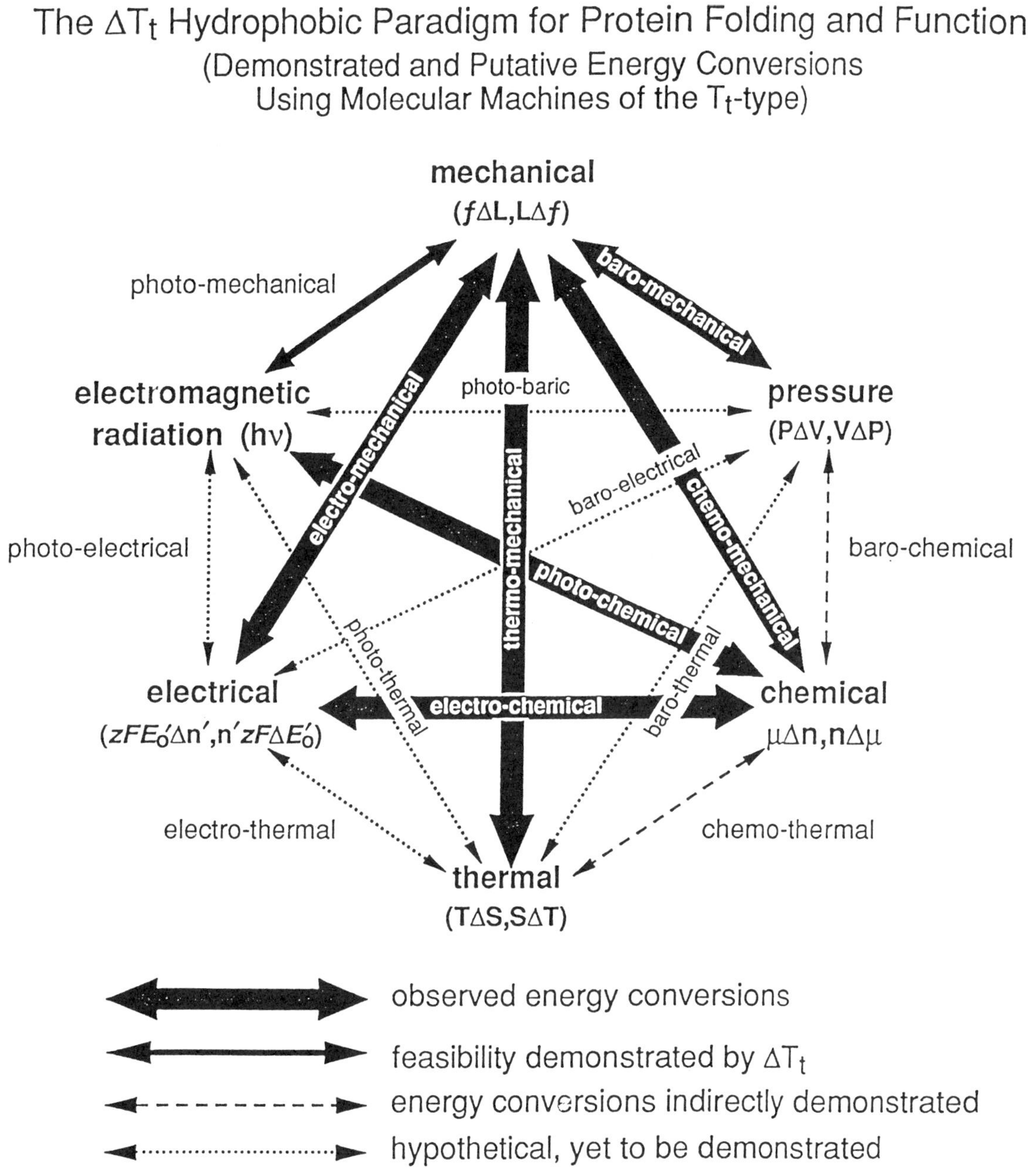

FIGURE 20.5. Hexagon with the six energies at the apices that can be interconverted by means of the inverse temperature transition of hydrophobic folding and assembly. (Reproduced with permission from reference 1. Copyright 1993 VCH.)

Postulate I: The input of thermal energy to a protein capable of hydrophobic folding and assembly on raising the temperature from below to above T_t of an inverse temperature transition can result in motion and the performance of mechanical work. The corollary is thermomechanical transduction.

Postulate II: Any energy input that changes T_t at which an inverse temperature transition occurs can be used to produce motion and perform mechanical work. The corollaries are as follows:

1. chemomechanical transduction,
2. electromechanical transduction,
3. baromechanical transduction, and
4. photomechanical transduction.

Postulate III: Different energy inputs, each of which can individually drive hydrophobic folding to produce motion and perform mechanical work, can be converted one into the other (transduced) by means of the inverse temperature transition with the correctly designed coupling and T_t value. The corollaries are as follows:

1. electrochemical transduction,
2. electrothermal transduction,
3. baroelectrical transduction,
4. photovoltaic transduction,
5. thermochemical transduction,
6. photothermal transduction,
7. barothermal transduction,
8. barochemical transduction,
9. photobaric transduction,
10. photochemical transduction, and
11. chemochemical transduction.

Molecular Machines for Effecting Drug Delivery

A molecular machine is a molecule, for our purposes a polymer, that is capable of converting energy from one form or location to another. A molecular engine is that class of molecular machines that can convert a given energy into useful mechanical work.

First-Order Molecular Machines

First-order molecular machines of the T_t-type are molecular engines in that they use the hydrophobic folding process directly to perform mechanical work. These molecular machines perform the energy conversions represented by the arrows that end at the mechanical apex of Figure 20.5. The control of matrix swelling or of coacervate dissolution to achieve the chemical work of drug release is the work of a first-order molecular machine of the T_t-type.

Second-Order Molecular Machines

When the properties of two functionalities are coupled by being a part of the same hydrophobic folding domain, the protein-based polymer of this composition is a second-order molecular machine of the T_t-type. For our purposes, an example might be the interaction between the more hydrophobic side chains of Phe(F), Ile(I), or Trp(W) residues and the ionizable functions of Glu(E), Asp(D), or Lys(K) residues that raise the value of T_t when the side chains are charged and that shift the pK_a of the ionizable function depending on the number, location, and intensity of the hydrophobic side chains. This second-order molecular machine of the T_t-type can be used to load drug into the matrix at concentrations related to the concentration of the ionizable function and with an affinity proportional to the hydrophobic-induced pK_a shift.

Hydrophobic-Induced pK_a Shifts

On the basis of the T_t-based hydrophobicity scale, any process that lowers T_t does so by increasing hydrophobicity, and any process that raises T_t does so by decreasing hydrophobicity. Replacing Val(V) residues by Phe(F) residues lowers T_t, but this increase in hydrophobicity causes an increase in the pK_a of the carboxyl moieties of Glu(E) and Asp(D) residues. This process is referred to as a hydrophobic-induced pK_a shift, and it has been demonstrated with Glu (*24, 25*), Asp (*26, 27*), and Lys (*28*) residues.

Relationship Between a ΔpK_a and Chemical Work

Chemical work, change in energy (ΔE), is defined as the product of the change in chemical potential, $\Delta\mu$, with the change in the number of moles of chemical species, Δn: $\Delta E = \Delta\mu \times \Delta n$. The chemical potential is defined as $\mu = RT \ln a$, where a is the activity of the chemical. Because the activity is the same as concentration for the very low proton concentrations of interest, $\mu = RT \ln[H^+]$, where $[H^+]$ is proton concentration in moles per liter. Also, because $pH = -\log[H^+] = -(\ln[H^+])/2.3$, $\mu = -2.3\ RT\ pH$. Accordingly, the change in chemical potential due to a change in hydrophobicity for the state at which the ionizable groups of the polymer are 50% ionized becomes $\Delta\mu = -2.3\ RT\ \Delta pK_a$. Therefore, per mole, the chemical work performed by a change in hydrophobicity effecting a change in pK_a is proportional to ΔpK_a. The chemical work performable by a matrix or coacervate containing a pK_a-shifted functional group can be the loading of an oppositely charged drug into the matrix or coacervate.

Nonlinear Hydrophobic-Induced pK_a Shifts

The change in hydrophobicity represented by the replacement of Val residues by more hydrophobic Phe residues, for the structures of Figure 20.6 (*29*), causes an increase in the pK_a values of Glu and Asp residues. As the number of Phe residues are increased at the indicated locations, the pK_a of the Asp and Glu residues are increased as indicated. However, the magnitude of the shift is not linear with the number of Phe residues (Figure 20.7) (*29*). When the change in hydrophobicity is equivalent to changing two Val residues to two Phe residues, the amount of chemical energy, ΔpK_a, obtained depends on how hydrophobic the polymer is before the change. If to begin with the polymer is of a hydrophobicity equivalent to no Phe residues, and two Val(V) residues were replaced by two Phe(P) residues, then the ΔpK_a would be small and the chemical work obtainable would be small. If, on the other hand, the polymer is of a hydrophobicity equivalent to having three Phe residues, and the increase in hydrophobicity is equivalent to two Val residues being replaced by two Phe residues, then the ΔpK_a would be large, and the chemical work obtainable would be large. This enhancement of the hydrophobicity of the folding domain to increase efficiency of energy conversion is called "poising", because the polymer is poised to perform work efficiently (*23*). The work that it can perform can be the loading of drug into the coacervate or the matrix.

Apolar–Polar Repulsive Free Energies of Hydration

As seen in Table 20.3, the T_t for poly(GEGVP) in the COOH state is 30 °C and in the COO^- state is 250 °C; the T_t for poly(GDGVP) in the COOH state is 45 °C and in the COO^- state is 170 °C. Analogously, the T_t for poly(GKGVP) in the NH_2 state is 35 °C and in the NH_3^+ is 120 °C. Even though qualitative differences exist, T_t always increases whenever the formation of charge occurs regardless of the sign, positive or negative. What charge requires, of course, is adequate hydration. It would seem that, in these polymers, the need for charged groups to obtain adequate hydration raises the value of T_t.

Stretch-Induced pK_a Shifts

With the polymers cross-linked to form a matrix by 20-Mrad γ-irradiation, as in X^{20}-poly[0.8-(GVGVP), 0.2(GEGVP)], in one particular experiment the pK_a in the unstretched state was 3.99, but on stretching of the elastic matrix with a 1-g load, the pK_a increased to 4.85 (*30*). On stretching, the forced unfolding of the hydrophobically folded state occurs with the consequences of exposure of hydrophobic side chains and an exothermic formation of water of hydrophobic hydration. Therefore, in spite of an increase in the amount of water present in the matrix, the free energy of the COO^- has increased; the charged moieties have not obtained adequate hydration. The proposed mechanism is a competition for hydration between the apolar (hydrophobic) and polar moieties in the matrix.

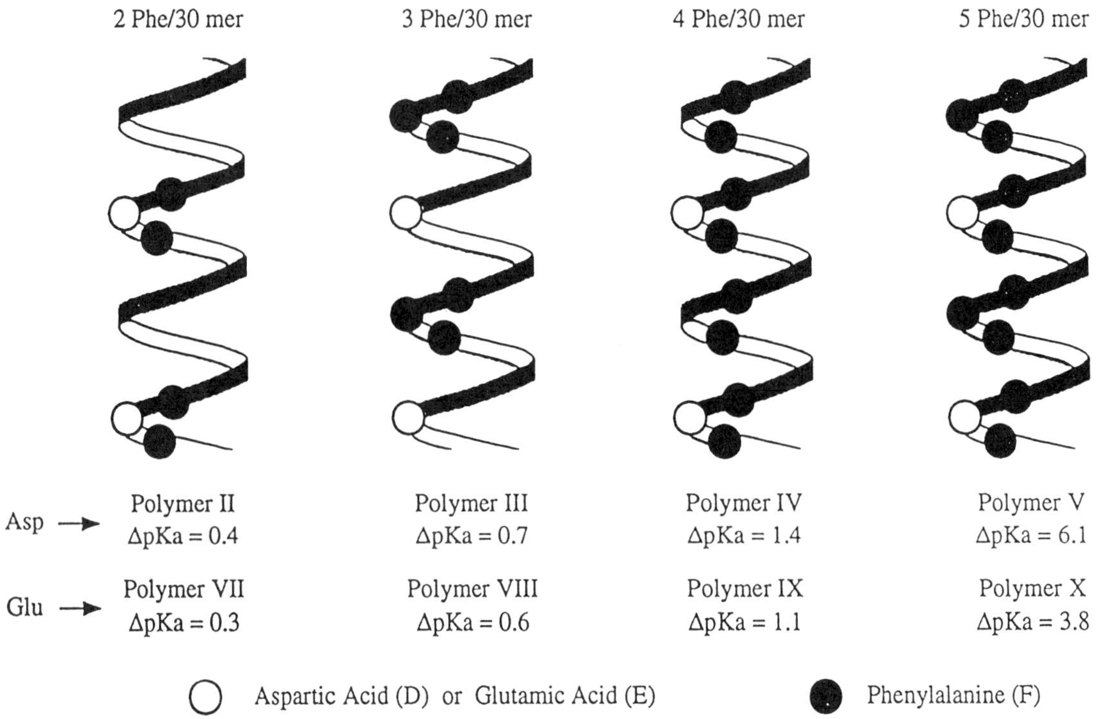

FIGURE 20.6. Schematic representation of poly(GVGVP)-based polymers with particular Val residues replaced by Phe residues and by either an Asp or a Glu residue per 30 residues (i.e., for every six pentamers or for two turns of the β-spiral). Data are plotted in Figure 20.7. (Reproduced with permission from reference 29. Copyright 1995 Elsevier.)

Charge Destructuring of Waters of Hydrophobic Hydration

The concept of competition for hydration between apolar and polar moieties should also be observable as a destructuring of the waters of hydrophobic hydration on formation of charged species. For the same polymer that was used in the stretch experiment, poly[0.8(GVGVP), 0.2(GEGVP)], the endothermic heat of the inverse temperature transition when in the COOH state is about 1 kcal/mol pentamers. This result is interpreted as the heat required for the destructuring of the waters of hydrophobic hydration that is necessary before the hydrophobic folding, and assembly of the inverse temperature transition can occur. After formation of two COO^- moieties per 100 residues, the heat required for the destructuring of the waters of hydrophobic hydration of the inverse temperature transition has been reduced to one-fourth (*31*). Accordingly, for the two COO^- moieties per 100 residues to form it appears that three-fourths of the water of hydrophobic hydration had to be destructured.

Hydrophobicity Effects

Increases in hydrophobicity increase the free energy of the charged state. Because of this result, the hydrophobic-induced pK_a shift increases the pK_a of carboxyls and decreases the pK_a of amino groups. The magnitude of ΔpK_a provides an estimate of the apolar–polar repulsive free energy of hydration; $\Delta G_{\text{a-p}} \approx 2.3\ RT\ \Delta pK_a$. The binding of charged drug can relieve this repulsive free energy of hydration. The $\Delta G_{\text{a-p}}$ value can be the driving force for loading positively charged drugs into carboxyl-containing coacervates and matrices and for

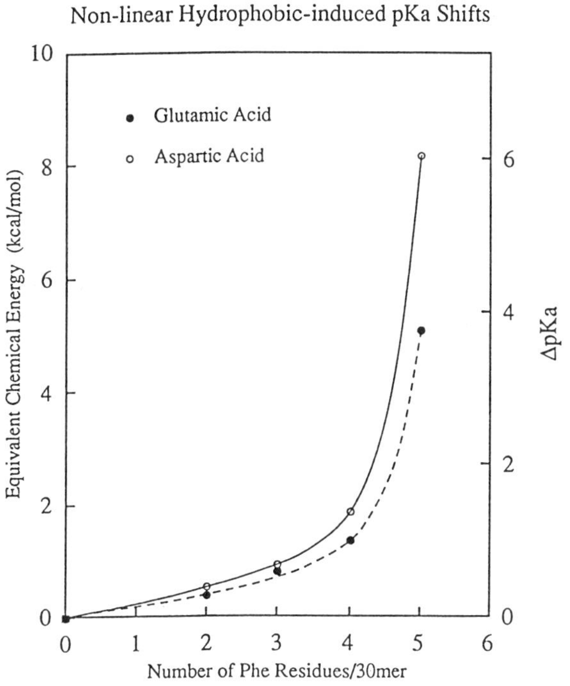

FIGURE 20.7. Plots of the ΔpK_a values for the sequences represented in Figure 20.6. (Reproduced with permission from reference 29. Copyright 1995 Elsevier.)

loading negatively charged drugs into amino-containing coacervates and matrices. The concentration of ionizable species within the polymer can be chosen to determine the maximal concentration of charged drug bound by ion pairing within the matrix, and the decrease in free energy due to ion pairing of polymer with the drug would be proportional to the magnitude of $\Delta G_{a\text{-}p}$; that is, to the ΔpK_a.

Transductional Control of Drug Loading and Release

Transductional control of drug loading and release uses the change of state achievable by the inverse temperature transition to achieve loading or release. A change in temperature can be used to change the state, as in the swell-doping approach to loading. In a particularly attractive approach, the protein-based polymer can be designed such that the drug itself provides the chemical energy to drive the assembly of the loaded vehicle and its release triggers the dispersal of the delivery vehicle. These approaches are considered in some detail.

Design of Protein-Based Polymer for Transductional Control of Loading and Release

The design of the protein-based polymer depends on the drug to be loaded into the coacervate or matrix and on the desired release rate. If the drug is neutral, that is, it does not have an ionizable functional group within several pH units of physiological pH, then increased polymer hydrophobicity or increased plastic nature of the matrix can be used to slow release rate. If the drug has an ionizable function within several pH units of 7.4, then a polymer with the opposite-signed ionizable function would be designed with the desirable number of ionizable functions to control the loading level and with a ΔpK_a to control binding constant and release rate. Clearly, limits exist between the number of ionizable functional groups (loading level) and the magnitude of ΔpK_a (binding constant and release rate). In general the fewer the ionizable functions the larger the ΔpK_a possible. The theoretical limits, however, appear to be quite broad, because the ΔpK_a for the Asp residue has been observed to be as large as 6 in Figure 20.7 with one Asp per 30 residues. This result would give a loading capacity of 0.18 M in the coacervate or matrix volume containing 400–500 mg of polymer per milliliter and, in a first-order approximation, a binding constant (K) of the order of 10^6, because $K \approx e^{-\Delta G\text{a-p}/RT} \approx e^{2.3\Delta pKa} \approx 10^{\Delta pKa}$. A drug-laden matrix with a sufficiently tight binding constant and with a low T_t value could be expected to reach the stage where a chemical clock would be required to facilitate release and to have a release profile dependent on the half-life of the chemical clock.

Transductional Loading of Drug into Matrix

Swell-Doping of Neutral Drugs

The inverse temperature transition in the role of a thermomechanical transducer can be used to load drugs into the matrix. A representative matrix is in its contracted state at 37 °C; a concentrated solution of drug, using a volume 5–10 times greater than the matrix, is added to the contracted matrix. The temperature is lowered below the transition temperature. The matrix swells and takes up the entire volume of concentrated drug solution; the drug-containing, swollen matrix is allowed to equilibrate for 24 h in order that the drug will become uniformly distributed. The temperature is raised to 37 °C, and the extruded volume of solution is removed. In the case of Biebrich scarlet red, a 0.1 M solution was used, and loading into the matrix was at a concentration of 0.3 M; that is, one molecule of Biebrich scarlet red for every three pentamers (*5*).

For the case of very hydrophobic drugs such as the steroids, which lack adequate solubility in water for swell-doping, the matrix can be swollen and lyophilized in the swollen state. The steroid is added to the lyophilized matrix in a solvent such as alcohol with 10% water that will wet the matrix, and the solvent swollen matrix is dried. Several cycles of addition and drying are usually sufficient. This approach has been taken for norgestrel, the progesterone-like steroid used for contraception, and dexamethasone, an anti-inflammatory glucocorticoid. The matrices were of poly(GVGVP), poly(GVGIP), and poly(AVGVP), and loadings were to average values for dexamethasone of 40%, 28%, and 11% and for norgestrel of 10%, 43%, and 8% by weight in the three matrices, respectively. Clearly, the steroids can be loaded into these matrices to high levels.

Loading of Ionizable Drugs

Ion-Pairing in a Hydrophobic Domain. The T_t-based hydrophobicity scale was determined in phosphate-buffered saline (PBS) (0.15 N NaCl, 0.01 M phosphate) to be as relevant to physiological conditions as possible, but also residues with charged side chains in the absence of NaCl would result in polymers having T_t values above 100 °C. The effect of NaCl is to lower the value of T_t for both COO^- and NH_3^+ moieties of poly[f_V-(GVGVP), f_E(GEGVP)] and poly[f_V(GVGVP), f_K(GKGVP)], respectively. On a molar basis, this lowering is greater with $CaCl_2$. From sodium NMR relaxation studies, the presence of the COO^- at a mole fraction, f_E, of 0.2 effects a sodium ion binding constant of about 10^{-1}/M. For the same polymer, the interaction is 10-fold greater for $CaCl_2$ (*32*). Of particular significance is that the effective binding constants for the ion pairing are made greater by increasing the hydrophobicity of the polymer as, for example, when using poly[f_V(GVGIP), f_E(GEGIP)] with an additional CH_2 moiety per pentamer. Our interpretation is that increased hydrophobicity decreases the effective dielectric constant surrounding the charges on the polymer due to the apolar–polar repulsive free energy of hydration: that is, due to a competition between the apolar and polar groups for hydration. The lower effective dielectric constant, ε, surrounding the charges increases the energy of interaction between the negative polymer charges and the divalent cations as given by the coulombic energy term, $\Delta E = q_1 q_2/\varepsilon r$, where q_1 and q_2 are the charges, and r is the distance between them.

Importance of Polymer Sequence Control. Just as the pK_a shifts increase as the hydrophobicity increases (*29*), so too will the affinity for an oppositely charged species. Similarly, just as the pK_a shifts are greater when there is control of the sequence (*14*), again, so too will be the affinity for oppositely charged species. Accordingly, the design of protein-based polymers for increasing the binding constants of ionizable drugs is directly parallel to the design of polymers for large pK_a shifts.

Use of Carboxylate To Effect Loading of Cationic Drug. The cationic drug of interest here is naltrexone; it is a narcotic antagonist of interest in drug addiction intervention (*33, 34*) and it contains a tertiary amine with a pK_a of 9.4. The protein-based polymer, poly(GFGVP GEGVP GFGVP), has a ΔpK_a of 1.1 at 37 °C at a concentration of 0.033 M Glu(E) in 0.01 M phosphate at pH 7.4. This polymer does not exhibit an inverse temperature transition in the accessible aqueous range: that is, the value of T_t is greater than 100 °C. On addition of 0.025 M naltrexone, however, the value of T_t appears at 60 °C. At equimolar polymer Glu and naltrexone concentrations (i.e., 0.033 M), the inverse temperature transition occurs at 34 °C. At a 50% molar excess of naltrexone, the value of T_t is 15 °C. When poly[0.7-(GFGVP), 0.3(GEGVP)] is studied, having the same mean composition but with a random sequence of pentamers and a lower ΔpK_a of 0.8 at 37 °C, the addition of an equimolar quantity of naltrexone would only reduce the value of T_t to 82 °C. The titrations of poly(GFGVP GEGVP GFGVP) and of poly[0.7(GFGVP), 0.3(GEGVP)] with naltrexone are given in Figure 20.8A as T_t versus [naltrexone]/[GEGVP]. Thus, the very addition of the cationic drug at 37 °C and pH 7.8, that is, the chemical energy input of increasing naltrexone concentration, results in the mechanical work of driving coacervate formation or of driving contraction of the cross-linked matrix as an integral part of loading with drug. In short, naltrexone has been titrated into two different solutions of Glu(E)-containing polymers, one with a fixed sequence having a Glu residue every 15 residues and the other with essentially the same composition but with a random arrangement of Glu-containing pentamers. Naltrexone, more effectively on a molar basis, lowers T_t for the fixed-sequence polypentadecapeptide as the polymer separates out to form a coacervate phase loaded with the naltrexone drug. The more effective interaction occurs with the polymer exhibiting the largest pK_a shift.

Similarly, when using the polytricosapeptide, poly(GEGVP GVGVP GVGFP GFGFP GVGVP GVGVP), the value of T_t is much greater than 100 °C at a pH above 6. On the addition of 0.5 equivalence of Leu-enkephalin amide, $^+$H-Tyr-Gly-Gly-Phe-Leu-NH_2 ($^+$YGGFL-NH_2), at a pH of 6.4, however, the T_t is lowered to 31 °C. By the addition of one equivalence of the cationic drug, the value of T_t is reduced to 12 °C. This result is shown in Figure 20.8B. For both the naltrexone and Leu-enkephalin amide examples of Figures 20.8A and 20.8B, the polymers are designed as molecular machines, actually as molecular engines, for the taking up of a cationic drug in the assembly of a drug-delivery vehicle.

Use of Charged Amino To Effect Loading of Anionic Drug. The anionic drug to be considered is Dazmegrel. It is a Pfizer drug that inhibits the enzyme, thromboxane synthetase, responsible in part for the tissue breakdown in the formation of pressure ulcers (*35, 36*), and it contains a carboxyl function with a pK_a of 4.9 and an imidazole function with a pK_a of 7.9. In this case the protein-based polymer contains the positively charged lysine (Lys, K) residue. In 0.01 M phosphate at pH 8.6, where Dazmegrel is negatively charged, poly[0.69(GVGIP), 0.31(GKGIP)] does not exhibit an inverse temperature transition in the aqueous range, but on adding a charge equivalence of Dazmegrel, the value of T_t is lowered to 21 °C. The titration is given in Figure 20.8C. Also included in Figure 20.8C is the titration for poly[0.86(GVGIP), 0.14(GKGIP)], where it is seen that the decrease in T_t is steeper for this polymer with a ΔpK_a of 0.8 at 37 °C, whereas the ΔpK_a is 0.33 at 37 °C for the polymer with f_K = 0.31. Also, the effect of increasing Dazmegrel concentration on the value of T_t for the polymer with $f_K = 0$ is to cause a small increase in T_t, as shown by the rising linear dotted curve of Figure 20.8C. Again the design of the polymer for the most effective molecular machine for loading of an anionic drug should include consideration of the ΔpK_a, but in this case it is the design of a molecular engine containing a pK_a-shifted positive charge for the loading of an anionic drug.

Transductional Release from Ion-Pair-Loaded Matrix

In the following section, two pair of protein-based polymers containing the COO^- function have

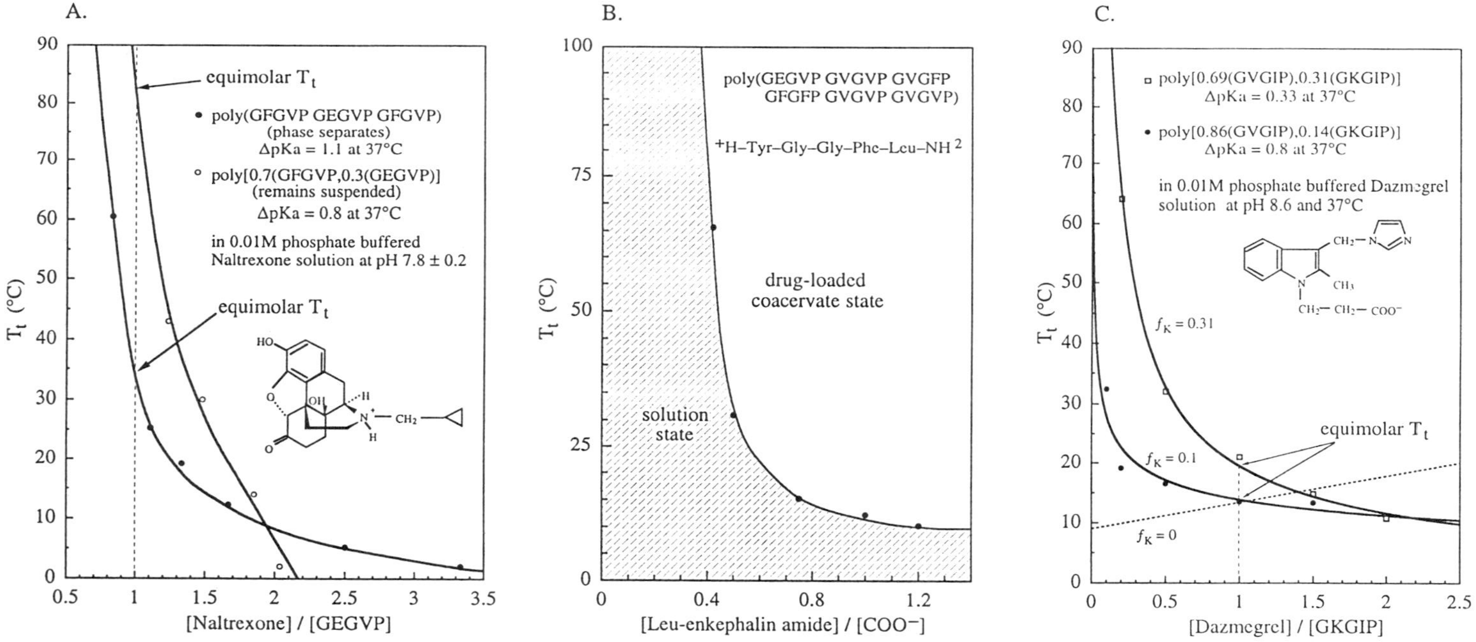

FIGURE 20.8. Titrations of charged drugs into solutions of protein-based polymers within which there are oppositely charged groups that have been hydrophobically poised as evidenced by the magnitude of their shifted pK_a values: A, naltrexone titrated into two different solutions of Glu(E)-containing polymers; B, Leu-enkephalin amide ($^+$Tyr-Gly-Gly-Phe-Leu-NH_2) loaded into poly(GEGVP GVGVP GVGFP GFGFP GVGVP GVGVP) (release from the loaded coacervate is shown in Figure 20.10 to be constant for a constant surface area); C, Dazmegrel titrated into two different polymer solutions composed of random polymers but in which the mole fraction of pentamers containing the positively charged Lys(K) amino side chain differs.

been loaded with the positively charged drug, naltrexone. One pair contains a single Glu residue per 30-mer and three and five Phe residues per 30-mer with ΔpK_a values of 0.6 and 3.8, respectively, at 20 °C, as shown in Figures 20.6 and 20.7. The other pair have the same positions for the five Phe residues per 30-mer, but in one case the Glu residue and in the other the Asp residue is the source of carboxylate, and they have different ΔpK_a values of 3.4 and 5.6, respectively, at 20 °C. The purpose of these comparisons is to determine whether an increase in the ΔpK_a improves the release profile as the result of an expected dependency of the ion-pair binding constant on ΔpK_a.

The drug was titrated into a 40-mg/mL polymer solution to a 50% excess of drug, thereby lowering T_t and driving coacervation by ion pairing. The coacervate was centrifuged in a small *conical* tube to provide a uniform surface to the coacervate. The overlying equilibrium solution was removed; its concentration of drug was determined, and it was replaced by 1 mL of PBS at pH 7.4 and 37 °C. The release of drug into the 1 mL was allowed to occur for 24 h. The 1 mL was removed; the drug concentration was determined; and a fresh 1 mL was added. The procedure was repeated for a sufficient number of days to define the release profile.

Naltrexone Bound to Coacervates with Differing Hydrophobicities and ΔpK_a

Following the described protocol, the release profiles from naltrexone-loaded coacervates of poly(GEGVP GVGVP GVGFP GFGFP GVGVP GVGVP) with a ΔpK_a of 0.6 at 20 °C and of poly(GEGFP GVGVP GVGFP GFGFP GVGVP GVGFP) with a ΔpK_a of 3.8 at 20 °C are given in Figure 20.9A for 37 °C. The polymers have one Glu residue per 30-mer and differ only in the number of Val residues that had been replaced by more hydrophobic Phe residues. Obviously, the drug-loaded coacervate with the larger ΔpK_a exhibits the most favorable release profile in that the concentration of drug released per day is maintained more nearly constant and is sustained over a longer period of time.

Naltrexone Bound to Coacervates Exchanging Glu for Asp and with Differing ΔpK_a

Again following the described protocol, the release profiles from naltrexone-loaded coacervates of poly(GEGFP GVGVP GVGVP GVGVP GFGFP GFGFP), ΔpK_a of 3.4 at 20 °C, and of poly(GDGFP GVGVP GVGVP GVGVP GFGFP GFGFP), ΔpK_a of 5.6 at 20 °C, are given in Figure 20.9B for 37 °C. The polymers have the same sequence with the exception that one contains the Asp residue, which has the more readily hydrophobically pK_a-shifted carboxyl, and the other contains the Glu residue, which does not exhibit as large a pK_a-shifted carboxyl. The polymer with the larger hydrophobic-induced pK_a shift exhibits the more sustained release profile.

These results indicate that the described rationale for the design of transductional matrices and coacervates does indeed provide a basis for developing more effective controlled-release devices. Transductional loading of an ionizable drug effectively uses ion pairing in a hydrophobic domain and a more sustained-release profile results from the matrix or coacervate with the larger hydrophobic-induced pK_a shift. Furthermore, more constant release profiles were expected to be obtained from matrices than were observed in Figure 20.9 by dissolving coacervates from the decreasing surface area in conical tubes.

Leu-Enkephalin Amide Bound to Coacervates with Differing Densities of Sites and ΔpK_a Values

Shown in Figure 20.10 is the release profile of Leu-enkephalin amide obtained by using a straight-walled tube that maintains a constant surface area. Two polymers were used: the first was poly(GEGVP GVGVP GVGFP GFGFP GVGVP GVGVP), and the second was poly(GFGFP GEGFP GFGFP). The ratio of Glu(E) to Phe(F) increased from 1:3 in the first to 1:5 for the second as the number of the more-polar COO^- moieties increases from one to two per 30-mer. The immediate result was the increase in the micromoles of sites from 95 to 167 for 250 mg of poly-

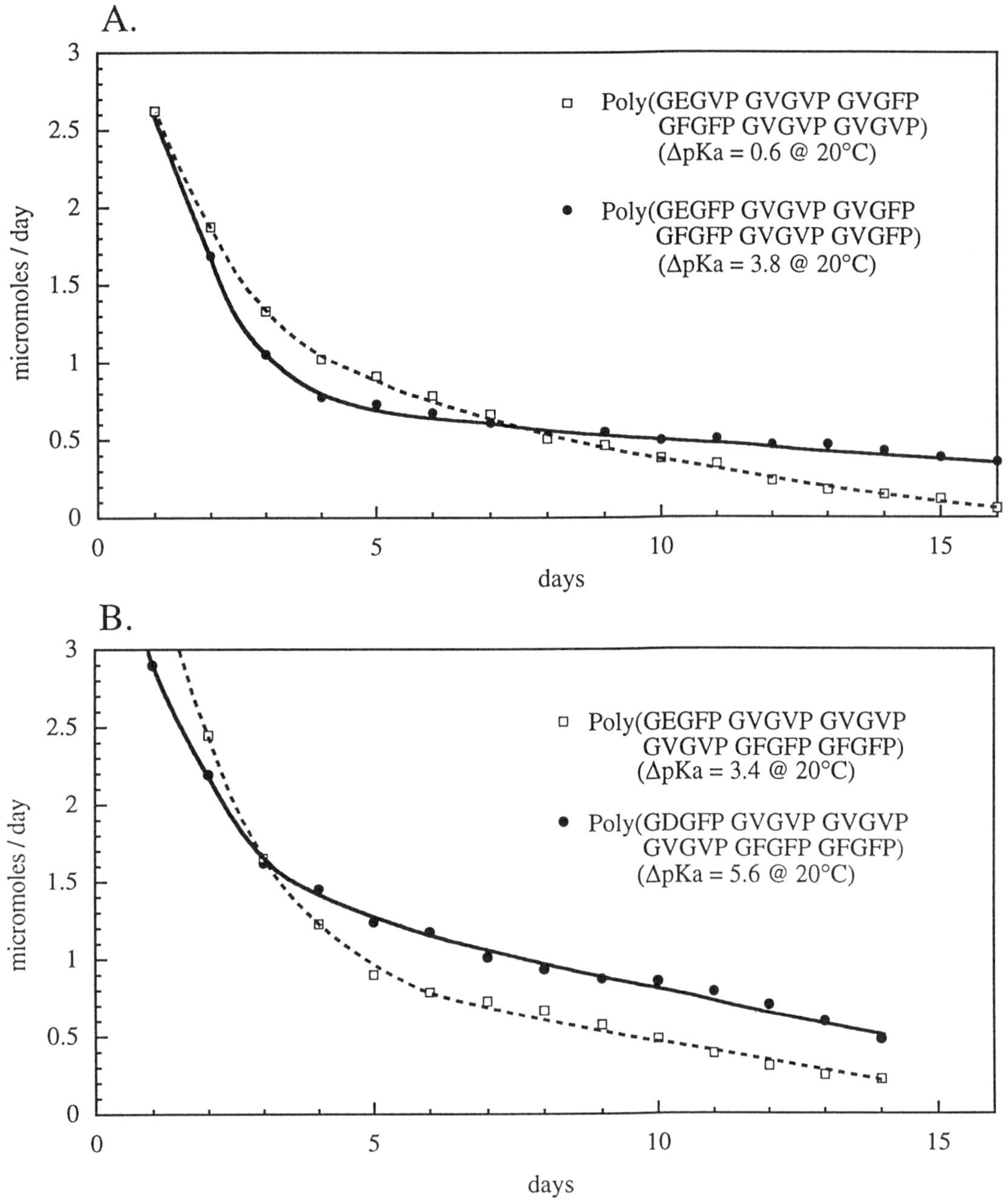

FIGURE 20.9. Profiles of naltrexone release from coacervates formed on titration of naltrexone into carboxylate-containing protein-based polymer solutions: A, two Glu-containing polytricosapeptides; B, two polytricosapeptides.

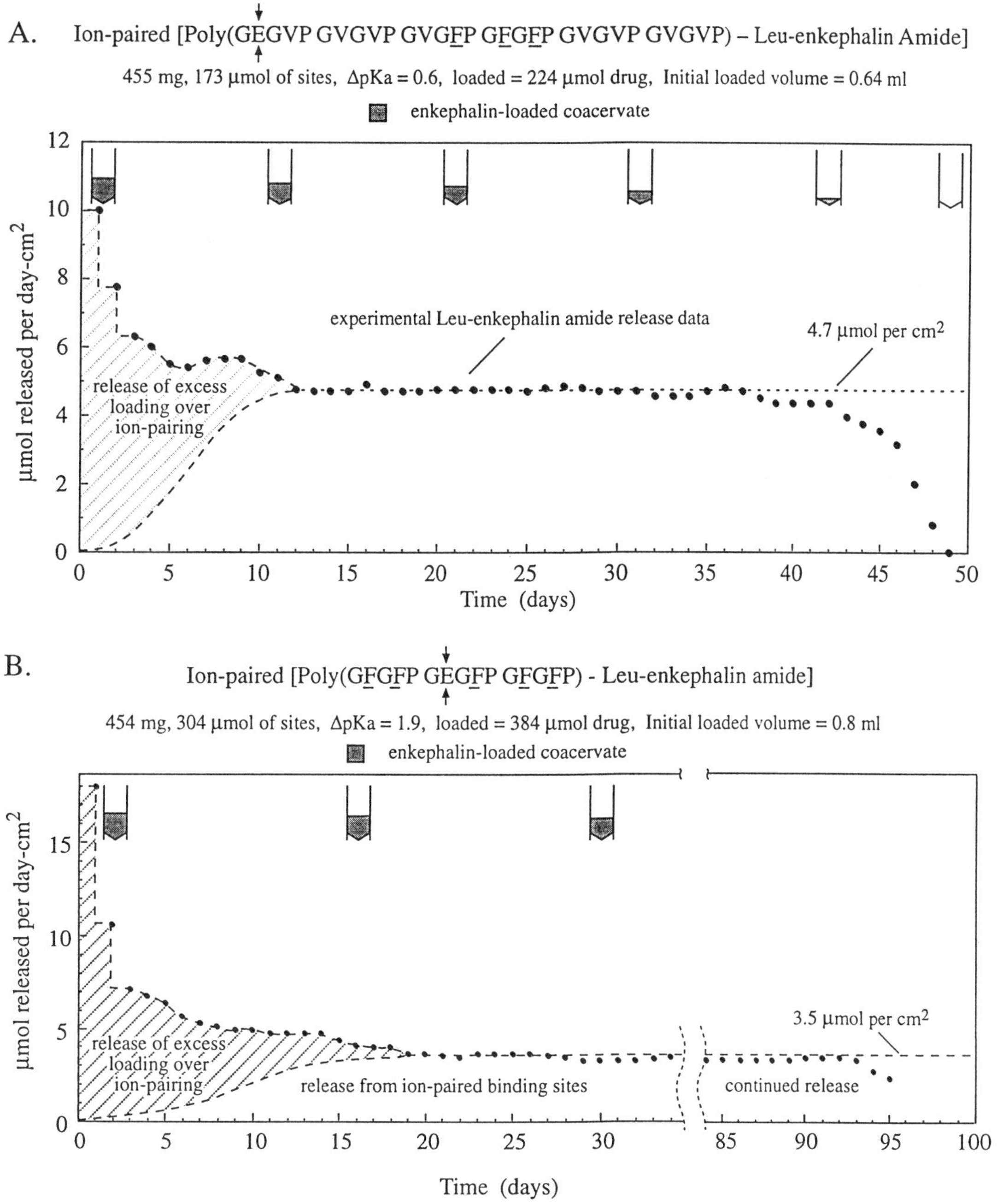

FIGURE 20.10. Release of Leu-enkephalin amide from two different Glu-containing polymers: A; profile from a 1 cm^2 surface area of ion-paired [poly(GEGVP GVGVP GVGFP GFGFP GVGVP GVGVP)–Leu-enkephalin amide]; B, profile from a 1 cm^2 surface area of ion-paired [poly(GFGFP GEGFP GFGFP)–Leu-enkephalin amide].

mer while increasing the ΔpK_a from 0.6 to 1.9. The expectation is that a higher concentration can be loaded into the second and that the release level would be lowered and sustained for a longer period of time. This expectation is demonstrated in Figure 20.10. In fact, the ion-pairing binding constant is sufficiently strong in both cases to result in a constant release for a constant surface area. For the case with the smaller ΔpK_a, the release level is higher and release continues for about 1.5 months; for the case with the larger ΔpK_a the release level is lower but continues at a constant level for the period of about 3 months.

Ion pairing in hydrophobic domains of designable intensity in transductional protein-based polymers may become an effective basis with which to design useful drug-delivery vehicles. In general, the release is proportional to an equivalent solubility product constant, K_{sp}: the smaller the K_{sp} the lower the release level. The magnitude of the K_{sp} is inversely proportional to the designed (i.e., the hydrophobic-induced) ΔpK_a.

Diffusional Release of Drugs from Swell-Doped, Neutral Matrices

Many drugs, for example, Biebrich scarlet red, naltrexone, Dazmegrel, and Leu-enkephalin and Leu-enkephalin amide, have been swell-loaded into neutral matrices such as X^{20}-poly(GVGVP), X^{20}-poly(GVGIP), and X^{20}-poly(AVGVP), and their diffusional release profiles have been determined. A representative profile is given in Figure 20.11 for Biebrich scarlet red diffusing from X^{20}-poly(GVGVP) after swell-loading into the matrix to a concentration of 0.3 M from a 0.1 M solution. Daily 1-mL volumes of PBS were added to the loaded matrix and removed after 24 h for determination of released drug. A plot of micromoles per day for 11 days showed a burst release, but for the

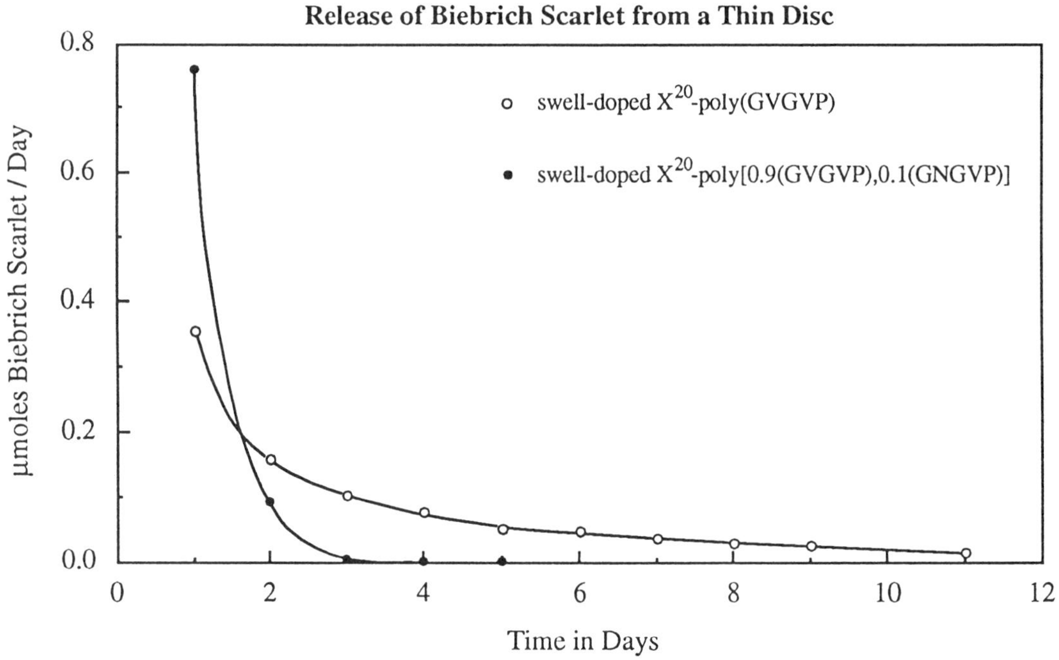

FIGURE 20.11. Release profiles for matrices swell-doped with the neutral Biebrich scarlet red model drug. (Reproduced from reference 5. Copyright 1994 American Chemical Society.)

week period from day 4 to day 11 a relatively sustained quantity dropped from 50 to 15 nmol/day from a small, thin, 8-μL, 0.035 × 0.545 cm disk (*5*). Given the size and shape of the disk without a reservoir depth from which to diffuse, this result is encouraging. A relatively sustained release was found but not one as favorable as those of Figure 20.10, where the drug had an enhanced binding to the condensed phase due to ion pairing under the influence of a hydrophobic domain sufficient to effect large pK_a shifts. Also shown in Figure 20.11 is the more rapid release from the matrix, X^{20}-poly[0.9(GVGVP), 0.1(GNGVP)], where the asparagine breakdown to aspartic acid results in the COO^--induced matrix swelling and disruption (*6*). This type of matrix is not where the chemical clock has its promise, but rather in a matrix that does not release until the formation of the carboxylates, such as a plastic matrix or a matrix in which the drug binding constant is too tight for release to occur.

Asparagine Chemical Clocks To Effect an Increase in T_t *and Control of Release Rate*

Two ways exist in which the release can be made entirely dependent on the rate of $CONH_2$ conversion to COO^-. Firstly, when the drug is trapped within the plastic state with little or no water, only the conversion of $CONH_2$ to COO^- at the surface of the drug-doped monolith is expected to increase the surface layer T_t above 37 °C, causing that surface layer to swell and to release its entrapped drug (*4*). Secondly, when the ion-pairing binding constant for the drug in the matrix is so strong as to prevent release, as previously considered, then a decrease in that binding constant can be achieved by creating an excess of COO^- moieties due to carboxamide hydrolysis with the consequence of an increase in T_t above 37 °C.

Time Course of Coacervate Dissolution at 37 °C

Three different polymer compositions were studied, each with four different mole fractions of asparagine chemical clocks: poly[f_V(GVGVP), f_N(GNGVP)], mole fractions f_N = 0, 0.11, 0.12, and 0.16; poly[f_V(GVGIP), f_N(GNGVP)], mole fractions f_N = 0, 0.07, 0.08, and 0.13; and poly[f_V(AVGVP), f_N(GNGVP)], mole fractions f_N = 0, 0.06, 0.09, and 0.12. The differential rates of coacervate dissolution were followed as T_t versus time in a 0.10 M phosphate solution at pH 7.4 (Figures 20.12A and 20.12B). In 0.10 M phosphate at pH 9.5, dissolution was complete in 2 days for the Val(V)-containing polymer, in 3 days for the Ile(I)-containing polymer, and in less than 2 days for the Ala(A)-containing polymer.

At pH 7.4 in 0.10 M phosphate, the elastic protein-based polymer, poly[f_V(GVGVP), f_N-(GNGVP)] exhibited a rate of increase of T_t that, as might be expected, was proportional to the mole fraction of pentamers containing Asn(N). In just over 8 days the coacervate of the polymer with f_N = 0.16 went completely into solution at 37 °C, whereas with f_N = 0.12 and 0.11, 11 and 12 days were necessary, respectively. As seen in the upper set of curves of Figure 20.12A, the profile is one of a nonlinear increasing ΔT_t per day. On the other hand, the more hydrophobic elastic protein-based polymer, poly[f_V(GVGIP), f_N(GNGVP)], just began to show a small increase in T_t by 14 days for the polymer with the highest mole fraction, 0.13, of pentamer containing the asparagine chemical clock, as shown in the lower set of curves of Figure 20.12A.

For the plastic protein-based polymer, poly-[f_V(AVGVP), f_N(GNGVP)], the value of T_t changed little up to 8 days, but after 8 days an abrupt dissolution occurred (Figure 20.12B). Apparently, the presence of the first few COO^- moieties catalyzed the rapid further breakdown of carboxamide.

Time Course for Disk Swelling and Disintegration in PBS at 37 °C

On 20-Mrad γ-irradiation, cross-linking of coacervate concentrations of six of the set of polymers—poly[f_V(GVGVP), f_N(GNGVP)], f_N = 0 and 0.11; poly[f_V(GVGIP), f_N(GNGVP)], f_N = 0 and 0.07; and poly[f_V(AVGVP), f_N(GNGVP)],

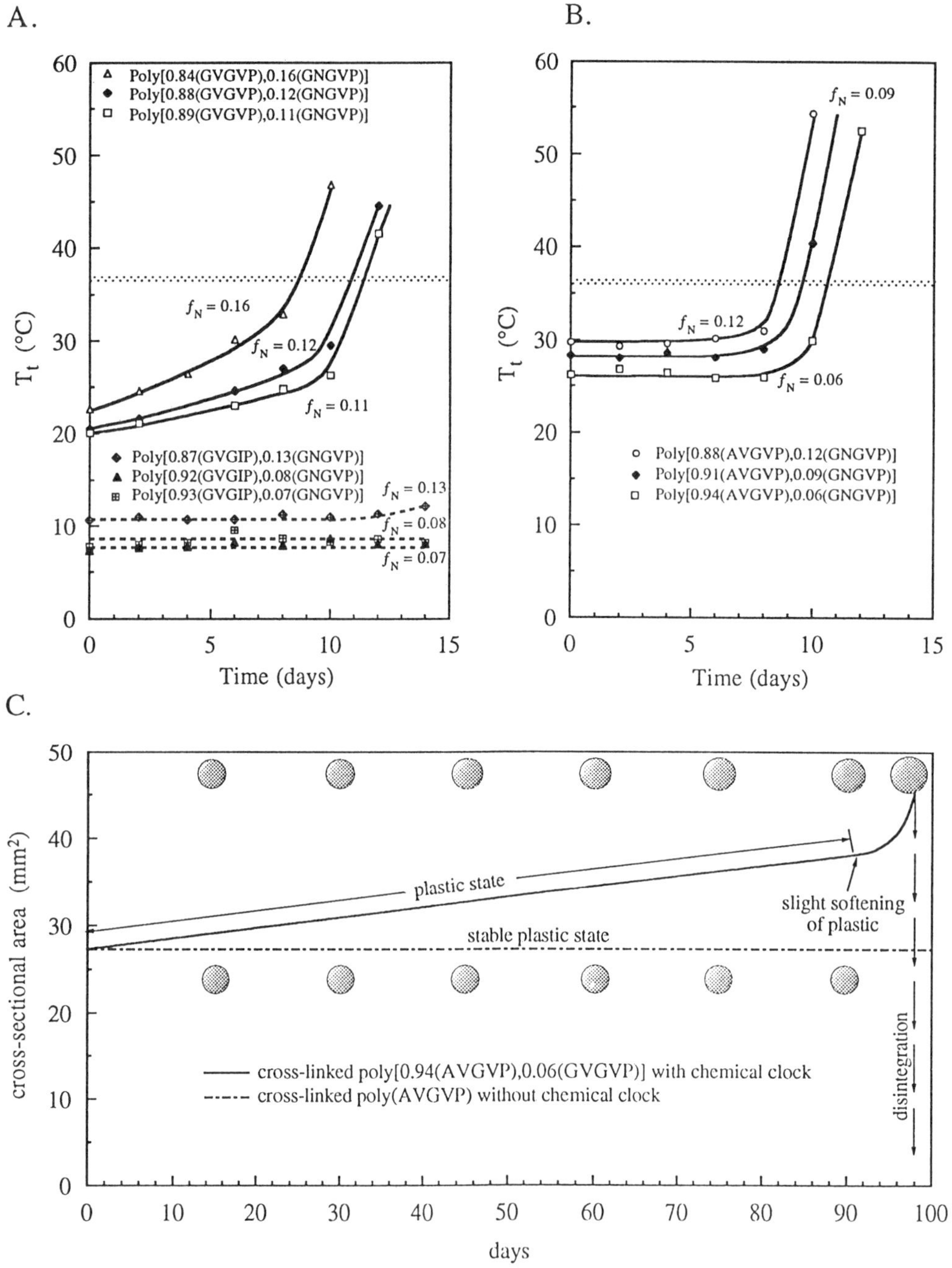

FIGURE 20.12. Time courses for the breakdown of three different parent polymer compositions each having different mole fractions of pentamers containing the asparagine (Asn, N) chemical clock: A, elastic protein-based polymers; B, plastic protein-based polymers; C, disks of the plastic state in which the presence of asparagine results in swelling and ultimate disintegration.

$f_N = 0$ and 0.06—matrices were obtained from which disks were punched. The disks were each placed in PBS at pH 7.4 and 37 °C. All of the disks with $f_N = 0$ remained unchanged for a period of time surpassing 5 months. For the asparagine-containing polymers, however, the disks slowly swelled and two ultimately disintegrated. The elastic disk, X^{20}-poly[0.89(GVGVP), 0.11(GNGVP)], swelled to twice its original area by 30 days and disintegrated to globs of hydrogel by 36 days. The different behaviors of plastic disks (X^{20}-poly[f_V(AVGVP), f_N(GNGVP)], $f_N = 0$ and 0.06) are represented in Figure 20.12C, where no change occurs in the plastic disk without asparagine. On the other hand, the asparagine-containing disk remained as a relatively hard plastic for 80 days while swelling about 30%; it then began to soften with more rapid swelling and disintegrated between 90 and 100 days. Finally, the more hydrophobic elastic disk, X^{20}-poly[0.93-(GVGIP), 0.07(GNGVP)], showed only a 10% swelling in 150 days.

The introduction of chemical clocks in combination with elastic and plastic states and with variable hydrophobicity can be expected to provide for a wide range of matrices with an equally wide range of release profiles.

Future Perspectives for Transductional Release Vehicles

Transductional protein-based polymers appear capable of performing the pairwise energy conversions, at least in one direction, that occur in living organisms. Because of this capability, these polymers can be expected, by suitable design, to provide for drug delivery in a particularly effective manner. Some specific considerations are given.

Coupling Release to the Diseased State: The Diagnostic–Therapeutic Pair

Transductional coacervates and matrices can be designed to function as integrated transducers in which they can both sense a particular energy input and actuate with the desired energy or work output. This characteristic means that a particular transductional vehicle could be designed to both sense the chemical anomaly of the diseased state and to use that chemical difference as the energy input to trigger the release of the desired dose of a therapeutic substance: that is, it could function as an integrated diagnostic–therapeutic pair.

By way of examples, a drug could be tightly bound in a nanosphere until a change in pH of the diseased tissue triggered its release, or until the high activity of an enzyme characteristic of the diseased tissue acted on the nanosphere to effect release, as could occur by the activity of a kinase to phosphorylate a sequence specific-site in the protein-based polymer. Also, release could be dependent on the change in the level of the oxidative state of an enzyme that was the result of a change in the level of a chemical such as glucose, and the enzyme could act on a redox couple attached to the protein-based polymer to achieve release of a therapeutic substance packaged within the nanosphere composed of the protein-based polymer. The list of such scenarios is endless, and the future holds much opportunity to design transductional polymers and matrices to function as diagnostic–therapeutic pairs.

External Control of Release

With the many energy inputs that could effect an energy output of a chemical release as represented in the hexagon of Figure 20.5, the drug could reside in a depot within the body until an external stimulus or even an internal stimulus controlled from without the body enhanced or slowed release. Examples of stimuli controlled from without the body could be electromagnetic radiation from various parts of the spectrum, applied electric fields or magnetic fields, applied pressure or a local change in temperature, or introduction of an otherwise innocuous chemical that would trigger release. Once there is design control of transductional coacervates and matrices, the potential for controlled release by numerous desired modalities becomes possible.

Concluding Overview

Inverse Temperature Transitions and the ΔT_t-Mechanism

The inverse temperature transition of hydrophobic folding and assembly on raising the temperature is observable in water for those polymers that have the correct balance of hydrophobic (apolar) and polar moieties. But this process is most effectively employed for controlled release of drugs by transductional means in protein-based polymers because of the available diversity of functional side chains and their precisely controllable composition and sequence. The temperature, T_t, at which the inverse temperature transition occurs becomes a measure of functional hydrophobicity and as such is the basis for a T_t-based hydrophobicity scale for protein-based polymer and protein engineering. Controlling the value of T_t by what is called the ΔT_t-mechanism becomes the basis for diverse energy conversions: that is, free-energy transductions involving the six energies—mechanical, pressure, thermal, chemical, electrical, and electromagnetic radiation. Of particular interest, these transductional capacities directly become the basis for the transductional loading of drug into coacervates and matrices composed of protein-based polymers that exhibit inverse temperature transitions and the transductional control of release of therapeutic substances from such constructs.

Swell-Doping and Coacervate Uptake

In general, coacervates dissolve and matrices swell as the temperature is lowered from above to below the transition temperature, T_t. On lowering the temperature of a matrix with an overlying concentrated drug solution, the matrix swells taking up drug and solvent; and on subsequently raising the temperature, the matrix contracts expelling solution, and many drugs are preferentially concentrated in the matrix. An analogous situation occurs below T_t with polymer and drug solutions: on raising the temperature above T_t, the polymer separates out of solution to form a more dense, viscoelastic, coacervate phase partitioning drug within it. Release from the matrix or the coacervate phase depends on the amount of water in the matrix or coacervate states and on the tightness of binding of drug, which can depend on the hydrophobicities of the polymer and the drug.

Transductional Polymers Effect Remarkable Loading and Release Potential

Where transductional polymers and matrices to date exhibit the most remarkable utility for control of loading and release is with drugs having the capacity to carry a charge. The protein-based polymer can be engineered to contain functional side chains that can be charged with the opposite sign. The number of charges or chargeable side chains in the polymer can be set to achieve the desired concentration of drug in the matrix or coacervate.

Most importantly, the ionizable side chain of the polymer, by proper design, can be hydrophobically primed to control the affinity of the drug for the matrix or coacervate. In short, increasing hydrophobicity decreases the effective dielectric constant surrounding the charged moieties of the polymer and thereby increases the free energy of interaction for ion pairing.

The loading process is simply one of titrating in the drug to form the ion pair; the formation of the ion pair causes a dramatic decrease in the value of T_t, causing the polymer to go from solution to coacervate or the matrix to go from swollen to contracted. Even though much basic work has yet to be done, binding is expected to be designed with ΔG values ranging from near zero to 8 kcal/mol (i.e., with binding constants for the ion pairing within the coacervates or matrices ranging from 1/M to 10^6/M or greater). With very tight binding constants, release could then fall under the control of chemical clocks.

Role of Chemical Clocks in Controlled Drug Release

Chemical clocks can be simply natural asparagine (Asn, N) and glutamine (Gln, Q) amino acid residues of the protein-based polymer sequence that hydrolyze to natural aspartic (Asp, D) and glutamic (Glu, E) acid residues. The half-lives for this conversion under physiological conditions can be as short as polymer preparation time allows or as long as a decade or more. The hydrolytic conversion of the carboxamide ($CONH_2$) side chain of N and Q residues to the COO^- moiety of E and D residues at the surface of the device dramatically raises the value of T_t for the surface layer of the matrix or coacervate and causes it to swell or dissolve with release of the drug contained in that layer. When the drug has been loaded into the plastic state containing little or no water or into an elastic state with a very tight ion-pair binding constant where diffusional release is blocked, release becomes dependent on the half-life of the chemical clock. Controlled release becomes the result of the effective pealing away of the drug-laden surface layer of the device, which, of course, is proportional to the half-life of the chemical clock.

Extensive Diversity of Transductional Control of Drug Release Portends a Challenging Future

In transductional polymers of the T_t-type, that is, polymers that exhibit inverse temperature transitions, each of the many ways in which the value of T_t can be shifted becomes a means for controlling the release of drug. These ways include each of the six generic energy inputs of Figure 20.5. Within each category of energy input, many potential examples exist. Of course, innumerable chemical energy inputs can be an integral part of the polymer design, such as in chemical clocks, in the control of ion-pair binding constants, and in the inclusion of enzymatic sites for proteolysis, for phosphorylation, etc. Also, the complete electromagnetic spectrum exists whereby the polymer design or certain of its intrinsic properties a particular range of the spectrum could be used to control, modulate, turn on, or turn off drug release. The possibilities for drug delivery using transductional polymers of the T_t-type and particularly using protein-based polymers of the T_t-type for controlled drug delivery are limitless.

Summary

Polymers designed to have the correct balance of hydrophobic and polar residues can be further designed to perform diverse free-energy transductions by means of inverse temperature transitions of hydrophobic folding and assembly. These transductional processes, which are most efficiently performed by polymers with precise sequence and with a substantial variety of functionalities available to their monomer side chains, can be very effective in achieving controlled release of therapeutic substances.

The strategies for loading of the matrices or of the polymer-rich (coacervate) phase depend on the properties of the drug. Drugs carrying a net charge can function as the chemical energy input for their own loading, because the polymer can be designed to contain the opposite charge such that the association with the drug can dramatically lower the temperature of the inverse temperature transition from above to below the operating temperature. Thus, when the polymer is in solution, the association with the drug drives the phase separation trapping the drug within; or when the polymer is cross-linked in the form of a swollen matrix, the drug drives the contraction of the swollen matrix as an integral process of loading the drug within the matrix. The tightness with which the drug is bound within the coacervate phase or within the matrix can be controlled by the degree to which the charged species within the matrix is poised: that is, by the extent to which the pK_a of the charged group of the polymer has been hydrophobically shifted.

Neutral drugs with adequate water solubility can be loaded into neutral matrices by swell-doping (i.e., by lowering the temperature of the matrix in the presence of a concentrated drug solution causing the matrix to swell and take up the drug). On raising the temperature above T_t, excess water is expelled leaving the drug preferentially associated within the matrix. If the neutral drug is too hydrophobic to have adequate solubility in water, the drug in organic solvents or in alcohol–water mixtures can be loaded into the lyophilized matrix obtained from the swollen state by successive addition and drying cycles. The rate of release of the drug from the matrix will depend on the hydrophobicity and the water content of the matrix.

Whether the drug is neutral or charged, it can be so tightly held within the matrix that release can be dependent on the hydrolysis of carboxamides of the naturally occurring amino acid residues (asparagines and glutamines) in the polymer to carboxylates to form the naturally occurring aspartic and glutamic acid residues. This breakdown occurs at the surface of the matrix causing the surface layer to swell and allowing for release of the drug from the surface layer with the subsequent degradation of the swollen state of the protein-based polymer. The polymers can be designed to contain carboxamides with half-lives that vary from days to decades.

Acknowledgments

This work was supported in part by NIDA Grant 1 R43 DA09511-01, NIH Grant RO1 HD31413-02, and Office of Naval Research Grant N00014-89-J-1970. We are pleased to acknowledge Auburn University Nuclear Science Center for carrying out the γ-irradiation cross-linking.

References

1. Urry, D. W. *Angew. Chem.* **1993**, *105,* 859–883; *Angew. Chem. Int. Ed. Engl.* **1993**, *32,* 819–841.
2. Urry, D. W. *Prog. Biophys. Mol. Biol.* **1992**, *57,* 23–57.
3. Robinson, A. B. *Proc. Natl. Acad. Sci. U.S.A.* **1974**, *71,* 885–888.
4. Urry, D. W. *Polym. Mater. Sci. Eng.* **1990**, *63,* 329–336.
5. Urry, D. W.; Gowda, D. C.; Harris, C. M.; Harris, R. D. In *Polymeric Drugs and Drug Administration;* Ottenbrite, R. M., Ed.; ACS Symposium Series 545; American Chemical Society: Washington, DC, 1994; pp 15–28.
6. Urry, D. W.; Nicol, A.; McPherson, D. T.; et al. In *Handbook of Biomaterials and Applications;* Marcel Dekker: New York, 1995; Vol. 2, pp 1619–1673.
7. Urry, D. W.; Jaggard, J.; Prasad, K. U.; Parker, T. M.; Harris, R. D. In *Biotechnology and Polymers;* Gebelein, C. G., Ed.; Plenum: New York, 1991; pp 265–274.
8. Urry, D. W.; Luan, C.-H.; Harris, C. M.; Parker, T. M. In *Proteins and Modified Proteins as Polymeric Materials;* McGrath, K.; Kaplan, D., Eds.; CRC Press: Boca Raton, FL, in press.
9. Urry, D. W.; Parker, T. M.; Reid, M. C.; Gowda, D. C. *J. Bioact. Compat. Polym.* **1991**, *6,* 263–282.
10. Urry, D. W.; McPherson, D. T.; Xu, J.; Gowda, D. C.; Parker, T. M. *Div. Polym. Mater. Sci. Eng.* **1995**, 259–281.
11. Nicol, A.; Gowda, D. C.; Parker, T. M.; Urry, D. W. *J. Biomed. Mater. Res.* **1993**, *27,* 801–810.
12. Urry, D. W.; Prasad, K. U. In *Biocompatibility of Tissue Analogues;* Williams, D. F., Ed.; CRC Press: Boca Raton, FL, 1985; pp 89–116.
13. Prasad, K. U.; Iqbal, M. A.; Urry, D. W. *Int. J. Pept. Protein Res.* **1985**, *25,* 408–413.
14. Urry, D. W.; McPherson, D. T.; Xu, J.; Daniell, H.; Guda, C.; Gowda, D. C.; Jing, N.; Parker, T. M. In *The Polymeric Materials Encyclopedia: Synthesis, Properties and Applications;* CRC Press: Boca Raton, FL, 1994; pp 2645–2699.
15. McPherson, D. T.; Xu, J.; Urry, D. W. *Protein Expression Purif.* **1996**, *7,* 51–57.
16. Urry, D. W.; Long, M. M.; Sugano, H. *J. Biol. Chem.* **1978**, *253,* 6301–6302.
17. Stackelberg, M. V.; Müller, H. R. *Naturwissenschaften* **1951**, *38,* 456.
18. Teeter, M. M. *Proc. Natl. Acad. Sci. U.S.A.* **1984**, *81,* 6014–6018.
19. Edsall, J. T.; McKenzie, H. A. *Adv. Biophys.* **1983**, *16,* 53–183.
20. Frank, H. S.; Evans, M. W. *J. Chem. Phys.* **1945**, *13,* 507–532.
21. Kauzmann, W. *Adv. Protein Chem.* **1959**, *14,* 1–63.
22. Urry, D. W.; Gowda, D. C.; Parker, T. M.; Luan, C.-H.; Reid, M. C.; Harris, C. M.; Pattanaik, A.; Harris, R. D. *Biopolymers* **1992**, *32,* 1243–1250.
23. Urry, D. W. *Int. J. Quantum Chem. Quantum Biol. Symp.* **1994**, *21,* 3–15.
24. Urry, D. W.; Gowda, D. C.; Peng, S. Q.; Parker, T. M.; Harris, R. D. *J. Am. Chem. Soc.* **1992**, *114,* 8716–8717.

25. Urry, D. W.; Peng, S. Q.; Parker, T. M. *J. Am. Chem. Soc.* **1993,** *115,* 7509–7510.
26. Urry, D. W.; Gowda, D. C.; Peng, S. Q.; Parker, T. M.; Jing, N.; Harris, R. D. *Biopolymers* **1994,** *34,* 889–896.
27. Urry, D. W.; Peng, S. Q.; Parker, T. M.; Gowda, D. C.; Harris, R. D. *Angew. Chem.* **1993,** *105,* 1523–1525; *Angew. Chem. Int. Ed. Engl.* **1993,** *32,* 1440–1442.
28. Urry, D. W.; Peng, S. Q.; Gowda, D. C.; Parker, T. M.; Harris, R. D. *Chem. Phys. Lett.* **1994,** *225,* 97–103.
29. Urry, D. W.; Gowda, D. C.; Peng, S. Q.; Parker, T. M. *Chem. Phys. Lett.* **1995,** *239,* 67–74.
30. Urry, D. W.; Peng, S. Q.; Hayes, L.; Jaggard, J.; Harris, R. D. *Biopolymers* **1990,** *30,* 215–218.
31. Urry, D. W.; Luan, C.-H.; Harris, R. D.; Prasad, K. U. *Polym. Prepr. (Am. Chem. Soc. Div. Polym. Chem.)* **1990,** *31,* 188–189.
32. Luan, C.-H.; Jing, N.; Parker, T. M.; Gowda, D. C.; Urry, D. W. "Lowering Hydrophobic Folding Temperature in Model Proteins by Charge Neutralization," in preparation.
33. Olsen, J. L.; Kinel, F. A. In *Naltrexone Research Monograph 28;* Willette, R. E.; Barnett, G., Eds.; National Institute on Drug Abuse: Rockville, MD, 1980.
34. Sharon, A. C.; Wise, D. L. In *Naltrexone Research Monograph 28;* Willette, R. E.; Barnett, G., Eds.; National Institute on Drug Abuse: Rockville, MD, 1980.
35. Wang, S.; Silberstein, E. B.; Lukes, S.; Robb, E.; Zou, W.-Z.; Bruno, L.; Heyd, T. J.; Waymack, J. P.; Alexander, J. W. *J. Int. Soc. Burn Inj.* **1986,** *12,* 312–317.
36. Heggers, J. P.; Robson, M. C.; Zachary, L. S. *J. Burn Care Rehabil.* **1985,** *6,* 466–468.

21

Synthetically Designed Protein-Polymer Biomaterials

Joseph Cappello

Materials science has yet to exploit the full properties potential of the vast array of natural protein-based structural materials that exist in nature. Previously, the availability of protein materials has been limited to only a few products that could be readily extracted from their natural sources. Now, technologies exist for not only manufacturing existing natural proteins using genetic engineering, but also for designing all new proteins with properties not available in nature. We have developed a technology for designing and producing "protein polymers", high-molecular-weight proteins composed of tandemly repeated sequential blocks of amino acids. These protein polymers display a variety of useful materials properties, including gelation at physiological conditions, film and fiber formation, adhesion to synthetic surfaces, controlled resorbability, and biological recognition.

A "protein polymer" is a polypeptide chain composed of amino acids which are arranged in a sequential manner and repeated tandemly to produce complex, high-molecular-weight repetitive proteins. The unit of repetition is a block of amino acids or a set of blocks of amino acids which define specific structural or functional units. This is in contrast to poly(amino acid)s where a single amino acid or a mixture of amino acids are polymerized producing homopolymers or random copolymers. The structure of poly(amino acid)s is defined statistically by determining the average composition and molecular weight of a mixture of polypeptide chains. Protein polymers also differ from typical sequential polypeptides which are produced by oligomerization of short sequences of relatively few amino acids. In sequential polypeptides the sequence of each peptide block is constant. However, the oligomerization of more than one block produces a mixture of protein chains whose composition, order of blocks, and molecular weight are not consistent. "Protein polymer" describes a set of sequential polypeptides whose exact amino acid sequence is specified at every amino acid position along the chain. The unit of repetition can be as long and complex as theory

Copyright © 1997 American Chemical Society

and current state-of-the-art chemistry and molecular biology allows, until the length of the repeating unit is equivalent to the molecular weight of the final product, in which case the product is not a protein polymer but a protein of unique sequence. Protein polymers also differ from natural repetitive proteins in that except for a few examples, the units of repetition of natural repetitive proteins are not exact. The repeating blocks conform to a consensus amino acid sequence, but any one block may differ from the consensus sequence by extensive substitution of specific amino acids within the block. In some cases, the lengths of the repeating blocks may also vary.

Some of these differences are illustrated by the following examples:

1. Silklike poly(amino acid) copolymer, $(Gly_3, Ala_2, Ser_1)_n$. Random copolymer of glycine (G), alanine (A), and serine (S) in an approximate compositional ratio of 3:2:1 depending on the relative reactivity of each amino acid: amorphous microstructure, no definable unit of repetition.
2. Silklike sequential polypeptide polymer, Beta-Silk, $(GlyAlaGlyAlaGlySer)_n$. Sequential block copolymer of glycine, alanine, and serine in an exact compositional ratio of 3:2:1 and an exact spatial sequence: highly crystalline silklike microstructure; however, no definable chain folding periodicity. Unit of repetition is 6 amino acids.
3. Silklike protein polymer with biologically active cell adhesion and chain folding, ProNectin F, $[(GlyAlaGlyAlaGlySer)_9GlyAlaAlaValThrGlyArgGlyAspSerProAlaSerAlaAlaGlyTyr]_n$. Sequential block copolymer of glycine, alanine, serine, valine (V), threonine, arginine (R), aspartate (D), proline (P), and tyrosine (Y) in an exact compositional ratio of 30:23:11:1:1:1:1:1:1. Highly crystalline silklike microstructure with uniform 12-nm periodic chain folding (*1*). Unit of repetition is 71 amino acids.

Life is dependent on the structure and function of proteins. Nature has chosen proteins as the macromolecules through which a wide variety of specific recognition, binding, and conversion of chemical species occurs. Proteins accomplish these functions by having the ability to attain unique three-dimensional conformations that allow them to control the spatial positions of diverse chemically functional groups. This control is accomplished by using 20 chemically distinct monomers, L-α-amino acids, and being able to precisely position them along the chain, which may extend for 1000 amino acids or more. Nature has empowered living cells with the molecular synthetic machinery to conduct this task and has packaged the machinery in such a way that it can be replicated and transferred almost unerringly to the individuals of a propagating cellular lineage.

A modest cost is associated with synthesizing proteins with this degree of precision. The expenditure of that cost is easy to rationalize for metabolic enzymes that accelerate chemical reactions several-thousandfold or proteins that are involved in fighting off disease. However, especially in higher animals, nature also has chosen to use this precision in the production of proteins that make up what seem to be bulk mechanical materials such as collagen and elastin for internal connective tissue; keratin for external body protection such as hair, nails, and hooves; and in insects and spiders, silks for prey capture, cocoon housing, etc. In fact, in some spiders, the production of a completely proteinaceous web is so taxing on its energy resources that it must supplement its daily nutritional intake by ingesting the web and reusing the amino acids for producing the next day's web. Is this just nature's folly or is there more sophisticated utility in these "bulk" materials than is currently understood? Clearly the roles of structural materials in the context of the complex biological systems in which they exist are only superficially understood and, unfortunately for biomaterials science, are insufficiently studied.

Regardless of our state of knowledge, materials are being introduced into biological systems, specifically as injectable or implantable devices, to improve the quality of life. These materials are chosen based on their existing profiles of properties or are synthetically modified or processed to obtain the properties that are thought to be desired. Very few materials are proactively designed with a set of desired properties in mind. Very few synthetic polymer systems allow one the capability to design novel properties. At best, existing properties can be modified, often at the expense of other properties or cost.

Chemical synthesis of DNA and modern biotechnology supply the means to produce large quantities of synthetically designed proteins. Even though a majority of the products of biotechnology are drugs, the prospect for producing materials from proteins is clearly demonstrated by natural protein-based materials such as silk, elastin, collagen, and keratin. Using proteins as an underlying foundation, we can design and build materials using the same amino acid components that nature uses. Natural proteins are our models. We have demonstrated that we can "borrow" segments of natural proteins that confer specific biological or chemical functions and incorporate them and their functions within a synthetically designed and produced protein polymer. The nature of these materials and their ability to be engineered by design present an opportunity for their application to, among other things, the biological delivery of therapeutic agents. The presentation that follows will introduce the technology for producing protein polymers and will present examples of how precise control over their structure can impact on their biological properties. Although not presented here, specific protein polymers have been processed into materials with varied forms and properties (*2*). These materials are currently in preclinical evaluation, testing their performance in wound healing, surgical repair, and drug delivery.

Production of Synthetically Designed Protein Polymers

For new, synthetically designed proteins to find utility as drug delivery materials, they must successfully compete with materials produced by other methods. The method of polymer production affects the final product in three ways:

- consistency of properties,
- quality assurance, and
- cost.

The two methods for producing sequence-controlled, high-molecular-weight proteins are chemical synthesis and biological production. In the chemical synthesis methods, amino acids can be sequentially linked into chains by stepwise chemical reactions (*3*). These methods are commonly used to produce peptides of 5–60 amino acids in length. Depending on the efficiency of each step in the reaction sequence, chemical synthesis produces a mixture of reaction products that must be eliminated from the final product by purification. Because these side-reaction products are all very similar in chemical composition to the desired product and may vary only insignificantly from each other in molecular weight, the purification is problematic. Currently, the purification of synthetic peptides is accomplished by high-performance liquid chromatography, which is an expensive and low-throughput method. The complexity and high cost of chemical peptide synthesis have precluded its use in the production of bulk protein-based materials.

By limiting the chain length to less than about 10 amino acids, solution-based chemical synthesis may be employed to produce sequence-controlled blocks, which may then be polymerized in a subsequent step to produce sequential polypeptides. Applied to the production of synthetic analogs of elastin protein, Urry et al. demonstrate the utility of these methods in Chapter 20. As for all chemical synthetic methods, the cost of production is strongly dependent on the length and complexity of the sequence block used for polymerization. If the polymer is to incorporate different sequence blocks, there is no way to precisely control the distribution of those blocks throughout the product. Even though useful protein-based materials have been prepared by polymerizing the products of chemical synthesis, the full extent of protein design cannot be exploited when specification of chain sequences is restricted only to short peptides.

Biotechnology offers an efficient means for producing proteins of precise chemical composition by the method of biological expression. This method commonly uses microorganisms for production, because they are already capable of synthesizing high-molecular-weight, sequence-controlled proteins. By using recombinant DNA technology, a microorganism can be genetically programmed to produce a new protein. The sequence of the new protein is specified by a chemically synthesized gene that is introduced and stably maintained within the microorganism. While the microorganism is fed inexpensive nutrients such as glucose and ammonium or nitrate salts, genetic and biochemical parameters are regulated to control production of the new protein. Once produced, the new protein is purified from a mixture of products contributed by the microorganism. Because the new protein may be chemically distinct from the other components of the mixture, its novel properties and composition can enable its purification using conventional protein separation methods. Many of these methods provide high throughputs. The scale of production is limited only by the size of the production vessel and by the capability of the processing equipment.

In contrast to the production cost of chemical synthesis, the production cost of biological expression is not dependent on the complexity of the design. Regardless of its amino acid composition, the cost of producing a protein by biological expression depends only on the cost of raw materials, scale of production, and productivity.

Several disadvantages apply to products produced by recombinant DNA methods as compared to chemical synthesis. Biological expression only efficiently uses the 20 natural α-amino acids (in the L-configuration) that are commonly found in natural proteins. However, Tirrell demonstrated that at least partial incorporation of some non-natural α-amino acids can be obtained in bacterial expression systems. Not every amino acid sequence will lead to an efficiently produced product. Early reports documented the failure, for example, to produce sequential polypeptides consisting of sequence blocks modeled on collagen repeating amino acid sequences (*4, 5*). Our laboratory has demonstrated the efficient production of five-protein polymers consisting of repetitive collagen-like blocks. Currently, more than 50 different protein polymers with ever-increasing diversity of amino acid compositions and sequences have been produced. Finally, biological production requires that the product be purified from a complex mixture of contaminants, some of which may be biologically active such as pyrogen. Even though the processes for purification of specific products will depend on their individual properties, the fact remains that a large number of biotechnology products have successfully passed purity and safety requirements and have been approved by the U.S. Food and Drug Administration for use in humans.

Programming Microorganisms To Produce Foreign Proteins

The programming of a microorganism to efficiently produce a new protein requires genetic manipulation of the microorganism, diverting energy from its normal functions to the production of the new protein. This programming can be accomplished only when the normal functioning of the microorganism is well characterized. Reprogramming requires knowledge of the nutrient requirements, optimal growth conditions, and organization of the genome, which includes information regarding the presence of DNA plasmids, the organism's chromosomal replication, and the configuration of its genes with regard to regulatory features such as promoters and translation signals. In addition, methods for manipulating and introducing foreign DNA into the cell must exist. Selectable genetic markers for detection and maintenance of the manipulated DNA must be available, as well as suitable gene promoters for the efficient transcription and translation of new genes. Fortunately, over the past 20 years, this information has accumulated for several microorganisms, plants, fungi, and animals and has enabled their use for the production of engineered proteins.

Escherichia coli strain K12, a nonpathogenic strain of a common colon bacterium, is the workhorse of engineered protein production. Although a variety of hosts are now available, *E. coli* is still the most widely used host for the expression of heterologous proteins in recombinant systems. Because the genetic structure of *E. coli* is well characterized, the genetic manipulations required to optimize protein expression are generally understood. Production using *E. coli* and other microorganisms is accomplished in large scale by the process of fermentation. Fermentation uses the biochemical processes occurring inside the microorganisms to convert raw materials into desired products.

Currently, fermentation procedures are used to produce materials for diverse industrial, medical, agricultural, and research applications. In the batch fermentation method, the first step is the addition of the nutrient media to the fermentor. The media is sterilized and the inoculum, a small liquid volume containing the actively growing microorganism, is added. Aerobic fermentations proceed under strict control of all physical and chemical parameters prespecified by standard operating procedures. Temperature and pH are monitored and maintained within a narrow range. The concentrations of carbon source, dissolved oxygen, nitrogen source, and essential nutrients are established at specified levels. Initially, the microorganisms experience a lag phase where they acclimate to the fermentor environment and do not increase in number. Following acclimation, cell division occurs and the microorganisms begin their exponential growth phase, reaching their maximum specific growth rate. This maximum specific growth rate results in the increase of biomass within the system. Because of either the depletion of nutrient sources or the increase in metabolic byproducts that inhibit the growth of the microorganisms, the microorganisms enter stationary phase, where they cease to increase in number. Ultimately, the microorganism progresses to death phase, where the rate of cell degradation is greater than the rate of new cell division. Schemes for suspending the growth cycle in the exponential growth phase or delaying the onset of the death phase have been employed that consist of continual dilution of the culture and continual harvest of product. The choice of fermentation strategy depends greatly on the specific properties of the product (i.e., whether the product is produced intracellularly or secreted).

Once the foreign product has been produced by fermentation, it must be separated and purified from the other products of the fermentation. A variety of procedures are used to purify protein polymers. In general, these procedures are the same ones used for the purification of other recombinant proteins from microorganisms.

Polymer Gene Design and Stability

Because of the peculiarity of their polymeric compositions, the biological production of protein polymers using recombinant DNA methods requires unique considerations. When large sections of foreign DNA are introduced into *E. coli*, investigators have observed that portions of clones containing natural sequences with tandem repeats are spontaneously deleted. Tandem repeats in DNA have been reported to be unstable, suffering loss or rearrangement through deletion and recombination (*6–9*). Doel et al. (*10*) reported the first attempt to create a synthetic gene encoding a polypeptide of repeating amino acid sequence. In this case, DNA plasmids containing repetitive genes of greater than 1,000 base pairs were introduced into *E. coli* cells, but the largest genes stably obtained were approximately 900 base pairs in length (encoding a maximum polypeptide length of 300 amino acids). Many researchers in the field of recombinant DNA technology have concluded that repeating DNA cannot be cloned and stably maintained in *E. coli*.

Because repetitive proteins must be encoded by repetitive genes, the production of protein polymers by biological expression requires that the genetic instability of repetitive genes be overcome. Genetic instability compromises the following:

- the ability to obtain repetitive genes greater than 1000 nucleotide base pairs in length (encoding proteins of 30 kDa or more) and
- the stable maintenance of the genes through the generations of exponential cell growth required for large-scale fermentation.

The most successful strategy has been one where the overall repetitiveness of the gene has been minimized (*11*). The DNA sequence of a repetitive gene can be designed to minimize direct tandem repetition by exploiting the degeneracy of the genetic code. One method for minimizing tandem repetition is to use a different codon for an amino acid when it is used more than once within an oligopeptide block. Another method is to increase the length of the monomer gene segment such that it encodes multiple oligopeptide blocks. In this way, adjacent, identical oligopeptide blocks can be encoded by nonidentical DNA sequences. To avoid or to reduce the deletion of tandem repeats, most researchers prefer to use microorganisms deficient in the deletional mechanisms of homologous recombination. Mutant strains of *E. coli* that are recombination deficient are available, as are strains deficient in a variety of other DNA modifying functions. However, it is still unclear whether using these strains has a definitive effect on gene stability (*4, 5, 11*).

Factors Influencing Microbial Production

After a desirable polymer gene has been introduced into a bacterial cell and has passed stability criteria, a culture is grown from this cell. After extensive characterization, this culture will serve as a master cell production stock and will be the source of the strain used for all subsequent experimentation and future productions. The master cell stock is split into multiple aliquots and can be stored in liquid nitrogen for many years. Thus, every batch of polymer produced from it will be consistent in identity over time.

The first step in the genetically controlled synthesis of protein polymers is to efficiently transcribe the polymer gene DNA into messenger RNA. Many transcriptional promoters are available that can direct gene transcription in *E. coli*, *Bacillus* sp., yeasts, and insect or mammalian cells in culture. To date, all protein-polymer expression has occurred in *E. coli* and *Saccharomyces cerevisiae* (*11–18*). Three promoters used extensively in *E. coli* originated from bacterial viruses, the phage T7 gene 10 promoter, and two phage lambda promoters, P_L and P_R. These three promoters have sufficient transcriptional activity to compete well with other cellular genes, thereby producing enough RNA to allow efficient production of the protein polymer.

Once transcribed, the messenger RNAs encoding the protein polymers must be stable. Also, these messenger RNAs must be able to bind ribosomes efficiently. In microbial systems, the stability of an RNA is often linked to its translation efficiency; the density of ribosomes progressing along its length may protect it from degradation by RNases. Any factors that slow down or halt the translation of that RNA into protein may encourage the premature disengagement of ribosomes from the RNA. For example, sequences such as inverted repeats, which can form internally base-paired "knots" in the RNA chain, can block or stall the progression of ribosomes down the chain. The polymer products of such events will have a shorter chain length than the predicted full-sized product. If the sequences responsible for premature disengagement of ribosomes from RNA occur at discrete locations within the repeating portion of the gene, a family of prematurely terminated products may be produced. A discrete ladder of products that increase in molecular weight in a stepwise fashion will be observed.

The high-level production of protein polymers composed of a large number of relatively few different amino acids may lead to debilitating metabolic consequences. The moment a cell begins expressing a protein-polymer gene, a competition for biosynthetic resources occurs between the polymer gene and all other genes of the cell.

Because the polymer gene is designed for high-level expression (it contains a strong transcriptional promoter, an efficient ribosome binding site, and is present in high copy number), it dominates these resources at the expense of other cellular functions. For instance, if a resource such as the supply of a particular amino acid is exhausted, not only is the production of the protein polymer limited, but the cell's ability to make the proteins or enzymes required to replenish the shortfall may be crippled as well. One example where gene design may influence this competition occurs in the DNA codons that are chosen to encode the polymer amino acids. Many organisms use one or a few codons to encode amino acids for which the genetic code is degenerate (more than one codon specifies the same amino acid). Preferred codons usually correspond to tRNAs that the organism produces in abundance. The overuse of rare codons may cause a critical depletion in a necessary tRNA substrate needed for continued protein synthesis. Codon depletion results in both the low expression of the desired product and a decrease in other cellular products also requiring that tRNA.

To minimize some of these problems, all current protein-polymer production has been accomplished by using expression promoters that are under inducible control (*10, 11, 13–19*). In inducible systems, the production cells can be grown and manipulated in the absence of expression. In this way, expression of the product will not usurp essential resources required for efficient growth. When the cell number in the culture has reached or is approaching a maximum, protein-polymer production is induced by providing an appropriate physical or chemical signal. Production will then continue until the accumulated product reaches maximum levels.

Polymer Purification

Methods for isolating a protein-polymer product from the surrounding cellular host components must necessarily be dependent on the properties of the product. Because protein polymers are so different in their amino acid composition and sequence from natural proteins, effective purifications have been accomplished by using surprisingly simple protein extraction and separation methods. For example, the purification scheme for silklike protein polymers is based on the extreme resistance of these products to solubilization in aqueous systems (*20*).

Protein polymers can also be purified by virtue of atypical charge distributions. The overall charge of a protein chain is determined by its negatively and positively charged amino acids. The ionization state of the charges varies depending on the pH of the solution in which they exist. The protein is likely to remain soluble as long as it contains a net charge. If a protein is placed in an aqueous environment where the pH approaches the isoelectric point of that protein (the pH at which the net charge of the protein is zero), it is likely to precipitate. Because most natural proteins contain both negative and positive charges, their isoelectric points will lie above pH 4, the pK_a of the most acidic side chain, glutamic acid. By decreasing the pH from neutral to near pH 4, many proteins will probably pass through their isoelectric point and become denatured. During this process, most proteins will expose hydrophobic domains to the exterior of their structure. These domains are inherently attractive to each other in aqueous solution and may induce other proteins, which may be marginally soluble in acidic conditions, to precipitate as well. Therefore, using acid precipitation, protein polymers that contain very few or no positively charged amino acids may remain in solution, whereas most other cellular proteins are precipitated.

Because the composition of a protein polymer is controlled at the DNA level, products can be designed to be purified easily by affinity methods (*21*). Affinity purification involves adsorption to a specific matrix for which the product displays selective binding. Examples of such systems have been reported for the purification of a variety of recombinantly expressed products. Many small segments of proteins have been identified that bind selectively to low-molecular-weight ligands, compounds such as sugars and metal ions. These protein segments can be fused directly to the product of interest by modifying the gene to include sequences encoding such a protein segment. The fusion product can then be isolated from the soluble cellular lysate by simply mixing it with a particulate matrix containing the immobilized ligand. Chromatography resins such as cross-linked dextran, agarose, or silica are commonly used. The sugar or metal ion is attached to the surface of the particles such that it may be efficiently bound by the fusion product. Nonspecifically bound proteins are washed away, and the fusion product is eluted from the resin either by competition with excess free ligand or by altering the conditions such that the binding function loses its activity. This process can be accomplished by changes in pH or ionic strength, or by the addition of denaturing agents such as urea or guanidine hydrochloride. Affinity purifications require that the fusion product be soluble. These methods are ideal for quickly purifying unknown products not yet sufficiently characterized for a purification scheme to be based on their properties. However, these methods are generally low in throughput and require the adsorption of product to a particulate resin with a limited binding capacity.

These examples illustrate how a variety of techniques conventionally used for protein purification are applicable to the purification of protein polymers, in spite of their atypical compositions. Moreover, the peculiarity of their compositions can be an advantage for simple separation of protein polymers from other natural cellular proteins. Microbially produced protein polymers have been purified to greater than 95% by using conventional, high-throughput methods. Purities in excess of 99% have been achieved by using more sophisticated bulk separation methods such as affinity adsorption.

Composition and Molecular Analysis

A number of parameters need to be assessed to determine the quality of a purified protein-polymer product. Both the identity of the product and the quality must be determined. Also, the identity and quantity of contaminants must be determined. Depending on the intended use of the product, acceptable criteria for these determinations must be established. The testing procedures used to assess these parameters must be validated to assure that results are statistically significant. Usually, test methods with appropriate reference standards are used for confirmation.

The identity of a protein polymer can be confirmed by amino acid composition, terminal amino acid sequence, and mass spectrometry. The nonprotein content of the product (the levels of compounds such as nucleic acids, lipids, and carbohydrates) can also be assessed by using standard assays. Water and other volatiles can be determined by performing mass spectrometry of the evolved gases.

If the protein chain has a high propensity for conformational order, determining the molecular weight of a novel protein polymer is not a trivial matter. Whether by gel permeation chromatography or sodium dodecyl sulfate–polyacrylamide gel electrophoresis, apparent molecular weights for some classes of protein polymers, as compared to typical reference standards, give values consistently different than theoretical. Recently, the exact molecular mass of a microbially produced protein polymer was determined by plasma desorption mass spectrometry (*22*). The experimental value was precise enough to allow the prediction of two amino acid changes in the polymer primary sequence. By nucleotide sequencing of the gene template, the two amino acid substitutions were subsequently confirmed. The corrected theoretical molecular weight agreed with the experimental value within two atomic units.

Control of Properties Through Sequence Specification

Even though the potential to create new protein compositions of varied characteristics through the design and production of protein polymers is indisputable, the utility of such products in drug delivery or any other application must be based on the actual properties that distinguish these materials from other polymers. Unfortunately, definition of the properties potential of protein-polymer materials is in its infancy. Examples of two families of protein polymers that we have produced and characterized are presented to illustrate the basic principals by which protein-polymer properties can be engineered.

ProNectin F

Protein-polymer design allows the direct incorporation of segments of natural proteins with known, well-characterized biological functions. Our laboratory created ProNectin F, the first protein polymer available for commercial use (*17*). We designed the ProNectin F polymer by using two oligopeptide blocks, a six amino acid block from silk fibroin (GAGAGS) and a 10 amino acid block from human fibronectin (VTGRGDSPAS). One oligopeptide block provided the structural properties of silk, whereas the other block provided the cell-attachment activity of fibronectin. Two short blocks of amino acid sequence were used between the silk blocks and the fibronectin block for structural continuity (GAA and AAGY). The blocks were configured into a repeating gene monomer such that one cell-attachment block would occur after every nine silklike structural blocks (*see* example 3). This string of blocks was repeated 13 times to yield a protein-polymer chain of 980 amino acids in length with an expected molecular weight of 72,738.

ProNectin F was designed to be used as a coating reagent for plasticware used for culturing animal cells in the absence of serum-supplemented media. The need for such reagents is now partially satisfied by natural cell-attachment proteins extracted from animal tissues or blood. While useful in this respect, natural proteins are not ideal in this application because they suffer from low specific activity, poor chemical stability, lot-to-lot variability, and potential viral contamination.

ProNectin F has the following key features, allowing it to function as an ideal cell culture coating:

- It efficiently presents the natural cell-attachment sequence of human fibronectin.
- It adheres to plastic surfaces such as polystyrene without denaturation.
- It remains shelf stable at room temperature for at least 1 year.
- It remains insoluble and adsorbed to the plastic in culture media at 37 °C.
- It withstands sterilization by autoclaving or irradiation without loss of activity.

Before producing the peptide blocks of ProNectin F, we generated predictions of the polymer properties through computer simulations of the possible conformations. The peptide blocks used in the design of ProNectin F were chosen on the basis of structural and functional studies of the peptides. The hexamer block GAGAGS from silk fibroin was chosen because of its tendency to produce stable, hydrogen-bonded beta sheets. Because of the thermal stability of protein structures containing this hexamer block, it could reliably confer both the capacity to be sterilized by autoclaving and the capability for long-term storage at room temperature both in the dry state and in aqueous solution. We determined the number of hexamers in a row to be at least four so that the solubility in water of the resulting polymer would be low and the number of alanine side chains (methyl groups) would be high. These characteristics would promote good adsorption to hydrophobic polystyrene surfaces and would maintain the protein on such surfaces, even in aqueous media.

To design an assembly of hexamer blocks into a chain that presented and configured the cell-attachment block properly, we made a set of assumptions. These assumptions allowed us to simulate the possible structures of ProNectin F by computer. Because the sequences of all monomer blocks making up the ProNectin F polymer were to be identical, they were treated as conformationally equivalent. Our modeling efforts could thus be confined to a sequence of less than 71 amino acids rather than the full-length polymer of 980 amino acids. This assumption facilitated the computer-modeling simulation of the cell-attachment sequence, which allowed us to make adjustments to the sequence to accommodate the block within the silklike protein scaffold. The computer simulations of the 17 amino acid cell-attachment block of ProNectin F demonstrated that while multiple conformations were possible, they all presented the cell-attachment block in an accessible configuration.

The amino acid sequence of the ProNectin F monomer was translated into a DNA sequence encoding the block structure shown in Figure 21.1. The amino acid blocks used in the design of ProNectin F and their strategy for assembly into the full length polymer are diagrammed. (Amino acids are abbreviated using the single letter code.) Oligonucleotides encoding this amino acid sequence were synthesized and used to construct a polymer gene. The polymer gene was introduced in *E. coli*, and isolated strains were shown to produce discrete ProNectin F products. One of these strains was selected for all future productions and analyses.

Once the chemical structure, purity, and identity of ProNectin F were confirmed, its physical, chemical, and biological properties were evaluated. Using natural fibronectin as a standard, we evaluated ProNectin F as a cell-attachment coating for polystyrene according to the performance criteria previously discussed. ProNectin F was shown to promote mammalian cell attachment at coating concentrations as low as 160 ng/cm^2. In aqueous solutions with polymer concentration of less than 1 μg/mL, ProNectin F adsorbed to polystyrene and promoted cellular attachment with a variety of cell types from different animal species. With some of these cells, ProNectin F exceeded the activity of natural cell-attachment proteins, even exceeding the activity of fibronectin, from which the ProNectin F cell-attachment sequence block was modeled. At room temperature, the polymer coating was optically clear and had a dry shelf life of at least 1 year. The ProNectin F powder was sufficiently thermally stable to withstand autoclaving at 120 °C with no loss of activity.

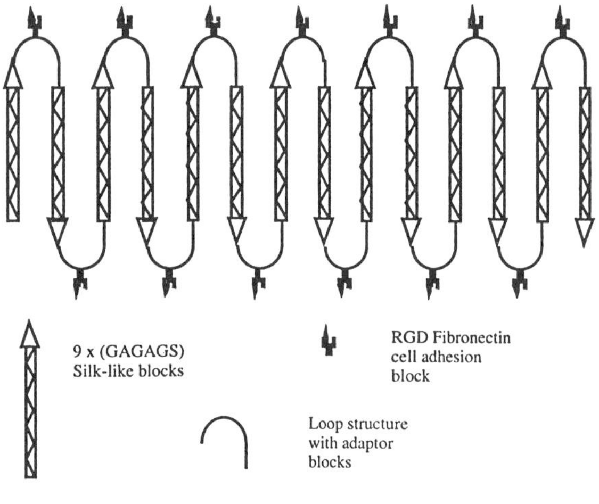

FIGURE 21.1. Diagram of ProNectin F block structure.

The ProNectin design has allowed the production of a variety of protein polymers with other biological activities simply by substituting the fibronectin cell adhesion block with appropriately configured blocks from other natural adhesion proteins. We have produced ProNectin polymers containing adhesive blocks from human laminin, which reproduce the biological activity of synthetic peptides containing the same laminin sequences. In all cases, the ProNectin backbone structure provides good surface coating properties and stable shelf life.

ProLastin Polymers Provide Controlled Resorption

The rate at which an implanted material resorbs or biodegrades within the body can be a major factor in determining its utility in drug delivery. Polylactide, polyglycolide, polyanhydrides, and polyorthoesters are examples of synthetic polymers that degrade in the body by hydrolysis. Collagen, glycosaminoglycans, and hyaluronic acid are examples of natural implantable materials that resorb at least partially by enzymatic degradation. Although, their physical structures influence their resorption, within the context of a single physical form, the rate of resorption of a material can only be modified by changing its chemical structure.

By being able to specify the entire amino acid sequence of a high-molecular-weight protein chain, we have shown that a set of protein-polymer materials can be created whose resorption rate can be systematically varied. They can be formed as both gels and solids, and in both states they have been compounded with exogenous drugs, cytokines, and growth factors. The materials can be processed from a variety of solvents including deionized water or saline; however, they are especially easy to formulate as dense films by evaporation from formic acid solution.

These protein polymers are called ProLastins, and they consist of silklike and elastin-like peptide (SELP) blocks in various block lengths and compositional ratios (*13, 23*). The silklike block consists of the six amino acid sequence GAGAGS (*24*). The elastin-like block consists of the five amino acid sequence, VPGVG (*25–27*). ProLastins are consistent in their periodic alternation of silk and elastin domains but they vary in the number of silk or elastin blocks within the domains (Figure 21.2). Hydropathy plots using the Kyte–Doolittle hydrophobic index algorithm (*28*) of five ProLastin polymers indicate the relative hydrophobicity of sliding windows of 11 amino acids along the entire length of each ProLastin polypeptide chain. The predicted amino acid sequence of each protein chain is graphically displayed as a function of its average hydrophobicity at each point along the chain. For each polymer, hydrophobicity increases in the vertical direction on the *y* axis. The segments at the beginning and end of each polymer chain that have negative hydrophobicity (below the *x* axis) are the short, nonrepetitive head and tail sequences that are produced as a consequence of the biological expression system and can be post-translationally cleaved, if desired. The plateaus on the *x* axis represent the silklike block domains of the polymers. The plateaus above the *x* axis represent the relatively more hydrophobic elastin-like block domains.

To facilitate the implantation and retrieval of specimens after implantation, ProLastin films of approximately 0.05–0.1-mm thickness were produced by solvent evaporation from formic acid and laminated to 1-cm diameter polyethylene discs. Specimens were produced from six ProLastin compositions and denatured collagen protein (DCP) produced identically as for the ProLastin films. Bovine collagen (fibrillar form, lot number 921101) was obtained from Colla-Tec, Inc. (Plainsboro, NJ). It was completely solubilized in 88% formic acid producing a clear but viscous solution.

Each disk was implanted subcutaneously in the back of male Fisher rats (CDF(344)/CrlBr: VAP Plus; 150–175-g body weight; from Charles River Laboratories, Raleigh, NC) such that the protein film was in direct contact with the muscle tissue. The specimens remained in the animals for 1 week, 4 weeks, and 7 weeks postimplantation. At each time interval six specimens per polymer group were retrieved for protein analysis. Additional specimens from each group were evaluated for tissue reaction by histology.

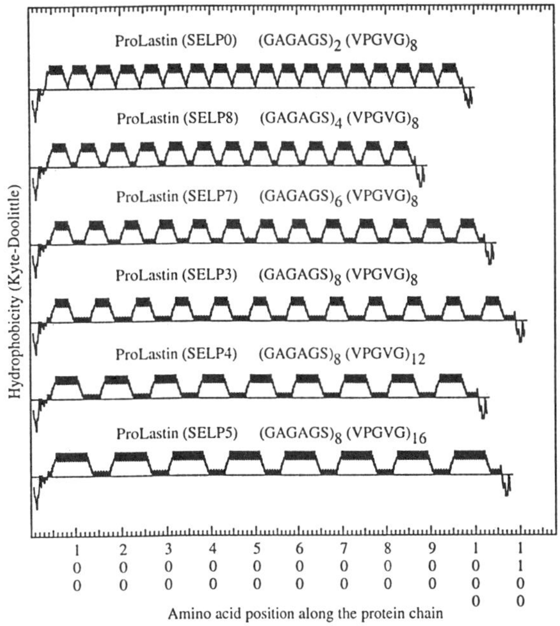

FIGURE 21.2. Diagram of ProLastin copolymer block structures.

Nonimplanted and retrieved specimens were analyzed to determine the mass of ProLastin film contained per specimen. Amino acid analysis was performed on each specimen. The mass of ProLastin film present on each specimen was determined. The amino acid contribution of the ProLastin protein was estimated based on the total content of the amino acids glycine, alanine, serine, valine, and proline, which for the pure polymers is >95%. Other amino acids potentially contributed by extraneous protein deposited onto the specimens during residence in the body were minimal and were excluded from these analyses. Average ProLastin film mass for nonimplanted specimens was determined from the same batch of specimens used for implantation. Average ProLastin film mass for retrieved specimens was similarly calculated except that replicates having values greater than 2 standard deviations from the mean were thrown out. Deviations were in many cases due to partial retrieval of specimens that had fragmented in the tissue after implantation and may not reflect true resorption.

Resorption analysis was conducted statistically by analyzing four specimen groups:

- nonimplanted,
- 1 week postimplantation,
- 4 weeks postimplantation, and
- 7 weeks postimplantation.

Six explant specimens were retrieved from each polymer group and control groups.

Figure 21.3 displays the remaining ProLastin film mass as a function of implant time. DCP indicates films produced from denatured collagen protein. Each bar represents the mean of at least four specimens. Error bars indicate one standard deviation around the mean. The results indicate that on implantation, SELP0 and DCP essentially resorbed by 1 week, falling below 4.3 and 2.3% of their nonimplanted masses, respectively. SELP8 and SELP3 resorbed by 7 weeks and had mean values of 18.1% and 58.2% remaining, respectively. SELP4 and SELP5 films showed no evidence of resorption by 7 weeks.

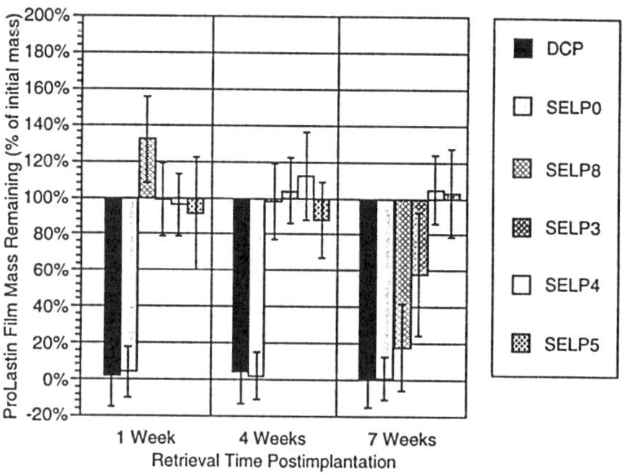

FIGURE 21.3. Diagram of resorption of ProLastin films.

Changes in the block composition of ProLastin polymers cause a variation in the resorption characteristics of films produced from these polymers. Faster resorption correlates with compositions containing domains of silklike blocks less than eight. SELP0 and SELP8 have repeating domains of 2 and 4 silklike blocks, respectively. SELP3, SELP4, and SELP5 all consist of domains containing 8 silklike blocks. The total content of silklike blocks in the copolymer composition does not correlate with resorption rate. The silklike block content of SELP8 is 35.3% of the total amino acids of the polymer. SELP5 and SELP4 silklike blocks account for 35.7% and 42.2% of the total amino acids of the polymers, respectively. SELP8 and SELP5 have identical elastin to silk block ratios of 2.0. Even though they are very similar compositionally, SELP8 films resorbed quickly, whereas SELP4 and SELP5 films did not resorb in 7 weeks. The lack of resorption of SELP4 and SELP5 films at 7 weeks postimplantation correlates with repeating domains containing greater than 8 elastin-like blocks. Although their silklike block lengths are identical, SELP4 and SELP5, with elastin-like block lengths of 12 and 16, resorb to a lesser degree than SELP3, which has an elastin-like block length of 8.

These data indicate that control of resorption rate of ProLastin polymers may be achieved by precisely controlling the number of tandem silklike and elastin-like blocks contained per repeating domain. Apparently, this control cannot be obtained simply by adjusting the compositional ratio of these blocks in the final copolymer. Faster resorption of ProLastins can be accomplished by decreasing the number of silklike blocks per repeating domain to less than eight. Slower resorption can be obtained by increasing the number of elastin-like blocks per repeating domain to greater than eight. Equivalent resorption and different material or biological properties may possibly be obtained by doing both. These data indicate that the specific amino acid sequence of the protein-polymer chains affects the rate of resorption to a greater extent than does their average amino acid content.

These studies do not address the mechanism by which ProLastin films resorb. The polyamide backbones of these proteins should not hydrolyze under physiological conditions. Our preliminary studies showed that except for SELP0, which dissolved in saline and serum at 37 °C, all ProLastin films were stable in vitro for at least 1 week in serum at 37 °C. More likely, enzyme-mediated mechanisms are involved. If digested by extracellular proteases, a process of surface erosion of the cleaved peptide fragments would most likely result, and a steady progression of resorption would occur from the exposed surfaces inward.

Discussion and Conclusions

Protein polymers cannot be categorized simply as a single new class of polymers. The properties that they can display are diverse. The number of distinct protein-polymer compositions that can be produced using the 20 naturally occurring amino acids is astronomical. For example, a relatively simple set of protein polymers composed of a single amino acid sequence block of five amino acids in length has 20^5, or 3.2×10^6, possible compositionally distinct members. If the block length doubles, the number of possible compositions increases to 1.02×10^{13}. If one combines different amino acid sequence blocks, for example one providing structural properties and one providing enzymatic cleavage, an endless number of these block copolymers can be produced that may vary in their bulk properties and their susceptibility to degradation.

Nature proves the utility of being able to stipulate the type and position of amino acids along a protein chain. A single change in one amino acid out of as many as a thousand can alter or destroy the structure and function of a natural protein. As protein designers, we have insufficient knowledge and means to design synthetic protein chains as nature does, specifying many hundreds, even thousands, of amino acids in a chain and achieving predictable properties. Fortunately, as protein-polymer designers, we do not have to. Our task is reduced to designing short blocks of amino acid sequence from 3 to 15 amino acids in length, which when combined into a polymer will provide desirable properties.

This task has been relatively manageable because all of the examples of protein polymers produced to date have exploited the sequences of natural proteins. Silk, elastin, collagen, and keratin proteins are the natural models on which the first generation of protein polymers has been based. Sequence blocks from these proteins, 3–15 amino acids in length, have been shown to be capable of reproducing specific characteristic structural features of the intact proteins from which they are derived. Every protein of nature is a collage of microproperties that emanate from its primary amino acid sequence. Assigning which blocks of amino acids are responsible for which properties of an intact protein is done with the help of computer modeling. With a certain degree of confidence, computer modeling can then also be used to simulate the structural changes that might occur when amino acids are substituted within a short peptide block.

We have recently begun to explore the potential utility of protein polymers in the delivery of drugs. Our current polymers have been designed to display simple properties. The ProLastins (silk–elastin block copolymers) are designed to be chemically deficient, containing no readily reactive side chains (the serine hydroxyl groups are the only marginally reactive group contained in the repetitive sequence blocks). They are readily soluble in aqueous solution and can be mixed with pharmaceutical agents and injected into the body. At body temperature, they form gels with varying molecular permeability without the need for chemical cross-linking. Depending on their precise arrangement of silk and elastin blocks, their susceptibility to degradation can be systematically adjusted providing resorption times from 1 week to many months after implantation. The materials are biocompatible and cause no undue tissue reaction. The ProLastins are an ideal framework on which to build more complex, functional materials.

We have demonstrated that biological recognition can be built into a protein polymer. The BetaSilk proteins consist exclusively of silklike peptide blocks and do not interact with animal cells in any specific way. When the RGD-containing cell attachment block from human fibronectin was introduced into it, ProNectin F was created. ProNectin F is recognized by cell adhesion receptors and promotes the efficient attachment of animal cells. We have similarly produced a number of additional ProNectin polymers by substituting the fibronectin RGD sequence block with blocks from other extracellular membrane proteins such as laminin. These polymers are currently under investigation.

The ability to combine structural properties with biological recognition and control over resorption simply as a design feature of the polymer chain is not available in any other synthetic polymer system. One of the biggest advantages of protein-polymer-based materials for drug delivery is that they can be designed to participate in the inherent protein recognition and processing mechanisms that already exist in the body. In many disease states, protein processing mechanisms are either directly or indirectly affected. In several instances, the actual mechanisms of protein degradation are known to be integral to the etiology of particular diseases such as osteoarthritis and some forms of cancer. We hope in the near future to demonstrate that protein polymers can effectively deliver drugs to diseased tissues where atypical protein processing or recognition mechanisms can be exploited.

References

1. Anderson, J. P.; Cappello, J.; Martin, D. C. *Biopolymers* **1994,** *34,* 1049–1058.
2. Cappello, J.; McGrath, K. P. In *Silk Polymers: Materials Science and Biotechnology;* Kaplan, D.; Adams, W. W.; Farmer, B.; Viney, C., Eds.; ACS Symposium Series 544; American Chemical Society: Washington, DC, 1994; pp 311–327.
3. Sarin, V. K.; Kent, S. B. H.; Tam, J. P.; Merrifield, R. B. *Anal. Biochem.* **1981,** *237,* 927.
4. Goldberg, I.; Salerno, A. J.; Patterson, T.; Williams, J. I. *Gene* **1989,** *80,* 305.
5. Salerno, A. J.; Goldberg, I. *J. Cell. Biochem.* **1992,** *Suppl. 16F,* 123.
6. Carlson, M.; Brutlag, D. *Cell* **1977,** *11,* 371.
7. Sadler, J. R.; Tacklenburg, M.; Betz, J. L. *Gene* **1980,** *8,* 279.
8. Gupta, S. C.; Weith, H. L.; Somerville, R. L. *Biotechnology* **1983,** *1,* 602.
9. Lohe, R. A.; Brutlag, D. L. *Proc. Natl. Acad. Sci. U.S.A.* **1986,** *83,* 696.
10. Doel, M. T.; Eaton, M.; Cook, E. A.; Lewis, H.; Patel, T.; Carey, N. H. *Nucleic Acids Res.* **1980,** *8,* 4575.
11. Ferrari, F. A.; Richardson, C.; Chambers, J.; Causey, J.; Pollock, S. C. World Office Patent Application WO88/03533, 1986.
12. McGrath, K. P.; Fournier, M. J.; Mason, T. L.; Tirrell, D. A. *J. Am. Chem. Soc.* **1992,** *114,* 727.
13. Cappello, J.; Crissman, J. W.; Dorman, M.; Mikolajczak, M.; Textor, G.; Marquet, M.; Ferrari, F. *Biotechnol. Prog.* **1990,** *6,* 198.
14. Creel, H. S.; Fournier, M. J.; Mason, T. L.; Tirrell, D. A. *Macromolecules* **1991,** *24,* 1213.
15. Urry, D. W.; Parker, T. M.; Minehan, D. S.; Nicol, A.; Pattanaik, A.; Peng, S. Q.; Morrow, C.; McPherson, D. T.; Gowda, D. C. *Proceedings of the American Chemical Society, Division of Polymeric Materials: Science and Engineering,* 203rd National

Meeting of the American Chemical Society, San Francisco, CA; American Chemical Society: Washington, DC, 1992; Vol. 66, p 399.
16. Masilamani, D.; Goldberg, I.; Salerno, A. J.; Oleksiuk, M. A.; Unger, P. D.; Piascik, D. A.; Bhattacharjee, H. R. *Biotechnology and Polymers;* Gebelein, C. G., Ed.; Plenum: New York, 1991; p 245.
17. Cappello, J.; Crissman, J. W. *Polym. Prepr. (Am. Chem. Soc. Div. Polym. Chem.)* **1990,** *31,* 193.
18. Tirrell, D. A.; Fournier, M. J.; Mason, T. L. *MRS Bull.* **1991,** *XVI*(7), 23.
19. Strausberg, R. L.; Anderson, D. M.; Filpula, D.; Finkelman, M.; Link, R.; McCandliss, R.; Orndorff, S. A.; Strausberg, S. L.; Wei, T. In *Adhesives from Renewable Resources;* Hemingway, R. W.; Conner, A. H., Eds.; ACS Symposium Series 385; American Chemical Society: Washington, DC, 1989; pp 453–464.
20. Cappello, J.; Ferrari, F. A.; Buerkle, T. L.; Textor, G. U.S. Patent 5,235,041, 1993.
21. Sassenfeld, H. M. *Trends Biotechnol.* **1990,** *8,* 88.
22. Beavis, R. C.; Chait, B. T.; Creel, H. S.; Fournier, M. J.; Mason, T. L.; Tirrell, D. A. *Proceedings of the American Chemical Society, Division of Polymeric Materials: Science and Engineering,* 203rd National Meeting of the American Chemical Society, San Francisco, CA; American Chemical Society: Washington, DC, 1992; Vol. 66, p 27.
23. Ferrari, F. A.; Richardson, C.; Chambers, J.; Causey, S. C.; Pollock, T. J.; Cappello, J.; Crissman, J. W. U.S. Patent 5,243,038, 1993.
24. Fraser, R. D. B.; MacRae, T. P.; Stewart, F. H. C.; Suzuki, E. *J. Mol. Biol.* **1965,** *11,* 706.
25. Urry, D. W.; Okamoto, K.; Harris, R. D.; Hendrix, C. F.; Long, M. M. *Biochemistry* **1976,** *15,* 4083.
26. Urry, D. W.; Okamoto, K. U.S. Patent 4,132,746, 1976.
27. Urry, D.W. *J. Protein Chem.* **1984,** *3,* 403.
28. Kyte, J.; Doolittle, R. F. *J. Mol. Biol.* **1982,** *157,* 105.

22 Biological Effects of Polymeric Drugs

Jun Liao and Raphael M. Ottenbrite

Polymeric drugs are the polymers that display a specific therapeutic activity of their own. This chapter provides a general review of the biological effects of two different types of polymeric drugs, polyanions and polycations. The biological effects include antitumor activity and antiviral activity, which occur through the immune system. The possible mechanisms for the biological activity are discussed.

Synthetic polymers have been used as biomaterials for biomedical applications and in drug delivery systems for many years. These polymers have been selected based on their physical properties, biocompatibility, and chemical inertness. However, the majority of synthetic polymers are not chemically inert, and many of them do elicit pronounced effects on biological systems. A number of these polymers display a wide range of therapeutic activities and are known as polymeric drugs (*1*).

This chapter is concerned with the biological effects of synthetic polymers that have a specific therapeutic activity of their own, without involving polymer–drug conjugates. Polymer–drug conjugates can be regarded as drug delivery systems that exhibit their therapeutic activities by means of releasing smaller therapeutic drug molecules from a polymer chain molecule. The polymeric drugs described here belong to the groups of polyanions and polycations.

Polyanions

Polyanions are polyelectrolytes with negative charges located along the polymer chain. They have a wide range of biological activities and have been investigated for use in the areas of oncology, virology, and immunology (*1–4*). The biological effects of the polyanions include antitumor and antiviral activities as well as immunological activity. The immune response of the host organism to the polyanions plays an essential role in their activities.

Specific polyanions can affect the biological functions of the host, and these polymers appear to

Copyright © 1997 American Chemical Society

behave similarly to specific proteins, glycoproteins, and polynucleotides that modulate a variety of biological responses. Among the most significant effects of the polyanions are their mitotic inhibitory effects and their role in antineoplastic process.

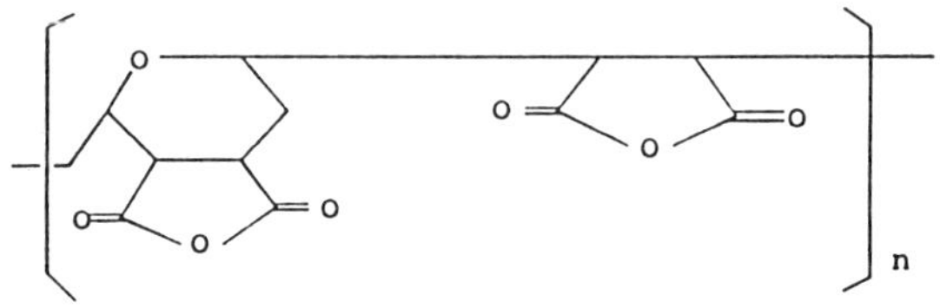

1, unhydrolyzed DIVEMA

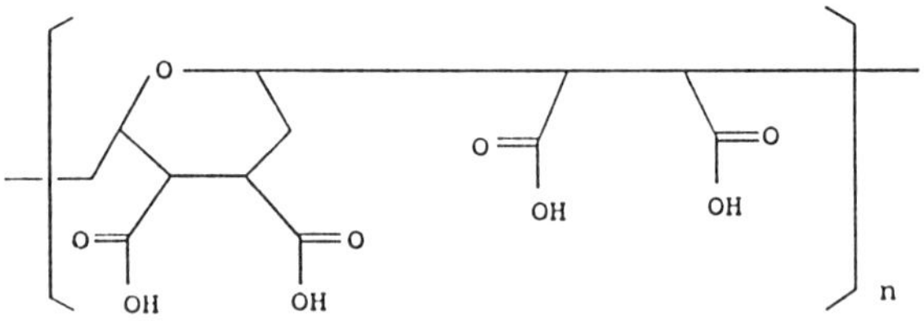

2, hydrolyzed DIVEMA

Antitumor Activity

The first clinical application of polyanions in cancer therapy was performed by Oestreich in 1910 (*5*) when he treated a cancer patient with chondroitin sulfate connective tissue extracts. Since then, a large number of the polyanions have been found to mediate antitumor effects. A wide variety of the polyanions have been studied, and a substantial number of reports were surveyed extensively in several reviews (*6–10*). Synthetic polyanions, such as pyran copolymer, poly(ethylene-*co*-maleic anhydride), poly(acrylic acid), and poly(maleic anhydride), have been studied for antitumor activity (*2, 6, 7*).

It has been suggested that the antitumor activity of polyanions is based on modulation of the immune system. Shown in Table 22.1 are some synthetic polyanions with immunostimulatory activities (*11*). Among the polyanions with antitumor activity, pyran copolymer and a low-molecular-weight copolymer of ethylene and maleic anhydride, carbetimer (NED 137), have been extensively studied.

Pyran copolymer, also known as DIVEMA (divinyl ether–maleic anhydride copolymer) (**1**, **2**), was first reported by Butler in 1960 (*12*) followed by Breslow (*13*). This copolymer immediately attracted the interest of the scientific community because of its pronounced antitumor activity. This antitumor activity appeared to be mediated via a number of mechanisms, such as the induction of interferon (IFN) production (*14*), macrophage activation (*15*), stimulation of antibody-dependent cellular cytotoxicity (ADCC) (*16*), and the cell function activation combined with the induction of IFN (*17*). The early development and initial evaluations of DIVEMA were reviewed extensively (*18–20*).

The in vitro toxicity of pyran-DIVEMA to tumor cells was low in comparison to the alkylating or other chemotherapeutic agents (*20*). However, pyran showed impressive antitumor activity against a broad range of animal models, including Lewis lung carcinoma (*21, 22*), B16 melanoma (*22*), Ehrlich carcinoma (*23*), Rauscher leukemia (*24, 25*), Friend leukemia (*25*), CH3 mammary tumor (*26*), Madison lung carcinoma (*27*), and L1210 leukemia (*28*). At the same time, pyran can also produce toxic side effects, such as anemia, hepatosplenomegaly, and leukocytosis. The initial clinical trials against human cancer revealed prohibitively high toxic thrombocytopenia (*29*), and the study was discontinued. The problem, however, was associated with the solubility of the administered dose, rather than the polymer itself.

The antitumor activity and toxicity of DIVEMA are related to its molecular weight (*2, 13, 30*). Breslow (*13*) showed that the acute toxicity of pyran in mice increased with increased molecular weight. The polymer fractions with molecular weight up to 15,000 stimulated the reticuloendothelial system (RES), whereas higher molecular weight fractions suppressed the RES, resulting in a biphasic response. On the other hand, pyran activities against Lewis lung carcinoma and Ehrlich ascites tumor were independent of molecular weight. Kaplan (*31*) found that a low-molecular-weight sample of narrow polydisperse (X18571-31) activated macrophages and was effective against Lewis

TABLE 22.1. Synthetic Polyanions with Immunostimulatory Activities

Polymer	*Activation of Cells In Vitro*	*Prolonged Survival In Vivo*	*Interferon Induction*		*Other*
			Suggested	*Proven*	
Maleic anhydride homopolymer		X			
Poly(acrylic acid)	X	X		X	X
Poly(ethacrylic acid)		X			
Poly(itaconic acid)					
Poly(methacrylic acid)		X			X
Poly(maleic acid)	X				
Poly(vinyl sulfate)					X
Dextran sulfate	X				X
Heparin		X			
Poly(xenyl phosphate)		X			
Poly(ethylene sulfonate)		X		X	X
Ethylene–maleic anhydride copolymers	X	X	X		X
Acrylic acid–maleic acid copolymer		X			
Styrene–maleic anhydride copolymer					X
Styrene–itaconic acid copolymer	X				X
Copolymers of bicyclo[2.2.1]heptene-2,3-dicarboxylic acid anhydride with					
Maleic anhydride		X			
Acrylic acid		X			
Vinyl acetate		X			
Vinyl alcohol					
Copolymers of maleic anhydride with					
Acrylic acid					X
Styrene					X
Methacrylic acid					X
Allyl phenol					X
Allyl succinic hydride					X
1,3-Dioxepin					X
Isobutenyl succinic anhydride					X
Poly(cyclohexyl-1,3-dioxepin-*co*-maleic anhydride)		X			
Poly(4-methyl-2-pentenone-*co*-maleic anhydride)		X			
1:2 Cyclopolymer of citraconic anhydride and divinyl ether		X			

lung carcinoma. A very low molecular weight sample (X19543-27) was not only nontoxic but inactive against these tumors, whereas very high molecular weight samples were active against Lewis lung tumor but showed a strong toxicity.

These findings were confirmed and elaborated on by Ottenbrite (*30*). Several fractions with different molecular weights of pyran, poly(acrylic acid-*alt*-maleic acid), poly(maleic acid), and poly(acrylic acid-*co*-3,6-endoxo-1,2,3,6-tetrahydrophthalic acid) were prepared and evaluated for activity against Lewis lung carcinoma and encephalomyocarditis virus. The results were reported and discussed elsewhere (*2, 30*). The biological toxicity of pyran increased with molecular weight: pronounced toxic effects were observed for molecular weights (M_w) above 15,000, and these effects were significantly aggravated over M_w of 50,000. Also, the pyran antitumor activity appeared to be independent of molecular weight. Therefore, by controlling the molecular weight and the molecular-weight distribution of pyran, many toxic effects were reduced while maintaining the antitumor activity (*13, 32*).

A narrow polydisperse pyran with M_w of <18,000, known as MVE-2 (*33*), showed an optimal antitumor activity and a lower toxicity. The animal studies showed an augmentation of natural killer (NK) cell activity (*34*), which is important for effective cancer therapy at the early stages of disease (*35*). MVE-2 was extensively investigated for its anticancer activity in preclinical and in clinical trials (*36*). The effectiveness of using MVE-2 in combination therapy with surgery (*37*) or other antineoplastic agents (*38, 39*) was also investigated.

The clinical pharmacology and the pharmacokinetics of MVE-2 were investigated thoroughly (*40*). The Phase I trials gave acceptable toxicity levels but no signs of significant anticancer activity against well-established cases (*41*). Proteinuria, one of the typical toxic effects observed, could be alleviated by adjusting the administered dose (*41*). In one of the Phase II clinical trial reports, MVE-2 was used as a single agent to treat metastatic malignant melanoma without encouraging results (*42*). A total of 16 patients were investigated and no significant clinical responses or immunologic responses were observed (*42*).

The specific chemical structures of the polyanions affect the biological activities. The effects of lipophilicity, chain rigidity, carboxylic acid strength, charge density, and charge distribution were evaluated (*2, 6, 43, 44*). The chemical structures of some of the more active polyanions are shown in Chart 22.1.

One of the polymers in Chart 22.1 is carbetimer, a low-molecular-weight (M_w = 1590) imide–amide derivative of a copolymer of maleic anhydride and ethylene (*43*). Carbetimer exhibits an antitumor activity, along with a remarkably low toxicity, against a variety of in vitro and in vivo tumor cell lines such as B16 melanoma, P815 mastocytoma, M5076 ovarian, colon 26, Madison 109 lung, and Lewis lung carcinoma (*43–49*).

The primary mechanism of the antitumor activity of carbetimer has not been clearly identified. However, several phenomena have been observed for this polymer:

1. Carbetimer does not exhibit any antitumor activity in the animals that have been immunosuppressed by whole-body irradiation or by being treated with cyclophosphamide. However, carbetimer does cause macrophage activation in vivo and does provide for a greater survival rate among the tumor-bearing animals than many other immunoadjuvants (*50*).
2. Carbetimer inhibits the growth of sensitive cells by inhibiting the uptake and metabolism of pyrimidine nucleosides both in vitro and in vivo (*51, 52*).
3. Carbetimer causes inhibition of cellular cyclic nucleotide phosphodiesterase (*53*).

Phase I trials have shown that carbetimer is well tolerated at doses up to 15,000 mg/m^3 given by infusion over a 1-h period (*54–57*). In several patients, carbetimer increased the T-helper:suppressor-cell ratios and induced interleukin-2 production (*56*). The major dose-limiting toxicity is reversible hypercalcemia. The minor toxicities at very high dose levels include visual disturbances, nausea, vomiting, and slight decreases in blood pressure (*57*).

Carbetimer was studied in a number of Phase II clinical trials (*58, 59*), and the results appear to be disappointing. In a trial by a Northern California Oncology Group (*59*), carbetimer was administered at a dose of 6,500 mg/m^3/day intravenously for 5 consecutive days to 14 patients with measurable metastatic or recurrent colorectal cancer. A total of 38 cycles of the therapy were administered, and nine patients completed at least three cycles of the treatment. No partial or complete tumor responses were observed. The toxicities observed were similar to those seen in Phase I trials.

Structure and abbreviation	Reference	Structure and abbreviation	Reference
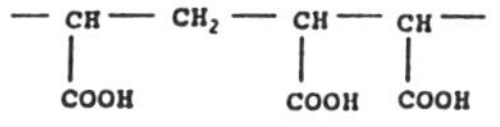 (AA-MA)	30, 45	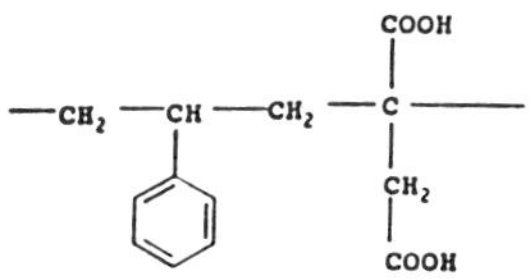 (IAS)	2, 47
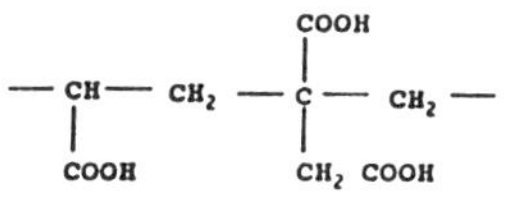 (AA-IT)	45	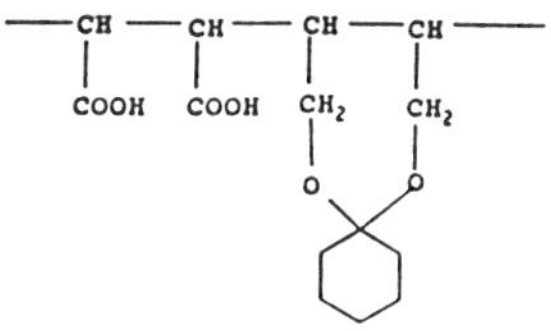 (MA-CDA)	2, 47
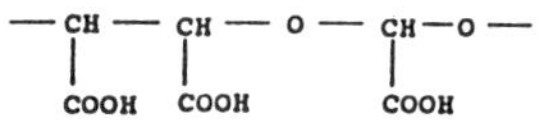 (COAM)	46	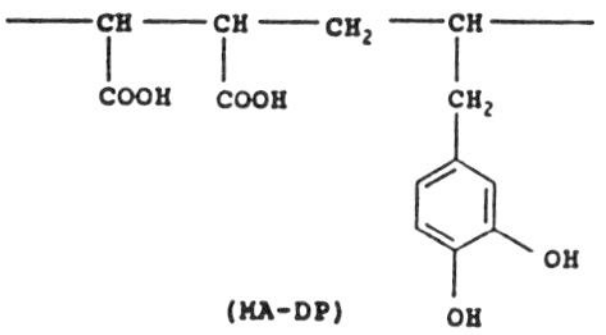 (MA-DP)	48
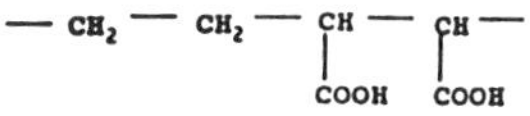 (E-MA)	6	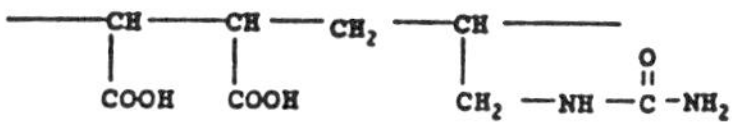 (MA-AU)	2, 47
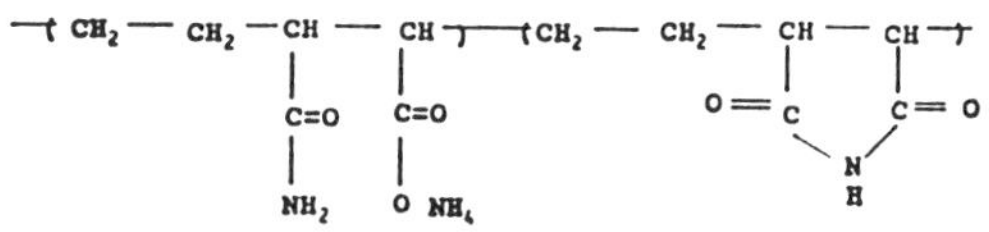 (Carbetimer)	43		

CHART 22.1. Structures of polyanions.

The initial interaction with the cell membrane appears to be the most important stage for the polyanions to exhibit biological activity. Normally, the interaction between the cell membranes and the polyanions is very weak, because both the cell surface and the polymers are negatively charged. Recently, we demonstrated (*60–62*) that the cell membrane affinity of the polyanions can be enhanced by grafting hydrophobic groups onto the polymer. Two series of polyanions were derived from poly(maleic anhydride-*alt*-cyclohexyl-1,3-dioxepin) (MA-CDA) and poly(maleic anhydride-*alt*-dihydroxyphenol) (MA-DP) by simple alkyl aminolysis of the anhydride moieties. Compared with MA-CDA and MA-DP, the modified polyanions had a higher affinity for cell membranes. This characteristic resulted in a significant increase of superoxide release from the dimethyl sulfoxide differentiated HL-60 cells. The long-term cytotoxicity on cultured J774 macrophage cells was determined for the polyanions with the improved cellular affinity.

Immunomodulation Effects

The immune system includes a heterogeneous group of lymphoid cells. The function of lymphoid cells is to protect against harmful invading substances and also to prevent or reduce tumor growth in otherwise normal tissues of the organism. Illustrated in Figure 22.1 are the basic principles of immune homeostasis (*63*).

The cellular immune responses are mediated through the thymus-derived T-lymphocytes. The antibody-dependent responses are mediated by the B-lymphocyte system. Macrophages and subpopulations of T-lymphocytes provide regulatory activities. Immunomodulators can directly or indirectly influence a specific immune reaction by affecting the regulatory system components.

Polyanions can function as chemical immunomodulators and produce a wide spectrum of effects on the immune system. Their biological activity involves the stimulation of the RES and the modulation of humoral or cell-mediated immune responses (*2, 64*). Poly(acrylic acid), pyran copolymer, and dextran sulfate enhanced the primary antibody response to sheep erythrocytes in vivo (*65–69*). Poly(acrylic acid) increased the amount of background plaques in unimmunized animals (*68*). On the other hand, administration of dextran sulfate or pyran copolymer in vivo did not stimulate the production of background plaques in unimmunized mice (*65, 69*). Inoculation of poly(acrylic acid) or dextran sulfate together with sheep red blood cells (sRBC) into thymectomized, irradiated, bone-marrow reconstituted (TxBM) mice resulted in an enhanced antibody response as measured by the plaque-forming cells and hemolysin titers, in comparison to the TxBM mice that had received sRBC alone (*70*). In contrast, pyran copolymer did not enhance specific activity in TxBM mice (*67*). The immunoadjuvant activity of many polyanions has been discussed in more detail elsewhere (*71*).

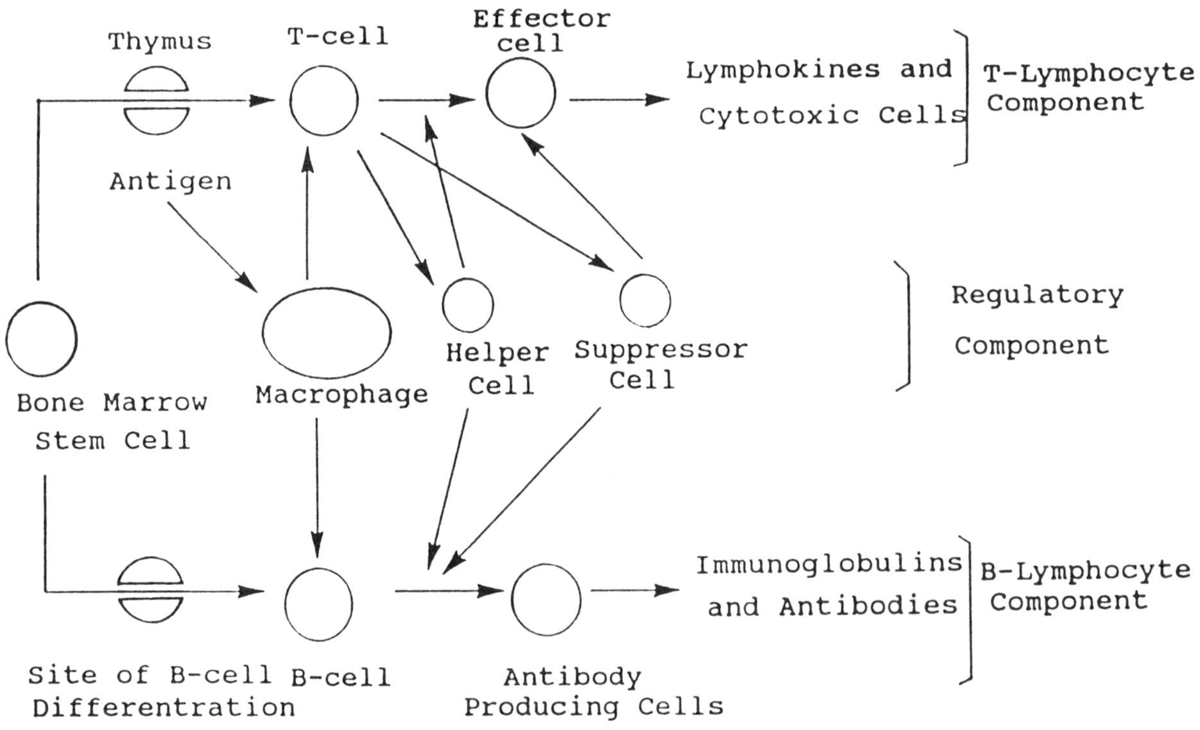

FIGURE 22.1. Basic principles of immune homeostasis.

The polyanions have the capacity to activate macrophages, a process that may relate to their effects on the plasma membrane lipids. On the other hand, in vitro studies suggested that polyanions may also provide immunomodulation through B-cell stimulation (*72, 73*). The antiviral, antibacterial, and antitumor activities of pyran may be mediated via the induction of IFN production (*74*), macrophage activation (*75*), and stimulation of ADCC (*16*).

The immunomodulation effects of the polyanions are related to their structural properties. A series of the polyanions that differ in molecular weight, lipophilicity, chain rigidity, and surface charge were evaluated for their ability to induce specific states of the macrophage activation (*2, 76*). The ability of the macrophages to become activated to tumoricidal capacity, elicited under the influence of the test polymers, was evaluated in the ^3H release and the morphological assays. The polymers with a lipophilic group, such as poly(itaconic acid-*alt*-styrene), MA-CDA, and poly(4-methyl-2-pentenone-*alt*-maleic anhydride) (MP-MA), exhibited the greatest capacity for the induction of tumoricidal macrophages in both the morphological and the ^3H release assays. The ectoenzyme profiles of the polymer-elicited macrophages were also evaluated. The 5′-N activity was elevated in all the test polymers that elicited peritoneal exudate cell populations when compared to thioglycolate, pyran, and *C. parvum*, respectively. The 5′-N levels detected in macrophage cells elicited by MA-CDA and MP-MA approached the level observed for resident populations.

Along with the elevated 5′-N activity observed in the polymer-elicited macrophages, these macromolecules also possessed a good tumoricidal activity. The survival rate studies were performed with mice that had received 2×10^4 Lewis lung cells subcutaneously. The results of these studies indicated that MA-CDA and MA-MP are effective antitumor agents and are more effective in inhibiting tumor growth than pyran or *C. parvum*. The results also indicated that there are several basic differences in the mechanisms of the immunomodulation by these structurally modified polyanions and pyran.

Sunamoto, Ottenbrite, and co-workers (*47*) encapsulated MA-CDA and some other polyanions in mannan derivative-coated liposomes. They found that the encapsulated MA-CDA induced the production of superoxide anion from the mouse peritoneal macrophages at a rate 3 times higher than unencapsulated MA-CDA during the first hour after administration. The free MA-CDA produced similar results after 24 h. These results indicate that the encapsulation of the polyanions in liposomes can enhance the efficiency of macrophage activation by the polyanions.

MVE-2 and carbetimer are the two polyanionic immunomodulators that have been clinically evaluated. The immunological target of MVE-2 (pyran copolymer) is the macrophage cell, and that of carbetimer is the B-cell (*77*). The activation of peritoneal macrophage cells by intraperitoneal administration of MVE-2 was demonstrated (*75, 78–82*) and is thought to be responsible, at least in part, for the observed antitumor effects. The MVE-2 activated macrophages are cytotoxic against M109, MBL-2, Lewis lung carcinoma, and B16 melanoma (*80–82*). The macrophage activation is involved in antibacterial activity of MVE-2. The intraperitoneal administration of MVE-2 protected mice against mortality induced by *Shigella flexneri* but not against that induced by *Diplococcus pneumoniae* (*83*). *Shigella flexneri* requires intracellular transport to be pathogenic, whereas *Diplococcus pneumoniae* does not require a macrophage transport system.

MVE-2 may exert its antimetastatic activity via the NK cells and the M109 target cells. Chirigos et al. (*82*) and Bartocci (*84*) demonstrated that MVE-2 does stimulate the NK cells. Thus, at least two cell types, the macrophages and the NK cells, are presumed to play an important role in the antitumor activity of MVE-2. The enhancement of the macrophages and the NK cells with immunomodulators may have the potential to provide protection against viral infections exacerbated during immunosuppression or immunodeficiency in cases such as acquired immunodeficiency syndrome (AIDS) and severe combined immunodeficiency (SCID) (*85, 86*).

Preliminary clinical trials revealed that MVE-2 moderately enhanced lymphocyte and monocyte ADCC, the NK activity, monocyte to macrophage maturation, and plasma levels of IFN (*41, 87, 88*).

The effect of carbetimer on the plaque-forming cells (PFC) response to T-dependent and T-independent cells was studied in normal and T-cell depleted Lewis (L) strain male rats (*89–91*). The intravenous administration of the antigen together with carbetimer induced a significant increase in the PFC response to sRBC in the L-rats. The administration of carbetimer also induced a significant increase in PFC response to deoxyribonucleoprotein (DNP)–ovalbumin and DNP-dependent antigen incorporated in $Al(OH)_3$ (*92*). The mechanism of this carbetimer activity is unclear.

Preclinical studies demonstrated that carbetimer had immunomodulatory activities, including augmentation of B-cell differentiation (*90*), increase of immunoglobulin M production in T-cell depleted animals in response to cellular antigens (*56*), and activation of alveolar macrophages (*93*). Growth in the percentage of the peripheral T-helper cells, an increased T-helper:T-suppressor-cell ratio, and induced interleukin (IL)-1 and IL-2 production were observed after administration of carbetimer in humans (*56*).

Antiviral Activity

Naturally occurring and synthetic polyanions were investigated for antiviral activity in a number of systems (*94, 95*). The early work concerning the antiviral activity of the polyanions was reviewed by Regelson (*96*) and Breinig et al. (*97*). The polyanions can also exert a potent antiviral effect against plant viruses (*98*). Shown in Table 22.2 is a broad spectrum of antiviral activities of pyran and other synthetic polyanion polymers (*7, 97*). The antiviral activity is molecular-weight dependent and appears to grow with the increase in the molecular weight of the polymer (*2*).

In contrast to standard antiviral chemotherapeutic agents, the polyanions can provide protection against viral infections in animals. The polyanion treatment protected mice from lethal inocula of either RNA or DNA cytopathic viruses (*99*). Treatment of mice with polyanions also inhibited tumor formation and delayed mortality after infection with RNA or DNA tumor viruses (*100*). Although polyanions usually elicit beneficial antiviral effects, treatment of animals with polyanions may also lead to the exacerbation of tumor formation by Moloney sarcoma (*101*) and Friend leukemia viruses (*100*). The exacerbation of Friend leukemogenesis and protection against it were related to the route of administration (*2*).

The mechanism of antiviral action of the polyanions has not been completely defined. The main proposed modes of antiviral action include:

- direct inactivation of the virus (*99, 102*),
- inhibition of virus replication (*22, 103*),
- interferon induction (*102, 104*),
- stimulation of phagocytosis and inflammation (*105–107*),
- specific immunoenhancement of humoral or cell-mediated immune responses against the virus (*108*), and
- enhancement of macrophage antiviral functions (*2, 65, 83, 109*).

In the protection regimen, polyanions appear to activate the macrophages, whereas in the adverse regimen, the drug appears to stimulate the target cells in which the virus replicates (*110*).

MVE-2 is a clinically used pleiotropic immunomodulator that can activate the NK cells and macrophages (*83*). Because some viral infections are exacerbated during immunosuppression or immunodeficiency related to the NK cell activity, such as AIDS and SCID, the enhancement of the NK cells with MVE-2 may provide protection against these viral infections. Recently, interest has increased in using immunomodulators to protect against viral infections (*85, 111–113*). A variety of

TABLE 22.2. Antiviral Activity In Vivo of Synthetic Polyanions

Virus	*Virus Class*	*Polyanion*	*Host*
RNA cytopathic viruses			
Encephalomyocarditis	Picornavirus	PCP	Mouse
MM	Picornavirus	PCP	Mouse
Mengo	Picornavirus	PCP, COAM, PAA	Mouse
Foot and mouth disease	Picornavirus	PCP, COAM, PAA	Mouse, guinea pig
Vesicular stomatitis	Rhabdovirus	PCP	Mouse
Influenza	Orthomyxovirus	COAM, PVS	Mouse
Semliki Forest	Togavirus	COAM	Mouse
Hog cholera	Togavirus	COAM, PAA	Swine
DNA cytopathic viruses			
Herpes simplex	Herpesvirus	PCP, COAM, CA	Mouse, rabbit
Vaccinia	Poxvirus	PCP, COAM, PAA, PVS, CA	Mouse, rabbit
RNA tumor viruses			
Friend leukemia	Retrovirus-C	PCP	Mouse
Rauscher leukemia	Retrovirus-C	PCP	Mouse
Gross leukemia	Retrovirus-C	PCP	Mouse
Moloney sarcoma	Retrovirus-C	COAM	Mouse
Mammary tumor	Retrovirus-B	COAM	Mouse
DNA tumor viruses			
Polyoma	Papovavirus	PCP	Mouse
Adenovirus 12	Adenovirus	PCP	Hamster

NOTE: PCP is pyran copolymer; COAM is chlorite-oxidized oxyamylose; PAA is poly(acrylic acid); PVS is poly(vinyl sulfate); and CA is carbopol.

immunomodulators, including MVE-2, provided antiviral protection to immunocompetent mice against murine cytomegalovirus (MCMV), herpes simplex virus (HSV-2), and a number of other viruses (*111, 112*). Further, the immunocompromised SCID mice were almost completely protected against the MCMV infection by MVE-2 (*113*).

Recently, MA-CDA polyanions with molecular weights of 2500 and 5400 were observed to exhibit antiviral activity against human immunodeficiency virus (HIV) in vitro in our laboratory (*114*). However, the MA-CDA polyanion with a molecular weight of 5400 was too rapidly cleared in vivo and was inactive against Rauscher leukemia virus. The mechanism of antiviral activity differs from that for the macrophage activation required for antitumor activity. Apparently, the lower molecular weight polymer is either cleared too readily by the RES, or the antiviral activity is elicited more effectively by larger macromolecules. Further studies on the anti-HIV activity of CDA-MA polyanions with higher molecular weight are in progress.

Polycations

Polycations—polyelectrolytes with electropositive charges located on the polymer chain—show a remarkable antitumor activity. Polycations are suggested to exert direct antitumor action against tumor cells both in vitro and in vivo. In the 1950s, Ambrose et al. (*115, 116*), using electromobility studies, observed an increase in the net negative charge of cells during carcinogenesis or during cell growth. The increased negative charge on tumor cell membranes is thought to be related to the tumor's invasive properties (*117*). Neutralization of the tumor cell negative charge may result in direct antitumor action.

Polyethyleneimine caused the agglutination of kidney tumor cells in vitro and exhibited antitumor activity against Ehrlich ascites carcinoma in vivo (*116*). Polylysine bound to and had an inhibitory activity against diploid Ehrlich ascites carcinoma (*118*), human epidermoid cancer cells (*119*), and polyoma virus-transformed fibroblasts in organ culture (*120*). Similar observations were reported for diethylaminoethyl (DEAE)–dextran (*121, 122*).

The polycations binding to the negatively charged surface of mammalian cells cause the charge neutralization (*123*) and the cell distortion, lysis, and agglutination (*124*). After cell inclusion by pinocytosis, the polycations may also bind to the cell nuclei (*125*). Furthermore, if the polycations bind to the internal membranes, they could interfere with cell functioning by affecting enzyme activity (*126*).

Polycations also exhibit immunomodulatory effects. With respect to the polycations' antitumor activity, these effects appear to enhance the direct cytotoxic action. In 1971, Moroson (*127*) reported a fourfold increase in survival of test animals after polyethyleneimine was administered for 9 days before the Ehrlich tumor transplantation, along with absence of the weight loss after the tumor transplantation. This result suggests that the polycations may render antitumor protection by stimulating the host immune response to transplanted cells as well as by direct antitumor action.

Polyvinylimidazoline enhanced the cellular uptake of antigen (*128*). Polycationic poly(amino acid)s, such as polylysine and polyarginine, and polyethyleneimine activated macrophages in a variety of systems (*129, 130*). Polylysine and polyarginine also exhibited an antibacterial activity, which was due to, to some degree, their immunomodulation effects (*129*).

The clinical use of polycations is limited due to their inherent toxicity to the animal species based on their destructive interaction with cell membranes (*131*). However, the toxicity can be relieved by using polycation preparations with lower molecular weight polymers (*131*).

References

1. Donaruma, L. G.; Vogl, O. *Polymeric Drugs;* Academic: New York, 1978.
2. (a) Ottenbrite, R. M.; Kuus, K.; Kaplan, A. M. In *Polymers in Medicine;* Chiellini, E.; Giusti, P., Eds.; Plenum: New York, 1983; p 3. (b) Kuus, K.; Ottenbrite, R. M.; Kaplan, A. M. *J. Biol. Response Modif.* **1985,** *4,* 46.
3. Fenichel, R. L.; Chirigos, M. A. *Immune Modulation Agents and Their Mechanisms;* Marcel Dekker: New York, 1984.
4. Turowski, R. C.; Triozzi, P. L. *Cancer Invest.* **1994,** *12,* 620.
5. Oestreich, R. *Berl. Klin. Wochenschr.* **1910,** *47,* 1698.
6. Regelson, W.; Kuhar, S.; Tunis, M.; Fields, J. E.; Johnson, J. J.; Glusenkamp, E. W. *Nature (London)* **1960,** *186,* 778.
7. Ottenbrite, R. M.; Regelson, W.; Kaplan, A.; Carchman, R.; Morahan, P.; Munson, A. In *Polymeric Drugs;* Donaruma, L. G.; Vogl, O., Eds.; Academic: New York, 1978; p 263.
8. Ottenbrite, R. M.; Kaplan, A. M. *Ann. N.Y. Acad. Sci.* **1985,** *446,* 160.
9. Donaruma, L. G. *Prog. Polym. Sci.* **1975,** *4,* 1.
10. Rihova, B.; Riha, I. *Crit. Rev. Ther. Drug Carrier Syst.* **1985,** *1,* 311.
11. Seymour, L. *J. Bioact. Compat. Polym.* **1991,** *6,* 178.
12. Butler, G. B. *J. Polym. Sci.* **1960,** *48,* 279.
13. Breslow, D. S. *Pure Appl. Chem.* **1976,** *46,* 103.
14. Merigan, T. C. *Nature (London)* **1967,** *214,* 416.
15. Dean, J. H.; Padarathsingh, L.; Keys, L. *Cancer Treat. Rep.* **1978,** *62,* 1807.
16. Tagliabue, A.; Mantovani, A.; Polentarutti, N.; Vecchi, A.; Spreafico, F. *JNCI J. Natl. Cancer Inst.* **1977,** *59,* 1019.
17. Papamatheakis, J. D.; Schultz, R. M.; Chirigos, M. A.; Massicot, J. G. *Cancer Treat. Rep.* **1978,** *62,* 1845.
18. Butler, G. B. *J. Macromol. Sci., Rev. Macromol. Chem. Phys.* **1982,** *C22,* 89.
19. Butler, G. B.; Xing, Y.; Gifford, G. E.; Flick, D. A. *Ann. N.Y. Acad. Sci.* **1985,** *446,* 149.

20. Niblack, J. F.; McCreary, M. B.; Stone, N. L. *Interferon Scientific Memoranda;* Aries Corporation (National Institute of Allergy and Infectious Diseases): McLean, VA, Memo 148/1, 1969.
21. *Modulation of Host Immune Resistance in Prevention or Treatment of Induced Neoplasms;* U.S. Government Printing Office: Washington, DC, 1977; p 277.
22. Morahan, P. S.; Munson, J. A.; Baird, L. B.; Kaplan, A. M.; Regelson, W. *Cancer Res.* **1974,** *34,* 506.
23. Little, A. P. *Report to Cancer Chemotherapy, National Service Center;* Little, Brown: Boston, MA, March 5, 1964; Contract Sa-43-ph-3789.
24. Hirsch, M. S.; Block, P. H.; Wood, M. C.; Monaco, A. P. *J. Immunol.* **1972,** *108,* 1312.
25. Chirigos, M. A.; Turner, W.; Pearson, J.; Griffin, W. *Int. J. Cancer* **1969,** *4,* 267.
26. Sandberg, J.; Goldin, G. *Cancer Chemother. Rep.* **1971,** *55,* 233.
27. Schultz, R. M.; Papamatheakis, J. D.; Luetzeler, J.; Ruiz, P.; Chirigos, M. A. *Cancer Res.* **1977,** *37,* 358.
28. Mohr, S. J.; Chirigos, M. A. In *Immune Modulation and Control of Neoplasia by Adjuvant Therapy;* Chirigos, M. A., Ed.; Raven: New York, 1978; p 415.
29. Regelson, W.; Munson, A. E. *Ann. N.Y. Acad. Sci.* **1970,** *173,* 831.
30. Ottenbrite, R. M.; Goodell, E.; Munson, A. *Polymer* **1977,** *18,* 461.
31. Kaplan, A. M. In *Dictionary of Scientific and Technical Terms;* Lapedes, D. N., Ed.; Mcgraw-Hill: New York, 1974; p 1257.
32. Kaplan, A. M.; Ottenbrite, R. M.; Regelson, W.; Cardman, R.; Morahan, P. S.; Munson, M. In *Handbook of Cancer and Immunology;* Waters, H., Ed.; Garland: New York, 1978; Vol. 5, p 135.
33. Breslow, D. S. In *Cyclopolymerization and Polymers with Chain-Ring Structures;* Butler, G.; Kresta, J. E., Eds.; ACS Symposium Series 195; American Chemical Society: Washington, DC, 1982; pp 1–9.
34. Lang, S. R. Z.; Salup, R. R.; Uriad, P. E.; Twilley, T. A.; Talmadge, J. E.; Herverman, R. B.; Wiltrout, R. W. *Cancer Immunol. Immunother.* **1986,** *21,* 19.
35. Talmadge, J. E.; Maluish, A. G.; Collins, M.; Schneider, M.; Herberman, R. D.; Oldham, R. K.; Wiltrout, R. H. *J. Biol. Response Modif.* **1984,** *3,* 634.
36. Carrano, R. A.; Luliucci, J. D.; Luce, J. K.; Page, J. A.; Imondi, A. R. In *Immune Modulation Agents and Their Mechanisms;* Fenichel, R. L.; Chirigos, M. A., Eds.; Marcel Dekker: New York, 1984; p 243.
37. Weese, J. L.; Gilbertson, E. M.; Syrjala, S. E.; Whitney, P. D.; Starling, J. R. *Dis. Colon Rectum* **1985,** *28,* 217.
38. Zaharko, D. S.; Corey, J. M. *Cancer Treat. Rep.* **1984,** *68,* 1255.
39. Yamamoto, H.; Miki, T.; Oda, T.; Hirano, Y.; Sera, M.; Akagi, M.; Maeda, H. *Eur. J. Cancer* **1990,** *26,* 253.
40. Rosenblum, M. G.; Rios, A. M.; Hersh, E. M. *Cancer Chemother. Pharmacol.* **1987,** *18,* 243.
41. Hainsworth, J. D.; Forbes, J. T.; Grosch, W. W.; Greco, F. A. *Cancer Immunol. Immunother.* **1986,** *22,* 68.
42. Rinehart, J.; LaFarge, J.; Gochnour, D.; Neidhart, J. *Cancer Immunol. Immunother.* **1987,** *24,* 244.
43. Fields, J. E.; Ascular, S.; Johnson, J. H.; Johnson, P. K. *J. Med. Chem.* **1982,** *25,* 1060.
44. Kaplan, A. M.; Kuns, K.; Ottenbrite, R. M. *Ann. N.Y. Acad. Sci.* **1985,** *446,* 169.
45. Miner, N. A.; Alfrey, T., Jr. U.S. Patent 2,803,302, 1974.
46. Claes, P.; Billiau, A.; De Clercq, E.; Desmyter, J.; Schonne, E.; Vanderhaeghe, H.; De Somer, P. *J. Virol.* **1970,** *5,* 313.
47. Sato, T.; Kojima, K.; Ihda, T.; Sunamoto, J.; Ottenbrite, R. M. *J. Bioact. Compat. Polym.* **1986,** *1,* 448.
48. Bey, P. S.; Ottenbrite, R. M.; Mills, R. R. *J. Bioact. Compat. Polym.* **1987,** *2,* 312.
49. Daniels, A. M.; Daniels, J. R. *Int. J. Cell Cloning* **1983,** *1,* 307.
50. Flak, R. E.; Makowka, L.; Nossal, N. A.; Rotstein, L. E.; Falk, J. A. *Surgery (St. Louis)* **1980,** *88,* 126.
51. Ardalan, B.; Kotarac, J.; Hrishikeshevan, H. J.; Kibbe, R.; Paget, G. E. *Proc. Am. Assoc. Cancer Res.* **1985,** *26,* 18.
52. Ardalan, B.; Paget, G. E. *Cancer Res.* **1986,** *46,* 5473.
53. Hrishikeshevan, H. J.; Ardalan, B.; Paget, G. E. *Cancer Res.* **1987,** *47,* 280.
54. Dodion, P.; DeValeriola, D.; Body, J. J.; Houa, M.; Noel, P.; Abrams, J.; Crespeigne, N.; Wery, F.; Kenis, Y. *Eur. J. Cancer Clin. Oncol.* **1989,** *25,* 279.
55. Fromm, M.; Berdel, W. E.; Schick, H. D.; Danhauser-Riedl, S.; Fink, U.; Remy, W.; Reichert, A.; Ankele, A.; Prauer, H. W.; Siewert, J. R.; Rastetter, J. *Invest. New Drugs* **1988,** *6,* 189.
56. Hanauske, A. R.; Melink, T. J.; Harman, G. S.; Clark, G. M.; Craig, J. B.; Koeller, J. M.; Boldt, D. H.; Kantor, B.; Kisner, D. L.; Orczyk, G.;

Anderson, D. W.; Paget, E.; Sarosy, G. A.; Von Hoff, D. D. *Cancer Res.* **1988,** *48,* 5353.
57. Body, J. J.; Margritte, A.; Cleeren, A.; Borkowski, A.; Dodion, P. *Eur. J. Cancer Clin. Oncol.* **1989,** *25,* 1831.
58. Keaton, M.; Brown, T.; Craig, J.; Fries, G.; Harmon, G.; Zaloznik, A.; Orczyk, G.; Von Hoff, D. D. *Invest. New Drugs* **1990,** *8,* 385.
59. Audeh, M. W.; Jocabs, C. D.; Davis, T. E.; Carlson, R. W. *Am. J. Clin. Oncol.* **1990,** *13,* 324.
60. Suda, Y.; Yamamoto, H.; Sumi, M.; Ito, N.; Yamashitâ, S.; Nadai, T.; Ottenbrite, R. M. *J. Bioact. Compat. Polym.* **1992,** *7,* 15.
61. Suda, Y.; Kusumoto, S.; Oku, N.; Yamamoto, H.; Sumi, M.; Ito, F.; Ottenbrite, R. M. *J. Bioact. Compat. Polym.* **1992,** *7,* 275.
62. Suda, Y.; Kusumoto, S.; Oku, N.; Ottenbrite, R. M. *Polym. Drugs Drug Adm.* **1994,** *545,* 149.
63. De Stevens, G. *Ann. N.Y. Acad. Sci.* **1992,** *685,* 430.
64. Ottenbrite, R. M.; Kuus, K.; Kaplan, A. M. *J. Macromol. Sci.* **1988,** *A25,* 873.
65. Morahan, P. S.; Regelson, W.; Munson, A. E. *Antimicrob. Agents Chemother.* **1972,** *2,* 16.
66. Brown, W.; Regelson, W.; Yajima, Y.; Ishizuko, M. *Proc. Soc. Exp. Biol. Med.* **1970,** *131,* 171.
67. Baird, L. G.; Kaplan, A. M. *Cell. Immunol.* **1975,** *20,* 167.
68. Diamantstein, T.; Wagner, B.; Beyse, I.; Odenwald, M. V.; Schultz, G. *Eur. J. Immunol.* **1971,** *1,* 335.
69. Diamantstein, T.; Wagner, B.; Beyse, I.; Odenwald, M. V.; Schultz, G. *Eur. J. Immunol.* **1971,** *1,* 340.
70. Diamantstein, T.; Wagner, B.; L'Age-Stehr, J.; Beyse, I.; Odenwald, M. V.; Schultz, G. *Eur. J. Immunol.* **1971,** *1,* 302.
71. Ottenbrite, R. M.; Regelson, W.; Kaplan, A.; Carchman, R.; Morahan, P.; Munson, A. In *Polymeric Drugs;* Donaruma, L. G.; Vogl, O., Eds.; Academic: New York, 1978; p 263.
72. Teodorczyk-Injeyan, J. A.; Makowka, L.; Falk, R. E.; Falk, J. A. *Scand. J. Immunol.* **1982,** *15,* 9.
73. Rotstein, L. E.; Makowka, L.; Falk, R. E.; Kirby, T. J.; Nossal, N.; Falk, J. A. *Transplantation* **1980,** *30,* 417.
74. Merigan, T. C. *Nature (London)* **1967,** *214,* 416.
75. Dean, J. H.; Padarathsingh, M. L.; Keys, L. *Cancer Treat. Rep.* **1978,** *62,* 1807.
76. Kuus, K.; Ottenbrite, R. M.; Kaplan, A. M. *J. Biol. Response Modif.* **1985,** *4,* 46.
77. Turowski, R. C.; Triozzi, P. L. *Cancer Invest.* **1994,** *12,* 620.
78. Morahan, P. S.; Barnes, D. W.; Munson, A. E. *Cancer Treat. Rep.* **1978,** *62,* 179.
79. Chirigos, M. A.; Stylos, W. A. *Cancer Res.* **1980,** *40,* 1967.
80. Barnes, D. W.; Morahan, P. S.; Loveless, S.; Munson, A. E. *J. Pharmacol. Exp. Ther.* **1979,** *208,* 392.
81. Pavlidis, N. A.; Schultz, R. M.; Chirigos, M. A.; Luetzeler, J. *Cancer Treat. Rep.* **1978,** *62,* 1817.
82. Chirigos, M.; Papademetriou, V.; Bartocci, A.; Read, E. *Adv. Immunopharmacol. Proc. Int. Conf. Immunopharmacol. 1st* **1981,** *1,* 217.
83. Carrano, R. A.; Iuliucci, J. D.; Luce, J. K.; Page, J. A.; Imondi, A. R. In *Immune Modulation Agents and Their Mechanisms;* Fenichela, R. L.; Chirigos, M. A., Eds.; Marcel Dekker: New York, 1984; p 247.
84. Bartocci, A.; Papademetriou, V.; Chirigos, M. A. *J. Immunopharmacol.* **1980,** *2,* 149.
85. Black, P. L.; McKinnon, K. M.; Wooden, S. L.; Ussery, M. L. *Ann. N.Y. Acad. Sci.* **1992,** *685,* 467.
86. Kunder, S. C.; Morahan, P. S. *Ann. N.Y. Acad. Sci.* **1992,** *685,* 618.
87. Rinehart, J. J.; Young, D. C.; Neidhart, A. *Cancer Res.* **1983,** *43,* 2358.
88. Rios, A.; Rosenblum, M.; Powell, M.; Hersh, E. *Cancer Treat. Rep.* **1983,** *67,* 239.
89. Falk, R. E.; Nossal, N. A.; Makowka, L.; Rotstein, L. E.; Falk, J. A. In *Regulation by T Cells;* Kilburn, D. G.; Levy, J. G.; Teh, H. S., Eds.; University of British Columbia: Vancouver, Canada, 1979; p 168.
90. Teodorczyk-Injeyan, J. A.; Filion, L.; Falk, J. A.; Falk, R. E.; Makowka, L. *J. Immunopharmacol.* **1983,** *5,* 147.
91. Makowka, L.; Falk, R. E.; Nossal, N. A.; Rotstein, L. E.; Falk, J. A. *Prog. Cancer Res. Ther.* **1981,** *6,* 295.
92. Lee, W. Y.; Sehon, A. H. *J. Immunol.* **1975,** *114,* 829.
93. Ardalan, B.; Hussein, A. M.; Shanahan, W. R., Jr.; Shields, M. J. *Cancer Treat. Rev.* **1991,** *18,* 73.
94. Brown, J. W.; Firshein, W. *Bacteriol. Rev.* **1967,** *31,* 83.
95. Regelson, W. *Adv. Cancer Res.* **1968,** *11,* 223.
96. Regelson, W. *Adv. Chemother.* **1968,** *3,* 303.
97. Breinig, M. C.; Munson, A. T.; Morahan, P. S. In *Anionic Polymeric Drugs;* Donaruma, L. G.; Ottenbrite, R. M.; Vogl, O., Eds.; John Wiley and Sons: New York, 1980; p 211.

98. Gianinazzi, S.; Kassanis, B. *J. Gen. Virol.* **1974,** *23,* 1.
99. Merigan, T. C.; Finkelstein, M. S. *Virology* **1968,** *35,* 363.
100. Schuller, G. B.; Morahan, P. S.; Snodgrass, M. *Cancer Res.* **1975,** *35,* 1915.
101. Gazdar, A. F.; Steinberg, A. D.; Spahn, G. F.; Baron, S. *Proc. Soc. Exp. Biol. Med.* **1972,** *139,* 1132.
102. McCord, R. S.; Breinig, M. K.; Morahan, P. S. *Antimicrob. Agents Chemother.* **1976,** *10,* 28.
103. Papas, T. S.; Pry, T. W.; Chirigos, M. A. *Proc. Natl. Acad. Sci. U.S.A.* **1974,** *71,* 367.
104. Merigan, T. C.; Regelson, W. *N. Engl. J. Med.* **1967,** *277,* 1283.
105. Regelson, W.; Morahan, P. S.; Kaplan, A. *Charged React. Polym.* **1975,** *2,* 131.
106. Breslow, D. S.; Edwards, E.; Newburg, N. *Nature (London)* **1973,** *246,* 160.
107. Declerq, E.; Desomer, P. *Infect. Immun.* **1973,** *8,* 669.
108. Hirsch, M. S.; Black, P. H.; Wood, M. L.; Monaco, A. P. *J. Immunol.* **1973,** *111,* 91.
109. Billiau, A.; Muyembe, J. J.; Desomer, P. *Nature (London)* **1971,** *232,* 183.
110. Morahan, P. S.; Schuller, G. B.; Snodgrass, M. J.; Kaplan, A. M. *J. Infect. Dis.* **1976,** *133,* A249.
111. Pinto, A. J.; Morahan, K. M.; Morahan, P. S. *Antiviral Res.* **1993,** *21,* 129.
112. Kunder, S. C.; Kelly, K. M.; Morahan, P. S. *Antiviral Res.* **1993,** *21,* 233.
113. Kunder, S. C.; Wu, L. X.; Morahan, P. S. *Antiviral Res.* **1993,** *21,* 233.
114. Ding, J. L.; Ottenbrite, R. M. In *Polymeric Drugs and Drug Administration;* Ottenbrite, R. M., Ed.; ACS Symposium Series 545; American Chemical Society: Washington, DC, 1994; pp 135–148.
115. Ambrose, E. J.; James, A. M.; Lowick, J. H. B. *Nature (London)* **1956,** *177,* 576.
116. Ambrose, E. J.; Easty, D. M.; Jones, P. C. T. *Br. J. Cancer* **1958,** *12,* 439.
117. Ambrose, E. J. In *The Biology of Cancer;* Ambrose, E. J.; Roe, F. J. C., Eds.; Van Nostrand: London, 1966; p 65.
118. Kornguth, S. E.; Stahmann, M. A.; Anderson, J. W. *Exp. Cell Res.* **1961,** *24,* 484.
119. Mehrishi, J. N. *Eur. J. Cancer* **1969,** *5,* 427.
120. Yarnell, M. M.; Ambrose, E. J. *Eur. J. Cancer* **1969,** *5,* 255.
121. Larsen, B.; Thorling, E. B. *Acta Pathol. Microbiol. Scand.* **1969,** *75,* 229.
122. Thorling, E. B.; Larsen, B. *Acta Pathol. Microbiol. Scand.* **1969,** *75,* 237.
123. Curtis, A. S. G. *The Cell Surface: Its Molecular Role in Morphogenesis;* Academic: New York, 1967; p 10.
124. Katchalsky, A. *Biophys. J.* **1964,** *4,* 9.
125. Mayhew, E.; Nordling, S. *J. Cell. Physiol.* **1966,** *68,* 75.
126. Fernandez-Moran, H.; Oda, T.; Blair, P. V.; Green, D. E. *J. Cell Biol.* **1964,** *22,* 63.
127. Moroson, H. *Cancer Res.* **1971,** *31,* 373.
128. Gall, D.; Knight, P. A.; Hampson, F. *Immunology* **1972,** *23,* 569.
129. Peterson, P. K.; Gekker, G.; Shapiro, R.; Freiburg, M.; Keane, M. F. *Infect. Immun.* **1984,** *43,* 561.
130. Moroson, H.; Rotman, M. In *Polyelectrolytes and Their Applications;* Rembaue, A.; Selegny, E., Eds.; D. Reidel: Boston, MA, 1975; p 187.
131. Arnold, L. U. J.; Dagan, A.; Gutheil, J.; Kaplan, N. O. *Proc. Natl. Acad. Sci. U.S.A.* **1979,** *76,* 3246.

23

New Biodegradable Polymers for Medical Applications

Elastomeric Poly(phosphoester urethane)s

Kam W. Leong, Zhong Zhao, and Basil I. Dahiyat

In developing biodegradable polymers for medical applications, we have incorporated a phosphoester group as the chain extender into polyurethanes, which we feel would afford a diversity of structures and physicochemical properties. Hydrolysis of the phosphoester confers degradability on the polymer, whereas the pentavalency of the phosphorus atom gives a pendant site for manipulation of properties and drug attachment. To maximize the chances that the polymer would decompose into nontoxic compounds, we converted putrescine and lysine into diisocyanates, expecting that after biochemical composition, the polyurethane synthesized from them would regenerate the starting diamines. The polymers dissolve readily in chlorinated solvents such as chloroform and can be cast into clear resilient films. The polymers typically have polydispersities over 20 and a bimodal molecular-weight distribution. These poly(phosphoester urethane)s are much weaker than the non-biodegradable polyurethanes based on aromatic diisocyanates. The complex dynamic moduli of these polymers range from 1.6 to 10.2 MPa, ultimate strength ranges from 0.62 to 3.07 MPa, and the elongation at break ranges from 36 to 180%. Degradation of the polymers in phosphate-buffered saline at 37 °C increases with the content of the biodegradable chain extender, spanning a mass loss rate of 10–40% in 100 days. The controlled-release capability of these polymers was also tested as both matrix and pendant delivery systems. Given the versatility of polyurethanes, which possess a range of physicochemical properties through manipulations of the soft segment and the chain extender, these poly(phosphoester urethane)s would make an interesting class of degradable and elastomeric biomaterials if they can pass the biocompatibility scrutiny.

Synthetic polymers have continued to find interesting applications in medicine. Not only can polymeric biomaterials afford tailor-made physicochemical properties, they can also serve as drug carriers. Integration of the structural support and controlled-delivery functions opens up new possibilities in biomaterial applications. The objective of this study was to design a biomaterial that is elastomeric, biodegradable, and capable of controlled drug delivery. The motivation was to develop a drug-eluting coating for metallic stents in combating re-stenosis.

Copyright © 1997 American Chemical Society

Ischemic heart disease, the leading cause of mortality in the Western world (*1*), designates pathological changes occurring in the myocardium from reduction of blood flow (ischemia) to reduction of oxygen supply (hypoxia). The primary factor contributing to ischemic heart disease is coronary arteriosclerosis, a general term for thickening and hardening of the arterial wall. Medical intervention is necessary when the arteriosclerotic process has occluded more than 70% of the cross-sectional luminal area of one of the coronary arteries. Angina pectoris and myocardial infarction are the two common clinical symptoms. Among the surgical procedures to relieve these symptoms is percutaneous transluminal coronary angioplasty (PTCA). It consists of percutaneous introduction of a dilating balloon catheter, via a brachial or femoral artery, into a stenosed coronary artery under a fluoroscopic monitor. When the part of the slender catheter containing the balloon straddles the stenosis, the balloon is briefly inflated with fluid to enlarge the coronary lumen. Even though PTCA is adequate for relieving acute symptoms, its overall success is hampered by re-stenosis. Between 20 to 30% of patients with an initially successful procedure experienced return of angina within 6 months of the procedure (*2*). Re-stenosis can be acute or delayed. Acute re-stenosis is caused by intimal dissection, and delayed re-stenosis is caused by elastic recoil of the vessel wall and neointimal smooth muscle hyperplasia at the PTCA site. One of the methods to prevent re-stenosis is the placement of a metallic stent at the site of angioplasty. Even though metallic stents prevent vascular occlusion, they show little improvement in arresting post-PTCA re-stenosis. Systemic administration of anticoagulants to prevent or reduce the incidence of re-stenosis has also been unsuccessful. Presumably, systemic administration is unable to provide a therapeutic concentration without causing systemic toxicity. Therefore, approaches have been proposed to deliver drugs to the PTCA site through a coating on the metallic stent or even through a stent that is entirely polymeric (*2*). An additional potential advantage, although still remaining to be proven, is that a polymeric surface that is less thrombogenic or less prone to incite neointimal smooth muscle hyperplasia than the metallic stents can be found.

The materials requirements for this application, therefore, include viscoelastic characteristics, biodegradability, biocompatibility, and controlled-release capability. The polymeric stent, or just as a coating to the metallic stent, should be able to expand 200–400% without rupture. Because the incidence rate of re-stenosis 6 months after the angioplasty is slim (<10%) (*3*), a biodegradable polymeric stent would be desirable. This stent would eliminate the lingering concern of long-term biocompatibility. An ideal polymeric stent would then be strong enough to hold the vessel patent for several months, drug-eluting to inhibit neointimal proliferation of smooth muscle cells, and biodegradable to breakdown into small non-toxic molecules without intravascular embolization. Such a stent would be difficult to realize, particularly in view of the prospect that a polymeric stent might disintegrate into particulates and embolize blood vessels downstream. The mechanical properties requirement is also demanding. The polymeric stent has to be stretched beyond its yield point and still maintain its radial strength. A compromise was made to apply a drug-eluting polymeric coating on the metallic stent. The materials requirements for this coating would remain the same.

After consideration of different classes of polymers, we opted for polyurethanes, which because of their inherent elastomeric nature possess the required properties of toughness and flexibility. This review summarizes our experience in developing these biodegradable elastomers for medical applications (*4–6*).

Design of Biodegradable Polyurethanes

Polyurethanes have been widely used to construct blood-contacting biomedical materials (*7*). The unique and excellent mechanical properties of polyurethane come from its two-phase microstructure, consisting of alternating blocks of hard and soft segments. The hard segment is usually composed of an oligomeric aromatic urethane or urethane-urea segment of molecular weight between 300–3000 resulting from the reaction of aromatic diisocyanates with a chain extender, which is typically a low-molecular-weight diol or diamine. The soft segment is typically a polyester-diol or polyether-diol with a molecular weight between 500 and 4000. The macroglycol and the urethane or urethane-urea together form an $(AB)_n$-type block polymer. Through variations of the soft and hard segments and the chain extenders, polyurethanes can exhibit diverse physicochemical properties. They can be hard and brittle, soft and tacky, and anywhere between those two extremes.

Used in a large number of clinical applications, polyurethanes are designed to be inert and stable in the body. Therefore, the first challenge is to incorporate a hydrolytically labile linkage into the backbone. Although recent studies show that polyurethanes, especially the aliphatic ones, undergo oxidative degradation in vivo (*8, 9*), the urethane and ether bonds are quite stable in a nonenzymatic aqueous environment. Using a bis(2-hydroxyethyl)phosphite as the chain extender, we introduced biodegradability by incorporating a phosphoester bond into the backbone of the polyurethane. Biodegradable polyurethanes have been proposed and studied before (*10*). The uniqueness of these biodegradable polyurethanes is the inclusion of a phosphoester linkage in the chain extender instead of a commonly used polyester component in the soft segment. The chemical structure of the repeating unit is shown in structure **1**.

A second challenge to the development of these poly(phosphoester urethane)s (PPU) is to ensure that the polymers break down into nontoxic residues. Intended to be stable, current biomedical polyurethanes are commonly synthesized from aromatic diisocyanates such as toluene diisocyanate (TDI) and methylene diphenyl isocyanate (MDI). When rendered biodegradable, however, polyurethanes derived from such diisocyanates might yield aromatic amines that are toxic. A plausible scenario is that when the polyurethane hydrolyzes, the biomaterial disintegrates into particulates, which would be phagocytozed. Under potent biochemical attack in the endolysosomes, the urethane bond would be decomposed to yield an amine. Efforts were then made to synthesize diisocyanates, which after polyurethane formation would decompose into nontoxic compounds. To achieve this synthesis, 1,4-diaminobutane and L-lysine were proposed as the precursors. 1,4-Diaminobutane is an intermediate in the urea cycle and goes by the biochemical name of putrescine. Although the toxic concentration of putrescine in humans is unknown, we thought it should have a chance of biocompatibility. L-Lysine, as one of the 20 basic amino acids, is nontoxic. We converted these two diamines to diisocyanates, expecting that after biochemical decomposition, the polyurethanes synthesized from them would regenerate the original diamines.

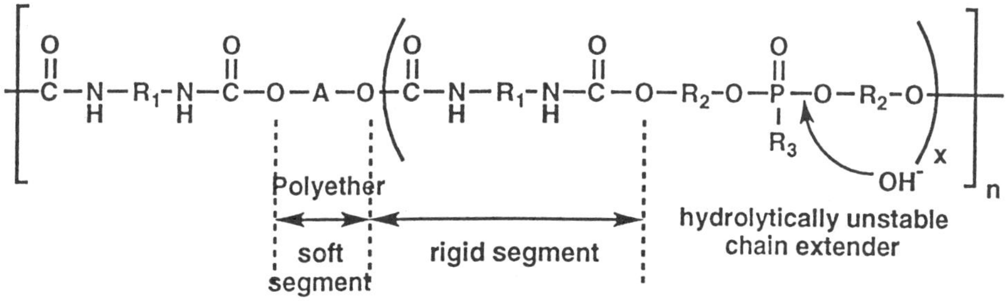

1, Basic unit of poly(phosphoester urethane).

The last requirement of controlled release is not difficult to satisfy. Like any other biodegradable polymers, these PPUs can be used for drug delivery via a diffusion-controlled or matrix degradation-controlled mechanism. However, when used in the form of a thin film, it might be difficult to obtain a sustained release because of the high surface-to-volume ratio, particularly when the drug is hydrophilic. One of the reasons we chose the phosphoester chain extender is because of the pentavalency of the phosphorus atom and a reactable side chain to which one can covalently link drugs for a pendant delivery system. We have evaluated these PPUs as both matrix and pendant delivery systems.

Synthesis

Monomers

As an attempt to understand the structure–property relationship of these PPUs, all three components of the polyurethane structure have been varied. Compositions of monomers used to synthesize PPUs are shown in Figure 23.1. In our earlier work using TDI and MDI as the diisocyanate, PPUs with molecular weights over one million were obtained (*4*). These PPUs were biodegradable in the presence of a phosphoester chain extender as expected. Concerned with the possible toxicity of the breakdown products, we focused our subsequent effort on the diisocyanates derived from putrescine and lysine (D-I and D-II) for a better chance of biocompatibility. The commonly used polyethers were used as the soft segments. For the chain extenders, C-I to C-III (Table 23.1) varying in hydrophobicity and pendant chain structure provided biodegradability, C-IV was a nonbiodegradable chain extender serving as a filler, and C-V was examined as a model drug that could be incorporated into the backbone of the polymer.

Bis(2-hydroxyethyl)phosphite (BGP) and bis(6-hydroxyhexyl)phosphite (BHP) were synthesized by adopting the procedures described by Borisov and Troev (*11*) (reaction 1). The product was a colorless viscous liquid. Even though the yield of this reaction was high, the product was difficult to purify. The transesterification of the dimethylphosphite and diol is likely to produce a number of side products. Purification of the products by vacuum distillation has proved to be difficult because of the high and close boiling points of the impurities. Proton NMR spectra of the products often showed more than one type of phosphite hydrogen, around 6.8 ppm, and total angular momentum quantum number ($\mathcal{J}_{P\text{-}H}$) of 694 Hz, suggesting side reactions of the side chains through either elimination or cyclization (*6*).

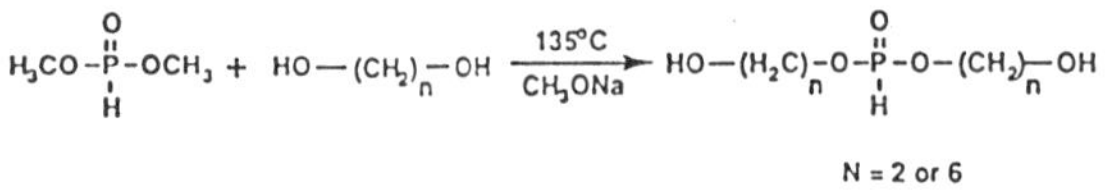

1

The diisocyanate BDI was synthesized via a Curtius rearrangement (*12*). Adipic dihydrazide was oxidized in aqueous nitrous acid forming the diacyl azide. Upon heating in benzene at 70 °C overnight, the azido compound eliminated two equivalents of nitrogen to generate BDI. The compound could be readily purified by vacuum distillation (reaction 2).

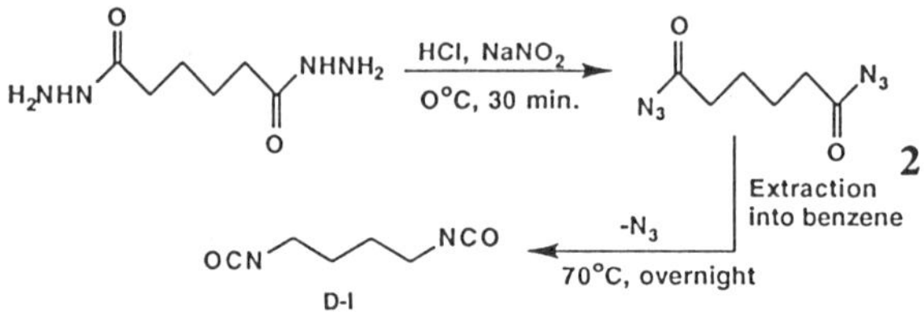

2

The synthesis of the diisocyanate derived from lysine (LDI) was more challenging. Initially we followed the procedures described in a patent, which transformed lysine to LDI by phosgenation in *o*-dichlorobenzene at 125 °C (*13*). However, after repeated experiments we failed to obtain reasonable yield in high purity by this method. LDI was then synthesized using a different scheme (reaction 3) (*14*). The lysine ethyl ester was converted to the disilazane intermediate by reacting it with hexamethyldisilazane at 120 °C for 24 h or

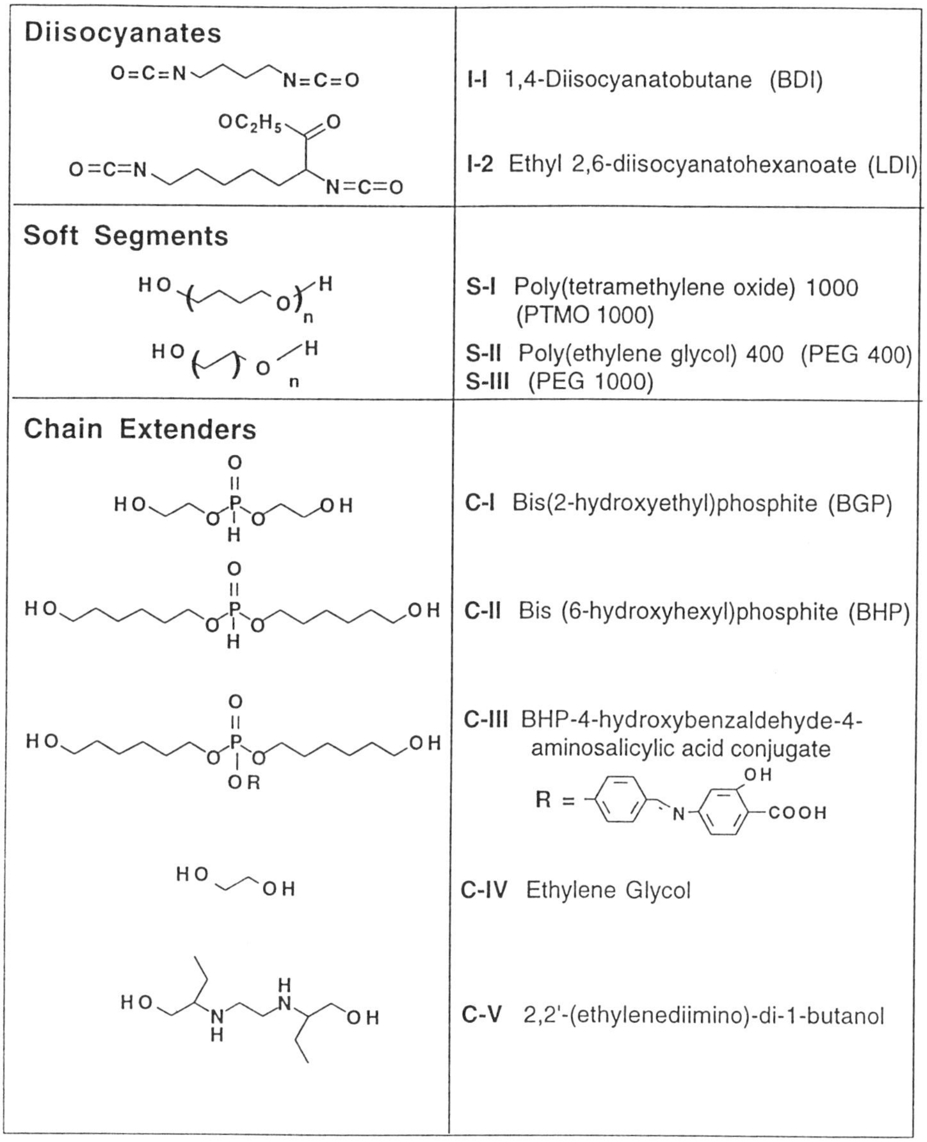

Diisocyanates
I-I 1,4-Diisocyanatobutane (BDI)
I-2 Ethyl 2,6-diisocyanatohexanoate (LDI)
Soft Segments
S-I Poly(tetramethylene oxide) 1000 (PTMO 1000)
S-II Poly(ethylene glycol) 400 (PEG 400)
S-III (PEG 1000)
Chain Extenders
C-I Bis(2-hydroxyethyl)phosphite (BGP)
C-II Bis (6-hydroxyhexyl)phosphite (BHP)
C-III BHP-4-hydroxybenzaldehyde-4-aminosalicylic acid conjugate
R =
C-IV Ethylene Glycol
C-V 2,2'-(ethylenediimino)-di-1-butanol

FIGURE 23.1. Chemical structures of monomers used in PPU synthesis.

more until the mixture turned clear. The excess hexamethyldisilazane was removed and the crude product purified by vacuum distillation at 100 °C and 0.1 mm Hg to yield a colorless liquid. Care was taken to exclude moisture from the system, and the bis(trimethylsilyl)-L-lysine was converted to the final product in anhydrous ethyl ether at −20 °C by using triethylamine as the acid acceptor. After the triethylamine hydrochloride was removed by filtration, the solution was evaporated to dryness. The residue was then purified by distillation at 120 °C and 0.1 mm Hg. The high purity of the compound was confirmed by Fourier transform (FT)IR and FT-NMR spectroscopies.

TABLE 23.1. Chemical Composition of PPU Studied

	Diisocyanate (mole ratio)		*Soft Segment (mole ratio)*			*Chain Extender (mole ratio)*				
Polymer	*BDI*	*LDI*	*PEG_{400}*	*PEG_{1000}*	*$PTMO_{1000}$*	*C-I*	*C-II*	*C-III*	*C-IV*	*C-V*
PPU-B1	4				2	1				1
PPU-B2	2				1	1				
PPU-B3	2				1				1	
PPU-B4	4				2	1			1	
PPU-B5	4		2			1				1
PPU-B6	2				1		1			
PPU-B7	2				1			1		
PPU-L1		2		1		1				
PPU-L2		2			1	1				
PPU-L3		2			1.75	0.25				
PPU-L4		2			1.40	0.60				

NOTE: *See* Figure 23.1 for chemical structures.

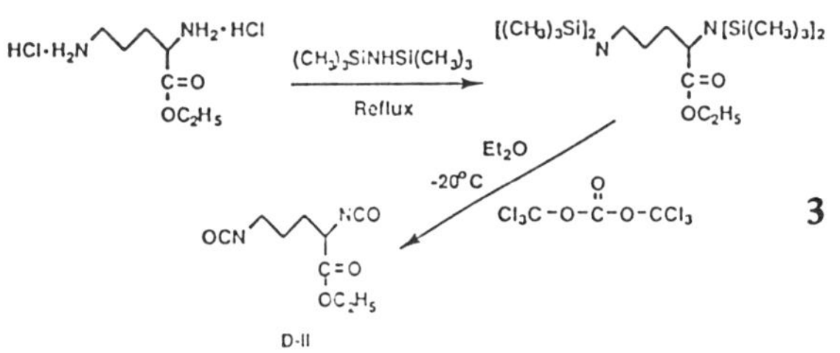

3

The drawback of this reaction scheme was the stringent requirement of anhydrous condition. The silylated intermediate also had to be distilled before the final phosgenation. Purification of the product was often hampered by presence of traces of triethylamine in the mixture, which catalyzed the unwanted side reactions of LDI at elevated temperatures. To search for a more practical and convenient procedure, the conventional phosgenation method was further investigated. After examining a number of solvents and reaction conditions, we found that LDI could be consistently produced in high purity by phosgenation of lysine ethyl ester in anhydrous toluene at 135 °C (reaction 4).

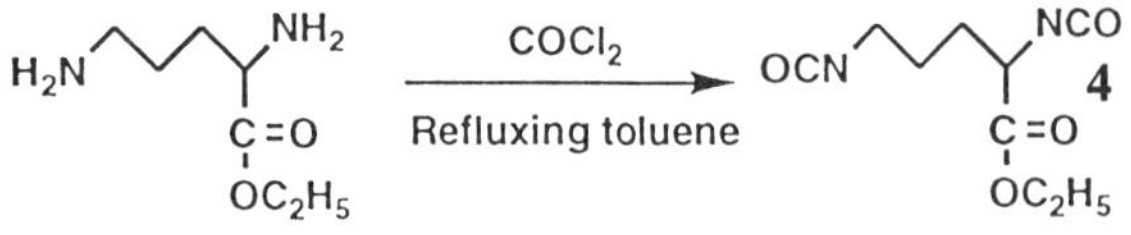

4

Polymerization

The standard two-step polymerization was used. Typically, a prepolymer was first prepared by reacting the diisocyanate with the various soft segments in dimethylformamide (DMF) or dimethylacetamide (DMA) in a two-to-one molar ratio at 100–110 °C for 2 h under nitrogen atmosphere. The polyethylene oxide (PEO) and poly(tetramethyl oxide) (PTMO) were vacuum-dried at 60 °C overnight before use. A solution of chain extenders containing BGP and other diols was then added to the prepolymer reaction mixture and further reacted for another 3 h. The polymer was isolated by quenching the reaction mixture in cold water or ethyl ether. Yields of over 90% could be obtained.

Solvent and Molecular-Weight Properties

For the BDI-based PPUs, polymerizations run at 110 °C gave PPUs with higher molecular weight that were stronger and more flexible than PPUs synthesized at less than 100 °C. This improvement in physical properties could have resulted from cross-linking due to allophanate or biuret bonds that only occurs at the higher temperature

(*15*). Further, the reaction of the phosphite hydrogen with isocyanates might have contributed to cross-linking. Using 20% excess BDI in the synthesis improved yields but did not result in any noticeable changes in material properties. PPUs based on $PTMO_{1000}$ were generally light colored, soft, and flexible, whereas PPUs based on PEG_{400} were hard, brittle, and tacky. All of the PEG_{400} polymers dissolved readily in water in addition to DMF, DMAc, alcohols, and chloroform. The $PTMO_{1000}$ polymers generally dissolved in DMF, DMAc, and chloroform but, as expected from the hydrophobic nature of the polyol, were insoluble in alcohols or water. The ready solubility of these PPUs in chlorinated solvents could be attributed to the solubilizing effect of the phosphoester group in the backbone. These polymers can be cast into clear resilient films and can be molded readily at 70 °C. The PPUs typically have polydispersities over 20 and bimodal molecular-weight distributions. One peak corresponds to chains in the 20-kDa range, and the other corresponds to a very high molecular weight polymer that often extended beyond the 3 million dalton exclusion limit of the gel permeation chromatography (GPC) columns used.

For the LDI-based PPUs, the molecular weight and polydispersity were similar to those of BDI-based PPUs. PPUs based on LDI often have molecular weights (M_w) over one million. The polymers were soluble in the reaction mixture of DMF or dimethyl acetamide (DMAc). However, once the polymers were cast into films at 50 °C, the polymers became insoluble. Cross-linking appears to have occurred during the drying process.

Thermal and Mechanical Properties

All the BDI-based PPUs showed a soft segment glass-transition temperature (T_g) between –60 °C and –65 °C due to relaxation of the soft segment, whereas the LDI-based PPUs showed a higher T_g around –40 °C. These values are lower than those reported for aromatic diisocyanate-based polyurethanes because of the greater flexibility of the aliphatic BDI and LDI. PPUs also showed several transitions between 30 °C and 140 °C that almost entirely disappeared if the samples were heated to 150 °C for 10 min and rapidly quenched to –120 °C immediately before differential scanning calorimetry (DSC) analysis. The annealing appeared to disrupt crystal structure that was slow to reorder. Letting annealed samples sit for several days resulted in reappearance of the affected peaks. Breakdown of PPUs as measured by thermal gravimetric analysis (TGA) began at 280 °C for all polymers tested with the exception of PPU-B3, which contained no BGP. Thus, the presence of phosphoester decreased the thermal stability of the polymer.

Dynamic mechanical testing was conducted to study stiffness and viscoelasticity as a function of time in vitro. One-percent offset and 0.3% strain amplitude were selected to stay within the linear ranges of the stress–strain curves for the materials and were the smallest values that gave reproducible results. Because varying the frequency from 0.001 to 0.1 Hz changed the dynamic modulus only slightly, 0.01 Hz was used for all measurements because it simplified data analysis. For PTMO-based PPUs, tan δ ranged from 0.07 to 0.21, depending on chain extender composition, whereas the magnitude of the complex dynamic modulus, E^*, varied from 1.6 MPa for PPU-B1 to 10.2 MPa for PPU-B2 (Table 23.2). For the BGP-based PPUs, phosphoester tended to increase both E^* and tan δ, whereas BHP-based polymers were usually softer than their BGP analogs, as can be expected from the longer hexamethylene chains in the monomer.

The ultimate tensile strengths (UTS) of BDI- and PTMO-based PPUs were in the 2–3 MPa range and were quite similar for all of the various compositions (Table 23.2). The elongations at break were more varied, and the trends could be correlated to the chain extender structures. PPU-B3 with ethylene glycol as its only chain extender was stiffer than the BGP-extended PPU-B2. PPU-B1, with half of its chain extender made up of the longer diol VII, is softer, and PPU-B6 is the softest yet. Elongations reflect the same tendency, and

TABLE 23.2. Mechanical Properties of PPU

Polymer	*Tan δ*	*Complex Dynamic Modulus, E* (MPa)*	*Ultimate Tensile Strength (MPa)*	*Elongation at Break (%)*
PPU-B1	—	1.6	2.50 ± 0.69	60 ± 18
PPU-B2	0.14	10.2	2.27 ± 0.39	36 ± 11
PPU-B3	0.08	7.4	3.07 ± 0.71	48 ± 13
PPU-B4	0.11	8.8	—	—
PPU-B6	0.12	6.5	1.89 ± 0.31	84 ± 18
PPU-L1	0.35	3.2	0.63 ± 0.04	146 ± 15
PPU-L2	—	—	1.18 ± 0.07	102 ± 7
PPU-L3	—	—	1.21 ± 0.08	101 ± 2
PPU-L4	—	—	0.62 ± 0.01	180 ± 8

PPU-B1 and PPU-B6 stretch more than PPU-B2 and PPU-B3. An analog to PPU-B2 was synthesized with hexamethylene diisocyanate (HDI) replacing BDI to gauge the effect of changing the isocyanate. Tensile properties were not significantly different. In contrast, the PPUs derived from LDI were all weaker and softer, having a lower UTS and higher elongation at break. The lower mechanical properties of these PPUs may be due to the asymmetric nature of lysine. This asymmetry of the hard segment might reduce the crystallizability of the polymer and render the final phase separation irregular and incomplete.

Effects of In Vitro Degradation on Physicochemical Properties

The effect of in vitro degradation on PPU-B2, PPU-B3, and PPU-B4, polymers with 25, 0, and 12.5 mol% BGP, respectively, was assessed. The thin, 300-μm samples reached swelling equilibrium in less than 3 days, and PPU-B2 and PPU-B4 swelled about 45% and remained nearly constant for 110 days. PPU-B3, without the phosphoester chain extender, swelled only about 10% and remained clear and translucent throughout the study in contrast to the other polymers, which turned opaque. In addition, PPU-B2 became appreciably softer and more difficult to handle until breaking up at around 100 days. Mass loss from the BDI-based PPUs is shown in Figure 23.2. Even though PPU-B3 showed no significant degradation for over 3 months, a rapid initial weight loss was seen for PPU-B2 and PPU-B4, of approximately 20% and 5%, respectively. The mass loss then leveled off for about 75 days before accelerating again. Degradation was still occurring during the plateau phase as verified by GPC. During the first 20 days of the study, molecular-weight distributions apparently increased as low-molecular-weight oligomers leached out and accounted for most of the early mass loss. As expected, PPU-B3, which contained no biodegradable chain extender, did not show any detectable change in molecular-weight distributions.

Although considerable mass loss and chain cleavage were seen, FTIR spectra of degraded samples showed no noticeable change from pristine polymers. DSC curves of PPU-B2 showed small shifts in the set of melting transitions above room temperature but not in any consistent or reproducible manner. Apparently, no change in chemical composition was occurring in the polymer samples during degradation. However, the identity of degradation products was not accounted for from these data. TGA did show differences after degradation. Mass loss began at 150 °C after 90 days in vitro compared with 280 °C for fresh polymers.

Changes in dynamic moduli during degradation are shown in Figure 23.3. Thicknesses of the

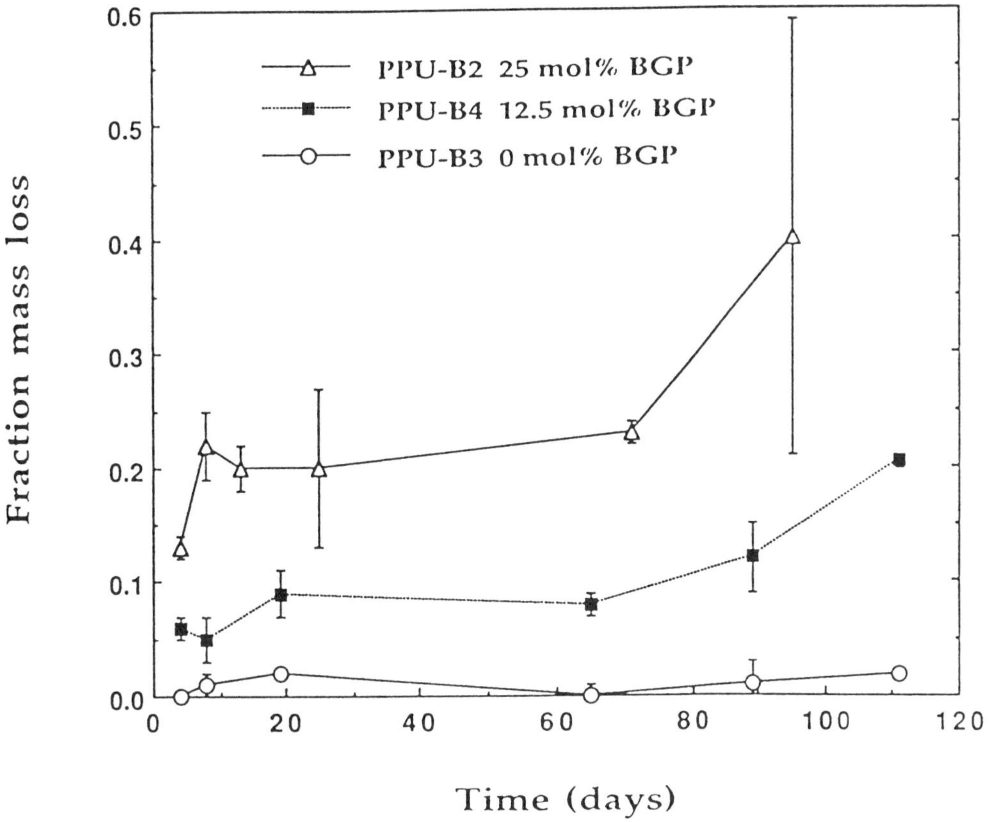

FIGURE 23.2. In vitro mass loss of PPU-B2, PPU-B3, and PPU-B4 as a function of time. All points are averages ± standard error of the mean (n = 3).

films were measured immediately before testing, and hence, moduli values are normalized to degraded sample dimensions. PPU-B2 and PPU-B4 showed initial increases in E^* followed by declines at later time points. PPU-B2 dropped off steeply and became too soft for the testing after 75 days, whereas PPU-B4 showed a more deliberate softening and PPU-B3 remained unchanged. The time course of E^* matched that of the mass loss and chain cleavage previously discussed, a result that suggests a possible explanation. The initial loss of low-molecular-weight oligomers, effectively raising the molecular weight, would be expected to stiffen the polymer. Then, as the high-molecular-weight polymer degraded, the modulus would be expected to drop. PPU-B2 showed a greater effect than PPU-B4 for both E^* and mass loss, and this result can be attributed to the difference in phosphoester contents, whereas the control PPU-B3 remained unchanged. Tan δ showed similar trends, but the structural features responsible for energy dissipation are not as apparent as those affecting the modulus.

For the LDI-based PPUs, the effects of the soft segment were first examined (Figure 23.4). The PTMO soft segment imparted a slightly higher resistance to hydrolysis than the PEO soft segment. The tensile strength, however, was twice as high with PTMO than PEO. The UTS declined by approximately 25% after 3 months, when mass losses of 18% and 13% were observed for PPU-L1 and PPU-L2, respectively. Disintegration of the

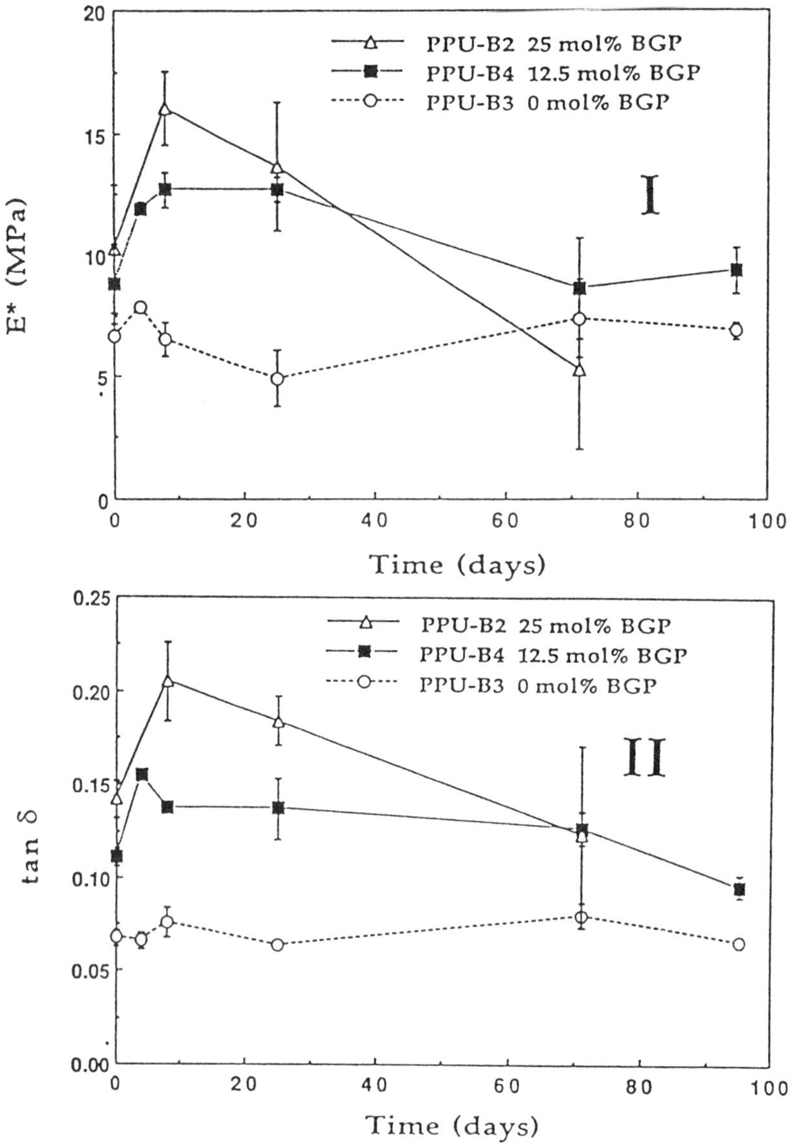

FIGURE 23.3. Change in complex modulus (I) and tan δ (II) of PPU-B2, PPU-B3, and PPU-B4 as a function of in vitro degradation. All points are averages ± standard error of the mean (n = 3).

UTS of LDI-based PPUs appears to be slower than the BDI-based PPUs. The UTS of PPU-B2 decreased by over 50% in 22 days, when only a 4% in mass loss was observed. The slower disintegration of the mechanical properties of the LDI-based PPUs might be attributed to some kind of cross-linking of the polymer, as evidenced by its change in solubility after film casting.

The effects of the molar content of BGP on the physicochemical properties of the LDI-based polymers are shown in Figure 23.5. When the BGP content was increased from 12.5 to 30

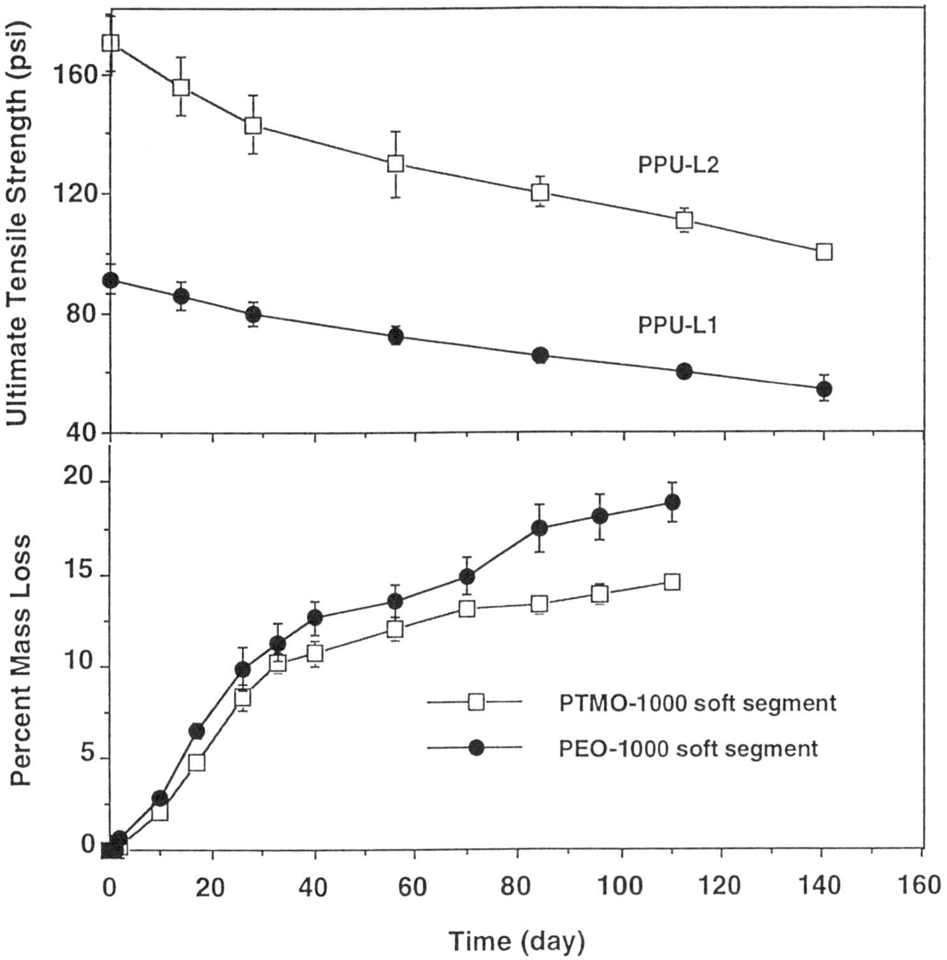

FIGURE 23.4. Effect of soft segment and in vitro degradation on UTS and mass loss of PPU-L1 and PPU-L2. All points are averages ± standard error of the mean ($n = 4$).

mol%, a twofold increase in degradation rate was observed. The corresponding equilibrium water uptake was also higher. The UTS of the two polymers was comparable, although the one with the lower BGP content had a significantly higher elongation at break.

Controlled Drug Delivery

Homogeneous DMA solution of PPU-L2 and dexamethasone (10 wt% drug loading) was cast into thin films of 0.8–1.0-mm thick. In vitro release kinetics studies showed a typical diffusion-controlled mechanism (Figure 23.6). The diffusion coefficient of dexamethasone in this polymer was estimated to be 5.3×10^{-6} cm^2/s. Complete drug release was observed in about 2 weeks. No substantial physical deterioration of the thin film was observed during the release period. To take advantage of the reactable side chain of the phosphoester chain extender, the covalent coupling of drugs to the PPU for a pendant drug delivery system was next examined.

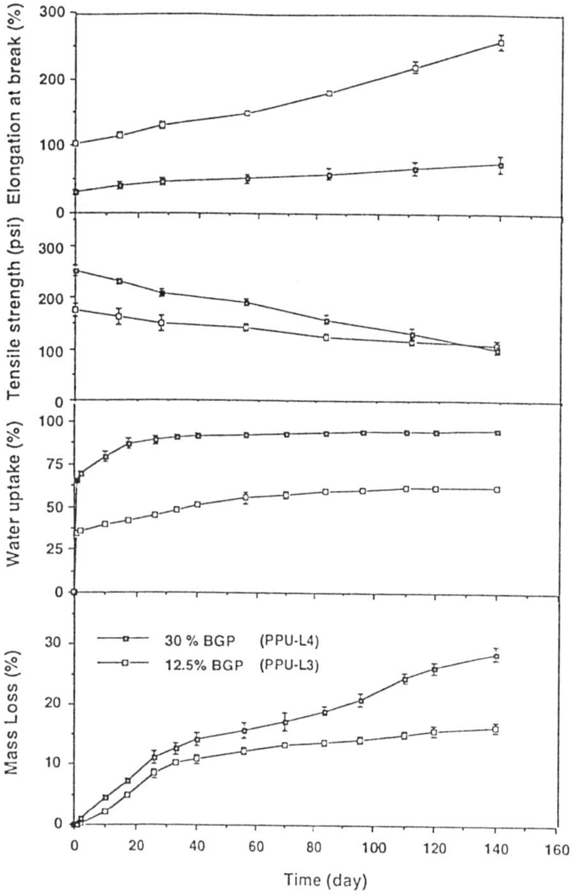

FIGURE 23.5. Effect of BGP content and in vitro degradation on physicochemical properties of PPU-L3 and PPU-L4. All points are averages ± standard error of the mean (n = 4).

Two approaches of linking drugs to the side chains of PPU were investigated. Benzocaine was attached after polymerization, whereas *p*-aminosalicylic acid (PAS) was coupled to the chain extender before polymerization. A spacer was also sometimes used to generalize the coupling chemistry. Coupling of a spacer 4-hydroxybenzaldehyde (HBA) to BHP via a P–O bond was achieved by adapting the method of Penczek and co-workers (*16*), and attachment of PAS to the spacer was achieved as described in reaction 5. Briefly, BHP was chlorinated in methylene chloride, followed by

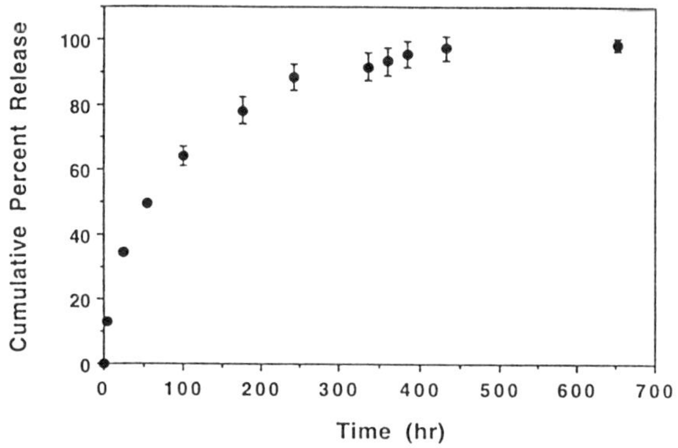

FIGURE 23.6. In vitro release of dexamethasone from 1-mm thick, 10-mm diameter PPU-L2 films with 10 wt% drug loading. Drug was dissolved in the PPU film.

reaction with a 2 molar excess of imidazole at room temperature. After removal of the excess imidazole, HBA was added and allowed to react for 2 days. After isolation, the BHP–HBA complex was reacted with an equimolar amount of PAS in absolute ethanol for 30 min. The product was a bright orange viscous liquid readily soluble in polar solvents. The monomer was used to form PPU-B7 (Table 23.1).

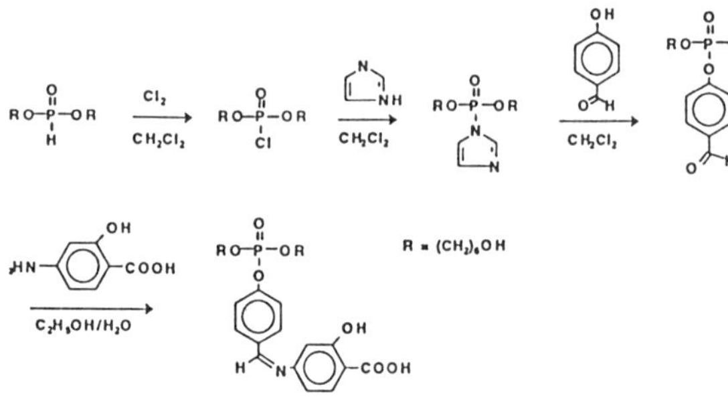

5

Benzocaine was coupled to PPU-B2 and PPU-B5 (reaction 6) by using the Todd reaction. For example, 67 mg of benzocaine, 760 mg of PPU-B5, and 42 mg of triethylamine were dissolved in 15 mL of CH_2Cl_2 and 10 mL of CCl_4 and stirred overnight at room temperature. The solvent was evaporated, and the product, a tacky, dark solid, was washed exhaustively with ether to remove unconjugated benzocaine.

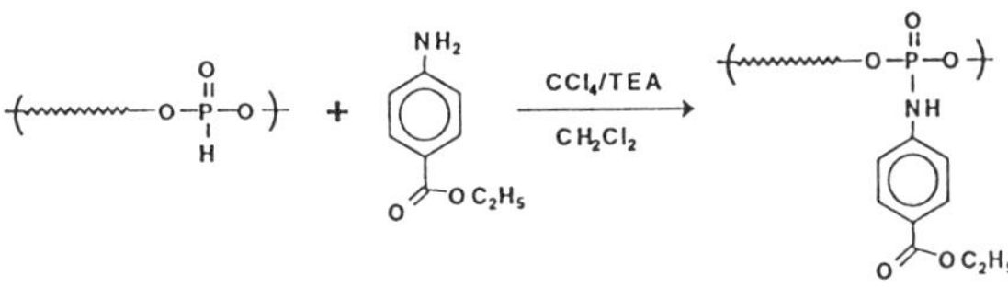

6

The drug-release kinetics of these two drugs are shown in Figure 23.7. In both cases, release was complete within hours. Complete release of PAS could be visualized as the samples changed from orange to colorless. The fast release of benzocaine from PPU-B2 and PPU-B5 indicates the lability of the phosphoroamidate bond. Interestingly there is little difference between benzocaine released from the soluble PPU-B5 and the solid PPU-B2 films. Chemical integrity of the released benzocaine was confirmed by UV spectroscopy.

In the case of PAS release from PPU-B7, the labile imine bond is responsible for the rapid release. Analysis by high-performance liquid chromatography (HPLC) showed that only free PAS was present in the buffer. Release of PAS could be realized through different pathways (reaction 7): for instance, cleavage of the polymer–spacer bond followed by detachment of the drug from the spacer, or even backbone scission preceding the side-chain disconnection. However, neither the spacer nor the spacer–drug conjugate was observed during the release experiment. The result attests to the higher hydrolytic reactivity of the imine bond compared with the phosphoester linkage but may also reflect the role of steric hindrance in protecting the spacer from cleavage.

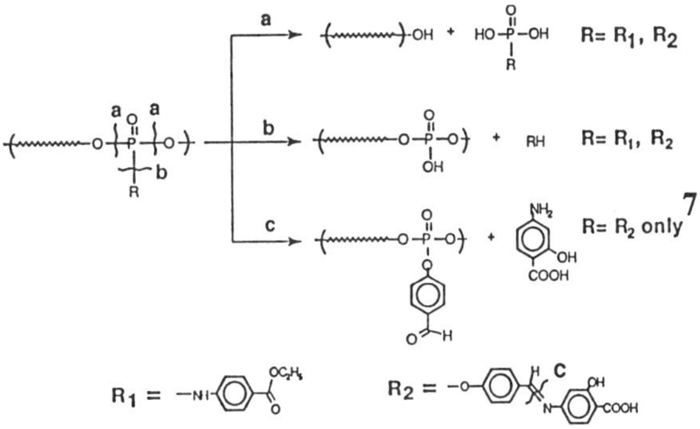

7

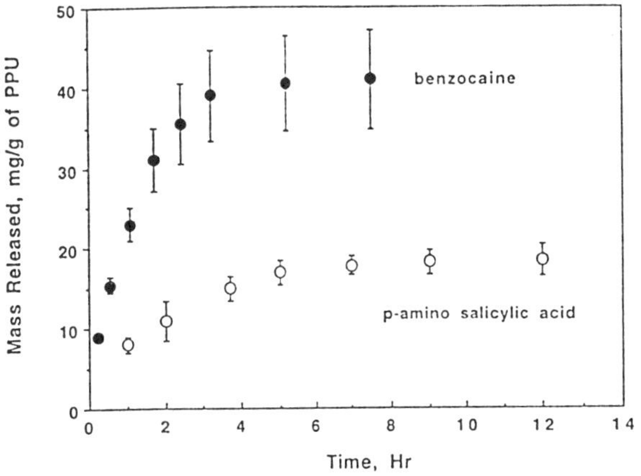

FIGURE 23.7. In vitro release of *p*-amino salicylic acid and benzocaine from PPU-B7 and PPU-B2 films, respectively. Drugs were covalently coupled to the side chain of PPU.

Each of the attachment schemes has its advantages and disadvantages. Coupling after polymerization is applicable to a wider range of drugs and is done on polymers having proven satisfactory properties. However, reaction of polymers is seldom complete, and the extent of drug attachment is difficult to control and predict. Starting with a drug-attached monomer can in principle lead to a better control over the final drug loading level. However, one must be careful that the drug has no functional groups that will interfere with the polymerization. This requirement precludes aliphatic hydroxyl and primary amino groups (other than the one attached to the spacer or the phosphoester diol) in the drug. Because no interference was observed in the polymerization of PAS-attached monomer, this study suggests that aromatic hydroxyl and carboxylic acid groups may not pose a problem, which is consistent with the knowledge that their reaction rates with isocyanate are much lower than that of aliphatic hydroxyl group.

We chose to couple PAS to the polymer by means of *p*-hydroxybenzaldehyde for two reasons. One is to include a spacer to facilitate the release. The other is to broaden the coupling chemistry available. The Todd reaction that links primary amino groups to the phosphite hydrogen has to be performed in chlorinated organic solvents, and hence is only applicable to lipophilic drugs. Attachment by means of the spacer HBA allows the use of more hydrophilic drugs as long as they are soluble in alcohols.

To take advantage of the facile reaction between an isocyanate and an alcohol, we also explored the possibility of directly incorporating the drug into the backbone of the polymer. Ethambutol, a diol drug, was added in conjunction with BGP during the chain-extension reaction. Release of ethambutol from PPU-B1 as determined by chemical assay is shown in Figure 23.8. The copper chelation assay did not distinguish free drug from drug still conjugated to polymer fragments. As illustrated by reaction 8, Figure 23.8 probably reflects the release of a combination of free drug and drug–oligomer conjugates. Because the release is more controlled by a matrix-degradation mechanism, a more sustained release could be obtained.

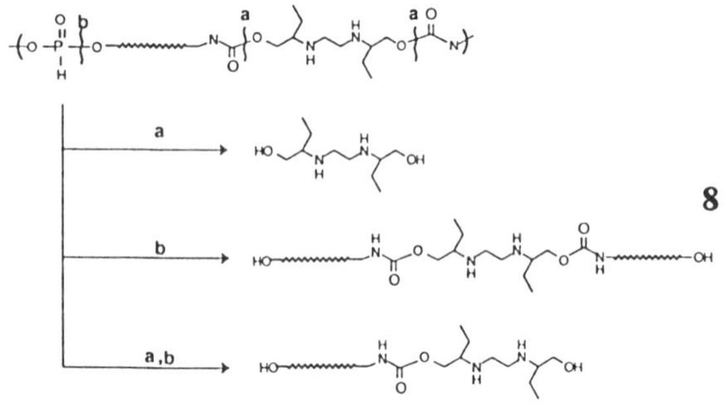

8

Conclusion

With regard to the original objectives of coating applications, the PPUs show promise of meeting the expectations. Several areas require further investigation. The most important ones are optimization of the polymer synthesis and biological evaluation. Both types of PPUs are much weaker than the nonbiodegradable polyurethanes based on aromatic diisocyanates. For instance, Biomer,

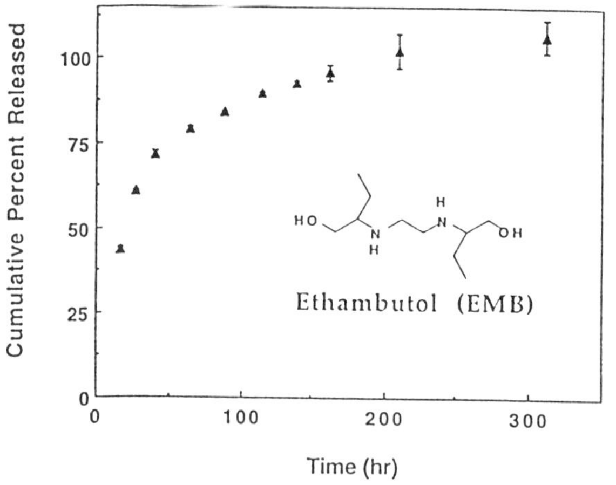

FIGURE 23.8. In vitro release of ethambutol from PPU-B5. Drug was incorporated into the backbone of the polymer.

built on MDI, PTMO, and a diamine chain extender, has a UTS around 40 MPa and an elongation of 600% (*15*).

The PPUs are expected to be softer because of the aliphatic nature of the hard segment, but there should be ample room for improvement. The PPUs have high polydispersities, and the low-molecular-weight fractions must have significantly weakened the mechanical strength of the polymer. The purity of BGP needs to be improved, probably through painstaking chromatographic separation. Systematic variations of the reaction conditions, notably ratio of diisocyanate to diol, reactant concentration, and polymerization temperature, also need to be carried out to narrow the molecular-weight distribution.

Through variation of the BGP content in the polymer, it is reasonable to expect that the PPUs can satisfy most biodegradation rate specifications. This result in turn would influence the controlled-release behavior. Lipophilic drugs probably can be released in a sustained manner from the stent coating. The pendant systems presented in this study, however, need to be modified if sustained release of a hydrophilic drug is desired. The hydrolytic lability of the phosphoroamidate bond is too high for this purpose. A phosphoester bond conjugating the drug to the polymer should be less unstable. It is envisioned that the hydroxyl group of a drug can be coupled to the imidazole of the phosphoester chain extender.

As with any other new biomaterials considerations, biocompatibility is the biggest unknown. We have designed the polymers from nontoxic monomers, but to predict biocompatibility strictly from chemical structures is impossible. One should also remember that polyurethane chemistry is extremely complicated. Other than the definite urethane, ether, and phosphoester bonds, there are also likely urea, allophanate, biuret, and acylurea bonds in these PPUs. The toxicology of these monomers and possible intermediates can only be evaluated in vivo.

In summary, given the versatility of polyurethanes, with high degrees of freedom to obtain the desired physicochemical properties through manipulations of the soft segment and the chain extender, the PPUs would make an interesting class of degradable and elastomeric biomaterials if they can pass the biocompatibility scrutiny.

References

1. Smith, J. J.; Kampine, J. P. *Circulatory Physiology: The Essentials,* 3rd ed.; Williams and Wilkins: Baltimore, MD, 1990.
2. Murphy, J. G.; Schwartz, R. S.; Huber, K. C.; Holmes, D. R., Jr. *J. Invasive Cardiology* **1991,** *3,* 144–148.
3. Botman, C. J.; el Gamal, M.; el Deeb, F.; Wesseling, F.; Bonnier, J.; Michels, R.; Landman, G. *Eur. Heart J.* **1989,** *10,* 112.
4. Shi, F. Y.; Wang, L. F.; Tashev, E.; Leong, K. W. In *Polymeric Drugs and Drug Delivery Systems;* Dunn, R. L.; Ottenbrite, R. M., Eds.; ACS Symposium Series 469; American Chemical Society: Washington, DC, 1991; pp 141–154.
5. Dahiyat, B. I.; Posadas, E. M.; Hirosue, S.; Hostin, E.; Leong, K. W. *React. Polym.* **1995,** *25,* 101–109.
6. Dahiyat, B. I.; Hostin, E.; Posadas, E. M.; Leong, K. W. *J. Biomater. Sci. Polym. Ed.* **1993,** *4,* 529.
7. Lelah, M. D.; Cooper, S. L. *Polyurethanes in Medicine;* CRC Press: Boca Raton, FL, 1986.
8. Smith, R. R.; Williams, D. F.; Oliver, C. J. *J. Biomed. Mater. Res.* **1987,** *21,* 1149.
9. Zhao, Q.; Topham, N.; Anderson, J. M.; Hiltner, A.; Lodven, G.; Payet, C. R. *J. Biomed. Mater. Res.* **1991,** *25,* 177.
10. Bruin, P.; Smedinga, J.; Pennings, A. J.; Jonkman, M. F. *Biomaterials* **1990,** *11,* 291–295.
11. Borisov, G.; Troev, K. *Eur. Polym. J.* **1973,** *9,* 1077.
12. Eckert, P.; Herr, E. *Chem. Abstr.* **1944,** 1564.
13. Garber, J. D. et al. U.S. Patent 1,351,368, 1964.
14. Katsarava, R. D.; Kartvelishvili, T. M. *Dokl. Akad. Nauk. UzSSR* **1985,** *281,* 591.
15. *High Performance Biomaterials: A Comprehensive Guide to Medical and Pharmaceutical Applications;* Szycher, M., Ed.; Technomic: Lancaster, PA, 1991.
16. Pretula, J.; Kaluzynski, K.; Penczek, S. *Macromolecules* **1986,** *19,* 1797.

24
"Intelligent" Polymers

Allan S. Hoffman

One can define "intelligent" polymers as those polymers that respond with large property changes to small physical or chemical stimuli. These polymers may be in various forms, such as in solution, on surfaces, or as solids. One may also combine intelligent aqueous polymer systems with biomolecules to yield a large family of polymer–biomolecule systems that respond intelligently not only to physical and chemical stimuli, but also to biological stimuli. This chapter overviews such interesting and versatile polymer systems.

The term "intelligent" polymers refers to soluble, surface-coated, or cross-linked polymer systems that exhibit relatively large and sharp physical or chemical changes in response to small physical or chemical stimuli. Although the well-known glass and melting transitions of solid polymers can fit within this definition, most of the recent interest in intelligent polymer systems focuses on aqueous polymer solutions, interfaces, and hydrogels. Intelligent polymers are also sometimes called smart, stimuli-responsive, or environmentally sensitive polymers. Figure 24.1 shows schematically such aqueous polymer systems in solution, on surfaces, or as hydrogels.

Many physical stimuli have been applied, such as changes in the following:

- temperature,
- electric fields,
- solvents,
- light,
- stress,
- sound, and
- magnetic fields.

Chemical or biochemical stimuli also have been used, such as changes in the following:

- reactants,
- pH,
- ions, and
- recognition.

The many potential responses to these stimuli are changes in the following:

- phase,
- shape and volume,
- optical transmission,
- mechanics,
- electrical signals,

Copyright ©1997 American Chemical Society

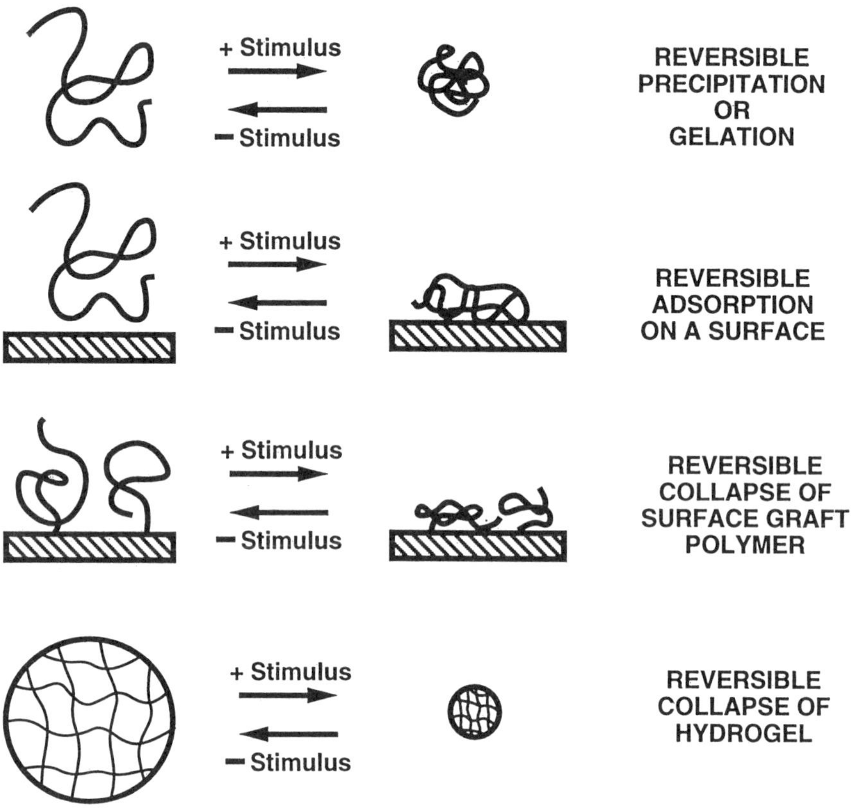

FIGURE 24.1. Schematic examples of intelligent polymer systems in solution, on surfaces, and as hydrogels.

- surface energies,
- reaction rates,
- permeation rates, and
- recognition processes.

Two typical examples of sharp responses are shown in Figure 24.2 for soluble polymers or hydrogels.

Many different properties of the polymer system may change when such sharp responses to stimuli occur. For example, when a soluble polymer is stimulated to precipitate, it will be selectively removed from solution, which will become cloudy. When such polymers are grafted or coated onto a solid support, then one may reversibly change the water absorption into the coated polymer, thus changing the wettability of the surface. When a hydrogel is stimulated to collapse, it will squeeze out its pore water, turn opaque, become stiffer, and shrink in size. One may take advantage of one or more of these *signals* for different end uses. A significant amount of research on these interesting systems has been carried out over the past 10–15 years, producing many publications and patents (*1–15*).

A number of possible molecular mechanisms can cause such sharp, sometimes discontinuous transitions in polymer systems, such as

- ionization or neutralization,
- ion exchange,
- ion–ion repulsion or attraction,
- release or formation of hydrophobically bound water,
- helix–coil transition,
- onset or inhibition of chain mobility,

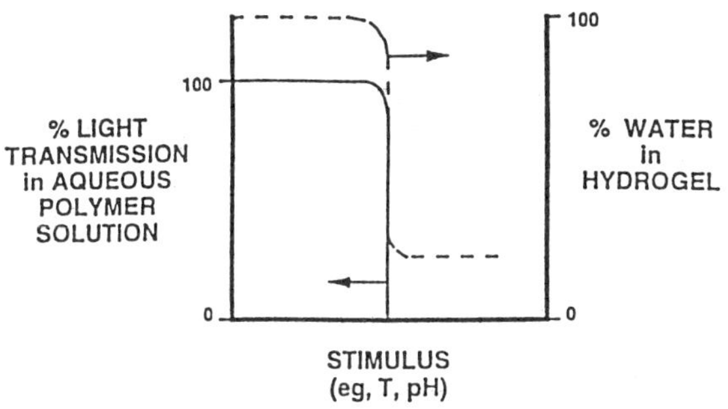

FIGURE 24.2. Typical responses to a stimulus for aqueous-based intelligent polymers in solution and as a hydrogel. The transition shown for the solution is called the cloud point (CP) or the lower critical solution temperature (LCST).

- crystallization or melting,
- isomerization between hydrophilic and hydrophobic forms,
- counterion movement in an electric field, and
- electron-transfer redox reaction.

Water is involved in most of these mechanisms. Numerous publications describe and discuss such mechanisms in both natural and synthetic polymers (*16–21*).

Temperature-Sensitive Polymers and Copolymers

Temperature-sensitive smart polymers have been studied extensively over the past 5–10 years (*1–3, 6–15, 22*). Many polymers exhibit a cloud point (CP) or lower critical solution temperature (LCST) in aqueous solutions (*12*). Some examples of thermally sensitive polymers with ether groups are as follows:

- poly(ethylene oxide) (PEO),
- poly[EO–propylene oxide (PO)] random copolymers,
- PEO–PPO–PEO triblock surfactants,
- alkyl-PEO block surfactants, and
- poly(vinyl methyl ether).

Examples of thermally sensitive polymers showing LCST behavior in aqueous solutions that have alcohol groups are as follows:

- hydroxypropyl acrylate,
- hydroxypropyl methylcellulose,
- hydroxypropyl cellulose,
- hydroxyethyl cellulose,
- methylcellulose, and
- poly(vinyl alcohol) derivatives.

Examples of thermally sensitive polymers showing LCST behavior in aqueous solutions that have *N*-substituted amide groups are as follows:

- poly(*N*-substituted acrylamides),
- poly(*N*-acryloyl pyrrolidine),
- poly(*N*-acryloyl piperidine),
- poly(acryl-L-amino acid amides), and
- poly(ethyl oxazoline).

One property that is common to these water-soluble or -insoluble polymers is that they each have a balance of hydrophilic and hydrophobic groups. The main mechanism of a thermally induced phase separation is the release of hydrophobically bound water. This is the mechanism of precipitation as well as of physical adsorption of a soluble LCST polymer onto a solid polymer substrate (*23–25*). If one increases or decreases the relative hydrophilic content of the temperature-sensitive polymer, such compositional changes will usually cause an increase or decrease, respectively, in the LCST, and it will have a similar effect on the tendency of the polymer to physically adsorb onto a particular solid polymer substrate (*25–27*).

We have been studying temperature-sensitive polymer systems with more than one stimulus-response sensitivity; these combination copolymers can exhibit very interesting properties, with many new and novel applications. If one combines temperature sensitivity with pH sensitivity in the same polymer, then the LCST of the copolymer may be especially sensitive to the pH because of the strong hydrophilic character of the ionized state of the pH-sensitive component. In a random vinyl copolymer containing both temperature-sensitive and pH-sensitive monomers, only a small fraction (e.g., 10 mol%) of a pH-sensitive monomer may be sufficient to completely eliminate the LCST phenomenon of the major, temperature-sensitive component when the pH is raised above the p*K* of the pH-sensitive component (*27–29*).

We recently extended these concepts to the synthesis of novel intelligent graft and block copolymer structures where the backbone polymer and grafted or block polymer chains each have a different stimulus-response sensitivity (*30*). In contrast to random copolymers, each of the responsive properties is retained over a wide range of conditions, because the individual, long block or graft segments in these types of copolymers tend to behave independently of each other. Figures 24.3 and 24.4 (*29–31*) illustrate this characteristic schematically for graft versus random copolymers prepared from monomers whose homopolymers have either temperature or pH sensitivities. Table 24.1 presents some typical data for homopolymers, random copolymers, and graft copolymers, all in solution (*29–31*). We found similar behavior of hydrogels based on these graft copolymers, and also for block copolymers in solution.

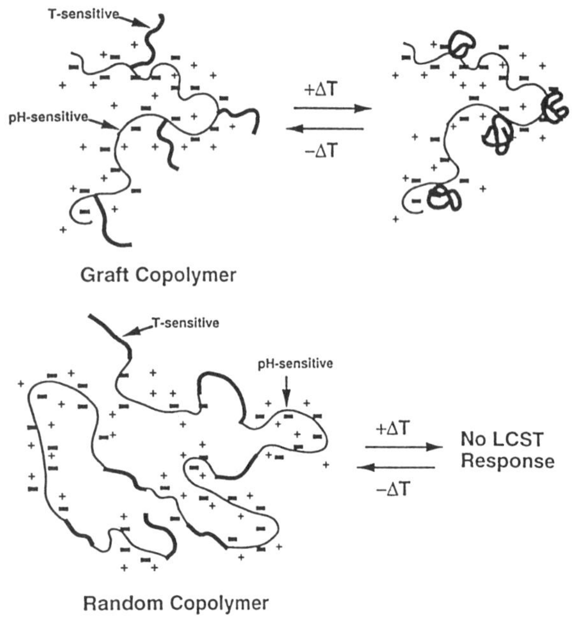

FIGURE 24.3. Temperature-induced responses of random vs. graft copolymers above the pK of the pH-sensitive component.

Biomolecules and Intelligent Polymers

A large number of biologically active molecules (biomolecules) may be combined with intelligent polymer systems. Table 24.2 lists many of these biomolecules, and a soluble intelligent polymer–biomolecule conjugate is shown schematically in Figure 24.5. The biomolecules may be conjugated to pendant groups along a polymer backbone or to one or both terminal ends of the polymer. In either case, the smart polymer may be a soluble polymer, a graft or block copolymer, a physically adsorbed polymer on a solid substrate, or a polymer chain segment within a hydrogel.

One of the important advantages of the conjugation of a biomolecule to a polymer molecule is the possibility of conjugating many of the same biomolecule to the same polymer molecule, thereby providing the opportunity for significant amplification of the biological activity (Figure 24.6).

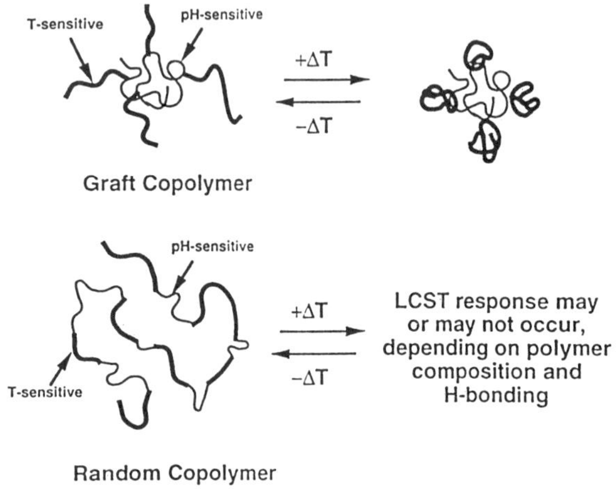

FIGURE 24.4. Temperature-induced responses of random vs. graft copolymers below the pK of the pH-sensitive component.

TABLE 24.1. Comparison of LCSTs of Random vs. Graft Copolymers of (*N*-isopropyl acrylamide) (NIPAAm) and Acrylic Acid (AAc)

Copolymer	*NIPAAm (wt%)*	*AAc (wt%)*	*LCST (°C), pH 4.0*	*LCST (°C), pH 7.4*
Random				
R-1	100	0	31	33
R-2	93	7	32	64
R-3	88	12	35	>95
R-4	79	21	38	>95
R-5	67	33	61	>95
R-6	57	43	>95	>95
R-7	46	54	>95	>95
Graft[a]				
AN-1	45	55	22	32
AN-2	28	72	22	32
G-20	19	81	16	32
G-25	24	76	16	32
G-30	29	71	16	32
G-50	49	51	16	32

[a]The AN samples were prepared by copolymerization of the macromonomer of NIPAAm with AAc. The G samples were prepared by coupling oligoNIPAAm onto PAAc through the reaction of the amino-terminal group of oligoNIPAAm with the carboxyl group of PAAc.

Another advantage is to be able to conjugate different biomolecules (e.g., an antibody and a drug) to the same polymer backbone. The biomolecule may also be physically entrapped within a hydrogel, either permanently as in the case of a large protein, or temporarily as in the case of a small drug molecule. There are many diverse biomedical and biotechnological applications of environmentally sensitive smart polymeric biomaterials, especially if they contain immobilized biomolecules. Table 24.3 lists examples of such applications.

Soluble Stimuli-Responsive Polymers

Soluble, environmentally sensitive polymers in aqueous solutions can be precipitated at specific environmental conditions. Such systems can be useful as temperature or pH indicators, or as on–off light transmission switches.

TABLE 24.2. Biomolecules Immobilized on or Within Intelligent Polymeric Biomaterials

Type	*Example*
Proteins and peptides	Enzymes
	Antibodies
	Antigens
	Cell adhesion molecules
	Blocking proteins
Saccharides	Sugars
	Oligosaccharides
	Polysaccharides
Lipids	Fatty acids
	Phospholipids
	Glycolipids
Other	Conjugates or mixtures of the above
	Labels
	Chromophores
	Therapeutic isotopes
	Stealth molecules
Drugs	Antithrombogenic agents
	Anticancer agents
	Antibiotics
	Contraceptives
	Drug antagonists
	Peptide and protein drugs
Ligands	Hormone receptors
	Cell surface receptors (peptides, saccharides)
	Avidin, biotin
Nucleic acids and nucleotides	Single- or double-stranded DNA, RNA (e.g, antisense oligonucleotides)

Three applications of intelligent polymer–biomolecule conjugates that we have been studying are illustrated schematically in Figure 24.7. In the first example in this figure, a biomolecule is conjugated to the polymer, and then it is selectively phase-separated from the solution by a small change in environmental conditions. In this way, an enzyme in a bioprocess may be easily phase-separated and recycled, also permitting easy recovery of the product at the same time (*32*).

In the second example in Figure 24.7, a recognition biomolecule or receptor ligand such as a

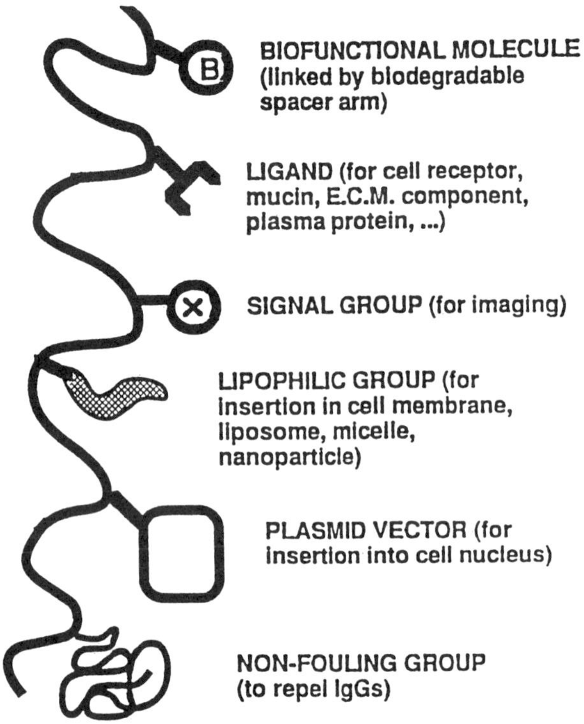

FIGURE 24.5. Schematic illustration of a variety of natural or synthetic biomolecules conjugated to an intelligent polymer backbone. In some uses only one type of biomolecule may be conjugated (although more than one of the same biomolecule may be conjugated), whereas in other uses more than one type of biomolecule is needed. Sometimes the intelligent polymer molecule may be conjugated by means of a reactive terminal group directly to the biomolecule.

cell-receptor peptide, protein A or G, streptavidin, or an antibody is conjugated to an intelligent polymer and used in a precipitation-induced affinity separation process. When mixed with a complex solution, the conjugate will selectively complex its binding partner and then it can be readily and cleanly separated by providing a stimulus (e.g., a small change in temperature), which causes the polymer–ligand conjugate or polymer–ligand–receptor complex or complex to precipitate. We have applied such a thermally induced affinity precipitation process to recover immunoglobulin G

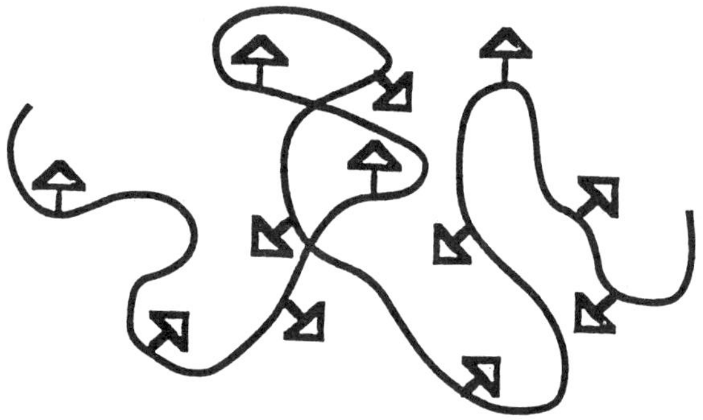

FIGURE 24.6. Illustration of the possibility of amplification of the action of a particular biomolecule (e.g., a drug, ligand, or enzyme) by conjugating many biomolecules to the same intelligent polymer molecule. The polymer may be in solution, on a surface, or in a hydrogel.

from solution by using a poly(NIPAAm)-protein A conjugate (*33*), and we are currently using this type of process to recover the CD-44 cell membrane receptor from cell membrane lysates.

In the third example in Figure 24.7, the affinity precipitation principle is extended to an immunoassay, where the affinity precipitate is the immune complex sandwich of polymer–antibody (no. 1) conjugate, which is coupled with its antigen, which is then complexed with another, labeled antibody (no. 2) (*34*). Such intelligent polymer-based affinity separations are potentially more efficient than traditional methods, such as affinity chromatography or the enzyme-linked immunosorbent assay (ELISA), both of which involve solid surfaces and the accompanying problems of antibody or ligand desorption and nonspecific adsorption.

We and others have conjugated lipids, such as fatty acids, fatty ethers, cholesterol, or phospholipids to temperature-responsive polymers (*35–38*). In one case, the conjugate of poly(NIPAAm) and a phospholipid was sonicated in an aqueous environment to simultaneously form liposomes and to phase-separate the conjugated polymer, causing a gel to form as the sonication caused the temperature to rise above the LCST (*38*). Presumably, the

TABLE 24.3. Applications of Intelligent Polymers and Combinations with Biomolecules

Type	*Application*
Polymers in solution	Reversible light transmission (sensors, switches) Precipitation separations Affinity precipitations (separations, sensors, and diagnostics) Phase-transfer catalysis Binding to and stimulating cells (cell separations, expression, endo- or exocytosis, lysis, etc.)
Polymers on surfaces[a]	Wettability changes Cell and protein attachment or detachment (implants, therapeutic devices) Bioactive surfaces (immobilized enzymes) Affinity separations Permeation switches in microporous membranes Optical indicators (sensors, switches)
Homogeneous or heterogeneous hydrogels	Separations (size or affinity) Drug delivery (pulsed, cyclic, controlled release) Immobilized enzymes, cells (bioprocesses, implants, therapeutic devices) Permeation switches (molecular pores) Robotics

[a]Polymers on surfaces are physically adsorbed or desorbed polymers, graft copolymers, or gels on surfaces.

conjugates had inserted themselves within different liposome membranes and acted as physical cross-links between the liposomes trapped within the gel network. This system was used to encapsulate insulin within the liposomes and then later to deliver it as the gel was cooled below the LCST of the polymer.

Another kind of intelligent polymer molecule in solution is based on the random incorporation along a water-soluble polymer backbone of the key chemical groups of a known recognition sequence from a natural biomolecule, in the same ratio as in the natural recognition molecule (*39*). If the polymer is a stimuli-responsive polymer, one might then be able to change its conformation by temperature change, enhancing the possibility of achieving transient conformations capable of recognition.

We have recently been conjugating intelligent polymers to specific sites on proteins by genetically engineering the protein to place a unique cysteine residue in a selected location and then conjugating the intelligent polymer to that site by reacting the cysteine –SH with a thiol-reactive end group on the polymer (*40, 41*). By this method, we intentionally conjugated poly(NIPAAm) conjugates away from the active site of cytochrome b5 (*40*), and more recently we placed them very close to the biotin binding pocket in streptavidin (SA) (*41*). In the SA case, we demonstrated that above the polymer collapse temperature the biotin was blocked from binding to the SA, whereas below that temperature the binding was complete. Such site-specific placement of intelligent polymers near the binding sites of recognition proteins can provide sensitive environmental control of the protein–ligand recognition process, which is involved in a wide variety of important applications in bioseparations, enzyme reactions, receptor-induced cell responses, drug delivery, and information processing (*41*).

Stimuli-Responsive Polymers on Surfaces

Stimuli-responsive polymers can also be chemically grafted (*42*) or physically adsorbed (*25*) onto

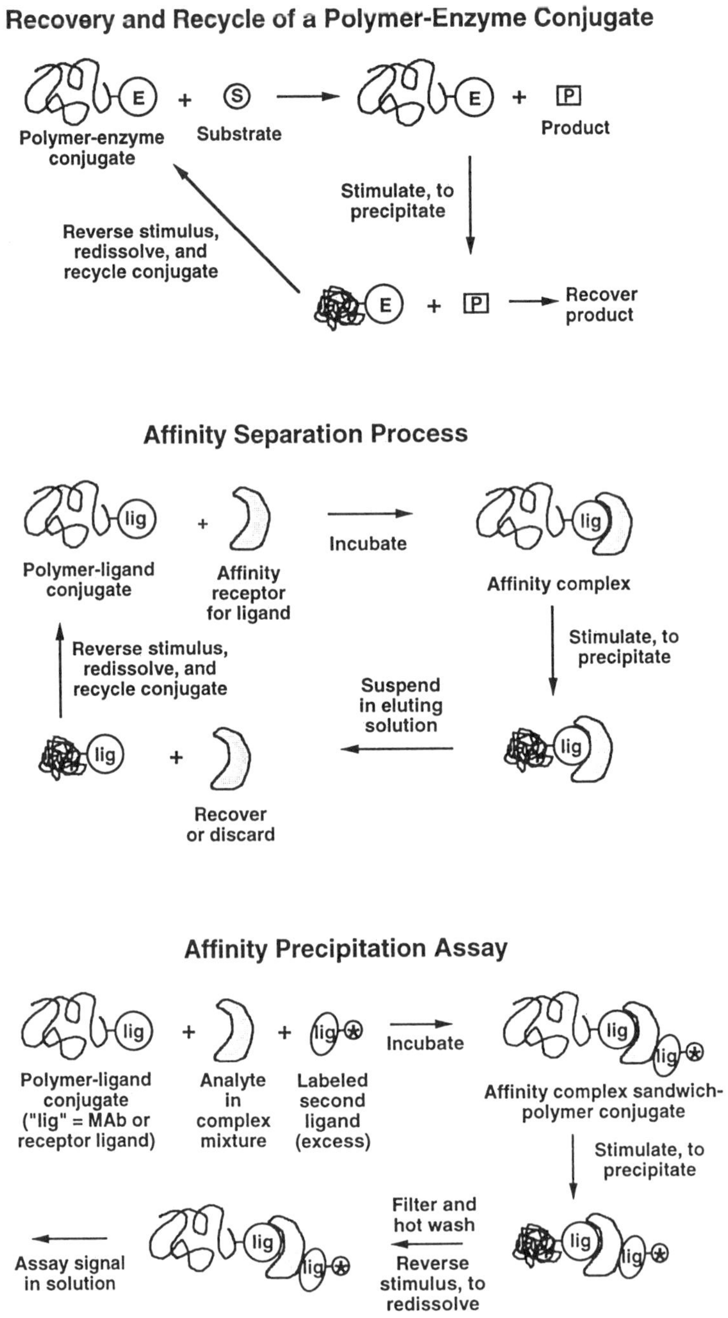

FIGURE 24.7. Three examples of intelligent polymer–biomolecule conjugates. The applications of these three examples include bioprocesses and downstream separations, affinity separations, sensors, diagnostics, and environmental water processing.

solid polymer supports, and then one can rapidly change surface film thickness, wettability, or surface charge in response to small changes in stimuli such as solution temperature, pH, or specific ionic concentrations. These responses can be much faster than for solids as hydrogels because the surface coatings can be very thin. Permeation switches can be prepared by depositing intelligent polymers onto the surfaces of pores in a porous membrane and stimulating their swelling (to block the pore flow) or collapse (to open the pore to flow) (*43–46*).

If proteins or cells are exposed to intelligent polymer surfaces that are in the swollen or collapsed state, they usually will preferentially adsorb on the more hydrophobic surface compositions (*47–50*) (Figure 24.8). If the temperature is cycled, fouling by adsorbed proteins or cells on surfaces may be reduced.

This concept led to another interesting application, wherein cells are reversibly cultured on a chemically grafted LCST polymer surface. After the cells have been cultured for a while, they can be detached without the use of trypsin, simply by reversing the stimulus, and converting the surface to the more hydrophilic condition (*48–50*). Specific cells may be physically separated on such LCST copolymer surfaces by adjusting the polymer composition to the desired cell surface character. If the smart polymer is only physically adsorbed to the surface and cells are cultured on top of it, when the cells are detached by lowering the temperature, they also remove the polymer, forming unusual shapes (*50*). We conjugated the peptide cell receptor ligand, RGD (arginine–glycine–aspartic acid) to special temperature-sensitive polymer compositions (having LCSTs close to ambient temperature) that have been matched to specific polymer support surface compositions. We have shown that cells will bind to surfaces having this conjugate physically adsorbed on them, whereas they will not bind to the LCST polymer-coated surfaces if RGD is not conjugated to the polymer (*51*). Surfaces coated with the RGD–LCST polymer conjugate could be used for thermally reversible cell cultures (Figure 24.9).

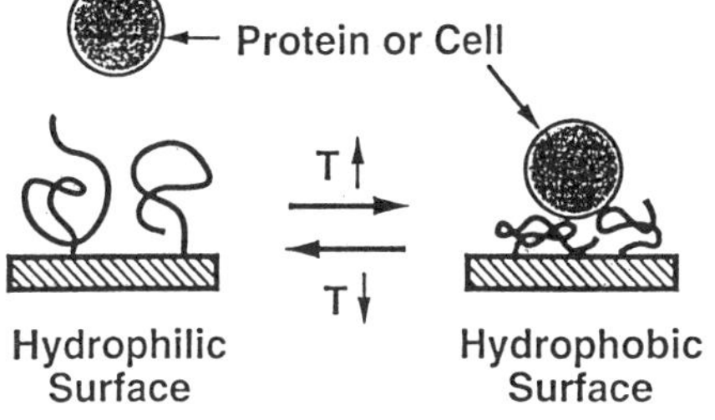

FIGURE 24.8. Use of a stimulus to convert a surface coated with an intelligent polymer from hydrophilic to hydrophobic, and thus from a protein- or cell-repelling surface to one that is more attractive to the protein or the cell.

In another study using such reversibly adsorbed LCST polymer–biomolecule conjugates, we conjugated a monoclonal antibody to an LCST polymer for use in a novel membrane-based immunoassay. When the assay solution is drawn through the microporous membrane after incubation, the temperature-sensitive polymer component of the polymer–antibody conjugate (which is complexed with the antigen) preferentially adsorbs on a protein-resistant cellulose acetate membrane surface at ambient temperature, both because its LCST is close to ambient temperature and also because its composition is designed to preferentially interact with the cellulose acetate (Figure 24.10) (*24*). This isolates the antigen on the membrane, since the rest of the solution is drawn through the membrane. The antigen is then assayed on the membrane surface.

Stimuli-Responsive Hydrogels

The most extensive work on intelligent polymers has been carried out on stimuli-responsive hydrogels, particularly those based on pH-sensitive monomers (e.g., *1, 7, 13, 28, 52–56*) and thermally sensitive monomers (e.g., *1–3, 6–14, 22*). In several interesting systems, the enzymes entrapped

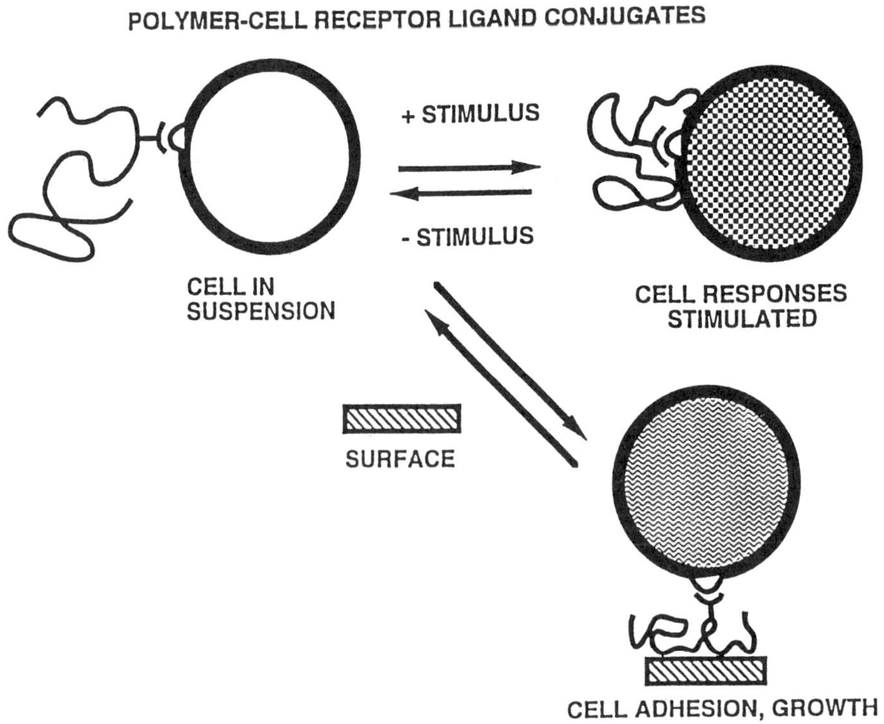

FIGURE 24.9. Cell receptor ligand–intelligent polymer conjugate binding to the cell membrane receptor and stimulating various cell responses, including reversible cell culture on a surface, when the polymer is stimulated to precipitate or redissolve.

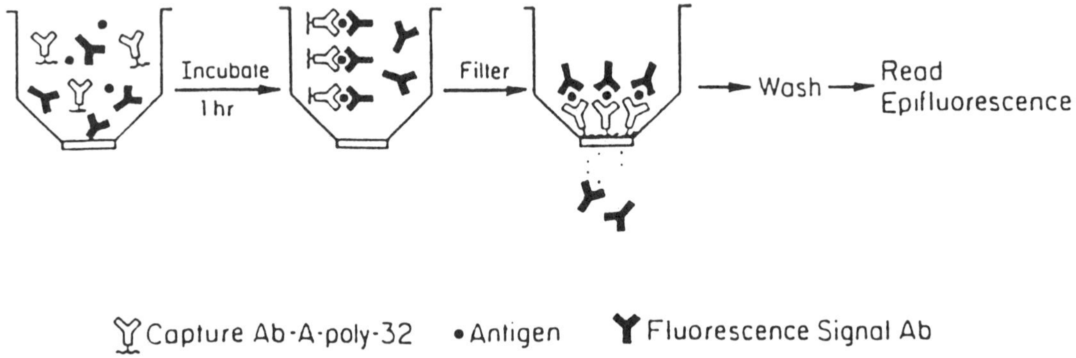

FIGURE 24.10. Novel membrane immunoassay based on selective adsorption of the polymer component of a polymer–antibody conjugate onto the surface of the membrane. The membrane itself is resistant to protein adsorption.

within the hydrogels create a local pH change when their substrate is converted to product, and this process can cause changes in swelling and pore sizes within the pH-sensitive gel (*4, 57*). Stimuli derived from changes in solvent mixtures, specific ions or solutes, pH, temperature, electric fields, or electromagnetic radiation have all been used to cause collapse or swelling of stimuli-responsive hydrogels. This action has led to applications such as desalting or dewatering of protein solutions (*2, 3*), microrobotics and artificial muscles (*13*), on–off immobilized enzyme reactors (*55, 58, 59*), and delivery of drugs (*4–6, 11–14, 28, 31, 38, 54, 56, 60*).

Delivery of drugs has been one of the most extensively studied application areas for stimuli-responsive hydrogels. Figure 24.11 illustrates many of these applications. Application of cyclic temperature or electric field stimuli has resulted in cyclic delivery of physically incorporated drug molecules (*11, 14, 61*). Figure 24.12 shows the new grafted hydrogel structures developed in our laboratories. These hybrid gels can be synthesized to display different combinations of sensitivities to different environmental stimuli (*30, 31*). Figure 24.13 shows how such novel structures may be used as pH- and temperature-sensitive carriers for controlled drug delivery.

When enzymes are immobilized within smart hydrogels, cyclic changes in environmental stimuli can lead to on–off activity of the enzyme due to the cyclic collapse and reswelling of the hydrogel pores. This action can also be used to enhance mass transport of substrate into and product out of immobilized enzyme hydrogels (*see* Bioreactions in Figure 24.11). We demonstrated significant enhancement of bioreactor productivity in such systems (*58, 59*).

If the enzyme is immobilized within a pH-sensitive gel, and the enzyme–substrate reaction produces a local microenvironmental pH change, then the actual stimulus for the resultant swelling or deswelling of the gel is the substrate concentration in the external solution. This type of gel can be used either as a biosensor or as a permeation switch: such as, to permit release of a drug such as

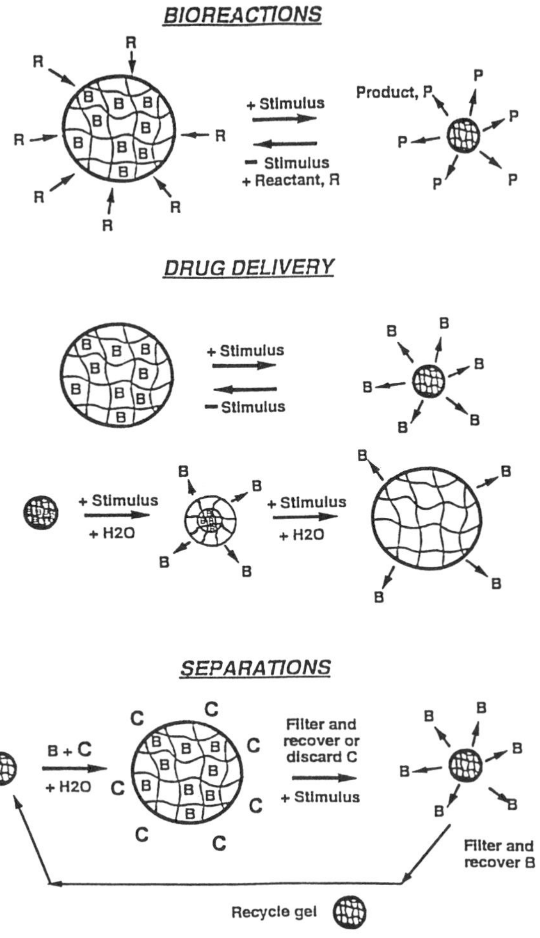

FIGURE 24.11. Various applications of intelligent hydrogels in medicine and biotechnology.

insulin through a swelling gel in response to an increase of a systemic metabolite such as glucose (*see* Drug Delivery in Figure 24.11) (*4, 57, 61*).

The pore sizes in an intelligent polymer hydrogel can be controlled by the environmental conditions and the hydrogel composition (*62*). Thus, such gels can be used to separate molecules on the basis of size (*2, 3, 61*) (*see* Separations in Figure 24.11). If a specific recognition ligand (e.g., an antibody) is immobilized within a stimuli-responsive gel, then its specific binding partner (e.g., antigen) can be selectively removed from

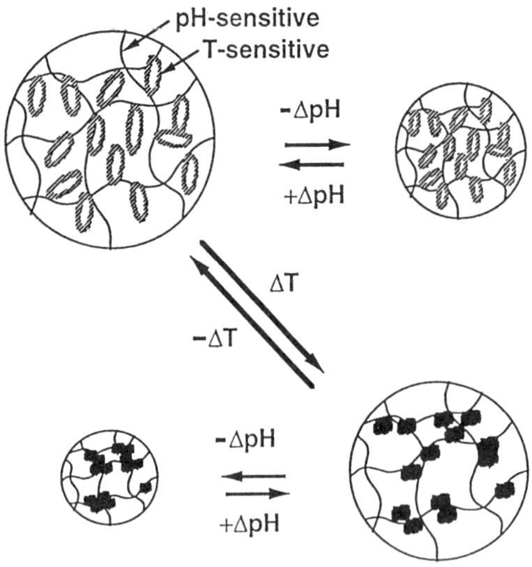

FIGURE 24.12. Individual responses of a dual-sensitivity hybrid intelligent hydrogel to individual temperature or pH stimuli.

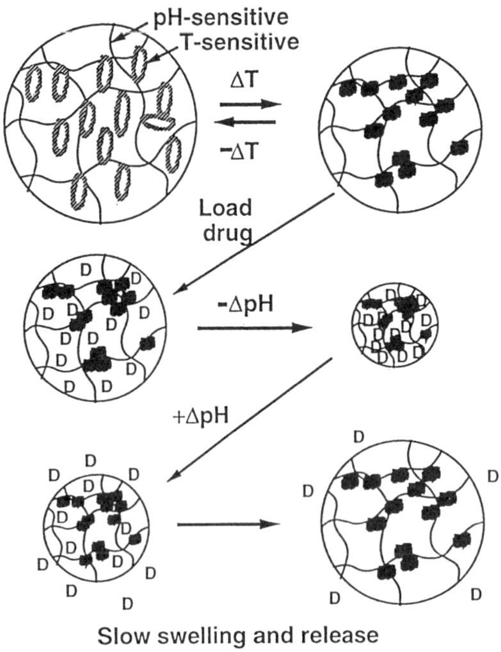

FIGURE 24.13. Loading and release of a drug from a dual-sensitivity, hybrid intelligent hydrogel. The release is stimulated by a rise in pH (e.g., from gastric to enteric conditions) ionizing the matrix of the hydrogel, while control of release rate is provided by the temperature-sensitive grafted chains that form hydrophobic domains above the LCST that resist rapid swelling of the ionized hydrogel.

solution if it is not too large to enter the gel. This process can be used to selectively recover a desired compound or to selectively remove an undesirable compound such as a toxin (*6, 63*). The binding alone could shrink the gel if there are multiple binding sites on the immobilized ligand or its binding partner (*63*).

Kinetics

The kinetics of the various intelligent responses described in this chapter will be very sensitive to the speed of the stimulus and the dimensions of the system being stimulated. Rates of polymer system responses are normally inversely proportional to the square of the smallest dimension of the system. The slowest systems will usually be solid hydrogels, which normally have greater dimensions than do hydrogel coatings. The coatings have greater dimensions than responsive polymers in solution, which should have the highest surface–volume ratio of all systems. Recently, a poly(NIPAAm) gel was synthesized to contain grafted side chains of poly(NIPAAm) oligomers (*64*). This gel collapsed much more rapidly than a conventional cross-linked poly(NIPAAm) gel. This result is presumably due to the initially rapid collapse of the grafted side chains, which then rapidly draw the gel network together by hydrophobic interactions. The speed of the stimulus is also important; it will vary according to the resistances it encounters in reaching the intelligent polymer molecules or segments. Clearly, electromagnetic radiation is the fastest stimulus, at least in systems transparent to the radiation, whereas those stimuli involving diffusion of molecules will be among the slowest.

Conclusions

One can see that the applications of such interesting smart polymer and hydrogel systems are varied and exciting. Also, they provide great opportunities for many diverse and novel applications in medicine and biotechnology.

Acknowledgments

I would like to acknowledge the stimulation I received from the many intelligent and creative contributions of my students and colleagues over the past 10–15 years while working in this exciting field. I would also like to note that the reference list is illustrative and not comprehensive, and I apologize for all omissions.

References

1. Tanaka, T. *Sci. Am.* **1981,** *244,* 124–138.
2. Cussler, E. L.; Stokar, M. R.; Vaarberg, J. E. *AIChE J.* **1984,** *30,* 578–582.
3. Cussler, E. L. U.S. Patent 4,555,344, 1985.
4. Ishihara, K.; Kobayashi, M.; Ishimaru, N.; Shinohara, I. *Polym. J.* **1984,** *16,* 625–631.
5. Heller, J. *Med. Device Diagn. Ind.* **1985,** *7,* 32–37.
6. Hoffman, A. S. *J. Controlled Release* **1987,** *6,* 297–305.
7. Peppas, N. A.; Korsmeyer, R. W. *Hydrogels in Medicine and Pharmacology;* CRC Press: Boca Raton, FL, 1987; Vols. 1–3.
8. Bae, Y. H.; Okano, T.; Hsu, R.; Kim, S. W. *Makromol. Chem. Rapid Commun.* **1987,** *8,* 481–485.
9. Hoffman, A. S.; Monji, N. U.S. Patent 4,912,032, 1990.
10. *Pulsed and Self-Regulated Drug Delivery;* Kost, J., Ed.; CRC Press: Boca Raton, FL, 1990.
11. Okano, T.; Bae, Y. H.; Jacobs, H.; Kim, S. W. *J. Controlled Release* **1990,** *11,* 255–265.
12. Hoffman, A. S. *Artif. Organs* **1995,** *19,* 458–467.
13. *Polymer Gels; Fundamentals and Biomedical Applications;* De Rossi, D.; Kajiwara, K.; Osada, Y.; Yamauchi, A., Eds.; Plenum: New York, 1991.
14. Kwon, I. C.; Bae, Y. H.; Kim, S. W. *Nature (London)* **1991,** *354,* 291–293.
15. Osada, Y.; Okuzaki, H.; Hori, H. *Nature (London)* **1992,** *355,* 242–244.
16. Dusek, K.; Patterson, D. *J. Polym. Sci. Part A* **1968,** *6,* 1209–1216.
17. Verdugo, P. *Biophys. J.* **1986,** *49,* 231.
18. Tanaka, T.; Sun, S. T.; Hirokawa, Y.; Katayama, S.; Kucera, J.; Hirose, Y.; Amiya, T. *Nature (London)* **1987,** *325,* 796–798.
19. Tanaka, T.; Fillmore, D.; Sun, S.-T.; Nishio, I.; Swislow, G.; Shah, A. *Phys. Rev. Lett.* **1980,** *45,* 1636–1639.
20. De Gennes, P. G. *Phys. Lett.* **1972,** *38,* 339–340.
21. Urry, D. W.; Harris, R. D.; Prasad, K. U. *J. Am. Chem. Soc.* **1988,** *110,* 3303–3305.
22. Schild, H. G. *Prog. Polym. Sci.* **1992,** *17,* 163–249.
23. Heskins, M.; Guillet, J. E. *J. Macromol. Sci. Chem.* **1968,** *2,* 1441–1455.
24. Monji, N.; Cole, C.-A.; Tam, M.; Goldstein, L.; Nowinski, R. C. *Biochem. Biophys. Res. Commun.* **1990,** *172,* 652–660.
25. Miura, M.; Cole, C.-A.; Monji, N.; Hoffman, A. S. *J. Biomater. Sci. Polym. Ed.* **1994,** *5,* 555–568.
26. Taylor, L. D.; Cerankowski, L. D. *J. Polym. Sci. Polym. Chem. Ed.* **1975,** *13,* 2551–2570.
27. Priest, J. H.; Murray, S. L.; Nelson, R. J.; Hoffman, A. S. In *Reversible Polymeric Gels and Related Systems;* Russo, P. S., Ed.; ACS Symposium Series 350; American Chemical Society: Washington, DC, 1987; pp 255–264.
28. Dong, L. C.; Hoffman, A. S. *J. Controlled Release* **1991,** *15,* 141–152.
29. Nabeshima, Y.; Chen, G. H.; Hoffman, A. S., University of Washington, Seattle, unpublished results, 1992.
30. Chen, G. H.; Hoffman, A. S. *Nature (London)* **1995,** *373,* 49–52.
31. Chen, G. H.; Hoffman, A. S. *Macromol. Chem. Phys.* **1995,** *196,* 1251–1259.
32. Chen, G. H.; Hoffman, A. S. *Bioconjugate Chem.* **1993,** *4,* 509–514.
33. Chen, J. P.; Hoffman, A. S. *Biomaterials* **1990,** *11,* 631–634.
34. Monji, N.; Hoffman, A. S. *Appl. Biochem. Biotechnol.* **1987,** *14,* 107–120.
35. Nightingale, J. A. S.; Grashin, J.; Hoffman, A. S. *Trans. Soc. Biomater. (Minneapolis, Minn.)* **1987,** *10,* 56.
36. Ringsdorf, H.; Venzmer, J.; Winnik, F. M. *Macromolecules* **1991,** *24,* 1678–1686.

37. Schild, H. G.; Tirrell, D. A. *Langmuir* **1991,** *7,* 1319–1324.
38. Wu, X. S.; Hoffman, A. S.; Yager, P. *J. Intell. Mater. Syst. Struct.* **1993,** *4,* 202–209.
39. Tardieu, M.; Gamby, C.; Avramoglou, T. *J. Cell. Physiol.* **1992,** *150,* 194–203.
40. Chilkoti, A.; Chen, G.; Stayton, P. S.; Hoffman, A. S. *Bioconjugate Chem.* **1994,** *5,* 504–507.
41. Stayton, P. S.; Shimoboji, T.; Long, C.; Chilkoti, A.; Chen, G.; Harris, J. M.; Hoffman, A. S. *Nature (London)* **1995,** *378,* 472–474.
42. Uenoyama, S.; Hoffman, A. S. *Radiat. Phys. Chem.* **1988,** *32,* 605–608.
43. Tirrell, D. A. *J. Controlled Release* **1987,** *6,* 15–21.
44. Iwata, H.; Oodate, M.; Uyama, Y.; Amemiya, H.; Ikada, Y. *J. Membr. Sci.* **1991,** *55,* 119–130.
45. Osada, Y.; Honda, K.; Ohta, M. *J. Membr. Sci.* **1986,** *27,* 327–338.
46. Yoshida, M. et al. *Radiat. Eff. Defects Solids* **1993,** *126,* 409–412.
47. Kawaguchi, H.; Fujimoto, K.; Mizuhara, Y. *Colloid Polym. Sci.* **1992,** *270,* 53–57.
48. Okano, T.; Yamada, N.; Sakai, H.; Sakurai, Y. *J. Biomed. Mater. Res.* **1993,** *27,* 1243–1251.
49. Yamada, N.; Okano, T.; Sakai, H.; Karikusa, F.; Sawasaki, Y.; Sakurai, V. *Makromol. Chem. Rapid Commun.* **1990,** *1,* 571–576.
50. Takezawa, T.; Mori, Y.; Yoshizato, K. *Bio/Technology* **1990,** *8,* 854–856.
51. Miura, M.; Cole, C.-A.; Monji, N.; Hoffman, A. S. *Trans. Soc. Biomater. (Minneapolis, Minn.)* **1991,** *14,* 130.
52. Lavsky, M. *Macromolecules* **1982,** *15,* 782–788.
53. Yu, H. Ph.D. Thesis, Oregon Graduate Institute, Beaverton, 1994.
54. Siegel, R. A.; Falamarzian, M.; Firestone, B. A.; Moxley, B. C. *J. Controlled Release* **1988,** *8,* 179–182.
55. Nakamae, K.; Miyata, T.; Hoffman, A. S. *Makromol. Chem.* **1992,** *193,* 983–990.
56. Dong, L. C.; Yan, Q.; Hoffman, A. S. *J. Controlled Release* **1992,** *19,* 171–178.
57. Albin, G.; Horbett, T. A.; Ratner, B. D. *J. Controlled Release* **1985,** *2,* 153–164.
58. Park, T. G.; Hoffman, A. S. *Appl. Biochem. Biotechnol.* **1988,** *19,* 1–9.
59. Park, T. G.; Hoffman, A. S. *Biotechnol. Bioeng.* **1990,** *35,* 152–159.
60. Bae, Y. H.; Kwon, I. C.; Kim, S. W. In *Polymeric Drugs and Drug Administration;* Ottenbrite, R. M., Ed.; ACS Symposium Series 545; American Chemical Society: Washington, DC, 1994; pp 98–110.
61. Horbett, T. A.; Ratner, B. D.; Kost, J.; Singh, M. In *Recent Advances in Drug Delivery_Systems;* Anderson, J. M.; Kim, S. W., Eds.; Plenum: New York, 1984; pp 209–220.
62. Park, T. G.; Hoffman, A. S. *Biotechnol. Prog.* **1994,** *10,* 82–86.
63. Kokufata, E.; Zhang, Y. Q.; Tanaka, T. *Nature (London)* **1991,** *351,* 302–304.
64. Yoshida, R.; Uchida, K.; Kaneko, Y.; Sakal, K.; Kikuchi, A.; Sakurai, Y.; Okano, T. *Nature (London)* **1995,** *374,* 240–242.

Modeling of Controlled Drug Delivery

25
Modeling of Self-Regulating Oscillatory Drug Delivery

Ronald A. Siegel

In recent years a number of classes of drug therapies have been shown to require a periodic, pulsatile regimen of delivery for efficacy or optimization. These therapies include drugs of tolerance, a wide spectrum of normally endogenous hormones, and cell cycle-specific drugs. Several delivery strategies have been proposed to respond to this need; however, these strategies almost always lead to only a few pulses of drug. For more prolonged periodic pulsatile delivery new design strategies are required. In this chapter one such design strategy is proposed, and the underlying theory of a simple device is discussed. A central feature of the device is a polymeric membrane whose permeability to the substrate of an enzyme-catalyzed reaction is inhibited by the product of that reaction. This negative feedback system can, under certain conditions, lead to oscillations in membrane permeability and in the levels of substrate and product in the device. Any one of these oscillating variables can then be used to drive a cyclic drug delivery process. The theoretical development demonstrates that the ability of the proposed device to oscillate depends critically on device parameters, as well as the nature of the underlying dynamics. The development of such a device will require an intimate interplay between theory and experiments.

The goal of controlled drug delivery is to provide patients with therapeutic and nontoxic drug levels over a desired period of time. For the majority of drugs, this goal can be ensured by maintaining the plasma drug concentration between a minimum effective concentration (MEC) and a minimum toxic concentration (MTC), MTC being higher than MEC for any realistic drug. When the difference between MTC and MEC is relatively small, it becomes advantageous to administer these drugs at a constant, or zero-order rate. By this means the large swings in drug concentration that may occur with multiple dosing are reduced or eliminated, and plasma drug levels are maintained within these bounds.

A diagram comparing zero-order with multiple dosing is shown in Figure 25.1a. Diagrams like this appear in many treatises on controlled release, and they represent a model that is deeply entrenched in the minds of workers in the field. A key underlying

Copyright © 1997 American Chemical Society

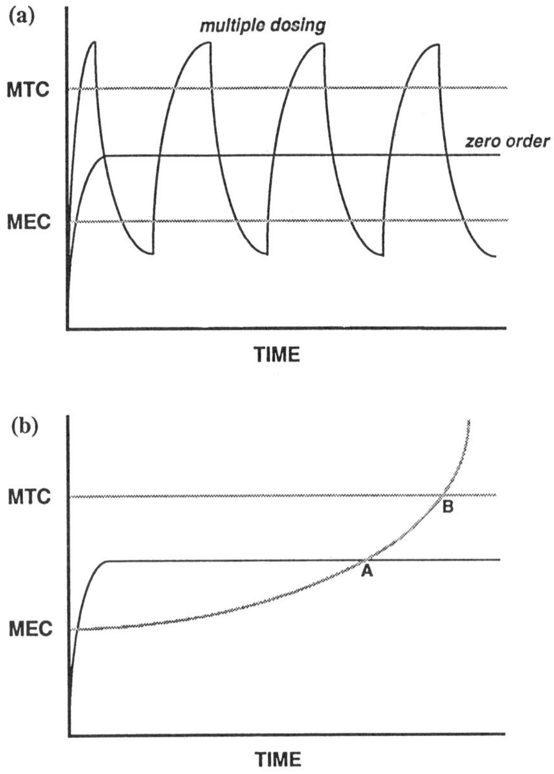

FIGURE 25.1. (a) Comparison of drug concentration profiles resulting from zero-order versus multiple dosing regimens for drugs to which no tolerance develops. Minimum effective concentration (MEC) and mean toxic concentration (MTC) do not change with time. (b) Response of system to zero-order delivery for a drug exhibiting tolerance. After prolonged exposure to drug, MEC starts to increase, crossing the steady-state drug level at point A. At point B, MEC crosses MTC, and no drug level will be efficacious.

assumption in the model embodied in Figure 25.1a is that there is a direct and constant pharmacodynamic relationship between plasma drug level and drug effect at steady state. This assumption is valid for many if not most drugs, which explains why zero-order release is optimal in most cases.

For certain classes of drugs, however, drug effect is determined not only by the present value of plasma drug concentration, but also by the history of drug exposure or the history of the effect itself. An obvious class fulfilling this description is the drugs of tolerance. By definition, a patient exposed to constant levels of such a drug will experience a diminished effect with the passage of time. Eventually the drug effect falls below some minimal level, at which point drug administration should be stopped until the patient can recover his or her sensitivity to the drug.

Figure 25.1b displays a conception of events that occur when a drug of tolerance is administered at constant rate. In this case MEC is not constant, but rather it increases with time due to drug exposure, eventually crossing over the steady-state drug level. Increasing drug input rate can maintain a minimal effect for a while, but eventually MEC will also cross MTC. At this point compensation becomes futile.

For drugs of tolerance, it appears that the most rational approach is to administer drug until tolerance develops to a significant extent, and then withdraw the drug for sufficient time to recover from tolerance. If drug must be administered over long periods of time, this process leads to a periodic dosing schedule. A well-known example is the cardiovascular drug nitroglycerin. Chronic, near zero-order administration of this drug by skin patches leads to development of tolerance to its beneficial effects after about 12 h. Nitroglycerin patches are currently recommended to be worn for 12 h at a time, alternated with 12 patch-free h (*1*).

A number of peptide hormones are released endogenously in a periodic, pulsatile manner (*2–7*). For many of these hormones replacement therapy based on zero-order administration is a failure, and best results occur when the hormones are delivered with a pulse frequency that closely mimics that occurring in a normal individual. A partial list of such hormones, which notably includes growth and sex hormones, is given along with the respective endogenous pulse frequencies in Table 25.1 (*2*).

A close analogy between these hormones and drugs of tolerance can be drawn. Hormones are

TABLE 25.1. Endogenous Pulsatile Secretion Frequencies of Various Hormones in Humans

Hormone	*Pulses/Day*
Growth hormone	9–16, 29
Prolactin	4–9, 7–22
Thyroid stimulating hormone	6–12, 13
Adrenocorticotropic hormone	15, 54
Luteinizing hormone	7–15, 90–121
Follicle stimulating hormone	4–16, 19
β-Endorphin	13
Melatonin	0, 18–24, 12–20
Vasopressin	12–18
Renin	6, 8–12
Parathyroid hormone	23, 24–139
Insulin	108–144, 120
Pancreatic polypeptide	96
Somatostatin	72
Glucagon	103, 144
Estradiol	8–16, 8–19
Progesterone	8–12, 6–16
Testosterone	13, 8–12
Aldosterone	6, 9–12
Cortisol	15, 39

NOTE: Extracted from a review article by Brabant et al. (*2*). Different ranges for a given hormone correspond to different primary references cited in the review article.

SOURCE: Reproduced with permission from reference 61. Copyright 1995 Elsevier.

released from glands to act on particular cells. However, in some cases the target cells eventually become desensitized (i.e., tolerant) to the hormones and must be relieved of exposure to recover their sensitivity (*8*). This result explains why in such a case glands as well as drug delivery systems must release hormones periodically.

Evidence has accumulated over the past few decades that certain cytostatic drugs are most successful when cells are in a particular phase of their cycle. Agents exist that, when administered with a period that *resonates* with the cycle of rapidly dividing (e.g., cancer) cells, can synchronize the cycles of a large fraction of such cells without having much effect on the slower dividing cells of the host. Periodic pulsatile administration of synchronizing agents along with agents whose cytotoxicity is phase-specific may significantly increase the efficiency and reduce toxicity of cancer chemotherapy (*9–12*).

All the examples listed previously point to the need for drug delivery systems that administer drug on a periodic and often pulsatile basis. A number of approaches that are variations on common drug delivery themes can be considered. Most simply, one might return to multiple oral or intravenous (iv) doses, which are known to lead to significant concentration swings when the interdose interval is comparable to or larger than the half-life of the drug in plasma. Even though this method involves no new delivery device development, it is not always feasible. If frequent pulses are required then neither oral nor iv multiple dosing is useful. The peaks and valleys in drug levels will be significantly attenuated if the dosing interval is smaller than the time constant associated with oral drug absorption. Moreover, variations in gastrointestinal transit time can degrade the periodic character of delivery. In the iv case frequent injections are not tolerable to the patient, although pulsing implantable or wearable pumps with percutaneous catheters might be considered.

Transdermal delivery may provide opportunities for periodic delivery. Skin patches can be applied and removed in any desired time sequence. Moreover, delivery across the skin from patches might sometimes be controlled by temporal patterns of thermal (*13*) or electrical (*14*) activation. For drugs that can be delivered across the skin, the efficacy of pulsatile delivery will depend on the characteristic time associated with percutaneous transport.

The problems with simple iv and oral multiple dosing have also spurred consideration of polymeric systems that can release drug in a periodic, pulsatile manner. Surface-bioerodible polymers containing alternating layers of drug and impermeable polymer would provide a straightforward mechanism for obtaining several pulses (*15*). Also, a number of strategies to release drug as a single pulse after a fixed delay exist. Drug can be encapsulated in a surface bioerodible layer (*16–18*) or a layer that will burst osmotically after a time that is determined by the thickness of the layer (*19*). Formulations containing many such single-pulse systems, programmed to release drug at different times, in principle can be produced to release drug in any desired pulse pattern.

The systems described in the last paragraph apparently will be limited for technical reasons to only several pulses: they are unlikely to be designed to produce, say, hundreds or thousands of periodically spaced pulses. (On the other hand, such formulations, if they could be eliminated from the body, could be administered at a much lower frequency than single dose formulations; and this characteristic could result in a significant improvement.) In this chapter we introduce a novel scheme for drug delivery that can oscillate forever, at least in principle. This scheme is inspired by work that has appeared in the past 40 years on chemical, biochemical, and membrane oscillators. At the time of this writing, the *drug delivery oscillator* exists only in theory, although experiments are in progress to construct such a device.

The present chapter is organized as follows. We begin with a brief review of the literature on artificial and natural chemical oscillators. Certain of these oscillators suggest a scheme for drug delivery oscillations, and the theory behind this scheme is outlined. Simplified mathematical descriptions that capture the important features of the scheme are then put forth and analyzed. The effect of varying the mathematical assumptions is examined in some cases. We conclude by discussing the feasibility of constructing a drug delivery oscillator based on the proposed scheme, using what has been learned from the modeling studies.

Mathematical modeling is essential in identifying device structural properties and parameters that are consistent with oscillatory behavior. To date, mathematical modeling has rarely been important in the design of controlled-delivery systems, although it has been useful in gaining insight into processes governing release behavior. We contend that for the proposed systems, modeling and design go hand-in-hand.

Chemical and Biochemical Oscillators

In closed systems, all chemical reactions eventually come to an equilibrium that represents a state of lowest Gibbs free energy. For decades after the original formulation of chemical thermodynamics it was generally accepted that reactions advance toward equilibrium in a monotonic, nonoscillatory fashion. Even though this assumption holds for a majority of known chemical reactions, a number of exceptions have been identified. In the 1920s Bray (*20*) discovered an inorganic reaction that displays several oscillations in component concentrations before finally reaching equilibrium (*20*). This work was generally dismissed as violating the second law of thermodynamics, although this law actually does not preclude such a path to equilibrium. A similar disbelief greeted Belousov's observations in the 1950s of oscillations in a reaction involving the cerium-catalyzed oxidation of citric acid by bromate ion (*21*). However, Zhabotinsky performed a number of mechanistic experiments with Belousov's reaction that established beyond doubt that oscillations are possible in chemical reactions (*22*). The Belousov–Zhabotinsky (BZ) reaction is presently the best studied chemical oscillator, although many other such reactions have been identified and studied (*23–27*).

Much of the interest in chemical oscillators is due to the belief that these reactions may serve as prototypes for oscillatory patterns seen in biological processes. Indeed, periodic processes are seen over numerous time scales in biology (*28*). Endogenous hormone release, as discussed previously, is one example. The literature on biochemical and biological oscillators is far too extensive to review here and readers are referred to several excellent reviews (*29–33*). However, we will discuss later a certain class of biochemical oscillators that provides hints as to how one might design oscillating drug delivery systems.

Even before Bray's initial discovery, Lotka (*34*) proposed a simple kinetic scheme for chemical oscillation. This work received little attention, although similar ideas by Lotka have been widely used in ecology. Later, Prigogine and co-workers proved by nonequilibrium thermodynamic analysis that if reactions are sufficiently far from equilibrium, then oscillatory behavior is indeed possible. They also proposed a reaction scheme, called the Brusselator, whose kinetic equations lead to sustained oscillations (*35*). (Prigogine was awarded the Nobel Prize in 1978 for this work).

A detailed reaction mechanism for the BZ reaction, along with a set of kinetic equations, was proposed by Field and Noyes in 1974 (*36*). Since then many theoretical and experimental studies of BZ have been published, and interesting phenomena such as wave propagation in spatially inhomogeneous systems have been observed and analyzed. At present the theory behind chemical oscillators is at a stage where conditions for oscillations have been identified, and this theory has allowed researchers to design new oscillators systematically (*26, 27*). Among these oscillators are pH oscillators (*37*). Recently, Giannos et al. (*38*) argued that the coupling of a pH oscillator with a drug whose flux through a membrane is pH-sensitive can lead to oscillations in delivery rate of that drug. This approach is an intriguing alternative to the strategy to be discussed below, although it appears to be limited to providing a finite number of pulses.

As already noted, oscillations cannot be sustained forever in a closed system, because such a system must eventually reach equilibrium. When such a system is replaced by an open or a partially open system, however, new possibilities arise. For example, if reactants are maintained at constant activity and reaction products are continuously removed, then equilibrium will never be achieved. In open or partially open systems the steady state, in which reactant inflow and consumption are just matched by product formation and removal, takes on the role that equilibrium plays in closed systems. Because reactants never deplete and products do not accumulate in open or partially open systems, reactions that exhibit only transient oscillations in closed systems can now exhibit sustained oscillations.

Two common ways to convert a closed system to an open or partially open system exist. The first is to use a continuously stirred tank reactor (CSTR) (*24–27*). Here, a reactor (e.g., a beaker) is connected to flow lines, and reactants are provided at the input port and reactor contents are removed at the output port. In the simplest and most common case volume inflow exactly matches volume outflow. Concentrations are made uniform in the reactor by stirring. In this setup, not only reactant concentrations at inflow, but also flow rate, are variables that determine system behavior. Because flow rate is readily controlled, the CSTR has become a popular vehicle for studying oscillating reactions, including enzyme reactions (*39*).

The second method for realizing an open or partially open system is to place a semipermeable membrane that is permeable to reactants and products between the reaction space and a reservoir containing reactants at constant activity (*40*). The reservoir also serves as a sink for reaction products. These membrane-based systems, even though less convenient than the CSTR for studying oscillating reactions, are more realistic in terms of modeling biological systems that exhibit oscillations as well as the potential drug delivery systems that are the main focus of this chapter. In addition, membrane-based partially open systems containing enzymes have been studied experimentally, and oscillations in reaction intermediate concentrations were demonstrated (*41, 42*).

In the systems described in the previous paragraph, the membrane plays a purely passive role in controlling the rate of entry of reactants and the rate of exit of products. The presence of a membrane, however, can lead to some interesting new possibilities. In the 1950s Teorell (*43, 44*) demonstrated that a steady electric current through a charged membrane separating two compartments at different ionic strengths can lead to oscillations in electric potential as well as volume flow between the two compartments. Subsequent experimental and theoretical studies showed that the behavior of Teorell's oscillator could be attributed to coupled electrical, chemical, and volume flows with the membrane itself. Thus, the membrane becomes an active participant in the oscillation (*45, 46*). Other membrane oscillators based on lipids have also been discovered (*47–51*). Again, an imposed electric current or field is needed for oscillations to occur. It appears that the ability of lipids to undergo phase transitions is important for the development of sustained oscillations.

Electric current-free membrane-based oscillators have been devised. One class of systems involves a membrane separating substrates from a compartmentalized enzyme or enzymes (*41–43*). With a proper combination of autocatalytic and inhibitory kinetics in the enzyme reactions, such systems have been shown to oscillate. The role of the membrane is to introduce mass-transfer limitations; without the membrane, oscillations could not occur in the systems studied. Another interesting system was studied by Caplan et al. (*52*) and later by Chay (*53, 54*). A hydrogel containing papain was shown theoretically to exhibit spatiotemporal oscillations in pH when exposed to the substrate benzoyl-L-arginine ethyl ester (BAEE). In this system, an interplay between the autocatalytic nature of the enzyme reaction and the mass-transfer kinetics of BAEE and the reaction product (H^+) was shown to be the source of the oscillatory behavior. Experimental confirmation of the existence of such oscillations was published by Naparstek et al. (*55*), but the pattern was difficult to maintain. Moreover, experimentally observed oscillation frequencies differed substantially from those predicted by theory.

An important analysis of membrane-based oscillating systems was published in 1968 by Katchalsky and Spangler (*56*). These authors considered a reactor containing an enzyme that communicates with the outside world through a membrane whose permeability to the enzyme product depends on the concentration of the product inside the reactor. Substrate for the enzyme reaction is assumed to be present at constant activity. (This condition can be met either by having the substrate present in the reactor above its solubility or by constant flux of substrate into the reactor.) If permeability to product increases with increasing product concentration, and if the relation between membrane permeability and product concentration exhibits hysteresis, then conditions can be found in which chemical oscillations will occur.

Another model for an oscillator that includes a responsive membrane was proposed by Hahn, Ortoleva, and Ross (*40*). These authors assumed that substrate was made available to the enzyme by permeation across the membrane from a constant-activity reservoir. In this case, permeation of substrate is assumed to be inhibited by product formation. Hahn et al. (*40*) showed that concentration oscillations were possible in such a system if the enzyme reaction could be essentially considered as two or more reactions in series, with roughly equal rate constants. Hysteresis in membrane permeability to substrate is not required for sustained oscillations to occur in this model, which is a starting point for the model for oscillatory drug delivery.

Before concluding this section, we discuss a class of models for biochemical oscillations that have, along with the last two references, inspired the present work. The regulation of cellular processes such as protein and de novo amino acid synthesis has been of considerable interest to cell biologists. Figure 25.2a is a schematic of a model, initially proposed by Goodwin, for regulation of protein synthesis (*32, 57*). In this model, an enzyme is produced by transcription of its gene's DNA into messenger RNA (mRNA), which is subsequently translated into the enzyme itself. The enzyme then catalyzes the conversion of some substrate to a product. Finally, the product inhibits the transcription process, thus providing negative feedback. Meanwhile the enzyme is degraded at a rate determined by its concentration. Under a variety of conditions, the scheme proposed by Goodwin can stabilize the steady-state enzyme concentration in the cell. However, if the rate constants associated with translation, degradation, and enzyme reaction are of similar order of magnitude, and the product inhibition exhibits sufficiently high cooperativity, then the steady state becomes unstable and enzyme concentrations will oscillate.

A generalization of the Goodwin model that was considered by a number of authors (*58–62*) is illustrated in Figure 25.2b. A chain of reactions converts an initial substrate to a final product, which inhibits the first reaction in the chain. The

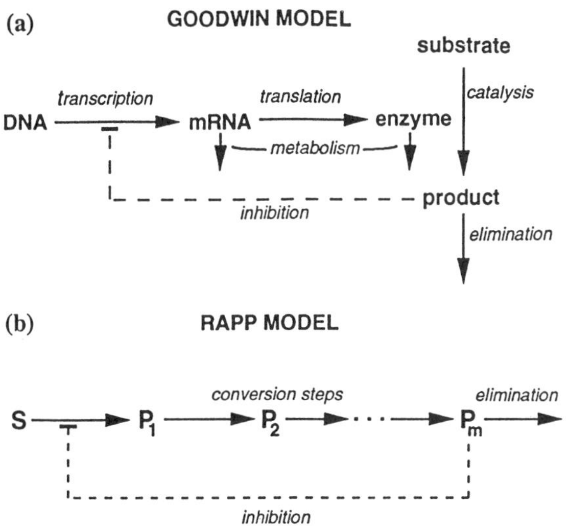

FIGURE 25.2. (a) Goodwin model for regulation of protein synthesis in cells based on product inhibition of gene activation. (b) Generalization of Goodwin model involving a cascade of reaction steps, with the final product inhibiting the first reaction step.

most interesting case is where the reactions in the chain all are first order with approximately the same rate constant. In this case the required degree of (negative) cooperativity for the product inhibition step decreases as the number of steps in the reaction chain increases. This important observation will figure in the discussion of the drug delivery scheme we will propose in the next sections.

Membrane-Based Drug Delivery Oscillator

Figure 25.3 illustrates a possible scheme for a drug delivery oscillator (*61, 62*). The device consists of a chamber containing the drug to be delivered (D), plus an enzyme (enz). The chamber is partially bounded by a membrane that controls solute influx and efflux. The substrate (S) for the enzyme is present externally at constant concentration. Substrate diffuses through the membrane into the chamber and is converted by the enzyme to a product (P). The product in turn diffuses out of the membrane. However, accumulation of product in the chamber inhibitorily affects the permeability of the membrane to the substrate. That is, increasing product concentration causes decreasing flux of substrate into the device.

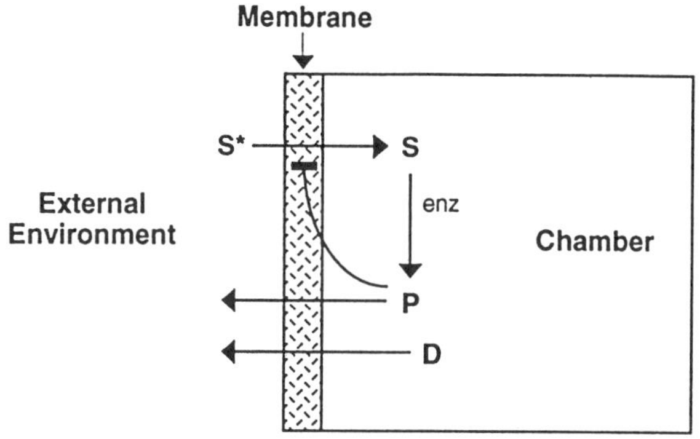

FIGURE 25.3. General scheme for proposed drug delivery oscillator.

The scheme in Figure 25.3 allows a number of means to control drug delivery. Drug solubility could be affected by substrate or product concentration, which will be seen to oscillate under favorable conditions. Alternatively, the permeability of the membrane to drug can oscillate with time along with the substrate permeability. In this chapter we will not be concerned with the particular mechanism of drug delivery; we will only investigate conditions for producing oscillations in those factors that affect drug delivery rate. We do note, however, that the rate of drug transport across the membrane must not be limiting if oscillations in those factors are to be translated into oscillating drug delivery (*38*).

Although in the present chapter we will be primarily concerned with the theory of oscillators based on Figure 25.3, it is worthwhile to discuss briefly a plausible realization of that scheme. Glucose is abundant in the bloodstream and is maintained at nearly constant levels in nondiabetic individuals. In the presence of oxygen glucose is converted to gluconic acid by the enzymes glucose oxidase (GluOx) and gluconolactonase (GluLac), according to the reactions

$$\text{Glucose} + O_2 + H_2O \xrightarrow{\text{GluOx}} \text{Gluconolactone} + H_2O_2 \xrightarrow{\text{GluLac}} \text{Gluconate}^- + H^+ + H_2O_2 \quad (1)$$

The peroxide product can be removed in the presence of excess catalase (Cat):

$$H_2O_2 \xrightarrow{\text{Cat}} H_2O + \tfrac{1}{2}O_2 \quad (2)$$

The net reaction, then, is

$$\text{Glucose} + O_2 \rightarrow \text{Gluconate}^- + H^+ + \tfrac{1}{2}O_2 \quad (3)$$

The dissociated form of gluconic acid is predominant at pH values well above its $pK = 3.6$.

In the scheme of Figure 25.3, then, we can identify glucose as the substrate S and hydrogen ion as the product P. Moreover, the enzyme reaction (eq 1) converting S to P in this case may be regarded as either a single first-order reaction or as the cascade of two first-order reactions, depending on whether the rate constants associated with the GluOx and GluLac steps are significantly different or comparable, respectively. In making this assertion we assume that the reactions are occurring in the linear, nonsaturated regimes.

To complete the realization of the proposed scheme, we require a membrane whose permeability to glucose is inhibited by increasing proton concentration. Polyacid gel membranes fit this requirement. In fact, very sharp decreases in membrane permeability to glucose have been shown to occur when the membrane consists of a copolymer of methacrylic acid and the lower critical solution temperature monomer *N*-isopropylacrylamide (*62*). Other polymer chemistries may also work as well or better.

Mathematical Modeling of Membrane Oscillator

To investigate the feasibility of the proposed scheme for a drug-delivery oscillator, we develop a mathematical model that summarizes many of the important factors in the scheme. Because the exact mathematical description will depend on the specific system design, and because experimental systems are as of this writing only beginning to be built, the present strategy is to make some simplifying assumptions to ease the modeling task. Later, effects of relaxing these assumptions will be explored.

In our model we assume that transport of substrate (S) into the device is due to passive diffusion with rate proportional to the gradient of S across the membrane. S is then assumed to be converted to product (P) by a first-order reaction. Thus we assume that the enzyme-catalyzed conversion is in its linear range. We also assume that only one enzyme is reaction limiting. Thus, in the GluOx–GluLac reactions (eqs 1 and 2), one of these reactions, usually eq 1, is rate determining. Finally, product is assumed to be eliminated by a first-order removal process, which could be diffusion across the membrane or a simple first-order reaction.

With these assumptions the following model equations may be written (*61*):

$$\frac{dC_S}{dt} = \gamma K(t)(C_S^\star - C_S) - \kappa C_S \quad (4)$$

$$\frac{dC_P}{dt} = \kappa C_S - (\gamma q)C_P \quad (5)$$

In these equations γ is the ratio of membrane area to chamber volume, κ is the first-order rate constant for conversion of S to P, C_S and C_P are intrachamber concentrations of S and P, respectively, $C_S^\star$ is the constant external substrate concentration, and K and q are the membrane permeabilities to S and P, respectively. Even though q is assumed to be a constant, K in eq 4 is taken to be a function of time, t, to reflect the dependence of K on C_P, which is expected to fluctuate in time. As will be seen, the specification of the effect of dynamics of product inhibition on membrane permeability is important in determining whether the system can oscillate.

Before proceeding, it is useful to note that the time scale of fluctuations of solutions of eqs 4 and 5 depends on the magnitude of the rate constants. We can nondimensionalize time by multiplying it by one of these rate constants. Choosing κ for this purpose, we define the dimensionless time $\tau = \kappa t$, and the previous equations become

$$\frac{dC_S}{d\tau} = K'(\tau)(C_S^\star - C_S) - C_S \qquad (6)$$

$$\frac{dC_P}{d\tau} = C_S - q'C_P \qquad (7)$$

where $K' = \gamma K/\kappa$ and $q' = \gamma q/\kappa$ are nondimensional permeabilities. Numerical results presented subsequently will be expressed in terms of dimensionless time. Translation back into real time is accomplished by setting $t = \tau/\kappa$.

We now consider a simple model linking permeability K' (i.e., K) to C_P. We assume that K' is a nonnegative, nonincreasing function of the *instantaneous value* of C_P. Thus, in eq 6, $K'(t) = K'(C_P)$, and the system is described completely by two equations corresponding to the dynamic variables C_S and C_P. Therefore, we say this system is two dimensional. In this case it can be shown rigorously that sustained oscillations are *not* possible. To show this result, we calculate the Jacobian matrix for eqs 6 and 7:

$$\mathbf{J} = \begin{bmatrix} \dfrac{\partial}{\partial C_S}\left(\dfrac{dC_S}{d\tau}\right) & \dfrac{\partial}{\partial C_P}\left(\dfrac{dC_S}{d\tau}\right) \\ \dfrac{\partial}{\partial C_S}\left(\dfrac{dC_P}{d\tau}\right) & \dfrac{\partial}{\partial C_P}\left(\dfrac{dC_P}{d\tau}\right) \end{bmatrix} = \begin{bmatrix} -(K'+1) & (C_S^\star - C_S)\dfrac{\partial K'}{\partial C_P} \\ 1 & -q' \end{bmatrix} \qquad (8)$$

A theorem called the Bendixon negative criterion (BNC) states that if, for a two-dimensional system, the trace of $\mathbf{J}$ (i.e., the sum of its diagonal elements) does not change sign for all possible combinations of C_S and C_P, then sustained oscillations are impossible (*32, 40*). Because the nondimensional permeabilities K' and q' are always positive, BNC is evidently met, and this particular system cannot oscillate indefinitely. This result is important because it indicates that systems designed according to the scheme of Figure 25.3, with product inhibition of substrate permeation, are not guaranteed to oscillate indefinitely. Other features must be added, as will be discussed.

The BNC is a rigorous criterion that is powerful when it can be used. However, BNC applies only to two-dimensional systems. Other mathematical results exist, such as the Poincaré–Bendixon (PB) theorem (*31, 32*), which give positive conditions for oscillatory behavior, but these results are also restricted to two-dimensional systems. Therefore, introducing an alternative analysis procedure is useful, and even though this procedure is less rigorous, it is more readily generalized to cases to be considered later for which the two-dimensional results cannot be applied (*31, 32*).

An important feature of any dynamical system such as that described by eqs 4 and 5 (or equivalently, eqs 6 and 7) is its steady-state solution, which is obtained by solving those equations with all time derivatives set to zero. Denoting the steady state by an overbar, we obtain an implicit solution for the steady-state values of C_S and C_P and the resulting reduced substrate permeability, K':

$$\overline{C}_S = \frac{C_S^\star}{1 + 1/\overline{K}'}; \quad \overline{C}_P = \frac{\overline{C}_S}{q'}; \quad \overline{K}' = K'(\overline{C}_P) \qquad (9)$$

Because K' is monotonically nonincreasing in $\overline{C}_P$, it can be shown that only one such steady-state solution exists. One may then ask whether this steady-state solution is locally stable. That is, if the system is started near steady state, will it decay back to steady state (local stability), or will the system move away from steady state permanently (local instability)? Often, local stability of the steady state indicates that sustained oscillatory behavior is precluded, whereas local instability is taken as evidence that sustained oscillations are possible.

To assess the local stability of the steady state, the dynamical equations are linearized around the steady-state solution. Defining $s = C_S - \overline{C}_S$ and $p = C_P - \overline{C}_P$, the linearized versions of eqs 6 and 7 are

$$\frac{d}{d\tau}\begin{pmatrix} s \\ p \end{pmatrix} = \overline{\mathbf{J}}\begin{pmatrix} s \\ p \end{pmatrix} \quad (10)$$

where $\overline{\mathbf{J}} = \mathbf{J}(\overline{C}_S, \overline{C}_P)$ is the Jacobian matrix defined in eq 8, evaluated at the steady-state condition. The local stability of the system is determined by the eigenvalues λ_1 and λ_2 of $\mathbf{J}$, which are the roots of the characteristic polynomial equation for $\mathbf{J}$,

$$\det \lambda\mathbf{I} - \overline{\mathbf{J}} = 0 \quad (11)$$

where $\mathbf{I}$ is the identity matrix and det denotes determinant. If the real parts of these eigenvalues are both negative, then the system is locally stable; whereas if either eigenvalue has a positive real part, then the system is locally unstable. If the eigenvalues contain imaginary parts then they will be complex conjugates of each other, and the local behavior near steady state will either involve oscillatory exponential decay to steady state if $Re\ \lambda_{1,2} < 0$, or oscillatory exponential flight away from steady state if $Re\ \lambda_{1,2} > 0$. The latter case is of particular interest. If the dynamical variables are bounded, then the oscillations cannot grow indefinitely, and exponential escape from the region near steady state should ultimately lead to a sustained oscillation of a finite amplitude: that is, a limit cycle (*31, 32, 40*).

We now evaluate the local stability of steady state for the system under current study. Using eqs 6–11, the characteristic equation is

$$\lambda^2 + (q' + K' + 1)\lambda + q'(K' + 1) - (C_S^* - \overline{C}_S)\left(\frac{dK'}{dC_P}\right)_{\overline{C}_P} = 0 \quad (12)$$

This equation can be rendered into a simpler and more illuminating form by considering the following relation, which can be obtained by rearranging eq 9:

$$C_S^* - \overline{C}_S = \frac{q'\overline{C}_P}{\overline{K}'} \quad (13)$$

Combination of eqs 12 and 13 yields

$$\lambda^2 + (q' + K' + 1)\lambda + q'\left[(K' + 1) - \left(\frac{d\ln K'}{d\ln C_P}\right)_{\overline{C}_P}\right] = 0 \quad (14)$$

The double logarithmic derivative term in eq 14 relates fractional changes in K' to fractional changes in C_P and will be called the *sensitivity term*. This term can be interpreted as the apparent order of product inhibition of substrate permeability (*63, 64*), as discussed in Appendix 25.1.

By definition, q' and K' are nonnegative. Moreover, K' is nonincreasing, so that $d\ln K'/d\ln C_P < 0$. Therefore eq 14 is of the form $\lambda^2 + b\lambda + c = 0$, with $b > 0$ and $c > 0$. The eigenvalues are given by

$$\lambda_1 = \frac{-b - \sqrt{b^2 - 4c}}{2}, \quad \lambda_2 = \frac{-b + \sqrt{b^2 - 4c}}{2} \quad (15)$$

Because $b > 0$ we can be sure that $Re\ \lambda_{1,2} < 0$, so the system is stable, and sustained oscillations are expected to be precluded, a result that is in line with the rigorous results obtained by using BNC. Local decay to steady state will be oscillatory when λ_1 and λ_2 are complex conjugates of each other, which occurs when $b^2 - 4c < 0$, or

$$(q' + K' + 1)^2 - 4q'[(K' + 1) - (d\ln K'/d\ln C_P)_{\overline{C}_P}] < 0 \quad (16)$$

which can be rearranged to the form

$$(d\ln K'/d\ln C_P)_{\overline{C}_P} < \frac{-[q' - (\overline{K}' + 1)]^2}{4q'} \quad (17)$$

When this condition holds the oscillatory decay to steady state will have period $T = 4\pi/\sqrt{4c - b^2}$. We now demonstrate how numerical behavior of the model can be anticipated by calculating eigenvalues. To this end a specific model for substrate permeability must be introduced. For illustrative purposes product inhibition of substrate permeability is assumed to be represented by an inhibitory Hill function: that is,

$$K' = \frac{K_0'}{1 + (C_P/C_{50})^n} \tag{18}$$

where $K_0' = \gamma K_0/\kappa$, K_0 being the maximal value of K achieved at $C_P = 0$, C_{50} is the product concentration required to reduce permeability to half its maximal value, and n is the Hill exponent, which is a measure of cooperativity of the inhibition process and also determines the sharpness of the transition between the high and low permeability regimes. Figure 25.4 illustrates the functional form of eq 18 for several values of n. This particular model is unlikely to be accurate for membranes based on polymer gels. It is being used here because of its simplicity and because it has been used in the past to model biochemical reaction networks with product inhibition.

In the following analysis we will simplify the notation by choosing concentration units such that $C_{50} = 1$. This definition is equivalent to nondimensionalizing all concentrations by C_{50}. Then, introducing K' from eq 18 into eq 9 we obtain a condition for steady state:

$$\overline{C}_P^{\,n+1} + (K_0' + 1)\overline{C}_P = q' K_0' C_S^\star \tag{19}$$

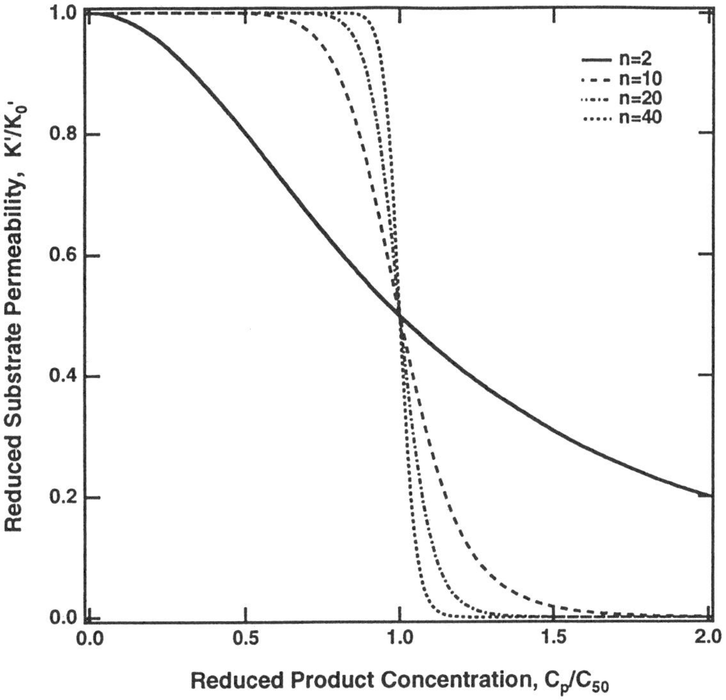

FIGURE 25.4. Plots of inhibitory Hill function given by eq 18 for several values of the cooperativity parameter (Hill exponent) *n*.

Once q' and K'_0 are specified, eq 19 is easily solved by using a root finder (*65*).

As an illustrative numerical example we set $q' = K'_0 = 1$, $C^\star_S = 20$, and $n = 10$. With these values we obtain for the steady-state substrate concentration, $\bar{C}_S = \bar{C}_P = 1.2966$ and the period of oscillatory approach to steady state, $T = 2.0596$. The results of a full numerical simulation of eqs 6 and 7 with these parameters are shown in Figure 25.5a. As initial conditions it was assumed that $C_S = 0$ and $C_P = 0$. Both C_S and C_P show the predicted decaying oscillations around the steady-state value. The inset in Figure 25.5a shows the deviation $s = C_S - \bar{C}_S$ at close approach to steady state for positive values of that deviation. As can be seen by comparing upward or downward zero crossings, the period of the decaying oscillation at close approach is essentially equal to the predicted value, T. However, oscillatory periods during the initial stage that is not near steady state can deviate significantly from T, as can be seen in the larger graph. This observation will figure later in the discussion. For completeness, Figure 25.5b shows the behavior of the membrane permeability to substrate, K'. This quantity fluctuates more strongly than C_S or C_P; this result is due to the sharpness of the relationship between C_P and K'.

On the basis of either the rigorous, global result using BNC, or the local stability analysis, we conclude that the two-dimensional system described previously cannot yield sustained oscillations. Therefore, features must be added. One possible modification is to assume that the membrane permeability to substrate does not respond immediately to product concentration inside the chamber. After all, the signaling process between the product and the membrane must involve some transport of the product and relaxation of the membrane, each of which will have associated time constants.

A simple way to represent delayed membrane response is to augment eqs 4 and 5 with a first-order equation relating the rate of change of membrane permeability to the difference between its current value and the value it would take at steady state given the current value of C_P: that is,

$$\frac{dK}{dt} = \alpha[K_\infty(C_P) - K] \quad (20)$$

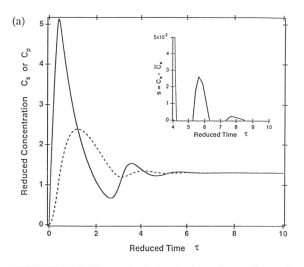

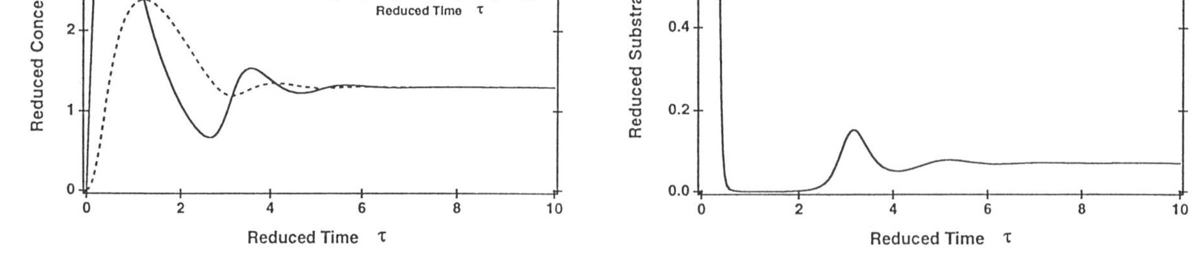

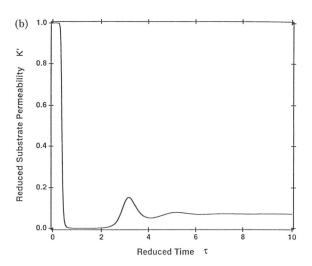

FIGURE 25.5. Numerical simulation of two-dimensional model with $q' = K'_0 = 1$, $C^\star_S = 20$, and $n = 10$. (a) Plot of reduced concentrations, C_S (—) and C_P (- - -) versus reduced time τ. Inset, plot of approach to steady state: $s = C_S - \bar{C}_S$ is plotted versus τ for positive values of s. (b) Plot of permeability $K'(C_P)$ versus τ.

where K_∞ is the steady-state permeability as a function of C_P. In this case, K is no longer a function of the instantaneous value of C_P; it depends as well on the past history of C_P. However, K_∞ depends only on the instantaneous value of C_P. Equation 20 may be interpreted as representing the membrane pursuing its proper steady-state permeability, which is continuously changing with time. Equation 20 can be recast in the form with dimensionless time:

$$\frac{dK'}{d\tau} = \alpha'[K'_\infty(C_P) - K'] \qquad (21)$$

where $\alpha' = \alpha/\kappa$ and $K'_\infty = \gamma K_\infty/\kappa$.

Equations 6, 7, and 21 constitute the system of governing equations, rendered dimensionless in time, for substrate concentration, product concentration, and membrane permeability, which are the dynamical variables for this augmented model. This system is three-dimensional, and the rigorous two-dimensional analysis techniques such as BNC and PB cannot be used. Therefore, we adapt the previously introduced local stability analysis techniques to the current model.

We first consider the nature of the steady state. From eq 21 it follows immediately that $\bar{K}' = K'_\infty(\bar{C}_P)$, and eq 9 holds with K'_∞ replacing K'. Thus, the steady-state substrate and product concentrations are unaltered by the presence of membrane relaxation.

The local stability analysis proceeds by linearizing eqs 6, 7, and 21. In addition to the perturbations s and p, defined previously, we define $k = K' - \bar{K}'$. We then have

$$\frac{d}{d\tau}\begin{pmatrix} s \\ p \\ k \end{pmatrix} = \bar{\mathbf{J}}\begin{pmatrix} s \\ p \\ k \end{pmatrix} \qquad (22)$$

where the Jacobian matrix is now given by

$$\bar{\mathbf{J}} = \begin{pmatrix} -[K'_\infty(\bar{C}_P)+1] & 0 & C_S^\star - \bar{C}_S \\ 1 & -q' & 0 \\ 0 & \alpha'[dK'_\infty/dC_P]_{\bar{C}_P} & -\alpha' \end{pmatrix} \qquad (23)$$

Once again, stability is determined by the eigenvalues of $\bar{\mathbf{J}}$, as determined from its characteristic equation (*see* eq 11):

$$\lambda^3 + (\alpha' + q' + K'_\infty + 1)\lambda^2 + [\alpha' q' + (\alpha' + q')(K'_\infty + 1)]\lambda + \left[\alpha' q'(K'_\infty + 1) - \alpha'(C_S^\star - \bar{C}_S)\frac{dK'_\infty}{dC_P}\right] = 0 \qquad (24)$$

where both K'_∞ and dK'_∞/dC_P are evaluated at $\bar{C}_P$. Again using eq 13 but with K'_∞ replacing $\bar{K}'$, we obtain a form of the characteristic equation that features the sensitivity term:

$$\lambda^3 + (\alpha' + q' + K'_\infty + 1)\lambda^2 + [\alpha' q' + (\alpha' + q')(K'_\infty + 1)]\lambda + \alpha' q'\left(K'_\infty + 1 - \frac{d\ln K'_\infty}{d\ln C_P}\right) = 0 \qquad (25)$$

Equation 25 is of the form

$$\lambda^3 + b\lambda^2 + c\lambda + d = 0 \qquad (26)$$

where the coefficients b, c, and d are all positive. According to Descartes' rule of signs (*32*), the roots (eigenvalues) of this cubic polynomial are either all real and negative, or one root is real and negative whereas the other two are complex conjugate pairs. In the former case where the roots are all real, the steady-state solution is locally stable, whereas in the latter case, stability depends on the real part of the complex eigenvalues. If the real part of these eigenvalues is negative, then the steady state is locally stable and oscillations around steady state are expected to decay. If the real part is positive, then the steady state is locally unstable and therefore cannot be reached. The system is then expected to settle eventually into an oscillatory, *limit cycle* behavior (*31, 32, 40*), because exponentially growing oscillating values of C_S, C_P, and K' will ultimately be bounded by limits on possible concentrations of substrate and product (e.g., $0 < C_S < C_S^\star$).

We now return to the specific model for membrane permeability, with K'_∞ replacing K' in eq 18. We assume that conditions can be set such that $\alpha' = q' = K'_0 = 1$. This assumption leaves two parameters, $C_S^\star$ and n, to be specified. Figure 25.6 shows a plot of the locus of the complex eigenvalue with positive imaginary part, $\lambda = \beta + i\omega$, and $C_S^\star$ is fixed and n is allowed to vary. At low values of n, β is less than 0, a result indicating local stability. However, with increasing n, β becomes positive, and the steady state becomes unstable. Also shown in Figure 25.6 is the case where n is fixed and $C_S^\star$ is varied; a crossover between the locally stable and locally unstable regimes can be seen with increasing $C_S^\star$. The loci for the different cases are virtually identical.

[The asymptotic equivalence of the eigenvalue loci in Figure 25.6 can be understood by noting that as long as $C_S^\star$ and n are sufficiently large, the solution of eq 19 is such that $\bar{C}_P$ must be larger than 1. Again provided n is large enough, this result means that K'_∞ is close to zero. The coefficients b and c (*see* eq 26) in eq 25 are then virtually constant, and only the sensitivity term appearing in the d coefficient has any bearing on the roots of the latter equation. Thus, the effect of $C_S^\star$ and n on the sensitivity term totally accounts for the joint effect of these parameters on the eigenvalues in the asymptotic regime, explaining why the root loci shown in Figure 25.6 all asymptotically collapse into a single curve.]

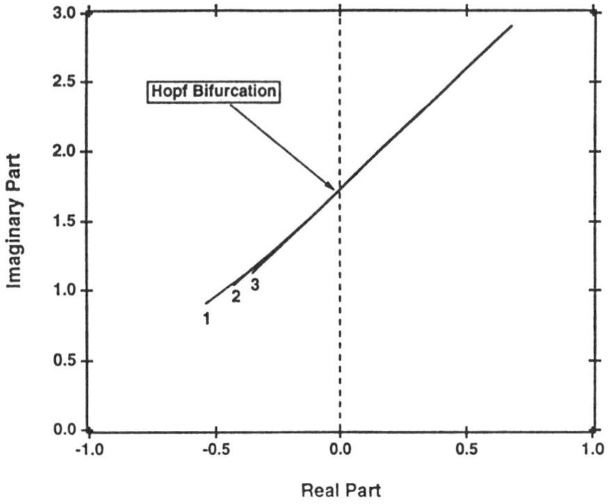

FIGURE 25.6. Locus of eigenvalue $\lambda = \beta + i\omega$ for $\alpha' = q' = K'_0 = 1$ for three-dimensional model: curve 1, $C_S^\star = 4$, n varying; curve 2, $C_S^\star = 20$, n varying; curve 3, $n = 20$, $C_S^\star$ varying. Hopf bifurcation occurs when locus crosses imaginary axis (real part = 0).

The crossover from local stability to local instability is called a *Hopf bifurcation* (*31, 32*). At the Hopf bifurcation point, λ is pure imaginary: that is, $\lambda = i\omega_C$. Considering that both the real and imaginary parts of the characteristic polynomial must vanish, the Hopf bifurcation condition for a three-dimensional system becomes (*see* eq 26):

$$bc - d = 0 \tag{27}$$

$$\omega_c^2 = c \tag{28}$$

If all but one of the model parameters are fixed (e.g., $C_S^\star$ or n), then eq 27 determines the value of that parameter at which the Hopf bifurcation occurs. Then, eq 28 gives the imaginary part of the eigenvalue.

[Note: The significance of the Hopf bifurcation point goes beyond its marking the threshold between local stability and instability. In describing local stability analysis, we have taken care to state that local instability is *often* a good predictor of sustained oscillatory behavior. Generally speaking there is no rigorous proof of this result. However, for parameter values sufficiently near the Hopf bifurcation point, a limit cycle does exist with period $T \approx 2\pi/\omega_C$. This result is the celebrated Hopf bifurcation theorem (*31, 32*). Although the range of parameter values for which this theorem is valid cannot be specified in general, and although the stability of the limit cycle is not always guaranteed by the theorem, the Hopf bifurcation theorem is a central mathematical result in the theory of nonlinear systems.]

The condition for Hopf bifurcation, given the model described by eqs 6, 7, and 21, is obtained by comparing eqs 25 and 26 and inserting the resulting values of b, c, and d into eq 27:

$$(\alpha' + q' + K'_\infty + 1)[\alpha' q' + (\alpha' + q')(K'_\infty + 1)] - \alpha' q' \left(K'_\infty + 1 - \frac{\mathrm{dln}\, K'_\infty}{\mathrm{dln}\, C_P} \right) = 0 \quad (29)$$

where K'_∞ and dln K'_∞/dln C_P are calculated at $\bar{C}_P$ as determined from eq 9. With these equations the threshold value of the Hill constant, n_{HB}, for local instability (Hopf bifurcation) can be determined for a given value of $C_S^\star$. A plot of n_{HB} versus $C_S^\star$, where K' is modeled according to eq 16 and with parameter values $\alpha' = q' = K'_0 = 1$, is shown in Figure 25.7. This plot, called the Hopf bifurcation curve, separates the parameter space into regions of local stability and local instability. Plots such as those in Figure 25.7 are called stability phase diagrams. For this particular model and set of parameter values, with increasing $C_S^\star$ the Hill constant n required for instability falls off rapidly, settling at a constant value of 8.

We now present numerical evidence that local instability implies sustained oscillatory behavior, whereas local stability precludes oscillatory behavior. Figure 25.8a shows the behavior of C_S, C_P, and K' as a function of time for $\alpha' = q' = K'_0 = 1$ and $C_S^\star = n = 20$. Initial conditions for these simulations are $C_S = C_P = 0$, $K' = 1$. Viewing the stability phase diagram in Figure 25.7, these parameter values indicate local instability of the steady state, and the numerical results in Figure 25.8a indeed show sustained oscillations. Figure 25.8b

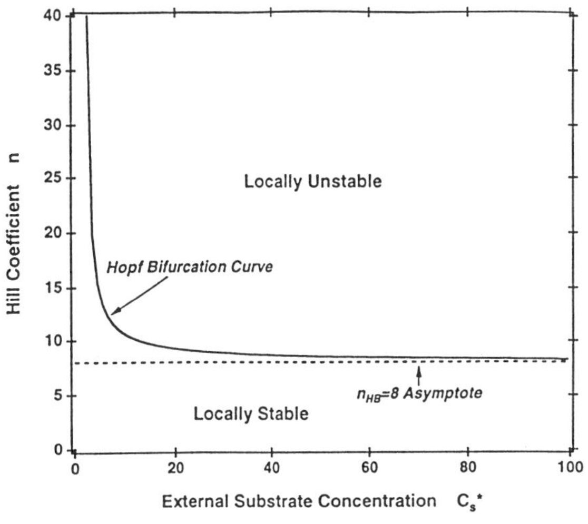

FIGURE 25.7. Hopf bifurcation diagram for simplest three-dimensional model with $\alpha' = q' = K'_0 = 1$. Hopf bifurcation curve $n_{HB}(C_S^\star)$ separates regions of local stability from regions of local instability. Dashed line is asymptote for n_{HB} given these conditions.

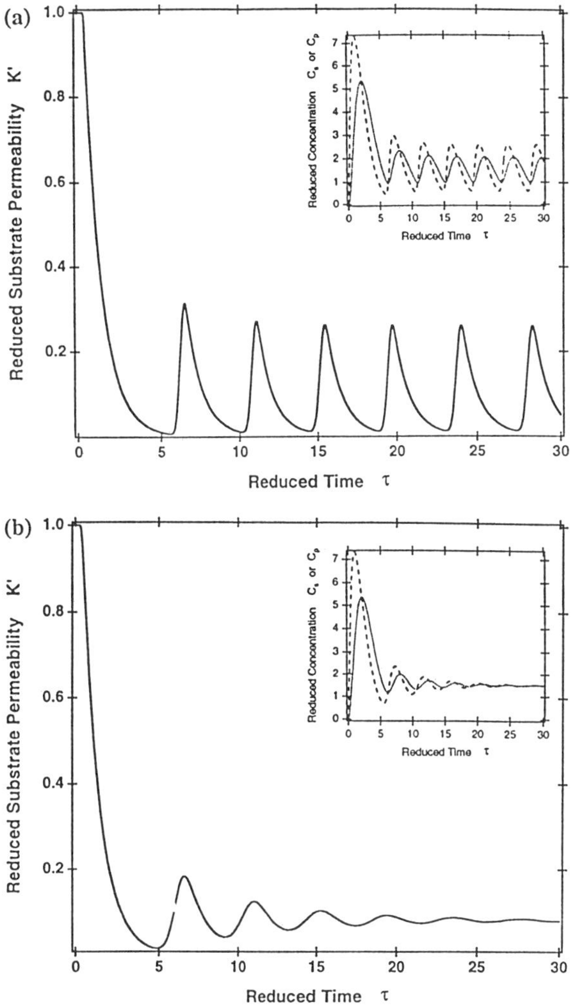

FIGURE 25.8. Numerical simulation of three-dimensional model with $\alpha' = q' = K'_0 = 1$, $C_S^\star = 20$: (solid line) K'. Insets: (dashed line) C_S; (solid line) C_P. (a) $n = 20$; (b) $n = 6$.

shows the dynamic behavior where C_S^* remains at 20 but n is reduced to 6. The same initial conditions were used as before. For this case the stability phase diagram indicates local stability, and the numerical results show decaying oscillations. Thus, the stability phase diagram appears to be a useful predictor of system behavior. Experience with the present system, as well as biochemical systems of related mathematical structure (*58–60*), indicates that stability phase diagrams provide an excellent means for picking parameter combinations leading to sustained oscillations. They also can be used in setting design criteria for components of a drug delivery oscillator, especially the gel membrane.

Figure 25.9a shows a phase plane plot of C_S versus C_P for the conditions of Figure 25.8a. Starting at the origin, this plot spirals around and converges to a closed curve, known as the limit cycle (*31, 32, 40*). For this system, this limit cycle will be approached regardless of the initial conditions. For example, Figure 25.9b shows the system behavior with the same parameter set, but with initial conditions near (but not exactly equal to) the steady-state values $(\bar{C}_S, \bar{C}_P, \bar{K}') = (1.15, 1.15, 0.06)$. Clearly, the same limit cycle is reached. Thus, the system will settle down to the same cyclic behavior regardless of its initial state. This alternative kind of stability is advantageous, because the system will return to its sustained oscillatory behavior even if it receives occasional jolts.

For the case $C_S^* = n = 20$, the complex eigenvalues have imaginary part $\omega = 2.302$, leading to an expected oscillation period of 2.73. This result differs from the actual period $T = 4.30$ seen in Figure 25.8a. Although the linearized equations can predict local instability and imply presence of limit cycle behavior, the actual limit cycle occurs sufficiently far from the steady-state solution that the linearized equations are no longer valid, and the limit cycle period generally cannot be predicted accurately from the eigenvalue. The exception to this statement is near the Hopf bifurcation curve, as previously discussed.

Returning to Figure 25.7, as C_S^* becomes large, n_{HB} appears to approach asymptotically the

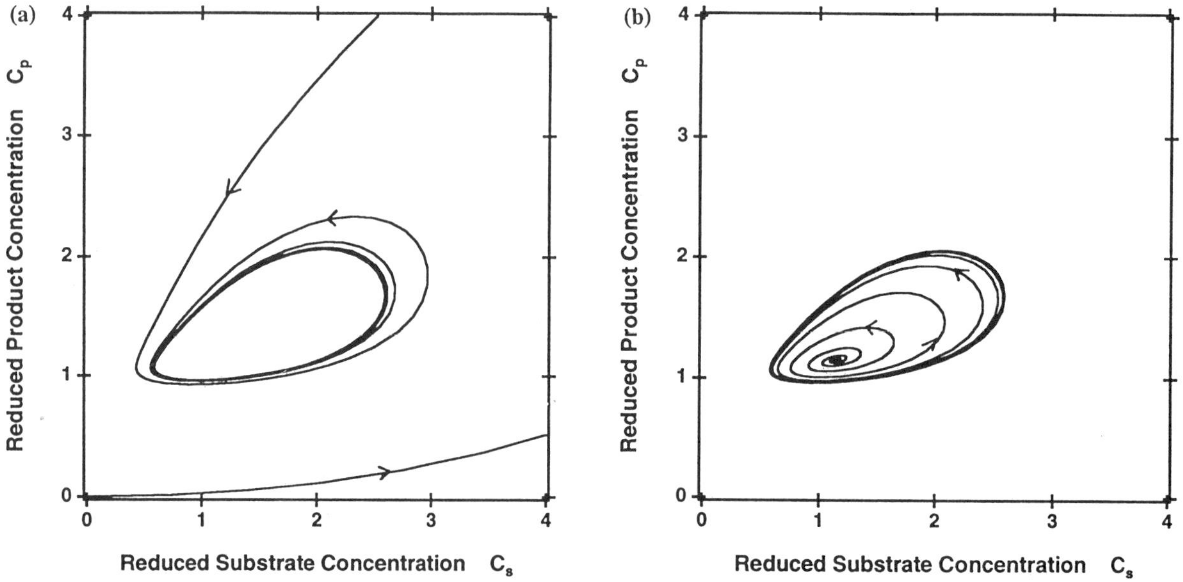

FIGURE 25.9. Phase plane plot of C_S versus C_P. $\alpha' = q' = K_0' = 1$, $C_S^* = 20$, $n = 20$. (a) Initial condition $C_P = C_S = 0$, $K' = 1$. (b) Initial condition $C_P = C_S = 1.15$, $K' = 0.06$ (near steady state).

value 8. In Appendix 25.2 we demonstrate the more general result that

$$\lim_{C_S^\star \to \infty} n_{HB} = 2 + \left(q' + \frac{1}{q'}\right) + \left(\alpha' + \frac{1}{\alpha'}\right) + \left(\frac{q'}{\alpha'} + \frac{\alpha'}{q'}\right) \quad (30)$$

The limiting value of 8 occurring when $\alpha' = q' = 1$ is a special case of eq 30. This asymptotic behavior is independent of the value of K_0', although K_0' can affect the value of n_{HB} at low values of $C_S^\star$.

The quantities in parentheses in eq 30 are all of form $(x + 1/x)$, which takes its minimal value when $x = 1$. Apparently, the smallest value of n that can be allowed in the limit of large $C_S^\star$ is precisely 8, and this result occurs when the first-order rate coefficients κ, q, and α are all equal. Clearly, if any of these rate constants goes to zero or infinity, then the asymptotic n_{HB} also becomes infinite, and the stability phase diagram shows only a region of local stability. This result is certainly consistent with the earlier result that infinitely fast relaxation of membrane permeability ($\alpha \rightarrow \infty$) precludes sustained oscillations.

Figure 25.10a shows Hopf bifurcation curves for the following combinations (q', α'): (2, 1), (1/2, 1), (1/10, 1), (1, 1/10), (1, 10). For the first two of these conditions, the asymptotic value of n_{HB}, calculated using eq 30, is 9, whereas for the latter three conditions the asymptotic value is 24.2. These asymptotic values are evidently approached by the corresponding Hopf bifurcation curves, although the rate of approach varies across conditions. For any of these conditions, local stability or instability is determined by whether the point ($C_S^\star$, n) lies above or below the corresponding Hopf bifurcation curve, respectively. Predictions based on the phase diagram are confirmed numerically in Figure 25.10b for ($C_S^\star$, n) = (20, 20). For both conditions where this point lies above the Hopf

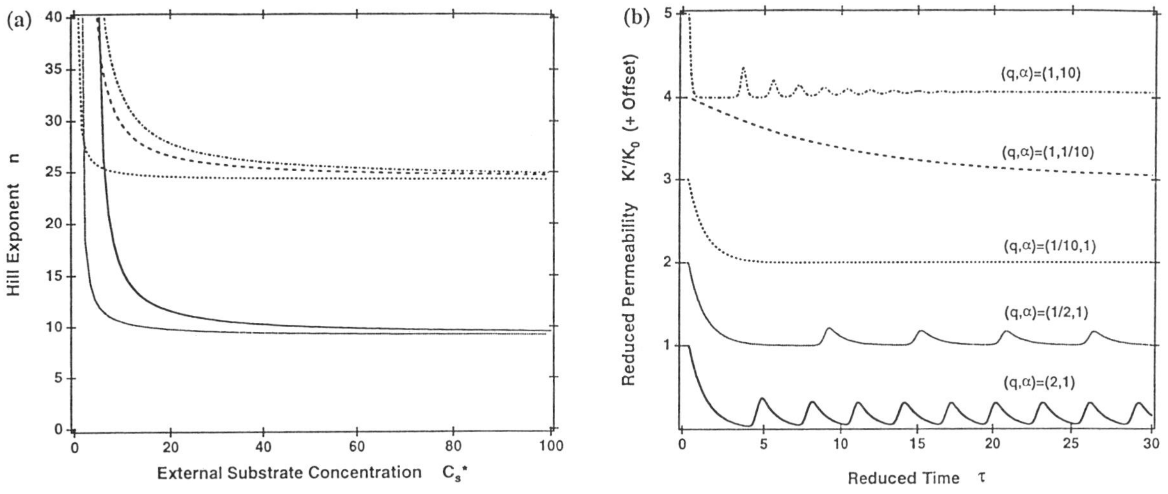

FIGURE 25.10. (a) Hopf bifurcation diagrams for the three-dimensional model with parameters q' and α' varied: $K_0' = 1$; (——) (q', α') = (2, 1); (••••••) (q', α') = (1/2, 1); (- - - -) (q', α') = (1/10, 1); (– – –) (q', α') = (1, 1/10); (– • – •) (q', α') = (1, 10). (b) Numerical results for the conditions giving rise to the Hopf bifurcation curves in part a with ($C_S^\star$, n) = (20, 20). Offsets are added to the curves for clarity.

bifurcation curve, sustained oscillatory behavior is seen, although the character and period of oscillation vary from condition to condition. Conversely, for conditions where this point lies below the Hopf bifurcation curve, the system decays to steady state. The nature of decay varies from case to case, however, differing in rate of decay as well as in the existence or absence of oscillations during the decay.

As already indicated, the Hill model for product effect on substrate permeability through the membrane is certainly incorrect. Aside from the fact that this model does not derive from any polymer physicochemical considerations, the model is flawed in that it permits permeability to decrease asymptotically to zero. This circumstance is unlikely, especially for a low-molecular-weight substrate such as glucose. More likely, a small but limiting degree of substrate permeability will persist even at the highest product concentrations. This result suggests a modified form for the permeability function:

$$K' = K'_\varepsilon + \frac{K'_0}{1+(C_P/C_{50})^n} \tag{31}$$

where $0 < K'_\varepsilon << K'_0$ represents the baseline permeability of the membrane. Figure 25.11a shows Hopf bifurcation curves for a series of values of K'_ε. Unlike the previous case, these curves do not asymptote to a constant value. Instead, after their initial descent, these curves turn upward and increase without limit with increasing $C_S^\star$. An important consequence is that for a fixed value of n, the range of $C_S^\star$ values that permit sustained oscillations becomes bounded both above and below. This situation was not true in Figure 25.6, which predicts only a lower bound in $C_S^\star$ for sustained oscillations. The permissible range of $C_S^\star$ values becomes narrower with increasing K'_ε. A qualitative explanation for this divergent behavior can be found in Appendix 25.2. Numerical examples that again confirm the predictive capability of the Hopf bifurcation curve analysis are presented in Figure 25.11b. The condition $C_S^\star = 20$, $n = 20$ is again simulated. This point lies in the unstable region for $K'_\varepsilon = 0.02$, but in the stable region when $K'_\varepsilon = 0.05$. In the $K'_\varepsilon = 0.02$ case sustained oscillations are observed, whereas for $K'_\varepsilon = 0.05$ the numerical simulation rapidly decays to steady state.

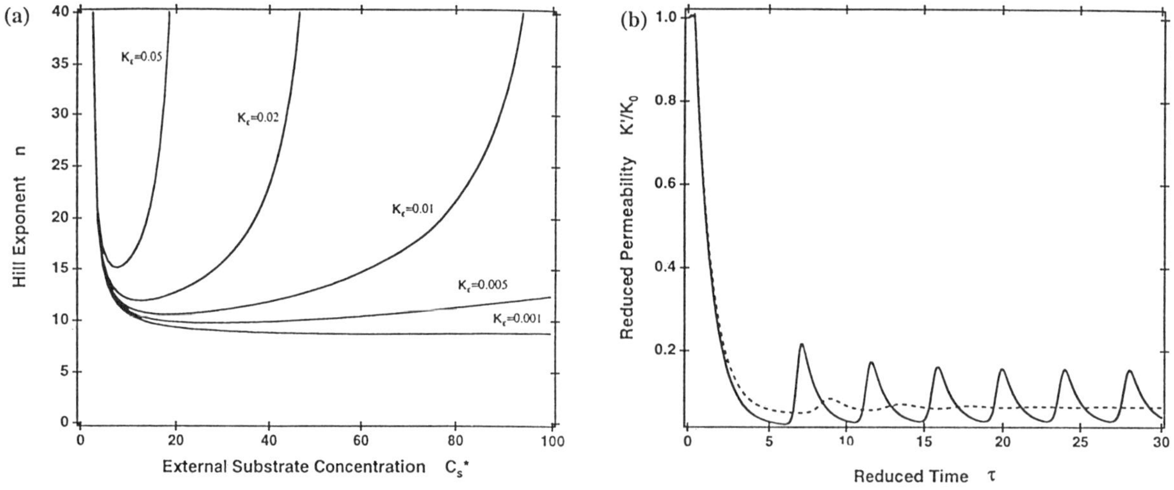

FIGURE 25.11. (a) Hopf bifurcation diagrams for the three-dimensional model with product effect on (reduced) substrate permeability described by an inhibitory Hill function with a small but nonzero baseline, K'_ε. Values of K'_ε appear next to corresponding Hopf bifurcation curves: $\alpha' = q' = K'_0 = 1$. (b) Numerical results for the condition $\alpha' = q' = K'_0 = 1$ and $(C_S^\star, n) = (20, 20)$ for (——) $K'_\varepsilon = 0.05$; (- - - -) $K'_\varepsilon = 0.02$.

Discussion

The foregoing analysis was based on a toy model representing the general scheme for a drug delivery oscillator illustrated in Figure 25.3. Toy models are often used to learn qualitative things about a system under study, without requiring a full and correct model of that system. In this spirit we can claim that we have ascertained a number of things about the class of drug delivery systems under consideration. First, product inhibition of substrate permeation by itself does not guarantee sustained oscillations. Therefore, haphazard design of a system based on this principle is unlikely to produce the desired result. Second, local stability analysis, along with the generation of stability phase diagrams and Hopf bifurcation curves, are useful means for identifying parameter sets in a given model that leads to sustained oscillatory behavior. Third, the character of the Hopf bifurcation curves and the phase diagrams can depend critically on the structure of the model.

The last point was made vivid when a seemingly trivial baseline substrate permeability was added to the initial model. It is particularly significant that major qualitative changes in behavior can occur with baseline levels that are nonzero but so small that experiments might ordinarily not be devised to measure them. This result is somewhat particular to the present toy model, which permits substrate permeability to decrease to zero with increasing product concentration in a way such that the order of inhibition (sensitivity term) remains large. This example teaches us the following lesson: one should be on the lookout to see whether the predictions are too good to be true, perhaps as a result of simplifications in the modeling. In addition, quantities such as the sensitivity term can give excellent insight into what kinds of problems might arise when perturbing a particular model.

Other variants of the toy model are possible, some of which we will now discuss. The enzyme reaction converting substrate to product is considered to be first order. If substrate concentrations inside the chamber become sufficiently large then one may need to use a Michaelis–Menten model to represent saturability of substrate to product conversion. As a second simple modification, we may wish to include the effect of product inhibition on *product* permeability of the membrane. After all, if the membrane is a collapsible gel, then its permeability to both substrate and product may be reduced in the collapsed state. Hopf bifurcation analyses of these modifications to the model show that they can significantly change the character of the stability phase diagrams, and hence the sets of parameters that will show sustained oscillations (*66*).

Previously we noted that in the toy model, the minimum value of n_{HB}, the value of n at which the Hopf bifurcation occurs when all other parameters are specified, is 8, and this result occurs in the case where there is no baseline permeability: that is, $K_\varepsilon' = 0$. In Appendix 25.2, this behavior is explained in terms of the sensitivity term. By considering now a realization based on glucose as substrate, H^+ as product, and a pH-sensitive swellable gel as the membrane (*see* reactions 1–3), the sensitivity term is interpreted as the slope of the graph of the $\log_{10}$ steady-state permeability of the membrane to glucose as a function of pH. Now a value of 8 of the sensitivity term corresponds to 8 log units (i.e., 10^8) in permeability change over one pH unit. Such behavior is, of course, preposterous. However, things look more reasonable when one considers a reduction in permeability by a factor $10^{0.8} \approx 6$ over a range of 0.1 pH unit. A change of this order was demonstrated recently in our laboratory with gels consisting of *N*-isopropylacrylamide and methacrylic acid (*62*). Clearly, the sharper the transition, the more likely the oscillatory behavior.

Unfortunately, a price must be paid for the increased sharpness in the permeability transition. According to the theory, conditions must be arranged such that the steady-state product concentration is in the transition range. In the present case, this statement means that pH in the chamber at steady state must lie within the range of the gel's swelling transition. As the transition becomes sharper, it will become more difficult to aim the steady-state pH into this range. (In the limiting case where the transition is a discrete jump, the precision in achieving the pH at which the jump occurs must be so perfect as to become unattainable.) From eq 9, moreover, the actual steady-state product concentration is determined by the parameters q' and C_S^*. It has already been demonstrated that q' should be chosen as close to unity as possible to guarantee oscillations, and this requirement tightly constrains the values of C_S^* that are compatible with sustained oscillations. However, the C_S^* parameter is a function of the external medium: that is, the physiologic milieu in which the device is placed. The designer seems to be stuck between a rock and a hard place, because the system parameters that are available to his or her control may not be compatible with the goal of sustained oscillations.

The conundrum described in the previous paragraph poses a challenge in the development of oscillating drug delivery systems. In the remainder of this section we discuss some approaches to overcome this problem. Only the main points will be discussed qualitatively, and more detail is reserved for other publications.

The first class of approaches would attempt to reduce the required sharpness in the steady-state substrate permeability characteristic: that is, the minimal value of n_{HB}. Thus far we have conceived of three ways of doing this. The first way is to consider the case, already described, where product permeability of the membrane is reduced with increasing concentration of product. Calculations of Hopf bifurcation curves indeed show that n_{HB} can be reduced below 8 by this means, although parameter ranges over which this reduction occurs appear to be rather narrow (*66*).

The second method to reduce the required sharpness is to increase the number of effective reaction steps between the substrate and the product. For example, reaction 1 shows that two enzymatic reactions in series are required to convert glucose to H^+, and GluOx first converts glucose to gluconolactone, and GluLac converts gluconolactone to a free proton. In the previous development it was assumed that the GluOx reaction is rate limiting; this is usually the case in practice because an excess activity of GluLac is present. However, if GluOx and GluLac concentrations are adjusted such that the two reaction steps have the same rate constant, then the reaction must be described as two first-order reactions in series. Moreover, the concentration of gluconolactone becomes a dynamical variable, and the system description is now four-dimensional. One could further increase the dimensionality of the description to five by having sucrose as the external substrate and including invertase, which catalyzes the hydrolysis of sucrose to glucose and fructose, in the chamber along with the other two enzymes.

Early in this chapter we mentioned that a close analogy exists between the proposed drug delivery oscillator and biochemical oscillators that consist of a chain of unidirectional reactions in series, the product at the end of the chain inhibiting the initial reaction on the chain (Figure 25.2b). Such reaction cascades, with negative retrograde feedback represented by an inhibitory Hill function, have been the subject of much investigation in the biomathematics literature. A key result is that if m is the number of reactions in the cascade (corresponding to the dimensionality of the system), then the minimal value of n, $n_{HB,c}$, compatible with sustained oscillations in this biochemical system is calculated by the formula $n_{HB,c} = \sec^m(1/m)$ (*32, 58–60, 64*). For example, when $m = 3$, 4, and 5, $n_{HB,c}$ takes the values 8, 4, and 2.89, respectively. As $m \to \infty$, $n_{HB,c} \to 1$. In the drug delivery oscillator system, one must also consider loss of intermediate across the membrane, as well as the fact that diffusion of substrate across the membrane is bidirectional, so the analogy to the biochemical cascade is not perfect. Nevertheless, one can expect that the effect of added reaction steps will be to reduce $n_{HB,c}$ and hence the sharpness required of the inhibition of the membrane's glucose permeability by product.

The third way to reduce the required sharpness is to interpose an inert gel layer between the product-sensitive membrane and the reaction space (Figure 25.12). The purpose of this layer is to create delays. It is well known from the theory

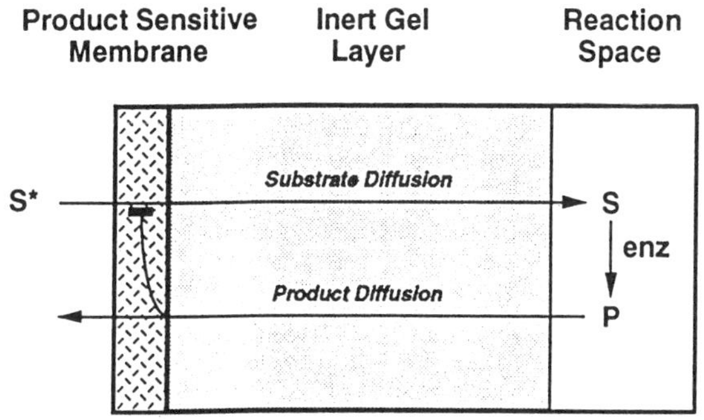

FIGURE 25.12. Schematic of a modified oscillating drug delivery oscillator featuring an inert gel layer separating the reaction space from the product-sensitive membrane.

of control systems that delays in feedback can create conditions for oscillatory instabilities. In fact, the effectiveness of added reaction steps in promoting oscillations, as discussed in the previous paragraphs, can be traced back to the extra delay incurred by each step. Even though diffusion processes are inherently bidirectional and therefore not exactly analogous to reaction cascades, it is reasonable to infer that the diffusional delays associated with the added inert gel layer should also promote oscillations. To treat this system mathematically, the lumped techniques previously described are no longer adequate. The concentrations of substrate and product at each point along the inert layer are dynamic variables, and there are an infinite number of them. Thus distributed techniques must be used. Description of these techniques lies beyond the scope of this chapter; however, suffice to say that diffusional delays can be shown to create conditions for oscillations, even when the reaction and membrane relaxation steps are so fast that they play no role in determining the temporal behavior of the system.

[A distinction between pure delays and diffusional delays should be kept in mind. A pure delay involves a single transit time for all molecules crossing a membrane. With diffusional delays, transit times for different molecules are distributed probabilistically between zero and infinity, but clustered around a mean value that depends on membrane structure. Because of this difference, diffusional delays are actually less efficient in causing instabilities than are pure delays (work to be submitted).]

In a rough sense, the steps of diffusion of substrate through the gel to the reaction space and diffusion of product from the reaction space through the gel to the variable membrane, kinetically replace the enzyme reaction and membrane relaxation steps, respectively.

The previous three methods serve to reduce the demand of sharpness of membrane sensitivity to product concentration, which is thought to be important because of the tight design constraints that accompany increasing sharpness. Sharpness is determined by the Hill parameter, n. When this parameter becomes infinite, the permeability to substrate is represented by a step function of the product concentration (Figure 25.4). Given the structure of this model, the step function would seem to be the limit, representing a first-order transition in gel swelling.

A property of the Hill function is that it is a true function: that is, it assigns a single value of (steady-state) membrane permeability given any value of product concentration. A mark of true first-order phase transitions in cross-linked polymer gels, however, is that some hysteresis be observed in swelling as a function of external conditions. Translated into the current discussion this would imply hysteresis in the (product concentration)–(membrane permeability-to-substrate) characteristic. The situation is illustrated in Figure 25.13. The permeability characteristic contains two branches, designated low and high, and two transition values of product concentration, $C_{P1} < C_{P2}$ Between C_{P1} and C_{P2}, substrate permeability is no longer a single-valued function of product concentration: C_{P1} depends on the recent history of C_{P2}. In Figure 25.13, the permeability curve actually has a backward "S" shape and includes a third branch in the range $C_{P1} < \bar{C}_P < C_{P2}$. This branch turns out to be unstable and generally will not be approached during the oscillations (*24, 25, 31, 32, 56*).

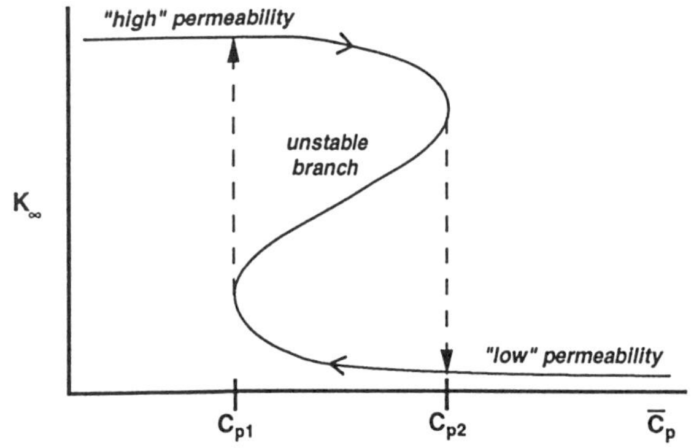

FIGURE 25.13. Permeability characteristic of a membrane that exhibits hysteresis in its steady-state permeability, K_∞. When $C_{P1} < \bar{C}_P < C_{P2}$, product concentration will oscillate between C_{P1} and C_{P2} along the trajectory containing the high and low permeability branches, plus the dashed jumps at the limiting values. The unstable branch of the characteristic curve is never reached.

Starting on the high branch and allowing C_P to increase, the permeability will remain on that branch until C_P reaches C_{P2}, at which point permeability will jump to the low branch. If on the other hand permeability is on the low branch and C_P is allowed to decrease, permeability will stay on the low branch until C_P reaches C_{P1}, at which point it jumps to the high branch. Noting that high permeability to substrate leads to increasing product concentrations, whereas low substrate permeability leads to decreasing product concentrations, it seems evident that hysteresis in membrane permeability will lead to the desired oscillations in membrane permeability, as well as oscillations in substrate and product concentrations.

Even though a full mathematical analysis of this situation will not be presented, we can state that oscillation frequency will depend on the ratio of high to low permeability, as well as the ratio C_{P2}/C_{P1}. Also, a requirement for oscillatory behavior is that $C_{P1} < \bar{C}_P < C_{P2}$; outside this range the system will simply come to rest at the appropriate steady state.

Production of membranes showing hysteresis (we will discuss evidence that this process is plausible) would be a major step in loosening the design constraints alluded to previously. Even though the band over which hysteresis occurs may be narrow, such tight conditions are no longer necessary on parameters associated with transport and reaction rates, as required for the Hill model, because the hysteresis itself causes the oscillation to occur. Thus, design parameters can be chosen to place the steady-state product concentration into the hysteresis band. We believe that this approach, perhaps combined with some of the other features described previously (particularly the inert interposed gel), provides the best chance for obtaining sustained oscillatory drug delivery.

The Hill model was introduced only for illustrative purposes, but it is obviously unable to describe membrane hysteresis; we therefore discard it. More appropriate are models that account for the physicochemical properties of cross-linked polyelectrolyte gels. A first-order model relating gel swelling to polymer hydrophobicity, cross-link density, and acidic group concentration, as well as pH and ionic strength, is obtained by combining Flory–Huggins theory with rubber elasticity theory and the Donnan theory for ionic osmotic forces (*67–70*). Such a model can exhibit either a one-to-one relation between gel swelling (and hence permeability) and hydrogen ion concentration, or a relation that shows hysteresis, depending on values of the model parameters. A model linking the swelling state of the gel to the gel's permeability to glucose will complete the specification of membrane behavior (*71*). This modeling approach is likely to lead to more accurate predictions of system behavior, and it will also enable the designer to build in specifications for a particular oscillator.

We conclude this section with a brief review of hysteresis phenomena that have been observed in cross-linked gels. To date measurements have only been made of equilibrium swelling properties. It is well established by now that gels that undergo discrete, first-order phase transitions do so in a manner that shows hysteresis (*72, 73*). The theory of first-order phase transitions actually predicts that such behavior can occur in many other systems

(*74*). For example, supercooling and superheating of a liquid around its freezing temperature can occur. Just as the supercooled and superheated states of liquids are metastable at the temperatures at which they are present, the swelling states seen in the hysteresis band may be metastable. In both liquids and in gels, nucleating events must occur to initiate transition to the globally stable state. Such events are apparently considerably more difficult to produce in a gel however, which explains why no extraordinary measures are required to observe the swelling hysteresis, whereas observations of supercooling and superheating in liquids must be carried out carefully (*75*).

First-order transitions with hysteresis have been observed as a function of temperature in (*N*-isopropylacrylamide)-*co*-(methacrylic acid) gels, the extent of transition increasing with degree of incorporation of methacrylic acid (*76*). Thus far the first-order phase transitions have been measured only in pure water, and it is not known whether addition of physiological concentrations of salts would eliminate the discrete phase transition. The same can be said for pH-induced transitions. We believe that incorporating comonomers that are even more hydrophobic than NIPA will enable the discrete transition to survive the inclusion of salt, but this hypothesis needs to be tested.

Conclusion

Self-sustained oscillatory drug delivery systems do not exist as of this writing, and their development represents a challenge.

We hope to have demonstrated that the construction and study of oscillatory systems will require a strong marriage between theory and experiment. For oscillating systems, modeling is critical, because the ability of a particular system to show sustained oscillatory behavior depends critically on design parameters. Hopefully, the present work will stimulate further investigations into the theory and practical design of oscillatory drug delivery systems.

Note Added in Proof

Since the original writing of this chapter, it has been established that gel membranes consisting of *N*-isopropylacrylamide and methacrylic acid can, at 37 °C and in the presence of added salt, undergo transitions in glucose permeability in response to changes in pH (*77*). These transitions show hysteresis with respect to pH. Therefore, such membranes are useful for studying oscillatory drug delivery based on the hysteresis mechanism discussed above, and work is proceeding in this direction. A mathematical model based on hysteresis has been sketched elsewhere (*78*) and will be the topic of future publications.

Acknowledgments

I acknowledge generous support from Amgen Inc., as well as helpful conversations with Colin G. Pitt, John P. Baker, Milos Dolník, and Michal Orkisz. Xiaoqin Zou is thanked for a thorough proofreading of the manuscript. Finally, John Ross, Irving R. Epstein, Kenneth Kustin, and Toyoichi Tanaka are thanked for hosting my sabbatical visits, during which many of the ideas presented here were formulated.

References

1. Thadhani, U. *Am. J. Cardiol.* **1992**, *70,* 43–53.
2. Brabant, G.; Prank, K.; Schöfl, C. *Trends Endocrinol. Metab.* **1992**, *3,* 183–190.
3. Knobil, E. *N. Engl. J. Med.* **1991**, *305,* 1582–1583.
4. Santoro, N.; Filicori, M.; Crowley, W. F., Jr. *Endocr. Res.* **1986**, *7,* 11–23.
5. Mathews, D. R.; Naylor, B. A.; Jones, R. G.; Ward, G. M.; Turner, R. C. *Diabetes* **1983**, *32,* 617–621.
6. Clark, R. G.; Jansson, J. O.; Isaksson, O.; Robinson, I. F. *J. Endocrinol.* **1985**, *104,* 53–61.
7. Giusti, M.; Cavagnaro, P. *J. Endocrinol. Invest.* **1991**, *14,* 419–429.
8. Li, Y.-X.; Goldbeter, A. *Biophys. J.* **1989**, *55,* 125–145.
9. Van Putten, L. M.; Keizer, H. J.; Mulder, J. J. *Eur. J. Cancer* **1976**, *12,* 79–85.

10. Dibrov, B. F.; Zhabotinsky, A. M.; Neyfakh, Y. A.; Orlova, M. P.; Churikova, L. I. *Math. Biosci.* **1983,** *66,* 345–347.
11. Dibrov, B. F.; Zhabotinsky, A. M.; Neyfakh, Y. A.; Orlova, M. P.; Churikova, L. I. *Math. Biosci.* **1985,** *73,* 1–31.
12. Churikova, L. I.; Krinskaya, A. V.; Dibrov, B. F.; Zhabotinsky, A.; Neyfakh, Y. A.; Gelfand, E. V. *Biofizika* **1984,** *101,* 746–749.
13. Mohr, J. M.; Schmitt, E. E.; Stewart, R. F. *Proc. Int. Symp. Controlled Release Bioact. Mater.* **1992,** *19,* 377–378.
14. Brand, R. M.; Guy, R. H. *Proc. Int. Symp. Controlled Release Bioact. Mater.* **1994,** *21,* 389–390.
15. Brannon-Peppas, L. *J. Controlled Release* **1992,** *20,* 201–208.
16. Sanders, L. M.; McRae, G. I.; Vitale, K. M.; Kell, B. A. *J. Controlled Release* **1986,** *2,* 187–195.
17. Wuthrich, P.; Ng, S. Y.; Fritzinger, B. K.; Roskos, K. V.; Heller, J. *J. Controlled Release* **1992,** *21,* 191–200.
18. Pozzi, F.; Furlani, P.; Gazzaniga, A.; Davis, S. S.; Wilding, I. R. *J. Controlled Release* **1994,** *31,* 99–108.
19. Kuethe, D. O.; Augenstein, D. C.; Gresser, J. D.; Wise, D. L. *J. Controlled Release* **1992,** *18,* 159–164.
20. Bray, W. C. *J. Am. Chem. Soc.* **1921,** *43,* 1262–1267.
21. Belousov, B. P. In *Oscillations and Travelling Waves in Chemical Systems;* Field, R. J.; Burger, M., Eds.; Wiley: New York, 1985; pp 605–613.
22. Zhabotinsky, A. M. *Biofizika* **1964,** *9,* 306–311.
23. Volkenshtein, M. V. *Biophysics;* MIR: Moscow, Russia, 1983.
24. Gray, P.; Scott, S. K. *Chemical Oscillations and Instabilities;* Clarendon: Oxford, England, 1990.
25. Scott, S. K. *Oscillations, Waves, and Chaos in Chemical Kinetics;* Clarendon: Oxford, England, 1994.
26. Epstein, I. R.; Kustin, K.; DeKepper, P.; Orbán, M. *Sci. Am.* **1983,** *248,* 112–123.
27. Epstein, I. R. *J. Phys. Chem.* **1984,** *88,* 187–198.
28. Rapp, P. E. *J. Exp. Biol.* **1979,** *81,* 281–306.
29. Rapp, P. E. In *Chaos;* Holden, A. V., Ed.; Princeton University: Princeton, NJ, 1986; pp 179–208.
30. Pavlidis, T. *Biological Oscillators: Their Mathematical Analysis;* Academic: New York, 1973.
31. Edelshtein-Keshet, L. *Mathematical Models in Biology;* McGraw-Hill: New York, 1988.
32. Murray, J. D. *Mathematical Biology;* Springer-Verlag: Berlin, Germany, 1989.
33. *Biological and Biochemical Oscillators;* Chance, B.; Ghosh, A. K.; Pye, E. K.; Hess, B., Eds.; Academic: New York, 1973.
34. Lotka, A. J. *Elements of Mathematical Biology;* Dover: New York, 1956.
35. Prigogine, I. *From Being to Becoming: Time and Complexity in the Physical Sciences;* W. H. Freeman: San Francisco, CA, 1980.
36. Field, R. J.; Noyes, R. M. *J. Chem. Phys.* **1974,** *60,* 1877–1884.
37. Rábai, G.; Orbán, M.; Epstein, I. R. *Acc. Chem. Res.* **1990,** *23,* 258–263.
38. Giannos, S. A.; Dinh, S. M.; Berner, B. *J. Pharm. Sci.* **1995,** *84,* 539–543.
39. Olsen, L. F.; Degn, H. *Biochim. Biophys. Acta* **1978,** *523,* 321–334.
40. Hahn, H.-S.; Ortoleva, P. J.; Ross, J. *J. Theor. Biol.* **1973,** *41,* 503–521.
41. Hervagault, J. F.; Thomas, D. *Eur. J. Biochem.* **1983,** *131,* 183–187.
42. Cook, L.; Larter, R.; Shen, P.; Geest, T. *J. Phys. Chem.* **1993,** *97,* 9060–9063.
43. Teorell, T. *J. Gen. Physiol.* **1959,** *42,* 831–845.
44. Teorell, T. *J. Gen. Physiol.* **1959,** *42,* 847–863.
45. Kobatake, Y.; Fujita, H. *J. Chem. Phys.* **1964,** *40,* 2219–2222.
46. Meares, P.; Page, K. R. *Proc. R. Soc. London Ser. A* **1974,** *339,* 513–532.
47. Larter, R. *Chem. Rev.* **1990,** *90,* 355–381.
48. Urabe, K.; Sakaguchi, H. *Biophys. Chem.* **1993,** *47,* 41–51.
49. Kim, J. T.; Larter, R. *J. Phys. Chem.* **1991,** *95,* 7948–7955.
50. Miyano, M.; Osada, Y. *Macromolecules* **1991,** *24,* 4755–4761.
51. Yagisawa, K.; Naito, M.; Gondaira, K.-I.; Kambara, T. *Biophys. J.* **1993,** *64,* 1461–1475.
52. Caplan, S. R.; Naparstek, A.; Zabusky, N. J. *Nature (London)* **1973,** *245,* 364–366.
53. Chay, T. R. *J. Theor. Biol.* **1979,** *80,* 83–99.
54. Chay, T. R. *Biophys. J.* **1980,** *30,* 99–118.
55. Naparstek, A.; Thomas, D.; Caplan, S. R. *Biochim. Biophys. Acta* **1973,** *323,* 643–646.
56. Katchalsky, A.; Spangler, R. *Q. Rev. Biophys.* **1968,** *2,* 127–175.
57. Goodwin, B. C. *Adv. Enzyme Regul.* **1965,** 425–438.
58. Viniegra-Gonzales, G. In *Biological and Biochemical Oscillators;* Chance, B.; Ghosh, A. K.; Pye, E. K.; Hess, B., Eds.; Academic: New York, 1973; pp 41–59.
59. Rapp, P. E. *J. Math. Biol.* **1976,** *3,* 203–224.
60. Tyson, J. J.; Othmer, H. G. *Prog. Theor. Biol.* **1978,** *5,* 1–62.

61. Siegel, R. A.; Pitt, C. G. *J. Controlled Release* **1995,** *33,* 173–186.
62. Baker, J.; Siegel, R. A. *Polymers in Medicine and Pharmacy;* Mikos, A. G.; Leong, K. W.; Yaszemski, M. J.; Tamada, J. A.; Radomsky, M. L., Eds.; Materials Research Symposium Proceedings; Materials Research Society: Pittsburgh, PA, 1995; Vol. 394, pp 119–130.
63. Higgins, J. *Ind. Eng. Chem.* **1967,** *59,* 18–62.
64. Thron, C. D. *Bull. Math. Biol.* **1991,** *53,* 383–401.
65. Press, W. H.; Flannery, B. P.; Teukolsky, S. A.; Vetterling, W. T. *Numerical Recipes;* Cambridge University: Cambridge, England, 1986.
66. Dolník, M., Brandeis University, personal communication, 1994.
67. Flory, P. J. *Principles of Polymer Chemistry;* Cornell: Ithaca, NY, 1953.
68. Siegel, R. A. In *Pulsed and Self-Regulated Drug Delivery;* Kost, J., Ed.; CRC Press: Boca Raton, FL, 1990; pp 129–157.
69. Harsh, D. C.; Gehrke, S. H. In *Absorbent Polymer Technology;* Brannon-Peppas, L.; Harland, R., Eds.; Elsevier: Amsterdam, Netherlands, 1991; pp 103–124.
70. Ohmine, I.; Tanaka, T. *J. Chem. Phys.* **1982,** *77,* 5725–5729.
71. Lustig, S. R.; Peppas, N. A. *J. Appl. Polym. Sci.* **1988,** *36,* 735–747.
72. Sato Matsuo, E.; Tanaka, T. *J. Chem. Phys.* **1988,** *89,* 1695–1703.
73. Kokofuta, E.; Tanaka, T. *Macromolecules* **1991,** *24,* 1605–1607.
74. Berry, R. S.; Rice, S. A.; Ross, J. *Physical Chemistry;* Wiley: New York, 1980.
75. Orkisz, M. Ph.D. Thesis, Massachusetts Institute of Technology, Cambridge, 1994.
76. Hirotsu, S.; Hirokawa, Y.; Tanaka, T. *J. Chem. Phys.* **1987,** *87,* 1392–1394.
77. Baker, J. P.; Siegel, R. A. *Macromol. Rapid Commun.* **1996,** *17,* 409–415.
78. Siegel, R. A.; Zou, X.; Baker, J. P. *Proc. Int. Symp. Controlled Release Bioact. Mater.* **1996,** *23,* 115–116.

Appendix 25.1. The Sensitivity Term

The sensitivity term, $\mathrm{dln}\ K'_\infty/\mathrm{dln}\ C_P$, will be shown to represent the apparent order of product inhibition of substrate flux. To do so it seems easiest to use analogies from chemical reaction kinetics (*63, 64*).

We first consider the n-molecular reaction

$$nX \xrightarrow{k} Y \tag{A}$$

which is described kinetically by the equation

$$R = \frac{dC_Y}{dt} = kC_X^n \tag{B}$$

Calculating the corresponding sensitivity term we find

$$\frac{\mathrm{dln}\ R}{\mathrm{dln}\ C_X} = n \tag{C}$$

which is precisely the order of the reaction.

As a second example we assume that X is converted to Y by an enzyme exhibiting Michaelis–Menten kinetics:

$$R = \frac{V_{max} C_X}{K_M + C_X} \tag{D}$$

In this case

$$\frac{\mathrm{dln}\ R}{\mathrm{dln}\ C_X} = \frac{K_M}{K_M + C_X} \tag{E}$$

Notice here that the sensitivity term depends on the concentration C_X, in contrast to the first example. In the present case the sensitivity term, or apparent reaction order, decreases continuously from one at low concentrations ($C_X << K_M$) to zero at high concentrations ($C_X >> K_M$), thus representing the transition of the enzyme reaction from its linear, first-order regime to its saturated, zero-order regime.

Returning to product effect on permeability, we find for the inhibitory Hill model represented by eq 18 (with $C_{50} = 1$) that

$$\frac{\mathrm{dln}\ K'_\infty}{\mathrm{dln}\ C_P} = \frac{-nC_P^n}{1 + C_P^n} \tag{F}$$

This function decreases continuously from 0 to $-n$ with increasing C_P. The negative sign indicates a negative order corresponding to inhibition of the permeation process.

For the modified inhibitory Hill model that includes a baseline permeability, eq 31, the sensitivity term, or apparent order of inhibition, is

$$\frac{\text{dln } K'_\infty}{\text{dln } C_P} = \frac{-nC_P^n}{\left(1+C_P^n\right)\left[1+\frac{K'_\varepsilon}{K'_0}\left(1+C_P^n\right)\right]} \qquad \text{(G)}$$

A plot of eq G is shown in Figure 25.14 for various values of K'_ε/K'_0 with $K'_\varepsilon << K'_0$ and n set to 10 and 20. The apparent order of inhibition decreases from 0 initially as in eq F. When C_P becomes of the order $(K'_0/K'_\varepsilon)^{1/n}$ the curves turn back up, approaching zero again with increasing C_P. For the large values of n such as are required for instability and oscillations, the turning point can occur very close to $C_P = 1$.

Appendix 25.2. Behavior of Hopf Bifurcation Curves at Large $C_S^\star$

In this appendix we first derive eq 30, that is $n_{HB} \to 2 + (q' + 1/q') + (\alpha' + 1/\alpha') + (q'/\alpha' + \alpha'/q')$ as $C_S^\star \to \infty$, when the effect of product on substrate permeability is given by the inhibitory Hill model, eq 18. Second, a qualitative discussion is given of the divergent behavior of n_{HB} when there is a small baseline permeability: that is, eq 31 is operative.

For the first case, $\overline{C}_P$ is calculated by using eq 19. For large $C_S^\star$ one may write asymptotically,

$$\overline{C}_P \sim (qK'_0C_S^\star)^{1/(n+1)} \qquad \text{(H)}$$

which approaches infinity, albeit rather slowly when n is large. Combining this expression with eq 16, K'_∞ is seen to vanish as $C_S^\star \to \infty$ (*see* eq G). Moreover, according to eq F the sensitivity term dln K'_∞/dln C_P approaches $-n$ as $C_S^\star \to \infty$. Introducing these results into eq 29, the condition for the Hopf bifurcation, we find in the limit of infinite $C_S^\star$.

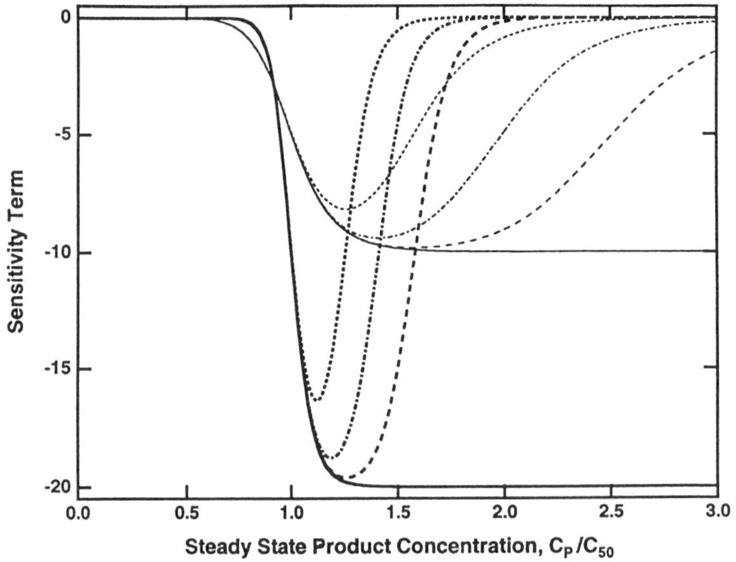

FIGURE 25.14. Plots of sensitivity term, dln K'_∞/dln C_P versus steady-state product concentration C_P (reduced by C_{50}) for $n = 10$ (thin lines) and $n = 20$ (thick lines): (——), $K'_\varepsilon = 0$; (– – –), $K'_\varepsilon = 10^{-4}$; (– • – •), $K'_\varepsilon = 10^{-3}$; (- - - -), $K'_\varepsilon = 10^{-2}$.

$$(\alpha' + q' + 1)(\alpha' q' + \alpha' + q') - \alpha' q'(1 + n_{HB}) \rightarrow 0 \quad \text{(I)}$$

which rearranges to eq 30.

In the presence of a small (nondimensionalized) baseline permeability, K'_ε, with $0 < K'_\varepsilon << K'_0$ (*see* eq 31), we no longer have the sensitivity term approaching $-n$ as $C^\star_S \rightarrow \infty$ (Figure 25.14), so eq I does not hold. In this case we are forced to return to the full form of eq 29, and we must find n sufficiently large that the sensitivity term, when calculated at steady state by using eq G, is of sufficient magnitude to compensate for the positive terms in eq 29. Will this result always be possible?

For this system, $\bar{C}_P$ is determined by a rearranged form of eq 9:

$$\frac{q'\bar{C}_P}{C^\star_S} = \frac{1}{1 + 1/K_\infty(\bar{C}_P)} \quad \text{(J)}$$

As n approaches ∞ the form of $K'_\infty(C_P)$ in eq 31 approaches that of a step function, taking the values $K'_\varepsilon + K'_0$ when $C_P < 1$ and K'_ε when $C_P > 1$. Two curves, representing each the left and right sides of eq J versus $\bar{C}_P$, are shown in Figure 25.15. The steady-state point lies at the intersection of these two curves. Evidently, when $1 + 1/(K'_\varepsilon + K'_0) < C^\star_S/q' < 1 + 1/K'_\varepsilon$ (curve A in Figure 25.15), the intersection lies in the steep part of the right-hand side of eq J, where the sensitivity factor has a large magnitude and $\bar{C}_P \approx 1$. For $C^\star_S/q' > 1 + 1/K'_\varepsilon$ (curve B in Figure 25.15), however, the intersection lies in the flat region of eq J, where $\bar{C}_P > 1$, and the sensitivity term is near zero. In the second case, then, it appears that eq 29 cannot be satisfied and no instabilities or sustained oscillations will be possible. This result is consistent with the predictions of Figure 25.10a, which shows the Hopf bifurcation curves diverging rapidly at finite concentrations.

Also, the foregoing analysis accounts for the divergence of n_{HB} for low values of $C^\star_S$, as seen in all of the stability phase diagrams. If $C^\star_S/q' < 1/(K'_\varepsilon + K'_0)$ (curve C in Figure 25.15), then $\bar{C}_P < 1$, and once again the intersection of curves lies in a flat region of the right-hand side of eq J, and the sensitivity term cannot ever be large enough to satisfy eq 29.

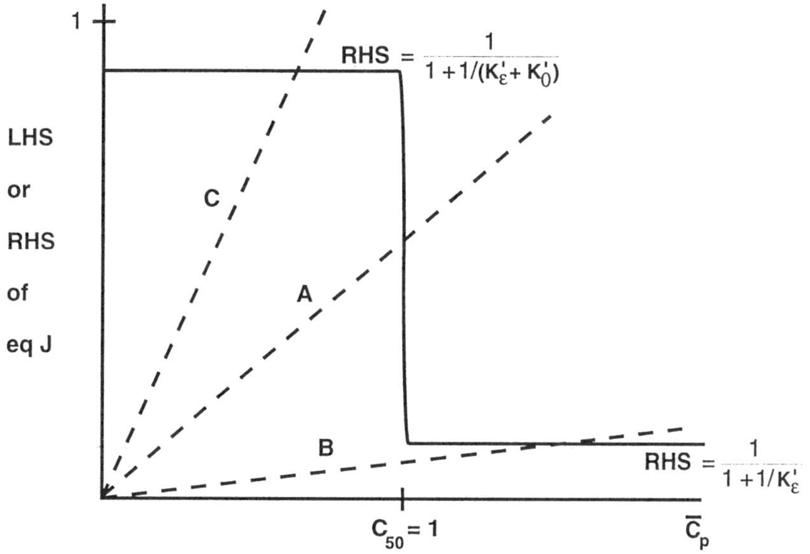

FIGURE 25.15. Plots of left-hand side (solid curve) and right-hand side (dashed lines) of eq J for the steady-state permeability model given by eq 31 with a large value of *n*. *See* Appendix 25.2 for definitions of curves A, B, and C.

26

The Role of Modeling Studies in the Development of Future Controlled-Release Devices

Balaji Narasimhan and Nicholas A. Peppas

We review and critically evaluate existing mathematical models for controlled drug release from polymeric systems. After a brief review of the fundamentals of solute diffusion through polymers, specific applications in controlled release are presented. These applications have been classified into diffusion-, swelling-, and chemically controlled systems. Diffusion (dissolution) limited models have been applied to study drug release from matrix (monolithic) and membrane (reservoir) devices. In swelling systems, relaxation phenomena compete with diffusional mechanisms, and these considerations form the basis of models that describe swelling-controlled devices. The chemically controlled systems consist of bioerodible and pendant-group polymer matrices. The important contributions in all of these systems are described. We conclude that an enhanced molecular understanding would render existing models applicable to a host of future controlled-release devices.

In the development of novel controlled-release systems, mathematical modeling of the release process plays a significant role because it establishes the mechanisms of drug or solute release and provides more general guidelines for the development of other systems. Undoubtedly, a number of successful controlled delivery systems have been developed as a result of an almost arbitrary selection of components, configurations, and geometries. Yet, development of advanced controlled-release systems is more and more dependent on judicious use of the fundamentals of solute diffusion through polymers.

The main objectives in the design of controlled-release systems for drug delivery are as follows:

- to design new systems based on general release expressions,
- to understand the mechanisms of bioactive agent release, and
- to optimize the release kinetics.

Copyright ©1997 American Chemical Society

These goals can be achieved by using accurate mathematical models that take into account the mechanistic aspects of the transport processes in drug delivery systems and the structural characteristics of the polymer. The model equations can be used to design new systems by selecting the optimal geometry, method of formulation, and size (*1–6*).

Modeling of such systems is not trivial and relies on careful representation of the physical situation. For example, proposed models must interpret the numerous problems that arise in testing of these controlled-release systems (e.g., in vitro or in vivo). The large drug loading and the solvent penetration into the polymer carrier could result in nonconstant diffusion coefficients. The polymer carrier could swell or deswell due to solvent transport. The occurrence of multicomponent transport instead of single drug diffusion complicates the analysis. Accurate mathematical models can overcome many of these shortcomings, and hence mathematical models play a pivotal role in the design of systems for controlled drug delivery.

Mathematical Modeling of Diffusion Processes

Fick's Law of Diffusion

The process of the transport of a drug or of a bioactive agent through a polymer or a controlled-release device usually can be described by Fick's law of diffusion. Exceptions to this mechanism exist, such as in solute transport from swellable release systems, and these exceptions will be discussed later.

Written in its differential form in one dimension, Fick's law can be expressed as

$$\mathcal{J}_1 = -D_{12}\frac{dc_1}{dz} \qquad (1)$$

Here, $\mathcal{J}_1$ is the molar flux of the drug in mol/(cm^2•s), D_{12} is the mutual diffusion coefficient of the drug in the polymer in cm^2/s, c_1 is the concentration of the drug in mol/cm^3 in the polymer, and z is the position in the device in centimeters.

By assuming that the molar flux of the drug and the mutual diffusion coefficient are constant, the above equation can be integrated to give what is called the integrated form of Fick's law:

$$\mathcal{J}_1 = D_{12}K\frac{\Delta c}{\delta} \qquad (2)$$

Here, K is a thermodynamic partition coefficient, and δ is the device thickness. The partition coefficient, K, is defined as

$$K = \frac{\text{drug concentration at interface}}{\text{drug concentration in bulk}} \qquad (3)$$

Several misconceptions related to the partition coefficient, its definition, and use are clarified by Lightfoot (*7*).

To determine the temporal evolution of the drug concentration (i.e., in unsteady-state problems), Fick's second law is used and is written as

$$\frac{\partial c_1}{\partial t} = D_{12}\frac{\partial^2 c_1}{\partial z^2} \qquad (4)$$

This equation is written in one dimension and assumes constant diffusion coefficients and constant boundaries.

Solutions to Fick's Law

Equations 1 and 4 can be solved to obtain drug concentration profiles for diffusion in polymer samples of various geometries. Crank (*8*) presented solutions of the diffusion equation for various initial and boundary conditions in infinite and semi-infinite media and for various geometries.

The most common analytical techniques for solving eq 4 used by most researchers working in the area of drug diffusion in polymers are separation of variables, Laplace transforms, and the method of reflection and superposition. For example, a solution of eq 4 can be obtained with the use of separation of variables by writing

$$c_1(z,t) = X(z)T(t) \tag{5}$$

and solving the two resulting ordinary differential equations to obtain a general solution of the form

$$c_1(z,t) = \sum_{n=1}^{\infty}(A_n \sin \lambda_n z + B_n \cos \lambda_n z) \exp(-\lambda_n^2 D_{12} t) \tag{6}$$

The parameters A_n, B_n, and λ_n (which are all constants) can be determined from the initial and boundary conditions of the specific problem.

Typical information obtained from these solutions includes

- fractional release of the drug, M_t/M_∞, where M_t is the mass of drug released at time t and M_∞ is the mass of drug released at infinite time; it is often assumed that M_∞ is the initial loading and that all of the drug is released;
- mass of drug released at time t per unit cross-sectional area A of the device (in mol/cm^2 or g/cm^2);
- drug release rate per unit cross-sectional area, $d(M_t/A)dt$ [in mol/(cm^2·s) or g/(cm^2·s)]; and
- drug concentration profiles in the polymer during release are obtained directly from the solution of the equations considered.

Usually, these profiles are reported in terms of the normalized concentration of drug $c_1/c_{1,0}$, where $c_{1,0}$ is the initial concentration of the drug as a function of dimensionless position x/δ for various dimensionless times $D_{12}t/\delta^2$.

These comments refer to analytical solutions of the Fickian equation. Such analytical solutions are often preferred in the pharmaceutical field because they provide a facile correlation between parameters. Yet, in modeling of general transport phenomena, numerical solutions are equally helpful. Numerical solutions of Fick's law allow for an easy presentation of modeling results and for evaluation of changing design parameters. Unfortunately, the field has been marred by the miscommunication and misinterpretation of relevant numerical information that is presented as if it is analytical in nature (*9*). Such analyses should be avoided.

Incorporation of Polymer Structure in Modeling Equations

Concentration-Dependent Diffusion Coefficients

Very rarely can the drug diffusion coefficient be assumed to be a constant value. This phenomenon occurs in the case of very dilute solutions and in systems where quasi-equilibrium approximations are valid. Typical values of the drug diffusion coefficients in polymers range from 10^{-6} to 10^{-7} cm^2/s for diffusion in rubbery polymers and from 10^{-10} to 10^{-12} cm^2/s for diffusion in glassy polymers.

Normally, the diffusion coefficient is concentration-dependent, and this characteristic modifies eq 4 as

$$\frac{\partial c_1}{\partial t} = \frac{\partial}{\partial z}\left[D_{12}(c_1)\frac{\partial c_1}{\partial z}\right] \tag{7}$$

Of interest is the exponential dependence of the drug diffusion coefficient on concentration proposed by Fujita (*10*) based on the free volume theory. This result is represented in eq 8.

$$D_{12}(c_1) = D_{10} \exp[-\beta_1(c_1 - c_0)] \tag{8}$$

Here, D_{10} and c_0 are surface diffusion coefficient and concentration, respectively, and β_1 is a constant that is system dependent. A similar expression can be written for the diffusion coefficient to show the influence of diluent concentration, as expressed in eq 9.

$$D_{12}(c_1) = D_{10} \exp[-\beta_2(c_{0,d} - c_1)] \tag{9}$$

Here, $c_{0,d}$ is the concentration of the diluent, and β_2 is a constant of the system.

Thermodynamic Considerations

Thermodynamic ideality is assumed when the flux J_1 in eq 1 is written in terms of the concentration gradient. For systems departing from ideality, Fick's law is expressed in terms of the chemical potential gradient, written as

$$\mathcal{J}_i = -cD_i \frac{x_i}{RT} \frac{\partial \mu_i}{\partial z} \quad (10)$$

Here, μ_i is the chemical potential of species i, x_i is the mole fraction of species i, R is the universal gas constant, and T is the temperature.

The chemical potential can be expressed in terms of the activity, a_i, as

$$\mu_i = RT \ln a_i + \mu_i^0 \quad (11)$$

Here, μ_i^0 is the chemical potential of the pure species, i. Relating the activity a_i to the activity coefficient, γ_i, we can finally write

$$\mathcal{J}_i = -cD_i \left(1 + \frac{\partial \ln \gamma_i}{\partial \ln x_i}\right) \frac{dx_i}{dz} \quad (12)$$

Comparing eqs 1 and 12, we can define an effective diffusion coefficient, D_{eff} as

$$D_{eff} = D_i \left(1 + \frac{\partial \ln \gamma_i}{\partial \ln x_i}\right) \quad (13)$$

In this analysis, D_i is an activity-based diffusion coefficient that is less dependent on concentration than D_{ij}.

Effect of Polymer Morphology

Proposed models must incorporate the influence of a number of structural parameters on D_{12}. These parameters include degree of crystallinity and crystallite size (for drug diffusion through semicrystalline polymers), degree of swelling (for swollen matrices and membranes), mesh size of cross-linked polymer networks, porous structure and tortuosity for porous polymers, and translational and relaxational behavior observed in swellable systems. All these parameters will be discussed in detail in the next section.

Modeling of Controlled-Release Devices

Having understood the mechanism of modeling diffusion processes in polymeric systems, we next focus on specific applications in modeling controlled-release systems. Specifically, we discuss diffusion, swelling, and chemically controlled systems *(11)*.

Diffusion-Limited Cases

Matrix (Monolithic) Devices

Matrix (monolithic) devices are available in many geometries. Usually the drug or bioactive agent is present either as a dispersion or a solution in the polymer.

Dispersed Drug, Nonporous Systems. These systems are where the matrix consists of a drug that is dispersed in the polymer at a concentration, c_d, that is greater than the drug solubility in the polymer, c_s. The model usually used in such cases invokes the pseudo-steady-state solution as introduced by Higuchi *(12)*.

The Higuchi model assumes that the solute (dry) particle size is much smaller in comparison to the polymer film thickness, and the model neglects boundary effects (Figure 26.1). Fick's law is used in the form of eq 1: that is, for thermodynamically ideal systems. Integrating eq 1 over the polymer film thickness $x^\star$, which contains only dissolved drug, we obtain

$$\frac{dM_t}{dt} = DA\frac{(c_s - 0)}{x^\star} \quad (14)$$

Here, D is the diffusion coefficient of the drug in the polymer, A is the cross-sectional area of the polymer film, and M_t is the amount of drug released at time t.

A mass balance over the dispersed drug region gives

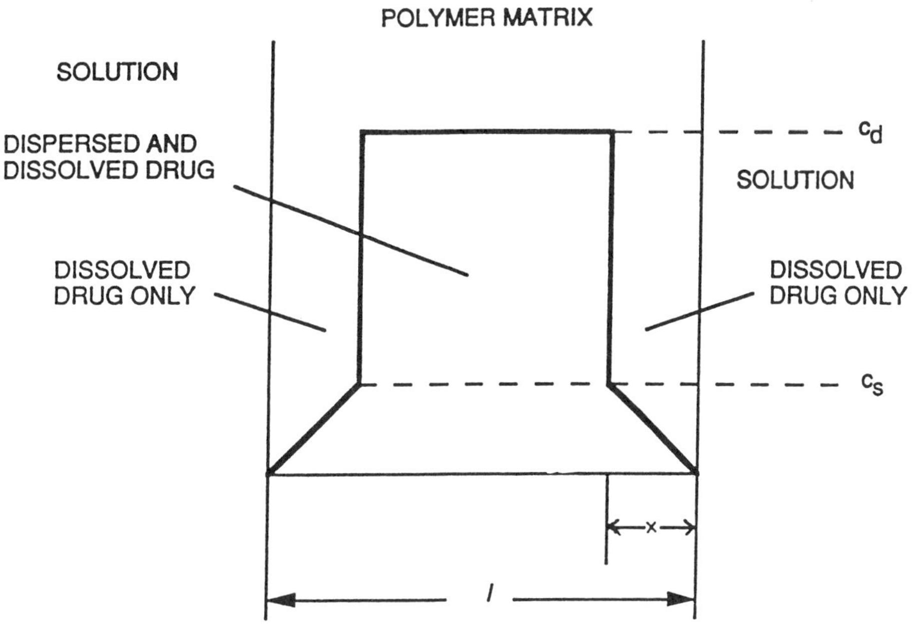

FIGURE 26.1. Parameter definition for drug delivery in nonporous systems. Derivation of the Higuchi model.

$$dM_t = \left(c_d - \frac{c_s}{2}\right)dx^* \qquad (15)$$

Equations 14 and 15 are solved for $x^\star$, which is then substituted in eq 15. Then, integration of dM_t over time yields

$$M_t = A\sqrt{Dc_s(2c_d - c_s)t} \qquad (16)$$

The Higuchi model predicts a square root of time dependence of the mass of drug released and an inverse square root of time dependence of the drug release rate. The effect of drug loading on the drug release rate is shown in Figure 26.2.

Paul and McSpadden (*13*) solved this problem of drug diffusion from dispersed matrix systems exactly. The exact solution is

$$c_1 = c_s \frac{\text{erf}\,\xi}{\text{erf}\,\xi^*} \qquad (17)$$

where erf means error function. The terms ξ and $\xi^\star$ are

$$\xi = \frac{x}{2\sqrt{Dt}} \quad \text{and} \quad \xi^* = \frac{x^*}{2\sqrt{Dt}} \qquad (18)$$

and $x^\star$ can be calculated from the roots of the following transcendental equation:

$$\sqrt{\pi}\xi^* \exp\left(\xi^{*2}\right) \text{erf}\,\xi^* = \frac{c_s}{c_d - c_s} \qquad (19)$$

The amount of drug released at time t can be obtained as

$$M_t = \frac{2c_s A}{\text{erf}\,\xi^*}\sqrt{\frac{Dt}{\pi}} \qquad (20)$$

This solution predicts significant deviations from the Higuchi equation, especially at early times.

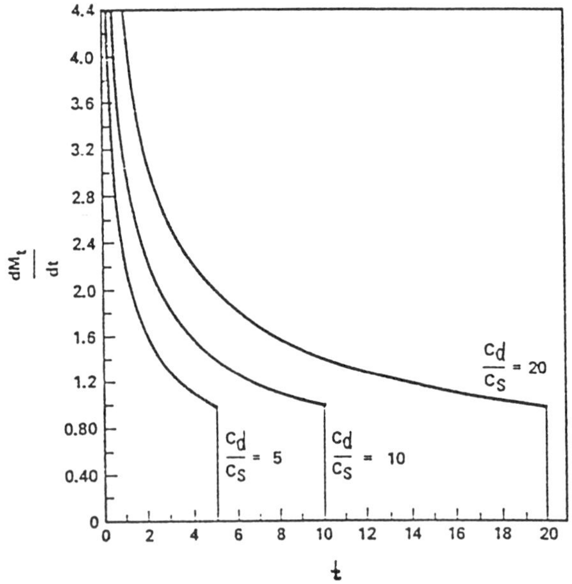

FIGURE 26.2. Effect of drug loading on the drug release rate (with $x^{\star 2}/8D = 1$ and $Ac_s x^\star/2 = 1$). (Reproduced with permission from reference 1. Copyright 1974 Plenum.)

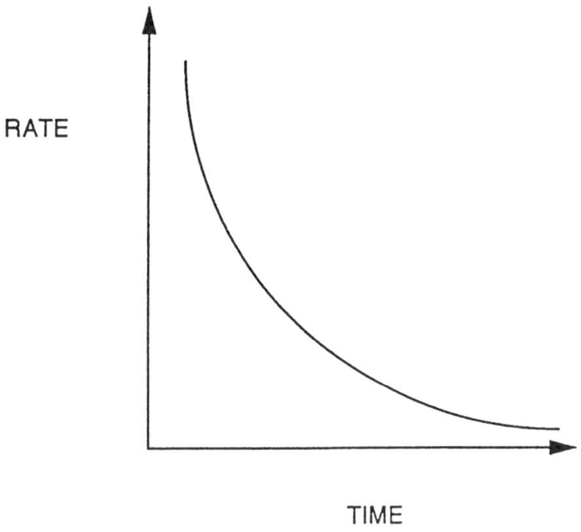

FIGURE 26.3. Drug release rates as a function of time for matrix systems with dissolved drug.

Dissolved Drug, Nonporous Systems. In this case, the drug is dissolved in the polymer below its solubility limit. For polymer films of thickness δ, with initial drug concentration $c_{1,0}$ and negligible boundary effects, eq 4 can be solved for the drug concentration and integrated over the film thickness to give

$$\frac{M_t}{M_\infty} = 1 - \sum_{n=0}^{\infty} \frac{8}{(2n+1)^2 \pi^2} \exp\left[\frac{-D(2n+1)^2 \pi^2}{\delta^2}\right] \quad (21)$$

For long times, the first term of this equation (n = 0) becomes dominant and eq 21 reduces to

$$\frac{M_t}{M_\infty} = 1 - \frac{8}{\pi^2} \exp\left(\frac{\pi^2 D}{\delta^2} t\right) \text{ for } \frac{M_t}{M_\infty} > 0.6 \quad (22)$$

For short times, the solution reduces to

$$\frac{M_t}{M_\infty} = 4\sqrt{\frac{Dt}{\pi\delta^2}} \text{ for } \frac{M_t}{M_\infty} < 0.6 \quad (23)$$

In conclusion, matrix controlled-release systems with dissolved drug (Figure 26.3) give

$$\frac{M_t}{M_\infty} \alpha \sqrt{t} \quad (24)$$

and

$$\frac{dM_t}{dt} \alpha \frac{1}{\sqrt{t}} \quad (25)$$

Similar expressions for the amount of drug released and for the drug release rate can be obtained for cylinders and spheres (Figure 26.4). The fractional release as a function of time with early and late time approximations is shown in Figure 26.5.

Drug Diffusion from Tablets. The model equations for drug release (*14*) from cylindrical tablets (Figures 26.6 and 26.7) can be written as

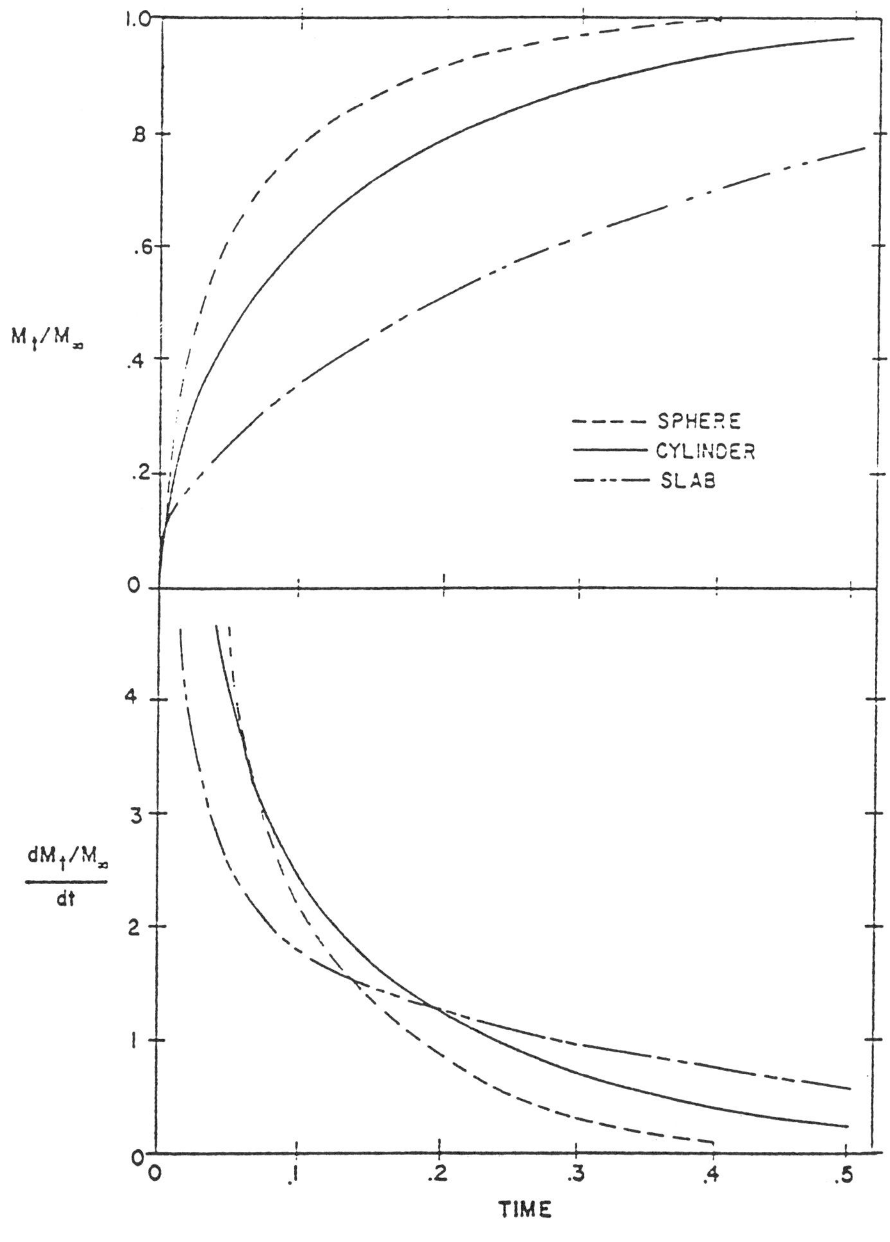

FIGURE 26.4. Fractional release and release rate as a function of time for slabs, cylinders, and spheres (with D/δ^2 or $D/r^2 = 1$). (Reproduced with permission from reference 1. Copyright 1974 Plenum.)

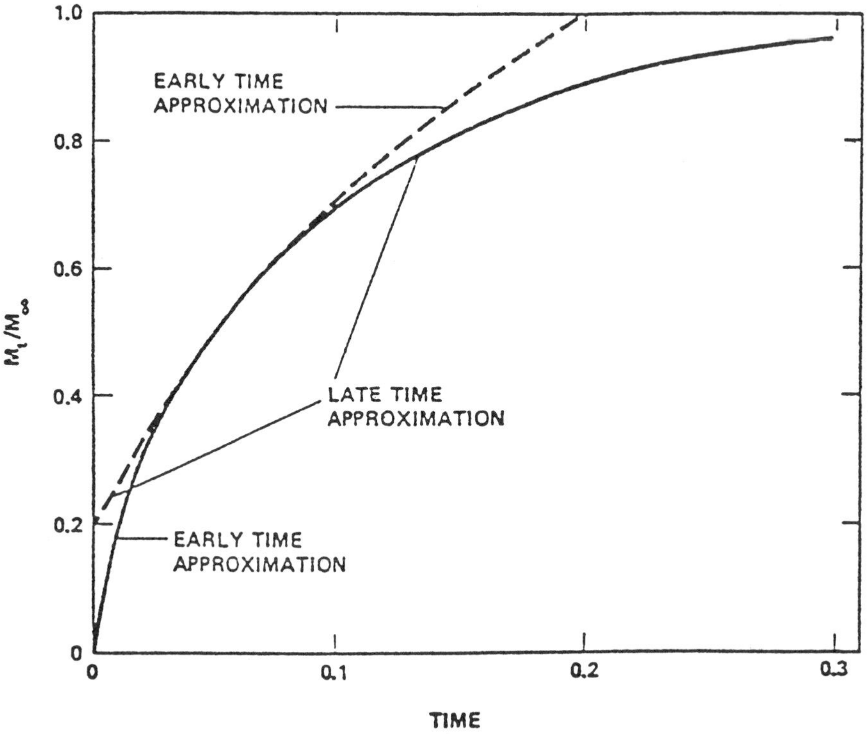

FIGURE 26.5. Fractional release as a function of time with early and late time approximations. (Reproduced with permission from reference 1. Copyright 1974 Plenum.)

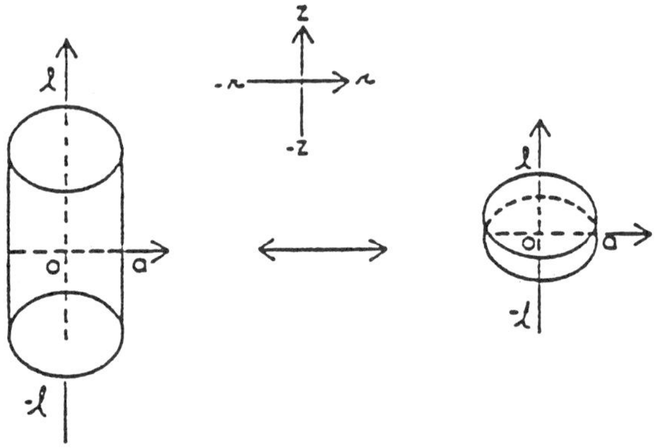

FIGURE 26.6. Schematic diagram of tablet dimensions represented in cylindrical coordinates. (Reproduced with permission from reference 14. Copyright 1976 Wiley.)

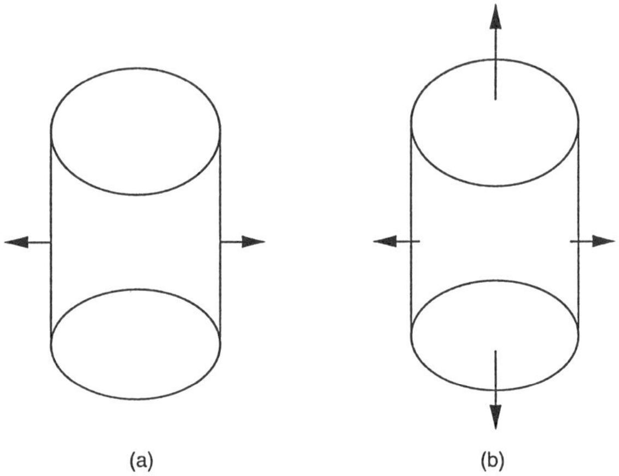

FIGURE 26.7. a, Previously proposed models with only drug from lateral surface accounted for and b, drug diffusion from all surfaces accounted for. (Reproduced with permission from reference 14. Copyright 1976 Wiley.)

$$\frac{\partial c_i}{\partial t} = D\left[\frac{\partial^2 c_i}{\partial r^2} + \frac{1}{r}\frac{\partial c_i}{\partial r} + \frac{\partial^2 c_i}{\partial z^2}\right] \quad (26a)$$

with

$$t = 0,\ 0 < r < R,\ c_i = c_0 \quad (26b)$$

and

$$t > 0,\ r = R,\ c_i = 0 \quad (26c)$$

The amount of drug released can be calculated as

$$\frac{M_t}{M_\infty} = 1 - \frac{8}{\delta^2 R^2}\sum_{m=1}^{\infty}\frac{\exp\left[-D\alpha_m^2 t\right]}{\alpha_m^2}\sum_{n=1}^{\infty}\frac{\exp\left[-D\beta_n^2 t\right]}{\beta_n^2} \quad (27)$$

where α_m are the roots of $\mathcal{J}_0(\alpha_m) = 0$ and

$$\beta_n = \frac{(2n+1)\pi}{2\delta} \quad (28)$$

The experimental and theoretical fractional drug releases from ethylene-vinyl alcohol (EVA)–hydrocortisone tablets from the work of Fu et al. (*14*) are shown in Figure 26.8. Very good agreement is observed between the model predictions and the experimental data.

A reasonable conclusion from this analysis is that zero-order release of drugs cannot be achieved by using matrix systems and simple geometries. However, exceptions exist (*15–18*). Geometries other than slabs, cylinders, or spheres can give zero-order release. For example, zero-order release is observed in the release of bovine serum albumin from coated ethylene-vinyl acetate copolymer (EVAc) hemispheres with an aperture for the release of the drug (Figure 26.9) as shown by Rhine et al. (*15*). The amount of drug released as a function of time is shown in Figure 26.10.

Porous Systems. Considerable research has been performed on the analysis of porous systems (*19*). These systems are usually characterized in terms of the porosity and tortuosity factors. Porous systems are produced either by compression of microparticles of the polymer and the drug or by dispersion of the drug in the polymer solution followed by subsequent evaporation of the solvent. The chaotic nature of the pore network and the continuous change of the pore structure

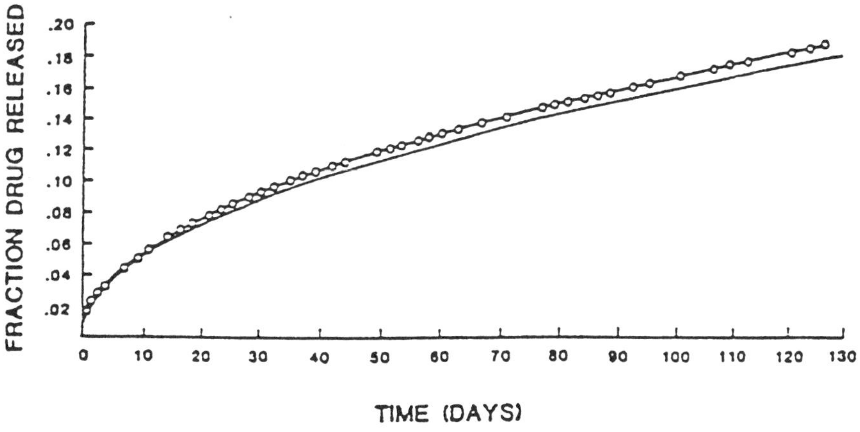

FIGURE 26.8. Experimental and theoretical drug release rate from EVA–hydrocortisone tablets. The points represent the data, and the line represents the model predictions. (Reproduced with permission from reference 14. Copyright 1974 Wiley.)

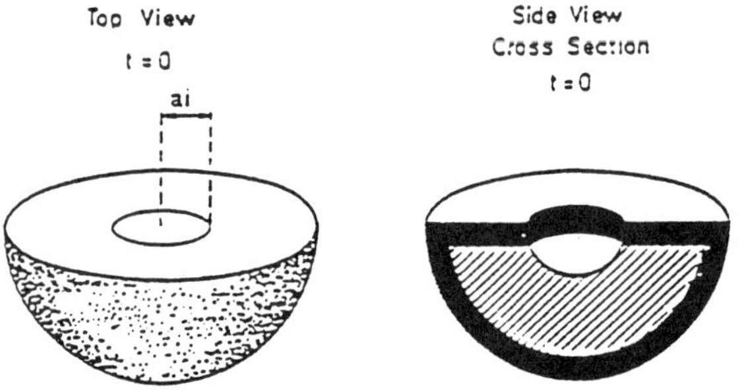

FIGURE 26.9. Schematic of hemisphere devices for zero-order release. (Reproduced with permission from reference 15. Copyright 1980 Academic.)

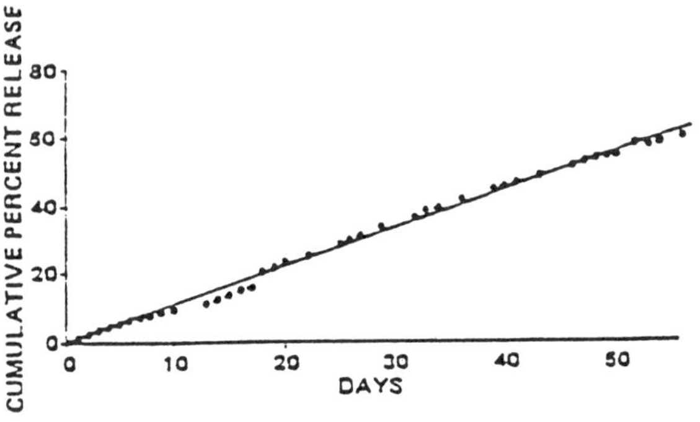

FIGURE 26.10. Release kinetics of bovine serum albumin from EVAc hemispheres. (Reproduced with permission from reference 15. Copyright 1980 Academic.)

during release render the mathematical modeling of the release kinetics quite difficult.

In semicrystalline polymers, drug diffusion is hindered because of the presence of crystallites. The volume fraction and the size of the crystallites affect the diffusion coefficient. The influence of these parameters is expressed in terms of "detour factors" (*20*). Figure 26.11 shows such effects for drug diffusion through a semicrystalline polymer.

Dispersed Drug, Porous Systems. For drug diffusion through a porous polymer film where the drug is dispersed throughout the polymer phase, Higuchi (*21*) developed a model that incorporates

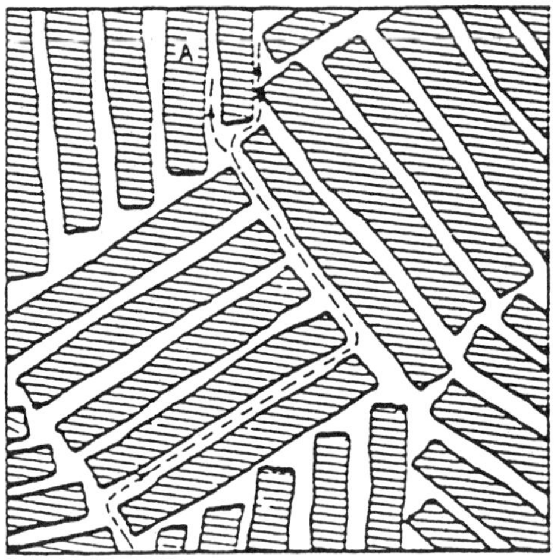

FIGURE 26.11. Detour and blocking effects in the amorphous phase of the crystalline polymer solid. The impermeable crystals are shadowed. The diffusion path marked by the broken line is blocked at A for large penetrant molecules. (Reproduced with permission from reference 19. Copyright 1989 VCH.)

the void fraction, ε, and tortuosity, τ, of the polymer in the diffusion coefficient D_{eff}. The model equation is solved by using the pseudo-steady-state approximation to yield the amount of drug released as

$$M_t = A\sqrt{D_{\text{eff}} c_s (2c_d - \varepsilon c_s) t} \tag{29}$$

where

$$D_{\text{eff}} = D_{\text{iw}} \frac{\varepsilon}{\tau} \tag{30}$$

and D_{iw} is the solute diffusion coefficient in water.

Similar models for other geometries were developed by Roseman and Higuchi (*22*) and Fessi et al. (*23*). Drug binding and release were treated by Desai et al. (*24*). The amount of drug released in this case is given by

$$M_t = A\sqrt{D_{\text{eff}} c_s \{2c_d - c_s[\varepsilon + K(1-\varepsilon)]\} t} \tag{31}$$

where K is a partition coefficient as defined in eq 3.

Dissolved Drug, Porous Systems. The analysis for this case is similar to that of the case of dissolved drug in nonporous polymeric systems. The difference arises in the definition of the diffusion coefficient in eqs 14, 16, and 20. In this case, an effective diffusion coefficient is defined by eq 30.

Diffusion/Dissolution Models. The release of drugs from porous systems can be treated as a dissolution-controlled phenomenon. To treat dissolution, a concentration term is added to the transport equation (eq 4) to give

$$\frac{\partial c_1}{\partial t} = D_{\text{eff}} \frac{\partial^2 c_1}{\partial x^2} + k(c_s - c_1) \tag{32}$$

Here, k is a dissolution constant. The solution to eq 32 under standard boundary conditions (*16*; *17*) is

$$c_1 = c_s \left[1 - \frac{\cosh \sqrt{\frac{k}{D_{\text{eff}}}} \left(x - \frac{l}{2} \right)}{\cosh \sqrt{\frac{k}{D_{\text{eff}}} \frac{l}{2}}} \right] \tag{33}$$

Here, l is the half-thickness of the polymer slab. The amount of drug released is given as

$$M_t = 2\varepsilon A c_s \left[\sqrt{k D_{\text{eff}}} \tanh \sqrt{\frac{k}{D_{\text{eff}}} \frac{l}{2}} + \frac{4}{l} \sum_{n=0}^{\infty} \frac{\alpha_n^2 D_{\text{eff}}^2}{\left(k + \alpha_n^2 D_{\text{eff}}\right)^2} \left\{ 1 - \exp\left[-\left(k + \alpha_n^2 D_{\text{eff}}\right) t \right] \right\} \right] \tag{34}$$

The release of drugs from nonswellable microparticles was studied by Harland et al. (*25*). For drug release from a spherical tablet of radius r, the transport equation is written as

$$\frac{\partial c_i}{\partial t} = D \left(\frac{\partial^2 c_i}{\partial r^2} + \frac{2}{r} \frac{\partial c_i}{\partial r} \right) + k(\varepsilon c_s - c_i) \tag{35}$$

where

$$D = \frac{D_0}{\tau} \exp\left[\alpha(\beta - c_i) \right] \tag{36}$$

Here, α and β are constants of the system. The solution was obtained by using standard boundary conditions. The amount of drug released is given by

$$M_t = 8\varepsilon c_s \pi R^3 \sum_{n=1}^{\infty} f\left(\frac{kR^2}{D}, t \right) \tag{37}$$

The dimensionless drug concentration (defined as $\psi = 1 - c/\varepsilon c_s$, where εc_s is the solubility of the drug) as a function of dimensionless radial position (defined as $\xi = r/R$) inside the sphere is shown in Figure 26.12. The mass of drug released as a function of time for various particle radii and for various values of the dissolution constant are shown in Figures 26.13 and 26.14.

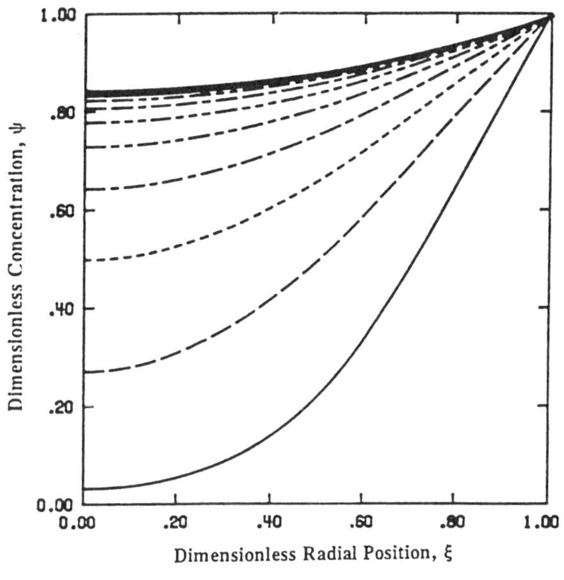

FIGURE 26.12. Dimensionless drug concentration as a function of dimensionless radial position in the sphere for D_i = 1.0. Curves from top to bottom are for τ = 0.5, 0.45, 0.4, 0.35, 0.3, 0.25, 0.2, 0.15, 0.1, and 0.05. (Reproduced with permission from reference 25. Copyright 1988 Elsevier.)

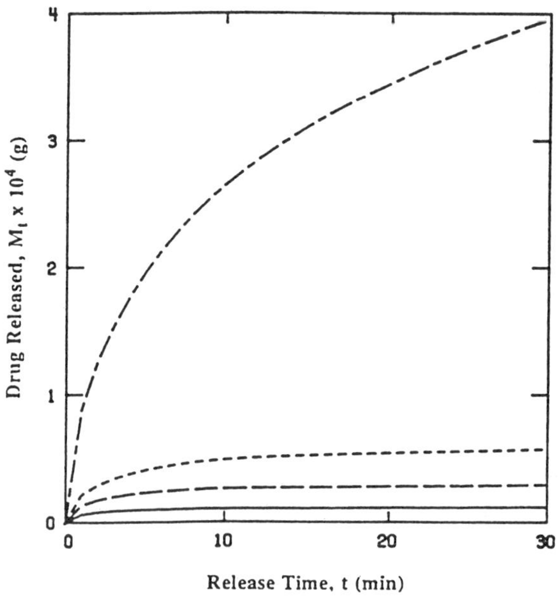

FIGURE 26.13. Amount of drug released as a function of release time. Drug diffusion coefficient, $D = 1 \times 10^{-6}$ cm^2/s; dissolution constant, $k = 1 \times 10^{-4}$ s^{-1}; $\varepsilon c_s = 0.1$ g/cm^3. Curves from top to bottom are for particle radius of $R = 0.1$, 0.05, 0.04, and 0.03 cm. (Reproduced with permission from reference 25. Copyright 1988 Elsevier.)

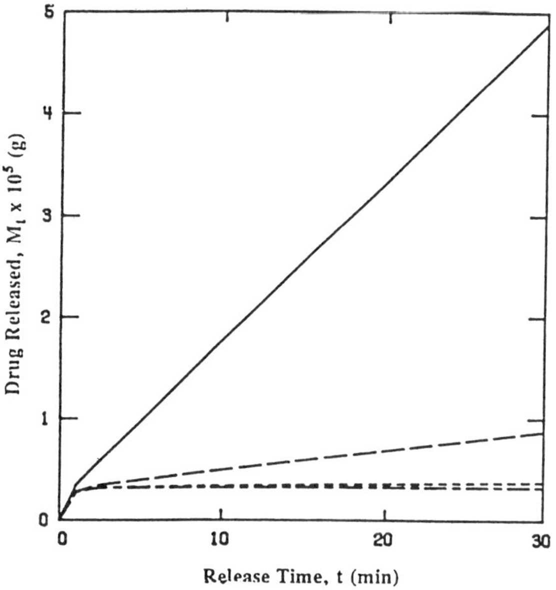

FIGURE 26.14. Amount of drug released as a function of release time. Drug diffusion coefficient, $D = 1 \times 10^{-6}$ cm^2/s, $\varepsilon c_s = 0.1$ g/cm^3; radius, $R = 0.02$ cm. Curves from top to bottom are for drug dissolution constants of $k = 10^{-2}$/s, 10^{-3}/s, 10^{-4}/s, and 10^{-5}/s. (Reproduced with permission from reference 25. Copyright 1988 Elsevier.)

Chang and Himmelstein (*26*) proposed a model for dissolution-controlled drug delivery. They noted that previous models could not provide exact information about the duration of the zero-order release. They also pointed out that previous models treated the drug release as if it occurred in a homogeneous system. They proposed a model that corrected the previous discrepancies by setting diffusion equations for the solid and the dissolved drug phases.

The equation for the undissolved drug was written as

$$\frac{\partial c_{sd}}{\partial t} = -K(c_s - c_d) \tag{38}$$

Here, c_{sd} is the concentration of the undissolved drug, c_d is the concentration of the dissolved drug, K is the dissolution rate, and c_s is the solubility of the drug.

The equation for the dissolved drug was written as

$$\frac{\partial c_d}{\partial t} = \frac{\partial}{\partial x}\left[D(c_d)\frac{\partial c_d}{\partial x}\right] - \frac{\partial c_{sd}}{\partial t} \tag{39}$$

Here, $D(c_d)$ is the diffusion coefficient of the drug in the polymer and is given by

$$D(c_d) = D_0\left[1 + M\left(c_{sd}^0 - c_{sd}\right)\right] \tag{40}$$

Here, c_{sd}^0 is the initial concentration of the drug, M is a system constant, and D_0 is the diffusion coefficient of the drug under conditions of initial loading.

The model was solved numerically. Drug release profiles and drug release rate profiles for

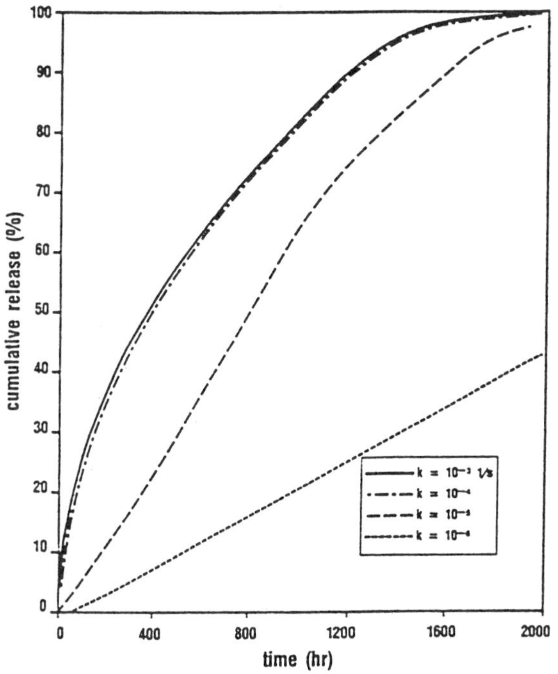

FIGURE 26.15. Drug release profile as a function of time from a slab device at various dissolution rates. Drug diffusion coefficient, $D = 10^{-9}$ cm^2/s and slab half thickness = 0.1 cm. (Reproduced with permission from reference 26. Copyright 1990 Elsevier.)

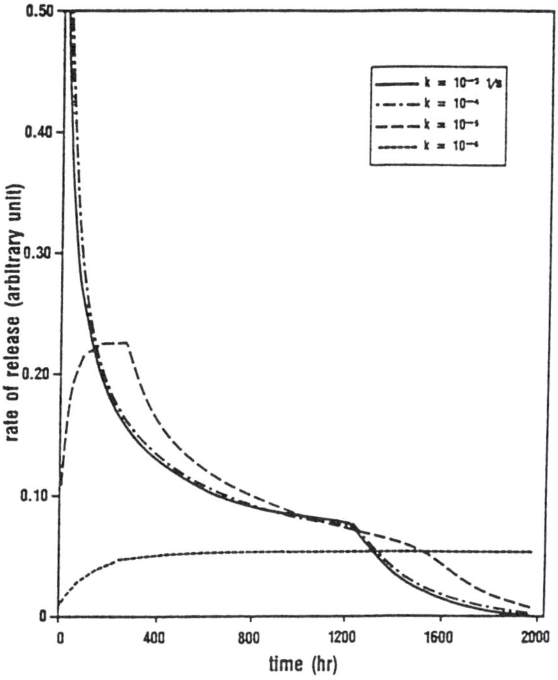

FIGURE 26.16. Drug release rate as a function of time from a slab device at various dissolution rates. All other parameters have the same values as in the plot represented by Figure 26.15. (Reproduced with permission from reference 26. Copyright 1990 Elsevier.)

various dissolution constants as a function of time for the case of drug release from a polymer slab are shown in Figures 26.15 and 26.16. The drug release profile as a function of time for the case of release from a spherical tablet is shown in Figure 26.17.

Membrane (Reservoir) Devices

In these systems, the drug or the bioactive agent is enclosed in relatively large quantities in a permeable synthetic membrane and is placed in contact with a fluid (usually water) at constant temperature. The drug may be present either in pure form or in solution. After an initial period of transient transport, steady states are reached. This process enables determination of the drug release rates. Mathematical models can be formulated for membrane devices either with perfect sink conditions or with release in finite volumes.

Perfect Sink Conditions. Modeling of these devices is done by applying Fick's law in its integrated form. Written for a planar membrane of thickness l (Figure 26.18), Fick's law becomes

$$\mathcal{J} = D\frac{\Delta c'}{l} \qquad (41)$$

The concentration of drug in the membrane, c', is related to that in the bulk phase, c by the partition coefficient, K (Figure 26.19). This relationship transforms the flux equation to

$$\mathcal{J} = \frac{1}{A}\frac{\mathrm{d}M_t}{\mathrm{d}t} = \frac{DK\Delta c}{l} \qquad (42)$$

or

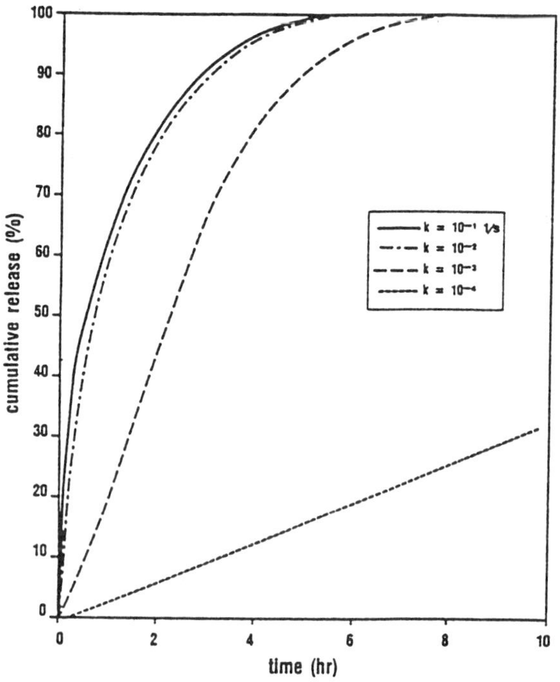

FIGURE 26.17. Cumulative release of drug from a sphere as a function of time. Drug diffusion coefficient, $D = 10^{-9}$ cm^2/s, and radius R of the sphere = 0.01 cm. (Reproduced with permission from reference 26. Copyright 1990 Elsevier.)

$$\frac{dM_t}{dt} = A\frac{DK\Delta c}{l} \tag{43}$$

Equation 43 shows that the drug release rate is independent of time, or in other words, the release kinetics are of zero-order. Hence, the amount of drug released per unit time from such devices is a function of (and can be controlled by) the area of the membrane A, the drug diffusion coefficient D, the partition coefficient K, the concentration Δc, and the thickness of the membrane l. The amount of drug released from such a device for the case of two specific drugs as a function of time is shown in Figure 26.20. Similar expressions can be obtained for release rates from spheres and cylinders.

The expression for the drug release rate from a sphere (Figure 26.21) is

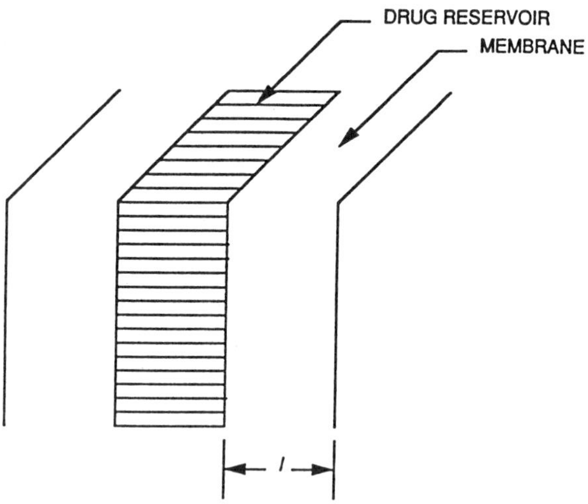

FIGURE 26.18. Planar membrane with drug reservoir under perfect sink conditions.

$$\frac{dM_t}{dt} = 4\pi\frac{DK\Delta c}{\left(\frac{r_0 - r_i}{r_0 r_i}\right)} \tag{44}$$

where r_0 is the radius of the membrane and r_i is the radius of the drug reservoir.

The expression for the drug release rate from a cylinder (Figure 26.22) is

$$\frac{dM_t}{dt} = A\frac{DK\Delta c}{\ln\frac{r_0}{r_i}} \tag{45}$$

where A is the area of the cylinder, r_0 is the radius of the membrane, and r_i is the radius of the drug reservoir. The ratio r_0/r_i is also an important parameter in these cases.

Release in Finite Release Volumes. Finite release volume devices are important in cases where the drug is available in solution, below its solubility limit. By using the Fickian equation and expressing the concentration in terms of amounts of released and remaining drugs $M_t^{(1)}$ and $M_t^{(2)}$,

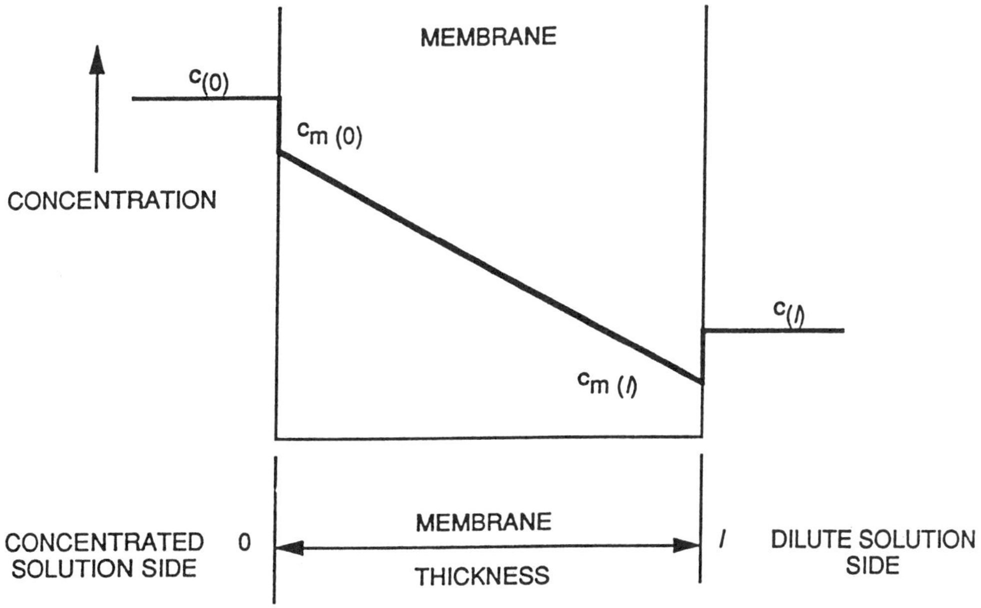

FIGURE 26.19. Parameter definition for the case of drug diffusion under perfect sink conditions.

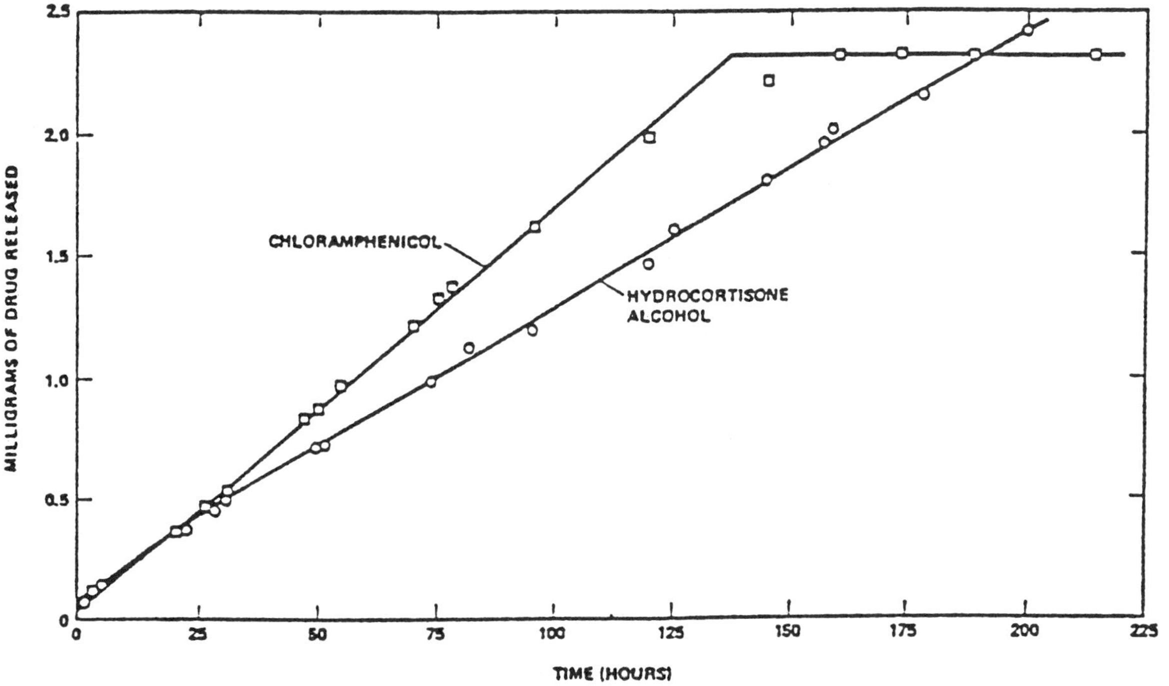

FIGURE 26.20. Zero-order release as obtained from a constant activity, reservoir source for chloramphenicol and hydrocortisone alcohol. (Reproduced with permission from reference 1. Copyright 1974 Plenum.)

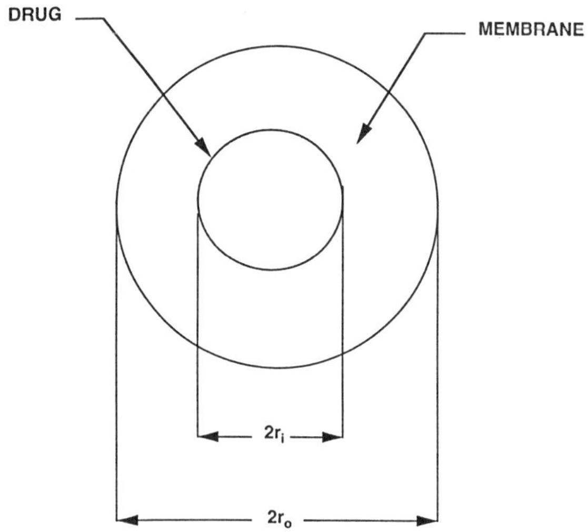

FIGURE 26.21. Spherical membrane with drug reservoir.

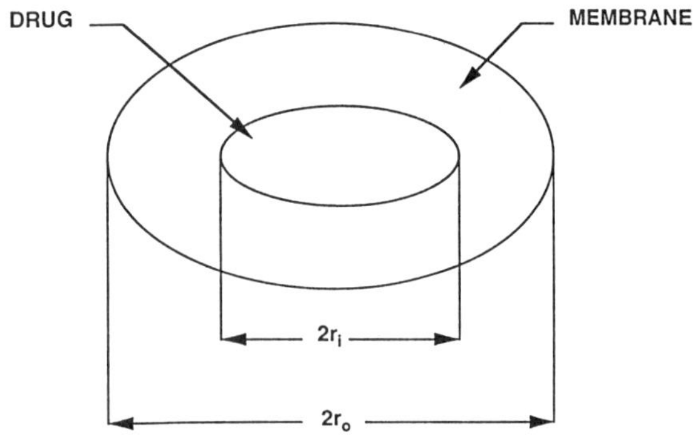

FIGURE 26.22. Cylindrical membrane with drug reservoir.

respectively, and volumes of reservoir and surrounding fluid V_1 and V_2 (Figure 26.23), we can write

$$\frac{dM_t^{(1)}}{dt} = \frac{ADK}{l}\left(\frac{M_t^{(2)}}{V_2} - \frac{M_t^{(1)}}{V_1}\right) \quad (46)$$

The total amount of drug can be expressed as

$$M_\infty = M_t^{(1)} + M_t^{(2)} \quad (47)$$

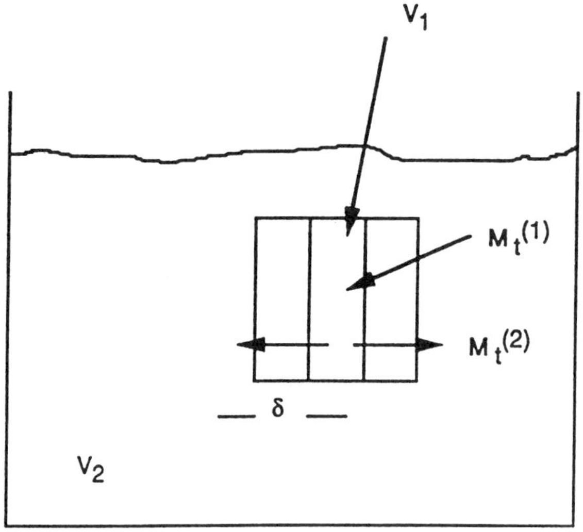

FIGURE 26.23. Diffusion of drug from a nonconstant activity reservoir system.

Initially, all the drug is in the device. Hence,

$$M_\infty = M_t^{(1)} \quad (48)$$

Thus, eq 46 is transformed as

$$\frac{dM_t^{(1)}}{dt} = -\frac{M_\infty ADK}{V_1 l}\exp\left[-\frac{ADK}{l}\left(\frac{1}{V_1}+\frac{1}{V_2}\right)t\right] \quad (49)$$

Hence, the drug release rate for these devices is of first order. This solution was proposed by Colton et al. (*27*).

Effect of System History on Initial Release Kinetics. The foregoing analysis was presented for the steady-state behavior of membrane devices. Experimental observations have shown that initial rate data are in disagreement with the previous model predictions. This phenomenon is explained by looking into the history of the device by considering the time-lag effect and the burst effect.

Time lag appears when a membrane device is used shortly after preparation. It is characterized by an induction period for the release of the drug. The corrected expression for the drug released is given by

$$M_t = A\frac{DK\Delta c}{l}\left(t - \frac{l^2}{6D}\right) \tag{50}$$

The burst effect appears when a membrane device is used long after preparation. It is characterized by a sudden fast release of the drug. The corrected expression for the drug released is given by

$$M_t = A\frac{DK\Delta c}{l}\left(t + \frac{l^2}{3D}\right) \tag{51}$$

The plots for the amount of drug released as a function of time incorporating the system history are shown in Figure 26.24.

Swelling-Controlled Devices

Numerous drug formulations are prepared by loading the drug onto the polymer matrix, usually in the glassy state, in a dissolved or dispersed phase. When the polymer is placed in contact with a thermodynamically compatible liquid, swelling occurs, the polymer begins to release its contents

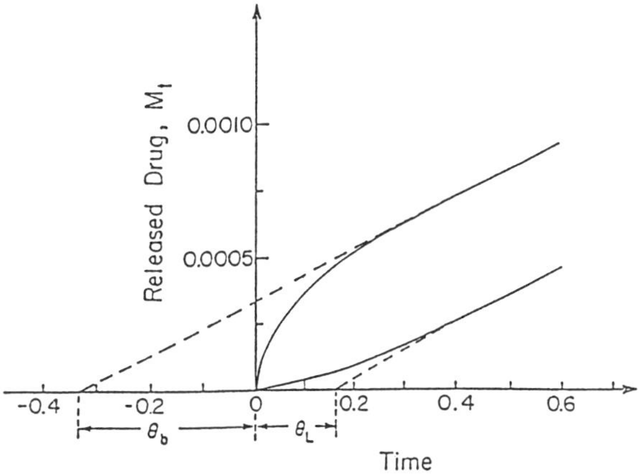

FIGURE 26.24. Released drug for reservoirs exhibiting burst and time-lag effects: $D = 10^{-6}$ cm²/s, $\delta = 10^{-3}$ cm, $c_{1,0} = 1$ mol/cm³.

to the surrounding fluid, and the drug diffuses through the relaxing gel-like polymer.

The transport of the drug through the polymer can be controlled either by the rate at which the macromolecular chains relax during the transition from a glassy to a rubbery state or by the diffusion of the drug through the rubbery polymer. Vrentas et al. (*28*) defined a dimensionless number called the Deborah number, *De*, to characterize the transport. The Deborah number is defined as

$$De = \frac{\lambda}{\theta} \tag{52}$$

Here, λ is the relaxation time, and θ is the diffusion time. When $De >> 1$ or when $De << 1$, Fickian behavior is observed. This result occurs when the transport is either completely relaxation controlled or completely diffusion controlled. When $De \sim 1$, anomalous diffusion behavior is observed. This result occurs when the relaxation time is of the order of the diffusion time.

Drug Release from Swellable Polymers

The modeling of these systems is complex owing to the number of phenomena occurring, and hence, the model equations are normally solved numerically. The transport of the solvent is modeled by

$$\frac{\partial c_1}{\partial t} = \frac{\partial}{\partial x}\left(D_1\frac{\partial c_1}{\partial x}\right) \tag{53}$$

or by

$$\frac{\partial c_1}{\partial t} = \frac{\partial}{\partial x}\left(D_1\frac{\partial c_1}{\partial x} - \upsilon c_1\right) \tag{54}$$

where υ is the convective velocity of the water.

The drug transport is modeled by

$$\frac{\partial c_2}{\partial t} = \frac{\partial}{\partial x}\left(D_2\frac{\partial c_2}{\partial x}\right) \tag{55}$$

Here, D_1 and D_2 are functions of c_1 and sometimes c_2. These values give rise to moving boundary problems that are solved numerically.

An important contribution to the modeling aspects of drug delivery systems was made by Lee (*2*). The model formulated by Lee provided simple but accurate solutions for the problem of drug release from swellable matrices by using a refined enthalpy balance method. It showed that swelling and erosion can be modeled within the same framework (Figure 26.25). The ideas of state and phase erosion were introduced in this work. Also, the importance of the parameter Ba/D (surface erosion over drug diffusion) was first established, where B is the surface erosion rate constant, a is the half thickness for a planar membrane (and the radius for a sphere), and D is the solute diffusion coefficient in the polymer. This parameter later became the swelling interface number, *Sw* (*29*). The importance of the phenomenon of front synchronization was first recognized by Lee. The fractional release of the drug as a function of time for a dispersed solute in a planar erodible polymer matrix for various solute loading levels is shown in Figure 26.26.

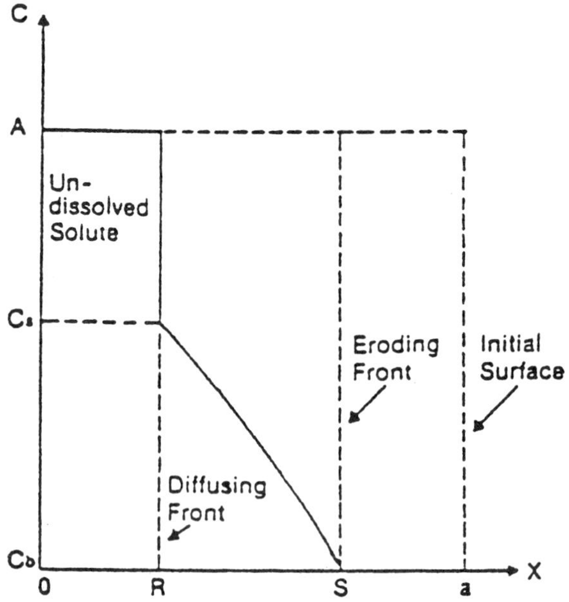

FIGURE 26.25. Schematic diagram of swelling and erosion. (Reproduced with permission from reference 2. Copyright 1980 Elsevier.)

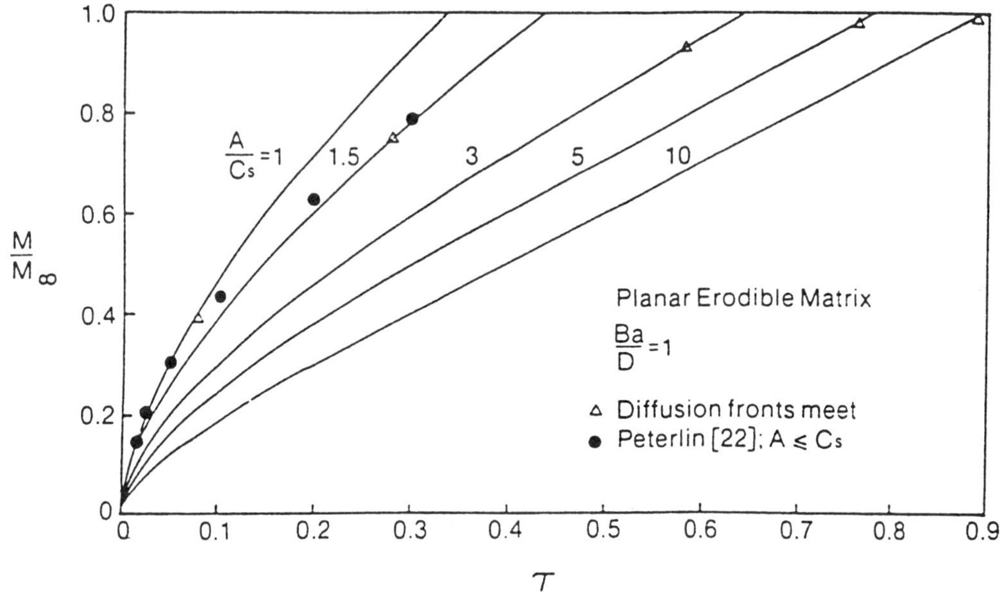

FIGURE 26.26. Fractional release as a function of time for a dispersed solute in a planar erodible polymer matrix (Ba/D = 1) with various solute loading levels. (Reproduced with permission from reference 2. Copyright 1980 Elsevier.)

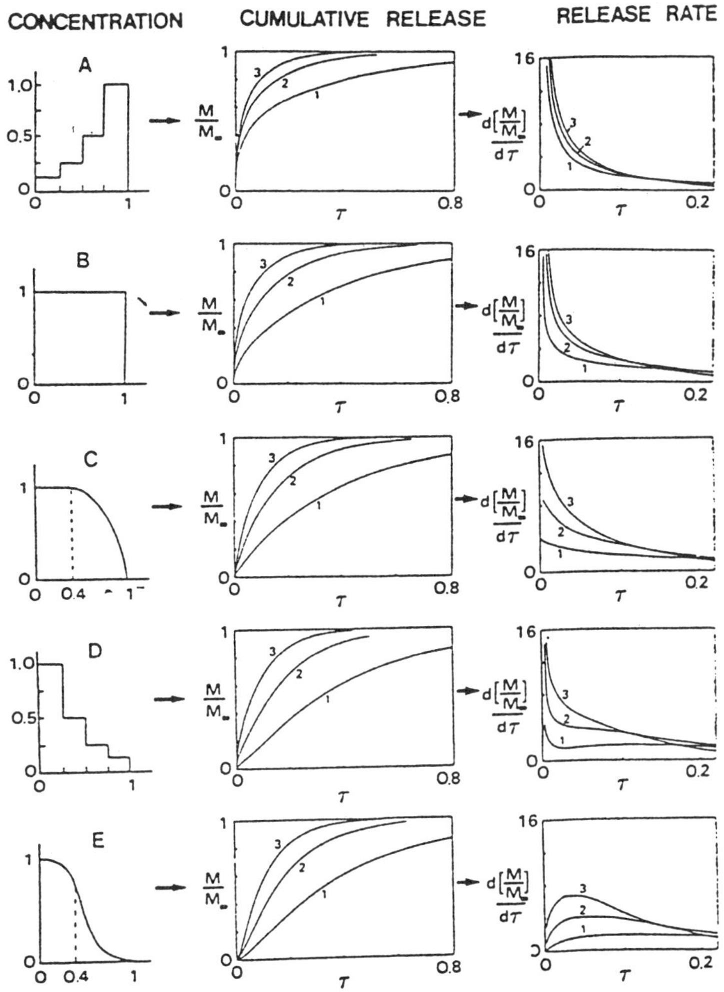

FIGURE 26.27. Characteristics of drug release from diffusion controlled matrix systems as a function of initial drug concentration distribution: 1, planar sheet; 2, cylinder; 3, sphere. (Reproduced with permission from reference 31. Copyright 1986 Elsevier.)

To obtain zero-order release from such erodible devices, the drug distribution was studied (Figure 26.27). A sigmoidal type of initial drug distribution gave almost zero-order release (*30, 31*). This effect is represented in Figure 26.28. Lee (*32*) also considered the effect of time-dependent diffusion coefficients on the release process. A time-dependent drug diffusion coefficient was defined as

$$D(t) = D_i + (D_\infty - D_i)[1 - \exp(-kt)] \quad (56)$$

Here, D_i is the drug diffusion coefficient initially, and D is that in the swollen polymer at long times. The model equations were exactly solved and the mass of drug released was obtained as

$$\frac{M_t}{M_\infty} = 1 - \sum_{n=0}^{\infty} \frac{8}{(2n+1)^2\pi^2} \exp\left(-(n+0.5)^2\pi^2\left\{\frac{D_\infty t}{l^2} + \frac{D_\infty}{kl^2}[1 - \exp(-kt)]\right\}\right) \quad (57)$$

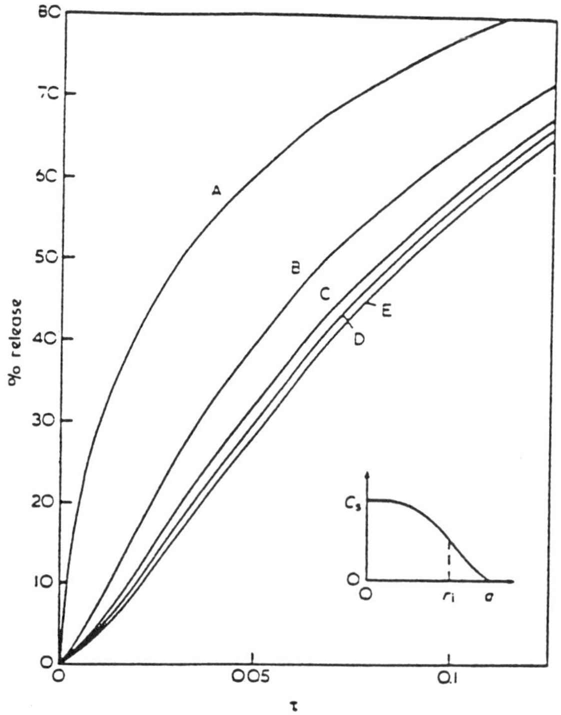

FIGURE 26.28. Effect of sigmoidal initial drug concentration distribution on the cumulative release from spherical matrices. Curves A through E represent different values of ξ_i. ξ_i is the initial position of the inflection point in the concentration profile. The distribution function is

$$f(\xi) = \frac{1 - \exp\left[-0.5\left(\frac{1-\xi}{1-\xi_i}\right)^2\right]}{1 - \exp\left[-0.5\left(\frac{1}{1-\xi_i}\right)^2\right]}$$

(Reproduced with permission from reference 30. Copyright 1984 Elsevier.)

Similar expressions were obtained for systems with dispersed drug. The fractional release as a function of square root of dimensionless time as a function of the Deborah number for a swellable polymer slab containing dissolved drug is shown in Figure 26.29.

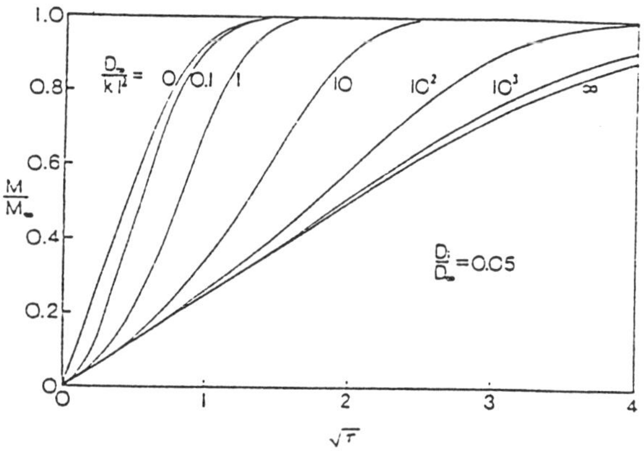

FIGURE 26.29. Fractional release as a function of the square root of dimensionless time as a function of the Deborah number for a swellable polymer sheet containing dissolved drug. (Reproduced from reference 32. Copyright 1987 American Chemical Society.)

Other Models

Kou et al. (*33*) proposed a model for drug diffusion through swellable cylinders with Fickian equations and concentration-dependent diffusion coefficients. Cohen and Erneux (*34, 35*) presented a treatment of free boundary problems that had fundamental ideas in solving drug delivery problems in swellable systems. They presented asymptotic solutions for the class of free boundary problems that model the controlled release of a drug from a swellable polymer.

Peppas and Franson (*29*) introduced the swelling interface number, *Sw*, by analogy to other dimensionless numbers previously discussed by Lee (*2*) and Hopfenberg and Hsu (*36*). This number describes the relative importance of water transport and drug diffusion. It is given by

$$Sw = \frac{\upsilon\delta(t)}{D} \qquad (58)$$

Here, υ is the convective velocity of the water, $\delta(t)$ is the gel layer thickness, and D is the drug diffusivity.

The first detailed model of drug transport with concentration-dependent diffusion coefficients and swelling was proposed by Korsmeyer et al. (*37*). Lustig and Peppas (*38*) proposed a free-volume-based model with three-dimensional swelling beyond the point where the fronts meet. Lee and Peppas (*39*) formulated a model for drug delivery from dissolution-controlled release systems. Lustig et al. (*40*) proposed a mathematical model for drug release based on rational thermodynamics, containing a full viscoelastic description of the polymer, concentration-dependent transport of the drug, and three-dimensional swelling.

Drug release from environmentally responsive polymeric systems has also been studied in great detail. Peppas and Brannon-Peppas (*41*) studied the drug release rates in a system that responded to changes in temperature. The time-dependent response of the drug was also studied by Brannon-Peppas and Peppas (*42*) and by Bell and Peppas (*43*). Models for drug delivery from polymeric systems that responded to changes in the pH and ionic strength of the surrounding medium were proposed by Hariharan and Peppas (*44, 45*).

In addition to these important modeling efforts, researchers have sought to analyze non-Fickian transport with relatively simple equations. One such effort relating the fractional release of the drug to time uses the so called swelling area (*Sa*) number (*46*). The swelling area number was defined as

$$Sa = \frac{1}{D}\frac{dA}{dt} \tag{59}$$

Here, A is the area available for drug release, and D is the drug diffusion coefficient. The Sa value was used to describe systems with varying area (Figure 26.30) due to swelling (specifically the Geomatrix systems).

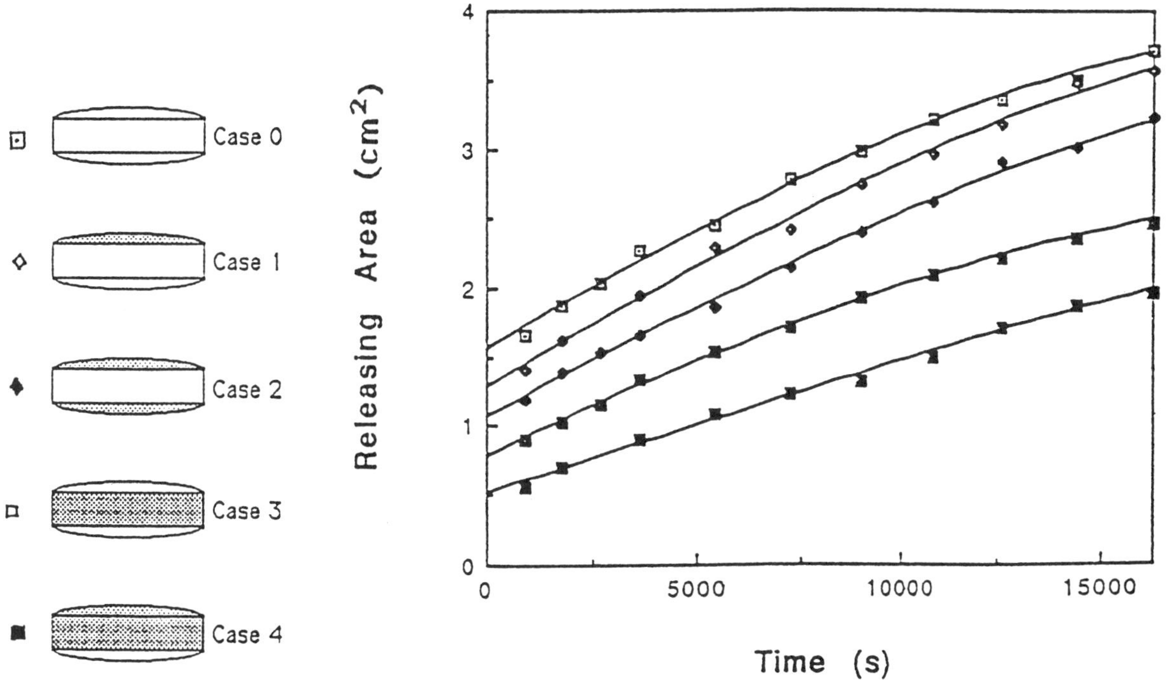

FIGURE 26.30. Releasing area as a function of time for the five systems shown. (Reproduced with permission from reference 46. Copyright 1992 Elsevier.)

$$\frac{M_t}{M_\infty} = \frac{c_b}{ac_{sd}}\left\{\sqrt{\frac{2\left[-D_s\left(c_s + c^\star - c_d - c_b\right) + D_d\left(c_s - c_b\right) + \left(D_d/c_b\right)\right]^2\left(1 - c^\star - c_s\right)t}{D_s\left(2 - c^\star - c_s\right)\left(c^\star + c_s - c_d - c_b\right) + D_d\left(c^\star + c_s\right)\left(c_s - c_b\right)}} + kc_b\right\} \quad (61)$$

Power Law Approach

A simple yet appealing way of describing controlled release and indeed modeling drug delivery is in terms of a power law equation given by

$$\frac{M_t}{M_\infty} = kt^n \quad (60)$$

Here, k and n are fitting parameters. Frisch, Hopfenberg, and others suggested the use of such equations to describe penetrant transport in glassy polymers. Korsmeyer and Peppas (*47*) were the first to propose that such equations can be used in controlled-release systems. Detailed analyses for nonswellable and swellable systems and for various geometries and size distributions based on the power law approach were provided by Ritger and Peppas (*48*).

Typically, the release data are fitted to the power law equation by plotting the logarithm of the dimensionless release as a function of the logarithm of the time. Then, from the slope the value of n is determined. At least a 95% confidence limit is essential for the nonlinear regression.

If the value of n obtained is 0.5, the mechanism of drug transport is Fickian and the drug release rate is time-dependent. If $0.5 < n < 1.0$, the transport is anomalous and the release rate is once again time-dependent. If $n = 1.0$, the transport is of the Case-II type and the release rate is time-independent; or in other words, zero-order release is obtained. If $n > 1.0$, Super-Case-II transport occurs and the resulting release rate is time-dependent. Detailed analyses of this form were performed by Ritger and Peppas (*48*).

Dissolution Model in Presence of Drug

Harland et al. (*49*) formulated a model for drug release in a dissolving polymer–solvent system. The transport was assumed to be Fickian and mass balances were written for the drug and the solvent at the glassy–rubbery interface and at the rubbery–solvent interface (Figure 26.31). The important parameters identified in the phenomenon were the polymer volume fraction $c^\star$ at the glassy–rubbery transition, the polymer volume fraction, c_d, for disentanglement of the chains, and the dissolution/mass-transfer coefficient, k. The expression for drug release as a function of time was obtained as in eq 61. Here, a is the half thickness of the polymer slab, D_s is the diffusion coefficient of the solvent in the polymer, D_d is the diffusion coefficient of the drug in the polymer, and c_b is the volume fraction of the drug in the bulk. A normalized gel layer thickness, δ, was defined as

$$\delta = \frac{S - R}{a} \quad (62)$$

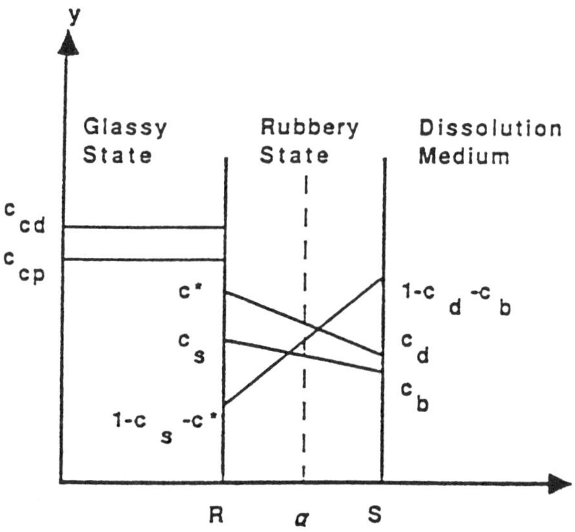

FIGURE 26.31. Parameter definition for drug/polymer tablet dissolution system. (Reproduced with permission from reference 49. Copyright 1988 Plenum.)

Here, S is the position of the rubbery–solvent interface, and R is the position of the glassy–rubbery interface. If front synchronization occurs, the normalized gel layer thickness depends on time as shown in Figure 26.32. Front synchronization leads to zero-order release in dissolution-controlled systems. The drug release and the gel layer thickness, both in normalized form and as a function of time, as predicted by the model are shown in Figures 26.33 and 26.34.

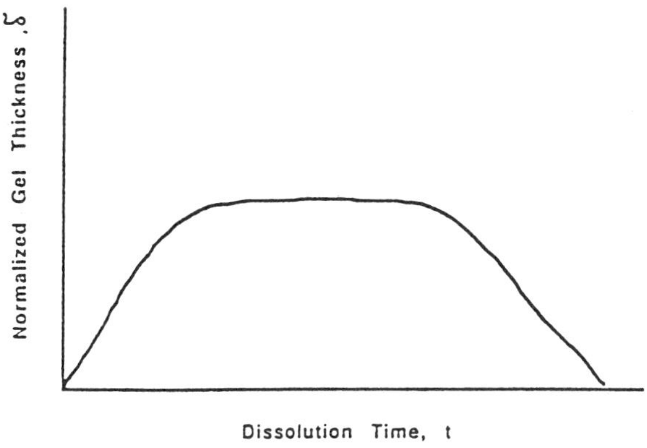

FIGURE 26.32. Dependence of normalized gel layer thickness, δ, on the dissolution time, *t*, for polymer tablets in the absence or presence of drugs. (Reproduced with permission from reference 49. Copyright 1988 Plenum.)

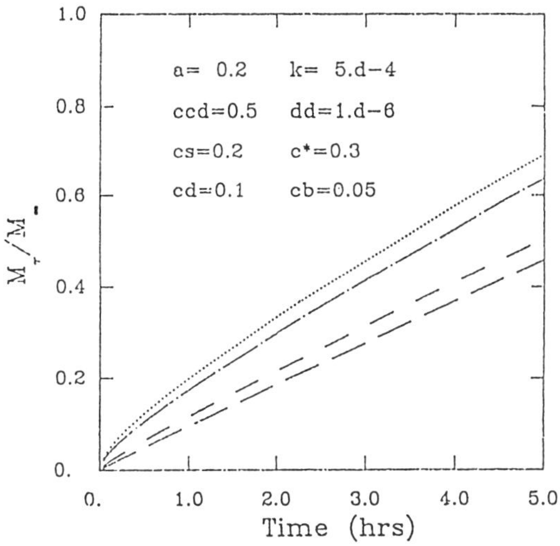

FIGURE 26.33. Normalized drug release as a function of time for drug/polymer tablet dissolution system. (Reproduced with permission from reference 49. Copyright 1988 Plenum.)

Osmotic Systems

Osmotic systems for drug delivery are designed by applying irreversible thermodynamics and the Kedem–Katchalsky (*50*) analysis. The total volume flow, $\mathcal{J}_\upsilon$, and the total exchange flow, $\mathcal{J}_D$, respectively, are given by

$$\begin{aligned} \mathcal{J}_\upsilon &= L_p \Delta p + L_{pD} \Delta \pi_s \\ \mathcal{J}_D &= L_{Dp} \Delta p + L_D \Delta \pi_s \end{aligned} \tag{63}$$

Here, L_i (i = p, pD, Dp, or D) represent Onsager coefficients, Δp is the hydrostatic pressure, and $\Delta \pi_s$ is the osmotic pressure of the solvent.

The systems under consideration may also be defined by the reflection coefficient, σ, and the permeability coefficient, L_p. The term σ is defined as

$$\sigma = -\frac{L_{pD}}{L_p} \tag{64}$$

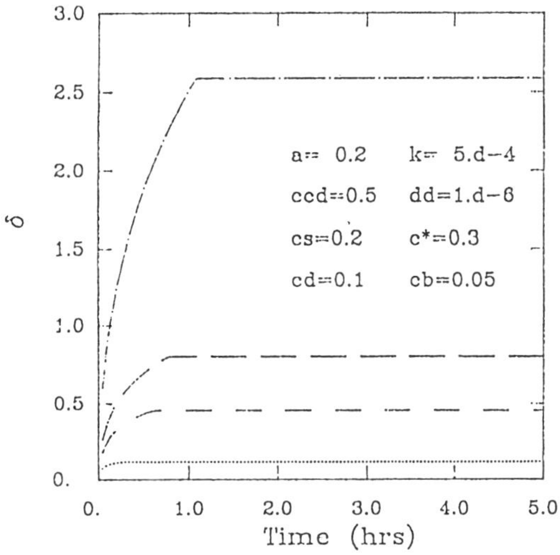

FIGURE 26.34. Normalized gel layer thickness as a function of time for drug/polymer tablet dissolution system. (Reproduced with permission from reference 49. Copyright 1988 Plenum.)

The osmotic pressure is given by

$$\Delta\pi_s = \frac{N}{V}RT = cRT \tag{65}$$

where N is the number of moles of the solvent, V is the solvent volume, R is the universal gas constant, and T is the temperature.

The volume flux, dV/dt for a membrane of thickness δ and cross-sectional area A is given by

$$\frac{dV}{dt} = \frac{A}{\delta}(\sigma\Delta\pi_s - \Delta p) \tag{66}$$

The value $\sigma = 0$ represents a course filter, and $\sigma = 1$ represents an impermeable solute. For large orifices,

$$\Delta\pi_s >> \Delta p \tag{67}$$

Hence,

$$\frac{dV}{dt} = \frac{A}{\delta}L_p\sigma\Delta\pi_s = \frac{A}{\delta}k\Delta\pi_s \tag{68}$$

Using

$$\frac{dM}{dt} = \frac{dV}{dt}c \tag{69}$$

we have

$$\frac{dM}{dt} = \frac{A}{\delta}k\Delta\pi_s c \tag{70}$$

Hence, the drug release rate in osmotically controlled systems can be tuned to achieve zero-order kinetics.

Chemically Controlled Devices

These systems include bioerodible and pendant-group polymer matrices.

Bioerodible Devices

Hopfenberg (*51*) derived expressions for drug release from erodible slabs, cylinders, and spheres. The erosion rate dM_t/dt was described in terms of an erosion constant, k, and the continuously changing available area for biodegradation, A_e as

$$\frac{dM_t}{dt} = kA_e \tag{71}$$

The expression for the drug release is

$$\frac{M_t}{M_\infty} = 1 - \left[1 - \frac{kt}{c_0 l}\right]^n \tag{72}$$

Here, $n = 1$ for a plane sheet of thickness $\delta = 2l$, $n = 2$ for a cylinder of radius $r = l$, and $n = 3$ for a sphere of radius $r = l$.

Heller and Baker (*52*) described polymer erosion rates in terms of the pH of the system. Figure 26.35 depicts their schematic for polymer erosion, and Figure 26.36 shows the calculated and observed erosion rates as a function of pH.

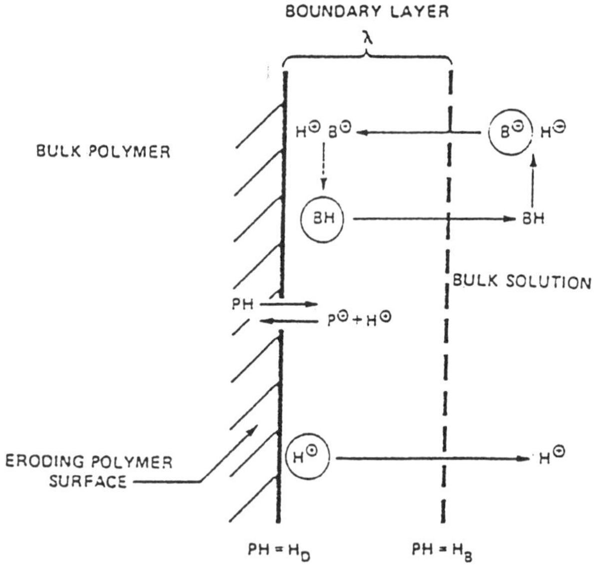

FIGURE 26.35. Model for erosion mechanism. (Reproduced with permission from reference 52. Copyright 1978 Wiley.)

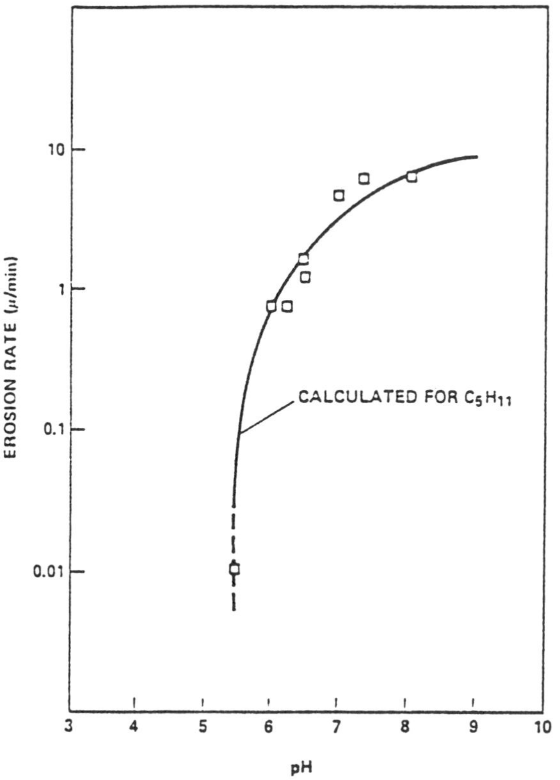

FIGURE 26.36. Calculated (solid line) and observed (broken line) erosion rates as a function of pH for *n*-pentyl half-ester of methyl ether-maleic anhydride copolymer. (Reproduced with permission from reference 52. Copyright 1978 Wiley.)

Thombre and Himmelstein (*53*) developed models to describe drug release from poly(ortho ester)s. A schematic of the system considered is shown in Figure 26.37. The basic form of the equations was

$$\frac{\partial c_i}{\partial t} = \frac{\partial}{\partial x}\left(D_i \frac{\partial c_i}{\partial x}\right) + r_i \qquad (73)$$

where i is any species, and r_i is the rate of generation of the ith species.

Typical boundary conditions used were

$$D_i \frac{\partial c_i}{\partial x} = k_i (c_{ib} - c_i) \qquad (74)$$

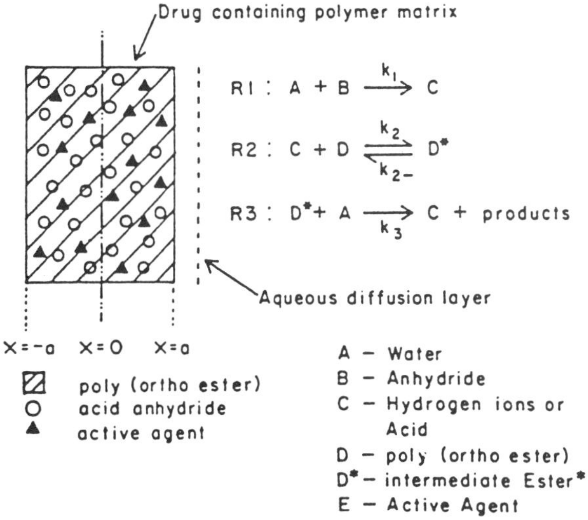

FIGURE 26.37. Schematic of drug release from poly(ortho ester)s. (Reproduced with permission from reference 53. Copyright 1985 AIChE.)

at the edge of the matrix and

$$c_A = K c_A^0 \qquad (75)$$

for water. Here, k_i is the rate constant for the generation of the ith species, c_{ib} is the concentration of the ith species in the bulk, K is a thermodynamic partition coefficient, and c_A^0 is the water concentration in equilibrium with the drug.

The diffusion-reaction model was solved in terms of typical Thiele moduli, ϕ_i, to describe the reaction-diffusion ratio, and Biot numbers, Bi_i, to describe the mass transfer–diffusion ratio. The solutions were reported as fractional amount of each species as a function of time as shown in Figure 26.38:

$$\phi_i = a\sqrt{\frac{k_i c_i}{D_i^0}} \qquad Bi_i = \frac{k_i a}{D_i^0} \qquad (76)$$

where D_i^0 is the diffusion coefficient of the ith species, and a is the half thickness of the polymer film.

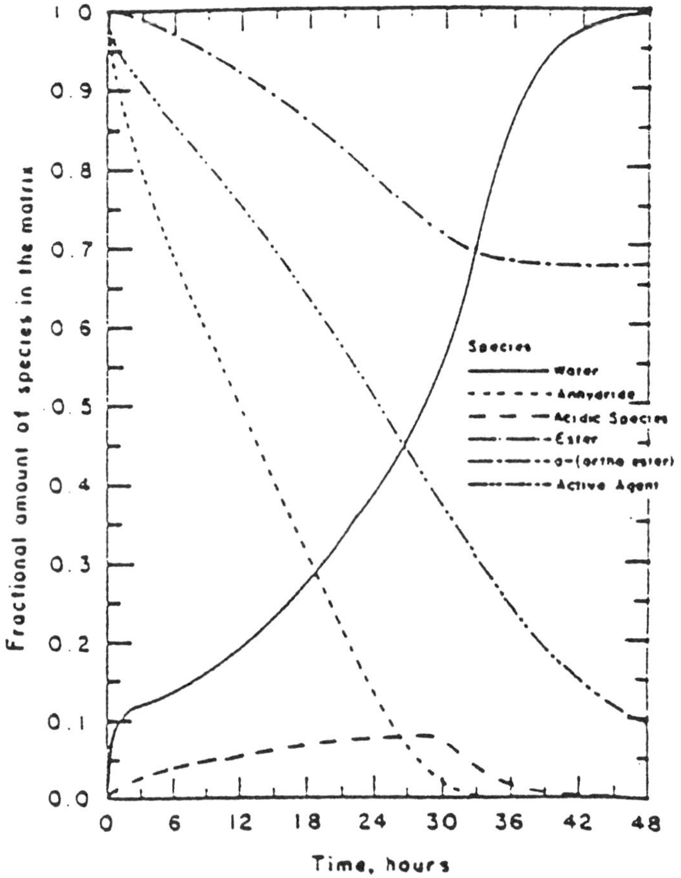

FIGURE 26.38. Fractional amount of species in the matrix as a function of time. (Reproduced with permission from reference 53. Copyright 1985 AIChE.)

Joshi and Himmelstein (*54*) applied the above analysis to bioerodible matrices and identified the water–polymer partition coefficient and the rate of hydrolysis as important parameters in regulating the release kinetics. The water solubility and the matrix component diffusivity as a function of the degree of hydrolysis were studied (Figure 26.39).

Gopferich and Langer (*55*) used Monte Carlo methods to model bioerosion. In their approach, the polymer matrix was represented as the sum of individual matrix parts and the erosion of each part was regarded as a random event. Comparisons with experimental results yielded good agreement. This analysis was later extended (*56*) to describe the release of a monomer from an eroding polymer. The model predicts parameters such as the porosity of the eroding polymer, the matrix weight, and amount of monomer released. The analysis could be extended to model drug release from such systems.

Pendant Chain Systems

Models have been proposed by Anderson and his associates (*57*) to describe cortisol hydrolysis from bound PGA systems. The hydrolysis starts at the surface giving hydrophilic polymer. The rate of water permeation, R_w, in the hydrophobic zone and the rate of hydrolysis, R_h, are important

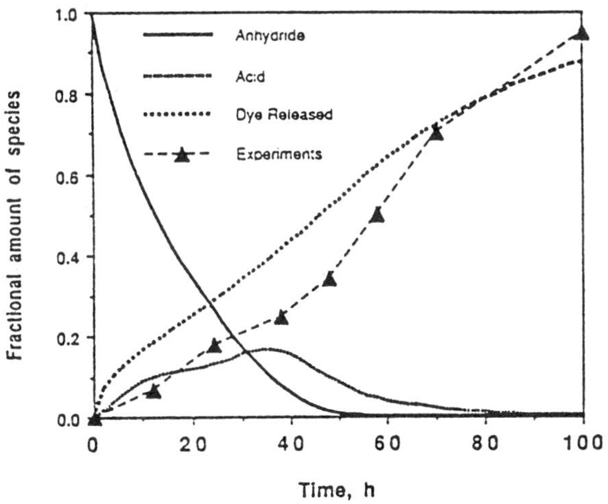

FIGURE 26.39. Cumulative release of active species from the polymer disk containing 0.25% of the anhydride. (Reproduced with permission from reference 54. Copyright 1991 Elsevier.)

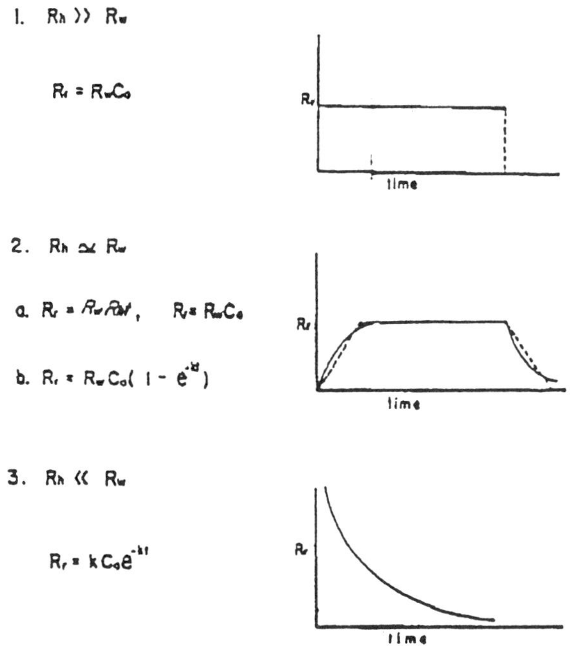

FIGURE 26.40. Release curves derived from the hydrophilic/hydrophobic models. (Reproduced with permission from reference 57. Copyright 1981 Plenum.)

parameters in tuning the release kinetics. The release curves derived from the hydrophilic/hydrophobic models are shown in Figure 26.40.

Conclusion

Drug release through controlled delivery polymeric systems has been modeled predominantly by steady-state and transient description of drug diffusion by use of Fick's law as in eqs 1 and 4. Drug release from matrix (monolithic) and membrane (reservoir) devices has been modeled, and the total amount of drug released and the drug release rate have been predicted for a variety of geometries.

A more accurate mathematical description is necessary in modeling drug release in swellable and porous polymeric systems. Consideration of aspects such as countercurrent solvent diffusion, chain disentanglement, polymer state transitions, degree of crystallinity, and porous structure will render existing models applicable to a variety of future controlled-release devices. The advent of codelivery polymeric systems has necessitated the need for models incorporating interacting and noninteracting multicomponent diffusion through polymers.

References

1. Baker, R. W.; Lonsdale, H. K. In *Controlled Release of Biologically Active Agents;* Tanquarry, A. C.; Lacey, R. E., Eds.; Plenum: New York, 1974; p 15.
2. Lee, P. I. *J. Membr. Sci.* **1980**, *7*, 255.
3. Peppas, N. A. In *Controlled Drug Bioavailability: Drug Product Design and Performance;* Smolen, V. F.; Ball, L. A., Eds.; Wiley: New York, 1984; Vol. 1, p 203.
4. Peppas, N. A. In *Medical Applications of Controlled Release Technology;* Langer, R. S.; Wise, D., Eds.; CRC Press: Boca Raton, FL, 1984; Vol. 2, p 169.
5. Fan, L. T.; Singh, S. K. *Controlled Release;* Springer: Berlin, Germany, 1989.

6. Peppas, N. A. In *Trends and Future Perspectives in Peptide and Protein Drug Delivery;* Lee, V. L.; Hashida, M.; Mizushima, Y., Eds.; Harwood Academic: Chur, Switzerland, 1995; p 23.
7. Lightfoot, E. N. *Transport Phenomena and Living Systems;* John Wiley: New York, 1974.
8. Crank, J. *The Mathematics of Diffusion,* 2nd ed.; Oxford University: New York, 1975.
9. Vergnaud, J. M. *Liquid Transport Processes in Polymeric Materials;* Prentice Hall: Englewood Cliffs, NJ, 1991.
10. Fujita, H. *Fortschr. Hochpolym. Forsch.* **1961,** *3,* 1.
11. Langer, R. S.; Peppas, N. A. *J. Macromol. Sci. Rev. Macromol. Chem. Phys.* **1983,** *23,* 61.
12. Higuchi, T. *J. Pharm. Sci.* **1961,** *50,* 874.
13. Paul, D. R.; McSpadden, S. K. *J. Membr. Sci.* **1976,** *1,* 33.
14. Fu, J. C.; Hagemeier, C.; Moyer, D. L. *J. Biomed. Mater. Res.* **1976,** *10,* 743.
15. Rhine, W. D.; Sukhatme, V.; Hseih, D. T.; Langer, R. S. In *Controlled Release of Bioactive Materials;* Baker, R., Ed.; Academic: New York, 1980; p 177.
16. Gurny, R.; Doelker, E.; Peppas, N. A. *Biomaterials* **1982,** *3,* 27.
17. Chandrasekaran, S. K.; Paul, D. R. *J. Pharm. Sci.* **1982,** *71,* 1399.
18. Peppas, N. A.; Korsmeyer, R. W. In *Hydrogels in Medicine and Pharmacy: Properties and Applications;* Peppas, N. A., Ed.; CRC Press: Boca Raton, FL, 1987; Vol. 3, p 109.
19. Siegel, R. A. In *Controlled Release of Drugs;* Rosoff, M., Ed.; VCH: New York, 1989; p 1.
20. Harland, R. S.; Peppas, N. A. *Colloid Polym. Sci.* **1989,** *267,* 219.
21. Higuchi, T. *J. Pharm. Sci.* **1963,** *52,* 1145.
22. Roseman, T. J.; Higuchi, W. I. *J. Pharm. Sci.* **1970,** *59,* 353.
23. Fessi, H.; Marty, J. P.; Puisieux, F.; Carstensen, J. T. *Int. J. Pharm.* **1978,** *1,* 265.
24. Desai, S. J.; Singh, P.; Simonelli, A. P.; Higuchi, W. I. *J. Pharm. Sci.* **1966,** *55,* 1224.
25. Harland, R. S.; Dubernet, C.; Benoit, J.-P.; Peppas, N. A. *J. Controlled Release* **1988,** *7,* 207.
26. Chang, N. J.; Himmelstein, K. J. *J. Controlled Release* **1990,** *12,* 201.
27. Colton, C. K.; Smith, K. A.; Merrill, E. W.; Farrell, P. C. *J. Biomed. Mater. Res.* **1971,** *5,* 459.
28. Vrentas, J. M.; Jarzebski, C. M.; Duda, J. L. *AIChE J.* **1975,** *21,* 894.
29. Peppas, N. A.; Franson, N. M. *J. Polym. Sci. Polym. Phys. Ed.* **1983,** *21,* 983.
30. Lee, P. I. *Polymer* **1984,** *25,* 973.
31. Lee, P. I. *J. Controlled Release* **1986,** *4,* 1.
32. Lee, P. I. In *Controlled-Release Technology: Pharmaceutical Applications;* Lee, P. I.; Good, W. R., Eds.; ACS Symposium Series 348; American Chemical Society: Washington, DC, 1987; pp 71–83.
33. Kou, J. H.; Fleisher, D.; Amidon, G. L. *J. Controlled Release* **1990,** *12,* 241.
34. Cohen, D. S.; Erneux, T. *SIAM J. Appl. Math.* **1988,** *48,* 1451.
35. Cohen, D. S.; Erneux, T. *SIAM J. Appl. Math.* **1988,** *48,* 1466.
36. Hopfenberg, H. B.; Hsu, K. C. *Polym. Eng. Sci.* **1978,** *18,* 1186.
37. Korsmeyer, R. W.; Lustig, S. R.; Peppas, N. A. *J. Polym. Sci. Polym. Phys. Ed.* **1986,** *24,* 395.
38. Lustig, S. R.; Peppas, N. A. *J. Appl. Polym. Sci.* **1987,** *33,* 533.
39. Lee, P. I.; Peppas, N. A. *J. Controlled Release* **1987,** *6,* 201.
40. Lustig, S. R.; Caruthers, J. M.; Peppas, N. A. *Chem. Eng. Sci.* **1992,** *47,* 3037.
41. Peppas, N. A.; Brannon-Peppas, L. *J. Membr. Sci.* **1990,** *48,* 281.
42. Brannon-Peppas, L.; Peppas, N. A. *Int. J. Pharm.* **1991,** *70,* 53.
43. Bell, C. L.; Peppas, N. A. *Polym. Prepr. (Am. Chem. Soc. Div. Polym. Chem.)* **1993,** *34,* 831.
44. Hariharan, D.; Peppas, N. A. *J. Membr. Sci.* **1993,** *78,* 1.
45. Hariharan, D.; Peppas, N. A. *J. Controlled Release* **1993,** *23,* 123.
46. Colombo, P.; Catellani, P. L.; Peppas, N. A.; Maggi, L.; Conte, U. *Int. J. Pharm.* **1992,** *88,* 99.
47. Korsmeyer, R. W.; Peppas, N. A. In *Controlled Release Delivery Systems;* Mansdorf, S. Z.; Roseman, T. J., Eds.; Marcel Dekker: New York, 1983; p 77.
48. Ritger, P. L.; Peppas, N. A. *J. Controlled Release* **1987,** *5,* 23.
49. Harland, R. S.; Gazzaniga, A.; Sangani, M. E.; Colombo, P.; Peppas, N. A. *Pharm. Res.* **1988,** *5,* 488.
50. Kedem, O.; Katchalsky, A. *Biochim. Biophys. Acta* **1958,** *27,* 229.
51. Hopfenberg, H. B. In *Controlled Release Polymeric Formulations;* Paul, D. R.; Harris, F. W., Eds.; ACS

Symposium Series 33; American Chemical Society: Washington, DC, 1976; pp 26–32.

52. Heller, J.; Baker, R.; Gale, R. M.; Rodin, J. O. *J. Appl. Polym. Sci.* **1978,** *22,* 1991.
53. Thombre, A. G.; Himmelstein, K. J. *AIChE J.* **1985,** *31,* 759.
54. Joshi, A.; Himmelstein, K. J. *J. Controlled Release* **1991,** *15,* 95.
55. Gopferich, A.; Langer, R. *Macromolecules* **1993,** *26,* 4105.
56. Gopferich, A.; Langer, R. *J. Controlled Release* **1995,** *33,* 55.
57. Tani, N.; Van Dress, M.; Anderson, J. M. In *Controlled Release of Pesticides and Pharmaceuticals;* Lewis, D. H., Ed.; Plenum: New York, 1981; p 79.

27

Computer Dynamics Simulation of Controlled Release

D. Robert Lu

Computer dynamics simulation of controlled release is a useful approach to study the controlling mechanism of drug release at a molecular level. Although to date the theories and applications are still at the early development stage, many studies have been carried out involving smaller molecules and various synthetic polymers, and the results were in general in agreement with those obtained experimentally. This chapter introduces the application of computer dynamics simulation of the behaviors of diffusive molecules in polymer matrix. We also explore the theoretical basis and general approaches of the simulation. The computer simulation tools and the associated experiments will also be discussed.

Development of controlled drug release systems involves complicated processes that include careful planning, detailed design, and subsequent experiments. With increasing power and widely ranging applications of modern computers, we can use computers not only to assist our data processing in the experiments but also to gain an insight to the molecular mechanism of controlled-release systems. The understanding of such a mechanism at the molecular level can ultimately develop a mechanism-based design strategy for new controlled-release systems.

The role of computer applications in the development of controlled-release systems can be briefly summarized in Figure 27.1. The first major application involves computer modeling and simulation of the controlled release at a molecular level that generate useful information for the rational design of the systems. The second application is in the area of in vivo and in vitro studies, such as modeling of drug release at a macroscopic level, in vivo–in vitro correlation of release profiles, and pharmacokinetic modeling of drug concentrations in the body. A third application is to assist the development stage in optimizing the device functions and manufactures. The computer simulation of controlled release at the molecular level, especially the molecular dynamics simulation, is a completely new area and little has been done. Therefore, this chapter will be solely devoted to

Copyright ©1997 American Chemical Society

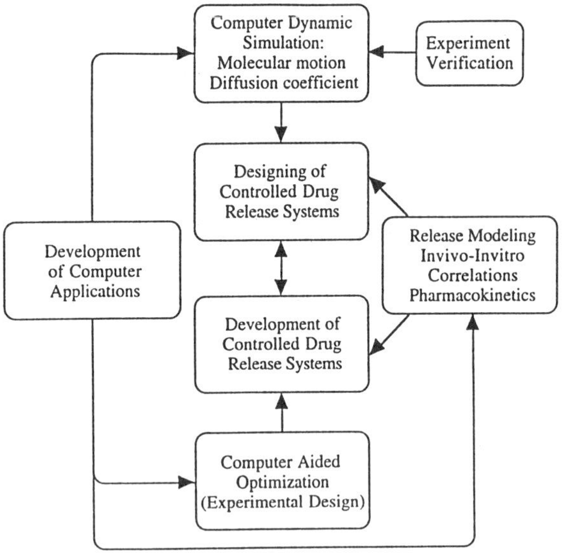

FIGURE 27.1. Role of computer applications in the development of controlled-release systems.

the discussion of the present studies and future applications in the controlled-release field. Because the polymer matrix system has been widely used for and is also the major unit of controlled-release systems, the chapter will mainly focus on those molecular dynamics simulations involving synthetic polymers. The computer applications in the in vitro and in vivo studies are relatively mature compared to the other applications, and excellent monographs can be found (*1–3*). The technique of computer-aided experimental design and optimization has been widely used in other areas, such as agriculture, chemistry, and traditional pharmaceutics, and has been proven useful and efficient (*4–6*). The application of the technique to the controlled-release area is important, owing to the nature of the release systems, which is much more complicated than the traditional dosage forms. Modeling of the response surfaces related to the device functions may provide a key role toward its optimization during the development process. Although these two aspects of the computer applications are important, they will not be covered in the content of this chapter because of the limit in length of this chapter. The relationship between these two applications and the molecular dynamics simulation of controlled release, however, can be clearly represented in Figure 27.1.

In summary, this chapter introduces the application of computer dynamics simulation of the behaviors of diffusive molecules in polymer matrix. The theoretical basis and general approaches will be discussed in detail to inspire the further development in this new and exciting area.

Controlled Release and Diffusion Coefficient

Computer simulation of a biochemical process is usually associated with the modeling studies of various systems involved in the process. For controlled-release systems, many different types of modeling studies have been carried out, and excellent correlations with experimental results were found (*7, 8*). Understanding of the mechanism of controlled release at a molecular level, however, is still in its early stage. Computer simulation corresponding to the existing models may provide insight on the mechanism and it is particularly interesting in the studies at the molecular level. The simulation may also provide useful information for the further design, preparation, and optimization of specific controlled-release devices.

As described in several chapters in this book, the use of various polymer materials has been revolutionary for the development of novel controlled drug release systems. A well-explored subject in controlled release of drug molecules from pharmaceutical dosage forms is the release profiles from polymer matrix devices. In many cases, we deal with a monolithic system in which the drug molecules are homogeneously dissolved or dispersed throughout the polymer mass. In general, when the drug loading is in the lower range, the release profile can be well described by the Higuchi model (*7*). The actual mathematical expression is dependent on the geometry of the release device. The rate of drug release from spherical matrices can be described by the following equation:

$$\frac{dF}{dt} = \frac{3DC_s}{r_0^2 C_0}\left[\frac{(1-F)^{1/3}}{1-(1-F)^{1/3}}\right] \tag{1}$$

where F is the fraction of drug released at any time, t, D is the diffusion coefficient of the drug molecules in the polymer matrix, C_s is the drug solubility in the polymer, r_0 is the radius of the device, and C_0 is the initial concentration of the drug in the polymer matrix. The integrated form of this equation, which is known as Baker and Lonsdale equation (*7*), is as follows:

$$\frac{3}{2}\left[1-(1-F)^{2/3}\right] - F = \frac{3DC_s}{r_0^2 C_0}t \tag{2}$$

The model and the corresponding equation have been well correlated with experimental results (*7, 8*). The equation is routinely used to fit the experimental data to generate a value for the constant portion at the right side of the equation. However, although release of drugs from polymer matrix can be described by these models, the actual mechanism of drug molecules diffusing through polymer matrix is not elucidated. Furthermore, the constant portion of the equation can only be used for fitting purposes. The theoretical calculation of its value solely based on information of the drug and polymer is not possible. From the equation, it is clear that the entire release profile is intrinsically controlled by the diffusion coefficient D, because C_s is usually a known value (at least the relative solubility) and the C_0 and r_0 are controllable variables in the manufacturing or experimental process. Obviously, predicting the values of diffusion coefficients solely based on the physicochemical and structural information of the drug and polymers is desirable. The ability to predict the diffusion coefficients will provide us with a useful tool for designing controlled-release systems. Given the information of a new drug and the desired release behaviors, one should then be able to calculate the diffusion coefficient for different polymers and hence choose the best candidate. In principle, the macroscopic release profile is actually controlled by the molecular behaviors of the drug and polymer molecules and the interactions between both. The diffusion coefficients may be calculated based on the classical molecular dynamics simulations by using the technique of center-of-mass mean-square displacement (*9, 10*). Molecular dynamics simulation calculates the self-diffusion coefficient of the diffusive molecules. In most cases, the mutual diffusion coefficient, however, can be approximated by the corresponding self-diffusion coefficient.

Attempts have been made to study the molecular diffusion in polymers by using molecular dynamics simulation and subsequently to calculate the diffusion coefficients solely based on the physicochemical and structural information (*10–15*). The studies were established based on the molecular interactions and motions of both diffusive molecules and polymer chains at the molecular level. The development of the techniques made it possible not only to predict the diffusion coefficients for the practical purposes, but also to generate more useful information for understanding the molecular behaviors of small molecules in polymer matrix. The work emerged from the field of physical chemistry at the beginning of this decade. Much of the progress appears to correspond with the development in computational power. Because of the limitation of the present computational power and relatively short period of development, the work is still at its early stage and, at the present time, only small gas molecules are employed in the studies. The scaling up of the research for pharmaceutical drug molecules will be of great interest and will significantly benefit the controlled-release studies. Because both gas molecules and larger drug molecules are dissolved in polymers and no solvent interaction is involved (in the cases discussed in this chapter), the scaling up appears to be feasible and the approaches similar in nature. The present development of molecular dynamics simulations for small gas molecules can well serve as an initial step for studies involving larger drug molecules.

Hopping Mechanism

Most of the simulations involve the so-called hopping mechanism (*10, 11*). In the polymer matrix, polymer chains are connected in a network based on the interactions among these chains. The network provides a structure in which a large number of cavities reside. The diffusive molecules that are in a dissolved state stay in these cavities (the polymer matrix containing solid particles is beyond the scope of this discussion). Because of the molecular motion, the diffusive molecules are constantly moving nondiscriminately in all directions, and the net result is an oscillation or a quasi-stationary state in the original cavities. However, within a certain period of time, some of the molecules experience a quick jump from one cavity to another resulting in the net movement of the molecules. The quasi-stationary period and the jump frequency depend on the structure and flexibility of the polymer matrix, the size of the diffusive molecules, and the interactions between polymer chains and diffusive molecules. During the quasi-stationary state, no net diffusion flow exists because the molecules are still located in the original cavities. The diffusion flow is a consequence of the molecular jump from one cavity to another. This process is called the hopping mechanism.

However, the polymer network is also in a dynamic state and is constantly reorganizing. The cavities where the diffusive molecules reside are consequently changing their locations in the polymer matrix. Nevertheless, no net diffusion flow is expected in the reorganizing process. The entire diffusion process involves several steps:

1. Dynamic movement and reorganization of the polymer chain network and the redistribution of the cavities occur.
2. Some of the diffusive molecules experience a quick jump from their original cavities to others leaving some empty cavities.
3. These empty cavities may disappear owing to the reorganization of polymer chains or refilling by other diffusive molecules.
4. The jump of the diffusive molecules is statistically in the direction of concentration gradient, partly due to larger number of empty cavities, resulting in the diffusion flow.

Model Setup

The model system usually contains a cubic molecular dynamics cell. A number of polymer chains (e.g., 20–100) with certain chain length (e.g., 20–200 monomer units) are placed in the molecular dynamics cell. The density of the polymer matrix is configured according to the polymer density determined experimentally. The size of the molecular dynamics cell, thus, is determined by the number of polymer units in the cell and the density required. A larger size of the cell (e.g., more polymer units) may provide some advantages but the this setup is sufficient for small molecules. To minimize the limitation due to its size, the molecular dynamics cell is configured with periodic boundary conditions on all three dimensions (described subsequently). The diffusive molecules are randomly configured into the cell and the maximal number of the diffusive molecules in a particular dynamics cell may be determined by the solubility of the molecule in the polymer. The initial generation of the molecular dynamics cell of amorphous polymers with experimental density can be made by the rotational isometric state (RIS) method of Theodorou and Suter (*16*). The method has been implemented in the Polymer/Amorphous_Cell modules of the INSIGHT II molecular modeling software (Biosym Technologies, San Diego, CA). The initial structure can then be energy-minimized and equilibrated for several hundred picoseconds by molecular dynamics at a constant temperature. Figure 27.2 shows a molecular dynamics cell of amorphous polypropylene generated by the RIS method with a minor refinement. For demonstration purposes, the cell contains only 30 polypropylene chains with 15 monomer units, each chain with a total number of 4080 atoms. As can be seen from the figure, the RIS method generates a rea-

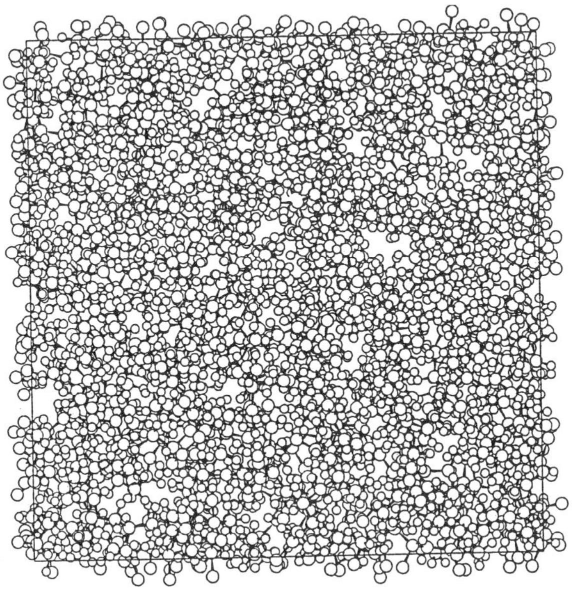

FIGURE 27.2. Demonstration of a molecular dynamics cell. The cell was generated by the RIS method with a minor refinement containing 30 amorphous polypropylene chains and 15 monomer units, each chain with a total number of 4080 atoms.

sonably good configuration. However, further energy refinement may be needed to set up an energy-minimized molecular dynamics cell with evenly distributed polymer units in space for subsequent simulation. The refinement may be done by using the Discover module (Biosym), which is interfaced with INSIGHT II.

Because of the large scale of computation involved and the limit of the current computer power, a "quasi-atoms" method may be used to reduce the extent of calculation. The hydrogen atoms of polymer chains may not be explicitly expressed. Instead, the atomic masses and the parameters defining the van der Waals interactions are applied to mimic individual functional groups. For example, a mass of 14.02 amu is used for methylene group and a mass of 15.02 amu is used for methyl group. On the other hand, depending on the extent of complexity of the setup, explicit hydrogen atoms may be involved in the simulation for more accurate results at the expense of computational time.

Several potential functions are used to characterize the polymer chain configurations in the molecular dynamics cell (*14*). The bond length, b, and bond angle, θ, are subjected to harmonic potentials V_b and V_θ, respectively:

$$V_b = \tfrac{1}{2}K_b(b - b_0)^2 \qquad (3)$$

$$V_\theta = \tfrac{1}{2}K_\theta(\cos\theta - \cos\theta_0)^2 \qquad (4)$$

where b_0 and θ_0 are the equilibrium bond length and angle, respectively, and K_b and K_θ are the force constants for bond length and bond angle, respectively. The internal rotations are evaluated by the potential V_ϕ:

$$V_\phi = K_\phi \Sigma\, a_n \cos^n\phi \qquad (5)$$

where ϕ is the dihedral angle of each unit on the polymer chain and K_ϕ and a_n are constants. The nonbonded interactions V_{nb}, including intramolecular as well as intermolecular interactions, are given by the truncated Lennard–Jones 12-6 potentials,

$$V_{nb,ij} = 4\varepsilon_s[(\sigma_s/r_{ij})^{12} - (\sigma_s/r_{ij})^6] \qquad r_{ij} < r_c \qquad (6)$$

$$V_{nb,ij} = 0 \qquad r_{ij} \geq r_c \qquad (7)$$

where r_{ij} is the distance between polymer unit i and j, and r_c is the cutoff distance. The values ε_s and σ_s are constants. Usually, the nonbonded potential is not considered when the units are separated by less than three bonds along the polymer chain. The values of the parameters in these equations are available in the literature (*11*). As an example, Table 27.1 lists the parameters used for polyethylene chains (*14, 17*).

Periodic Boundary Condition

As mentioned previously, because of the limitation of the computer power the molecular dynamics

TABLE 27.1. Parameters Used To Characterize Polyethylene Chain Configurations in Molecular Dynamics Cell

Parameter	*Value*
b_0	0.15 nm
$\cos\theta_0$	–0.33
K_b	3.46×10^7 J nm^{-2} mol^{-1}
K_θ	5.00×10^5 J mol^{-1}
K_ϕ	9.00×10^3 J mol^{-1}
a_n (n = 0–5)	1.00; 1.31; –1.41; –0.33; 2.83; –3.39
ε_s	500 J mol^{-1}
σ_s	0.38 nm
r_c	0.80 nm

cell contains only a relatively small number of polymer chains. To make the simulation resemble a realistic situation, a periodic boundary condition is implemented in the configuration of the cell (*16*). The cell is treated as part of an infinite system and is surrounded by the displaced images of the polymer chains in the cell. In this infinite system, if the tailmost backbone atoms of a polymer chain locate in the simulation cell, the chain is considered as the parent molecule, and otherwise as the image molecule. Figure 27.3 shows a simulation cell with all the eight surrounding neighboring cells based on the periodic boundary condition. All the polymer chains in the surrounding cells are composed of the displaced images of the parent cell. The arrangement of the parent molecule and the image molecules in each cell is made in a way that each cell contains the exact number of atoms intended to be placed in the cell. This configuration ensures the elimination of surface effects on those atoms located in the periphery of the simulation cell. To demonstrate the periodic boundary condition, only one polypropylene chain composed of 100 monomer units is packed into the simulation cell (Figure 27.3). The configuration was generated without using the RIS method. The uneven distribution of the atoms in the cell can therefore help us to examine the packing pat-

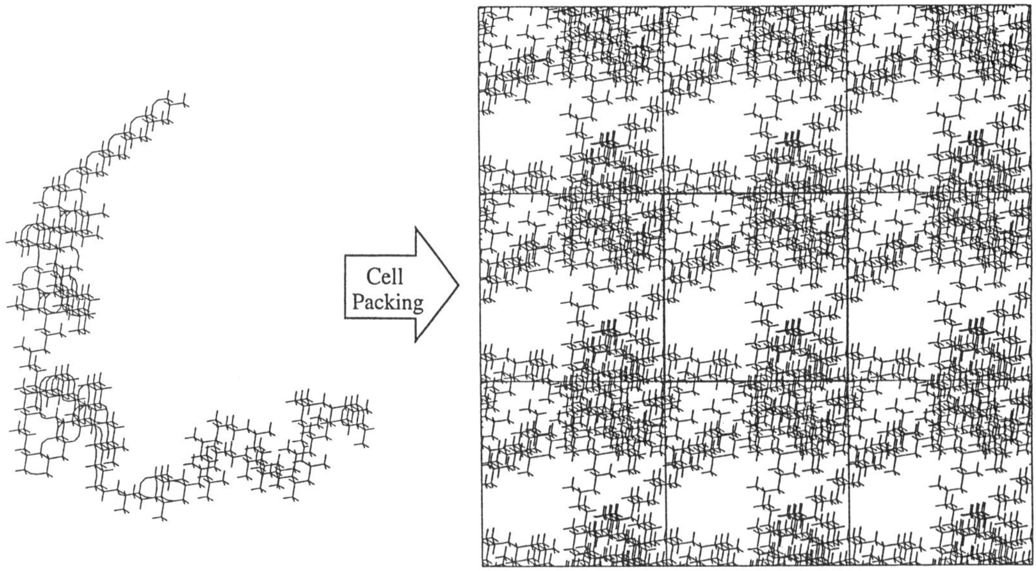

FIGURE 27.3. A demonstration of a dynamics simulation cell with all eight surrounding neighboring cells based on the periodic boundary condition. Only one polypropylene chain composed of 100 monomer units was packed into the simulation cell, and the configuration was generated without using the RIS method.

tern and can better demonstrate the periodic boundary condition.

Simulation Steps

After the setup of the simulation cell, the molecular dynamics calculations can be carried out using the Verlet algorithm or its modified version (*9*). The most popular one for dynamics simulation is the leapfrog algorithm, which is a modified version of the original Verlet algorithm. The algorithms are briefly described as follows. The technique deals with the classical equations of motion for a system composed of a number of molecules interacting with each other based on the potential interaction energies. For each atom, i, with Cartesian coordinate of $\mathbf{r}_i$, the motion equation can be written as follows:

$$m_i \frac{d^2 r_i}{dt^2} = f_i \qquad (8)$$

where $\mathbf{m}_i$ is the mass of the atom and $\mathbf{f}_i$ is the force on that atom associated with the kinetic and potential energies. The equation also applies to the center of mass motion of a molecule and can be employed for our simulation purpose. The general simulation step of the molecular dynamics is that, given the molecular positions, velocities, and other dynamic information at time t, we attempt to obtain the positions, velocities, and new dynamic information at a later time $t + \delta t$, to a sufficient degree of accuracy. In most cases, the new position can be evaluated by Taylor expansion about $\mathbf{r}(t)$:

$$\mathbf{r}(t + \delta t) = \mathbf{r}(t) + \delta t\mathbf{v}(t) + 1/2\ \delta t^2\mathbf{a}(t) + 1/6\ \delta t^3\mathbf{b}(t) + \ldots \qquad (9)$$

where $\mathbf{r}$, $\mathbf{v}$, $\mathbf{a}$, and $\mathbf{b}$ are the positions, velocities, accelerations, and third time derivatives of $\mathbf{r}$, respectively. By truncating the expansion to retain only up to second derivatives of $\mathbf{r}$, a reasonable accuracy may be obtained for the calculation of the new position. The Verlet algorithm can further eliminate the velocity term by addition of the equations of the Taylor expansion with respect to $\mathbf{r}(t + \delta t)$ and $\mathbf{r}(t - \delta t)$:

$$\mathbf{r}(t + \delta t) = \mathbf{r}(t) + \delta t\mathbf{v}(t) + \tfrac{1}{2}\delta t^2\mathbf{a}(t) + \ldots \qquad (10)$$

$$\mathbf{r}(t - \delta t) = \mathbf{r}(t) - \delta t\mathbf{v}(t) + \tfrac{1}{2}\delta t^2\mathbf{a}(t) - \ldots \qquad (11)$$

The resulting Verlet algorithm can be written as:

$$\mathbf{r}(t + \delta t) = 2\mathbf{r}(t) - \mathbf{r}(t - \delta t) + \delta t^2\mathbf{a}(t) \qquad (12)$$

and the missing velocity term can be evaluated by:

$$\mathbf{v}(t) = \frac{\mathbf{r}(t + \delta t) - \mathbf{r}(t - \delta t)}{2\delta t} \qquad (13)$$

The leapfrog algorithm is a modified version of the Verlet algorithm and is widely used because of its better handling of velocities. The equations are as follows:

$$\mathbf{r}(t + \delta t) = \mathbf{r}(t) + \delta t\mathbf{v}(t + \tfrac{1}{2}\delta t) \qquad (14)$$

$$\mathbf{r}(t + \tfrac{1}{2}\delta t) = \mathbf{v}(t - \tfrac{1}{2}\delta t) + \delta t\mathbf{a}(t) \qquad (15)$$

The main difference between the leapfrog algorithm and the Verlet algorithm is that the velocities leap over the coordinates to obtain the next midstep values in the leapfrog algorithm. The selection of the δt term is based on the computational power, the number of atoms in the dynamics cell, and the simulation scale (e.g., the overall number of time steps performed in the simulation). A small δt term may improve the accuracy of the results to some extent. However, the trade-off for a smaller δt term is the reduced simulation scale owing to the limitation of the computational power. The usual time increment δt is about 10^{-15} to 10^{-14} s. At present, a typical simulation run has about 10,000–100,000 time steps, which represents the behavior of the simulation cell in a duration of 100–1,000 ps.

During the simulation run, the new coordinates of the molecules can be written to a file at predetermined time increments (e.g., 10^{-12} s) to keep track of the trajectory of the molecular motion. Another important measurement during the simulation of the molecular motion is the mean-square displacement of the center of mass $<|\mathbf{R}_i(t) - \mathbf{R}_i(0)|^2>$ of the molecules from the initial positions. In this mean-square displacement expression, $\mathbf{R}_i(t)$ and $\mathbf{R}_i(0)$ represent the position of molecule i at time t and time $t = 0$, respectively. The average denoted by angle brackets is evaluated over all diffusive molecules and all possible time origins. Specifically, a time origin is selected and after a number of time steps (e.g., 150, 170, 200 time steps) the positions of the molecules are recorded and compared with those at the time origin to calculate the square of the displacement of the collective center of mass of the molecules. The time origin is then moved 75 time steps and the process is repeated. A large amount of data can be collected in this manner, and the mean-square displacement of the center of mass $<|\mathbf{R}_i(t) - \mathbf{R}_i(0)|^2>$ can be calculated (*18*).

Molecule Motion and the Diffusion Coefficient

On the basis of the coordinates of the molecules at all time increments that were saved in a file during simulation, the molecular motion of both diffusive molecules and polymer chains can be examined. The trajectories of the molecular motions in a period of time provide useful information in the study of molecular behaviors of the diffusive molecules in the polymer matrix. The hopping patterns usually can be seen by constructing a curve of the molecular displacement as a function of time. The mean-square displacement of all the diffusive molecules averaged over all the molecules and all possible time origins can also be calculated from the trajectory data.

The diffusion coefficients can then be calculated based on the center-of-mass mean-square displacement of the diffusion molecules. With a large amount of the sampling time, the center-of-mass mean-square displacement becomes linear in time.

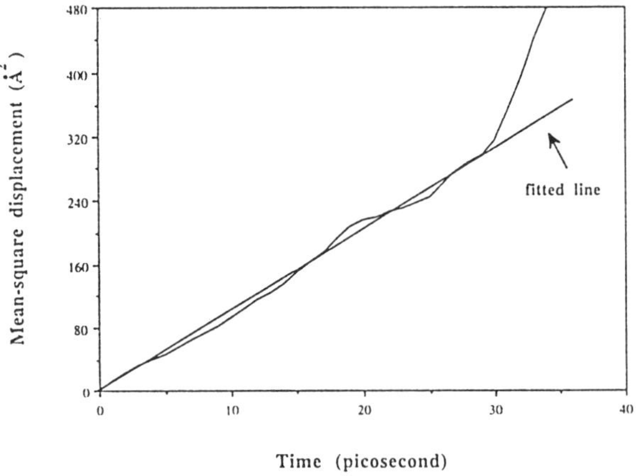

FIGURE 27.4. Center-of-mass mean-square displacement of N_2 molecules in a polydimethylsiloxane matrix as a function of time.

Figure 27.4 shows the mean-square displacement of N_2 molecules in a polydimethylsiloxane matrix as the function of time. The graph was directly generated by using the files of n2_pdms.car and n2_pdms.his in the INSIGHT II package. The displacement was almost linear with time at the early portion. After that, statistical errors caused the curve to diverge (*11*). The diffusion coefficient D can be calculated from the slope of the linear line based on the Einstein relation:

$$\mathbf{D} = <|\mathbf{R}_i(t) - \mathbf{R}_i(0)|^2> / 6t \qquad (16)$$

At present, a number of simulations have been carried out to examine the diffusion coefficients of small molecules in different polymer matrices. Table 27.2 summarizes the results from these studies (*10–14, 19–21*). The predicted diffusion coefficients were generally in a certain degree of agreement with the values obtained experimentally. However, direct comparison of the calculated values with the experiment values is sometimes difficult because the simulation is usually conducted in an "ideal" condition that is different from the experiment condition. For example, real polyethylene is not entirely amorphous but con-

TABLE 27.2. Simulation Studies for Examination of Diffusion Coefficients of Small Molecules in Different Polymer Matrices

Polymer	*Diffusive Molecule*	*Diffusion Coefficient ($cm^2\ s^{-1}$)*	*Temperature (K)*	*Ref.*
Polyethylene	CH_4	7.7×10^{-5}	300	10
Polyethylene	CH_4	1.2×10^{-5}	298	12
Polyethylene	O_2	$1.3–1.6 \times 10^{-5}$	300	14, 19
Polyethylene (glassy)	O_2	3.8×10^{-7}	300	20
Polypropylene	H_2	4.3×10^{-5}	300	11
Polypropylene	CH_4	4.8×10^{-7}	300	11
Polypropylene	O_2	4.0×10^{-6}	300	11
Polyisobutylene	He	3.7×10^{-5}	300	21
Polyisobutylene	CH_4	6.2×10^{-6}	298	12
Polydimethylsiloxane	He	1.8×10^{-4}	300	13
Polydimethylsiloxane	CH_4	2.1×10^{-5}	300	13

tains a significant amount of crystalline material. Therefore, a smaller value of diffusion coefficient is expected for the experiments (*10*). In many cases, experiment values were not available for the direct comparison. The detailed information regarding the relative comparison may be seen in references 10–14 and 19–21.

Important Factors for Molecular Dynamics Simulation

Many factors need to be considered for the setup of molecular dynamics simulation. Furthermore, as indicated earlier, only very small diffusive molecules have been employed in the simulation studies at present. For the application of the simulation to controlled drug release system, many more additional factors need to be considered. The diffusive molecules are drug molecules and the size of which is certainly much larger than that of gas molecules. Also, the type of polymer systems involved in drug delivery is more complicated. The structures of the commonly used polymers for controlled-release systems are much more complicated than simple polymers such as polyethylene or polypropylene. Nevertheless, the present simulations may be used as the initial studies for drug release simulations.

Polymer Chain Length

On the construction of molecular dynamics cell, polymer chains with different length may be used to pack the cell with desired density. Because of the limitation of computational power, a chain length of 20–200 unit segments was typically used for each of the polymer chains. However, when short chain length was used, the diffusion rate calculated from the molecular dynamics approach tended to be larger than that obtained from experiments. This phenomenon may be due to the large number of loose ends of the polymer chains, which makes the redistribution of free cavities easier. An attempt was made to address the effect of polymer chain length on the simulation results (*19*). Two cubic molecular dynamics cells for polyethylene were constructed with periodic boundary conditions, and each contained 600 CH_2 segments and 10 O_2 molecules. The first cell contained 30 polymer chains each with 20 segments, and the second one had a model for a noncyclic chain with no free ends. The simulation result indicated that the shorter polymer chain length in the first molecular dynamics cell resulted in a faster diffusion as well as faster relaxation of internal rotations to a certain extent compared with that from the second molecular dynamics cell. Therefore, an appropriate chain length may give a better result.

Glass-Transition Temperature

The temperature factor involved in the simulation model may largely influence the results. Temperature may affect the simulation because it changes the molecular energies and also the density of polymer packing in the molecular dynamics cell. However, a critical point that has a direct impact is the polymer glass-transition temperature (T_g). In most cases, the molecular diffusion in glassy polymer involves a more complicated process than that in rubbery polymer. On lowering the temperature below T_g, the polymer changes from a viscous rubbery state to a glass state. A sudden change occurs in the motion status of the polymer chains rendering the molecules difficult to diffuse through the polymer. Molecular dynamics simulation indicated that below T_g, most of the molecules are trapped by cages created by the polymer chain. Molecular jump motion, however, was observed in some cases (*20*). Compared to the simulation results above T_g, the diffusion is more difficult in the glass polymer chain than in the rubbery polymer. Because most polymers have a lower T_g than the body temperature, the studies involving rubbery polymer would be sufficient in our field. The studies on the glass polymer, however, may generate useful information for some special situations.

Many polymers do not have a completely amorphous structure above T_g. They always contain a substantial fraction of crystalline state (semicrystallinity), which may cause a slower diffusion of the drug molecules. Neglecting this phenomenon in the molecular dynamics simulation may result in a faster diffusion rate than that obtained from experiments. Because the size of microcrystallites is much larger than that of a typical molecular dynamics cell, the modeling of semicrystallinity is difficult with the present computational power.

Nature of Diffusive Molecules

The nature of the diffusive molecules greatly influences the value of the diffusion coefficient, especially for the studies related to controlled drug release from polymer matrix. At present, only very small molecules are involved in the simulation studies. One of the reasons is the limit of the current computational power, and another reason is the complexity of larger molecules compared with these small molecules. Because this area is new, more experience and results need to be accumulated before applying the simulation technique to larger molecules. For a larger molecule, not only are more atoms involved in the calculations, but the shape and orientation of the molecule in the polymer matrix also play an important role. A specific orientation of the larger diffusive molecules may greatly affect the direction of the molecular motion.

The solubility of the diffusive molecules in the polymer matrix is also important for simulation. A consideration during the model setup stage should be to pack an appropriate number of the diffusive molecules in the simulation cell. A concentration of the diffusive molecules in the simulation cell that exceeds the corresponding solubility would result in an unrealistic value of the diffusion coefficient.

Simulation of Diffusion Through Liposomes

Liposomes have been widely studied as controlled-release devices. The diffusion of drug molecules through the liposome bilayer is an interesting subject for investigation to understand the release and transport of drug from liposomes. In dipalmitoyllecithin liposomes, the translational motion of small nitroxide solutes appears to be faster in the center of the bilayer than that near the polar interface, as indicated by ^{13}C relaxation studies (*22*). On the basis of NMR experiments the surface density of the dimyristoylphosphatidylcholine (DMPC) and dipalmitolyphosphatidylcholine lipid chains affects the partitioning of benzene into the lipid bilayers (*23*).

To further understand the diffusion process in liposomes, molecular dynamics simulations of solute diffusion in a lipid bilayer may be performed. Bassolino-Klimas et al. (*24*) constructed a lipid bilayer model consisting of 36 DMPC lipid molecules arranged into two monolayers. An all-atom representation of the molecules including the lipids, benzene (the diffusive molecules), and solvating water molecules was employed, and the study involved a large scale of computation. In agreement with experimental evidence, the results indicated that small solutes did not affect bilayer thickness but they caused slight perturbations in the ordering of the hydrocarbon chains. The rate of translational diffusion was faster in the center than near the head groups. Agreement was also found between the simulation results and those obtained from experiments, in the rotational correlation times and diffusion coefficients of similar solutes.

Simulation of Protein–Polymer Interactions

Controlled-release devices for protein drugs are of great interest, as indicated in several chapters in this book. The understanding of behaviors of proteins in the polymer matrix would be of great help for the design of the corresponding controlled-release devices. The molecular dynamics simulation of proteins in polymer matrix is even more complicated and not feasible at present. However, the knowledge of the interactions between proteins and polymers could be the first step toward this molecular dynamics simulation and could serve as the framework for the future simulation of protein diffusion in the polymer matrix. A number of computer simulation studies examined the behaviors of proteins on contact with polymer surfaces, and these studies are worthy of mention in this chapter.

Protein molecules are tightly bound with surfaces of polymer matrix because of the strong interactions between the protein and polymer. Computer simulations have examined the nature of these interactions and the corresponding conformational changes of protein molecules. Lu and Park (*25, 26*) studied the interactions between several proteins and several polymers. The proteins were lysozyme, trypsin, immunoglobulin F_{ab}, and hemoglobin and the polymers were polystyrene, polyethylene, polypropylene, poly(hydroxyethyl methacrylate), and poly(vinyl alcohol). In agreement with experimental observations in the literature, the studies showed that stronger interactions existed between proteins and more hydrophobic polymers. In addition to the potential energy interaction, the solvation interaction (the hydrophobic force) also played an important role. Also, the orientation of the protein molecules with respect to the polymer surfaces resulted in a larger difference in the overall interaction. The conformational change of protein molecules was another factor affecting the interactions and should be heavily considered (*27*). These studies suggest that in the future studies of diffusion of protein molecules in polymer matrix, the orientational and conformational effect of protein molecules should be considered (the effect is less important for small molecules).

Computer Simulation Tools

Several molecular modeling software programs are available for large-scale molecular modeling and may be used to perform the molecular dynamics simulations to examine the diffusion behaviors in polymer matrix. Two particular packages are mentioned for reference.

INSIGHT II

INSIGHT II is a molecular modeling computer program from Biosym cooperation (San Diego,

CA). In conjunction with the molecular mechanics/dynamics package Discover (Biosym), INSIGHT II can be used to build, manipulate, and simulate virtually any class of molecules or molecular systems. Molecular properties can also be studied through the INSIGHT II interface with other Biosym products. The programs can be run on Silicon Graphics IRIS (SGI) series workstation or on IBM RISC6000 workstations.

Most of the computational functions are associated with each module implemented in the INSIGHT II. These modules can interact with each other to perform different tasks. The core module is the Viewer module. From the Viewer module the user can have access to other modules, such as Builder (molecule construction), Biopolymer (peptides and proteins), Polymerizer (polymer construction), Ampac/Mopac (electron configuration), Delphi (electrostatic potentials), Discover (minimization and dynamics calculation), DMol (quantum chemistry), Docking (nonbond energy calculation), Amorphous_Cell (amorphous polymers), Crystal_Cell (crystalline polymers), RIS (rotational isomeric state), Analysis (molecular conformation), Networks (polymer network), QSPR (quantitative structure–property relationships), and many other modules. For our simulation purposes, the most useful ones are Polymerizer, Discover, Amorphous_Cell, and RIS. The polymer chains of interest can be constructed by using the Polymerizer module. The user can have the choice of specified tactility (atactic, isotactic, or syndiotactic) or copolymers (alternating, diblock, or random) in construction of the polymer chains. Combined with the RIS module, the Amorphous_Cell module can be used to pack the polymer chains into the molecular dynamics cell with appropriate periodic boundary conditions. Discover can then be used to refine the polymer configuration in the molecular dynamics cell and perform the dynamics simulations. The trajectory of molecular motion, mean-square displacement, and diffusion coefficient can be directly calculated by using the Amorphous_Cell/Discover modules.

SYBYL

Developed by Tripos Associates, Inc., SYBYL is molecular modeling software that can be used to build, study, and manipulate molecules. The program also runs on Silicon Graphics IRIS (SGI) series workstation or on IBM RISC6000 workstations. A number of modules have been implemented in SYBYL for different computational purposes. These modules includes Advanced Computations (conformational analysis and quantum chemistry), Biopolymer (large biomolecules—polypeptides and polynucleotides), Polymer (synthetic polymers), Dynamics (the motions and the configuration space of molecular systems), QSAR (quantitative structure–activity relationships), and QSPR (quantitative structure–property relationships). The Polymer and Dynamics modules are particularly useful for molecular dynamics simulations of polymers. The Polymer module provides basic building and modeling tools to handle the modeling of synthetic polymers. The polymer can be partially crystalline, such as polyethylene and polyethylene terephthalate, or noncrystalline, such as polystyrene and polymethyl methacrylate. The module can be used to examine the conformation and configuration of polymer chains; perform energy minimization; and examine the physical properties such as vibrational, thermodynamic, electrostatic and polarizability, and mechanical. The Dynamics module allows the user to perform molecular dynamics simulations by using various force fields and difference algorithms, such as the Verlet method or the leapfrog method.

Experimental Verifications of Simulation

To verify the simulation results, particularly the diffusion coefficient, a number of experimental methods may be used. This aspect is important to examine the assumptions involved in the simulation and the accuracy of the results. However, many of the experimental methods for diffusion study in the literature (28) are only useful for liquid and not suitable for studying the diffusion in polymer matrix. Three main methods may be useful for examining the drug diffusion in polymer matrix.

Radiotracer Technique

Radiolabeled molecules can be used to study the molecular diffusion through polymer matrix. Several possible configurations are based on this technique. One is to measure the rate of diffusion of labeled molecules from one region of the polymer matrix into another, which originally is a radioactive-free region. The radioactive molecules may be placed on or incorporated into the top layer of the polymer matrix, and the diffusion of the molecules is monitored by directly measuring the attenuation of the emitted radiation as a function of time. The diffusion coefficient can then be calculated from the attenuation curve (*29*). Another possible configuration is to construct a two-compartment apparatus. The compartments are divided by a polymer film of interest. One compartment is filled with a solution containing radiolabeled molecules, and the second is filled with the solution containing no radiolabeled molecules. The diffusion can be measured by counting the radioactivities of the second compartment (*30*).

Infrared Absorption Technique

Because infrared (IR) light penetrates many polymer films, it can be used to directly measure the diffusion of IR-sensitive molecules in those polymer matrices. One of the possible procedures is to prepare two cylindrical pellets of the polymer. One of the pellets contains a dispersion of the IR-sensitive molecules and the other does not. The two pellets are then joined by heat or by a solvent. After a certain period of diffusion time, a slice (around 100 μm in thickness) is microtomed across the joined pellet. The slice is traversed across the slit of an IR microdensitometer and the transmitted intensity is measured. A diffusion concentration with respect to the position on the slice can then be plotted and the diffusion coefficient can be calculated (*29*).

Pulsed-Gradient Spin Echo NMR Spectroscopy

Pulsed-gradient spin echo NMR (PGSE–NMR) spectroscopy may be used to examine the diffusion of small molecules through the polymer matrix (*31*). The requirement of PGSE–NMR spectroscopy is the capability to generate at least two-pulse radio-frequency sequences, typically the 90°-τ-180° sequence. The method is based on the principle that a large gradient of magnitude is applied in the measurement only twice for a small duration, after the 90° pulse and after the 180° pulse (*32*). A steady gradient of much smaller magnitude may also be used to narrow the echo sufficiently for the measurement of the signal baseline and to enhance the time and phase stability of the echo. The molecular motion of the system may be studied by measuring characteristic nuclear relaxation times, the spin–spin relaxation time T_2, or the spin–lattice relaxation times T_1 or T_{1r}. The nature of the molecular motions that can be examined by the PGSE–NMR technique includes vibrations and rotations of all or part of a molecule and also the spatial translations relative to the neighboring molecules, which provides useful information for diffusion studies.

Conclusions

Synthetic polymers have been widely used in the development of various controlled-release devices. For a large portion of these devices, drugs or bioactive compounds are encapsulated in the polymer matrix in a homogeneous dispersion form. The interactions between drug molecules and polymer chains as well as the dynamic motions of these two components provide the controlling mechanism of drug release at a molecular level. Computer simulation of such a system involving the interactions and dynamic motions may help us to better understand the controlling mechanism. The applications of molecular dynamics simulation of these systems have emerged only in recent years owing to the enormous computational power required in the simulation. Theories and strategies for these simulations are still in the development stage and the progress seems in parallel with the development in computer power. As an example, a generally accepted and widely used model for a polypropylene system is the RIS model by Theodorou and Suter (*16*). The model provides a good representation of the polymer chains in a molecular dynamics cell with the three-dimensional periodic boundary conditions. In the original article, the backbone carbon atoms and pendant hydrogen atoms were expressed explicitly. The methyl units, however, were grouped together in single "quasi-atoms" of appropriate size. The treatment can save tremendous amount of computational time, and the results have been satisfactory. The major limitation of explicit treatment of the atoms in the methyl unit was the computational power. With the fast-increasing computer power, the simplified treatment may not be necessary in future calculations. Each atom in the unit can then be treated individually to provide better accuracy of the simulation (*21*). For controlled-release systems, the molecular dynamics simulation may provide the detailed information of the polymer–drug interactions and dynamic motion at the molecular level. The simulation may also be used to calculate, on theoretical bases, the diffusion coefficient of molecules through a polymer matrix, which is an important piece of information in designing controlled-release systems.

The present simulation work mainly involves small gas molecules. Although a larger molecule of benzene has been used in the molecular dynamics simulation of lipid bilayer system (*24*), further studies and development are needed to apply the simulation technique to controlled drug release systems. With the rapidly increasing computational power, it can be reasonably predicted that a large amount of research will be conducted in this dynamic simulation area over the next 10 years. The direct applications of the simulation results to practical development of various controlled-release systems will also be largely explored.

References

1. Baker, R. W. In *Controlled Release of Biologically Active Agents*; Tanquary, A. C.; Lacey, R. E., Eds.; Plenum: New York, 1974; pp 15–71.
2. Welling, P. G.; Dobrinska, M. R. In *Controlled Drug Delivery: Fundamentals and Applications,* 2nd ed.; Robinson J. R.; Lee, V. H. L., Eds.; Marcel Dekker: New York, 1987; pp 253–291.
3. Shargel, L.; Yu, A. B. C. *Applied Biopharmaceutics and Pharmacokinetics,* 2nd ed.; Appleton: Norwalk, CT, 1985; pp 347–372.
4. Bayne, C. K.; Rubin, I. B. *Practical Experimental Designs and Optimization Methods for Chemists;* VCH: Deerfield Beach, FL, 1986; pp 171–186.
5. Khuri, A. I.; Cornell, J. A. *Response Surfaces;* Marcel Dekker: New York, 1987.
6. Schwartz, J. B. In *Modern Pharmaceutics,* 2nd ed.; Banker, G. S.; Rhodes, C. T., Eds.; Marcel Dekker: New York, 1990; pp 803–828.
7. Baker, R. W. In *Controlled Release of Biologically Active Agents;* John Wiley & Sons: New York, 1987; pp 50–75.
8. Dubernet, C.; Benoit, J. P.; Peppas, N. A.; Puisieux, F. *J. Microencapsulation* **1990,** *7,* 555–565.
9. Allen, M. P.; Tildesley, D. J.*Computer Simulation of Liquids;* Oxford: New York, 1987.
10. Muller-Plathe, F. *J. Chem. Phys.* **1991,** *94,* 3192–3199.
11. Muller-Plathe, F. *J. Chem. Phys.* **1992,** *96,* 3200–3205.
12. Boyd, R. H.; Pant, P. V. K. *Macromolecules* **1991,** *24,* 6325–6331.
13. Sok, R. M.; Berendsen, H. J. C.; van Gunsteren, W. F. *J. Chem. Phys.* **1992,** *96,* 4699–4704.
14. Takeuchi, H. *J. Chem. Phys.* **1990,** *92,* 5643–5652.
15. Muller-Plathe, F.; Laaksonen, L.; van Gunsteren, W. F. *J. Mol. Graphics* **1993,** *11,* 118–126.
16. Theodorou, D. N.; Suter, U. W. *Macromolecules* **1985,** *18,* 1467–1478.
17. Rigby, D.; Roe, R. J. *J. Chem. Phys.* **1987,** *87,* 7285–7292.

18. Jolly, D. L.; Bearman, R. J. *Mol. Phys.* **1980,** *41,* 137–147.
19. Takeuchi, H. *J. Chem. Phys.* **1990,** *93,* 4490–4491.
20. Takeuchi, H. *J. Chem. Phys.* **1990,** *93,* 2062–2067.
21. Muller-Plathe, F.; Rogers, S. C.; van Gunsteren, W. F. *Chem. Phys. Lett.* **1992,** *199,* 237–243.
22. Dix, J. A.; Kivelsion, D.; Diamond, J. M. *J. Membr. Biol.* **1978,** *40,* 315–342.
23. DeYoung, L. R.; Dill, K. A. *Biochemistry* **1988,** *27,* 5281–5289.
24. Bassolino-Klimas, D.; Alper, H. E.; Stouch, T. R. *Biochemistry* **1993,** *32,* 12624–12637.
25. Lu, D. R.; Park, K. *J. Biomater. Sci. Polym. Ed.* **1990,** *1,* 243–260.
26. Lu, D. R.; Lee, S. J.; Park, K. *J. Biomater. Sci. Polym. Ed.* **1991,** *3,* 127–147.
27. Lu, D. R. *J. Biomater. Sci. Polym. Ed.* **1993,** *4,* 323–335.
28. Cussler, E. L. In *Diffusion, Mass Transfer in Fluid Systems;* Cambridge: New York, 1984; pp 132–145.
29. Klein, J. *Contemp. Phys.* **1979,** *20,* 611–629.
30. Barson, C. A.; Dong, Y. M. *Eur. Polym. J.* **1990,** *26,* 329–332.
31. Kim, D.; Caruthers, J. M.; Peppas, N. A.; von Meerwall, E. *J. Appl. Polym. Sci.* **1994,** *51,* 661–668.
32. von Meerwall, E. D. *Rubber Chem. Technol.* **1985,** *58,* 527–560.

Regulatory Issues

28

Food and Drug Regulations for Controlled-Release Dosage Forms

Garnet E. Peck

In 1906 the Food and Drug Act of the U.S. Department of Commerce was put into place to protect the consuming public from drug products that were not properly controlled as far as quality standards, including potency. As time moved on, the act was changed to cover safety and effectiveness of drug products that entered the marketplace. Some of these changes were made in response to adverse effects of drugs that caused illness and, in numerous cases, death. Also over the years there has been a greater emphasis on manufacturing methodology, and that emphasis has generated the current Good Manufacturing Practices as well as the current Good Laboratory Practices. Today regulations clearly define what a new product is and the requirements for establishing a new drug product as far as safety and effectiveness are concerned. In considering controlled-release systems, frequently various polymeric substances are used to delay the drug's release within the body. Sufficient data must be given to demonstrate that the product does have a controlled-release profile that is satisfactory for therapeutic purposes. It is also important to recognize regulatory concerns in developing suitable controlled-release products. The ultimate goal is to produce delivery systems that are beneficial to society.

The original purpose of the Food and Drug Act of 1906, in response to a problem noted by the U.S. Commerce Department, was to prevent the distribution of adulterated materials to the consuming public (*1*). Numerous materials were in commerce that were not of appropriate quality or standards, and thus deficiencies were seen in the various therapies in use at that time, which were primarily drugs derived from natural sources. This material was difficult to control because it was being imported to the United States from various parts of the world. This basis for the regulations soon became limited, as indicated by the various revisions to the Food and Drug Regulations. In 1938 a major revision took place because of the serious tragedy that occurred concerning the for-

Copyright ©1997 American Chemical Society

mulation of sulfanilamide elixir. This solubilized, very potent drug substance and the product caused numerous deaths when introduced into the marketplace. The revisions to the act became known as the Food, Drug, and Cosmetic Act of 1938. The act had major new provisions, which included the following:

- extension of coverage to cosmetics and devices;
- requirement of predistribution clearance for safety of new drugs;
- provision for tolerances for unavailable poisonous substances;
- authorization of standards of identity, quality, and fill for foods;
- authorization of factory inspections; and
- addition of the remedy of court injunctions to previous remedies of seizure and prosecution.

These additions to the act became highly significant. They introduced various concepts that concerned the formulation and manufacturing of pharmaceutical drug products as well as foods, cosmetics, and devices. Regulations within the act have indirect impact on the formulation of pharmaceutical products, and these regulations include the addition to the act of the Color Additive Amendments in 1960. This addition allowed the U.S. Food and Drug Administration (FDA) to have full control over these particular colorants, to ensure safety (*2*).

In 1962, another major provision, known as the Kefauver–Harris Drug Amendments, arose from the tragedy involving thalidomide, which had been marketed in various countries of the world but was still undergoing clinical testing in the United States. The amendments required a greater degree of safety assurance of any new drug substance before it could be cleared for the marketplace. The amendments increased the amount of testing, for example, to ensure that no substance would harm or have the potential of causing any detrimental effects to the patient or to those who might be affected by the ingestion of the drug, which potentially could cause harmful effects over and above the therapeutic effect.

At this time increased interest was placed on new drug testing and the methods by which the material would be evaluated in clinical studies. These elements are discussed in another section of this chapter. An addition to the regulations that affect the development of new products came in 1963, when the Good Manufacturing Practices (GMPs) and regulations were established. These processes were later changed in 1978 to increase the impact on the drug products that were to be distributed by various manufacturers. About that time another set of regulations was generated that became guidelines. These regulations involved current Good Laboratory Practices (GLPs). With these two sets of documents, the FDA and was able to implement control over the operations of most pharmaceutical firms as well as numerous research centers involved in new drug development and new product development. The overall purpose to this day is to protect the consuming public and to ensure a quality product for all those taking drug substances either for immediate treatment or long-term therapy.

General Considerations for New Drugs

When the Food, Drug, and Cosmetic Act of 1938 was passed, a new era of drug product development began. It was the beginning of a requirement for the preclearance of drug product before its marketing. The act required the assurance of safety and stated minimum requirements for manufacturing quality control. However, it required only 60 days for review by the FDA before the distribution of any new drug product. This requirement was changed in 1962 when increased information concerning safety, ethicality, and manufacturing controls of the drug product were initiated. For the first time, a drug firm or other organization interested in a new drug substance had to inform the FDA of the desire for testing of the new drug in humans. The document to initiate a human study involving a new drug (which includes not only new drug substances but also new dosage forms for existing products) is known officially as the Notice of Claimed Investigational Exemption for a New Drug (abbreviated IND). From the *Code of Federal Regulations*, a new drug is one not presently recognized by experts in the field of clinical pharmacology to be safe and effective based on currently available clinical evidence. The definitions of a new drug substance or product are as follows (*3*):

1. "New-drug substance" means any substance that when used in the manufacture process or packaging of a drug causes the drug to be a new drug but does not include intermediates used in the synthesis of such substances.

2. The newness of a drug may arise by reason (among others) of a drug use of

   - any substance that composes such drug in whole or in part whether it is an active substance or an excipient carrier coating or other component;
   - a combination of two or more substances, none of which is a new drug;
   - the proportion of a substance in a combination even though such combination containing such substance in other proportion is not a new drug;
   - such drug in diagnosis, curing, mitigating, treating, or preventing a disease or to affect a structure or function of the body even though such drug is not a new drug when used in other disease or to affect another structure or function of the body; and
   - a dosage or method or duration of administration or application or other conditions of use prescribed, recommended, or suggested in the labeling of such drug even though such drug when used in other dosage or other method or duration of administration or application or different condition is not a new drug.

Thus, one is not simply talking about a new medicinal agent with newly established therapeutic activity, but one may be also talking about modifying substances to produce a new method by which we can deliver the drug and, in particular, alteration of the drug delivery over time. This area must be investigated before a new product is distributed. In particular, controlling the release of the drug substance over time is an example of newness.

To determine whether the filing of an IND (in human testing) is reasonable for a new substance, preclinical studies must be conducted that include the following:

- acute toxicity (to determine the safety of the drug in animals),
- preformulation studies covering the physical and chemical characterization of the new substance, and
- pharmacological screening and evaluation (to verify that a new drug substance has some potential for therapeutic activity).

In many instances, when a controlled-release system is being designed, adequate data may be available already for the drug that satisfy the previously stated requirements. In many instances the development of a controlled-release system is considered to be an additional way by which the product may be used in the treatment of either individuals or animals with a suitable system that will perform over an extended period of time that is different from the product that may have been prepared as an immediate-release system. However, a place for the evolution of new polymer systems hopefully will occur that will produce different release profiles or profiles that may be more easily reproduced.

Elements of Claimed Exemption

To initiate studies that would demonstrate that a system is valuable in the treatment of particular disease states or will perform in a particular manner, a document and a plan of study may be necessary that would require some fundamental work within the area of clinical investigation. To do this work, a document entitled Notice of Claimed Investigational Exemption for a New Drug (IND) must be prepared. This process is done when a number of factors require the collection of information for safety and ethical purposes through clinical evaluation. This document puts together information that may be available from various sources, including a product that has been in the market as an immediate-release product and now is proposed as a controlled-release system. The following elements are necessary for a complete IND (*2*):

- best available descriptive name of the drug product and how it is to be administered;
- complete list of components of the drug product;
- quantitative composition of the drug product;
- description of the source and preparation of any new drug substance used as a component;
- description of the methods, facilities, and controls used to prepare and distribute the drug product;
- information available from preclinical investigations and any clinical studies that may have been conducted;
- copies of any labels or labeling to be used in the study;
- scientific training and experience to qualify the investigators;
- complete information on each investigator, including training and experience;
- outline of any phase or phases of the planned clinical investigation;
- assurance that the FDA will be notified if the studies are discontinued;
- assurance that investigators will be notified if a new drug application is approved or the studies discontinued;
- explanation of why a drug product must be sold during a study; and
- agreement not to start a study for 30 days before a notice of receipt of the IND by the FDA.

This information is rather complete as far as what is required to proceed into a clinical evaluation. Many elements involve the physical and chemical properties of the drug substance itself. In the case of controlled-release systems, the methods by which the drug product is to be manufactured and the controls used to ensure that the product will perform as claimed must be explained. These studies have been at times difficult to analyze, especially the release profiles of drugs coming from extended-release or controlled-release systems. This area also causes some degree of discussion as to the in vitro evaluation of drug products and in particular the controlled-release systems and the desire to match the in vitro–in vivo information so that one can attain a correlation. Although other details are given in the previous list, it does indicate the degree of complexity that might be necessary for the introduction of a new system. Depending on the clinical evaluation of a product the IND could be fairly straightforward. Another significant point that is raised in the IND application is the outline of all of the phases that will be needed for the clinical investigation. This outline includes Phase I, which is a study on healthy humans that is also referred to as clinical pharmacology. Phase II is the evaluation of the drug product on a few individuals that have the disease state for which the product is being investigated. Finally, Phase III is a larger study to ensure that the product will in fact perform as desired or claimed. This area does cause some problems in the evaluation of controlled-release data with regard to the pharmacokinetics of the various systems. The methods by which these systems are evaluated are discussed in other sections of this text.

To assist the investigator or the individual preparing an IND, various guidelines and other documents are available from the FDA. The purpose of these regulations, especially in the IND stage, is to ensure that satisfactory data are collected to demonstrate safety and effectiveness of a new compound or new system through carefully controlled clinical studies and then to guide the submitter in the preparation of the data that related to the new compound in a manner that will allow a reasonable period of time for approval.

Once adequate information has been gathered and it has been judged that the claims can be supported for the use of this system, the clinical information gathered during the investigational stage is summarized, analyzed, and then used to prepare the New Drug Application (NDA). This document would ensure that the product will perform as indicated. In other words, once a new drug substance or anything that is defined as a new drug has been thoroughly investigated in Phase III clinical trial, the NDA is prepared. The data should be obtained under a specific IND that has permitted the gathering of the following information:

- acute and chronic toxicity,
- extended clinical pharmacology,
- full-scale clinical evaluation,
- complete product design,
- complete package design, and
- complete label information.

Because of today's myriad regulations, the NDA submission has become a compilation of information that could be compared in size to any one of the well-known encyclopedias. The components of the NDA for human use are as follows (*2*):

- table of contents;
- summary;
- evaluation of safety and effectiveness;
- copies of the label and all other labeling to be used for the drug;
- statement as to whether the drug is (or is not) limited in its labeling and by this application to use under the professional supervision of a practitioner licensed by law to administer it;
- full list of the articles used as components of the drug;
- full statement of the composition of the drug;
- full description of the methods used in, and the facilities and controls used for, the manufacture, processing, and packing of the drug;
- samples of the drug and articles used as components;
- full reports of preclinical investigations that have been made to show whether or not the drug is safe for use and effective in use;
- list of investigators;
- full reports of clinical investigations that have been made to show whether the drug is safe for use and effective in use; and
- if this application is supplemental, full information on each proposed change concerning any statement made in the approved application.

These components vary slightly when the application is for veterinarian purposes.

The information must be generated so that sufficient instructions can be given for the use of the drug substance or pharmaceutical product in order that optimal outcomes can be produced. Certain systems must be taken in a very specific manner, which is part of the labeling information needed. The clinical studies may have indicated that the product would need to be taken at a particular time of day, with or without food, to ensure proper therapy. This fact becomes important with controlled-release systems when during the clinical trials the manner in which the patient is dosed is very important. Studies should establish, if at all possible, what type of diet is used and, as closely as possible, the best time of day to administer such a system. Many factors discussed in this text influence the performance of a controlled-release system. From a regulatory standpoint these factors are important to establish and to demonstrate in the NDA that the therapeutic outcome desired can be obtained. What becomes very important is the description of the method of manufacturing including the types of excipients, the exact processing, and the equipment being used to produce the product, and the limits that are permissible to attain a delivery system that is accurate. This description could require extensive specifications if polymeric materials are used to control the release of a particular medicinal agent. As stated earlier, the processing of this system may also be very important. Processing variables may be significant as far as release profiles are concerned. Depending on the experience of the reviewers within the FDA, numerous questions could be raised about the performance of a system and the processing method used to prepare the delivery system.

In the case of immediate-release systems, extensive preclinical studies in animals may have been required to establish the safety and the approximate dose level of the system. Usually in the controlled-release area these data have been collected for the immediate-release system and may not need to be repeated for the controlled-release system. Generally, the studies are done after assurance of safety in the human subject fairly early in the investigation. The extent of studies in animals will depend on the particular drug under consideration.

Once the NDA has been prepared it is submitted to the FDA for review and comment. The initial time period is 180 days before they are required to respond; however, this period may vary depending on the material and the manner in which the NDA is received by the reviewers. Depending on their experience, certain additional information will be required. At times the reviewing process is somewhat time consuming and predictable for new drugs to reach the U.S. market. Currently, attempts are being made to reduce the time needed for NDA approvals, and this process has been expedited by the fee system used for the review of NDAs. This fee allows the FDA to increase the personnel doing such activity. However, a very new system to the U.S. marketplace may still require extensive review within the agency. The NDA is reviewed by chemists, pharmacologists, medical officers, statisticians, and, when necessary, by those scientists and professionals who have the understanding and expertise for the evaluation of a particular drug. The total review process is represented by Table 28.1, which indicates the time required to get approval of the NDA and also what is required postmarketing of a particular product (*4*). This part of the process is called Phase IV for a new product. The time estimates given in the table are somewhat current but may change from time to time, depending on the newness of the system that has been proposed as a drug product.

Approval of New Controlled-Release Systems

Previously I described what constitutes a new drug product as far as the FDA is concerned. Any judgment, as far as a drug delivery is concerned, that fits into any one of the categories would constitute a new drug product. Thus, no current separate method or category exists by which a controlled-release system would be approved. It is a modification generally of a currently available drug, and it thus contains new substances and new claims that would place the system into the new drug area. Once the decision is made to delay the release of the material, change the site of release, and control the release of the system (which would include prolonging the release by current definition), this product becomes a new drug and would require a new drug application.

Generally, this type of system is developed for a specific reason. Some of these reasons may include the following: avoid patient compliance problems, employ less total drug, and improve efficiency in treatment, which may include improved bioavailability of some drugs. Also, the ability of controlling symptoms while the patient is asleep may be included. Finally, some economic gain may occur by developing such a dosage form (*5*). These reasons are important as far as generating a system that would control the release of the

TABLE 28.1. Evaluation of New Drug Product and Associated Regulatory Considerations

Events	*Years for Studies*
Preclinical testing	1–2
Laboratory and animal studies	
Safety considerations	
Component compatibility	
Investigation of a new drug filing	Variable
Proposal dosage form and supporting data (including stability evaluation)	
Phase I studies	1
Healthy humans	
Limited length	
Dose finding	
Phase II studies	1–2
Limited studies in ill patients	
Phase III studies	About 3
Expanded to large population of ill patients	
Filing of new drug application	Approximately 6 months to prepare
Summary and reporting of clinical results	
Establishing final dosage form and manufacturing procedures	
FDA review	2–3
Postapproval evaluation	At least 1

drug substance. All of these possible potential advantages of the system in part could place the material in the new drug area.

The mechanism by which a controlled-release system would need governmental approval, thus, would obviously follow the development of any new drug product. However, data would be available to support any preclinical information as from the aspect of what the drug may do and its safety. However, it may be necessary to establish a safety level should the dosage form contain more than was previously administered by a conventional dosage form. This higher level would obviously not have had a safety determination done and thus may require such studies. The studies at a higher dose level, however, may or may not be a complicating factor. The establishment of an appropriate release profile as far as toxicity is concerned would be of concern. Numerous examples in the literature cover the drug theophylline (*5*). This interest is due to the narrow therapeutic window of the particular drug substance and thus a requirement for appropriate studies to establish that controlled-release administration of this particular substance was justified. This process remains in the area of clinical pharmacology of an active moiety.

As with most drug substances, an adequate demonstration of the release profile in vitro should be established within the appropriate limits being chosen for the controlled-release system. For claims that are being made, this profile must be adequate to demonstrate sufficient release of the material in vitro. Retention by the controlling system or polymeric system if being used should not be excessive approaching 100% release. A reasonable release profile must match with the claims of the particular product. Once established, adequate in vivo studies should be conducted to demonstrate the capability of the system used for controlling the release. As is explained in other chapters, the in vivo data can be interpreted in a number of ways to establish the claims that will be made under a new drug application. This portion of the studies may become complicated and detailed if there is an unusual mechanism by which the profile needs to be judged. In a recent publication concerning the scale-up of extended-release products, the methods by which a judgment should be made are outlined (*6*).

Once sufficient data have been gathered to demonstrate that the product is both safe and effective, an NDA for this particular system would be filed. Establishing this NDA as a supplement to another NDA, which would be a misunderstanding, would be very difficult. The new delivery system is definitely in the new drug area and would need to be appropriately reviewed by the agency before approval could be obtained. However, the length of time for approval may be shorter than the original approval of the new drug substance.

Manufacture and Control of New Controlled-Release Dosage Forms

Because controlled-release systems or extended-release products have unique properties and characteristics, additional manufacturing parameters and control requirements may be necessary to ensure that the products will perform as outlined in any study of the system. In particular, if currently available polymers are used, some requirements need to be observed as far as the material is concerned.

An example of a material that might be used in the controlled-delivery area is ethylcellulose. It currently has generally recognized as safe status, which indicates that it may be used as a food additive in a number of different products. Thus, it would allow easy use in a pharmaceutical system for whatever purpose that is being considered. This material has a number of specifications that include the degree of substitution of various groups on the cellulosic backbone. Other standard specifications are used to judge this material from the standpoint of use in a controlled-release system. Sufficient data are available, for example, in the solubility area for judgment of its utility. A number of references are in the literature as to specific applications of this substance, which would support its use in a controlled-release system. But what is important is that the manufacture of the material is clearly defined as well as having appropriate specifications of the basic polymer substance. Once this material is selected for a system then how it is used to prepare the controlled-release device would have to be outlined with appropriate processing parameters given. This process becomes important from the regulatory standpoint to ensure reproducibility of the system, which is meant to delay the release of a medicinal agent. This point may become important with the regulatory agency to ensure that performance will be sufficiently equivalent from batch to batch and to also have the appropriate release profile over time. Thus, adequate long-term dissolution stability data must be determined. In most cases, a method will be used taken from the U.S. Pharmacopeia/National Formulary (USP/NF) that may be appropriately modified for the system under consideration. However, pH levels that are normally experienced in the gastrointestinal tract would be used in any in vitro test procedure. The degree of reproducibility of these tests is an important part of the control package that would be presented for the agency. It has been observed over time that there may be a long-term dissolution problem with a polymer, depending on the material into which it is incorporated. Data must be sufficient to support whatever stability claims made for the finished dosage form. Concerns thus may be present about the stability of the plasticizer that is normally used with most polymeric systems in the controlled-release product design (*7*).

As with all dosage forms that are being marketed, the establishment of adequate controls of the manufacturer of the finished dosage form as well as its control and stability evaluation are of extreme importance. From time to time products have failed, in particular, based on their changes in the dissolution profiles. Some feeling for the stability of the system is important to establish early in product formulation so that no surprises happen at a later date. The stability also allows the formulator to attempt to modify the system so that appropriate product specifications are in place to reflect long-term expiration dating of a product.

Current Good Laboratory and Good Manufacturing Practices

My intention in this section of the chapter is not to detail the regulations involving current Good Laboratory Practices and Good Manufacturing Practices (*8, 9*). What are presented are the elements of these regulations and how they affect general research activities, as well as the development of products that may be available for either clinical trial or are expected to be ready for commerce.

Even though the researcher within an organization may have motives to produce products that could be of life-saving capabilities, in this time span the FDA may request certain information that involves basic and product development information that may be in the laboratory notebooks or files. Knowing this, researchers involved in pharmaceutical and medical device products often accurately record information and document the activities within the different types of laboratory environments. To this end, there needs to be a sensitivity about certain issues and an understanding of how some of these practices may be implemented by the FDA.

In the case of GLPs, in the mid-1970s there were deficiencies in the record keeping of information pertaining to animal studies, chemical and physical testing within laboratories, and the certification of instruments calibration and accuracy and precision. The reasons behind these concerns were the need to support the idea that the data generated would confirm the safety and efficacy of the product from the data generated to support the research concerning new drug substances or the data to support the evaluation of new medical devices. Numerous examples were cited at that time that involved the deficiencies of animal studies. These examples included poor record keeping of short- and long-term dosing of animals for efficacy studies. Some deficiencies also noted were general record keeping for human clinical studies, such as drug response. For chemical assays, preparation and expiration dates were not given for many reagents. Also, stability testing on finished dosage forms and products was not supported by adequate information as to the validity of the test methods.

The deficiencies extended not only into medicinal agents but also those tests required for food products. A concern existed about tests on diagnostic products and the ability to certify that these tests were valid and had been validated. One major concern was the application of quality control procedures to all events taking place in the laboratory to support products that would be entered into either the IND stage or to examine the data that was to support NDA applications. These regulations were felt to create testing facilities that could be easily monitored by a federal agency to be certain that appropriate data were being collected and were truly valid. Also, an element exists of either certification or ensuring that the individuals doing the testing in the laboratories are adequately trained. One of the final elements of testing of the GLP regulations is the assurance that responsible individuals are in place who could be held accountable for the activities of the laboratory in general. As with many regulations, an inspection provision may be used by the FDA to ensure compliance with the regulations.

As we look at the harmonization of various activities within research institutes or groups responsible for the initiation of studies concerning new pharmaceutical products or medical devices, the laboratory practices must be certifiable and have certain procedures in place that will ensure this certification. The important point that should be made now is that the current regulations should be reviewed by a responsible individual to ensure compliance at all times.

In the case of the current GMPs, when the revisions to the FDA regulations for NDAs were implemented in 1962, a deficiency with regard to currently manufactured products was observed. The following year the current GMPs were put in place and have since been revised to reflect the current situation within the industry. In 1979 major changes were made that still have an ongoing impact and are still being reviewed with regard to the ability of a group to comply with these particular regulations. These regulations control the manufacture of pharmaceutical dosage forms and can be applied to other materials used in humans. Slightly modified regulations exist for medical devices. Current GMPs appeared in the *Code of Federal Regulations* sections 210 and 211. As in the case of GLPs, it would be necessary to review the current regulations for any modifications. What is paramount in these regulations is the ability to reproduce a product on a continuing basis and thus the introduction of the concept of validation of process and process design. A number of elements are covered in these regulations, including organization scheme, buildings and facilities, equipment, laboratory facilities, components, packaging, and distribution of pharmaceuticals. Other elements of these regulations could be reviewed by appropriate individuals who specialize in regulatory affairs.

What is important is the control of the components used in the preparation of dosage forms. Components require specifications that would ensure that when used in a product, the performance of that product would be as claimed. In many of the dosage forms involving controlled drug delivery, various polymeric substances are used. These substances must have adequate specifications to ensure reproducibility. As has also been noted over the years, some systems contained trace contaminants that must be eliminated before approval of these particular materials. If there is a possibility of a contaminant, methods for testing for their presence must exist. This concept is a new one as far as raw materials are concerned and comes from the harmonization efforts of various agencies that began in 1991. There will be further concern about the amount of trace toxic substances in the near future. A sensitivity to this particular issue is important. The methacrylates were suspect as far as trace monomers were concerned for a number of years. They are now used extensively in the preparation of controlled-release products.

Through the current GMP regulations, the FDA can ensure to a great degree that adequate manufacturing processes are in place for the preparation and distribution of pharmaceutical dosage forms. The regulations that are used for products sold in commerce are also applied to those products used in clinical research. Thus, an identical system must be in place with supporting groups to ensure compliance to these regulations. As far as the pharmaceutical scientist is concerned, the important elements from early on to later considerations have to do with the specifications of the raw materials and the finished dosage forms. The understanding and the implementation of good processing methodology and the assurance of adequate stability for the protection of the consumer are essential.

By means of both current GLPs and current GMPs, the FDA can attempt to ensure adequate controls both in the laboratory and in the manufacturing environment so that products of high quality are distributed within commerce. Many believe that these practices are aids and guidelines and when used will help to produce quality products.

Drug Manufacturing Regulatory Responsibilities

To attempt to produce products without recognizing the regulatory climate that is around the industry would not be acceptable. Although efforts have been made to decrease the number of regulations on the industry, this decrease will not take place to any great extent. It is necessary to accept the situation of control over the operations and distribution of pharmaceutical products and to appreciate the efforts of regulators, not only the FDA but also other regulatory bodies around the world. In recent years efforts have been made to harmonize these regulations on an international basis, and this process is ongoing. Much has been accomplished between the European agencies, the agency in Japan, and also the FDA. With this in mind and a sensitivity to what is required of products entering the marketplace, the scientist working in developing new products will do it in a manner that will allow compliance as products evolve over a period of time. It has become commonplace for the FDA, for example, to ask for the justification as to why a particular product was developed in the manner it was, a review of the components required for the controlling of the release of a drug substance, and the rationale behind the use of the particular components. This would draw thus on the scientific background of the individuals responsible for the development of a new product. It is necessary for the scientist to explain to the regulator what was involved in the particular research efforts. This concept is fairly new, and it is a review process that will require in some cases an adjustment to the manner in which research is conducted. The responsibility is basically to produce drugs that are safe and effective and have an adequate stability so that the consumer at time of use has a product that performs under the stated claims and prerequisites.

Recently there has been increased interest in the development of generic dosage forms. Because drug products are no longer covered by patents, it is possible to develop a generic equivalent product by filing an abbreviated NDA (*3*). The product that is developed must have an identical in vivo release profile as the innovator product. This requirement is especially true if a different polymeric system is used to control the release. Methods for the judgment of equivalency have been described in the literature (*6*).

References

1. *Federal Food, Drug, and Cosmetic Act as Amended and Related Laws;* Superintendent of Documents. U.S. Government Printing Office: Washington, DC, 1992.
2. Peck, G. E. In *Modern Pharmceutics,* 3rd ed.; Banker, G. S.; Rhodes, C. T., Eds.; Marcel Dekker: New York, 1996; pp 753–772.
3. *Code of Federal Regulations, Title 21, Foods and Drugs,* Part 310, New Drugs; U.S. Government Printing Office: Washington, DC, 1995; pp 6–70.
4. Mathieu, M. *New Drug Development: A Regulatory Overview*; Parexel International: Cambridge, MA, 1990.
5. Chiao, C. S.; Robinson, J. R. In *Pharmaceutical Sciences,* 18th ed.; Gennaro, A., Ed.; Mack: Easton, PA, 1995; pp 1660–1675.
6. Skelly, J. P. et al. *Pharm. Res.* **1993,** *10,* 1800–1805.
7. *Handbook of Pharmaceutical Excipients;* Wade, A.; Weller, P. J., Eds.; American Pharmaceutical Association: Washington, DC, and the Royal Pharmaceutical Society of Great Britain: London, 1994.
8. *Good Laboratory Practice Regulations;* Hirsch, A. F., Ed.; Drugs and the Pharmaceutical Sciences Series 38; Marcel Dekker: New York, 1989.
9. Willig, S. H.; Tuckerman, M. M.; Hitchings, W. S. *Good Manufacturing Practices for Pharmaceutics,* 2nd ed.; Drugs and the Pharmaceutical Sciences Series 16; Marcel Dekker: New York, 1982.

29
Pharmacodynamic and Pharmacokinetic Considerations in Controlled Drug Delivery

Philip R. Mayer

Previously, pharmacokinetics has been the most important bridge to therapeutic response when characterizing a controlled drug delivery system. Now, beyond the mathematical relationship between plasma concentration and clearance, steady-state pharmacodynamic correlations must be understood to appropriately design a controlled-release product. This approach, whether for a marketed drug or a new chemical entity, should decrease product development time and lead to faster regulatory approvals for new dosage forms.

For many years, pharmacokinetic input into controlled drug delivery products involved a fairly straightforward thought process. Given the mean plasma clearance of drug A in humans, find an in vitro release rate for a dosage form that will result in the in vivo absorption rate needed to provide the plasma concentration needed for efficacy. The emphasis was placed on clearance and absorption rate constant, and much less thought was given to the concentration of drug A necessary for activity. Even though disposition and absorption pharmacokinetic parameters are still of great importance, today a more complete understanding of the concentration–response relationship is needed to most appropriately design a controlled drug delivery product. Research efforts to quantify the pharmacodynamics of a drug will enhance the development of a controlled drug delivery system and, importantly, should lead to faster regulatory approval for a novel dosage form.

Pharmacodynamic and Pharmacokinetic Principles

Pharmacodynamics

The goal of drug treatment is to elicit a desired pharmacologic response without an undesired toxicologic response. But the process by which a drug entity is released from the dosage form and delivery system and provides this response is not a simple process, due to physicochemical properties,

Copyright ©1997 American Chemical Society

physiologic constraints, and biochemical principles. This process (Figure 29.1) most frequently has been described on a chronological basis with release and absorption of the drug molecule as the first steps. Then, the drug in a central compartment or plasma may distribute to the biophase where the receptor, enzyme, or other active site is located. But, as the importance of pharmacodynamics has become apparent, the final step may actually be the key one for understanding and designing a controlled drug delivery system.

The importance of the pharmacologic effect, of course, is necessarily associated with a quantitative measure of that effect and the plasma concentration or other readily measured fluid in equilibrium with the active site. The effect is typically not an all-or-none phenomenon, but it is a graded response such that

$$E = PC + E_0 \qquad (1)$$

where E is the measured effect, P is a proportionality constant, C is the plasma concentration, and E_0 is the baseline effect. In many cases, a more satisfactory relationship is observed when the logarithmic measure of drug concentration is used:

$$E = P \log C + E_0 \qquad (2)$$

A graph of pharmacologic response versus concentration or the log of concentration will provide a measure of drug efficacy and the plasma concentration needed to achieve the desired response (Figure 29.2). The slope of the linear response will denote the steepness of this concentration–response relationship and the range of concentrations that will result in activity. Simultaneous measures of a graded toxicologic response will create a window for desired efficacy with a tolerable frequency of adverse events. The farther that this curve is displaced to the right, the easier it is to design a dosage form to balance the inherent therapeutic and toxicity aspects of the drug entity. However, when this therapeutic index is narrow, the need for a controlled delivery system to maintain concentrations within a desired, tight range is more crucial.

The relationship between response and concentration is linear within a given range but is frequently more complex, with a less pronounced increase in effect at low concentrations and a maximum effect plateau observed at high concentrations. This maximal effect, E_{max}, has been incorporated into a pharmacologic response model, the E_{max} model. In this model (*1*), the effect more closely simulates classic biochemical or enzymatic relationships. Again using a baseline effect to chart any changes, the sigmoid E_{max} model can be defined as:

$$E = E_0 + \frac{E_{max} C^n}{E_{50}^n + C^n} \qquad (3)$$

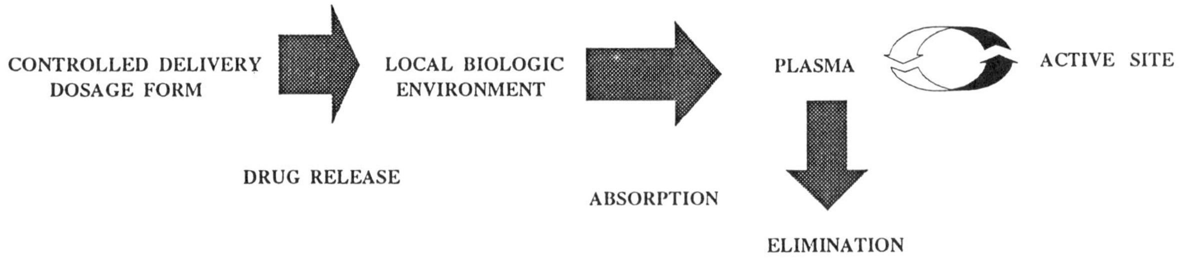

FIGURE 29.1. Processes involved in drug delivery from the dosage form to the pharmacologic effect provided at the active site.

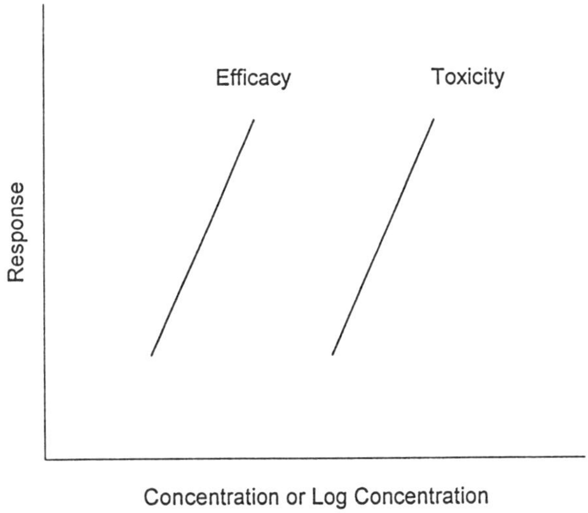

FIGURE 29.2. Relationship between pharmacologic and toxicologic responses and concentration. The relative distance between efficacy and toxicity is the therapeutic index of the drug substance.

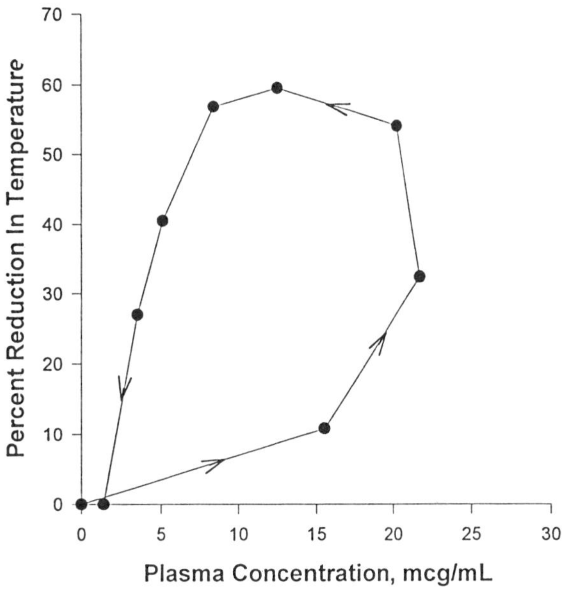

FIGURE 29.3. Antipyretic effect versus plasma concentration for a nonsteroidal antiinflammatory drug. The arrows show the time course of the pharmacodynamic–pharmacokinetic relationship.

where E_{50} is the drug concentration resulting in 50% of the E_{max} response. The actual shape of the curve is also a function of *n*, a constant that affects the hyperbolic nature of the concentration–response relationship. When this constant is 1, a simpler E_{max} model results. The sigmoid E_{max} equation, often called the Hill Equation because of its application to physiologic processes by A. Hill in 1910 (*2*), has been used to model drug effects directly or when linked to plasma concentrations by a separate effect compartment (*3*).

Even though a simple hyperbolic function may predict the concentration–response relationship for some drugs, a lag time is also quite likely to result in a delayed response or for a response to linger following a decline in plasma concentration. In these cases, a hysteresis loop may occur during a dosing interval as concentrations rise and fall (Figure 29.3). Here, an anticlockwise hysteresis loop is observed, indicating that the effect lags behind the plasma concentration. Thus, for the same concentration, the effect is greater at a later time, during the elimination phase of the concentration–time profile. This process must be investigated for a controlled-release system even though the range of drug concentrations will not be as great as when the drug is administered by a first-order release dosage form. This result will examine whether a pulsed input or other formulation variables are preferred to incorporate intrinsic or chronobiologic alterations of response.

Recently, more physiologic-based pharmacodynamic models have been proposed by Dayneka, Garg, and Jusko (*4*) to characterize drugs that result in pharmacologic activity through an indirect mechanism. These new models may be more satisfactory than effect-compartment models when there is a disparity or time lag between plasma concentrations and pharmacologic effect.

The characterization of pharmacologic response is most appropriate during chronic or steady-state conditions. Although one may investigate pharmacodynamic relationships in humans or animals following single doses, longer term therapy frequently results in an altered response due to adaptation, tolerance, or other mechanisms. Because most controlled-release dosage forms are developed for chronic therapy, the therapeutic window observed with concentrations at steady state are a more accurate depiction of the concentrations needed for efficacy in a controlled-release dosage form. Therefore, effects and concentrations (in eqs 1–3) are most appropriately steady-state measures, if possible.

Pharmacokinetics

The most useful pharmacokinetic parameter for determining dosing strategies is total plasma clearance (CL). The CL is meaningful because of its relationship with dosing rate and plasma concentration at steady state (C_{ss}). For a zero-order input (R_0):

$$C_{ss} = \frac{R_0}{\mathrm{CL}} \tag{4}$$

Therefore, the plasma concentration, which in turn produces a desired effect, is determined solely by the input rate for drug delivery and the plasma clearance. This equation is most routinely used for a constant intravenous infusion, but it would also hold for other methods of drug delivery if a true zero-order release is present for the drug substance. In this case, a bioavailability term (F) would indicate the fraction of the dose actually absorbed ($F \times R_0$).

Thus, in a rearrangement of eq 4, the product of drug clearance (or CL/F, if CL is not determined following an intravenous dose) and the effective steady-state plasma concentration are needed to determine the input rate necessary for the controlled drug delivery system. This straightforward calculation assumes that the drug obeys linear pharmacokinetic principles: that is, the clearance is constant and first-order elimination is observed at several doses. Also, the concentration must be chosen based on pharmacodynamic principles and steady-state human data. For each of these measures, mean CL and C_{ss} are the preferred parameters, but one must realize that inherent intersubject variability exists for each parameter. A range of input rates may be calculated based on a range of expected parameters. For example, theophylline has a relatively narrow range of effective plasma concentrations (10–20 μg/mL). Even though a mean C_{ss} value of 15 μg/mL could be used for calculation purposes, one should appreciate the entire target range, and if serious or numerous toxicities occur frequently at the upper end of this range, the input rate calculation must be weighted toward the lower end of the therapeutic range.

The plasma elimination half-life ($t_{1/2}$), though a widely used pharmacokinetic parameter, is not directly needed for determining the input rate for a controlled drug delivery system. The half-life will help provide an initial estimate of the usefulness of a controlled delivery dosage form. A compound with a relatively long elimination half-life (greater than 12–24 h) for the active drug moiety would be a poor candidate for a controlled-release dosage form, whereas treatment with a drug exhibiting a shorter half-life, but not extremely short, would be improved by delivery in a controlled-release system. The half-life is primarily needed to determine a dosing interval for a regular release dosage form; the dosing interval for a controlled-release system is more a function of the maximum quantity of drug available in and the release rate from the dosage form.

Similarly, the volume of distribution is not a major consideration in the development of a drug delivery device. Although a large volume of distribution will necessitate a larger dose to provide a specific plasma concentration than a smaller volume of distribution, the actual potency of the drug is more critical to determining the feasibility of a drug quantity or amount able to be administered in a controlled-release system.

The in vitro release rate for a dosage form, though not strictly a pharmacokinetic parameter, is clearly quite critical to the eventual regulatory approval and the effectiveness of the controlled-release delivery system. As observed in Figure 29.1, the rate of release from the dosage form is the first step in the sequence leading to the desired response. From a pharmacokinetic perspective, this process is also important because it is the rate-determining step leading to drug in the plasma and at the active site. Therefore, the release rate from the drug delivery system is most often the only variable available in drug development to achieve the expected pharmacologic activity. This release rate must be reproducible, be unaffected as much as possible by physiologic factors, and, as a final result, provide the therapeutic plasma concentration.

Development of Controlled-Release Product for Marketed Drug

Although immediate release dosage forms are accompanied by a multitude of formulation difficulties, it is generally easier to develop a solution, capsule, tablet, or other dosage form with normal-release characteristics than it is to develop a controlled-release dosage form. Many immediate release products are currently under investigation or are marketed as approved drugs. However, because of a low therapeutic index or improved patient compliance, a controlled-release dosage form can contribute positively to patient drug therapy.

For many drugs, a narrow therapeutic window places limits on the dosage form and resulting plasma concentrations, making it difficult to provide the needed pharmacologic activity without unnecessary toxicity. For example, theophylline was an older drug with notorious difficulty in dosing because of peak concentrations (generally, >20 μg/mL) that led to increased heart rate, headache, dizziness, and other more serious toxic responses at higher concentrations. Also, trough concentrations below 8–10 μg/mL might exacerbate breathing problems at the end of a dosing interval. A controlled-release dosage form is ideally suited to decrease this inherent fluctuation in plasma concentrations for theophylline or other drugs with this low therapeutic index and relatively wide swings in concentration during a dosing interval.

As a measure of the peak-to-trough difference during a dosing interval, the fluctuation index (FI) may be calculated:

$$\mathrm{FI} = \frac{C_{max} - C_{min}}{C_{avg}} \tag{5}$$

where C_{max}, C_{min}, and C_{avg} are the maximum, minimum, and average concentrations, respectively, within a dosing interval. As a matter of principle, the fluctuation index should be smaller for a controlled-release than for an immediate-release dosage form, and therefore, probability of providing plasma concentrations within the therapeutic window will be improved.

Because frequent drug administration, especially three or four times daily, is related to patient noncompliance (5), decreasing the dosing schedule to once or twice daily has clear advantages. The ideal oral controlled-release dosage form would be administered once daily without regard to meals, which should be more amenable to varied patient lifestyles and to improved compliance.

These attributes for a controlled-release dosage form can be observed in data obtained from a bioequivalence study for a carbonic anhydrase inhibitor. In drug concentration–time profiles obtained from a representative subject (Figure 29.4), an immediate-release dosage form administered twice daily is compared with a controlled-release product administered once daily as a twofold larger dose. After dosing to steady state, administration of the same daily dose clearly provides equivalent areas under the plasma concentration–time curve (AUC) for both dosage forms. Therefore, the average concentration (AUC/24 h) is also the same over the sampling interval. However, the peak plasma concentration for the controlled-release dosage form is lower by approximately 30% compared with the immediate-release product. As a consequence, the fluctuation index is correspondingly decreased. These data also demonstrate the importance of developing an oral controlled-release dosage form that is not affected by coadministration with food. In this instance, a dose taken with a high-fat meal does not alter the

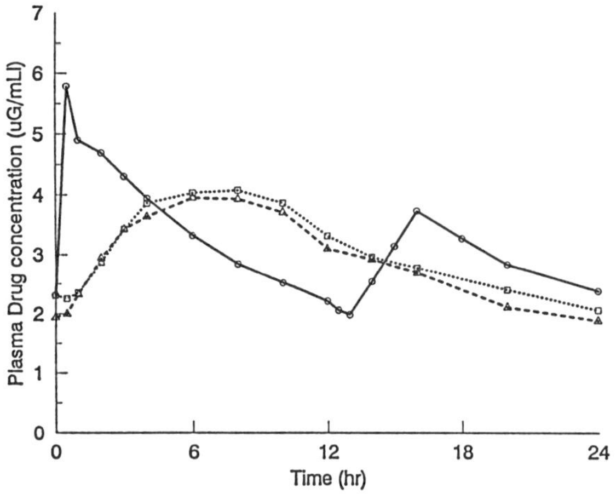

FIGURE 29.4. Plasma concentration versus time profile for drug being developed as an oral controlled-release product: immediate-release dosage form (○) and controlled-release dosage form under fasting (□) and fed (△) conditions.

rate (as measured by C_{max} and the time needed to reach the peak concentration) or extent (AUC) of drug absorption from the controlled-release dosage form (Figure 29.4). It is especially critical for an oral dosage form with a full daily dose of drug not to *dose dump* when administered into a gastric environment that has been altered by a meal.

From a purely pharmacokinetic perspective, a controlled-release dosage form offers improvements over a more frequently dosed immediate-release product by providing more consistent plasma concentrations. Also, a less frequently dosed drug product will lead to a more compliant patient in most cases, although that benefit may be overstated when an immediate-release dosage form is dosed only twice daily. But, the overall improvement in clinical outcome for a patient may be a more important measure of the utility of a controlled-release product. In the 1980s, demonstrating the near plateau in a concentration–time profile and statistical bioequivalence for AUC measures might have been sufficient. Now, analogous to the enhanced importance of pharmacodynamics, clinical trial data will be necessary for Board of Health approval in many cascs. These data should document the improvement in efficacy with a concurrent decrease in adverse events. In some cases, pharmacokinetic data alone may be sufficient for approval, but probably only when strong supporting pharmacodynamic data exists: that is, one must be able to demonstrate that the concentrations are indeed correlated with pharmacologic activity.

Development of Controlled-Release Product for New Chemical Entity

The ideal progression for a controlled-release dosage form may very well be simultaneous with the discovery and overall development of a drug substance. Following positive findings in an in vitro or early pharmacology screening model, one must make the conversion from the effective in vitro concentration or in vivo dose to an effective concentration in humans. If the in vitro data is obtained in a buffer environment, then a correction for protein binding (total drug concentration times free fraction) may be necessary because only free drug is active. For an animal model, plasma concentrations should be measured to begin to estimate a concentration–response relationship. Typically, substantial interanimal variability exists such that an initial attempt at constructing a pharmacodynamic model is possible without a large range of doses due to the natural range of data obtained in early animal studies.

Concurrent with early pharmacology studies, range-finding and shorter, pivotal toxicology studies will be performed. This process allows an early decision to pursue a controlled-release dosage form based on two considerations. First, toxicokinetic parameters for exposure, C_{max} and AUC, should be compared with the corresponding estimates for human efficacy to obtain an initial estimate of therapeutic index. Although many assumptions are necessary to predict behavior in humans, a narrow therapeutic index may provide a rationale for a controllcd-release dosage form. Second, preformulation data, including physicochemical characterization of the drug, are required to make a judgment about the feasibility of developing a controlled-release dosage form given the desired route of administration and an early estimate of dose requirements.

Although the initial safety studies in humans are more likely to be performed with a solution or other immediate-release dosage form, a bioequivalence study will allow comparison of the safety data already accumulated. This study need not have rigorous bioequivalence testing, but it certainly must provide measures of the in vivo disposition of the new dosage form. This dosage form should be viewed as a first attempt to formulate a controlled-release dosage form; other variables will be altered to finally provide a dosage form that can be administered in pivotal Phase III clinical trials.

The criteria on which to judge the success and eventual marketability of a controlled-release dosage form are threefold. Most importantly, it must provide plasma concentrations that are acceptable given the pharmacodynamics of the drug. Because the concentration–response relationship has been initially defined in animals, an even more accurate characterization is needed in humans. The dosage form must result in drug input rates consistent with the plasma concentrations needed. Second, the rate of drug input should be or at least approach zero-order release in humans. The kinetics of drug absorption can be estimated by deconvolution methods (*6*) or Wagner–Nelson (*7*) or Loo–Riegelman (*8*) plots, but they should also be apparent by a near plateau in plasma drug concentrations. Finally, the dosage form should not be affected by coadministration of food, if an oral product, and be consistent and reproducible in its delivery. The final dosage form to be marketed must be bioequivalent with the one used in clinical trials to link the pivotal clinical efficacy and safety data. With adequate understanding and controls on each of the steps involved in the manufacturing of the product and in the studies to be conducted, the final bioequivalence study will become a less stressful situation.

Conclusion

Quite simply, the pharmacodynamics of the drug itself must be characterized and the pharmacokinetics of the dosage form must be optimized as a controlled-release dosage form is developed. The key link for these formulation processes and pharmacokinetic parameters is the plasma drug concentration. Because drug concentration is proportional to pharmacologic and toxicologic responses, determination of the pharmacodynamics is the first step involved in designing a controlled drug delivery product and leading to improved patient therapy.

References

1. Holford, N. G.; Sheiner, L. B. *Pharmacol. Ther.* **1982,** *16,* 143–166.
2. Hill, A. V. *J. Physiol. (London)* **1910,** *40,* 4–7.
3. Sheiner, L. B.; Stanski, D. R.; Vogel, S.; Miller, R. D.; Ham, J. *Clin. Pharmacol. Ther. (St. Louis)* **1979,** *25,* 358–371.
4. Dayneka, N. L.; Garg, V.; Jusko, W. J. *J. Pharmacokinet. Biopharm.* **1993,** *21,* 457–478.
5. Greenberg, R. N. *Clin. Ther.* **1984,** *6,* 592–599.
6. Veng-Pedersen, P. *J. Pharm. Sci.* **1980,** *69,* 298–305.
7. Wagner, J. G.; Nelson, E. *J. Pharm. Sci.* **1963,** *52,* 610–611.
8. Loo, J.; Riegelman, S. *J. Pharm. Sci.* **1968,** *57,* 918–928.

Index

Copy editing: Scott Hofmann-Reardon
Production: Paula M. Bérard and Susan Drake Fisher
Acquisition: Anne Wilson
Jacket design: Amy O'Donnell

Text designed and typeset by Betsy Kulamer, Washington, DC
Printed and bound by Maple Press Company, York, PA